AF900190

Active Volcanoes of the World

Series Editors

Corrado Cimarelli, Section for Mineralogy, Petrology and Geochemistry, Ludwig-Maximilians-University Munich, Department of Earth and Environmental Sciences, München, Germany

Sebastian Muller, Mineralogisches Museum, Marburg, Hessen, Germany

About 500 active volcanoes presently exist on the Earth's surface, of which around 50 erupt each year. Volcanoes played a crucial role in the evolution of the planet and early life, and are constantly reshaping the morphology of Planet Earth. Many active volcanoes are located in dense settlement areas, with over 500 million people living in close proximity of still active or dormant volcanoes.

On one side, volcanoes provide valuable soil and rock basis for agriculture, but often the "mountains of fire" cause disastrous societal and economical disasters caused by ash clouds, lahars, lava flows, and pyroclastic flows. Eruptions are still difficult to predict, although volcanologists around the world are constantly working on new ways to understand the character and behavior of volcanoes.

Active Volcanoes of the World is an official book series of the International Association of Volcanology and Chemistry of the Earth's Interior (IAVCEI). The series aims to be a scientific library of monographs that provide authoritative and detailed reviews of state-of-the art research on individual volcanoes or a volcanic area that has been active in the last 10.000 years, e.g. the Teide Volcano or the Chiapas Region. The books in the series cover the geology, eruptive history, petrology and geochemistry, volcano monitoring, risk assessment and mitigation, volcano and society, and specific aspects related to the nature of each described volcano.

The Active Volcanoes of the World series contains single and multi-authored books as well as edited volumes. The Series Editors, Dr. Corrado Cimarelli and Dr. Sebastian Müller are currently accepting proposals and a proposal document can be obtained from the Publisher, Dr. Annett Buettner (annett.buettner@springer.com).

More information about this series at https://link.springer.com/bookseries/10081

Giovanni Orsi • Massimo D'Antonio •
Lucia Civetta
Editors

Campi Flegrei

A Restless Caldera in a Densely Populated Area

Springer

Editors
Giovanni Orsi
Dipartimento di Scienze della Terra dell'Ambiente e delle Risorse
Università degli Studi di Napoli Federico II
Naples, Italy

Massimo D'Antonio
Dipartimento di Scienze della Terra dell'Ambiente e delle Risorse
Università degli Studi di Napoli Federico II
Naples, Italy

Lucia Civetta
Emeritus Professor of Geochemistry and Volcanology
Università degli Studi di Napoli Federico II
Naples, Italy

ISSN 2195-3589 ISSN 2195-7029 (electronic)
Active Volcanoes of the World
ISBN 978-3-642-37059-5 ISBN 978-3-642-37060-1 (eBook)
https://doi.org/10.1007/978-3-642-37060-1

© Springer-Verlag GmbH Germany, part of Springer Nature 2022
This work is subject to copyright. All rights are reserved by the Publisher, whether the whole or part of the material is concerned, specifically the rights of translation, reprinting, reuse of illustrations, recitation, broadcasting, reproduction on microfilms or in any other physical way, and transmission or information storage and retrieval, electronic adaptation, computer software, or by similar or dissimilar methodology now known or hereafter developed.
The use of general descriptive names, registered names, trademarks, service marks, etc. in this publication does not imply, even in the absence of a specific statement, that such names are exempt from the relevant protective laws and regulations and therefore free for general use.
The publisher, the authors and the editors are safe to assume that the advice and information in this book are believed to be true and accurate at the date of publication. Neither the publisher nor the authors or the editors give a warranty, expressed or implied, with respect to the material contained herein or for any errors or omissions that may have been made. The publisher remains neutral with regard to jurisdictional claims in published maps and institutional affiliations.

This Springer imprint is published by the registered company Springer-Verlag GmbH, DE part of Springer Nature.
The registered company address is: Heidelberger Platz 3, 14197 Berlin, Germany

Foreword

Campi Flegrei is a location of rich human history that is intertwined with volcanism. As a volcanologist, I am drawn to the area for the wide range of eruptive activity that has occurred there over the past several tens of thousands of years. This ranges from the super-eruption that produced the widespread Campanian Ignimbrite, to the small (by volcanological standards) cone-forming eruption of Monte Nuovo in the sixteenth century. The two large nested calderas (related to the Campanian Ignimbrite and to the Neapolitan Yellow Tuff) that host the Campi Flegrei have experienced unrest —indicative of a living volcanic system—throughout the human habitation of the area, with multiple episodes of uplift and subsidence (inflation and deflation). Many volcanologists have lodged at the campground in Solfatara Crater, with its fumaroles that display another sign of the living volcano—Campi Flegrei's active hydrothermal system.

Volcanic activity has benefitted humans in many ways at Campi Flegrei, for example with coastal craters forming beautiful protected harbours for trading and military sea vessels, the rich volcanic soils, and abundant tuffs that form excellent building materials. At the same time, volcanism poses a major risk to the dense population of Naples that spills into the Campi Flegrei and to the towns within it, especially since previous eruptions have been dominated by dangerous explosive phenomena, such as pyroclastic density currents and widespread tephra fall. This large human population, and the living nature of the volcanic system, make the Campi Flegrei an area with one of the highest volcanic risks on Earth.

Tremendous research has been conducted over the years to understand the volcanic and human history of the area. Painstaking work has led to the development of a robust geologic framework of the eruptive history of the volcanic field. This framework facilitates detailed studies of the past and current magmatic–hydrothermal processes, using a range of approaches that include petrology, heat flow, geophysics, deformation monitoring, fumarole and water chemistry, and numerical and experimental modelling. Distal studies of Campi Flegrei tephra layers provide important Pleistocene geochronologic makers over a wide portion of Europe, aiding in interpretation of Quaternary geology and prehistoric human activity. All of this provides necessary constraints for assessments of volcanic hazards as well as geomorphic hazards such as landslides, which are common in rugged volcanic areas with abundant altered deposits (and exasperated by urbanisation).

This book integrates the huge body of existing research on the Campi Flegrei. The focus is on volcanism and volcanic hazards, but chapters on the practical uses of volcanic materials in the area and the history of human habitation nicely round out the work. The book is an invaluable resource for all those conducting research on the Campi Flegrei volcanic system and as a launching point for future researchers in the coming decades.

Greg A. Valentine
Department of Geology
and Center for Geohazards Studies
University at Buffalo
Buffalo, NY, USA

Introduction

In the history of the European culture, Campi Flegrei represent the myth of Rome celebrated in the Aeneid by Virgil (70 BC–19 BC), who made Aeneas, the Trojan hero of the Greco-Roman mythology and progenitor of the family of Emperor Augustus, disembark in Cuma. Travellers of all times have been looking for the myth and not for the history on the Phlegraean shores. Among them, Francesco Petrarca (AD 1304–1374), who wrote about his journey to the Campi Flegrei in AD 1343: "I have seen all the places of Virgil".

The Campi Flegrei are a magical place within the Mediterranean where striking volcanic phenomena, inexplicable to the ancients, have fuelled the myth for millennia. Local history, stories, legends, literature form an Intangible Cultural Heritage of the humanity that the Campi Flegrei preserve. However, for this heritage to be well understood and interpreted today and in future, we are called to take care of the natural features from which it was generated.

The Campi Flegrei are also one of the best places in the world where to find the effects of the encounter-clash between nature and Man, the species presently dominant on Earth. Since millennia, they are a privileged scenery for this encounter-clash with phases of strong anthropisation and slow but unrestrainable volcanic events. Among these, the best known and somehow specific is the bradyseism that testifies "the breath of the Earth" and causes ground uplift and subsidence with birth or disappearance of pieces of land and environmental changes.

Despite the presence of three active Volcanoes (Campi Flegrei, Somma-Vesuvio, Ischia) with the direct and related potential hazards they pose, the Neapolitan area has been inhabited by humans since Neolithic times. The earliest organised society developed with the settlement of Hellenic colonies in the eighth century BC at Ischia, Cuma in the Campi Flegrei, and Megaride islet close to the coast of the city of Naples, and in the sixth century BC at Pozzuoli. Presently, the area hosts a population of more than two million people for a variety of reasons, among which soil fertility, temperate climate, beautiful scenery, abundance of thermo-mineral waters and easily extractable volcanic rocks suitable for being used for engineering and architectural purposes. Furthermore, the area, located right in the middle of the Mediterranean, the cradle of the Western Culture, has been always in a very favourable position from a political, commercial and strategic point of view. The millenary coexistence of volcanic phenomena and human's activity has determined the basic conditions for the development and conservation of an extraordinary biodiversity on the outskirts of Naples, one of the most densely populated European urban areas.

Calderas are depressions that occur in all volcanic environments and geodynamic settings on Earth as well as on other planets, with a diameter larger than that of the explosive vents or craters. Topics such as formation and evolution of calderas, including pre-collapse tumescence, collapse and post-collapse resurgence, are still debated in the scientific literature. Caldera collapses repetitively occurring in the same area after several eruptions form nested collapse structures. Presently, 97 out of the 446 calderas known on Earth are in unrest. Volcanic unrest is the deviation of a given volcano, in particular of its manifestations, such as ground deformation, degassing, seismicity and gravity acceleration, relatively to its baseline. It is likely generated by pressure and/or volume increase in the magmatic or hydrothermal system, variation in the local stress field or by a combination of more than one of them.

The Campi Flegrei, to the west of and partially including the city of Naples, give name to the Campi Flegrei volcanic field which dominant feature is the Campi Flegrei caldera, a nested, resurgent and restless structure in the densely inhabited Neapolitan volcanic area, within the Campanian Plain. The earliest studies on the volcanism of the Neapolitan area date back to the eighteenth century. The book "Campi Phlegraei. Observations on the volcanos of the Two Sicilies" by Sir William Hamilton, British Ambassador to the Kingdom of Naples from 1764 to 1800 and volcanologist, was published in 1776 in French and English because it was intended for the travellers of the Grand Tour. Actually, it is the first comprehensive document on the volcanoes of Southern Italy. Scipione Breislak, an Italian geologist and naturalist, presented the results of his studies in the book "Topografia Fisica della Campania" published in 1798. The Neapolitan volcanic area has also attracted the interest of other outstanding scientists such as Johnston Lavis (1889), Dell'Erba (1892), De Lorenzo (1904) and Rittmann (1950). The latter was the first to envisage the presence of a caldera in the Phlegraean area. Since then, timing, geometry and evolution of the Campi Flegrei caldera and its magmatic feeding system have been variably interpreted by many authors. The most shared idea among the scientific community is that it is a complex, nested and resurgent structure resulting mainly from two major collapses related to the catastrophic Campanian Ignimbrite ($\sim$40 ka) and the Neapolitan Yellow Tuff ($\sim$15 ka) eruptions. The youngest of these calderas is the portion still active of the entire system. One part of this caldera is submerged. This makes both basic research and monitoring more difficult. Only since few years, researches on the submerged portion of the caldera have been intensified by deploying sea-bottom, multi-parametric stations for monitoring and carrying out offshore seismic and morphological surveys. The youngest caldera has been the site of an intense volcanism and deformation in the past 15 kyrs. The volcanism has generated not less than 70 eruptions, most of which have been explosive, with phreatomagmatic dominating over magmatic explosions. The deformation has mostly consisted in resurgence of the floor of the younger caldera with a maximum net uplift of about 100 m. The caldera floor is affected by short-term deformation episodes, also known as bradyseismic events. The earliest evidence of this phenomenon prior to the installation of a monitoring system in the area has been recognised at the Serapeo in Pozzuoli, the remnant of a Roman market which oldest part was built in the first century AD and reported by Antonio Parascandola in a landmark paper published in 1947. The most recent major events have occurred in 1969–1972 and in 1982–1984 totalling a maximum net uplift of 350 cm at Pozzuoli. These events, which stimulated the interest of many scholars on the caldera, have been followed by some minor episodes known as miniuplifts. The last of these episodes began in late 2004 and is still ongoing. These bradyseismic events have been interpreted as transient short-term deformation events within the long-term deformation. The magmatic feeding system has a complex architecture with melts stagnating, differentiating, mingling and mixing at various depths. It includes a widespread reservoir at about 8–10 km depth that feeds smaller magma chambers at depths varying between 2 and 8 km that in turn feed the eruptions. At shallower depth is an active hydrothermal system that generates diffuse soil degassing, as well as fumarole and hot-water spring activity, especially at Solfatara and Pisciarelli localities.

The persistent state of activity, testified by the current unrest, and the explosive character of volcanism result in a high level of volcanic hazard potential of the caldera. The dense urbanisation, with a population of not less than 350,000 people in the most active portion of the caldera, determines a very high risk. In this situation, the Italian Dipartimento della Protezione Civile (www.protezionecivile.it) in 2012 has increased the alert level from Base to Attention, the second of 4 levels provided by the Campi Flegrei National Emergency Plan.

This book aims at presenting the current and most widely shared knowledge of the scientific community engaged in the study of the various aspects of the Campi Flegrei caldera. In particular, the volcanic and deformation history, as well as the present setting of the caldera defined through geological and volcanological (Chap. "Volcanic and Deformation History of the Campi Flegrei Volcanic Field, Italy") and geophysical (Chap. "Seismic and Gravity Structure of the Campi Flegrei Caldera, Italy") studies, is summarised. Another topic extensively treated concerns the evolution through time, architecture and current state of the magmatic feeding system (Chaps. "An Evolutionary Model for the Magmatic System of the Campi Flegrei Volcanic Field (Italy) Constrained by Petrochemical Data", "Origin and Differentiation History of the Magmatic System Feeding the Campi Flegrei Volcanic Field (Italy) Constrained by Radiogenic and Stable Isotope Data", and "Tephrochronology and Geochemistry of Tephra from the Campi Flegrei Volcanic Field, Italy"). The dynamics of the magma chamber is highlighted through experimental (Chap. "Rheological Properties of the Magmas Feeding the Campi Flegrei Caldera (Italy) and Their Influence on Mixing Processes") and modelling (Chap. "Magma Chamber Dynamics at the Campi Flegrei Caldera, Italy") approaches. The permanent monitoring system of the volcano is presented (Chap. "The Permanent Monitoring System of the Campi Flegrei Caldera, Italy"). The current knowledge on the very active hydrothermal system that has developed within the caldera and plays an important role in the current evolution of the volcano is reviewed (Chap. "The Hydrothermal System of the Campi Flegrei Caldera, Italy"). The characteristics of the recent unrest episodes, including the ongoing one, and their causes hypothesised by modelling data collected through the monitoring system, are also presented (Chaps. "Historic Unrest of the Campi Flegrei Caldera, Italy" and "Source Modelling from Ground Deformation and Gravity Changes at the Campi Flegrei Caldera, Italy"). The evolution of the knowledge on the volcanic and related hazards of the caldera has been reviewed in detail (Chap. "Volcanic Hazard Assessment at the Campi Flegrei Caldera, Italy"). An overview of landslide hazard and risk is the subject of a specific chapter (Chap. "Landslide Hazard and Risk in the Campi Flegrei Caldera, Italy"). The building and technological characteristics of the typical volcanic rocks of the caldera and their use in architecture are discussed (Chap. "Building Stones and Technological Materials of the Campi Flegrei Caldera, Italy"). The book ends with a chapter devoted to the urban development of the caldera (Chap. "The Urban Development of Campi Flegrei, Italy").

Giovanni Orsi
Massimo D'Antonio
Lucia Civetta

Contents

Volcanic and Deformation History of the Campi Flegrei Volcanic Field, Italy

Giovanni Orsi

Abstract

This chapter reviews the current state of knowledge on the volcanic and deformation history of the Campi Flegrei caldera. This caldera, a nested and resurgent collapse structure in a current state of unrest, is the world's best example of a restless volcano in a densely populated area. It formed through two major collapses related to the Campanian Ignimbrite ($\sim$39 ka) and the Neapolitan Yellow Tuff ($\sim$15 ka) eruptions. Each caldera collapse mostly occurred through activation of portions of northeast-southwest and northwest-southeast regional faults. The younger caldera is the currently active portion of the entire structure and has been the site of intense volcanism and an ongoing deformation. This volcanism produced not less than 70 eruptions, largely explosive with very few effusive events, concentrated in three clusters of activity, called epochs, and separated by two periods of quiescence during which two widespread paleosols formed. During each epoch, the eruptions occurred at mean time intervals of tens of years. The last event produced the Monte Nuovo tuff cone in AD 1538, after $\sim$3,000 years of quiescence. Deformation mostly consists of a resurgence process that has generated a net uplift of about 90 m mainly through portions of regional faults probably active also during caldera collapses. Within this process, a deformation lasting since the caldera collapse called "long-term deformation" and one that takes place over years or decades called "short-term deformation" can be distinguished. Vent locations through time provide a good tracer of the long-term deformation. The vents of both epochs 1 and 2 were located along the structural boundary of the Neapolitan Yellow Tuff caldera, while those of epoch 3 were mostly concentrated in the north-eastern sector of the resurgent portion of the caldera floor. Short-term ground deformation episodes, also known as bradyseism, are transient events within the long-term resurgence process and can be reconstructed over the past 2,000 years. In particular, levelling data available since 1905 highlight a general subsidence interrupted by three major and many minor uplifts, with the ongoing one beginning in late 2004. Some geological and volcanological aspects of the Campi Flegrei caldera are still under debate within the scientific community. One of these aspects concerns the Campanian Ignimbrite eruption and related caldera collapse, namely the relations among architecture of the magmatic feeding system, timing and dynamics of the eruption, type (central, fissure) and areal

G. Orsi (✉)
Dipartimento di Scienze della Terra, dell'Ambiente e delle Risorse, Università degli Studi di Napoli Federico II, Napoli, Italy
e-mail: giovanni.orsi@unina.it

© Springer-Verlag GmbH Germany, part of Springer Nature 2022
G. Orsi et al. (eds.), *Campi Flegrei*, Active Volcanoes of the World,
https://doi.org/10.1007/978-3-642-37060-1_1

distribution of the eruption vents, and volume of caldera collapse versus volume of erupted magma. A very preliminary hypothesis aimed at providing a still missing unifying explanation for all the mentioned aspects of the Campanian Ignimbrite eruption is proposed. This hypothesis argues that the eruption was fed by vents located on a network of regional faults that controlled the collapse of a caldera in the Neapolitan-Phlegraean area. Further crucial scientific points that need to be addressed in order to increase our knowledge of the volcano are summarised in conclusions.

1 Introduction

Calderas are a type of volcano with a large summit depression wider than explosive vents or craters, occurring in all volcanic environments and geodynamic settings on Earth (Williams 1941), as well as on other planets (e.g., Sigurdsson et al. 2015).

The processes by which collapse calderas form and evolve, including pre-collapse tumescence, collapse and post-collapse resurgence, are still debated in the scientific literature (Cole et al. 2005; Branney and Acocella 2015). The geometry resulting from the variable dynamics of their collapse has resulted in five end-member caldera types (e.g., Lipman 1997; Cole et al. 2005). Among these, piecemeal collapse calderas are characterised by numerous floor blocks and/or multiple collapse centres (e.g., Moore and Kokelaar 1998). Collapses that repeat in the same area form nested calderas after several eruptions (Lipman 1997 and references therein). Deformation preceding and following caldera collapse is very common. It is due to magmatic overpressure, although hydrothermal activity may also play an important role (Martí et al. 2008). Post-collapse caldera resurgence (Marsh 1984) is generally interpreted as the result of overpressure re-establishment in the magma chamber after collapse (Smith and Bailey 1968). Resurgence dynamics vary from doming (Smith and Bailey 1968) to disjointing in differentially uplifted blocks (Orsi et al. 1991b) of the magma chamber roof rocks, and likely depend upon the aspect ratio of the chamber (Acocella et al. 2001).

Reconstruction of the volcanic and deformation history of a caldera is complicated by coeval and intense volcanism and deformation. The information that can be acquired varies in quantity and quality with the age of the event, such that older events are generally less well defined than younger ones. Geological, morpho-structural, and petrological investigations help image the subsurface structural setting of volcanoes (Heiken et al. 1990; Druitt et al. 1999). The quality of such images depends upon availability of exposures of the internal part of the volcano, which is related to erosion depth. Advances in understanding both structure and collapse dynamics of a caldera have resulted from geological studies of deeply eroded (e.g., Branney and Kokelaar 1994) and geophysical investigations of poorly eroded calderas (e.g., Hill 1976; Finlayson et al. 2003; Husen et al. 2004; Zollo et al. 2008). Furthermore, studies of magmatic volatiles permit assessment of both the physico-chemical state and the thermobarometric conditions of the major magma storage systems through time (e.g., Spilliaert et al. 2006; Arienzo et al. 2010; Moretti et al. 2013b, 2019).

Currently, 97 out of the 446 calderas known on Earth are in unrest (Geyer and Martí 2008; Siebert et al. 2010; Acocella et al. 2015). Unrest of a caldera, as in any type of volcano, includes appearance or increase of phenomena such as ground deformation, degassing, seismicity and gravity changes relative to their baseline. These are likely generated by increase of pressure and/or temperature, mass of melt or fluids in the magmatic and/or hydrothermal system, variation of the local stress field, or by a combination of these.

The Campi Flegrei caldera (CFc) is a nested, resurgent, and restless structure that derives its name from Campi Flegrei (CF), a morpho-structural depression to the west of the centre of the city of Naples, in the Campanian Plain (southern Italy) (Figs. 1 and 2) (Orsi et al. 1996). The CFc is the dominant feature of the CF volcanic field, one of the three volcanic fields, together with those of Ischia and Procida islands, that form the Phlegraean volcanic district (Orsi et al. 1996). This volcanic district and the neighbouring Somma-Vesuvio volcanic complex have developed within the Campanian volcanic area (Fig. 2). The CFc results from at least two collapses related to catastrophic eruptions. The floor of the youngest of these caldera depressions, dated at about 15 ka, has been the site of intense volcanism and deformation. The volcanism has been dominated by explosive events and generated its last eruption in AD 1538, while the deformation is testified by an ongoing unrest, also known as bradyseism, a word coined by Issel (1883) to mean slow vertical ground movement. In 2012, the values and trends of the monitored physical and chemical parameters of the ongoing unrest, such as ground deformation, seismicity, gravity changes and compositional variation of fluid effluents (for reviews see Chaps. The Permanent Monitoring System of the Campi Flegrei Caldera, Italy; The Hydrothermal System of the Campi Flegrei Caldera, Italy; Historic Unrest of the Campi Flegrei Caldera, Italy and Source Modelling from Ground Deformation and Gravity Changes at the Campi Flegrei Caldera, Italy), induced the Italian Dipartimento della Protezione Civile to increase the alert level from "base" to "attention" (www.protezionecivile.it). The volcanic hazards posed by this caldera and the related risk are extremely high because of the current unrest state, the explosive character of its volcanism, and the 350,000 people living within its active portion, contiguous to and partly including the city of Naples (for a review see Chap. Volcanic Hazard Assessment at the Campi Flegrei Caldera, Italy). In addition to volcanic hazards, the area is exposed to a variety of natural hazards related to its geological setting and evolution (e.g., seismic, hydrogeological, tsunami) (Orsi et al. 2003; Chap. Landslide Hazard and Risk in the Campi Flegrei Caldera, Italy). Furthermore, the use of the territory and volcanic rocks by the inhabitants for fulfilment of their needs has resulted in an increase of the hydrogeological hazards.

CFc may be the world's best example of an unrest caldera within a densely populated area. This area, despite the occurrence of active volcanoes, has been the site since at least late Neolithic times ($\sim$6.2 ka cal. BP) of human habitation and activity (Albore Livadie 2002; Saccoccio et al. 2013; Chap. Building Stones and Technological Materials of the Campi Flegrei Caldera, Italy). Development of an organised society in the area began in the eighth century BC with the settlement of Hellenic colonists between Cuma and the present city of Naples, which served as a strategic location along the Mediterranean trade routes, and habitation has uninterruptedly continued since (Orsi et al. 2013; Chap. The Urban Development of Campi Flegrei, Italy and references therein) (Figs. 1 and 2). These colonists named the area Burning Fields (Campi Flegrei in Italian) due to the occurrence of volcanic cones and craters, intense fumarole and hot-water spring activity, and volcanic eruptions at Ischia (Figs. 1 and 2).

The volcanic and deformation history of the CFc will be summarised in this chapter. Geological investigations have been carried out by many scientists since the eighteenth century (for an exhaustive list of references see Orsi et al. 1996, 2004, 2009; Di Vito et al. 1999; De Vivo et al. 2001; Rolandi et al. 2003; Perrotta et al. 2006, 2010; Isaia et al. 2009, 2019; Acocella 2010; Fedele et al. 2011; Smith et al. 2011; Capuano et al. 2013; Scarpati et al. 2013; Vitale and Isaia 2014). Stratigraphic sequences penetrated by deep and shallow boreholes (e.g., Penta 1954; Rosi et al. 1983; Rosi and Sbrana 1987; Orsi et al. 1996; Di Vito et al. 1999) and geophysical investigations (e.g., De Bonitatibus et al. 1970; Finetti and Morelli 1974; Florio et al. 1999; Bruno 2004; Judenherc and Zollo 2004; Berrino et al. 2008; Zollo et al. 2008; Capuano et al. 2013; Steinmann et al. 2018; Sacchi et al. 2019) have also contributed to the definition of its geological and structural setting.

2 Geological Outlines of the Campanian Volcanic Area, Phlegraean Volcanic District and Somma-Vesuvio Volcanic Complex

The Tyrrhenian margin of the Southern Apennines thrust belt has been affected by extensional processes resulting from its eastward migration related to a progressive retreat of the subducting

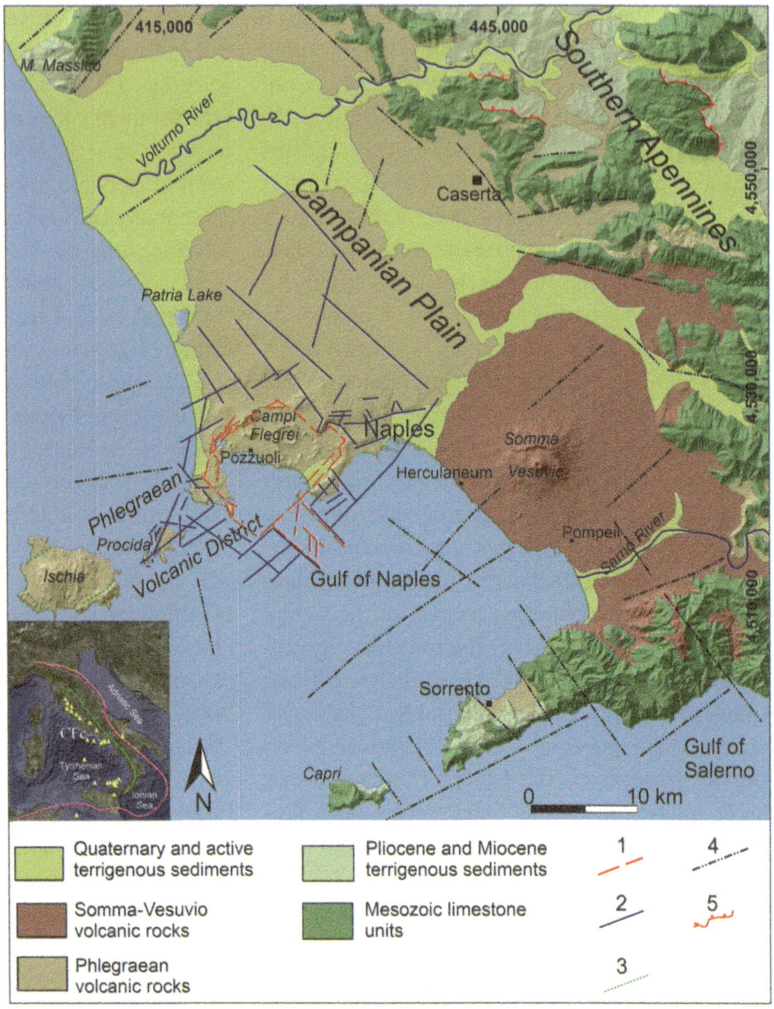

Fig. 1 Geological sketch map of the Neapolitan volcanic area. **1** Features active during the Neapolitan Yellow Tuff eruption. **2** Features active during the Campanian Ignimbrite eruption. **3** Features active during the Mount Epomeo Green Tuff. **4** Regional faults. **5** Overthrusts (modified after Cameli et al. 1975; Orsi et al. 1996). In the inset: traces of the subduction-related compression (purple line) and extension (green line) fronts along the Apennine chain (modified after Acocella and Funiciello 2006), and main Quaternary volcanic centres (yellow triangles). For details on the Campi Flegrei caldera, see text and Figs. 4 and 14

Ionian lithosphere since Miocene time (Patacca et al. 1990; Jolivet et al. 2009) (Fig. 1). Extension has occurred along mainly northwest-southeast normal and subordinate northeast-southwest transverse faults (Acocella and Funiciello 2006 and references therein). This extension generated the graben of the Campanian Plain, a large Plio-Pleistocene tectonic depression (Cinque et al. 1993) within which conditions for magma ascent were established with the generation of the Campanian volcanic area. This volcanism developed over the area comprising the Campanian Plain and its extension to the southwest into the Tyrrhenian Sea, where it included the islands of Procida and Ischia (Fig. 1). The geological setting and landscape of this volcanic area are presently dominated by four volcanoes: Somma-Vesuvio, CF, Procida and Ischia (e.g., Santacroce et al. 2003) (Fig. 1). The timing of the onset of volcanic activity in the Campanian volcanic area is still poorly constrained. A stratigraphic sequence, about 1,300 m thick, including calc-alkaline basalt to andesite as old as 2 Ma, was bored in the Campanian Plain below the CF potassic rocks (Di Girolamo 1978). Distal tephra similar in major and trace element composition to the recent products of the CFc and spanning in age from early/middle to late Pleistocene, have been found embedded in lacustrine (Wulf et al. 2004; Munno and Petrosino 2007; Petrosino et al. 2014a, b; Wutke et al. 2015; Giaccio et al. 2014, 2017) and

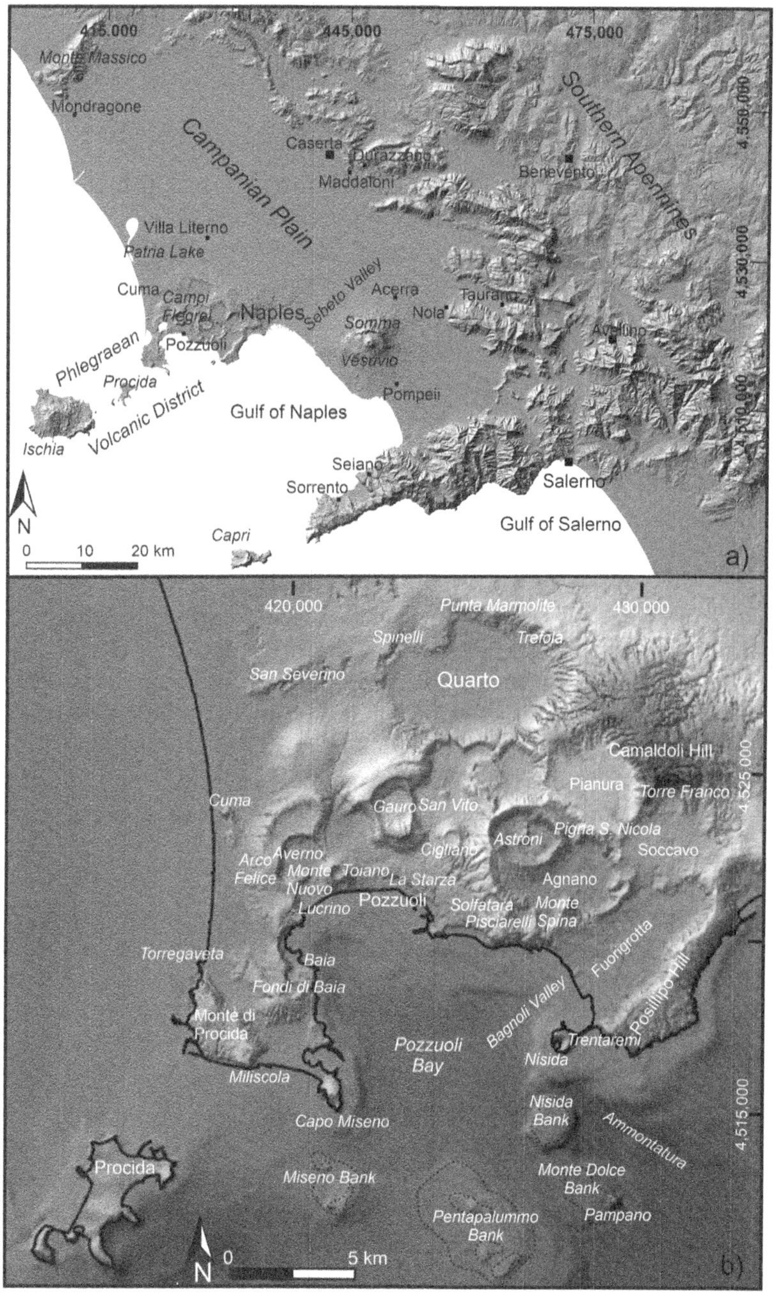

Fig. 2 a DTM of the Neapolitan volcanoes, Campanian Plain and surrounding Southern Apennines; **b** Digital Terrain Marine Model of the Campi Flegrei and Pozzuoli Bay. Figure modified after Capuano et al. (2013)

marine sequences cored in the Tyrrhenian and Ionian seas (Paterne et al. 1986, 2008; Insinga et al. 2014; Bourne et al. 2015; Petrosino et al. 2016) (Fig. 1). Several of these sedimentary successions are located thousands of kilometres from the Campanian volcanic area (Leicher et al. 2016; Vakhrameeva et al. 2019), testifying to the occurrence of high volcanic explosivity index (VEI; Newhall and Self 1982) eruptions during at least the past 700 kyrs. Finding the terrestrial counterparts of these tephra, however, is a difficult task due to the almost total absence of proximal outcrops. The occurrence of at least five ignimbrites older than ~100 ka along the

Southern Apennines thrust belt bordering the Campanian Plain, both on the Sorrento Peninsula and at the foothills of the Nola and Caserta mountains (Fig. 2), is reported by De Vivo et al. (2001) and Rolandi et al. (2003). These ignimbrites, found in very scattered outcrops, are described by the authors as mainly represented by up to several metres thick, deeply altered, sand to coarse-sand deposits embedding altered juvenile fragments. Two ignimbrites, identified at Seiano on the Sorrento Peninsula (Fig. 2), yield $^{40}Ar/^{39}Ar$ ages of 289.6 ± 1.9 and 245.9 ± 3 ka, respectively. Another ignimbrite, dated at 157 ± 1 ka by $^{40}Ar/^{39}Ar$ methods, crops out in the Taurano area and has recently been found also in the Calore River valley (Amato et al. 2018). The largest of these ignimbrites yields an $^{40}Ar/^{39}Ar$ age of 116.1 ± 4.4 ka and is best exposed in the Durazzano-Maddaloni area, where it is composed of a brown ash matrix engulfing cm-sized pumice lapilli (Fig. 2). The youngest of these units crops out at Piano di Sorrento and has been dated at 104.83 ± 0.79 ka ($^{40}Ar/^{39}Ar$; Rolandi et al. 2003). Rolandi et al. (2003) hypothesised that these ignimbrites were erupted through one or more feeding fissures located along northeast-southwest, northwest-southeast and east–west trending regional faults in a wide volcanic area comprising the entire Campanian Plain and the Roccamonfina volcano. These ignimbrites could represent the product of a fissure activity of the Campanian volcanic area predating the centralisation of activity of the Somma-Vesuvio volcanic complex and the Phlegraean volcanic district (Figs. 1 and 2).

The Somma-Vesuvio volcanic complex includes an older cone (Monte Somma) affected by a multicyclic, nested caldera, and a younger one (Vesuvio) that grew inside the caldera (Cioni et al. 1999, 2008; Santacroce et al. 2008; Sbrana et al. 2019) (Fig. 1). Its volcanic activity began after 40 ka and sporadically continued with its most recent eruption in AD 1944. The Phlegraean volcanic district is a northeast-southwest aligned structure including the three volcanic fields of CF, Procida and Ischia (Orsi et al. 1996). The beginning of volcanism at the Ischia volcanic field, although not precisely defined, must have occurred earlier than 150 ka and its most recent eruption occurred in AD 1301–02 (Vezzoli 1988; de Vita et al. 2010; Sbrana et al. 2018). This volcanic field has been affected by a caldera collapse related to the Mount Epomeo Green Tuff eruption (55 ka; Brown et al. 2008, 2014 and references therein). The age of the beginning of volcanism at the CF volcanic field is also unknown, while its last eruption occurred in AD 1538. The timing of the inception of volcanism in the Procida monogenetic volcanic field is unknown but it last erupted at 23,624 ± 330 cal. years BP (Morabito et al. 2014 and references therein). The Phlegraean volcanic district is associated with a dense network of northeast-southwest and northwest-southeast, with subordinate north–south and east–west trending, tectonic and volcano-tectonic faults and fractures. These features correspond to the regional tectonic lineaments characterising the Campanian Plain and the adjacent Southern Apennines thrust belt (Orsi et al. 1996; Di Vito et al. 1999; Acocella 2010; Capuano et al. 2013; Vitale and Isaia 2014) (Fig. 1). Somma-Vesuvio volcanic complex, and CF and Ischia volcanic fields, the latter two affected by resurgent calderas, are still active (Santacroce et al. 2003). Moretti et al. (2013b) argued that similarities and differences among the Neapolitan volcanoes depend upon a common causal link, which is the interplay between the northwest-southeast normal and northeast-southwest transverse regional faults and a widespread deep volatile source derived from subducted metasediments. This hypothesis was based on geochemical and structural data, and on the fact that all these volcanoes are underlain by a major region of magma stagnation and gas fluxing at 8–10 km depth (Auger et al. 2001; Zollo et al. 2008; Dallai et al. 2011; Pappalardo and Mastrolorenzo 2012; Fedi et al. 2018). The different amounts of magma erupted over time likely result from the flow-rate of deep slab-derived fluids or melts through the mantle wedge. The efficiency of fluid transfer varies with the permeability of the overlying lithosphere, which increases at the intersection of the two main regional fault systems (Moretti et al. 2013b). This is the case of the

CF and Ischia volcanic fields, and the Somma-Vesuvio volcanic complex where ascent and accumulation of fluids at crustal levels is favoured, with consequent production and storage of large amounts of magma. These conditions have never been established at the Procida volcanic field, as it is located along the northeast-southwest transverse fault system.

3 Volcanic History of the Campi Flegrei Volcanic Field

3.1 Pre-Campi Flegrei Caldera Volcanism

The timing and frequency of the CF volcanic field activity prior to the CI eruption are not well defined, due to scarcity of exposures (Figs. 3 and 4). The deposits of this activity are discontinuously exposed mostly along the scarps bordering the CFc depression, in particular along the scarps of Monte di Procida and Cuma hills, the northern slopes of the Quarto and Soccavo plains, and the southern flanks of the Camaldoli and Capodimonte hills (Fig. 4). The discontinuity of exposures and scarcity of absolute age determinations do not permit the definition of the evolution of volcanism in time and space. Therefore, the products of this volcanism cannot be presented following a stratigraphic order. They include phonolitic to alkali-trachytic lava and pyroclastic sequences (Di Girolamo et al. 1984; Rosi and Sbrana 1987; Orsi et al. 1996; Pappalardo et al. 1999; Perrotta et al. 2006, 2010; Pabst et al. 2008; Scarpati et al. 2013; Tomlinson et al. 2012). Lavas crop out between the Acquamorta and Torregaveta beaches and along the western slopes of Monte di Procida, and also constitute the Punta Marmolite (Quarto) and Cuma (Fig. 4) lava domes. The Punta Marmolite dome was dated by K/Ar method at about 47 ka by Cassignol and Gillot (1982). Buried lavas have been encountered in drill cores and diggings for the underground railway network under the San Martino Hill in the city of Naples. Pyroclastic successions mostly include loose deposits intercalated with paleosols, varying from fine- to coarse-grained and lithified sequences. The loose deposits, mostly occurring in the Neapolitan-Phlegraean area, either represent distal facies or have been erupted from local vents, such as the two tuff cones older than CI, recently identified by Perrotta et al. (2010) at Miliscola (Fig. 4) and named Vitafumo and Miliscola volcanoes. Pyroclastic units, likely erupted from vents located at Ischia and Procida, are exposed at Monte di Procida, Cuma and San Severino (Orsi et al. 1996; Pappalardo et al. 1999). Sequences of lithified pyroclastic beds collectively known as the Ancient Tuffs (Tufi Antichi in Italian; Scherillo 1957) occur in the city of Naples, in particular in the San Martino-Capodimonte area. The proximal imprint of the sedimentological characteristics of these tuffs (presence of ballistic blocks, prominent sandwave structures) led Scarpati et al. (2013) to hypothesise that they were erupted from local vents. The age of these volcanoes is stratigraphically constrained by the $^{40}Ar/^{39}Ar$ age of 78.29 ± 0.35 ka yielded by the San Martino Lower Tephra, a pyroclastic fall that drapes the youngest Ancient Tuffs (San Sepolcro volcano; Scarpati et al. 2013). A several-metre-thick sequence of pyroclastic units, intercalated with paleosols, partly erupted from vents located to the north of the CFc, occurs along the sector of the caldera wall north of Quarto (Orsi et al. 1996) (Fig. 4). Its upper portion is correlatable with the Tufi di Torre di Franco (^{14}C ages > 42 ka; Alessio et al. 1973) described by Rittmann (1950b), cropping out at the base of the pyroclastic sequence of the Camaldoli Hill (Naples). A pyroclastic unit occurring in the lower portion of a sequence exposed along the northern scarps delimiting the CF depression yielded a $^{40}Ar/^{39}Ar$ age of about 60 ka (Pappalardo et al. 1999). Some of the units of this sequence relate to volcanism extending beyond the northern limits of the present CFc. A sequence of 11 pyroclastic units older than CI has been cored in the Ponti Rossi area and in the northern and eastern portions of the city of Naples (Pappalardo et al. 2002a).

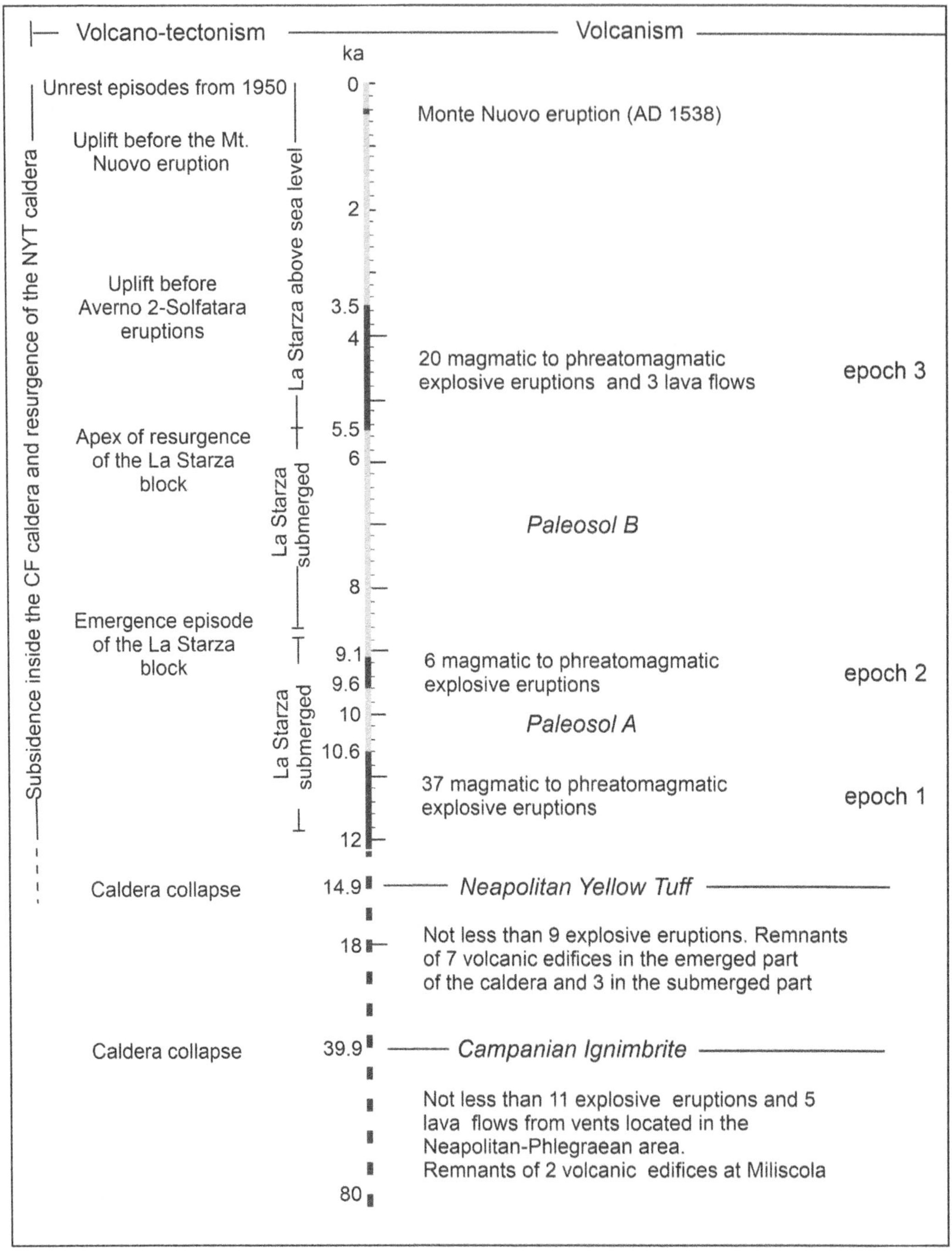

Fig. 3 Chronogram of the volcanic and deformation history of the Campi Flegrei volcanic field over the past 80 kyrs. Dark grey: periods of volcanic activity; light grey: periods of quiescence. Figure modified after Di Vito et al. (1999) and Orsi et al. (2004)

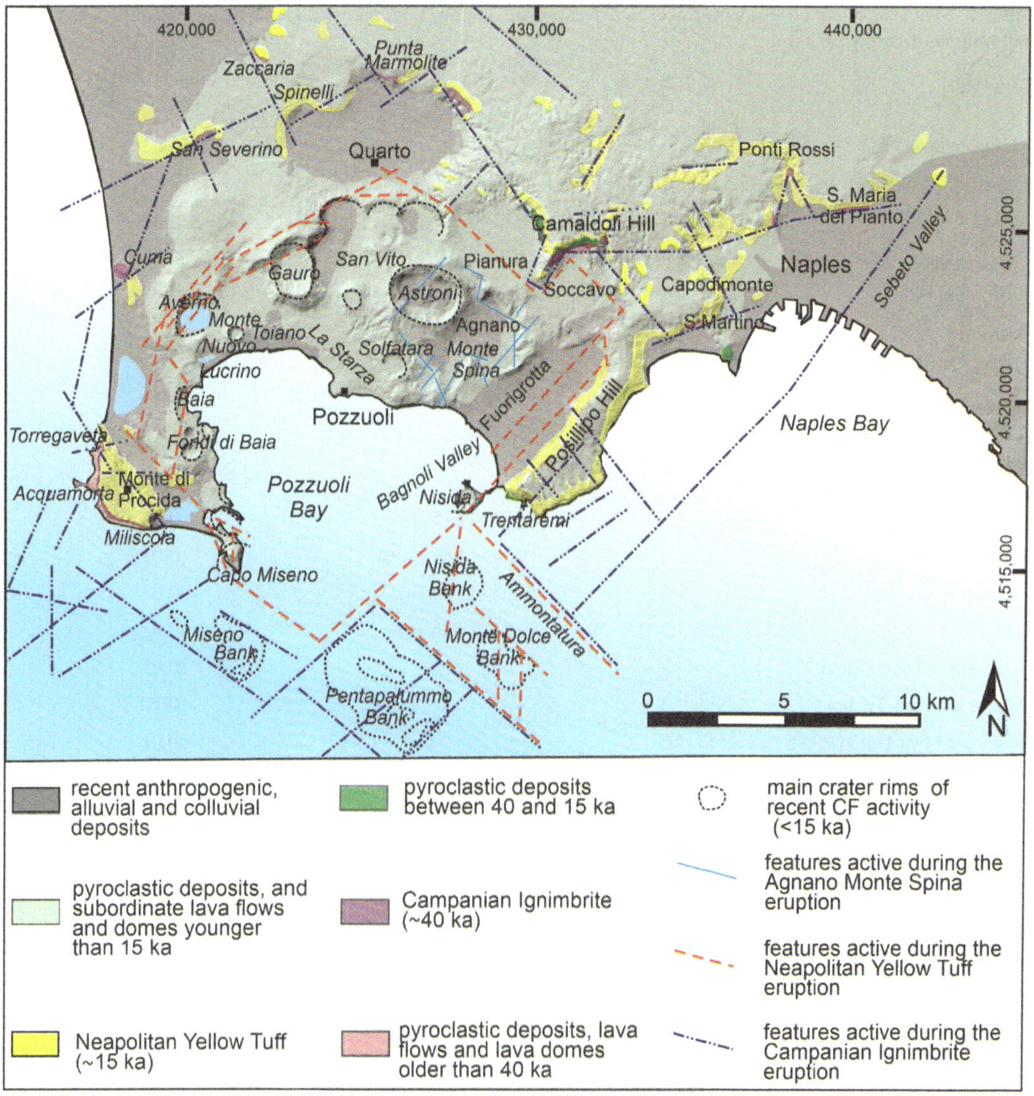

Fig. 4 Geological sketch map of the Campi Flegrei caldera

3.2 Volcanism of the Campi Flegrei Caldera

The present CFc formed through two major collapses related to the trachytic-to-phonolitic Campanian Ignimbrite (CI; ~40 ka) and the latitic-to-phonolitic Neapolitan Yellow Tuff (NYT; ~15 ka) eruptions (Orsi et al. 1992, 1996, 2004, and references therein). A third collapse between these two has been hypothesised by Albert et al. (2019). Orsi et al. (1996) and Di Vito et al. (1999) highlighted that, after each collapse, volcanism was confined within the collapsed area. These authors subdivided the volcanic and deformation history of the CFc according to the occurrence of the CI and NYT catastrophic eruptions and related caldera collapses, and post-NYT caldera volcanism and deformation (Fig. 3). The activity younger than NYT (Table 1) has been subdivided into three epochs of activity also based on geochronological constraints and the occurrence of two widespread paleosols. The most recent eruption took place as a single event in AD 1538 after about

Table 1 Synthesis of the eruptions younger than Neapolitan Yellow Tuff, with indication of the epoch of activity, age, and volume of erupted magma

Eruption	Epoch	Age (cal. years BP) References	Magma volume (DRE) km^3
Monte Nuovo		AD 1538	0.03
Epoch 3: Volcanism between 5.5 and 3.5 ka			
Nisida	3	3,213–4,188 (e)	0.02
Fossa Lupara	3	3,978–4,192 (c)	0.02
Astroni 7	3	4,098–4,297 (c)	0.07
Astroni 6	3		0.12
Astroni 5	3		0.10
Astroni 4	3		0.14
Astroni 3	3		0.16
Astroni 2	3		0.02
Astroni 1	3	4,153–4,345 (c)	0.06
Capo Miseno	3	3,259–4,886 (c, e)	0.02
Averno 2	3	3,895–4,205 (h)	0.07
Solfatara	3	4,181–4,386 (d)	0.03
Accademia Lava Dome	3		<0.01
Monte Olibano Tephra	3		<0.10
Solfatara Lava Dome	3		<0.01
Paleo-Astroni 3	3		0.02
M.te Olibano Lava Dome dome	3		<0.01
Santa Maria delle Grazie	3	4,382–4,509 (d)	<0.10
Agnano-Monte Spina	3	4,482–4,625 (b)	0.85
Paleo-Astroni 2	3	4,712–4,757 (c, f)	<0.30
Paleo-Astroni 1	3	4,745–4,834 (b, f)	0.05
Monte Sant'Angelo	3	4,832–5,010 (c)	<0.30
Pignatiello 2	3		<0.02
Cigliano	3		0.05
Agnano 3	3		0.19
Averno 1	3	5,064–5,431 (c)	<0.10
Agnano 2	3		<0.01
Agnano 1	3	5,266–5,628 (c)	0.02
Epoch 2: Volcanism between 9.6 and 9.1 ka			
San Martino	2	9,026–9,370 (c)	0.05
Sartania 2	2		<0.10
Pigna San Nicola	2	9,201–9,533 (a)	<0.30
Costa San Domenico	2		<0.10
Monte Spina Lava Dome	2		<0.01
Sartania 1	2	9,500–9,654 (c)	<0.10
Fondi di Baia	2	9,525–9,695 (c)	0.04
Baia	2		<0.01

(continued)

Table 1 (continued)

Eruption	Epoch	Age (cal. years BP) References	Magma volume (DRE) km^3
Epoch 1: Volcanism between 15 and 10.6 ka			
Porto Miseno	1	10,347–12,860 (e)	<0.10
Bacoli	1	11,511–14,154 (e)	0.20
Casale	1		<0.10
Pisani 3	1	10,516–10,755 (c)	<0.10
Pignatiello 1	1		<0.10
Montagna Spaccata	1		<0.02
Concola	1		<0.01
Fondo Riccio	1		<0.01
Pisani 2	1		<0.30
Pisani 1	1		<0.30
Soccavo 5	1		<0.10
Minopoli 2	1		<0.10
Paleo-San Martino	1		<0.10
Soccavo 4	1		<0.30
S4s3_2	1		<0.10
S4s3_1	1		<0.30
Soccavo 3	1		<0.10
Soccavo 2	1		<0.10
Paleo-Pisani 2	1		<0.30
Paleo-Pisani 1	1		<0.10
Pomici Principali	1	11,915–12,158 (g)	0.85
Gaiola	1		<0.10
Soccavo 1	1		0.50
Paradiso	1		<0.10
Minopoli 1	1		<0.10
Archiaverno	1	12,555–12,721 (c)	<0.10
Torre Cappella	1		<0.10
La Pigna 2	1		<0.10
La Pigna 1	1	12,749–13,110 (c)	<0.10
La Pietra	1		<0.10
Santa Teresa	1		<0.10
Gauro	1	12,721–15,511 (e)	0.50
Mofete	1		<0.10
Bellavista	1		<0.10

All eruption ages but that of Monte Nuovo are from Smith et al. (2011) and Bevilacqua et al. (2016). Original ages and dating methods are reported in: (a) Scandone et al. (1991), (b) de Vita et al. (1999), (c) Di Vito et al. (1999), (d) Isaia et al. (2009), (e) Di Renzo et al. (2011), (f) Passariello et al. (2010), (g) Lane et al. (2011), (h) Chap. Origin and Differentiation History of the Magmatic System Feeding the Campi Flegrei Volcanic Field (Italy) Constrained by Radiogenic and Stable Isotope Data

3,000 years of quiescence. The younger caldera is the portion still active of the entire CFc structure and its floor has been the site of intense volcanism and an ongoing deformation. The latter and its accompanying seismicity and variation in fumarole effluents, generate the bradyseismic events and the present unrest state of the volcano with the consequent high level of volcanic hazards (see Chap. Volcanic Hazard Assessment at the Campi Flegrei Caldera, Italy).

The volcanism of the CFc has erupted magmas varying in composition from shoshonite, to latite, to trachyte, to phonolitic trachyte. It has been fed by a complex system including a long-lasting deep reservoir at about 8–9 km depth and shallow magma chambers formed between 2 and 6 km depth before each eruption (Di Girolamo et al. 1984; Melluso et al. 2012; Orsi et al. 1995; Pappalardo et al. 1999, 2002a; Judenherc and Zollo 2004; Tonarini et al. 2009; D'Antonio et al. 2007; Pabst et al. 2008; Zollo et al. 2008; Arienzo et al. 2009, 2010, 2011, 2016; De Siena et al. 2018; Di Renzo et al. 2011, 2016; Pappalardo and Mastrolorenzo 2012; Forni et al. 2018a, b). Mantle-derived primitive magmas stagnate, differentiate and are partially contaminated by crustal rocks in the deep reservoir and then ascend to the shallow chambers where they are subject to differentiation processes such as fractional crystallisation, mixing and mingling. Details on the composition of the erupted magmas and the architecture of the magmatic feeding system are reported in Chaps. An Evolutionary Model for the Magmatic System of the Campi Flegrei Volcanic Field (Italy) Constrained by Petrochemical Data; Origin and Differentiation History of the Magmatic System Feeding the Campi Flegrei Volcanic Field (Italy) Constrained by Radiogenic and Stable Isotope Data; Tephrochronology and Geochemistry of Tephra from the Campi Flegrei Volcanic Field, Italy; Rheological Properties of the Magmas Feeding the Campi Flegrei Caldera (Italy) and Their Influence on Mixing Processes; Magma Chamber Dynamics at the Campi Flegrei Caldera, Italy.

3.2.1 Campanian Ignimbrite Eruption

The CI eruption is the largest eruption within the Mediterranean area over the past 200 kyrs and the first cataclysmic caldera-forming event of the CFc. The age of this eruption, due to its magnitude and the important role that its deposits play as a stratigraphic marker bed on a more than continental scale, has been determined by several methods on samples collected in various environments in both proximal and distal areas. ^{14}C and K/Ar determinations yielded ages ranging from 26,900 ± 700 to 42,000 ± 4,000 years (Alessio et al. 1971, 1973, 1974; Deino et al. 2004) and from 30.0 ± 3.5 to 38.5 ± 1.8 ka (Cassignol and Gillot 1982), respectively. $^{40}Ar/^{39}Ar$ ages range from 36.0 ± 1.0 to 39.85 ± 0.14 ka (Ramrath et al. 1999; De Vivo et al. 2001; Ton-That et al. 2001; Deino et al. 2004; Giaccio et al. 2017). The most recently defined $^{40}Ar/^{39}Ar$ dating result (39.85 ± 0.14 ka) is taken here as the age of the CI eruption.

The products of the CI eruption and their stratigraphic sequence as well as source area and occurrence of a syn-eruptive caldera collapse are the subjects of a more than two-century-long scientific debate begun with the article by Breislack (1778). Currently, the most accepted hypotheses within the scientific community are that the eruption took place in the CF, generated a complex pyroclastic sequence and was accompanied by a caldera collapse. The stratigraphic sequence includes a basal Plinian fallout (Rosi et al. 1999) overlain by various types of pyroclastic density current (PDC) deposits, breccias (e.g., Breccia Museo) and spatter agglutinates (e.g., Piperno) (Rosi et al. 1983, 1996; Rosi and Sbrana 1987; Orsi et al. 1996; Perrotta et al. 2006, 2010; Fedele et al. 2008). Fisher et al. (1993) estimated that the PDC deposits with a thickness up to 5 m were laid down over an area of about 3,000 km^2 (Fig. 5a). The fallout deposits covered an area of at least 2×10^6 km^2, reaching a distance in excess of 2,500 km from the source (Pyle et al. 2006; Marti et al. 2016; Smith et al. 2016) (Fig. 5b). Silleni et al. (2020), in calculating the magnitude

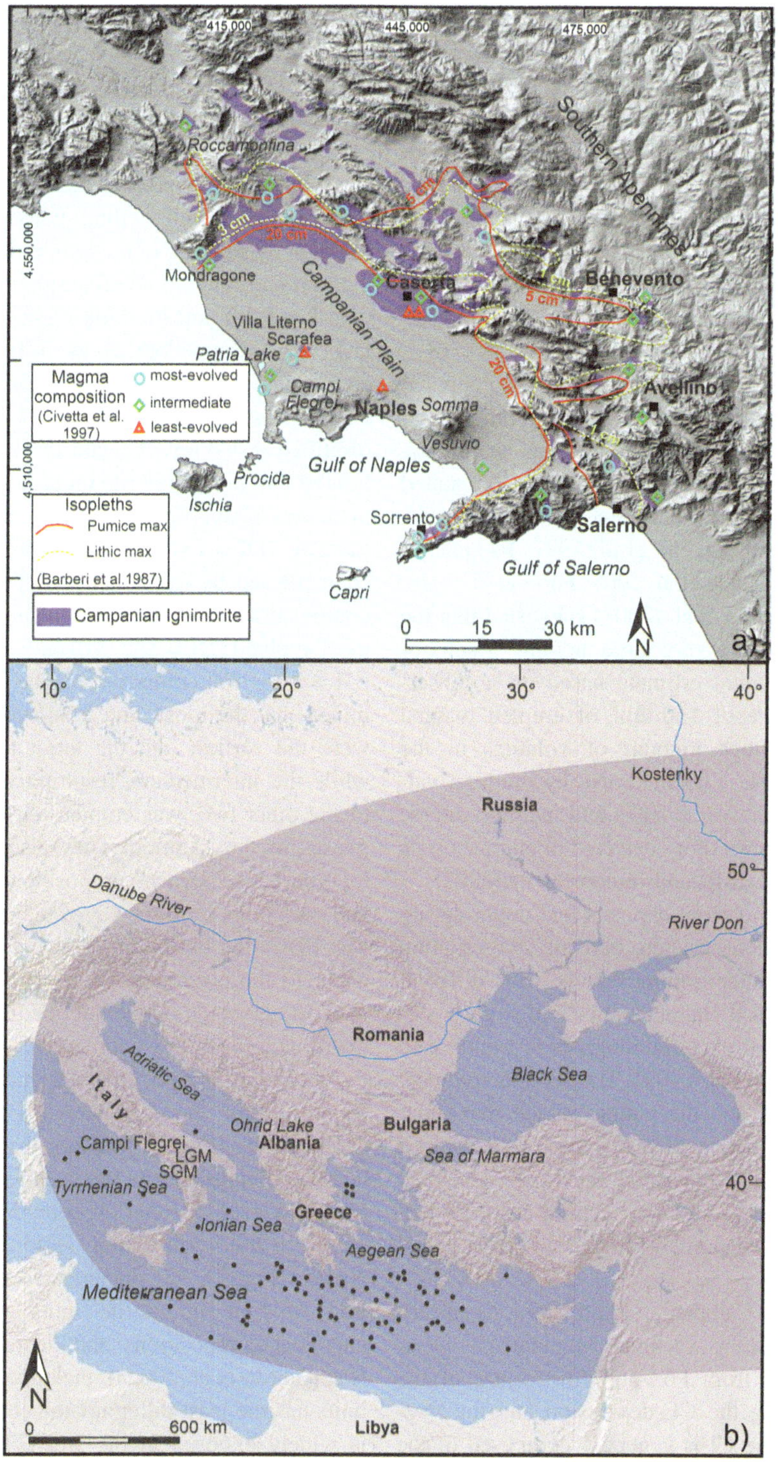

Fig. 5 Distribution of the Campanian Ignimbrite: **a** in the Campania area (modified after Barberi et al. 1978 and Civetta et al. 1997), **b** in distal settings. LGM: Lago Grande di Monticchio, SGM: San Gregorio Magno. Black dots locate deep-sea sediment cores including Campanian Ignimbrite distal tephra

of the CI eruption, estimated that the 0 m isopach for the ignimbrite encloses an area of about 6,000 km^2, similarly to what suggested by Thunell et al. (1979). The total volume of erupted magma has been estimated at different times using diverse methods, with estimates varying from the earliest evaluation of 60 km^3 dense rock equivalent (DRE) (Thunell et al. 1979) to the more recent values of 300 km^3 DRE (Fedele et al. 2007) and the similar ranges of 180–280 km^3 DRE (Costa et al. 2012) and 181–265 km^3 DRE (Silleni et al. 2020). The chemical composition of the erupted deposits varies from trachytic to trachyphonolitic and has been interpreted as resulting from a vertically zoned magma chamber (Di Girolamo et al. 1970; Barberi et al. 1978; Civetta et al. 1997; Pappalardo et al. 2002a; Fedele et al. 2008; Forni et al. 2016, 2018a, b). Fedele et al. (2007) suggested that the eruption emitted a very large amount of sulphur (10^{15} g; minimum estimate based on a conservative estimate of 150 km^3 of erupted magma DRE). The huge amount of volatiles in the erupted magma is testified also by the high values of void fraction (>70%) and the high degree of vesicles connectivity detected in pumice clasts (Moretti et al. 2019 and references therein).

The Plinian fallout deposits laid down during the earlier reconstructed part of the eruption include two superimposed units with different sedimentological characteristics and slightly different dispersal axes, although both trending to the east (Rosi et al. 1999). The sedimentological characteristics and the pattern of the maximum lithic and pumice isopleths of these beds suggest that the column from which they were deposited was erupted by a unique central vent located within the Neapolitan-Phlegraean area (Rosi et al. 1999).

Within the Campanian Plain, along its eastern margin, and in the Apennine Mountains as far as about 80 km from Pozzuoli (the centre of the area including the CF depression and the Pozzuoli Bay), the CI is generally composed of the basal Plinian deposits, if within its axis of dispersal, overlain by ignimbrite units. Di Girolamo (1970) and Barberi et al. (1978) carried out an analysis of the maximum pumice and lithic clasts of the PDC deposits in the Campanian Plain and the surrounding Apennine Mountains. Their results show a roughly northwest-southeast elongation of the isopleths (Fig. 5a). Fisher et al. (1993) recognised in these units the classical ignimbrite sequence described by Sparks et al. (1973), with a thin, fines-poor layer 1, overlain by the bulk of the ignimbrite called layer 2 and represented by the finer grained basal zone (sub-layer 2a) and the bulk of the deposit (sub-layer 2b). Cappelletti et al. (2003), although concurring with this subdivision, named the ensemble of layers 1 and 2a as Unconsolidated Stratified Ash Flow. Civetta et al. (1997) highlighted that the ignimbrite units as a whole range in composition from trachyte to phonolitic-trachyte (DI = 75 − 90) and each of them is characterised by one of three different chemical compositions: a least evolved (DI = 75 − 83), a most evolved (DI = 88 − 90) and an intermediate (DI = 84 − 87) composition. The authors determined that the most- and least-evolved magmas were the earliest and the latest to be erupted, while the intermediate, resulting from mingling of the other two, was erupted during the central phases of the eruption. They also found a correlation between each composition and distances and directions from the CF of the units (Fig. 5a). The PDCs fed by the most-evolved magma only occur to the north and south of the CFc, not to the east, those fed by the intermediate magma are distributed over the entire area, and those fed by the least-evolved magma have only been found within the Campanian Plain. Only at Mondragone (Figs. 2 and 5a), the intermediate-composition ignimbrite has been found overlying the most-evolved one. A core drilled at Ponti Rossi, in the north-eastern sector of the city of Naples (Fig. 2), has documented the superposition of four ignimbrite units with different chemical compositions and degree of welding (Pappalardo et al. 2002a); the uppermost of these units has the least differentiated composition and is widely exposed in the area. Flow directions determined by means of anisotropy of magnetic susceptibility studies on the ignimbrites show that, at the time of deposition, they were not flowing outward from the vent area, but, instead, were moving as gravity flows even down slopes facing the vent area (Fisher et al. 1993; Ort et al. 2003). The authors explained this apparent contradiction by assuming that the transport system had flow properties distinct from the deposition system. The PDCs may have been initially extremely diluted and turbulent, moving in an expanded state over the landscape, and as they transported particles to the site of deposition, they became density stratified, forming the depositional system consisting of gravity currents. The distribution of the ignimbrites with various composition in distal areas together with anisotropy of magnetic susceptibility flow direction data imply that the two earlier PDCs travelled long distances, passing over topographic barriers >1,000 m high and flowed $\sim$35 km over sea water, while the later and least differentiated ones only travelled within the Campanian Plain (Fisher et al. 1993; Civetta et al. 1997; Ort et al. 2003). The very low particle concentration within the two earlier erupted currents, together with the huge amount of extruded magma, its high sulphur content and the high values of void fraction and degree of vesicles connectivity of the pumice were taken by Moretti et al. (2019) as evidence that the eruption dynamics was dictated by a magma reservoir highly enriched in volatiles, rather than fluid entrainment from hydrothermal bodies or seawater.

Scattered outcrops of a sequence including a basal fallout bed (occurring only in a few exposures mostly at the eastern periphery of the CF) overlain by variably welded ash deposits, spatter beds, coarse lithic breccias, and large scoria block deposits can be found in various locations. In particular, they occur along the slopes delimiting the Neapolitan-Phlegraean lowland, including CF and the southern part of the city of Naples, in the walls of some quarries excavated in the Campanian Plain north of CF, and at the island of Procida (Fig. 5a). The rocks of these scattered exposures, attributed by some authors (Lirer et al. 1991; Scandone et al. 1991) to local eruptions that occurred at various times, are generally interpreted as being correlated to each other and representing the proximal facies of the CI eruption deposits, although their general stratigraphic sequence has been variably reconstructed (Rosi et al. 1983, 1996, 1999; Rosi and Sbrana 1987; Orsi et al. 1996; Perrotta et al. 2006, 2010; Fedele et al. 2008). The outcrops in the CF are located along slopes bordering the Phlegraean lowland toward the southwest (Monte di Procida), the west (Cuma), and the north (San Severino, Spinelli, Punta Marmolite, northeast of the Quarto Plain, between the Pianura Plain and Camaldoli, north of the Soccavo Plain) (Fig. 2). Those in the Naples area are exposed along the slopes delimiting to the north the morphologically lower part of the city (San Martino, Montesanto, Fontanelle, Ponti Rossi, and Santa Maria del Pianto at Poggioreale) (Fig. 4). Within the Campanian Plain are outcrops at about 15 km from Pozzuoli, namely at Patria Lake and Scarafea (Fig. 5a). Next to Patria Lake, a PDC bed is topped by a breccia layer interpreted by Rosi et al. (1983) as the likely product of a phreatic phase, according to the abundance of lithic clasts. In a quarry at Scarafea, a breccia overlies a sintered ignimbrite and likely filled a channel radial to the CF that was scoured into the underlying still-soft ignimbrite by the flow from which it was deposited (Rosi et al. 1991, 1996; Ort et al. 1999, 2003). The sequence contains particularly large and poorly vesiculated juvenile spatter clasts, up to 150 cm in size, in a non-graded flow unit. At Procida, proximal facies of the CI sequence including lithic breccias and spatter beds are exposed between 9 and 14 km southwest of Pozzuoli (Fedele et al. 2008) (Fig. 5a).

Ash correlatable to CI occurs in diverse geological settings at many distances up to more than 2,500 km from the source area (Fig. 5b). In southern Italy, it has been cored in lacustrine sequences at Lago Grande di Monticchio (Wulf et al. 2004; Tomlinson et al. 2012) and San Gregorio Magno basin (Munno and Petrosino 2007) (Fig. 5b). CI ash extensively occurs in sediments of the Eastern Mediterranean Sea (Tyrrhenian, Adriatic, Ionian, and Aegean seas), where it forms the Y-5 tephra marker (e.g., Keller et al. 1978; Thunell et al. 1979; Cornell et al. 1983; Paterne et al. 1988; Lowe et al. 2012; D'Antonio et al. 2016) and of the Sea of

Marmara (Makaroglu et al. 2020) (Fig. 5b). It has also been found in Albania, at Lake Ohrid (Caron et al. 2010), in peat basins in Greece (Seymour et al. 2004; Lowe et al. 2012), in caves in Bulgaria, Greece and Libya (Vitaliano et al. 1981; Douka et al. 2014; Lowe et al. 2012; Ivanova et al. 2016), in southeast Romania (Lower Danube basin) (Veres et al. 2013), and in archaeological sites in southern Russia (Kostenky along the Don river) and in central Europe (Pyle et al. 2006; Lowe et al. 2012).

Vent geometry and evolution of the CI eruption, as well as detailed correlation between proximal and distal sequences, are still unresolved topics. Di Girolamo (1970) stressed the uniqueness of the PDC deposits of the CI and left open the possibility that they had been generated by a fissure- or a central-vent eruption. The latter might have generated a caldera collapse in the area of Acerra or around the Patria Lake, both in the Campanian Plain north of Naples (Fig. 2). Barberi et al. (1978) and Di Girolamo et al. (1984), based on the maximum lithic and pumice clast isopleths pattern (Fig. 5a), suggested that the PDC deposits of the CI sequence were erupted through a northwest-southeast fracture system extending between the city of Naples and Villa Literno in the Campanian Plain (Fig. 2). Scandone et al. (1991) proposed that the CI was the result of multiple eruptions in the Acerra area (Fig. 2). A large number of authors, based on volcanological, sedimentological, petrological and flow direction data, locate the CI vent area in the CF and imply that the eruption was the largest known caldera-forming event in the CF volcanic field (e.g., Rosi et al. 1983, 1996, 1999; Rosi and Sbrana 1987; Barberi et al. 1991; Fisher et al. 1993; Orsi et al. 1996; Civetta et al. 1997; Pappalardo et al. 2002a; Ort et al. 1999, 2003; Perrotta et al. 2006, 2010; Fedele et al. 2008; Scarpati et al. 2013; Silleni et al. 2020). It is suggested here that, although the earlier Plinian phase was erupted through a central vent in the CF, the subsequent phases were fed through faults and fractures and/or discrete vents along them. These features, portions of regional structures in the Neapolitan-Phlegraean area, acted as the faults along which the CI caldera collapse occurred, and extended toward the north into the surrounding plain and toward the southwest into the area of the present Procida Island. This hypothesis contributes to the understanding of both facies and component-size distribution and, as will be discussed later in the section devoted to caldera collapses, of areal distribution of the PDCs of different chemical composition, and of discrepancy between the volume estimates of erupted magma and caldera collapse.

The effects of the CI eruption on climate and hominid life have also been investigated. Fedele et al. (2007) highlighted that the eruption took place at the beginning of one of the Heinrich Events cold periods, namely Heinrich Event 4. It emitted a very large amount of ash and sulphur, comparable to the amounts emitted by the Toba and Bishop Tuff super-eruptions, and should have induced a global cooling of 3–4 °C for at least 2–3 years. The coincidence of Heinrich Event 4 with both widespread ash deposition and probable dramatic cooling related to the CI eruption has been interpreted in terms of effects on hominid life. Golovanova et al. (2010) suggested that Neanderthals left portions of Europe that were inhabited by anatomically modern humans coming from Africa. Fedele et al. (2007) hypothesised that the CI eruption induced more gradual cultural and evolutionary changes that allowed the anatomically modern humans to outcompete and finally supplant Neanderthals. Lowe et al. (2012) proposed that the extinction of the Neanderthals in Europe was not related to the CI eruption, but rather to anatomically modern humans with whom they had already interacted prior to the eruption. Neanderthals, earlier saved because they lived in small groups with a high mobility, were later supplanted by anatomically modern humans.

3.2.2 Volcanism Between Campanian Ignimbrite and Neapolitan Yellow Tuff Eruptions

Volcanism between the CI and NYT eruptions was confined within the CI caldera floor and characterised by explosive, mainly phreatomagmatic eruptions, fed by magmas varying in composition from trachyte to phonolitic trachyte

(Pabst et al. 2008; Pappalardo et al. 1999). A stratigraphic sequence of the products of this activity is far from being reconstructed due to scarcity and incompleteness of proximal exposures, despite some insights arise from both drilled sequences and distal tephra. Therefore, this activity and its timing, as well as the age of resumption of volcanism after the CI eruption is not well defined.

Volcanic rocks dated between the CI and NYT eruptions mostly occur along the scarps of the CI caldera topographic rim, in the central and eastern part of the city of Naples, and along the north-western and south-western slopes of the Posillipo Hill and the western slopes of the Monte di Procida morphological high (Fig. 4).

Among the pyroclastic sequences, generally separated by paleosols, the best known is the one occurring at the mouth of the Verdolino Valley, to the southeast of the Camaldoli Hill, and named the Tufi Biancastri (Italian for Whitish Tuffs) by Rittmann (1950b) and later detailed by Orsi et al. (1996). Pappalardo et al. (1999) reported $^{40}Ar/^{39}Ar$ ages of 30.3 ± 0.2 and 14.6 ± 0.6 ka for the lowermost and uppermost units of the sequence, respectively. Calcaterra et al. (2007) named a pyroclastic sequence, exposed along the western slopes of the Camaldoli Hill facing the Pianura Plain, the Masseria del Monte Tuff. This tuff, which lies above the CI proximal sequence, is composed of a poorly lithified, strongly laminated greyish-white sequence of plane-parallel to cross-stratified ash beds with pumice layers and lenses. The sequence, attaining a maximum thickness of 50 m, was interpreted as the remnant of a tuff cone whose vent was located in the area of Soccavo. Albert et al. (2019) hypothesised that a large-magnitude eruption with an associated caldera collapse occurred at 29.3 ± 0.7 ka and produced a voluminous pyroclastic deposit, which they named Masseria del Monte Tuff. Based on chemical composition and age determinations, the authors define the tuff by correlating one of the units forming the Tufi Biancastri sequence as the proximal deposit and the widespread Y-3 marine tephra layer widely occurring in the Mediterranean sediments (Keller et al. 1978; Thunell et al. 1979) as the distal counterpart. Lack of correlation by Albert et al. (2019) of their Masseria del Monte Tuff with the homonymous unit previously defined by Calcaterra et al. (2007) raises a stratigraphic problem that still needs to be solved, despite the first attempt to do so presented in Chap. Tephrochronology and Geochemistry of Tephra from the Campi Flegrei Volcanic Field, Italy. Albert et al. (2019) used mathematical modelling to estimate that the eruption ejected about 17 km^3 (DRE) of magma and had a VEI of 6. The lack of both thick and widespread pyroclastic deposits and physical evidence of a caldera were explained by the authors as due to unusual dynamics of the eruption and to the later collapse within the area of the caldera related to the NYT eruption, respectively. The remnants of the Trentaremi volcano, an edifice built in two different phases separated by a hiatus, whose products were recently $^{40}Ar/^{39}Ar$ dated at 21.64 ± 0.28 ka (Scarpati et al. 2013), are exposed along the Posillipo cliff facing the Fuorigrotta Plain (Figs. 2 and 4). Dark scoriaceous products likely erupted by the neighbouring Torregaveta volcano and belonging to the same time interval are exposed along the Monte di Procida coast. The age of the Solchiaro volcano, the latest eruptive centre of the Procida volcanic field, has been recently assessed at 23,624 ± 330 cal. years BP by calibrating a previous ^{14}C age (Morabito et al. 2014 and references therein).

Many inland boreholes (Orsi et al. 1996; Pappalardo et al. 1999) have also penetrated rocks of this age. The rocks exposed outside the CI caldera are above sea level, while those inside the structure are now below sea level. The majority of them (Figs. 2 and 4) were produced by explosive eruptions dominated by hydromagmatic phases with subordinate magmatic events. Their sedimentological characteristics suggest that they were erupted by local vents within the CI caldera floor and deposited in subaerial environments (Figs. 2 and 4).

The very rich distal tephra record of the time interval comprised between the CI and NYT eruptions implies that CI caldera activity during this period was much more intense than the very scattered proximal outcrops might suggest. At least ten Phlegraean tephra layers have been found in the Lago Grande di Monticchio lacustrine sequence (Wulf et al. 2008, 2012) located at about 130 km east of Pozzuoli. Their proximal counterparts have not been clearly identified, apart from TM-15 and TM-10 layers that have been correlated to the Masseria del Monte Tuff, also related to the marker Y-3 by Albert et al. (2019; ~29 ka), and the Lagno Amendolare pyroclastic sequence, respectively (Tomlinson et al. 2012). Therefore, this lacustrine sequence shows no evidence of Phlegraean tephra blowing in that direction, which is roughly the dominant wind direction in the area, between the CI and Y-3. This could imply that for about 9 kyrs after the CI catastrophic event, the CF system did not produce large eruptions. Lagno Amendolare is a Phlegraean explosive event named after the Vesuvian Lagno Amendolare valley where these products have been recognised, with a ^{14}C age of 15,439 ± 190 cal. years BP (see Damaschke et al. 2013). Distal tephra possibly related with this event is widespread and has been found in the Adriatic Sea (Bourne et al. 2010) about 250 km northeast of Pozzuoli. Other distal tephra, generically ascribed to the Tufi Biancastri sequence, has been identified at Lake Accesa (Magny et al. 2006) about 350 km northwest of Pozzuoli.

3.2.3 Neapolitan Yellow Tuff Eruption

The NYT eruption, the second largest eruption of the CF volcanic field, is the last cataclysmic caldera-forming event of the CFc (Orsi and Scarpati 1989; Orsi et al. 1991a, 1992, 1995, 1996). The age of this eruption has been defined by various authors and with different methods because, similarly to the CI, its widely dispersed products are an important stratigraphic marker bed at continental scale. The earliest ages were determined by ^{14}C and K/Ar methods and are synthesised in Deino et al. (2004). Bronk Ramsey et al. (2015) performed a Bayesian modelling of the ^{14}C ages that returned a best estimate of 14,194 ± 172 cal. years BP. The most recent ^{40}Ar/^{39}Ar age determinations yielded similar values of 14.9 ± 0.4 ka (Deino et al. 2004) and 14.5 ± 0.5 ka (Galli et al. 2017). The genesis of the NYT deposits has long been debated in the literature. They were interpreted by some authors as the fallout products of a single eruption (Scherillo and Franco 1960; Lirer and Munno 1975; Di Girolamo et al. 1984), while by others as resulting from different events that occurred in shallow marine water (Rittmann 1950a, b; Rosi et al. 1983; Rosi and Sbrana 1987). Orsi and Scarpati (1989) first interpreted the NYT as the products of a complex eruption with varying dynamics. Orsi et al. (1991a, 1992, 1995) and Wohletz et al. (1995) presented a more detailed reconstruction of the complex stratigraphic and chemostratigraphic NYT sequence. These authors defined some parameters of the NYT eruption that will be discussed in the following and include chemical composition and volume of the erupted magma, architecture of the magmatic system, eruption dynamics and timing and relations among eruption dynamics, magma chamber withdrawal, and timing of a syn-eruptive caldera collapse.

The complex pyroclastic sequence generated by the NYT eruption includes two members (Orsi and Scarpati 1989; Orsi et al. 1991a, 1992, 1995; Wohletz et al. 1995). The Lower Member and Upper Member have different textural and sedimentological characteristics, as well as dispersal and magma composition, suggesting that the eruption dynamics and thus the depositional mechanisms changed during the course of the event. The eruption was phreatoplinian-to-phreatomagmatic and a major change in its dynamics occurred in conjunction with caldera collapse. The boundary between the two members, defined by a sharp change in textural characteristics and attitude of the deposits, shows no evidence of a long time-break between their emplacements. The cumulative eruption duration is estimated at one to several days by Wohletz et al. (1995).

The Lower Member, comprising a sequence of interlayered phreatomagmatic and minor magmatic tephra layers, was generated by the largest known trachytic phreatoplinian eruption (Orsi et al. 1992). It includes 13 layers whose characteristics vary with distance from the vent.

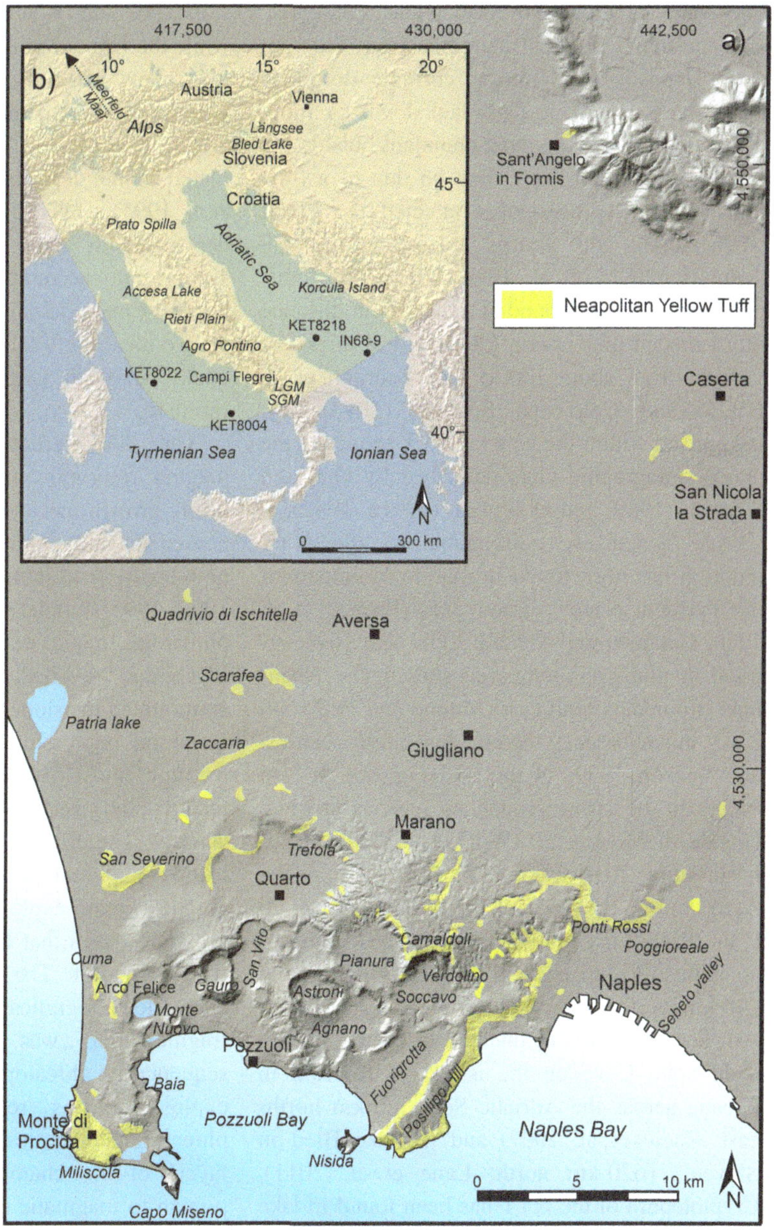

Fig. 6 Distribution of the Neapolitan Yellow Tuff: **a** in the Campania area (modified after Orsi et al. 1992), **b** in distal settings. LGM: Lago Grande di Monticchio; SGM: San Gregorio Magno; black dots locate deep-sea sediment cores including Neapolitan Yellow Tuff distal tephra

The fine-grained phreatoplinian deposits are very rich in accretionary lapilli, implying the presence of a large amount of steam, and were laid down by upstream-erosive PDCs in proximal areas and ash fallout in both proximal and distal areas. In contrast, the coarse-grained fallout beds were generated by magmatic explosions forming sustained columns and downwind-dispersed plumes containing pumice with subordinate fine ash and steam that facilitated formation of accretionary lapilli. The Lower Member has been found at a maximum distance of 35 km from the vent area, at Sant'Angelo in Formis at the foot of the Apennine Mountains (Fig. 6a). Its thickness varies from 11 to 1.25 m.

The Upper Member is composed of a sequence of deposits from variably diluted PDCs produced by phreatomagmatic explosions.

The characteristics of these deposits vary laterally and along different azimuths from the vent area. Deposits from denser currents dominate inside the CFc, while beds laid down by more dilute currents are the most abundant outside the caldera depression. This may be due to a flow transformation process that affected the PDCs when hitting the scarps bordering the CF depression (Orsi et al. 1992). Deposits of this member have been found at a maximum distance of 14 km, in the Caserta Plain, with a thickness varying from about 100 to 7 m. Scarpati et al. (1993) and Cole and Scarpati (1993) also recognised within the general NYT sequence the major stratigraphic units described by Orsi and Scarpati (1989) and Orsi et al. (1991a, 1992).

Ash genetically correlatable to the NYT eruption has been found in marine sediments of the Tyrrhenian and Adriatic seas (Paterne et al. 1988; Calanchi and Dinelli 2008) and is recognised in numerous terrestrial sites in the Apennine Mountains both east (Munno and Petrosino 2007 and references therein) and north (Narcisi and Vezzoli 1999) of the CF (Fig. 6b). It also occurs in lacustrine sequences such as at Lago Grande di Monticchio (Wulf et al. 2004; Tomlinson et al. 2012 and references therein), in the Agro Pontino (Eisner et al. 1986) and in the Rieti Plain (Ricci Lucchi and Vigliotti 1998) (Fig. 6b) at about 130 km east, 130 km northwest, and 200 km north of the source area, respectively. Ash layers 10–30 cm thick have been found in Vela Spila Cave on the island of Korcula in Croatia across the Adriatic Sea (320 km northeast; Radic et al. 2007) and at Lake Bled in Slovenia (620 km north; Lane et al. 2011). Cryptotephra of the NYT has been found in lake sediments at Meerfelder Maar in the Eifel region of Germany ($\sim$1,200 km north; Lane et al. 2015) at Längsee in Austria ($\sim$660 km north; Schmidt et al. 2002), at Prato Spilla "C" site ($\sim$500 km north-northwest; Davies et al. 2002) and Lake Accesa ($\sim$360 km northwest; Magny et al. 2006) in northern Italy (Fig. 6b).

Not less than 10 cm of Lower Member covered an area of about 1,000 km^2, including the Pozzuoli and Naples bays, the Campanian Plain between Naples and Caserta and the western slopes of the Apennine Mountains (Fig. 6a), while the Upper Member deposits were dispersed over about 250 km^2 (Wohletz et al. 1995). The volume of erupted magma, ranging in composition from alkali-trachyte to latite, has been estimated at not less than 40 km^3 DRE (Wohletz et al. 1995). This estimate was based only on the deposits with a minimum thickness of 10 cm. Taking into account also the NYT ash layers found in many localities up to 660 km from the source area, the total volume of erupted magma could be much larger than, even double, this estimate.

The NYT eruption tapped a three-layer magma reservoir that had undergone a step-filling growth mechanism by repeated arrivals of new melts, and was likely triggered by influx of a new trachytic-to-latitic magma batch (Orsi et al. 1992, 1995; Wohletz et al. 1995). It began with phreatomagmatic explosions generating dilute PDCs that travelled as far as about 35 km and surmounted the slopes bordering the CF morpho-structural depression. These explosions tapped the uppermost volatile- and likely bubble-rich magma and were followed by a pause in the activity. The eruption then resumed with significantly different dynamics. Highly energetic phreatoplinian explosions were produced with pressure waves that likely flushed the water out of the conduit. The following magma did not encounter a significant amount of water and its fragmentation was mostly magmatic. This sequence of phreatoplinian to mostly magmatic explosions was repeated six times. Each phreatoplinian explosion tapped all three magma layers of the chamber, whereas the phreatoplinian to magmatic explosions only drained the intermediate trachytic layer. Increase in withdrawal depth has been attributed to a dramatic pressure decrease on the magma beneath the conduit and/or a large increase in eruptive mass flux, both consequences of the efficiency of the water/magma interaction during phreatoplinian explosions. A time break occurred after this central-vent phase of the eruption and deposition of the Lower Member. Activity resumed, probably in relation to the beginning of the caldera collapse, through multiple vents located along the caldera faults. During this phase, explosions generated variably dilute PDCs that laid down the Upper Member sequence. Cole and Scarpati (1993) favoured the idea that the Upper Member was erupted by a single vent rather than by a ring fracture and multiple vents.

3.2.4 Post-Neapolitan Yellow Tuff Volcanism

Very intense volcanism followed the NYT eruption and was mostly controlled by the deformation resulting from both the NYT caldera collapse and the ongoing resurgence of its floor (Orsi et al. 1996, 1999a, b, 2004; D'Antonio et al. 1999; Di Vito et al. 1999, 2016; Isaia et al. 2009, 2019) (Table 1; Figs. 3 and 4). Many authors have presented diverse hypotheses on its evolution through time. Di Vito et al. (1999) first subdivided this volcanism in three epochs of clustered activity separated by two significant periods of quiescence during which two widespread paleosols developed, and followed after a long hiatus by the AD 1538 Monte Nuovo eruption (D'Oriano et al. 2005; Piochi et al. 2005; Guidoboni and Ciuccarelli 2011; Arzilli et al. 2016; Di Vito et al. 2016). This reconstruction was later improved with recognition of more eruptions, second-order pauses in the volcanic activity and determination of new absolute ages (Orsi et al. 2004, 2009; Isaia et al. 2009; Di Renzo et al. 2011; Fedele et al. 2011; Smith et al. 2011). The most recent review on this volcanic activity (Smith et al. 2011) presents a synthesis of the chronostratigraphic sequence highlighting the still controversial points. The inferred location of the vents active during each of the recognised epochs of activity are shown in Fig. 7.

During each epoch of the post-NYT activity (Table 1; Fig. 3), eruptions occurred at mean time intervals of tens of years. They were mostly explosive and only subordinately effusive, although it is possible that poorly dispersed products of some effusive eruptions could have been partially hidden or dismantled by the following activity. The latter produced small-volume lava domes and lava flows. The explosive events were dominated by phreatomagmatic phases and characterised by alternation of phreatomagmatic and magmatic explosions, although occurrence of contemporaneous phreatomagmatic and magmatic fragmentation dynamics has also been documented (Dellino et al. 2004). They generated dilute and turbulent PDCs and particles fallout. Some events that had peculiar characteristics or can be taken as representative of recurrent phenomena, will be illustrated below in detail. These events also merit particular attention in a volcanic hazard assessment. Among them, the Averno 2 eruption (Di Vito et al. 2011) and the growth of the Astroni volcano (Isaia et al. 2004; Tonarini et al. 2009) were characterised by migration of the vent over a limited area. Similarly, Pistolesi et al. (2017) suggested a syn-eruptive vent migration for the Baia-Fondi di Baia eruptions at the onset of epoch 2. Averno 2 and Solfatara eruptions likely were contemporaneous events with vents located several kilometres apart (Isaia et al. 2009; Pistolesi et al. 2016). The Astroni volcano was built through 7 eruptions in a short time span (Di Vito et al. 1999; Isaia et al. 2004; Tonarini et al. 2009). The Agnano-Monte Spina eruption, the largest magnitude event of epoch 3 of the post-NYT activity, was accompanied by the volcano tectonic collapse of the present Agnano Plain (de Vita et al. 1999) (Figs. 3 and 4). Decades of ground deformation prior to the Monte Nuovo eruption are well documented (Parascandola 1946; Guidoboni and Ciuccarelli 2011; Di Vito et al. 2016).

One open problem is whether the Monte Nuovo eruption is a precursor of a new epoch of activity. According to the current knowledge of the post-NYT activity, there is no evidence of any temporal outliers like the Monte Nuovo event during the quiescent periods between epochs 1, 2 and 3. Furthermore, during each of the three epochs, the eruptions occurred at mean time intervals of tens of years, but about 500 years of quiescence have followed the Monte Nuovo event. Therefore, it is likely that the Monte Nuovo eruption is an isolated, but unique, event.

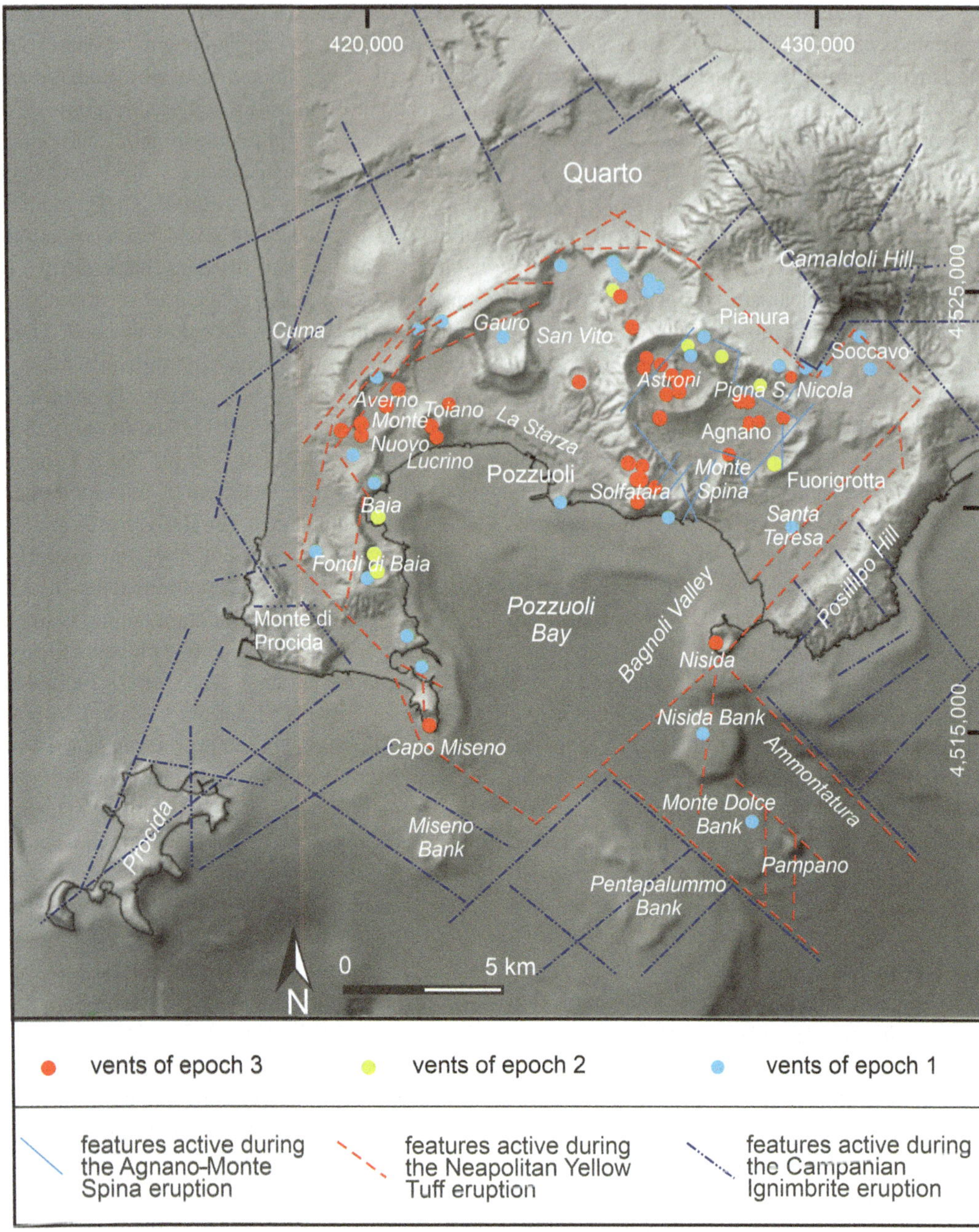

Fig. 7 Vents active during each of the three epochs of activity over the past 15 kyrs at the Campi Flegrei caldera. Figure modified after Orsi et al. (2009)

Epoch 1

Epoch 1 of activity began after the NYT eruption (~15 ka) and lasted until ~10.6 ka (Table 1). The oldest dated eruption (Gauro), whose deposits are not the stratigraphically lowermost of the post-NYT sequence, yielded a ^{14}C age of 12,721–15,511 cal. years BP (Table 1 and references therein; Fig. 3). At that time, all but the northern portion of the floor of the NYT caldera was invaded by the sea, which was about

90 m lower than at present (Lambeck et al. 2011). This epoch generated not less than 30 trachybasaltic-to-trachytic, magmatic to phreatomagmatic explosive eruptions that emitted from 0.001 to 0.5 km^3 of magma DRE, with the only exception being the Pomici Principali (Lirer et al. 1987b) that erupted 0.85 km^3 of magma DRE (Table 1 and references therein; Fig. 3). The majority of the eruptive vents were located along the marginal faults of the NYT caldera (Fig. 7). The pyroclastic deposits of these eruptions were dispersed within the floor of the caldera and out to a few kilometres beyond its rim (Fig. 8). The Pomici Principali eruption has been investigated by many authors (for an exhaustive reference list see Di Vito et al. 1999). It occurred at 11,915–12,158 cal. years BP and extruded 0.85 km^3 DRE of magma (Smith et al. 2011) (Table 1; Fig. 8). The eruption began with a phreatomagmatic vent-opening phase, continued with mainly magmatic and subordinate phreatomagmatic explosions, and ended with dominantly phreatomagmatic explosions. The magmatic explosions generated Plinian columns that laid down fallout beds dispersed to the north and northeast with a thickness of 20 cm at distances of more than 30 km (Fig. 8; Orsi et al. 2004), and PDCs that surmounted the CFc marginal scarps to the north. Only the PDCs of the Soccavo 1 Tephra were dispersed in the plain north-westward of the caldera margin. Many of the later eruptions of this epoch laid down PDC deposits over the central part of the present city of Naples and fallout deposits over its western part (Orsi et al. 2004, 2009).

A quiescent period, which lasted about 1.0 kyr, followed this epoch of intense volcanic activity. During quiescence, the earlier of the two widespread paleosols (Paleosol A) formed in the emerged portion of the caldera floor (Fig. 3) (Di Vito et al. 1999).

Epoch 2

Epoch 2 of activity lasted between ~9.6 and ~9.1 ka and generated 8 explosive eruptions that emitted volumes of trachybasaltic-to-trachytic magmas varying between <0.1 and 0.3 km^3 DRE (Table 1 and references therein; Fig. 3). The vents of all these eruptions were located along portions of the marginal faults of the NYT caldera, with the majority in the north-eastern sector, and only those of the Baia and Fondi di Baia eruptions in the western sector (Fig. 7). The deposits of these eruptions were mostly distributed across the caldera floor (Fig. 9) (Orsi et al. 2004). The largest event of this epoch was the Pigna San Nicola eruption at 9,201–9,533 cal. years BP and extruded 0.1–0.3 km^3 DRE of magma (Di Vito et al. 1999; Orsi et al. 2004, 2009; Smith et al. 2011) (Table 1). The eruption was characterised by alternation of magmatic and phreatomagmatic explosions that generated a sequence of pumice lapilli and subordinate ash fallout beds, surmounted by fine- to coarse-ash sandwave PDC deposits (Di Vito et al. 1999).

At least 3.5 kyrs of quiescence followed this epoch of activity. During this period, the sea level rose, invaded the area presently occupied by the Fuorigrotta, Agnano, San Vito and Toiano plains (Fig. 4), and partly eroded the volcanoes formed around the plains. Contemporaneously the thick and widespread Paleosol B formed in the emerged part of the CFc (Di Vito et al. 1999) (Fig. 3).

Epoch 3

This epoch of activity lasted from ~5.5 to ~3.5 ka, covering Eneolithic and part of the Bronze Age. It included 25 explosive and 3 effusive latitic-to-phonolitic eruptions, all within the NYT caldera, with a mean frequency of one every 50 years (Table 1 and references therein; Fig. 3). The explosive eruptions varied from magmatic to phreatomagmatic and produced fallout and PDC deposits mostly dispersed in the north-eastern part of the CFc and a few kilometres beyond the caldera rim (Orsi et al. 2009) (Fig. 10). Orsi et al. (2009), based on the magnitude of these eruptions as well as on the parameters upon which it depends, subdivided them according to a threefold (small, medium, large) classification. The large class only includes the Agnano-Monte Spina eruption. Most of the eruptive vents were located in the north-eastern sector of the NYT caldera floor; conversely, the vents of the two Averno and

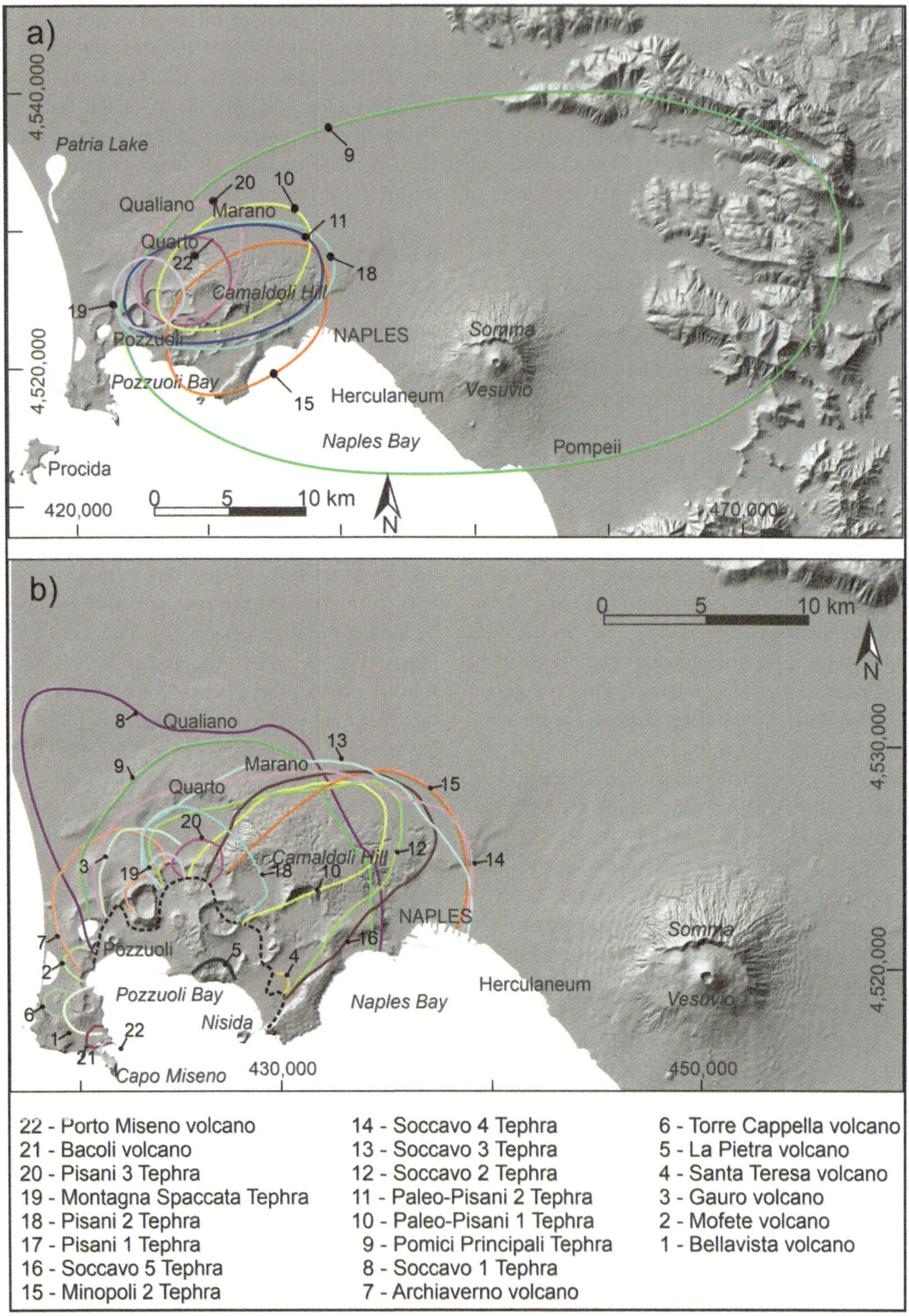

Fig. 8 Areal distribution of the pyroclastic deposits of the eruptions of epoch 1: **a** pyroclastic fallout, **b** pyroclastic density currents. Black dotted line in (**b**) is the limit of the maximum sea ingression between epochs 2 and 3. Figure modified after Orsi et al. (2004)

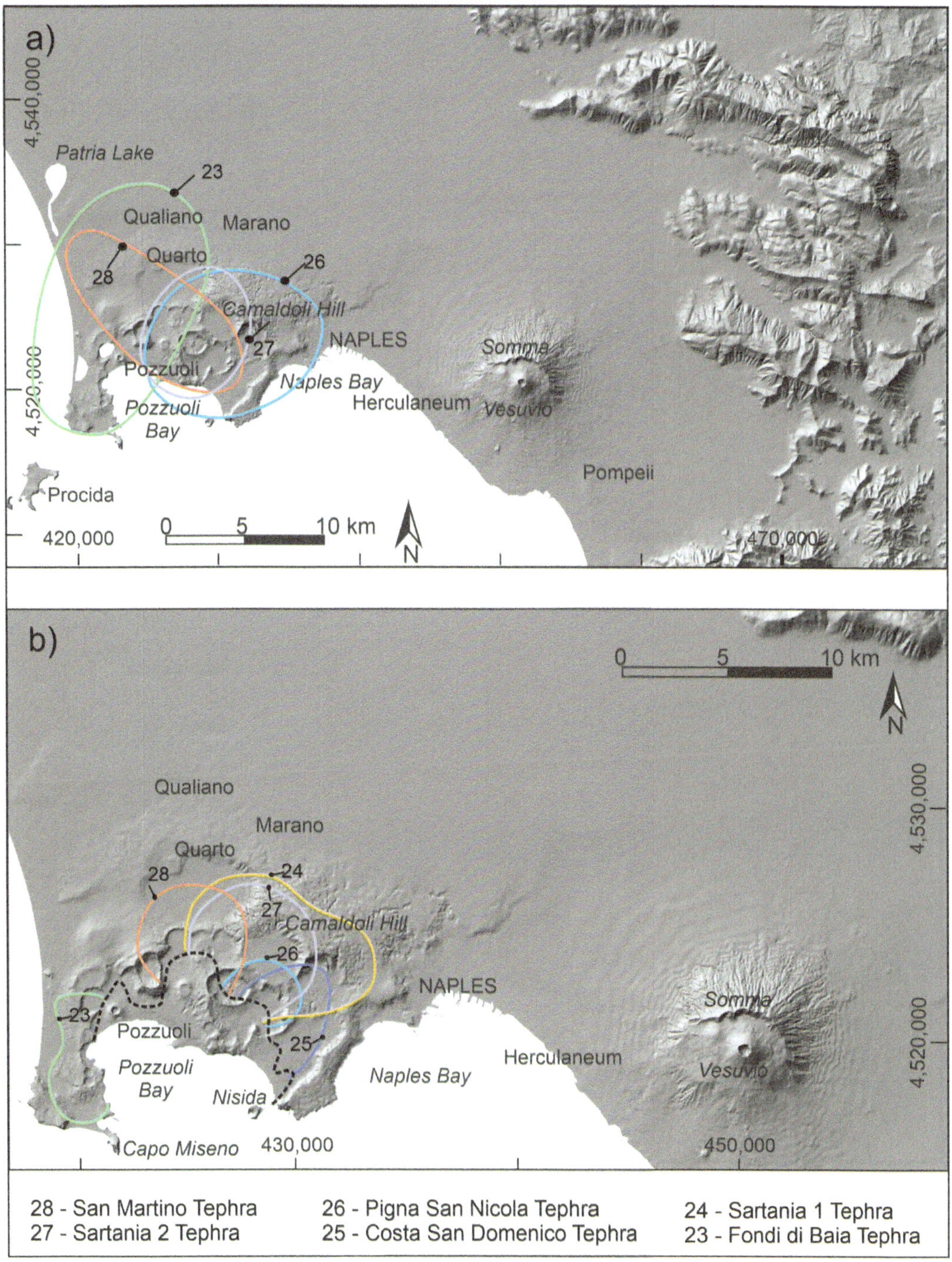

Fig. 9 Areal distribution of the pyroclastic deposits of the eruptions of epoch 2: **a** pyroclastic fallout, **b** pyroclastic density currents. Black dotted line in (**b**) is the limit of the maximum sea ingression between epochs 2 and 3. Figure modified after Orsi et al. (2004)

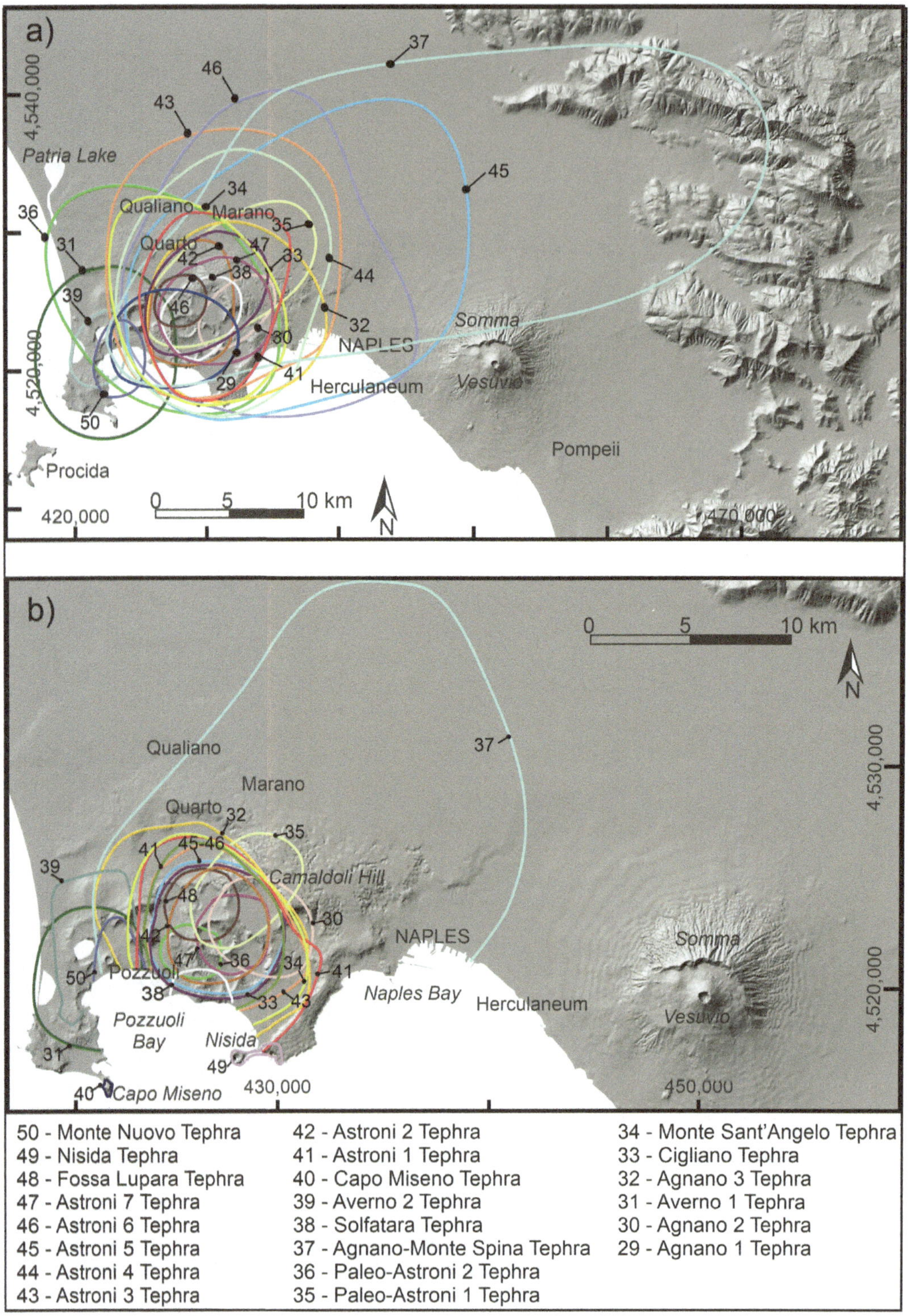

Fig. 10 Areal distribution of the pyroclastic deposits of the eruptions of epoch 3: **a** pyroclastic fallout, **b** pyroclastic density currents. Figure modified after Orsi et al. (2004)

the later Monte Nuovo events were in the north-western sector (Fig. 7). The Nisida and Capo Miseno eruptions occurred at the eastern and western edges of the horseshoe-shaped inland portion of the caldera (Fig. 7). Nisida, Monte Dolce and Pampano banks were located within the Monte Dolce-Ammontatura morpho-structural depression (Fig. 7) controlled by northwest-southeast and northeast-southwest faults likely reactivating portions of structures active during the CI eruption and caldera collapse. Isaia et al. (2009), having recognised a period of quiescence of about 200 years following the Agnano-Monte Spina eruption, subdivided the eruptions of this epoch of activity into two sub-groups. The Agnano-Monte Spina, Averno 2 and Astroni eruptions will be described here because they are representative of the types of volcanic events that occurred in this epoch and generally in the post-NYT activity. Therefore, they are important also in terms of volcanic hazard assessment (see Chap. Volcanic Hazard Assessment at the Campi Flegrei Caldera, Italy).

The Agnano-Monte Spina eruption has been reconstructed in great detail by de Vita et al. (1999 and references therein). It occurred in the north-eastern sector of the NYT caldera floor (Fig. 11) at 4,482–4,625 cal. years BP (Smith et al. 2011) (Table 1) and was accompanied by a volcano-tectonic collapse. It was the largest magnitude event of this epoch, extruded a volume of 0.85 km^3 of magma (DRE) and distributed its deposits over a very large area (Orsi et al. 2009) (Fig. 11). The eruption was characterised by phreatomagmatic explosions generating PDCs, and two Plinian events that produced sustained columns up to 30 km high. The umbrella clouds of these two columns were north-easterly dispersed and laid down fallout deposits not thinner than 10 cm at distances of up to 50 km. The PDCs flowed both within and outside the CFc depression where they travelled at least 15 km over the surrounding plain. The eruption was interrupted by a pause that is presently testified by an erosional unconformity. After this pause, the activity resumed through variable vents and was characterised by vent migration along fractures. Tephra layers associated with this eruption are at Lago Grande di Monticchio (Wulf et al. 2004), at Accesa Lake (Magny et al. 2006), and in the Adriatic (Matthews et al. 2015 and references therein) and Tyrrhenian (Crocitti et al. 2019 and references therein) seas.

The Averno 2 eruption has been reconstructed in detail by Di Vito et al. (2011) (Table 1;

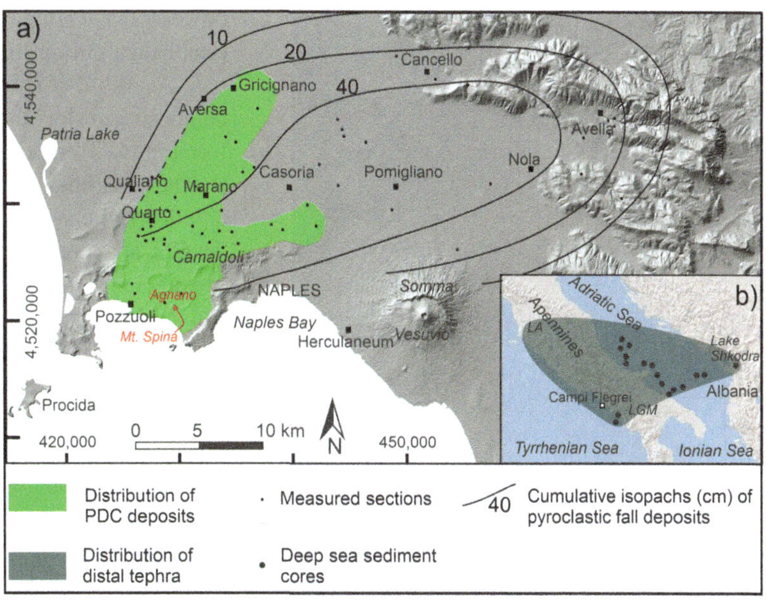

Fig. 11 Areal distribution of the pyroclastic deposits of the Agnano-Monte Spina eruption: **a** in the Campanian area, **b** in distal settings. LGM: Lago Grande di Monticchio; LA: Lake Accesa. Figure modified after de Vita et al. 1999 and Orsi et al. 2009)

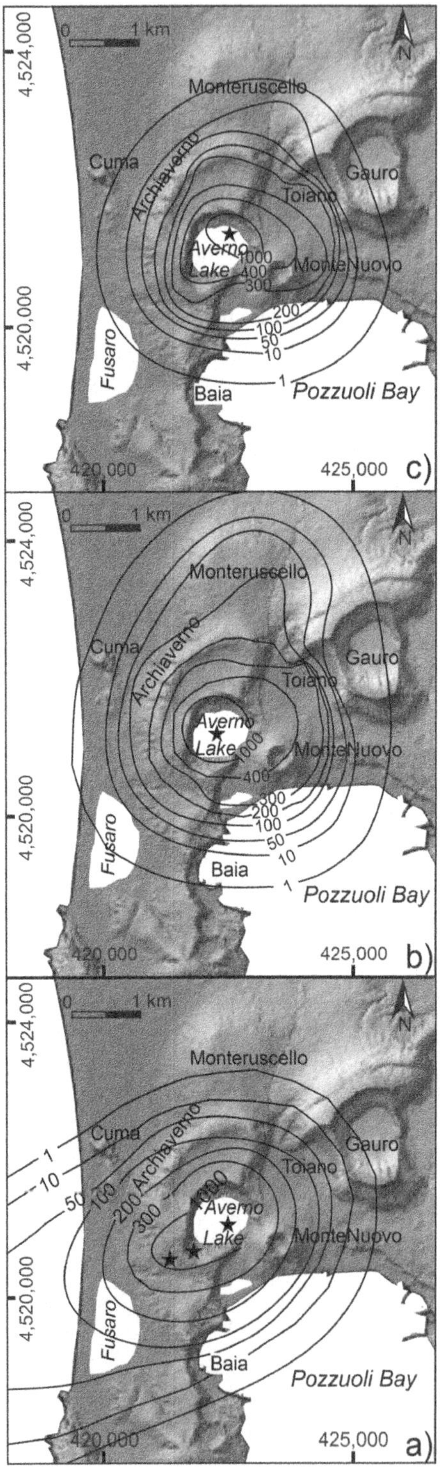

Fig. 12 Total thickness distribution of the three members of the Averno 2 pyroclastic sequence: **a** Member A, **b** Member B and **c** Member C. Figure modified after Di Vito et al. (2011)

Fig. 12). Pistolesi et al. (2016) suggested, based on stratigraphic and stratimetric data, that this eruption was coeval with the Solfatara event, dated at 4,181–4,386 cal. years BP (Smith et al. 2011) (Table 1). The Averno 2 eruption and the Astroni volcano described below are good examples of Phlegraean volcanic events characterised by syn-eruptive vent migration along fracture systems. The Averno 2 eruption extruded 0.07 km^3 DRE of alkali-trachytic magma and was triggered by the arrival of a less-evolved magma into a shallow reservoir and its mingling/mixing with an already present more-evolved melt. The magmatic system included a dyke-shaped shallow reservoir intruded into a 2-km-long portion of a northeast-southwest-trending fault system, bordering the La Starza resurgent block. The eruption was characterised by two phases. The first phase was dominated by magmatic explosions with formation of columns that reached a maximum height of 10 km, while the second produced explosions driven by water-magma interaction generating wet and dry surge deposits. The activation of a 2-km-long portion of a northeast-southwest-trending fault system controlled the change from the magmatic to the phreatomagmatic explosive phases. It also induced vent migration from southwest to northeast during the course of the eruption. The close correlation among structural setting, magmatic processes, and eruption and withdrawal dynamics established during the course of the low magnitude Averno 2 eruption is similar to that already demonstrated for the high-magnitude caldera-forming NYT eruption (Orsi et al. 1992, 1995).

The Astroni volcano is a well-preserved elliptical edifice in the north-eastern sector of the CFc at the north-western edge of the Agnano-Monte Spina volcano-tectonic collapse (Figs. 4 and 7). It has been investigated for more than a century, with the last comprehensive studies carried out by Isaia et al. (2004) and Tonarini et al. (2009). This volcano grew through seven eruptions that extruded a total of 0.7 km^3 of magma (DRE) and deposited their products over a large area (Isaia et al. 2004; Orsi et al. 2009) (Fig. 13). The activity began with the Astroni 1 eruption at 4,153–4,345 cal. years BP and ended

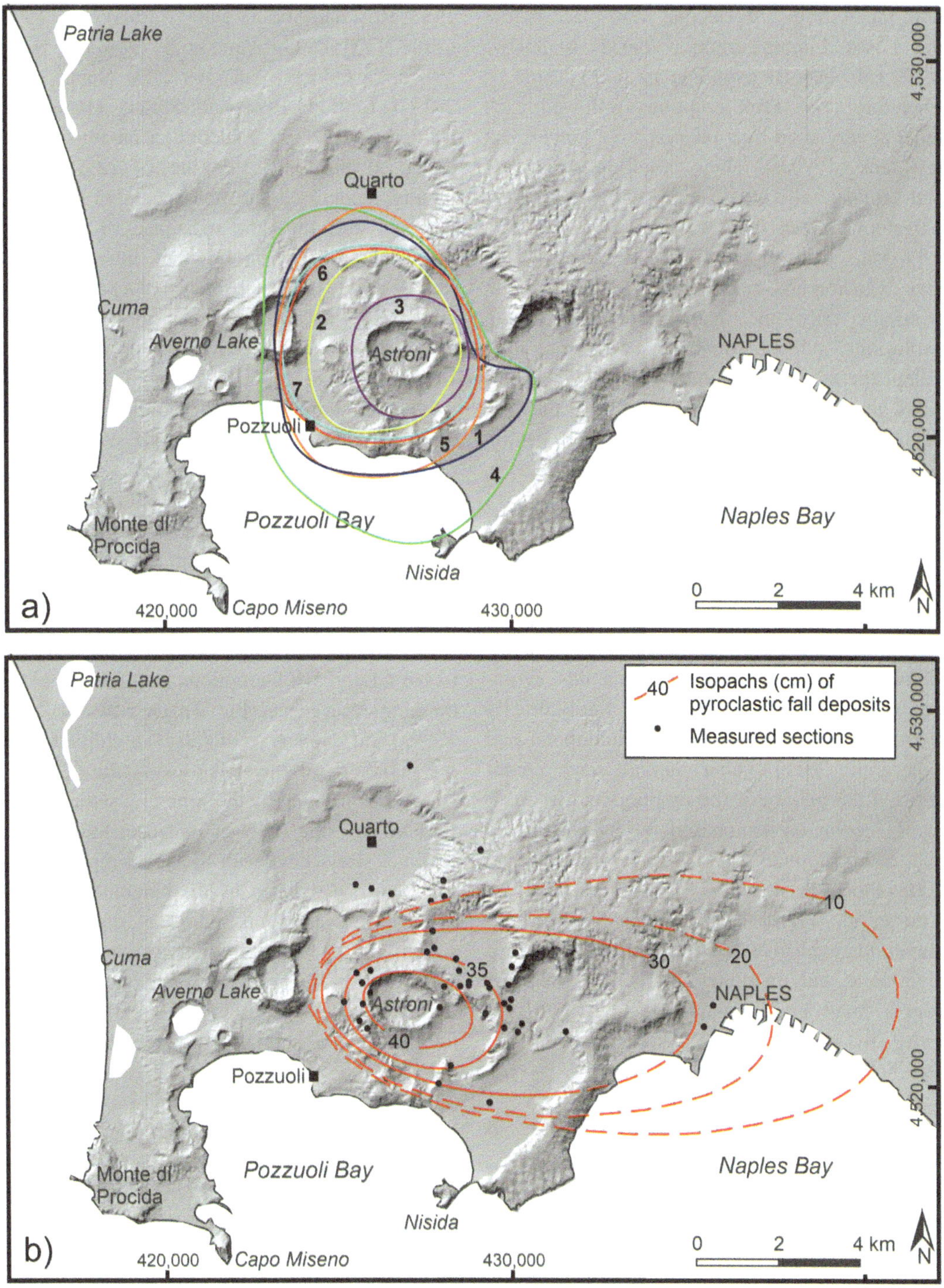

Fig. 13 **a** Total distribution of the pyroclastic density current deposits of each of the 7 eruptions of the Astroni volcano. **b** Isopach map of the Astroni 6 pyroclastic fall deposits; black dots locate measured sections. Figure modified after Isaia et al. (2004)

with the Astroni 7 event that took place before the Fossa Lupara eruption dated at 3,978–4,192 cal. years BP (Smith et al. 2011) (Table 1). Therefore, the seven eruptions followed each other at very short time intervals, not longer than few tens of years. They varied in magnitude and almost all were explosive, dominantly phreatomagmatic. Only eruption 5 generated a lava dome and eruption 7, the last, ended with lava extrusion and spatter cone activity. Eruption 6 is the only one that produced magmatic explosions, although minor, with generation of fallout deposits. The phreatomagmatic explosions generated PDCs that spread over the CFc depression and the surrounding plain laying down sequences of surge beds.

Monte Nuovo Eruption

The AD 1538 Monte Nuovo eruption (Figs. 3 and 4) occurred about 3,000 years after the end of epoch 3 and is the only historical eruption of the caldera. Therefore, it is the only event for which detailed descriptions of dynamics, evolution and precursory phenomena were directly recorded (Parascandola 1946; Guidoboni and Ciuccarelli 2011). Other authors have reconstructed the history of the eruption and its magmatic feeding system based on volcanological and compositional data (e.g., D'Oriano et al. 2005; Piochi et al. 2005; Arzilli et al. 2016; Di Vito et al. 2016). The eruption was a low-magnitude explosive event that formed a 130-m-high cone and distributed pyroclastic deposits over a small area around the vent. It was preceded by significant ground deformation and seismicity that lasted for tens of years.

4 Deformation History of the Campi Flegrei Caldera

The CF depression is widely interpreted as that of a caldera, although a few authors deny the existence of the caldera (e.g., Milia and Torrente 2003; Milia et al. 2006). Various hypotheses have been proposed on the genesis and architecture of the CFc. Some have related it to a single volcanic event, either the CI (Rittmann 1950a; Rosi et al. 1983; Rosi and Sbrana 1987; Barberi et al. 1991) or the NYT (Di Girolamo et al. 1984; Lirer et al. 1987a; Scandone et al. 1991; De Natale et al. 2017) eruption. Rosi and Sbrana (1987) and Barberi et al. (1991) identified an inner, more collapsed portion inside the CI caldera. Not having recognised the uniqueness of the NYT deposits and thus the existence of the high-magnitude NYT eruption, they interpreted this feature as due to the subsidence of a portion of the CI caldera floor during the volcanism from the CI eruption to about 10 ka. Many later authors (e.g., De Natale et al. 2006; Sacchi et al. 2014; Steinmann et al. 2016, 2018) refer to the NYT caldera as the inner structure. Scandone et al. (1991) related a gravity minimum in the Acerra area north of the CF (Fig. 2) to a caldera and suggested that this structure was the result of several collapses accompanying multiple CI eruptions. Orsi et al. (1992, 1996) first interpreted the CFc as a nested, resurgent and restless structure resulting from two different major nested collapses related to the CI and NYT eruptions and affected by an ongoing resurgence (Fig. 4). Orsi et al. (1996, 1999a) and Capuano et al. (2013) highlighted that the geometry of the collapses of the CFc was strongly controlled by the structural setting of the area at the time of each collapse, thus by the regional stress field with the superimposition of the local one induced by the magmatic activity. The present structural setting of the area also suggests that regional fault systems played a major role in the definition of the boundaries of both caldera collapses and disjointing of their floors. It also supports a piecemeal collapse mechanism, at least for the NYT caldera (Capuano et al. 2013). In light of this, the hypothesis that the "inner caldera ring fault" is the sum of portions of faults likely activated during both CI and NYT caldera collapses within the collapsed areas is favoured here. The hypothesis that the caldera is a nested structure (Orsi et al. 1996) has been accepted by all later authors sharing the idea of its existence, although with diverse interpretations of the boundary of each collapse (e.g., Scarpati et al. 1993; Di Vito et al. 1999; Perrotta et al. 2006; Acocella 2008; Fedele et al. 2008; Capuano et al. 2013; Vitale and Isaia

2014). Acocella (2008), based on a re-interpretation of the geological profiles constructed by Orsi et al. (1996) and new structural data, suggested that the CFc consists of two nested tectonic depressions active during both CI and NYT eruptions although with different amounts of collapse. Steinmann et al. (2016, 2018) suggested that the faults of the NYT caldera in Pozzuoli Bay resulted from re-activation of the marginal faults of the CI caldera. De Natale et al. (2016), by interpreting the core of a single drillhole in the Bagnoli area, disputed the previously proposed reconstructions of the caldera structural setting. In particular, they proposed a smaller collapsed area and a small volume for the infilling caldera deposits, particularly those of the CI. They also hypothesised a complex collapse mechanism with a progressive deepening toward the centre of the structure. In addition to the two major and provable caldera events, a minor volcano-tectonic collapse occurred during the Agnano-Monte Spina eruption and formed the Agnano Plain depression (de Vita et al. 1999).

The definition of the caldera structure has greatly benefitted also from the interpretation of the results of geophysical investigations and borehole drillings (Penta 1954; Rosi and Sbrana 1987; Aster and Meyer 1988; Barberi et al. 1991; Orsi et al. 1996; Di Vito et al. 1999; Florio et al. 1999; Bruno et al. 2003; Bruno 2004; Judenherc and Zollo 2004; Vanorio et al. 2005; Berrino et al. 2008; Zollo et al. 2008; Capuano et al. 2013; Piochi et al. 2014; De Siena et al. 2018). CFc is filled to 3 km depth by a sequence of pyroclastic deposits with intercalated marine and continental sediments; the denser rocks located between 2 and 3 km depth (Capuano et al. 2013) have been interpreted as diagenetically consolidated volcanic and non-volcanic sediments, affected by hydrothermal alteration (e.g., Mormone et al. 2011 and references therein). Seismic reflection studies (Judenherc and Zollo 2004; Vanorio et al. 2005; Zollo et al. 2008; Chap. Seismic and Gravity Structure of the Campi Flegrei Caldera, Italy) suggest the likely occurrence of thermometamorphic rocks bearing water and/or gas at 3–4 km depth, and of a 1-km-thick, low velocity layer at 8 km depth. The former overlies the Mesozoic carbonate sequence, in turn overlying a crystalline basement whose top is at about 6 km depth, while the latter is interpreted as a partial melting zone. Temperature profiles measured in geothermal wells within the caldera give maximum values of about 350 °C at about 2.5 km depth (Rosi and Sbrana 1987).

The overall CFc structure is subsiding, but the floor of the NYT caldera has been affected by an ongoing resurgence (Gunter 1903; Cinque et al. 1985; Rosi and Sbrana 1987; Orsi et al. 1996, 1999a; Di Vito et al. 1999; Capuano et al. 2013; Marturano et al. 2018; Isaia et al. 2019). The most prominent feature of this resurgence is the morphological height of the La Starza marine terrace (Cinque et al. 1985) in its subaerial portion. Orsi et al. (1996, 1999a) distinguished a deformation operating within the resurgence process since the NYT caldera collapse, called "long-term deformation", and one that takes place over short periods of time (years to decades), called "short-term deformation", also known as bradyseism. Based on the similarity of the geometry of both deformations, Orsi et al. (1999a) concluded that short-term deformation episodes are transient events within the long-term deformation. Both syn- and post-collapse deformation patterns of the NYT caldera have been interpreted as indicative of piecemeal collapse and block resurgence mechanisms (Orsi et al. 1996; Di Vito et al. 1999; Capuano et al. 2013). Similar deformation dynamics can be hypothesised for the CI caldera collapse on the basis of the areal distribution of the proximal facies and of the few morpho-structural elements relative only to some marginal faults of the structure. Sacchi et al. (2019 and references therein) have suggested that the NYT caldera floor has been deformed by the growth of a resurgent dome. The geometry of all collapses, as well as of the resurgence within the NYT caldera, has been mostly dictated by partial reactivation of portions of pre-existing features, mainly northwest-southeast and northeast-southwest and subordinate north–south and east–west regional faults (Fig. 14) (Orsi et al. 1996, 1999a, b; Di Vito et al. 1999; Acocella et al. 2004; Capuano et al. 2013; Vitale and Isaia 2014). The present

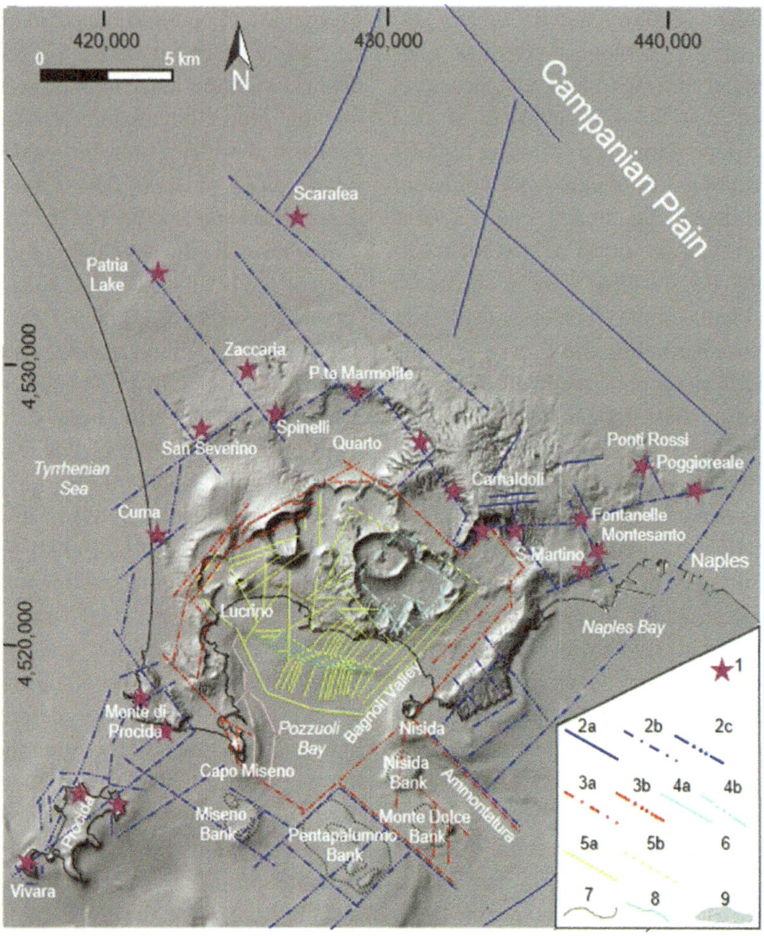

Fig. 14 Structural sketch map of the Campi Flegrei caldera. (1) outcrops of CI proximal deposits, (2a) faults probably active during the CI eruption, (2b) faults of the Campanian Ignimbrite caldera, (2c) faults and fractures active during the Campanian Ignimbrite eruption along which there was magma extrusion, (3a) faults likely active during the Neapolitan Yellow Tuff eruption, (3b) faults active during the Neapolitan Yellow Tuff eruption, (4a) faults likely active during the Agnano-Monte Spina eruption, (4b) faults active during the Agnano-Monte Spina eruption, (5a) faults active during resurgence of La Starza block, (5b) faults likely active during resurgence of La Starza block, (6) faults active during subsidence of the Neapolitan Yellow Tuff caldera floor, (7) rim of La Starza marine terrace, (8) rim of the post-Würmian level surface, (9) post-Würmian level surface

structural setting of the entire caldera is dominated by these faults and fractures and morpho-structural elements related to volcano-tectonism, as highlighted also by geological, morpho-structural and geophysical data (Cinque et al. 1985; Di Vito et al. 1999; Florio et al. 1999; Bruno 2004; Judenherc and Zollo 2004; Orsi et al. 2009; Sacchi et al. 2009, 2019; Acocella 2010; Capuano et al. 2013; Vitale and Isaia 2014). The main structural lineaments, which have controlled both genesis and evolution of the caldera, have also been preferential pathways for migration of magmas, aqueous fluids and gases. Therefore, the areal distribution of the active vents through time (Fig. 7) is a tracer of the structures that have acted as feeding features for volcanism and that have controlled the dynamics of the post-NYT deformation (Orsi et al. 1996, 1999a; Di Vito et al. 1999; Rivalta et al. 2019) (Fig. 14). Magmatism, volcanism, and local and regional deformation are intimately related. Such a relation can be verified at any scale from caldera collapse, to caldera resurgence, to areal distribution of volcanic vents through time and composition of the erupted magmas, to direction of vent migration during a single eruption (Orsi et al. 1996, 1999a, 2004; D'Antonio et al. 1999; Di Vito et al. 1999, 2011).

4.1 Caldera Collapses

CFc is an active, nested and resurgent structure including the CF depression, the Pozzuoli Bay, and the southern portion of the city of Naples. The CF morpho-structural depression mostly results from alternating, sometimes coeval, constructive and destructive volcanic and/or volcano-tectonic events, as well as time-space variations of the relations between sea and earth-surface level. It is composed of a subaerial and a submerged portion (Figs. 2 and 4). The subaerial portion is horseshoe shaped and includes diverse volcanic landforms, the flat-topped height of the La Starza marine terrace, and several lowlands. It is bordered by high-angle scarps to the north-northeast, Monte di Procida to the southwest, and Posillipo Hill to the east. Its south-central portion includes the La Starza marine terrace (40 m a. s.l.) cut on the south by a fossil marine cliff (Cinque et al. 1985). The volcanic landforms are monogenetic volcanoes, mostly tuff-rings and tuff-cones with very few lava domes. The oldest of these landforms, partially dismantled by erosion and/or volcano-tectonics, are typically covered by younger deposits and volcanic edifices, while the youngest are commonly well preserved. The submerged portion of the CF depression, forming the floor of Pozzuoli Bay, has been subdivided by Capuano et al. (2013) into a north-eastern and a south-western sector. The two sectors are divided by the roughly northwest-southeast-trending morpho-structural lineament that connects Lucrino to Bagnoli Valley, named the "Lucrino-Bagnoli Valley lineament" (Figs. 2 and 4). This lineament is well evidenced by the distribution of many earthquakes of the 1982–1984 bradyseismic crisis and some of the more recent mini-uplift episodes (Orsi et al. 1999a, b; Chap. Historic Unrest of the Campi Flegrei Caldera, Italy and references therein). It is composed of several sectors formed by scarps with different slopes and directions. The eastern and western flanks of the deeper sector of Pozzuoli Bay are high-angle evolved flanks of volcanic edifices. This sector extends south-eastward into the rectangular Monte Dolce-Ammontatura morpho-structural low composed of a series of faulted blocks whose downthrow increases north-eastward. This morpho-structural feature is delimited by the Pentapalummo Bank to the southwest and a northwest-southeast high-angle scarp to the northeast (Figs. 2 and 4).

4.1.1 Campanian Ignimbrite Caldera Collapse

The collapse of a caldera within the CF volcanic field over the course of the CI eruption was first proposed by Rittmann (1950a) and then by Thunnel et al. (1979). Rosi et al. (1983) defined the north-eastern, northern, western and south-western portions of the CI caldera rim, to which Rosi and Sbrana (1987) added a southern submerged portion passing through the Nisida and Pentapalummo banks. Barberi et al. (1991) corroborated this definition of the caldera margin, partly subaerial and partly submerged, with gravity and magnetic data. Orsi et al. (1996) extended the collapsed area suggested by Rosi and Sbrana (1987) along a regional Antiapenninic (northeast-southwest) fault system to the central and eastern portions of the city of Naples and to the northernmost portion of the Bay of Naples. This hypothesis was later corroborated by Perrotta et al. (2006), who detailed the inland north-eastern margin of the structure. Vitale and Isaia (2014) accepted the inland caldera rim suggested by Rosi and Sbrana (1987), and hypothesised a margin towards the southeast beneath the present Posillipo Hill. The subaerial portion of the CI caldera margin (Fig. 14) is characterised by high-angle surfaces cut into the CI, locally including older sequences, and uncomformably overlain by younger rocks. According to Orsi et al. (1996), the caldera margin is exposed at Monte di Procida, Cuma, San Severino, Punta Marmolite, Trefola, Soccavo, and at the Vomero-San Martino and Ponti

Rossi-Poggioreale alignments of topographic highs (Figs. 2 and 14). The authors associated the submerged portion of this margin to two structural features. One is the alignment of the Pentapalummo Bank and Miseno Bank volcanoes, considered younger than CI and likely located along the caldera structural boundary, towards the south. The other is the Anton Dohrn Canyon-Sebeto Valley northeast-southwest structural trend, towards the southeast (Fig. 4). According to more recent marine geology data, Pentapalummo Bank is older than CI, Miseno Bank is aged between CI and NYT, while Nisida, Monte Dolce and Pampano banks are younger than NYT (e.g., Sacchi et al. 2019). Therefore, improving the conclusions of Orsi et al. (1996) and Capuano et al. (2013), it is suggested here that the submerged sector of the CI caldera collapse is delimited by portions of northeast-southwest and northwest-southeast regional faults, parts of which were later re-activated during the NYT caldera collapse, as illustrated in Fig. 14.

In order to put forward any hypothesis on the deformation related to the CI eruption, some volcanological and magmatic aspects, presented in Sect. 3.2.1, have to be taken into account and are summarised here. They include the isopachs and the maximum lithic and pumice clast isopleths for the Plinian fallout beds, the maximum lithic and pumice clast isopleths for the PDC deposits, the correlation between areal distribution of the various PDCs and chemical composition of the feeding magmas, the areal distribution of proximal (e.g., breccias and Piperno) deposits, the relations between the volume estimates of erupted magma and collapsed rocks, and the current knowledge on the architecture of the CF magmatic feeding system. Stratigraphic and stratimetric data suggest that the earliest reconstructed part of the eruption generating a Plinian column was fed by a central vent located within the Neapolitan-Phlegraean area (Rosi et al. 1999). The hypothesis that this Plinian phase was followed by the beginning of the collapse of a caldera and that the subsequent eruption phases generated high-volume PDCs through the marginal caldera faults is commonly accepted.

The semi-elliptic pattern of the maximum pumice and lithic isopleths of the PDCs (Fig. 5a) was interpreted by Barberi et al. (1978) as evidence that the eruption occurred through a northwest-southeast regional fracture system located between Naples and Villa Literno.

The relation between areal distribution of the PDC deposits and chemical composition of the feeding magmas (Fig. 5a) suggested to Civetta et al. (1997) that the caldera collapse was not synchronous. During the earliest stages of the collapse, after the Plinian explosions, a series of fractures opened in the northern and southern parts of the CF. In subsequent stages, PDCs were produced by the general collapse of the caldera. All these PDCs were highly diluted and expanded. The last stages generated poorly expanded PDCs that were fed by the least-evolved magma and only travelled within the Campanian Plain.

The areal distribution of the proximal facies of the CI deposits, previously discussed and illustrated in Fig. 14, is considered here as a good tracer of the location of the active vents during the course of the eruption. It clearly highlights that vents were located both along and far from the marginal faults of the caldera collapse.

The estimates of the volume of magma erupted during the course of the CI eruption and the size of the related caldera collapse presently available in the literature do not agree. The CI caldera reconstructed by Orsi et al. (1996), the largest of the proposed outlines of the structure, covers an area of about 230 km^2. Such an area would be enclosed within a theoretical circle of about 17 km in diameter, although the proposed shape of the caldera is much more irregular. Assuming an average downthrow of 700–800 m as evaluated by Barberi et al. (1991) by interpreting the results of many deep boreholes, the collapsed volume could be about 170 km^3. This value is slightly more than half of the volume of erupted magma estimated by Fedele et al. (2007) and smaller than the minimum values of the ranges estimated in more recent papers (Costa et al. 2012; Silleni et al. 2020). This puts a focus on the importance of having a good estimate of the erupted volume of magma. Assuming a downthrow of about 100 m as suggested by De

Natale et al. (2016) based on one single drill hole, the volume of the collapse would be around 23 km^3, much smaller than any recently suggested estimate of the volume of the CI erupted magma (180–300 km^3). To account for these estimated volumes of magma, a circular area of 18–23 km in diameter should have collapsed for 700–800 m. This is impossible in the Neapolitan-Phlegraean area due to the constraints posed by its geological configuration, in particular by the areal distribution of the exposed rocks older than CI (Fig. 4). Various hypotheses have been formulated on the architecture and dynamics of magmatic reservoirs feeding caldera-forming eruptions and recently synthesised by Cashman and Giordano (2014).

It is hypothesised here that the whole CI sequence was not erupted from vents located only within the area of the present CF morphological depression, where a caldera collapse took place. It was also erupted through fissure vents and/or discrete central vents along structures trending through the Campanian Plain and the area of the present Procida Island. A still missing very important piece of information is the ratio between the volume of magma erupted within the caldera collapse area and that emitted along the structural features outside this area. The activity

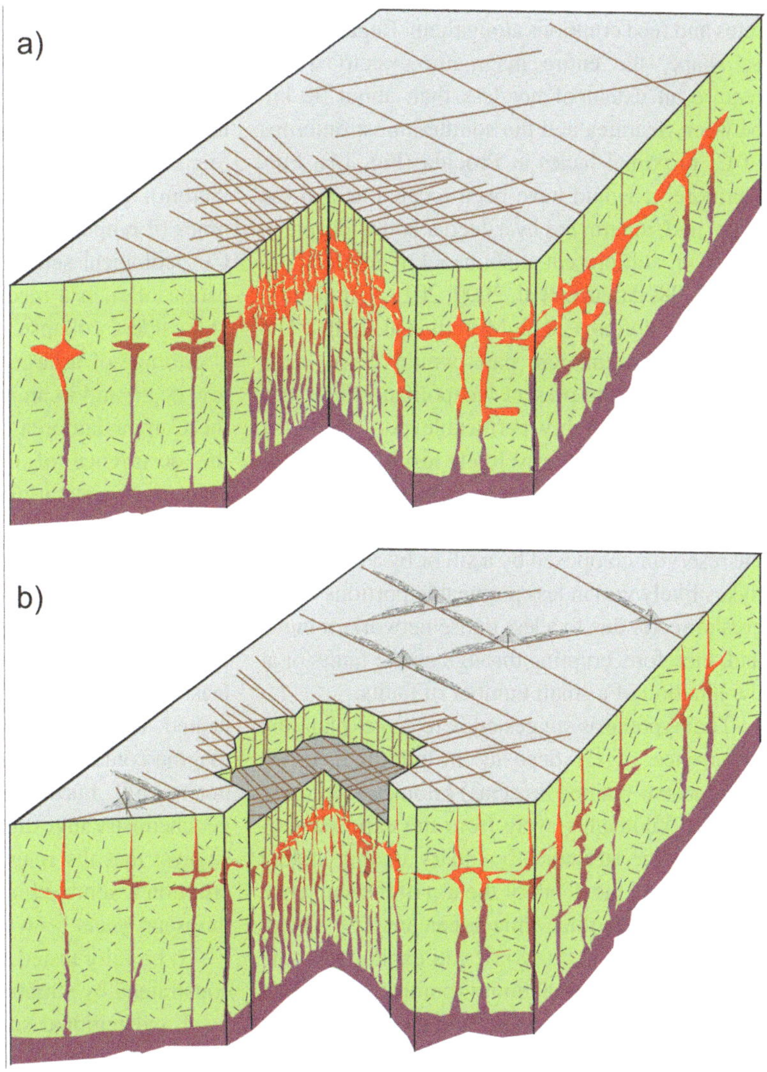

Fig. 15 Artistic restoration of a possible widely dispersed magmatic reservoir feeding the large-volume, caldera-forming Campanian Ignimbrite eruption through multiple vents. **a** Before eruption. **b** After eruption and caldera collapse

through scattered central/fissure vents along fault and fracture systems, even at long distance from the CF, and production of large-volume, highly-dilute PDCs associated with a syn-eruption deformation, need a complex magma reservoir. This reservoir could have had at least two different types of architecture. It could have been shaped as a sill with a thicker portion or as a network of dykes with a zone in which they were more intensely interconnected. Either the thicker portion of the sill or the zone with a more intense dykes interconnection was the site of magma accumulation, thus corresponding to the magma chamber of a typical magma reservoir (Fig. 15). The dykes departing from the magma accumulation zone would intrude regional faults and fractures and feed eruptions along them. Regardless of its shape, the entire magmatic system would require an extent of not less than about 30 km between Scarafea and the southernmost outcrops of CI proximal facies at Procida (Fig. 14). Furthermore, it would have been located at 3–6 km depth and, in turn, fed by a deeper and widespread reservoir, still present at about 8 km depth (Zollo et al. 2008; Arienzo et al. 2009, 2011; Pappalardo and Mastrolorenzo 2012; Moretti et al. 2019). The magma accumulation part would have been located under the present Neapolitan-Phlegraean area, within a highly permeable portion of the Campanian shallow crust likely resulting from the occurrence of a dense network of faults (Moretti et al. 2013b and references therein). The portion of the reservoir composed by a sill or by a number of dykes likely was in less permeable portions of the crustal sector due to a less dense network of faults and, therefore, erupting through single faults or at the junction of a small number of faults.

The hypothesis suggested here can explain the occurrence of CI deposits in proximal facies (e.g., coarse-grained breccias) kilometres to more than ten kilometres away from the exposed segments of the structural boundary of the caldera (Fig. 14) and the depth of the top of the sequence within the Neapolitan-Phlegraean area (Orsi et al. 1996; Fedele et al. 2008). It can explain also the geometry of the maximum pumice and lithic clast isopleths (Fig. 5a) (Barberi et al. 1978), and justify the asynchronous collapse of the caldera suggested by Civetta et al. (1997). Furthermore, it can explain the difference between size of the caldera collapse and volume of the erupted magma, notwithstanding the uncertainty on these two volumes that could vary from few to tens of cubic kilometres.

To justify the differences between the volume of erupted magma and a smaller volume of caldera collapse during the course of an eruption, variable hypotheses have been proposed. The stacked sills/lenses hypothesis suggested for the Villa Senni mafic eruption at Colli Albani (30 km^3 DRE) (Cashman and Giordano 2014 and references therein), seems not suitable for the CI eruption and caldera collapse because of differences in volume, composition and timing of the erupted magmas. Also a mechanism of caldera collapse kilometres away from the eruption vent with horizontal magma displacement, as in the case of the 1912 Novarupta (Alaska) eruption (e.g., Hildreth and Fierstein 2000 and references therein), can justify a difference between volumes of erupted magma and of caldera collapse. CF geological setting does not permit the invocation of such a mechanism for the CI eruption and caldera collapse.

Storage of a large volume of magma at shallow depth and in an areally extended reservoir in the Campanian area could also be supported by the occurrence of the earlier magmatism and volcanism generating both voluminous fissure-fed ignimbrites (De Vivo et al. 2001; Rolandi et al. 2003) and thick pyroclastic sequences from central vents located outside the Neapolitan-Phlegraean depression (Orsi et al. 1996). The long-lasting magmatic activity in the area (at least 2 Ma; Di Girolamo 1978) could have heated the local shallow crust, allowing the magma contained in a sill or a network of dykes not to chill. Furthermore, the results of isotope-geochemistry investigations (Arienzo et al. 2011) suggest that the magma extruded during the CI eruption assembled in the reservoir in only a few thousand years ($\sim$6 kyrs) before the event.

To be fully tested, this hypothesis needs to be constrained by structural/geophysical evidence at local and regional scale, petrological data interpreted in light of the spatial and structural position of the eruption vents of the analysed rocks, and more refined estimates of the volumes of collapsed rocks and erupted magma, both total and intra-caldera versus extra-caldera ratio.

4.1.2 Neapolitan Yellow Tuff Caldera Collapse

Orsi and Scarpati (1989), Orsi et al. (1991a, 1992, 1995) and Wohletz et al. (1995) highlighted the relations among eruption dynamics, magma withdrawal and caldera collapse during the NYT eruption by integrating stratigraphic, sedimentologic and compositional data. Orsi et al. (1996) used surface and sub-surface geological data as well as geophysical and geochronological evidence to relate the pseudo-circular high-density feature encompassing an inhomogeneous low-density crustal sector, which they linked to the structural boundary of the NYT caldera (Fig. 14). This geophysical feature had been interpreted previously as the inner, more depressed portion of the collapse accompanying the CI eruption (Rosi and Sbrana 1987; Barberi et al. 1991). Additional structural data presented by Acocella (2010) and Vitale and Isaia (2014) have supported the subaerial reconstructed NYT caldera margin. Capuano et al. (2013) have detailed the structural features and present architecture of the NYT caldera previously outlined by Orsi et al. (1996) by integrating the results of the first fully 3D inversion of gravity data with geological and morpho-structural data and re-evaluation of the available literature. The structural setting of the NYT caldera presented here and synthesised in Fig. 14 is a step forward from the work of Capuano et al. (2013), based on a new synthesis of morpho-structural data and recent literature (e.g., Sacchi et al. 2014, 2019; Steinmann et al. 2016, 2018). The pseudo-circular high-density feature was likely generated by high-density shallow intrusions and/or dykes mostly feeding the volcanism of epochs 1 and 2 of the post-NYT activity and solidified within the marginal faults of the caldera (Figs. 4 and 14) (Orsi et al. 1992, 1996, 2009; Capuano et al. 2013). The portion encompassed by this gravity feature corresponding to the caldera fill has been subdivided into a north-eastern and a south-western sector, separated by the scarps of the northwest-southeast Lucrino-Bagnoli Valley lineament (Figs. 4 and 14) (Capuano et al. 2013). The south-western (deeper) and the north-eastern (shallower) sectors represent the "undeformed-to-subsiding portion" and the "resurgent portion" of the caldera, respectively (Fig. 14). The collapse also involved part of the rectangular Monte Dolce-Ammontatura morpho-structural low (Figs. 4 and 14) along northwest-southeast faults with formation of a series of blocks with increasing downthrows to the northeast. This feature was intersected to the northwest by a northeast-southwest fault system that dropped the main part of the caldera floor. Occurrence of blocks of different thickness and their displacement at different depths support a piecemeal caldera-collapse type, with the Pozzuoli Bay more downthrown than the Monte Dolce-Ammontatura morpho-structural low.

The pseudo-circular portion of the NYT caldera covers an area of about 80 km^2 and its collapse was estimated at about 800–1,000 m by extrapolating the values measured in deep boreholes in the San Vito area to the whole caldera floor (Rosi and Sbrana 1987; Barberi et al. 1991). Such an extrapolation, despite the piecemeal type collapse, gives an estimate of the volume of the collapse of this portion of about 70 km^3. The volume of the collapse of the Monte Dolce-Ammontatura feature, as well as the contribution of its dynamics to the withdrawal of the NYT magma reservoir, is difficult to estimate, due to the current poor knowledge of its geometry. Nevertheless, despite all these uncertainties, the volume estimates of both caldera collapse and magma extruded during the NYT eruption (40–80 km^3 DRE) are in fairly good agreement.

4.1.3 Agnano-Monte Spina Volcano-Tectonic Collapse

The Agnano-Monte Spina eruption was accompanied by a volcano-tectonic collapse (Figs. 4 and 14) whose main features and timing in relation to the eruption dynamics and evolution have been defined on the basis of stratigraphic, stratimetric, sedimentologic, componentry, structural, surface and sub-surface geologic, and sea-level variation data by de Vita et al. (1999). The collapse began during the pause in the eruption and continued after its end. The collapsed area that includes the Agnano Plain has a northwest-southeast-elongated roughly rectangular shape, defined mainly by northwest-southeast and northeast-southwest marginal faults, likely portions of pre-existing structures. It is about 6 km^2 and was affected by a total downthrow of about 60 m, yielding a volume of the collapse of about 0.4 km^3, less than half of the 0.85 km^3 (DRE) volume of extruded magma estimated by Orsi et al. (2009). Such a discrepancy could arise from an underestimate of the amount of the collapse as it was evaluated extrapolating the results of the study of a single core to the entire collapsed area (de Vita et al. 1999). Furthermore, the amount of the collapse could have been affected also by the vertically extended architecture of the Agnano-Monte Spina magmatic feeding system, comprising at least two connected and superimposed sub-reservoirs (Arienzo et al. 2010).

4.2 Neapolitan Yellow Tuff Caldera Resurgence

4.2.1 Long-Term Deformation

The post-15-ka resurgence of the NYT caldera floor has generated a net maximum uplift of 85–100 m detected at the La Starza marine terrace (Cinque et al. 1985; Rosi and Sbrana 1987; Orsi et al. 1996, 1999a, b; Di Vito et al. 1999; Capuano et al. 2013; Sacchi et al. 2014; Marturano et al. 2018; Isaia et al. 2019). The eruptions of epochs 1 and 2 were fed by magmas rising through portions of the NYT caldera marginal faults. During epoch 3, the areal distribution of the active vents significantly changed. Most of them were located in the eastern sector, with only those of the two Averno and the Monte Nuovo eruptions occurring in the western sector of the resurgent portion of the caldera, northeast of the Lucrino-Bagnoli Valley lineament. The Nisida and Capo Miseno events occurred outside the resurgent portion of the caldera floor erupting from vents located on northwest-southeast and northeast-southwest trending features. The vents of the two Averno and the Monte Nuovo eruptions were located along the northeast-southwest San Vito-Averno fault system that drops the La Starza block to the northwest, where it likely intersects the northwest-southeast Lucrino-Bagnoli Valley lineament (Orsi et al. 1996; Di Vito et al. 1999). Nisida erupted at the intersection of the northeast-southwest and northwest-southeast structures of the caldera margin, at the northern edge of the Monte Dolce-Ammontatura morpho-structural low, while Capo Miseno erupted along the structures that connect Monte di Procida to Pentapalummo Bank. Agnano-Monte Spina volcano-tectonic collapse occurred prevailingly along northwest-southeast and northeast-southwest-trending faults. These fault systems (northwest-southeast, northeast-southwest, east–west, and north–south) characteristic of the post-collapse deformation are still active and affect ground deformation patterns during the recent bradyseismic episodes (Orsi et al. 1999a).

During epochs 1 and 2 of activity, the central part of the NYT caldera was invaded by a shallow sea and the pyroclastic deposits were reworked to form sediments with constant marine-transitional characteristics. As sea level was rising during the deposition of these sediments, the caldera floor must also have been uplifted to maintain the sedimentary facies. Such caldera resurgence is also demonstrated by the emergence of the La Starza marine terrace topping the La Starza block at the end of epoch 2, shortly before 8.6 ka. During the quiescence between epochs 2 and 3, the caldera floor subsided to reach the maximum depth (80/100 m b.s.l.). Such a subsidence was followed by a resurgence that produced the definitive emergence of the La Starza block before the beginning of epoch 3, even while the sea level was still rising. During the 200 years of quiescence following the Agnano-Monte Spina eruption, while the magmatic system refilled with new magma, a general uplift of few tens of metres affected the caldera floor (Isaia et al. 2009). Also, after the end of epoch 3, the caldera floor subsided but an uplifting event occurred before the AD 1538

Monte Nuovo eruption. The caldera floor appears to have generally subsided during the periods of quiescence and resurged just before and during the 3 epochs of activity.

The resurgent portion of the NYT caldera has a complex structural setting (Fig. 14) dominated by faults and fractures highlighted by structural data, morpho-structural lineaments, alignments of eruption vents, fluid emissions at the surface, and the results of geophysical and geochemical investigations. Orsi et al. (1996) suggested that the resurgence occurs through a simple-shearing mechanism as highlighted for similar resurgent blocks at Pantelleria and Ischia (Orsi et al. 1991b), and interpreted the La Starza marine terrace as the top of the resurgent La Starza block. This hypothesis was based on the structural setting of the resurgent portion of the NYT caldera floor and its relations with the areal distribution of the vents active through time. The resurgent block can be subdivided into two sectors, separated by a north–south fault system called the Pozzuoli north–south lineament (Capuano et al. 2013). This is the most prominent of the north–south structures affecting the caldera floor and is highlighted by a complex sequence of short-wavelength, low-density anomalies extending from 200 to 1,800 m depth. In addition to the northeast-southwest and northwest-southeast trending fault systems, the eastern and western sectors have been significantly affected also by east–west and north–south trending structures, respectively. The distribution of the eruption vents through time is a tracer of the dynamics of the caldera deformation and supports the block resurgence mechanism for the caldera floor deformation (Orsi et al. 1996, 1999a; Di Vito et al. 1999) (Fig. 7). A resurgent dome would have generated an annular caldera moat that would have hosted the later eruption vents such as at Valles (Smith and Bailey 1968) and Long Valley (Bailey et al. 1976) calderas. This is not the case at the NYT caldera (Orsi et al. 1996, 2004; Di Vito et al. 1999; Selva et al. 2012). Sacchi et al. (2019 and references therein) support the hypothesis of a resurgent dome with the occurrence of a set of closely spaced faults within the submerged part of the resurgent portion of the NYT caldera floor. The authors interpret these features as the markers of the apical graben of the dome. A Digital Terrain Marine Model of the entire NYT caldera (20 × 20 m cell) with an overage grid resolution of 1 m and maximum elevation error of 1 m presented by Capuano et al. (2013) (Fig. 14) highlights that the faults are northeast-southwest oriented and affect the entire south-eastern part of the resurgent portion of the caldera floor, not only its apical part. Furthermore, they extend north-eastward, connecting with those cutting through the sector of the caldera in which most of the eruptions of epoch 3, including the Agnano-Monte Spina event and related volcano-tectonic collapse, have taken place. The Digital Terrain Marine Model also shows that the south-western flank of the resurgent feature is cut by a marine level surface, post-Würmian in age (De Pippo et al. 1984) and affected by the same faults.

4.2.2 Short-Term Deformation

Short-term ground deformation is ongoing within the NYT caldera floor. It has been documented by Parascandola (1947) at the Serapeo of Pozzuoli, the remnant of a Roman marketplace, built in the first century BC. Measurements of the vertical ground deformation along a levelling network were performed for the first time in 1905, and then repeated in 1919, 1922, 1953, 1968 and between 1970 and 1972. In 1975 a dense levelling network was established by the Osservatorio Vesuviano. Since then, it has been integrated with other monitoring networks to form the present monitoring system for the active Neapolitan volcanoes and is managed by the Istituto Nazionale di Geofisica e Vulcanologia - Osservatorio Vesuviano (see Chap. The Permanent Monitoring System of the Campi Flegrei Caldera, Italy).

Reviews of the vertical ground deformation of the NYT caldera for the period between 1905 and 2019 are presented in Del Gaudio et al. (2010), Ricco et al. (2019) and in Chaps. Historic Unrest of the Campi Flegrei Caldera, Italy; Source Modelling from Ground Deformation and Gravity Changes at the Campi Flegrei Caldera, Italy. Figure 16 is a synthetic representation of the ground deformation over the past 2,000 years using archaeological and paleontological data for paleo-height indicators before 1905, and levelling data afterwards. The monitoring system highlighted a general subsidence interrupted by uplift events in 1950–52, 1969–72, and 1982–84. After 1984, minor uplifts occurred in 1989, 1994 and 2000, followed by a new ongoing uplift that began in late 2004/early 2005 (Chaps. Historic Unrest of the Campi Flegrei Caldera, Italy; Source Modelling from Ground Deformation and Gravity Changes at the Campi Flegrei Caldera, Italy). All the ground deformation events have been accompanied by seismicity and gravity changes, as well as variations in geochemical parameters of gas and water effluents. The data and their proposed interpretations are extensively reviewed in Chaps. The Hydrothermal System of the Campi Flegrei Caldera, Italy; Historic Unrest of the Campi Flegrei Caldera, Italy; Source Modelling from Ground Deformation and Gravity Changes at the Campi Flegrei Caldera, Italy. Various hypotheses have been suggested for the triggering mechanisms of the phenomenon. They include intrusion of small magma batches at shallow depth (3–4 km) (e.g., Giudicepietro et al. 2017 and references therein), injection of hot magmatic fluids from shallow magmatic intrusions into the hydrothermal system (e.g., Chiodini et al. 2016), and magma intrusion at shallow depth until 1984, its subsequent cooling, and the ascent through it of

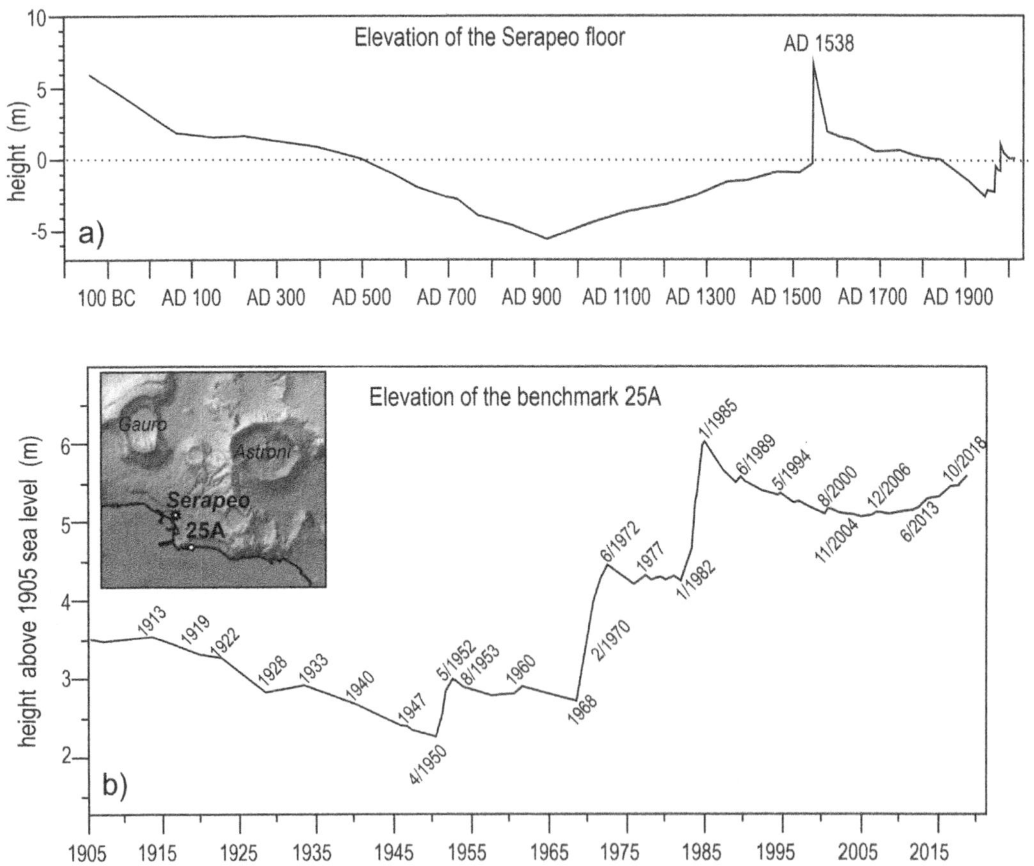

Fig. 16 Ground level deformation of the Neapolitan Yellow Tuff caldera floor recorded at: **a** Serapeo of Pozzuoli, since the first century BC, and **b** benchmark 25A, the site of the measured maximum vertical displacement, since 1905. Figure modified after Del Gaudio et al. (2010) and Ricco et al. (2019). For details on the Permanent Monitoring System of the Campi Flegrei caldera, see Chap. The Permanent Monitoring System of the Campi Flegrei Caldera, Italy

magmatic fluids from depth that interact with the geothermal system (e.g., Moretti et al. 2017).

5 Conclusive Remarks

The CFc, part of the CF volcanic field, which is in turn part of the Phlegraean volcanic district within the Campanian volcanic area, is one of the best examples worldwide of a densely inhabited unrest caldera with which humans have experienced a multi-millennial coexistence. It has attracted the attention of many earth scientists over the past 3 centuries. In recent decades, research on its geological and volcanological evolution has been fostered also because the Italian political authorities, urged by the scientific community, have realised the high level of volcanic and related (e.g., earthquakes, landslides) hazards to which the area is prone. Such hazards, due to the very dense urbanisation of the area and, more in general, the very high exposed value, result in extremely high risks.

The post-Miocene structural evolution of the Tyrrhenian margin of the Southern Apennines thrust belt has been characterised by extensional processes along mainly northwest-southeast normal and northeast-southwest transverse faults. This has strongly affected both genesis and evolution of the Phlegraean volcanic district. Volcanism and deformation of the CF volcanic field have been dominated by the two catastrophic eruptions of the CI and NYT and related caldera collapses, as well as the resurgence of the NYT caldera. Accordingly, their evolution has been subdivided based upon these events. The volcanism following each of the two cataclysmic events was confined within the collapsed sectors. The area active before the CI eruption and caldera collapse extended beyond the morpho-structural borders of the CF into the Campanian Plain (e.g., Orsi et al. 1996), in the area of the present city of Naples (e.g., Perrotta et al. 2010; Scarpati et al. 2013), and in the area of the Pentapalummo Bank (e.g., Bruno et al. 2003; Sacchi et al. 2019). Scarcity of exposures of rocks aged between the two caldera-forming eruptions prevents a detailed reconstruction of the volcanic and deformation history in that time interval. This lack of data results from the concentration of volcanic activity within the CI caldera floor that was largely involved in the later NYT caldera collapse.

Currently, there is a general consensus among the scientific community, with very few exceptions, that the CF volcanic field is dominated by the CFc, a nested, resurgent and restless structure mainly resulting from two major collapses related to the CI and NYT eruptions and affected by an ongoing resurgence. Both collapses occurred through movement along faults, most of which were portions of pre-existing regional structures. It is suggested here that the CI eruption was fed by a reservoir extending well behind the CFc and began with a central-vent Plinian phase venting from within the CF morpho-structural depression. During the course of the eruption, there was, in addition to the caldera collapse within the Neapolitan-Phlegraean area, also activation/re-activation of portions of regional faults outside the collapsed area along which magma was extruded. Some of these faults trended north, within the Campanian Plain, and southwest, into the area of the present Procida Island. The CI magma reservoir was either a sill located at 3–6 km depth with a thick portion, or a network of dykes with a zone in which they were more intensely interconnected. In the thicker portion of the sill or in the zone of more intense dyke interconnections, there was an accumulation of magma that can be considered as the magma chamber of a typical magma reservoir. Despite its shape, the reservoir extended from the Campanian Plain, north of the CF, to the area now occupied by the Procida Island. Furthermore, its greatest magma accumulation zone was located below the Neapolitan-Phlegraean area, where the crust was more permeable due to intersection of many fault systems, and its withdrawal caused the caldera collapse. The NYT eruption was accompanied by a piecemeal collapse caldera through new faults, and activation and/or re-activation of portions of regional faults, some of which were already active during the CI caldera collapse.

Since its formation, the NYT caldera has been the site of intense and closely correlated volcanism and deformation. Volcanism has been concentrated in three periods of activity characterised by clusters of eruptions called epochs, separated by periods of quiescence (Di Vito et al. 1999; Orsi et al. 2004, 2009; Smith et al. 2011). During the three epochs, which occurred between ~15 and 10.6, 9.6 and 9.1, and 5.5 and 3.5 ka, eruption frequency was on the order of 50–70 years. The two periods of quiescence lasted 1,000 and 3,600 years. D'Antonio et al. (1999) and Di Vito et al. (1999) highlighted that the dynamics and the geometry of the deformation related to both the NYT caldera collapse and the long-term resurgence contemporaneous to the post-NYT volcanism affected the areal distribution of the active vents through time and chemical composition of the erupted magmas. This long-term deformation is the sum of short-term deformation episodes, also called bradyseism, representing transient events within the long-term process (Orsi et al. 1999a). The present unrest phase of the caldera (see Chaps. The Permanent Monitoring System of the Campi Flegrei Caldera, Italy; The Hydrothermal System of the Campi Flegrei Caldera, Italy; Historic Unrest of the Campi Flegrei Caldera, Italy; Source Modelling from Ground Deformation and Gravity Changes at the Campi Flegrei Caldera, Italy) represents one of these events.

The volcanism of the CFc, despite the paucity of exposures of rocks aged between CI and NYT, is dominated by explosive activity with phreatomagmatic prevailing over magmatic fragmentation. Phreatomagmatic explosions generated dilute PDCs that deposited sequences of surge beds. Magmatic explosions formed sustained columns that produced tephra fallout and only very rarely collapsed to give rise to dense and laminated PDCs. The magmas feeding the CI eruption that generated the most voluminous and likely most dilute PDCs did not interact efficiently with external water. The dilution of the PDCs was interpreted as due to exsolution of huge amount of volatiles in the magma (Fedele et al. 2007; Moretti et al. 2019). The earliest phase of the NYT eruption, which laid down the Lower Member, was the largest trachytic phreatoplinian event ever recognised (Orsi et al. 1992). At least once, two contemporaneous eruptions (Averno 2 and Solfatara) occurred in different sites within the NYT caldera floor (Isaia et al. 2009). Both the growth of the Astroni volcano (Isaia et al. 2004; Tonarini et al. 2009) and the Averno 2 eruption (Di Vito et al. 2011) experienced vent migration over a limited area. The Astroni volcano was built through seven eruptions, five explosive, one effusive, and a last one poorly explosive to effusive, following each other at very short time intervals.

The volcanic and deformation history of the CFc clearly highlights a close relation among magmatic, volcanic and deformation processes at all scales, such as different size caldera collapses, caldera floor resurgence, areal distribution of active vents through time, direction of vent migration during a single eruption, areal distribution of magmas with different chemical composition erupted through time.

In light of the current knowledge of the volcanic and deformation history of the CFc, and also taking into account the present state of the volcano and the high level of both hazards and risk (Chap. Volcanic Hazard Assessment at the Campi Flegrei Caldera, Italy), some crucial topics of the volcanic system should be further investigated. One of these topics is the relations among eruption dynamics, magma withdrawal and structural deformation at regional and local (caldera collapse) scale prior to, during and after the CI eruption. Another topic concerns the dynamics of the NYT caldera resurgence through time by means of integrated inland and marine geological, volcanological, petrological, geochemical and geophysical investigations. A further topic is the relations between short-term and long-term deformation dynamics.

The results of these investigations would increase our understanding of both the evolution and present state of the volcano, thus the capability of the scientific community to interpret its signals in order to put forward well constrained hypotheses on its future behaviour, and finally to improve the current volcanic hazard assessment.

Acknowledgements I feel deeply indebted to Paola Petrosino for her invaluable contribution to the preparation of this chapter. Without her contribution, I would have not completed my work. I warmly thank Raffaello Cioni and Michael Ort for very thorough reviews of the first version of the manuscript.

References

Acocella V (2008) Activating and reactivating pairs of nested collapses during caldera forming eruptions: Campi Flegrei (Italy). Geophys Res Lett 35:L17304. https://doi.org/10.1029/2008GL035078

Acocella V (2010) Evaluating fracture patterns within a resurgent caldera: Campi Flegrei, Italy. Bull Volcanol 72:623–638. https://doi.org/10.1007/s00445-010-0347-x

Acocella V, Funiciello R (2006) Transverse systems along the extensional Tyrrhenian margin of central Italy and their influence on volcanism. Tectonics 25, TC2003. https://doi.org/10.1029/2005TC001845

Acocella V, Cifelli F, Funiciello R (2001) The control of overburden thickness on resurgent domes: insights from analogue models. J Volcanol Geotherm Res 111:137–153. https://doi.org/10.1016/S0377-0273(01)00224-4

Acocella V, Funiciello R, Marotta E, Orsi G, de Vita S (2004) The role of extensional structures on experimental calderas and resurgence. J Volcanol Geotherm Res 129:199–217. https://doi.org/10.1016/S0377-0273(03)00240-3

Acocella V, Di Lorenzo R, Newhall C, Scandone R (2015) An overview of recent (1988 to 2014) caldera unrest: knowledge and perspectives. Rev Geophys 53:896–955. https://doi.org/10.1002/2015RG000492

Albert PG, Giaccio B, Isaia R, Costa A, Niespolo EM, Nomade S, Pereira A, Renne PR, Hinchliffe A, Mark DF, Brown RJ, Smith VC (2019) Evidence for a large-magnitude eruption from Campi Flegrei caldera (Italy) at 29 ka. Geology 47:1–5. https://doi.org/10.1130/G45805.1

Albore Livadie C (2002) A first Pompeii: the early bronze age village of Nola-Croce del Papa (Palma Campania phase). Antiquity 76(294):941–942. https://doi.org/10.1017/S0003598X00091675

Alessio M, Bella F, Improta S, Belluomini G, Cortesi C, Turi B (1971) University of Rome carbon-14 dates IX. Radiocarbon 13(2):395–411. https://doi.org/10.1017/S0033822200008511

Alessio M, Bella F, Improta S, Belluomini G, Calderoni G, Cortesi C, Turi B (1973) University of Rome carbon-14 dates X. Radiocarbon 15(1):165–178. https://doi.org/10.1017/S0033822200058677

Alessio M, Bella F, Improta S, Belluomini G, Calderoni G, Cortesi C, Turi B (1974) University of Rome carbon-14 dates XII. Radiocarbon 16:358–367. https://doi.org/10.1017/S0033822200059658

Amato V, Aucelli PPC, Cesarano M, Filocamo F, Leone N, Petrosino P, Rosskopf CM, Valente E, Casciello E, Giralt S, Jicha BR (2018) Geomorphic response to late quaternary tectonics in the axial portion of the Southern Apennines (Italy): a case study from the Calore River valley. Earth Surf Proc Land 43 (11):2463–2480. https://doi.org/10.1002/ESP.4390

Arienzo I, Civetta L, Heumann A, Wörner G, Orsi G (2009) Isotopic evidence for open system processes within the Campanian Ignimbrite magma chamber. Bull Volcanol 71:285–300. https://doi.org/10.1007/s00445-008-0223-0

Arienzo I, Moretti R, Civetta L, Orsi G, Papale P (2010) The feeding system of Agnano-Monte Spina eruption (Campi Flegrei caldera, Italy): dragging the past into present activity and future scenarios. Chem Geol 270:135–147. https://doi.org/10.1016/j.chemgeo.2009.11.012

Arienzo I, Heumann A, Wörner G, Civetta L, Orsi G (2011) Processes and timescales of magma evolution prior to the Campanian Ignimbrite eruption (Campi Flegrei, Italy). Earth Planet Sci Lett 306(3–4):17–228. https://doi.org/10.1016/j.epsl.2011.04.002

Arienzo I, Mazzeo FC, Moretti R, Cavallo A, D'Antonio M (2016) Open-system magma evolution and fluid transfer at Campi Flegrei caldera (Southern Italy) during the past 5 ka as revealed by geochemical and isotopic data: the example of the Nisida eruption. Chem Geol 427:109–124. https://doi.org/10.1016/j.chemgeo.2016.02.007

Arzilli F, Piochi M, Mormone A, Agostini C, Carroll MR (2016) Constraining pre-eruptive magma conditions and unrest timescales during the Monte Nuovo eruption (1538 AD; Campi Flegrei, Southern Italy): integrating textural and CSD results from experimental and natural trachy-phonolites. Bull Volcanol 78:72. https://doi.org/10.1007/s00445-016-1062-z

Aster RC, Meyer RP (1988) Three-dimensional velocity structure and hypocenter distribution in the Campi Flegrei caldera, Italy. Tectonophysics 149:195–218. https://doi.org/10.1016/0040-1951(88)90173-4

Auger E, Gasparini P, Virieux J, Zollo A (2001) Seismic evidence of an extended magmatic sill under Mt. Vesuvius. Science 294:1510–1512. https://doi.org/10.1126/science.1064893

Bailey RA, Dalrymple GB, Lanphere MA (1976) Volcanism, structure, and geochronology of Long Valley Caldera, Mono County, California. J Geophys Res Solid Earth 81:725–744. https://doi.org/10.1029/JB081i005p00725

Barberi F, Innocenti F, Lirer L, Munno R, Pescatore T, Santacroce R (1978) The Campanian Ignimbrite: a major prehistoric eruption in the Neapolitan area (Italy). Bull Volcanol 41:10–31. https://doi.org/10.1007/BF02597680

Barberi F, Cassano E, La Torre P, Sbrana A (1991) Structural evolution of Campi Flegrei caldera in light of volcanological and geophysical data. J Volcanol Geotherm Res 48:33–49. https://doi.org/10.1016/0377-0273(91)90031-T

Berrino G, Corrado G, Riccardi U (2008) Sea gravity data in the Gulf of Naples. A contribution to delineating the

structural pattern of the Phlegraean volcanic district. J Volcanol Geotherm Res 175:241–252. https://doi.org/10.1016/j.jvolgeores.2008.03.007
Bevilacqua A, Flandoli F, Neri A, Isaia R, Vitale S (2016) Temporal models for the episodic volcanism of Campi Flegrei caldera (Italy) with uncertainty quantification. J Geophys Res Solid Earth 121:7821–7845. https://doi.org/10.1002/2016JB013171
Bourne AJ, Lowe JJ, Trincardi F, Asioli A, Blockley S, Wulf S, Matthews IP, Piva A, Vigliotti L (2010) Distal tephra record for the last ca 105,000 years from core PRAD 1–2 in the central Adriatic Sea: implications for marine tephrostratigraphy. Quat Sci Rev 29(23–24):3079–3094. https://doi.org/10.1016/j.quascirev.2010.07.021
Bourne A, Albert P, Matthews I, Trincardi F, Wulf S, Asioli A, Blockley S, Keller J, Lowe J (2015) Tephrochronology of core PRAD 1–2 from the Adriatic Sea: insights into Italian explosive volcanism for the period 200–80 ka. Quat Sci Rev 116:28–43. https://doi.org/10.1016/j.quascirev.2015.03.006
Branney MJ, Kokelaar P (1994) Volcanotectonic faulting, soft-state deformation, and rheomorphism of tuffs during development of a piecemeal caldera, English Lake District. Geol Soc Am Bull 106(4):507–530. https://doi.org/10.1130/0016-7606(1994)106%3c0507:VFSSDA%3e2.3.CO;2
Branney M, Acocella V (2015) Calderas. In: Sigurdsson H, Houghton B, McNutt S, Rymer H, Stix J (eds) Encyclopaedia of volcanoes, 2nd edn. Academic Press, San Diego (CA), USA, pp 249–315. ISBN: 9780123859389
Breislack S (1778) Topografia Fisica della Campania. Brazzini, Firenze, Italy, pp 1–368
Bronk Ramsey C, Albert PG, Blockley SPE, Hardiman M, Housley RA, Lane CS, Lee S, Matthews IP, Smith VC, Lowe JJ (2015) Improved age estimates for key Late Quaternary European tephra horizons in the RESET lattice. Quat Sci Rev 118:18–32. https://doi.org/10.1016/j.quascirev.2014.11.007
Brown RJ, Orsi G, de Vita S (2008) New insights into Late Pleistocene explosive volcanic activity and caldera formation on Ischia (southern Italy). Bull Volcanol 70(5):583–603. https://doi.org/10.1007/s00445-007-0155-0
Brown RJ, Civetta L, Arienzo I, D'Antonio M, Moretti R, Orsi G, Tomlinson EL, Albert PG, Menzies MA (2014) Assembly, evolution and disruption of a magmatic plumbing system before and after a cataclysmic caldera-collapse eruption at Ischia volcano (Italy). Contrib Mineral Petrol 168:1035. https://doi.org/10.1007/s00410-014-1035-1
Bruno PP (2004) Structure and evolution of the Bay of Pozzuoli (Italy) using marine seismic reflection data: implications for collapse of the Campi Flegrei caldera. Bull Volcanol 66:342–355. https://doi.org/10.1007/s00445-003-0316-9
Bruno PPG, Rapolla A, Di Fiore V (2003) Structural settings of the Bay of Naples (Italy) by seismic reflection data: implications for the Campanian volcanism. Tectonophysics 372:193–213. https://doi.org/10.1016/50040-1951(03)00327-5
Calanchi N, Dinelli E (2008) Tephrostratigraphy of the last 170 ka in sedimentary successions from the Adriatic Sea. J Volcanol Geotherm Res 177:81–95. https://doi.org/10.1016/j.jvolgeores.2008.06.008
Calcaterra D, Orsi G, Parise M, de Vita S, Di Vito MA, Palma B, Coppin D (2007) Slope processes in weathered volcaniclastic rocks of the Naples area: the Camaldoli Hill case. Geomorphology 87:132–157. https://doi.org/10.1016/j.geomorph.2006.03.040
Cameli GM, Rendina M, Puxeddu M, Rossi A, Squarci P, Taffi L (1975) Geothermal research in Western Campania (Southern Italy) geological and geophysical results. In: Lawrence Berkeley Laboratory, University of California (ed) Second United Nations symposium on the development and use of geothermal resources. Proceedings, Library of Congress, vol 1, pp 315–328
Cappelletti P, Cerri G, Colella A, de' Gennaro M, Langella A, Perrotta A, Scarpati C (2003) Post-eruptive processes in the Campanian Ignimbrite. Mineral Petrol 79:79–97. https://doi.org/10.1007/s00710-003-0003-7
Capuano P, Russo G, Civetta L, Orsi G, D'Antonio M, Moretti R (2013) The Campi Flegrei caldera structure imaged by 3-D inversion of gravity data. Geochem Geophys Geosyst 14:4681–4697. https://doi.org/10.1002/ggge.20276
Caron B, Sulpizio R, Zanchetta G, Siani G, Santacroce R (2010) The Late Holocene to Pleistocene tephrostratigraphic record of Lake Ohrid (Albania). Comp Rend Geosci 342:453–466. https://doi.org/10.1016/j.crte.2010.03.007
Cashman KV, Giordano G (2014) Calderas and magma reservoirs. J Volcanol Geotherm Res 288:28–45. https://doi.org/10.1016/j.jvolgeores.2014.09.007
Cassignol C, Gillot PY (1982) Range and effectiveness of unspiked potassium-argon dating: experimental ground work and application. In: Odin GS (ed) Numerical dating in stratigraphy. Wiley, New York (NY), USA, pp 159–179
Chiodini G, Paonita A, Aiuppa A, Costa A, Caliro S, De Martino P, Acocella V, Vandemeulebrouck J (2016) Magmas near the critical degassing pressure drive volcanic unrest towards a critical state. Nat Commun 7:13712. https://doi.org/10.1038/ncomms13712 | https://www.nature.com/ncomms/
Cinque A, Rolandi G, Zamparelli V (1985) L'estensione dei depositi marini olocenici nei Campi Flegrei in relazione alla vulcano-tettonica. Boll Soc Geol It 104 (2):327–348
Cinque A, Patacca E, Scandone P, Tozzi M (1993) Quaternary kinematic evolution of the Southern Apennines. Relationships between surface geological features and deep lithospheric structures. Ann Geophys 36:249–260
Cioni R, Santacroce R, Sbrana A (1999) Pyroclastic deposits as a guide for reconstructing the multi-stage evolution of the Somma-Vesuvius caldera. Bull Volcanol 60:207–222. https://doi.org/10.1007/s004450050272

Cioni R, Bertagnini A, Santacroce R, Andronico D (2008) Explosive activity and eruption scenarios at Somma-Vesuvius (Italy): towards a new classification scheme. J Volcanol Geotherm Res 178:331–346. https://doi.org/10.1016/j.jvolgeores.2008.04.024

Civetta L, Orsi G, Pappalardo L, Fisher RV, Heiken GH, Ort M (1997) Geochemical zoning, mixing, eruptive dynamics and depositional processes—the Campanian Ignimbrite, Campi Flegrei, Italy. J Volcanol Geotherm Res 75:183–219. https://doi.org/10.1016/S0377-0273(96)00027-3

Cole PD, Scarpati C (1993) A facies interpretation of the eruption and emplacement mechanisms of the upper part of the Neapolitan Yellow Tuff, Campi Flegrei, southern Italy. Bull Volcanol 55:311–326. https://doi.org/10.1007/BF00301143

Cole JW, Milner DM, Spinks KD (2005) Calderas and caldera structures: a review. Earth-Sci Rev 69:1–26. https://doi.org/10.1016/j.earscirev.2004.06.004

Cornell W, Carey S, Sigurdsson H (1983) Computer simulation of transport and deposition of Campanian Y-5 ash. J Volcanol Geotherm Res 17:89–109. https://doi.org/10.1016/0377-0273(83)90063-X

Costa A, Folch A, Macedonio G, Giaccio B, Isaia R, Smith VC (2012) Quantifying volcanic ash dispersal and impact of the Campanian Ignimbrite super-eruption. Geophys Res Lett 39:L10310. https://doi.org/10.1029/2012GL051605

Crocitti M, Sulpizio R, Insinga DD, De Rosa R, Donato P, Iorio M, Zanchetta G, Barca D, Lubritto C (2019) On ash dispersal from moderately explosive volcanic eruptions: examples from Holocene and Late Pleistocene eruptions of Italian volcanoes. J Volcanol Geotherm Res 385:198–221. https://doi.org/10.1016/j.jvolgeores.2018.07.009

Dallai L, Cioni R, Boschi C, D'Oriano C (2011) Carbonate-derived CO_2 purging magma at depth: influence on the eruptive activity of Somma-Vesuvius, Italy. Earth Planet Sci Lett 310(1–2):84–95. https://doi.org/10.1016/j.epsl.2011.07.013

Damaschke M, Sulpizio R, Zanchetta G, Wagner B, Bohm A, Nowaczyk N, Rethemeyer J, Hilgers A (2013) Tephrostratigraphic studies on a sediment core from Lake Prespa in the Balkans. Clim Past 9:267–287. https://doi.org/10.5194/cp-9-267-2013

D'Antonio M, Civetta L, Orsi G, Pappalardo L, Piochi M, Carandente A, de Vita S, Di Vito MA, Isaia R (1999) The present state of the magmatic system of the Campi Flegrei caldera based on the reconstruction of its behaviour in the past 12 ka. J Volcanol Geotherm Res 91(2–4):247–268. https://doi.org/10.1016/S0377-0273(99)00038-4

D'Antonio M, Tonarini S, Arienzo I, Civetta L, Di Renzo V (2007) Components and processes in the magma genesis of the Phlegrean Volcanic District, southern Italy. In: Beccaluva L, Bianchini G, Wilson M (eds) Cenozoic volcanism in the Mediterranean area. Boulder, CO, USA, Geological Society of American Special Paper, vol 418, pp 203–220. https://doi.org/10.1130/2007.2418(10)

D'Antonio M, Mariconte R, Arienzo I, Mazzeo FC, Carandente A, Perugini D, Petrelli M, Corselli C, Orsi G, Principato MS, Civetta L (2016) Combined Sr-Nd isotopic and geochemical fingerprinting as a tool for identifying tephra layers: application to deep-sea cores from Eastern Mediterranean Sea. Chem Geol 443:121–136. https://doi.org/10.1016/j.chemgeo.2016.09.022

Davies SM, Branch NP, Lowe JJ, Turney CSM (2002) Towards a European tephrochronological framework for Termination 1 and the Early Holocene. Phil Trans R Soc Lond A 360:767–802. https://doi.org/10.1098/rsta.2001.0964

De Bonitatibus A, Latmiral G, Mirabile L, Palumbo A, Sarpi E, Scalera A (1970) Rilievi sismici per riflessione: strutturali, ecografici (fumarole) e batimetrici nel Golfo di Pozzuoli. Boll Soc Nat Napoli, Italy 79:97–115

Deino AL, Orsi G, Piochi M, de Vita S (2004) The age of the Neapolitan Yellow Tuff caldera-forming eruption (Campi Flegrei caldera—Italy) assessed by $^{40}Ar/^{39}Ar$ dating method. J Volcanol Geotherm Res 133:157–170. https://doi.org/10.1016/S0377-0273(03)00396-2

Del Gaudio C, Aquino I, Ricciardi GP, Ricco C, Scandone R (2010) Unrest episodes at Campi Flegrei: a reconstruction of vertical ground movements during 1905–2009. J Volcanol Geotherm Res 195:48–56. https://doi.org/10.1016/j.jvolgeores.2010.05.014

Dellino P, Isaia R, La Volpe L, Orsi G (2004) Interaction between particles transported by fallout and surge in the deposits of the Agnano-Monte Spina eruption (Campi Flegrei, Southern Italy). J Volcanol Geotherm Res 133(1–4):193–210. https://doi.org/10.1016/S0377-0273(03)00398-6

De Natale G, Troise C, Pingue F, Mastrolorenzo G, Pappalardo L, Boschi E (2006) The Campi Flegrei caldera: unrest mechanisms and hazards. In: Troise C, De Natale G, Kilburn CRJ (eds) Mechanisms of activity and unrest at large calderas vol 269. Geological Society London Special Publication, pp 25–45. https://doi.org/10.1144/GSL.SP.2006.269,01.03

De Natale G, Troise C, Mark D, Mormone A, Piochi M, Di Vito MA, Isaia R, Carlino S, Barra D, Somma R (2016) The Campi Flegrei Deep Drilling Project (CFDDP): new insight on caldera structure, evolution and hazard implications for the Naples area (Southern Italy). Geochem Geophys Geosyst 17:4836–4847. https://doi.org/10.1002/2015GC006183

De Natale G, Troise C, Kilburn CRJ, Somma R, Moretti R (2017) Understanding volcanic hazard at the most populated caldera in the world: Campi Flegrei, Southern Italy. Geochem Geophys Geosyst 18:2004–2008. https://doi.org/10.1002/2017GC006972

De Pippo T, Di Cara A, Guida M, Pescatore T, Renda P (1984) Contributi allo studio del Golfo di Pozzuoli: lineamenti di geomorfologia. Mem Soc Geol It 27:151–159

De Siena L, Sammarco C, Cornwell DG, La Rocca M, Bianco F, Zaccarelli L, Nakahara H (2018) Ambient seismic noise image of the structurally controlled heat and fluid feeder pathway at Campi Flegrei caldera.

Geophys Res Lett 45:6428–6436. https://doi.org/10.1029/2018GL078817

de Vita S, Orsi G, Civetta L, Carandente A, D'Antonio M, Di Cesare T, Di Vito M, Fisher RV, Isaia R, Marotta E, Ort M, Pappalardo L, Piochi M, Southon J (1999) The Agnano-Monte Spina eruption (4.1 ka) in the resurgent, nested Campi Flegrei caldera (Italy). J Volcanol Geotherm Res 91:269–301. https://doi.org/10.1016/S0377-0273(99)00039-6

de Vita S, Sansivero F, Orsi G, Marotta E, Piochi M (2010) Volcanological and structural evolution of the Ischia resurgent caldera (Italy) over the past 10 ka. In: Groppelli G, Viereck-Goette L (eds) Stratigraphy and geology of volcanic areas, vol 464. Geological Society American Special Paper, pp 193–241. https://doi.org/10.1130/2010.2464(10)

De Vivo B, Rolandi G, Gans PB, Calvert A, Bohrson WA, Spera FJ, Belkin HE (2001) New constraints on the pyroclastic eruptive history of the Campanian volcanic Plain (Italy). Mineral Petrol 73:47–65. https://doi.org/10.1007/s007100170010

Di Girolamo P (1970) Differenziazione gravitativa e curve isochimiche nella "Ignimbrite Campana" (Tufo Grigio Campano auct.). Rend Soc It Mineral Petrol 26:1–45

Di Girolamo P (1978) Geotectonic settings of Miocene Quaternary volcanism in and around the eastern Tyrrhenian Sea border (Italy) as deduced from major element geochemistry. Bull Volcanol 41:229–250. https://doi.org/10.1007/BF02597225

Di Girolamo P, Ghiara MR, Lirer L, Munno R, Rolandi G, Stanzione D (1984) Vulcanologia e petrologia dei Campi Flegrei. Boll Soc Geol It 103:349–413. https://doi.org/10.1007/BF02597225

Di Renzo V, Civetta L, D'Antonio M, Tonarini S, Di Vito MA, Orsi G (2011) The magmatic feeding system of the Campi Flegrei caldera: architecture and temporal evolution. Chem Geol 281:227–241. https://doi.org/10.1016/j.chemgeo.2010.12.010

Di Renzo V, Wohletz K, Civetta L, Moretti R, Orsi G, Gasparini P (2016) The thermal regime of the Campi Flegrei magmatic system reconstructed through 3d numerical simulations. J Volcanol Geotherm Res 328:210–221. https://doi.org/10.1016/j.jvolgeores.2016.11.004

Di Vito MA, Isaia R, Orsi G, Southon J, de Vita S, D'Antonio M, Pappalardo L, Piochi M (1999) Volcanism and deformation since 12,000 years at the Campi Flegrei caldera. J Volcanol Geotherm Res 91:221–246. https://doi.org/10.1016/S0377-0273(99)00037-2

Di Vito MA, Arienzo I, Braia G, Civetta L, D'Antonio M, Di Renzo V, Orsi G (2011) The Averno 2 fissure eruption: a recent small-size explosive event at the Campi Flegrei caldera (Italy). Bull Volcanol 73:295–320. https://doi.org/10.1007/s00445-010-0417-0

Di Vito MA, Acocella V, Aiello G, Barra D, Battaglia M, Carandente A, Del Gaudio C, de Vita S, Ricciardi GP, Ricco C, Scandone R, Terrasi F (2016) Magma transfer at Campi Flegrei caldera (Italy) before the 1538 AD eruption. Sci Rep 6:32245. https://doi.org/10.1038/srep32245

D'Oriano C, Poggianti E, Bertagnini A, Cioni R, Landi P, Polacci M, Rosi M (2005) Changes in eruptive style during the AD 1538 Monte Nuovo eruption (Phlegrean Fields, Italy): the role of syn-eruptive crystallization. Bull Volcanol 67(7):601–621. https://doi.org/10.1007/s00445-004-0397-z

Douka K, Jacobs Z, Lane C, Grün R, Farr L, Hunt C, Inglis RH, Reynolds T, Albert P, Aubert M, Cullen V, Hill E, Kinsley L, Roberts RG, Tomlinson EL, Wulf S, Barker G (2014) The chronostratigraphy of the Haua Fteah cave (Cyrenaica, northeast Libya). J Hum Evol 66:39–63. https://doi.org/10.1016/j.jhevol.2013.10.001

Druitt TH, Edwards L, Mellors RM, Pyle DM, Sparks RSJ, Lanphere M, Davies M, Barreiro B (1999) Santorini volcano. Geol Soc London Mem 19:1–176. https://doi.org/10.1144/GSL.MEM.1999.019.01.12

Eisner W, Kamermans H, Wymstra AT (1986) The Agro Pontino survey: results from a first pollen core in Prospettive storico-antropologiche in Archeologia preistorica. Seconda parte. 5-Metodi della ricerca sul territorio e modelli di insediamento. Dialoghi di archeologia 4(2):145–153

Fedele FG, Giaccio B, Isaia R, Orsi G, Carroll M, Scaillet B (2007) The Campanian Ignimbrite factor: towards a reappraisal of the Middle to Upper Palaeolithic "Transition". In: Grattan J, Torrence R (eds) Living under the shadow: the cultural impacts of volcanic eruptions. One World Archaeology Series 53. Left Coast Press, Walnutt Creek, CA, USA, pp 19–41. ISBN-13:978-1-59874-268-8

Fedele L, Scarpati C, Lanphere M, Melluso L, Morra V, Perrotta A, Ricci G (2008) The Breccia Museo formation, Campi Flegrei, southern Italy: geochronology, chemostratigraphy and relationship with the Campanian Ignimbrite eruption. Bull Volcanol 70:1189–1219. https://doi.org/10.1007/s00445-008-0197-y

Fedele L, Insinga DD, Calvert AT, Morra V, Perrotta A, Scarpati C (2011) $^{40}Ar/^{39}Ar$ dating of tuff vents in the Campi Flegrei caldera (southern Italy): toward a new chronostratigraphic reconstruction of the Holocene volcanic activity. Bull Volcanol 73:1323–1336. https://doi.org/10.1007/s00445-011-0478-8

Fedi M, Cella F, D'Antonio M, Florio G, Paoletti V, Morra V (2018) Gravity modeling finds a large magma body in the deep crust below the Gulf of Naples, Italy. Sci Rep 8:8229. https://doi.org/10.1038/s41598-018-26346-z

Finetti I, Morelli C (1974) Esplorazione sismica a riflessione dei Golfi di Napoli e Pozzuoli. Boll Geofis Teor Appl 16(62–63):175–222

Finlayson DM, Gudmundsson O, Itikarai I, Nishimura Y, Shimamura H (2003) Rabaul volcano, Papua New Guinea: seismic tomographic imaging of an active caldera. J Volcanol Geotherm Res 124:153–171. https://doi.org/10.1016/S0377-0273(02)00472-9

Fisher RV, Orsi G, Ort M, Heiken G (1993) Mobility of large-volume pyroclastic flow—emplacement of the Campanian Ignimbrite, Italy. J Volcanol Geotherm Res 56:205–220. https://doi.org/10.1016/0377-0273(93)90017-L

Florio G, Fedi M, Cella F, Rapolla A (1999) The Campanian Plain and Phlegrean Fields: structural setting from potential field data. J Volcanol Geotherm Res 91:361–379. https://doi.org/10.1016/S0377-0273(99)00044-X

Forni F, Bachmann O, Mollo S, De Astis G, Gelman SE, Ellis BS (2016) The origin of a zoned ignimbrite: insights into the Campanian Ignimbrite magma chamber (Campi Flegrei, Italy). Earth Planet Sci Lett 449:259–271. https://doi.org/10.1016/j.epsl.2016.06.003

Forni F, Degruyter W, Bachmann O, De Astis G, Mollo S (2018a) Long-term magmatic evolution reveals the beginning of a new caldera cycle at Campi Flegrei. Sci Adv 4(11):eaat9401. https://doi.org/10.1126/sciadv.aat9401

Forni F, Petricca E, Bachmann O, Mollo S, De Astis G, Piochi M (2018b) The role of magma mixing/mingling and cumulate melting in the Neapolitan Yellow Tuff caldera-forming eruption (Campi Flegrei, Southern Italy). Contrib Mineral Petrol 173:45. https://doi.org/10.1007/s00410-018-1471-4

Galli P, Giaccio B, Messina P, Peronace E, Amato V, Naso G, Nomade S, Pereira A, Piscitelli S, Bellanova J, Billi A, Blamart D, Galderisi A, Giocoli A, Stabile T, Thil F (2017) Middle to Late Pleistocene activity of the northern Matese fault system (southern Apennines, Italy). Tectonophysics 699:61–81. https://doi.org/10.1016/i.tecto.2017.01.007

Geyer A, Martí J (2008) The new worldwide collapse caldera database (CCDB): a tool for studying and understanding caldera processes. J Volcanol Geotherm Res 175(3):334–354. https://doi.org/10.1016/j.jvolgeores.2008.03.017

Giaccio B, Galli P, Peronace E, Arienzo I, Nomade S, Cavinato GP, Mancini M, Messina P, Sottili G (2014) A 560–440 ka tephra record from the Mercure Basin, southern Italy: volcanological and tephrostratigraphic implications. J Quat Sci 29(3):232–248. https://doi.org/10.1002/jqs.2696

Giaccio B, Hajdas I, Isaia R, Deino A, Nomade S (2017) High-precision ^{14}C dating and $^{40}Ar/^{39}Ar$ dating of the Campanian Ignimbrite (Y-5) reconciles the time-scales of climatic-cultural processes at 40 ka. Sci Rep 7:45940. https://doi.org/10.1038/srep45940

Giudicepietro F, Macedonio G, Martini M (2017) A physical model of sill expansion to explain the dynamics of unrest at calderas with application to Campi Flegrei. Front Earth Sci 5:54. https://doi.org/10.3389/feart.2017.00054

Golovanova LV, Doronichev VB, Cleghorn NE, Koulkova MA, Sapelko TV, Shackley MS (2010) Significance of ecological factors in the Middle to Upper Paleolithic transition. Curr Anthropol 51:655–691. https://doi.org/10.1086/656185

Guidoboni E, Ciuccarelli C (2011) The Campi Flegrei caldera: historical revision and new data on seismic crises, bradyseisms, the Monte Nuovo eruption and ensuing earthquakes (twelfth century 1582 AD). Bull Volcanol 73:655–677. https://doi.org/10.1007/s00445-010-0430-3

Gunter RT (1903) Contributions to the study of earth-movements in the Bay of Naples. Nichols JB and Sons, Parliament Mansions Westminster

Heiken G, Goff F, Gardner JN, Baldridge WS, Hulen JB, Nielson DL, Vaniman D (1990) The Valles/Toledo caldera complex, Jemez Volcanic Field, New Mexico. Ann Rev Earth Planet Sci 18:27–53. https://doi.org/10.1146/annurev.ea.18.050190.000331

Hildreth W, Fierstein J (2000) Katmai volcanic cluster and the great eruption of 1912. Geol Soc Am Bull 112(10):1594–1620. https://doi.org/10.1130/0016-7606(2000)112%3c1594:KVCATG%3e2.0.CO;2

Hill DP (1976) Structure of Long Valley caldera, California, from a seismic refraction experiment. J Geophys Res Solid Earth 81(5):745–753. https://doi.org/10.1029/JB081i005p00745

Husen S, Smith RB, Waite GP (2004) Evidence for gas and magmatic sources beneath the Yellowstone volcanic field from seismic tomographic imaging. J Volcanol Geotherm Res 131(3–4):397–410. https://doi.org/10.1016/S0377-0273(03)00416-5

Insinga DD, Tamburrino S, Lirer F, Vezzoli L, Barra M, De Lange GJ, Tiepolo M, Vallefuoco M, Mazzola S, Sprovieri M (2014) Tephrochronology of the astronomically-tuned KC01B deep-sea core, Ionian Sea: insights into the explosive activity of the Central Mediterranean area during the last 200 ka. Quat Sci Rev 85:63–84. https://doi.org/10.1016/j.quascirev.2013.11.019

Isaia R, D'Antonio M, Dell'Erba F, Di Vito MA, Orsi G (2004) The Astroni volcano: the only example of close eruptions within the same vent area in the recent history of the Campi Flegrei caldera (Italy). J Volcanol Geotherm Res 133:171–192. https://doi.org/10.1016/S0377-0273(03)00397-4

Isaia R, Marianelli P, Sbrana A (2009) Caldera unrest prior to intense volcanism in Campi Flegrei (Italy) at 4.0 ka B. P.: implications for caldera dynamics and future eruptive scenarios. Geophys Res Lett 36:L21303. https://doi.org/10.1029/2009GL040513

Isaia R, Vitale S, Marturano A, Aiello G, Barra D, Ciarcia S, Iannuzzi E, D'Assisi Tramparulo F (2019) High-resolution geological investigations to reconstruct the long-term ground movements in the last 15 kyr at Campi Flegrei caldera (southern Italy). J Volcanol Geotherm Res 385:143–158. https://doi.org/10.1016/j.jvolgeores.2019.07.012

Issel A (1883) Le oscillazioni lente del suolo o bradisismi. Atti R Univ Genova, Italy IV:1–210

Ivanova S, Gurova M, Spassov N, Hristova L, Tzankov N, Popov V, Marinova E, Makedonska J, Smith V, Ottoni C, Lewis M (2016) Magura Cave, Bulgaria: a multidisciplinary study of Late Pleistocene human palaeoenvironment in the Balkans. Quat Int

415:86–108. https://doi.org/10.1016/j.quaint.2015.11.082
Jolivet L, Faccenna C, Piromallo C (2009) From mantle to crust: stretching the Mediterranean. Earth Planet Sci Lett 285:198–209. https://doi.org/10.1016/j.epsl.2009.06.017
Judenherc S, Zollo A (2004) The Bay of Naples (southern Italy): constraints on the volcanic structures inferred from a dense seismic survey. J Geophys Res Solid Earth 109:B10312. https://doi.org/10.1029/2003JB002876
Keller J, Ryan WBF, Ninkovich D, Altherr R (1978) Explosive volcanic activity in the Mediterranean over the past 200,000 yr as recorded in deep-sea sediments. Geol Soc Am Bull 89:591–604. https://doi.org/10.1130/0016-7606(1978)89%3c591:EVAITM%3e2.0.00:2
Lambeck K, Antonioli F, Anzidei M, Ferranti L, Leoni G, Scicchitano G, Silenzi S (2011) Sea level change along the Italian coast during the Holocene and projections for the future. Quat Int 232(1–2):250–257. https://doi.org/10.1016/j.quaint.2010.04.026
Lane CS, Andri M, Cullen VL, Blockley SPE (2011) The occurrence of distal Icelandic and Italian tephra in the Lateglacial of Lake Bled, Slovenia. Quat Sci Rev 30:1013–1018. https://doi.org/10.1016/j.quascirev.2011.02.014
Lane CS, Brauer A, Martín-Puertas C, Blockley SPE, Smith VC, Tomlinson EL (2015) The Late Quaternary tephrostratigraphy of annually laminated sediments from Meerfelder Maar, Germany. Quat Sci Rev 122:192–206. https://doi.org/10.1016/j.quascirev.2015.05.025
Leicher N, Zanchetta G, Sulpizio R, Giaccio B, Wagner B, Nomade S, Francke A, Del Carlo P (2016) First tephrostratigraphic results of the DEEP site record from Lake Ohrid (Macedonia and Albania). Biogeoscience 13:2151–2178. https://doi.org/10.5194/bg-13-2151-2016
Lipman PW (1997) Subsidence of ash-flow calderas: relation to caldera size and magma chamber geometry. Bull Volcanol 59:198–218. https://doi.org/10.1007/s004450050186
Lirer L, Munno R (1975) Il tufo giallo napoletano (Campi Flegrei). Period Mineral 44:103–118
Lirer L, Luongo G, Scandone R (1987a) On the volcanological evolution of Campi Flegrei. EOS, Trans Am Geophys Union 68:226–234. https://doi.org/10.1029/E0068i016p00226
Lirer L, Mastrolorenzo G, Rolandi G (1987b) Un evento pliniano nell'attività recente dei Campi Flegrei. Boll Soc Geol It 106(3):461–473
Lirer L, Rolandi G, Rubin M (1991) ^{14}C age of the "Museum Breccia" (Campi Flegrei) and its relevance for the origin of the Campanian Ignimbrite. J Volcanol Geotherm Res 48(1–2):223–227. https://doi.org/10.1016/0377-0273(91)90044-Z
Lowe J, Barton N, Blockley S, Bronk Ramsey C, Cullen VL, Davies W, Gamble C, Grant K, Hardiman M, Housley R, Lane CS, Lee S, Lewis M, MacLeod A, Menzies M, Muller W, Pollard M, Price C, Roberts AP, Rohling EJ, Satow C, Smith V, Stringer C, Tomlinson EL, White D, Albert P, Arienzo I, Barker G, Carandente A, Civetta L, Ferrier C, Gaudelli JL, Karkanas P, Koumouzelis M, Muller UC, Orsi G, Pross J, Rosi M, Shalamanov-Korobar L, Sirakov N, Tzedakis PC (2012) Volcanic ash layers illuminate the resilience of Neanderthals and early modern humans to natural hazards. P Natl Acad Sci 109(34):13532–13537. https://doi.org/10.1073/pnas.1204579109
Magny M, De Beaulieu JL, Drescher-Schneider R, Vanniere B, Walter-Simonnet AV, Millet L, Bossuet G, Peyron O (2006) Climatic oscillations in central Italy during the Last Glacial-Holocene transition: the record from Lake Accesa. J Quat Sci 21(4):311–320. https://doi.org/10.1002/jqs.999
Makaroglu O, Nowaczyk NR, Eris KK, Namik Cagatay M (2020) High-resolution palaeomagnetic record from Sea of Marmara sediments for the last 70 ka. Geophys J Int 222(3):2024–2039. https://doi.org/10.1093/gji/ggaa281
Marsh BD (1984) On the mechanics of caldera resurgence. J Geophys Res Solid Earth 89(B10):8245–8251. https://doi.org/10.1029/JB089iB10p08245
Martí J, Geyer A, Folch A, Gottsmann J (2008) A review on collapse caldera modelling. In: Gottsmann J, Marti J (eds) Caldera volcanism; analysis, modelling and response, vol 10. Elsevier, Amsterdam, The Netherlands, Develop Volcanol, pp 233–284. https://doi.org/10.1016/S1871-644X(07)00006-X
Marti A, Folch A, Costa A, Engwell S (2016) Reconstructing the plinian and co-ignimbrite sources of large volcanic eruptions: a novel approach for the Campanian Ignimbrite. Sci Rep 6:21220. https://doi.org/10.1038/srep21220
Marturano A, Isaia R, Aiello G, Barra D (2018) Complex dome growth at Campi Flegrei Caldera (Italy) in the last 15 ka. J Geophys Res Solid Earth 123(9):8180–8197. https://doi.org/10.1029/2018JB015672
Matthews IP, Trincardi F, Lowe JJ, Bourne AJ, MacLeod A, Abbott PM, Andersen N, Asioli A, Blockley SPE, Lane CS, Oh YA, Satow CS, Staff RA, Wulf S (2015) Developing a robust tephrochronological framework for Late Quaternary marine records in the Southern Adriatic Sea: new data from core station SA03-11. Quat Sci Rev 118:84–104. https://doi.org/10.1016/j.quascirev.2014.10.009
Melluso L, De' Gennaro R, Fedele L, Franciosi L, Morra V (2012) Evidence of crystallization in residual, Cl-F-rich, agpaitic, trachyphonolitic magmas and primitive Mg-rich basalt-trachyphonolite interaction in the lava domes of the Phlegrean Fields (Italy). Geol Mag 149(3):532–550. https://doi.org/10.1017/S0016756811000902
Milia A, Torrente M (2003) Late-Quaternary volcanism and transtensional tectonics in the Bay of Naples, Campanian continental margin, Italy. Mineral Petrol 79(1–2):49–65. https://doi.org/10.1007/s00710-003-0001-9

Milia A, Torrente MM, Giordano E, Mirabile L (2006) Rapid changes of the accommodation space in the Late Quaternary succession of Naples Bay, Italy: the influence of volcanism and tectonics. In: De Vivo B (ed) Volcanism in the Campania Plain: Vesuvius, Campi Flegrei and Ignimbrites, vol 9. Elsevier, Amsterdam, The Netherlands, Develop Volcanol, pp 53–68

Moore I, Kokelaar P (1998) Tectonically controlled piecemeal caldera collapse: a case study of Glencoe volcano, Scotland. Geol Soc Am Bull 110(11):1448–1466. https://doi.org/10.1130/0016-7606(1998)110%3c1448:TCPCCA%3e2.3.00;2

Morabito S, Petrosino P, Milia A, Sprovieri M, Tamburrino S (2014) A multidisciplinary approach for reconstructing the stratigraphic framework of the last 40 ka in a bathyal area of the eastern Tyrrhenian Sea. Glob Planet Change 123(A):121–138. https://doi.org/10.1016/j.gloplacha.2014.10.005

Moretti R, Arienzo I, Civetta L, Orsi G, Papale P (2013a) Multiple magma degassing sources at an explosive volcano. Earth Planet Sci Lett 367:95–104. https://doi.org/10.1016/j.epsl.2013.02.013

Moretti R, Arienzo I, Orsi G, Civetta L, D'Antonio M (2013b) The deep plumbing system of the Ischia island (southern Italy): a physico-chemical and geodynamic window on the fluid-sustained and CO_2-dominated magmatic source of Campanian volcanoes. J Petrol 54(5):951–984. https://doi.org/10.1093/petrology/egt002

Moretti R, De Natale G, Troise C (2017) A geochemical and geophysical reappraisal to the significance of the recent unrest at Campi Flegrei caldera (Southern Italy). Geochem Geophys Geosyst 18:1244–1269. https://doi.org/10.1002/2016GC006569

Moretti R, Arienzo I, Di Renzo V, Orsi G, Arzilli F, Brun F, D'Antonio M, Mancini L, Deloule E (2019) Volatile segregation and generation of highly vesiculated explosive magmas by volatile-melt fining processes: the case of the Campanian Ignimbrite eruption. Chem Geol 503:1–14. https://doi.org/10.1016/j.chemgeo.2018.10.001

Mormone A, Piochi M, Bellatreccia F, De Astis G, Moretti R, Della Ventura G, Cavallo A, Mangiacapra A (2011) A CO_2-rich magma source beneath the Phlegraean Volcanic District (Southern Italy): evidence from a melt inclusion study. Chem Geol 287:66–80. https://doi.org/10.1016/j.chemgeo.2011.05.019

Munno R, Petrosino P (2007) The late quaternary tephrostratigraphical record of the San Gregorio Magno basin (southern Italy). J Quat Sci 22(3):247–266. https://doi.org/10.1002/jqs.1025

Narcisi B, Vezzoli L (1999) Quaternary stratigraphy of distal tephra layers in the Mediterranean—an overview. Glob Planet Change 21:31–50. https://doi.org/10.1016/S0921-8181(99)00006-5

Newhall CG, Self S (1982) The volcanic explosivity index (VEI) an estimate of explosive magnitude for historical volcanism. J Geophys Res Solid Earth 87:1231–1238. https://doi.org/10.1029/JC087iCO2p01231

Orsi G, Scarpati C (1989) Stratigrafia e dinamica eruttiva del Tufo Giallo Napoletano. Boll GNV 2:917–930

Orsi G, Civetta L, Aprile A, D'Antonio M, de Vita S, Gallo G, Piochi M (1991a) The Neapolitan Yellow Tuff: eruptive dynamics, emplacement mechanism and magma evolution of a phreatoplinian-to-plinian eruption. In: Orsi G, Rosi M (eds) Large Ignimbrite eruptions of the Phlegraean Fields caldera: the Neapolitan Yellow Tuff and The Campanian Ignimbrite. IAVCEI—Commission on Explosive Volcanism, Workshop on Explosive Volcanism 1–8 Sept, Napoli, Italy, Guidebook, pp 76–115

Orsi G, Gallo G, Zanchi A (1991b) Simple shearing block resurgence in caldera depressions. A model from Pantelleria and Ischia. J Volcanol Geotherm Res 47:1–11. https://doi.org/10.1016/0377-0273(91)900974

Orsi G, D'Antonio M, de Vita S, Gallo G (1992) The Neapolitan Yellow Tuff, a large-magnitude trachytic phreatoplinian eruption: eruptive dynamics, magma withdrawal and caldera collapse. J Volcanol Geotherm Res 53:275–287. https://doi.org/10.1016/0377-0273(92)90086-S

Orsi G, Civetta L, D'Antonio M, Di Girolamo P, Piochi M (1995) Step-filling and development of a three-layers magma chamber: the Neapolitan Yellow Tuff case history. J Volcanol Geotherm Res 67:291–312. https://doi.org/10.1016/0377-0273(94)00119-2

Orsi G, de Vita S, Di Vito M (1996) The restless, resurgent Campi Flegrei nested caldera (Italy): constraints on its evolution and configuration. J Volcanol Geotherm Res 74:179–214. https://doi.org/10.1016/S0377-0273(96)00063-7

Orsi G, Civetta L, Del Gaudio C, de Vita S, Di Vito MA, Isaia R, Petrazzuoli SM, Ricciardi G, Ricco C (1999a) Short-term ground deformations and seismicity in the nested Campi Flegrei caldera (Italy): an example of active block-resurgence in a densely populate area. J Volcanol Geotherm Res 91:415–451. https://doi.org/10.1016/S0377-0273(99)00050-5

Orsi G, Petrazzuoli S, Wohletz K (1999b) The interplay of mechanical and thermo-fluid dynamical systems during an unrest episode in calderas: the Campi Flegrei caldera (Italy) case. J Volcanol Geotherm Res 91:453–470. https://doi.org/10.1016/S0377-0273(99)00051-7

Orsi G, de Vita, Di Vito M, Isaia R, Nave R, Heiken G (2003) Facing volcanic and related hazards in the Neapolitan area. In: Heiken G, Fakundiny R, Sutter J (eds) Earth sciences in the cities: a reader. American Geophysical Union Special Publication Series, vol 56, pp 121–170. ISBN 0-87590-299-5

Orsi G, Di Vito MA, Isaia R (2004) Volcanic hazard assessment at the restless Campi Flegrei caldera. Bull Volcanol 66:514–530. https://doi.org/10.1007/s00445-003-0336-4

Orsi G, Di Vito M, Selva J, Marzocchi W (2009) Long-term forecast of eruption style and size at Campi Flegrei caldera (Italy). Earth Planet Sci Lett 287:265–276. https://doi.org/10.1016/j.epsl.2009.08.013

Orsi G, Cioni R, Di Renzo V (2013) The Campanian Plain during the Bronze Age: development of volcanism and impact of the Vesuvius Avellino eruption in a densely populated area. In: Meller H et al (eds) Proceedings of the 4th archaeological conference of central Germany "1600 - cultural change in the shadow of the Thera-Eruption?", pp 117–134

Ort MH, Rosi M, Anderson CA (1999) Correlation of deposits and vent locations of the proximal Campanian Ignimbrite deposits, Campi Flegrei, Italy, based on natural remanent magnetization and anisotropy of magnetic susceptibility characteristics. J Volcanol Geotherm Res 91:167–178. https://doi.org/10.1016/S0377-0273(99)00034-7

Ort M, Orsi G, Pappalardo L, Fisher RV (2003) Anisotropy of magnetic susceptibility studies of depositional processes in the Campanian Ignimbrite, Italy. Bull Volcanol 65:55–72. https://doi.org/10.1007/s00445-002-0241-2

Pabst S, Wörner G, Civetta L, Tesoro R (2008) Magma chamber evolution prior to the Campanian Ignimbrite and Neapolitan Yellow Tuff eruptions (Campi Flegrei, Italy). Bull Volcanol 70:961–976. https://doi.org/10.1007/s00445-007-0180-z

Pappalardo L, Civetta L, D'Antonio M, Deino A, Di Vito M, Orsi G, Carandente A, de Vita S, Isaia R, Piochi M (1999) Chemical and Sr—isotopical evolution of the Phlegraean magmatic system before the Campanian Ignimbrite (37 ka) and the Neapolitan Yellow Tuff (12 ka) eruptions. J Volcanol Geotherm Res 91:141–166. https://doi.org/10.1016/S0377-0273(99)00033-5

Pappalardo L, Civetta L, de Vita S, Di Vito M, Orsi G, Carandente A, Fisher RV (2002a) Timing of magma extraction during the Campanian Ignimbrite eruption (Campi Flegrei Caldera). J Volcanol Geotherm Res 114:479–497. https://doi.org/10.1016/S0377-0273(01)00302-X

Pappalardo L, Piochi M, D'Antonio M, Civetta L, Petrini R (2002b) Evidence for multi-stage magmatic evolution during the past 60 ka at Campi Flegrei (Italy) deduced from Sr, Nd and Pb isotope data. J Petrol 43:1415–1434. https://doi.org/10.1093/petrology/43.8.1415

Pappalardo L, Mastrolorenzo G (2012) Rapid differentiation in a sill-like magma reservoir: a case study from the Campi Flegrei caldera. Sci Rep 2:712. https://doi.org/10.1038/srep00712

Parascandola A (1946) Il Monte Nuovo e il Lago Lucrino. Boll Soc Nat Napoli, Italy 55:151–312

Parascandola A (1947) I fenomeni bradisismici del Serapeo di Pozzuoli. Genovese, Naples, Italy

Passariello I, Talamo P, D'Onofrio A, Barta P, Lubritto C, Terrasi F (2010) Contribution of the radiocarbon dating to the chronology of Eneolithic in Campania (Italy). Geochronometria 35:25–33. https://doi.org/10.2478/v10003-010-0008-2

Patacca E, Sartori R, Scandone P (1990) Tyrrhenian basin and Apenninic arcs: kinematic relations since late Tortonian times. Mem Soc Geol It 45:425–451

Paterne M, Guichard F, Labeyrie J, Gillot PY, Duplessy JC (1986) Tyrrhenian Sea tephrochronology of the oxygen isotope record for the past 60,000 yrs. Mar Geol 72:259–285. https://doi.org/10.1016/0025-3227(86)90123-4

Paterne M, Guichard F, Labeyrie J (1988) Explosive activity of the South Italian volcanoes during the past 80,000 years as determined by marine tephrochronology. J Volcanol Geotherm Res 34:153–172. https://doi.org/10.1016/0377-0273(88)90030-3

Paterne M, Guichard F, Duplessy JC, Siani G, Sulpizio R, Labeyrie J (2008) A 90,000–200,000 yrs marine tephra record of Italian volcanic activity in the Central Mediterranean Sea. J Volcanol Geotherm Res 177:187–196. https://doi.org/10.1016/j.jvolgeores.2007.11.028

Penta F (1954) Ricerche e studi sui fenomem esalativo-idrotermali e il problema delle "forze endogene." Ann Geofis 3:317–409

Perrotta A, Scarpati C, Luongo G, Morra V (2006) The Campi Flegrei caldera boundary in the city of Naples. In: De Vivo B (ed) Volcanism in the Campania Plain: Vesuvius, Campi Flegrei and Ignimbrites, vol 9. Elsevier, Amsterdam, The Netherlands, Develop Volcanol, pp 85–96

Perrotta A, Scarpati C, Luongo G, Morra V (2010) Stratigraphy and volcanological evolution of the southern sector of Campi Flegrei and Procida Island, Italy. In: Groppelli G, Viereck-Goette L (eds) Stratigraphy and geology of volcanic areas, vol 464. Geological Society American Special Paper, pp 171–191.https://doi.org/10.1130/2010.2464(09)

Petrosino P, Jicha BR, Mazzeo FC, Russo Ermolli E (2014a) A high resolution tephrochronological record of MIS 14–12 in the Southern Apennines (Acerno Basin, Italy). J Volcanol Geotherm Res 274:34–50. https://doi.org/10.1016/j.jvolgeores.2014.01.014

Petrosino P, Russo Ermolli E, Donato P, Jicha B, Robustelli G, Sardella R (2014b) Using Tephrochronology and palynology to date the MIS 13 lacustrine sediments of the Mercure Basin (Southern Apennines—Italy). Ital J Geosci 133:169–186. https://doi.org/10.3301/IJG.2013.22

Petrosino P, Morabito S, Jicha BR, Milia A, Sprovieri M, Tamburrino S (2016) Multidisciplinary tephrochronological correlation of marker events in the eastern Tyrrhenian Sea between 48 and 105 ka. J Volcanol Geotherm Res 315:79–99. https://doi.org/10.1016/j.jvolgeores.2014.01.014

Piochi M, Mastrolorenzo G, Pappalardo L (2005) Magma ascent and eruptive processes from textural and compositional features of Monte Nuovo pyroclastic products, Campi Flegrei, Italy. Bull Volcanol 67:663–678. https://doi.org/10.1007/s00445-005-0410-1

Piochi M, Kilburn CRJ, Di Vito MA, Mormone A, Tramelli A, Troise C, De Natale G (2014) The volcanic and geothermally active Campi Flegrei caldera: an integrated multidisciplinary image of its buried structure. Int J Earth Sci 103(2):401–421. https://doi.org/10.1007/s00531-013-0972-7

Pistolesi M, Isaia R, Marianelli P, Bertagnini A, Fourmentraux C, Albert P, Tomlinson E, Menzies M, Rosi M, Sbrana A (2016) Simultaneous eruptions from multiple vents at Campi Flegrei (Italy) highlight new eruption processes at calderas. Geology 44 (6):487–490. https://doi.org/10.1130/G37870.1

Pistolesi M, Bertagnini A, Di Roberto A, Isaia R, Vona A, Cioni R, Giordano G (2017) The Baia–Fondi di Baia eruption at Campi Flegrei: stratigraphy and dynamics of a multi-stage caldera reactivation event. Bull Volcanol 79:67. https://doi.org/10.1007/s00445-017-1149-1

Pyle DM, Ricketts GD, Margarib V, van Andela TH, Sinitsync AA, Praslovc ND, Lisitsyn S (2006) Wide dispersal and deposition of distal tephra during the Pleistocene 'Campanian Ignimbrite/Y5' eruption, Italy. Quat Sci Rev 25:2713–2728. https://doi.org/10.1016/j.quascirev.2006.06.008

Radic D, Lugovic B, Marjanac L (2007) Neapolitan Yellow Tuff (NYT) from the Pleistocene sediments in Vela Spila on the island of Korcula: a valuable chronostratigraphic marker of the transition from the Palaeolithic to the Mesolithic. Opus Archaeol 31:7–26

Ramrath A, Zolitschka B, Wulf S, Negendank JFW (1999) Late Pleistocene climatic variations as recorded in two Italian maar lakes (Lago di Mezzano, Lago Grande di Monticchio). Quat Sci Rev 18:977–992. https://doi.org/10.1016/S0277-3791(99)00009-8

Ricci Lucchi MA, Vigliotti L (1998) Facies e paleoambienti di depositi fluvio-lacustri tardo-quaternari (Piana di Rieti, Appennino centrale). Giorn Geol 60 (1):321–323

Ricco C, Petrosino S, Aquino I, Del Gaudio C, Falanga F (2019) Some investigations on a possible relationship between ground deformation and seismic activity at Campi Flegrei and Ischia Volcanic Areas (Southern Italy). Geosciences 9(5):222. https://doi.org/10.3390/geosciences9050222

Rittmann A (1950a) Sintesi geologica dei Campi Flegrei. Boll Soc Geol It 69:117–177

Rittmann A (1950b) Rilevamento geologico della Collina dei Camaldoli nei Campi Flegrei. Boll Soc Geol It 69:129–178

Rivalta E, Corbi F, Passarelli L, Acocella V, Davis T, Di Vito MA (2019) Stress inversions to forecast magma pathways and eruptive vent location. Sci Adv 5(7): eaau9784. https://doi.org/10.1126/sciadv.aau9784

Rolandi G, Bellucci F, Heizler MT, Belkin HE, De Vivo B (2003) Tectonic controls on the genesis of ignimbrites from the Campanian Volcanic Zone, southern Italy. Mineral Petrol 79:3–31. https://doi.org/10.1007/s00710-003-0014-4

Rosi M, Sbrana A (eds) (1987) Phlegrean Fields. Quad Ric Sci 114(9):1–175, CNR Rome, Italy. ISSN 0556-9664

Rosi M, Sbrana A, Principe C (1983) The Phlegrean Fields: structural evolution, volcanic history and eruptive mechanisms. J Volcanol Geotherm Res 17:273–288. https://doi.org/10.1016/0377-0273(83)90072-0

Rosi M, Vezzoli L, Aleotti P, De Censi M (1991) Campanian Ignimbrite eruption: proximal deposits. In: Orsi G, Rosi M (eds) Large Ignimbrite eruptions of the Phlegraean Fields caldera: the Neapolitan Yellow Tuff and The Campanian Ignimbrite. IAVCEI—Commission on Explosive Volcanism, Workshop on Explosive Volcanism, Napoli, 1–8 Sept, Guidebook

Rosi M, Vezzoli L, Aleotti P, De Censi M (1996) Interaction between caldera collapse and eruptive dynamics during the Campanian Ignimbrite eruption, Phlegraean Fields, Italy. Bull Volcanol 57:541–554. https://doi.org/10.1007/BF00304438

Rosi M, Vezzoli L, Castelmenzano A, Greco G (1999) Plinian pumice fall deposit of the Campanian Ignimbrite eruption (Phlegrean Fields Italy). J Volcanol Geotherm Res 91:179–198. https://doi.org/10.1016/S0377-0273(99)00035-9

Sacchi M, Alessio G, Aquino I, Esposito E, Molisso F, Nappi R, Porfido S, Violante C (2009) Risultati preliminari della campagna oceanografica CAFE_07—Leg 3 nei Golfi di Napoli e Pozzuoli, Mar Tirreno Orientale. Quad Geofis 64:1–25. ISSN 1590-2595

Sacchi M, Pepe F, Corradino M, Insinga DD, Molisso F, Lubritto C (2014) The Neapolitan Yellow Tuff caldera offshore the Campi Flegrei: Stratal architecture and kinematic reconstruction during the last 15 ky. Mar Geol 354(1):15–33. https://doi.org/10.1016/j.margeo.2014.04.012

Sacchi M, Caccavale M, Corradino M, Esposito G, Ferranti L, Hamori Z, Horvath F, Insinga D, Marino C, Matano F, Molisso F, Natale J, Passaro S, Pepe F, Toth T (2019) The use and beauty of ultra-high-resolution seismic reflection imaging in Late Quaternary marine volcaniclastic settings, Napoli Bay, Italy. Foldtani Kozlony 149(4):371. https://doi.org/10.23928/foldtkozl.2019.149.4.371

Saccoccio F, Marzocchella A, Vanzetti A (2013) The field system of Gricignano d'Aversa (Southern Italy) and the agrarian impact in the Piana Campana, ca. 3900 cal BP. Quat Int 303:8292. https://doi.org/10.1016/j.quaint.2013.02.021

Santacroce R, Cristofolini R, La Volpe L, Orsi G, Rosi M (2003) Italian active volcanoes. Episodes 26(3):227–234

Santacroce R, Cioni R, Marianelli P, Sbrana A, Sulpizio R, Zanchetta G, Joron JL (2008) Age and whole rock-glass compositions of proximal pyroclastics from the major explosive eruptions of Somma-Vesuvius: a review as a tool for distal tephrostratigraphy. J Volcanol Geotherm Res 177:1–18. https://doi.org/10.1016/j.volgeores.2008.06.009

Sbrana A, Marianelli P, Pasquini G (2018) Volcanology of Ischia (Italy). J Maps 14(2):494–503. https://doi.org/10.1080/17445647.2018.1498811

Sbrana A, Cioni R, Marianelli P, Sulpizio R, Andronico D, Pasquini G (2019) Volcanic evolution of the Somma-Vesuvius Complex (Italy). J Maps 16(2):137–147. https://doi.org/10.1080/17445647.2019.1706653

Scandone R, Bellucci F, Lirer L, Rolandi G (1991) The structure of the Campanian plain and the activity of the

Neapolitan Volcanoes. J Volcanol Geotherm Res 48:1–31. https://doi.org/10.1016/0377-0273(91)90030-4
Scarpati C, Cole P, Perrotta A (1993) The Neapolitan Yellow Tuff—a large volume multiphase eruption from Campi Flegrei, Southern Italy. Bull Volcanol 55:343–356. https://doi.org/10.1007/BF00301145
Scarpati C, Perrotta A, Lepore S, Calvert A (2013) Eruptive history of Neapolitan volcanoes: constraints from ^{40}Ar-^{39}Ar dating. Geol Mag 150(03):412–425. https://doi.org/10.1017/S0016756812000854
Scherillo A (1957) I "tufi antichi" tra S. Maria Apparente e via Parco Grifeo in Napoli. Boll Soc Natur Napoli 66:69–84
Scherillo A, Franco E (1960) Rilevamento stratigrafico del territorio comunale di Napoli. Boll Soc Nat Napoli 69:255–262
Schmidt R, van den Bogaard C, Merkt J, Müller J (2002) A new Lateglacial chronostratigraphic tephra marker for the south-eastern Alps: The Neapolitan Yellow Tuff (NYT) in Längsee (Austria) in the context of a regional biostratigraphy and palaeoclimate. Quat Int 88(1):45–56. https://doi.org/10.1016/51040-6182(01)00072-6
Selva J, Orsi G, Di Vito M, Marzocchi W, Sandri L (2012) Probability hazard map for future vent opening at the Campi Flegrei caldera, Italy. Bull Volcanol 74:497–510. https://doi.org/10.1007/s00445-011-0528-2
Seymour KS, Christanis K, Bouzinos A, Papazisimou S, Papatheodorou G, Moranc E, Denes G (2004) Tephrostratigraphy and tephrochronology in the Philippi peat basin, Macedonia, Northern Hellas (Greece). Quat Int 121:53–65. https://doi.org/10.1016/j.quaint.2004.01.023
Siebert L, Simkin T, Kimberly P (2010) Volcanoes of the World, 3rd edn. Smithsonian Institution, University of California Press, Berkeley (CA), USA, pp 1–551. ISBN 978-0-520-26877-7
Sigurdsson H, Houghton B, McNutt SR, Rymer H, Stix J (eds) (2015) Encyclopedia of volcanoes, 2nd edn. Academic Press, San Diego (CA), USA, pp 1–1456. ISBN: 9780123859389
Silleni A, Giordano G, Isaia R, Ort MH (2020) The magnitude of the 39.8 ka Campanian Ignimbrite Eruption, Italy: method, uncertainties and errors. Front Earth Sc-Switz 8:543399. https://doi.org/10.3389/feart.2020.543399
Smith RL, Bailey RA (1968) Resurgent Cauldrons. Mem Geol Soc Am 116:613–662. https://doi.org/10.1130/MEM116-p613
Smith VC, Isaia R, Pearce NJG (2011) Tephrostratigraphy and glass compositions of post-15 kyr Campi Flegrei eruptions: implications for eruption history and chronostratigraphic markers. Quat Sci Rev 30:3638–3660. https://doi.org/10.1016/j.quascirev.2011.07.012
Smith VC, Isaia R, Engwell SL, Albert PG (2016) Tephra dispersal during the Campanian Ignimbrite (Italy) eruption: implications for ultra-distal ash transport during the large caldera-forming eruption. Bull Volcanol 78(6):45. https://doi.org/10.1007/s00445-016-1037-0
Sparks RSJ, Self S, Walker GP (1973) Products of ignimbrite eruptions. Geology 1(3):115–118. https://doi.org/10.1130/0091-7613(1973)1%3c115:POIE%3e2.0.00:2
Spilliaert N, Allard P, Metrich N, Sobolev AV (2006) Melt inclusion record of the conditions of ascent, degassing, and extrusion of volatile-rich alkali basalt during the powerful 2002 flank eruption of Mount Etna (Italy). J Geophys Res Solid Earth 111:B04203. https://doi.org/10.1029/2005JB003934
Steinmann L, Spiess V, Sacchi M (2016) The Campi Flegrei caldera (Italy): formation and evolution in interplay with sea-level variations since the Campanian Ignimbrite eruption at 39 ka. J Volcanol Geotherm Res 327:361–374. https://doi.org/10.1016/j.jvolgeores.2016.09.001
Steinmann L, Spiess V, Sacchi M (2018) Post-collapse evolution of a coastal caldera system: insights from a 3D multichannel seismic survey from the Campi Flegrei caldera (Italy). J Volcanol Geotherm Res 349:83–98. https://doi.org/10.1016/j.jvolgeores.2017.09.023
Thunell R, Williams D, Tappa E, Rio D, Raffi I (1979) The age, origin and volcanological significance of the Y-5 ash layer in the Mediterranean. Quat Res 12:241–253. https://doi.org/10.1016/0033-5894(79)90060-7
Tomlinson EL, Arienzo I, Wulf S, Smith VC, Carandente A, Civetta L, Hardiman M, Orsi G, Rosi M, Thirlwall M, Muller W, Menzies MA (2012) Campi Flegrei Italy: geochemistry of the proximal sources for major Mediterranean tephras (C-1, C-2, Y-3 & Y-5). Geochim Cosmochim Acta 93:102–128. https://doi.org/10.1016/j.gca.2012.05.043
Tonarini S, D'Antonio M, Di Vito MA, Orsi G, Carandente A (2009) Geochemical and isotopical (B, Sr, Nd) evidence for mixing and mingling processes in the magmatic system feeding the Astroni volcano (4.1–18 ka) within the Campi Flegrei caldera (South Italy). Lithos 107:135–151. https://doi.org/10.1016/j.lithos.2008.09.012
Ton-That T, Singer B, Paterne M (2001) 40Ar/39Ar dating of latest Pleistocene (41 ka) marine tephra in the Mediterranean Sea: implications for global climate records. Earth Planet Sci Lett 184(3–4):645–658. https://doi.org/10.1016/S0012-821X(00)00358-7
Vakhrameeva P, Wulf S, Koutsodendris A, Tjallingii R, Fletcher WJ, Appelt O, Ludwig T, Knipping M, Trieloff M, Pross J (2019) Eastern Mediterranean volcanism during marine isotope stages 9 to 7e (335–235 ka): insights based on cryptotephra layers at Tenaghi Philippon, Greece. J Volcanol Geotherm Res 380:31–47. https://doi.org/10.1016/j.jvolgeores.2019.05.016
Vanorio T, Virieux J, Capuano P, Russo G (2005) Three-dimensional seismic tomography from P wave and S wave microearthquake travel times and rock physics characterization of the Campi Flegrei Caldera. J Geophys Res Solid Earth 110:B03201. https://doi.org/10.1029/2004JB003102
Veres D, Lane CS, Timar-Gabor A, Hambach U, Constantin D, Szakacs A, Fulling A, Onac BP (2013) The

Campanian Ignimbrite/Y5 tephra layer—a regional stratigraphic marker for isotope Stage 3 deposits in the Lower Danube region, Romania. Quat Int 293:22–33. https://doi.org/10.1016/j.quaint.2012.02.042

Vezzoli L (ed) (1988) Island of Ischia. Quad Ric Sci 114 (10):1–122. CNR, Rome, Italy. ISSN 0556-9664

Vitale S, Isaia R (2014) Fractures and faults in volcanic rocks (Campi Flegrei, southern Italy): insight into volcano-tectonic processes. Int J Earth Sci (geol Rundsch) 103:801–819. https://doi.org/10.1007/s00531-013-0979-0

Vitaliano CJ, Taylor SR, Farrand WR, Jacobsen TW (1981) Tephra layer in Franchthi cave, Peloponnesos, Greece. In: Self S, Sparks RSJ (eds) Tephra studies. Nato Advanced Study Institutes Series, Reidel, Dordrecht, The Netherlands, C 75, pp 373–379

Williams H (1941) Calderas and their origin. Univ Calif Publ Bull Depart Geol Sci 25:239–346

Wohletz K, Orsi G, de Vita S (1995) Eruptive mechanisms of the Neapolitan Yellow Tuff interpreted from stratigraphic, chemical and granulometric data. J Volcanol Geotherm Res 67:263–290. https://doi.org/10.1016/0377-0273(95)00002-C

Wulf S, Kraml M, Brauer A, Keller J, Negendank JFW (2004) Tephrochronology of the 100 ka lacustrine sediment record of Lago Grande di Monticchio (southern Italy). Quat Int 122:7–30. https://doi.org/10.1016/j.quaint.2004.01.028

Wulf S, Kraml M, Keller J (2008) Towards a detailed distal tephrostratigraphy in the Central Mediterranean: the last 20,000 yrs record of Lago Grande di Monticchio. J Volcanol Geotherm Res 177:118–132. https://doi.org/10.1016/j.jvolgeores.2007.10.009

Wulf S, Keller J, Paterne M, Mingram J, Lauterbach S, Opitz S, Sottili G, Giaccio B, Albert PG, Satow C, Tomlinson EL, Viccaro M, Brauer A (2012) The 100–133 ka record of Italian explosive volcanism and revised tephrochronology of Lago Grande di Monticchio. Quat Sci Rev 58:104–123. https://doi.org/10.1016/j.quascirev.2012.10.020

Wutke K, Wulf S, Tomlinson EL, Hardiman M, Dulski P, Luterbacher J, Brauer A (2015) Geochemical properties and environmental impacts of seven Campanian tephra layers deposited between 40 and 38 ka BP in the varved lake sediments of Lago Grande di Monticchio, southern Italy. Quat Sci Rev 118:67–83. https://doi.org/10.1016/j.quascirev.2014.05.017

Zollo A, Maercklin N, Vassallo M, Dello Iacono D, Virieux J, Gasparini P (2008) Seismic reflections reveal a massive melt layer feeding Campi Flegrei caldera. Geophys Res Lett 35:L12306. https://doi.org/10.1029/2008GL034242

Seismic and Gravity Structure of the Campi Flegrei Caldera, Italy

Francesca Bianco, Paolo Capuano, Edoardo Del Pezzo, Luca De Siena, Nils Maercklin, Guido Russo, Maurizio Vassallo, Jean Virieux, and Aldo Zollo

Abstract

We present a comprehensive review of seismic and gravity observations and tomographic models produced over the past four decades in order to understand the structure of the crust beneath the Campi Flegrei caldera. We describe the main lithological and structural discontinuities defined through these observations, illustrate their geophysical responses, and discuss the constraints they give to the understanding of magmatic and volcanic processes. Micro-seismic crises related to caldera unrest, and ambient seismic noise measurements provide comprehensive seismic data to local earthquake and ambient noise tomography. In combination with reflection data from onshore and offshore active seismic experiments, velocity tomography reconstructs the elastic properties of the caldera between surface and ~4 km depth. Active experiments also define the depth of lithological interfaces and deep (~7.5 km) partially molten bodies. Seismic attenuation tomography provides information complementary to velocity tomography, defining lateral lithological changes and the geometry of onshore and offshore fluid and magma bodies down to 4 km depth. Once compared with seismic analyses, gravity data highlight lateral changes in the offshore caldera structures. During the deformation and seismo-geochemical unrest (1982–1984), they permitted to reconstruct a minor (<1 km lateral extent) melt volume related to the point of maximum uplift measured at the caldera. Seismic coda-wave amplitude inversions depict the caldera rim limits in analogy to velocity tomography and map the lateral extension of ~4-km-deep deformation source. Once combined with the results from velocity tomography and gravity inversions, they reconstruct the feeding systems that connect deep deformation source and shallow vents across the eastern caldera, capped by a seismic horizon around a depth of 2 km.

F. Bianco (✉) · E. Del Pezzo
Istituto Nazionale di Geofisica e Vulcanologia, Sezione Osservatorio Vesuviano, Napoli, Italy
e-mail: francesca.bianco@ingv.it

P. Capuano
Dipartimento di Fisica "Eduardo R. Caianiello", Università di Salerno, Fisciano, SA, Italy

E. Del Pezzo
Instituto Andaluz de Geofisica, Universidad de Granada, Granada, Spain

L. De Siena
Institute of Geosciences, Johannes Gutenberg University, Mainz, Germany

N. Maercklin · G. Russo · A. Zollo
Dipartimento di Fisica, Università degli Studi di Napoli Federico II, Napoli, Italy

M. Vassallo
Istituto Nazionale di Geofisica e Vulcanologia, Sezione Roma1, Roma, Italy

J. Virieux
Institut des Science de la Terre, Université Grenoble Alpes, Grenoble, France

© Springer-Verlag GmbH Germany, part of Springer Nature 2022

G. Orsi et al. (eds.), *Campi Flegrei*, Active Volcanoes of the World,
https://doi.org/10.1007/978-3-642-37060-1_2

1 Introduction

Active volcanoes undergo human-scale, time-variable ground deformation caused by both the ascent of magmas and their volatiles through the crust until eruption and the shallow circulation of hydrothermal fluids. The volcanic systems show strong heterogeneity in the physical rock properties and a large variety of non-linear thermo-mechanical processes preceding and accompanying the eruptions. Such a complexity calls for a multi-disciplinary and integrated approach for data acquisition, analysis and modelling. A reliable model of the three-dimensional distribution of the physical rock properties is necessary to understand and simulate magmatic and hydrothermal processes, including melts and fluids upwelling and eruption, and to detect and track changes in the volcanic medium properties. Such a model has been constructed for the Campi Flegrei caldera (CFc), the target volcano of this volume.

Geophysical monitoring has improved dramatically from its first steps in the 1980s (Malone et al. 1983). Since 2005, the massive deployment of digital telemetry and reliable and dense broad-band seismic networks has enabled high-quality data recording in volcanic regions, allowing an improvement in the interpretation of seismogram complexities (Hellweg 2000). This enormous technological effort has been driven by the search for increasingly refined subsurface volcano images (with spatial resolution of a few hundred meters to kilometres) mostly using local earthquake (Benz et al. 1996; Nakamichi et al. 2002; Monteiller et al. 2005; Mora et al. 2006; Patané et al. 2006) and, more recently, ambient seismic noise (Brenguier et al. 2007) tomography. The use of broad-band seismometers has opened new perspectives (Neuberg et al. 1994; Rowe et al. 1998, 2000) about the identification and physical interpretations of the large variety of typical volcanic seismic signals, such as tremors, low frequency events and tornillos, with the aim of inferring physical models of their source and relating them to underlying volcanic processes (Chouet 1988; Ferrazzini and Aki 1987; Hellweg 2000; Jousset et al. 2003; Neuberg et al. 2006). To understand volcanic processes and improve monitoring capabilities, researchers have investigated the relations between the various seismic signals and the changes in the volcano micro-scale features, such as increased heat flow or magma and hydrothermal fluids migration and chemical transformations.

Delay-time tomography of longitudinal (P) wave and shear (S) wave first-arrivals from both local micro-earthquakes and active seismic surveys is the most widely used technique to obtain subsurface images of active volcanoes (Nercessian et al. 1984, Mount Dole; Thurber 1984, Kilauea; Achauer et al. 1988, Newberry Volcano). These techniques have been extensively applied to the CFc (e.g., Battaglia et al. 2008) allowing the determination of smooth layered models. These models lack details on small interfaces, discontinuities, magma or gas accumulation zones. The velocity spatial variation mirrors the distribution of lithological and structural properties of the rocks, while the ratio between P-wave and S-wave velocities (V_P/V_S) maps rock defects, pores and cracks, and their fluid contents. Fluid-filled pores in matrix rocks result in a decrease in both V_P and V_S, while variation in V_P/V_S depends on both fluids type and pores shape (Nakajima and Hasegawa 2003; Schmeling 1985; Takei 2002).

An accurate imaging of complex, inland volcanic structures, characterised by rough topography and a highly heterogeneous propagation medium requires deeper analysis and modelling of seismograms with full waveform inversions. During the last decades, the massive data availability and the experience gained from industrial seismic exploration has led to the development and application of advanced data processing, analysis and modelling techniques to the Somma-Vesuvio volcanic system (The Internal Structure of Mount Vesuvio—TomoVes Project) and to the CFc (SEismic Reflection/Refraction Acquisition Project for Imaging complex volcanic Structures —SERAPIS project) (e.g., Gasparini et al. 1998; Zollo et al. 2003). More recently, onshore active seismic experiments have produced extensive datasets for small volcanoes, allowing to model small-scale structures that control gas emission at the CFc (De Landro et al. 2017). These techniques permit the use of unconventional data acquisition geometries (sparse, irregular distribution of sources/receivers) and complex waveforms in the complete range of incidence angles from near-vertical to wide-angle (Auger et al. 2003; Operto et al. 2004). Both analysis and modelling of volcano seismic data have been enhanced by exploiting the whole information carried out by the direct, reflected and converted wave field through modelling of reflection amplitudes versus offset (or incidence angle). This can give important insights into the physical and lithological nature of subsurface discontinuities (Zollo et al. 2008; Maercklin and Zollo 2009). In addition, the spatial variations of both P-wave and S-wave velocity fields, as represented by tomography images, can be analysed and interpreted in terms of lithological and rheological features for rock characterisation. This interpretation can be performed by using theoretical modelling of seismic velocities in biphasic structures based on field data as well as on laboratory experiments (Vanorio et al. 2005).

The attenuation of elastic waves is quantified through the quality factor Q (the ratio between the total energy lost by a wave cycle and the energy associated with the cycle itself). Seismic attenuation tomography (Schurr et al. 2003; Nakajima and Hasegawa 2003; Hansen et al. 2004; Eberhart-Phillips et al. 2005; De Gori et al. 2005; De Siena et al. 2014a, b; Prudencio et al. 2015) has shown significant sensitivity to fluid reservoir and melt bodies at the CFc once interpreted in combination with velocity tomography results (De Siena et al. 2014a, b, 2017a, b; Serlenga et al. 2016; Akande et al. 2019). Q depends mainly on both rock temperature and presence of fluid-permeated fractures (Eberhart-Phillips et al. 2005). In addition, as Q for P-waves (Q_P) is different from that for S-waves (Q_S), the joint measure of Q for both waves is crucial for a correct characterisation of the physical state of the rocks inside a volcano. As an example, Q_p^{-1} can be used to discriminate between water-filled media and gas reservoirs (Hansen et al. 2004), while high Q_S^{-1} are a strong clue for the presence of melt-filled inclusions, which are always characterised by high V_P/V_S values (Takei 2002). Ambiguous interpretations of the causes of seismic velocity anomalies can be overcome by interpreting them jointly with the measures of Q_p^{-1} and Q_S^{-1} (De Siena et al. 2009, 2010, 2014a, b).

Scattering transfers the high-frequency energy of direct P-waves and S-waves into the coda of the seismograms (Sato et al. 2012). Due to anelasticity (seismic energy lost through heat) and scattering, the total quality factor (QT) can be separated in an anelastic part (intrinsic-Q, Qi) and in a scattering part (scattering-Q, Qsc) (Sato et al. 2012) as

$$\mathrm{QT} = \mathrm{Qi} \cdot \mathrm{Qsc}/(\mathrm{Qi} + \mathrm{Qsc})$$

It is challenging to separate scattering attenuation from intrinsic dissipation in single path estimates, necessary for 3D imaging (Del Pezzo et al. 2006). This is a limit for unequivocal interpretations of attenuation imaging, as the focussing-defocussing effects become indistinguishable from the scattering attenuation effects (De Siena et al. 2014b). An alternative is the use of seismic coda waves, which comprise energy scattered by direct waves in the heterogeneous volcanic medium. Both single scattering tomography (Tramelli et al. 2006) and diffusive coda imaging (De Siena et al. 2017a; Akande et al. 2019) have complemented direct wave velocity and attenuation imaging at the caldera. Scattering is also a primary ingredient in ambient noise tomography in volcanoes (Brenguier et al. 2007). This technique has provided the first onshore structural image of the CFc after the 1982–1984 unrest (De Siena et al. 2018).

Non-seismic geophysical methods help to better recognise volcano characteristics, complementing the information from seismic data and providing an integrated, more robust structural picture. Gravity investigations offer images of rock density distribution inside the Earth (e.g., Thorpe et al. 1981; Campos Enriquez and Arredondo-Fragosso 1992; Locke et al. 1993; Gailler et al. 2009; Yokoyama and Mena 1991;

Sanders et al. 1995; Davy and Caldwell 1998; Masturyono et al. 2001). If repeated through time, the study of the gravity field provides information on the evolution of the volcanic structure (e.g., Rymer 1994; Yokoyama 1989; Fournier et al. 2004). Several attempts have been made to extract useful information from gravity data collected at the CFc between 1979 and 1984 in the framework of a geothermal project and constrained by lithological information from boreholes (AGIP 1987). The data have been analysed through 2.5D modelling via the equivalent prism density method and boundary analysis (e.g., Cassano and La Torre 1987; Fedi et al. 1991; Florio et al. 1999; Berrino et al. 2008). These analyses related a gravity low centred on the town of Pozzuoli to a pyroclastic sequence with intercalated marine and continental sediments down to 2–3 km depth and overlying a thermo-metamorphic horizon. A fully 3D gravity inversion (Capuano et al. 2013) provided new insights into the internal caldera structure down to 3 km depth, defining features related to regional or volcano tectonic lineaments and dynamics. The model is parameterised on a $0.8 \times 0.8 \times 0.4$ km grid with a sequence of layers from sea bottom (0.2 km b.s.l.) down to 3 km b.s.l. with 0.4 km spacing, besides the subaerial and sea layers.

The CFc is one of the Neapolitan volcanoes that include the Somma-Vesuvio volcanic complex and the Ischia and Procida volcanic fields. All these volcanoes are located in the Campanian Plain, a Plio-Pleistocene tectonic depression (Cinque et al. 1993) formed along the Tyrrhenian margin of the Southern Apennines thrust belt in response to extensional processes (e.g., Patacca et al. 1990; Jolivet et al. 2009). The Southern Apennines belt is composed of Mesozoic carbonate and Mio-Pliocene terrigenous sequences, overlain by Quaternary continental deposits generated by the general building up of the chain and an intense volcanism. The Campanian Plain includes a 2–3,000 m thick succession of continental and marine sediments intercalated with volcanic deposits. The Mesozoic carbonate sequence of the adjacent Apennines mountains downthrown during extension, forms the basement of this sedimentary succession. The basement underlying the CFc includes a carbonate sequence with top at ~ 3 km depth, as inferred from seismic reflection investigations (Finetti and Morelli 1974; Bruno et al. 1998). The deepest rocks bored in the area (AGIP 1987) are calc-silicate metacarbonate rocks possibly resulting from metamorphic processes involving limestones and dolostones of the Mesozoic sequence (Vanorio and Kanitpanyacharoen 2015). The metacarbonate rocks would undergo decarbonation reactions that release lime-rich fluids promoting the formation of a seismic horizon characterised by a fibril-rich, fragile matrix in the 1–2 km depth range. This shallow interface is identified as a "caprock" by rock physics investigations (Vanorio and Kanitpanyacharoen 2015) and is analogue to those developing in geothermal reservoirs, delays deformation signals and dampens seismic and electromagnetic signals (Vanorio et al. 2005; De Siena et al. 2017a, b; Calò and Tramelli 2018; Siniscalchi et al. 2019; Akande et al. 2019).

The CFc is a nested, resurgent and restless structure in the densely populated Neapolitan area (Orsi et al. 2003; Chaps. Volcanic and Deformation History of the Campi Flegrei Volcanic Field, Italy; The Urban Development of Campi Flegrei, Italy). It results from two major nested collapses occurred during the Campanian Ignimbrite (CI; ~ 40 ka) and the Neapolitan Yellow Tuff (NYT; ~ 15 ka) eruptions. After each caldera event, volcanism was concentrated within the collapsed area. Therefore, the NYT caldera is the portion still active of the entire structure (Capuano et al. 2013 and references therein). It has been the site of intense volcanism and a long-term deformation related to the resurgence of its floor (Orsi et al. 1996, 1999, 2004; Di Vito et al. 1999). The post-NYT volcanism has generated not less than 70 eruptions grouped in three epochs of activity separated by quiescence (see Orsi et al. 1992, 1996, 2004; Chap. Volcanic and Deformation History of the Campi Flegrei Volcanic Field, Italy and references therein). The last eruption occurred in AD 1538, after 3 kyrs of quiescence, and produced the Monte Nuovo tuff cone (Di Vito et al. 2016 and references therein). The long-term deformation of the NYT caldera floor has generated a maximum net uplift of ~90 m of La Starza resurgent block (Cinque et al. 1985). This caldera is historically characterised by recurrent, large-amplitude vertical ground displacements, known as bradyseisms, consisting of subsequent episodes of uplift and subsidence (Casertano et al. 1976) accompanied by seismicity, variation of the gravity field and increase in the fumarole and thermal spring activity. These unrest events have been interpreted as transient short-term episodes within the long-term deformation related to resurgence (Orsi et al. 1999). Oliveri del Castillo and Quagliariello (1969) and Bonafede (1991) were the first to interpret and model the Campi Flegrei ground uplift events in terms of pressure and temperature perturbations in a porous fluid-saturated medium. Such a perturbation can be generated by magmatic degassing below a hydrothermal system (Chiodini et al. 2003, 2016, 2021). The subsidence has been interpreted as the effect of lateral diffusion of hydrothermal fluids after their ascent (Gaeta et al. 1998), or as the combination of their cooling and lateral diffusion (Todesco et al. 2003). The geochemical unrest has been primarily monitored at the Solfatara crater, to the east of Pozzuoli (Chiodini et al. 2021). This area is affected by secondary deformation and seismic source paired with the primary source of uplift during unrest (Amoruso et al. 2014; De Siena et al. 2017a, b; Pepe et al. 2019). Solfatara is the surface expression of a wide hydrothermal system, identified by many researchers as the primary source of recent unrest at the caldera (Chap. The Hydrothermal System of the Campi Flegrei Caldera, Italy and references therein).

The CFc is among the best-studied volcanoes in the world due to its historical and current activity, its location in a densely populated area and the consequently high potential volcanic hazard and risk (see Chap. Volcanic Hazard Assessment at the Campi Flegrei Caldera, Italy). This chapter reviews the main observations and models inferred from the analysis of massive seismic and gravity datasets collected over the past two decades, allowing for a comprehensive geophysical hypothesis on the structural setting of the crust beneath the CFc.

2 Data and Observables

2.1 Local Earthquake Seismic Data

Since 1969 the seismicity of the caldera has been monitored by a seismic network managed by the Osservatorio Vesuviano and, for a limited period of time, by the Italian Oil Company AGIP. Most of the seismicity, considering both released seismic energy and number of seismic events, has been concentrated during two main uplift episodes that occurred in 1970–1972 and 1982–1984 (Chaps. The Permanent Monitoring System of the Campi Flegrei Caldera, Italy; The Hydrothermal System of the Campi Flegrei Caldera, Italy; Historic Unrest of the Campi Flegrei Caldera, Italy). Generally, seismicity increased when the ground uplift velocity increased. Seismicity during 1970–1972 uplift episode was characterised by a $M_L = 2.5$ maximum magnitude and several seismic swarms consisting of tens of micro-earthquakes. In the 1982–1984 episode more than 10,000 micro-earthquakes were detected; about 6,000 of them had a magnitude greater than 0.1 and included about 20 events with magnitude comprised between 2.8 and the maximum recorded value of 4.2. This episode was also characterised by seismic swarms, the largest of which occurred on April 1, 1984 and produced 513 recorded events in about 6 h. From January to June 1984 the University of Wisconsin and the Osservatorio Vesuviano jointly deployed a temporary portable digital seismic network to record the intense micro-earthquake activity. The network consisted of 13 stations equipped with three-component sensors and placed in a total of about 20 inland sites. This was the first passive seismic experiment in Italy in which digital 12 bit data-loggers were used for earthquake recording. The entire earthquake dataset from that deployment has been recovered, organised in a database of three-component waveforms, and has been completely manually re-picked to avoid possible inconsistencies merging new picks and data already used in previous studies (Capuano et al. 2006). The waveform archive consists of more than 1,000 events of which about 850 recorded by more than 3 stations. Since the end of the 1982–1984 bradyseismic episode up to the end of 2005, the CFc experienced a period of seismic quiescence, interrupted only by very few earthquakes, hence this archive represents a unique collection of waveforms produced by seismic sources in the complex volcanic area that can be very useful to better understand the post-bradyseism caldera dynamics.

The picking operation of the University of Wisconsin dataset resulted in 4,583 P-phase and 3,321 S-phase picks. Data are weighted following four classes of reading accuracy from weight 0 (best) to 3 (worst). The weight assignment takes into account the reading time error to give an idea of the pick quality (picking accuracy is in the range 0.02–0.05 s for P-waves and 0.02–0.10 s for S-waves). About 80% of the P-phase picks have a weight $p \leq 1$ and 70% of the S-phase picks a weight $p \leq 2$ (Capuano et al. 2006).

2.2 Active Seismic Data

The SERAPIS project was a dense and extended marine active seismic survey carried out in the bays of Naples and Pozzuoli in September 2001, utilising offshore seismic sources, and data acquisition on the sea bottom and inland. It was aimed at providing new insight on the caldera structure and investigating its magmatic feeding system (Zollo et al. 2003). During the course of the experiment, the vessel Nadir of IFREMER, equipped with 12, 16-L airguns, produced about 5,000 shots at 125 m distances, which were recorded by an array of 72 ocean bottom seismographs (OBS) and 62 onshore three-component stations (Fig. 1). All the seismic lines were sampled at least twice using a staggered configuration, resulting in a smaller source spacing (less than 65 m). SERAPIS provided a dense 3-D seismic coverage of the Pozzuoli Bay, the coastal portion of the CFc and the western sector of the Naples Bay. The denser 2D network of 35 OBS within the Pozzuoli Bay was designed with the aim of detecting reflected and converted waves from possible shallow to deep geological discontinuities beneath the CFc.

For most analyses all SERAPIS waveforms were band-pass filtered between 5 and 15 Hz. P-wave first-arrival travel-times were manually picked on the pre-processed traces, arranged as Common Receiver Gathers (CRG). The SERAPIS dataset processing consisted in examining 300,000 OBS waveforms and 400,000 waveforms from inland stations, providing 65,000 and 25,000 P-wave first-arrival times, respectively. These travel-time data have been used to infer 3D images of the structures of the CFc and the Naples Bay (Zollo et al. 2003; Judenherc and Zollo 2004), and were also jointly inverted with travel-times from the micro-earthquake dataset to obtain both P-wave and S-wave velocity images (Battaglia et al. 2008). Manually picked reflection travel-times and amplitudes from SERAPIS data lead to a 1D structural model of main discontinuities (Zollo et al. 2008). An iterative tomographic analysis based on automatic refined picking has been proposed and developed by Satriano et al. (2008), providing a first application of automated seismic imaging techniques and a feasibility study for the continuous-time monitoring of the space–time changing volcanic structure.

Besides the SERAPIS dataset, older marine, 2D multi-channel seismic reflection data (Finetti and Morelli 1974) were reprocessed (Bruno et al. 2003), and a high-resolution multichannel seismic survey focussed on mapping the very shallow geological features was carried out in the Pozzuoli Bay in January 2008 (Sacchi et al. 2009).

2.3 Gravity Dataset

Gravity data acquired by various institutions through several surveys performed inland and offshore the CFc and its surroundings have been collected by Capuano et al. (2013) who merged measurements on the mainland (Maino and Tribalto 1971; Cassano and La Torre 1987; Servizio Geologico d'Italia unpublished data), on the nearby island of Procida (Imbò et al. 1964) and offshore in the Gulf of Naples (Berrino et al. 1998). To obtain a better data coverage in the southern part of the study region,

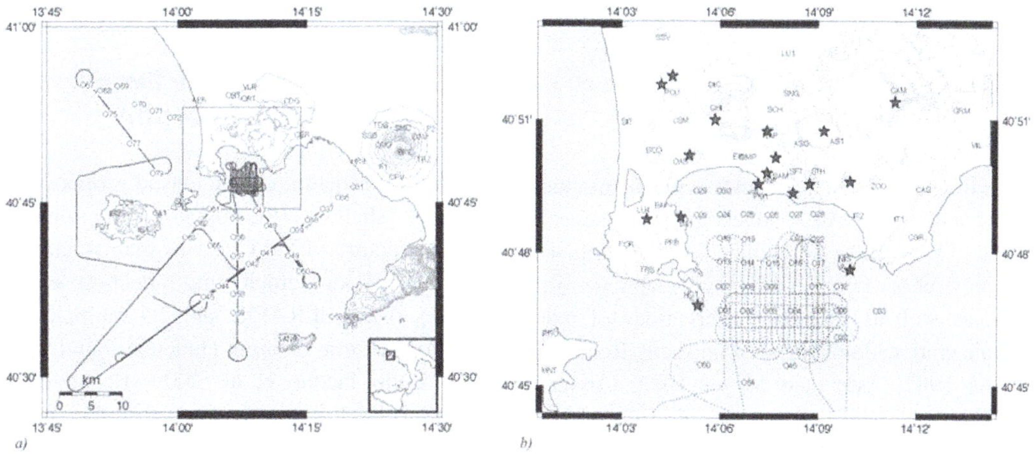

Fig. 1 **a** Map of the area investigated during the SERAPIS experiment; **b** details at Campi Flegrei. Black lines trace the path of the vessel during the survey. Dark and light circles with ID display the position of seismographs and OBS during the experiment. Stars show the locations of the UW stations deployed during 1984

the collected data were complemented with the Gravity Map of Italy. After a homogenisation process and the removal of inconsistencies, the gravity dataset consisted of 1,077 gravity measurements covering an area of about 440 km^2 (Fig. 2).

The effect of the water column and the regional gravity field were subtracted from the gravity anomaly. The sea water effect was evaluated using a large-scale bathymetric map and a water density of 1,030 kg/m^3 (Stabile et al. 2007). For the regional field correction, Capuano et al. (2013) considered that the most important regional feature of the area is the Mesozoic limestone basement that underlies the whole region at few kilometres depth. To retrieve the shape of the interface of the basement they used the results of the seismic tomography survey by Judenherc and Zollo (2004). Capuano et al. (2013) also assumed a value of 2,600 kg/m^3 for the density of the carbonate basement according to the results of other gravity investigations on the area (e.g., Cassano and La Torre 1987; Berrino et al. 2008). The local gravity anomaly used for the density inversion was obtained by subtracting all these contributions. The resulting gravity anomaly map shows a minimum associated with the central portion of the CFc, together with relative maxima along its margin and in the Camaldoli area (Fig. 2).

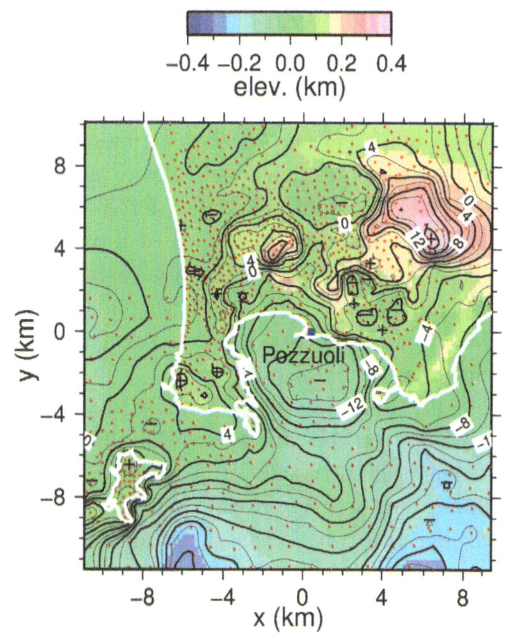

Fig. 2 Gravity anomaly map, contoured at 2 mGal intervals, obtained merging several gravity measurements datasets (see main text for details). The colour scale refers to the elevation model. Figure modified after Capuano et al. (2013)

3 The Campi Flegrei Caldera Structure Deduced by Velocity Tomography and Gravity Modelling

Since the late 1980s, the shallow crustal structure of the caldera has been investigated by several −1 to −3 km deep boreholes, local earthquake seismic tomography, gravity and magnetic surveys, as well as sporadic observations of teleseismic and wide-angle seismic data (Rosi and Sbrana 1987; Aster and Meyer 1988; Cassano and La Torre 1987; Ferrucci et al. 1992). High temperature gradients have been measured at rather shallow depths (450 °C at 3 km depth; Agip 1987). The caldera appears to be filled by a few kilometres thick sequence of volcanic and non-volcanic deposits, characterised by low V_P, high V_P/V_S and high P-wave attenuation whose geometry is consistent with the negative gravity anomaly (de Lorenzo et al. 2001). The relocated seismicity contemporaneous to the 1982–1984 ground uplift episode shows that most of the events were confined in a low-velocity layer with maximum depths of 3–4 km (Aster and Meyer 1988). The possible occurrence of a magmatic reservoir at about 4–5 km depth is based mainly on extrapolation at depth of temperature data and teleseismic observations (de Lorenzo et al. 2001; Ferrucci et al. 1992).

In the following we summarise the current knowledge on the shallow crustal structure as inferred from the results of the massive seismic velocity and gravity investigations carried out over the past two decades, referring to the 1D layered model of the CFc structure proposed by Zollo et al. (2008). In particular, we present and describe the main lithological and geological discontinuities beneath the caldera, and the geophysical characteristics of the embedded medium, from the Earth surface downward through the crust.

3.1 Shallow Unconsolidated, Water-Saturated Marine Sediments (from the Earth Surface to 900 m Depth)

Detailed information on the elastic properties of the very shallow (the uppermost kilometre) crustal structure of the CFc was mainly inferred from velocity tomography and reflection seismic analyses using SERAPIS and 2D multichannel reflection seismic datasets (Judenherc and Zollo 2004; Dello Iacono et al. 2009; Bruno et al. 2003; Sacchi et al. 2009). The subsurface of the caldera is very complex, displaying an extremely variable seismic response, especially in correspondence of the Pozzuoli Bay, where zones with excellent reflectivity and areas of a very low signal to noise ratio (S/N) alternate at relatively short distances. The signal decay is mainly due to the massive presence of volcanic rocks and edifices interlayered with Plio-Quaternary non-volcanic sediments that produce strong scattering of the seismic energy (Bruno et al. 2003). This effect is also highlighted by independent direct measurements of the separated intrinsic and scattering attenuation (Del Pezzo and Bianco 2013a). The interpreted seismic sections acquired during a multichannel seismic survey performed in the Pozzuoli Bay in January 2008, identify the well-known volcanic edifices (Pentapalummo Bank, Miseno Bank, Nisida Bank) and other buried features representing the remnant of volcanic edifices covered by marine and volcanic sediments (Sacchi et al. 2009). In such a complex environment, the determination of a velocity image of the shallow structure of the caldera requires a joint analysis of velocity tomography images and shallow seismic reflection data. This analysis was performed using an ad-hoc processing and unconventional methods applied to the data acquired within the caldera during the SERAPIS project (Zollo et al. 2003).

A refined 3D tomographic velocity model up to 1 km depth beneath the Pozzuoli Bay was obtained by the inversion of about 45,000 P-wave first-arrival travel-times, using the inversion method described in Judenherc and Zollo (2004). The obtained velocity model (Fig. 3) shows an area of low velocities (mean value of ~2,200 m/s) in the central part of the Pozzuoli Bay bordered to the south by a complex arc-shaped anomaly (Dello Iacono et al. 2009). The deeper part of the anomaly (beneath 700 m, with Vp > 3,500 m/s) correlates with rock sequences made up of agglomerate tuffs and interbedded lavas (AGIP 1987), which form the southern edge of the caldera (Barberi et al. 1991; Orsi et al. 1996). The shallower part of the anomaly that tends to split into two parallel arcs is correlated with dikes, volcanic mounds and hydrothermal alteration zones (De Bonitatibus et al. 1970; Pescatore et al. 1984; Milia and Torrente 2003). The presence of these structures suggests the existence of a highly fractured area, through which fluids (mainly gases) may have been able to rise towards the surface.

1D velocity profiles extracted from the 3D tomography model show a steep change in the P-wave velocity at about 600 m depth (Dello Iacono et al. 2009). At shallower depth, V_P values range from 1,500 to 2,000 m/s and indicate the presence of a thick, low-velocity sequence of

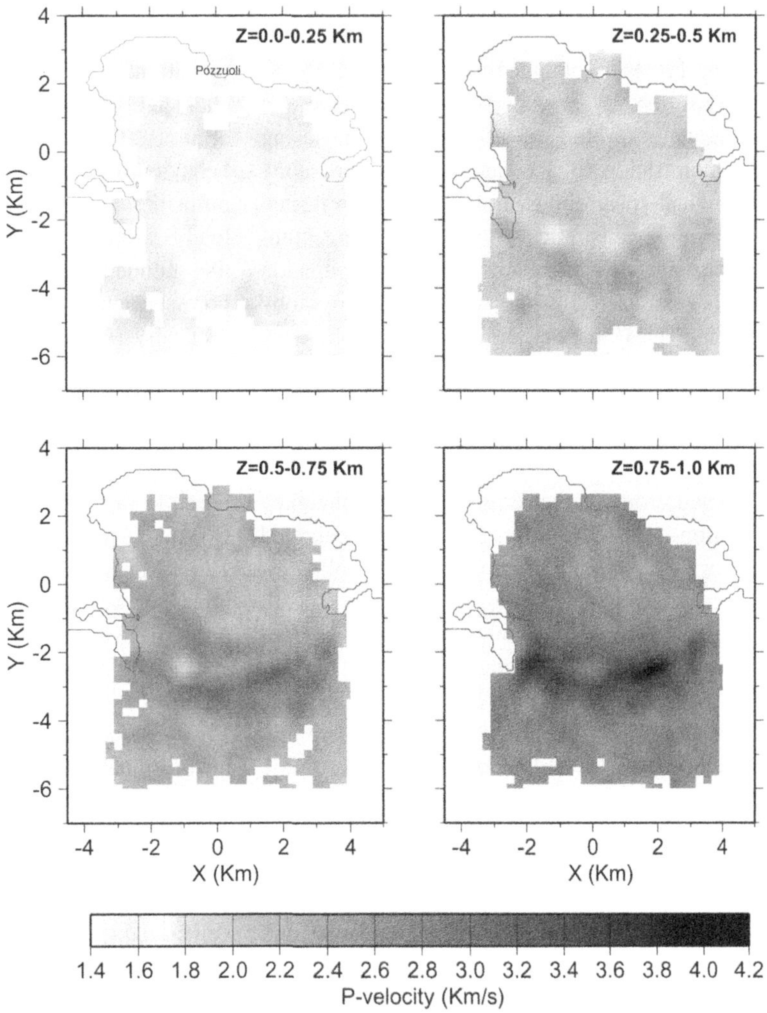

Fig. 3 Plane views of the tomography velocity model. Light grey indicates low velocities, and dark grey indicates high velocities. The top and bottom depth of each layer is indicated. Land topography is indicated by a solid line. Peripheral light grey indicates not resolved area for insufficient or absent ray coverage. Note the double-arc shaped P-wave velocity anomaly in the section 0.5–0.75 km, stretching between Nisida and Capo Miseno.
Figure modified after Dello Iacono et al. (2009)

sediments in the shallowest part of the caldera fill. In contrast, from 600 to 900 m depth, the P-wave velocity sharply increases from 2,000 to 2,600 m/s. This velocity discontinuity is also confirmed by a strong seismic reflector located at about 0.6–0.7 s two-way time on seismic sections, with data organised in Common Mid-point gathers as described in Vassallo et al. (2008). A total of 1,108 travel-time readings of P-to-P (PP) reflected waves have been manually picked on vertical-component sections corrected for a move-out velocity of 1,600 m/s. The distribution of picked travel-times has a mean value of 0.64 s with a standard deviation of 0.06 s. Assuming the mean two-way time of the PP phase as the zero-offset time for the identified reflector, and using the mean P-velocity profile inferred from the seismic tomography model of the Pozzuoli Bay area, an interface depth of 600 ± 120 m is obtained. The value of V_P/V_S ratio of the layer was computed assuming a horizontally layered subsurface model and using a maximisation of a coherency function along theoretical travel-times of P-to-S (PS) reflected phases. The theoretical travel-times were computed using the information on the propagation media extracted by the velocity tomography structure previously determined. The coherency function is computed for all Common-Midpoint gathers using the amplitudes of seismic traces in a window along the theoretical travel-times, and V_P/V_S ratio with the related uncertainty are obtained from the maximum of the coherency function (Vassallo et al. 2008). A weighted average of $V_P/V_S = 3.7 \pm 0.9$ for the shallowest layer was computed using the estimates for all Common Midpoint gathers and the inverse of the uncertainties as weighting factors. This relatively high value is an average estimate of the whole shallow layer that could be strongly influenced by the saturation conditions of the rocks in the shallowest hundreds of metres. Theoretical rock physical modelling of the V_P/V_S values as a function of porosity suggests that the shallow layer is likely formed by a sequence of incoherent, water saturated, volcanic and marine sediments that was deposited on the caldera floor (Dello Iacono et al. 2009).

A layer with such physical characteristics could have important impact on the understanding of the processes that rule the ongoing deformation of the caldera. Therefore, to test the hypothesis of a large V_P/V_S value, Dello Iacono et al. (2009) modelled the variation of the V_P/V_S ratio as a function of porosity in unconsolidated sediments, withstanding pressure conditions appropriate to describe a layer above 600 m depth. The elastic properties of the rocks depend upon the properties of the solid frame, such as composition and grain contact stiffness, and pore fluid and porosity. High V_P/V_S values are characteristic of shallow water-flow sediments at depth between 400 and 2,000 m below seabed (Huffman and Castagna 2001). At such a burial depth, these sediments are poorly consolidated (i.e., low grain contact stiffness) and withstand low effective stresses. Thus, they are near a transition between loose and consolidated rocks characterised by a critical porosity (Φc) (Nur et al. 1991, 1995). Φc separates the mechanical behaviour of rocks into two distinct domains: for porosities lower than Φc, the mineral grains are load-bearing, whereas at porosities greater than Φc, sediments fall apart and behave as a mineral-pore suspension. Dello Iacono et al. (2009) theoretically reproduced the effects of poorly consolidated sediments lying at 500 m depth on the elastic properties, via modelling as proposed by Dvorkin et al. (1999). The lower Hashin–Shtrikman bound is the tightest minimum bound possible from range of composite moduli for a two-phase material (e.g., Hashin and Shtrikman 1963; Mavko et al. 1998). It was adopted as the most appropriate model to simulate the elastic properties of these unconsolidated sediments and, therefore, looking for the softest rock arrangement. Figure 4 shows the variation of the V_P/V_S value as a function of porosity for an unconsolidated quartz-sand having a critical porosity of 38%. Figure 4a shows that the V_P/V_S value increases with increasing porosity, as in fluid-bearing sediments shear-wave velocity approaches zero while the compression-wave velocity does not fall below the velocity of a suspension of sand in water. Zimmer (2003) showed a similar trend in the V_P/V_S value

(Fig. 4b) for a set of reconstituted, unconsolidated sand and glass-bead samples under low pressure conditions. The modelling results and the experimental measurements reported for unconsolidated sediments (Zimmer 2003) suggest that the large changes predicted in the V_P/V_S value at low effective pressures are consistent with the hypothesis of poorly consolidated, fully water-saturated sediments lying above 600 m depth.

The density structure (Fig. 5) has been retrieved by inverting gravity data searching for a Tichonov regularised solution of the linear discrete inverse problem having the smallest possible weighted norm (Capuano et al. 2013). The application of regularisation theory to gravimetric and electromagnetic data is extensively discussed by Zdhanov (2002). The key point of the theory is that a stable solution of the inverse problem can be obtained requiring that the model fits the observed data within a given accuracy δ, usually linked to data and model errors. The weighting matrix associated to the model parameters is chosen to maximise the sensitivity of model parameters to data changes (Zdhanov 2002). To evaluate the goodness of the 3D density anomalies distribution at depth, the density transitions from positive to negative have been positively compared with the stratigraphic sequences drilled at Mofete and San Vito (AGIP 1987), showing that they well match the changes from tuffites and sedimentary material to lavas (flows and domes) (Rosi et al. 1983). The final gravity model obtained by Capuano et al. (2013) has a root mean square of 0.16 mGal, the misfit respect to the observed data is almost everywhere very small and the resolution is good down to

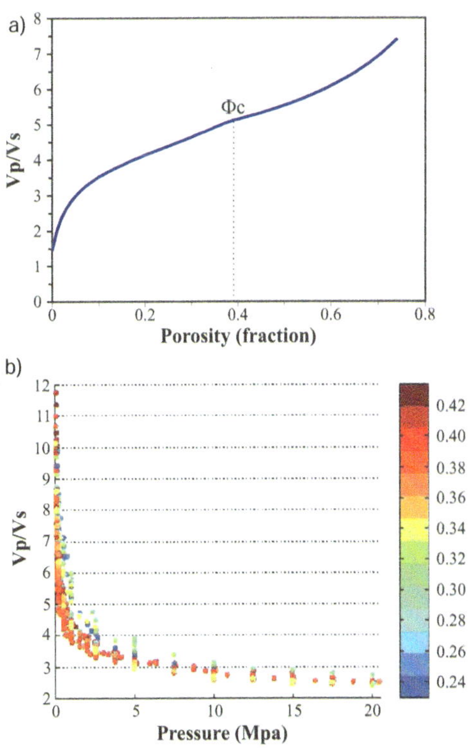

Fig. 4 **a** Variation in the V_P/V_S ratio as a function of porosity in unconsolidated, water-saturated sediments. The critical porosity is 0.38 and the number of contacts per grain is 8.5. **b** V_P/V_S ratio data for water-saturated glass-bead samples plotted against the effective pressure. Data are in colour scale as a function of porosity. Figure modified after Zimmer (2003)

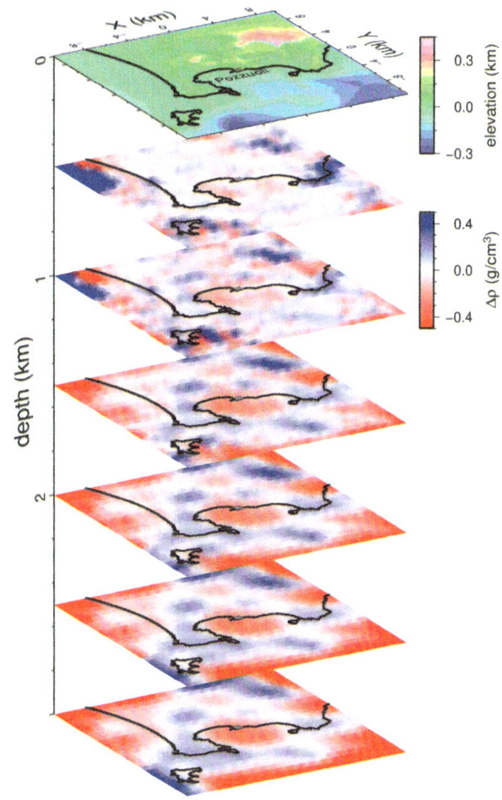

Fig. 5 3D imaging of the density model at different depths (0–3 km). The uppermost slice shows the topography. Density model after Capuano et al. (2013)

about 3 km depth. Between 0 and 200 m depth, the resulting density model is characterised by alternation of positive and negative anomalies mostly in correspondence of rock bodies of different characters (e.g., lava flows, lava domes, pyroclastic deposits). Between 200 and 600 m depth (Fig. 6) the caldera is characterised mostly by negative density anomalies. The Pozzuoli Bay morphological low is dominated by a density low about 6 km wide. Its western portion includes a large-wavelength, roughly north–south aligned and irregularly shaped structure. This density low is compatible with low velocity inferred by the seismic analysis and ascribed to the caldera fill composed of volcanic and non-volcanic sediments.

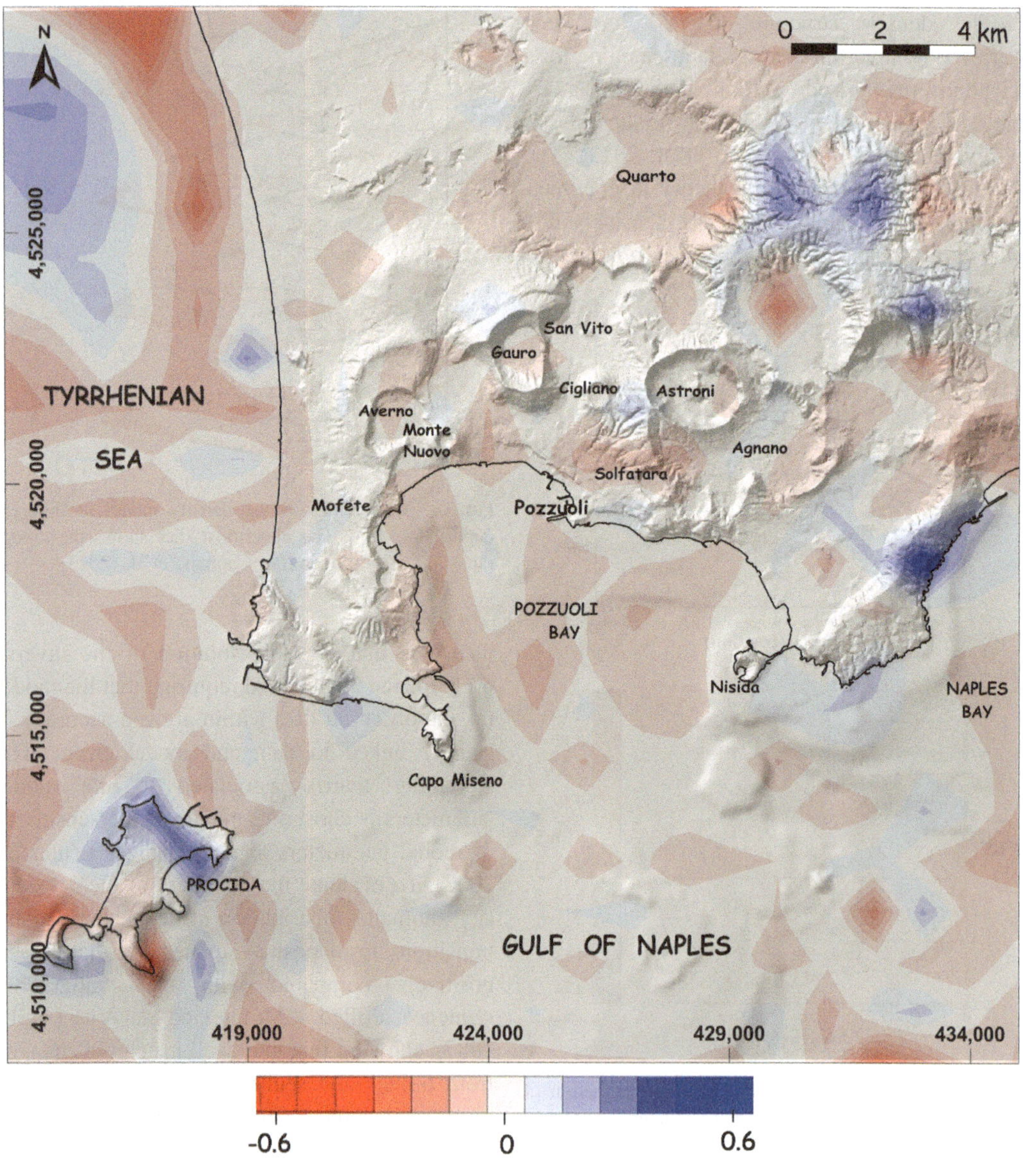

Fig. 6 Density horizontal slice between 200 and 600 m b.s.l. Colour scale represents density contrasts (g/cm^3). Figure modified after Capuano et al. (2013)

3.2 Volcanic and Non-Volcanic Deposits, Buried Caldera Rim and Fluid-/Gas-Bearing Thermometamorphic Rocks (from 500 m to ~4 km Depth)

The highest resolution seismic image of the caldera buried rim was obtained by the inversion and modelling of 77,000 arrival times of P-wave first-arrivals recorded during the SERAPIS experiment by the OBS deployed in the Pozzuoli Bay (Zollo et al. 2003). A delay-time inversion technique has been applied, based on a linearised, perturbative approach, with the use of smoothness constraint equations to regularise the solution and to stabilise the inversion procedure (Hole 1992; Benz et al. 1996).

The 3D tomography images show the presence of an arc-like, vertically-extended high P-velocity anomaly (V_P = 3.5–4.0 km/s) within the southern portion of the Pozzuoli Bay, interpreted as the expression of the caldera margin (Fig. 7). This anomaly has its top at about 800 m depth, extends to about 2,000 m depth and follows a pattern nearly concentric to the coastal line. The shape of the reconstructed caldera rim is well consistent with and closely matches the density model inferred by gravity data which have been integrated with the existing data set (Berrino et al. 1998) and re-processed by Capuano and Achauer (2003), providing with an updated Bouguer anomaly map. Litho-stratigraphic and sonic log data from several 2–3 km deep inland boreholes, drilled for geothermal exploration purposes, indicated that the caldera margin is formed by a sequence of compacted tuffs, tuffs with interbedded lavas and thermo-metamorphic rocks, with the latter at about 2–2.5 km depth. P-wave log velocities in boreholes vary with depth from 2.7–3 km/s at 0.8–1 km depth to 3.6–4.2 km/s at 2–2.2 km depth (Agip 1987), which are consistent with tomography velocities in the same depth range.

de Lorenzo et al. (2001) and De Siena et al. (2010) found evidence for an anomaly of low P-wave attenuation (low Qp) and high temperature between 2 and 3 km depth, occurring along the eastern onshore sector of the caldera rim. These observations led the authors to suggest that this anomalous zone could be the site where intense fracturing may have occurred during caldera collapse and resurgence, thus representing a preferential pathway for magma to migrate toward the surface and then erupt. This hypothesis is also supported by the presence of eruption

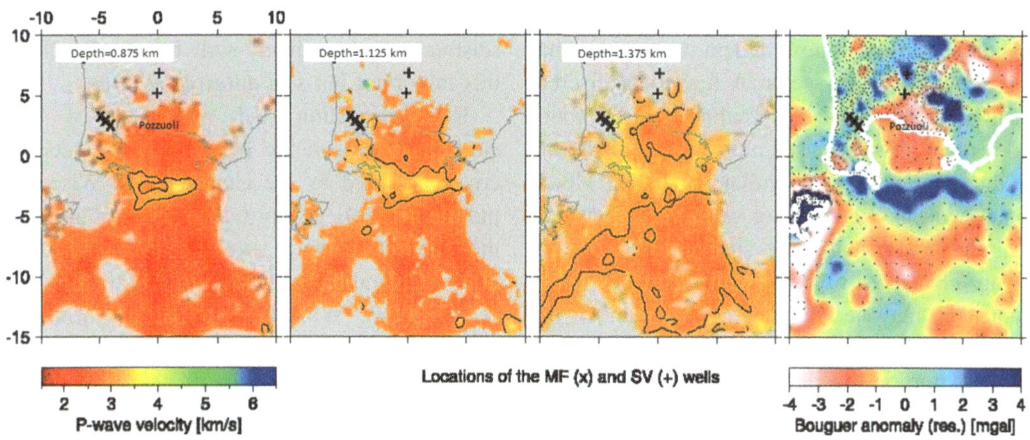

Fig. 7 P-wave velocity and Bouguer gravity anomaly images. The tomography P-wave velocity images are displayed from the left to the right at three different depths (875, 1,125 and 1,375 m). Symbols "+" and "×" indicate the location of S. Vito and Mofete wells, respectively, drilled by AGIP in the late 1970s for geothermal exploration purposes. A low-pass filtered map of the Bouguer gravity anomaly (reduction density: 2.4 g/cm^3) is shown on the rightmost panel (after Zollo et al. 2003)

vents along or nearby the onshore sector of the eastern caldera rim.

Consistently with seismic results, the density model locates below 600 m depth the top of the caldera rim, marking a density discontinuity. The caldera top is not flat, but deepens from north to south of about 800 m. The caldera low density has a diameter of ~6 km, even if it is characterised by a complex structural pattern of blocks that have undergone differential movements, are located at different depths and are separated by structural lineaments (Fig. 5). In fact, the 3D density anomalies distribution sketches a pseudo-circular high-density feature, encompassing an inhomogeneous low-density crustal sector. The pseudo-circular feature shows a complexity in its south-eastern portion, likely related to activation in that area of portions of northwest-southeast trending regional faults during the CI and NYT collapses (Capuano et al. 2013; Chap. Volcanic and Deformation History of the Campi Flegrei Volcanic Field, Italy).

The complex arc-shaped velocity anomaly has a close counterpart in the density model. In fact, a high-density pseudo-circular feature encloses the caldera both inland and offshore (Fig. 8). The continuity of this feature is well established except in the north-eastern sector of the caldera, where it is less marked and interrupted by structural discontinuities. In this part of the caldera, the high-density margin may be partly obliterated by the post-caldera volcanism that concentrated in this sector. A few small discrete density highs located at shallow depth (about 1 km) may be recognised along this feature, one of these being closely correlated with the Monte Gauro volcano and another, with very small lateral dimension and more similar to a conduit, being closely correlated with Astroni. These anomalies may be related to the remnant of the feeding system of the post-NYT eruptions.

The portion encompassed by the pseudo-circular high-density feature can be subdivided into a north-eastern and a south-western sector, according to the complex distribution of anomalies. This subdivision is corroborated by the trending of morphostructural lineaments. The root of this pseudo-circular anomalies' alignment appears to extend down to about 2,500 m. In fact, at 3 km depth the density contrast abruptly decreases and fades, likely in correspondence of the bottom of the caldera fill, consistently with the seismic results. The density distribution, the caldera structural boundary and the post-collapse deformation pattern support piecemeal collapse and block resurgence mechanisms (Capuano et al. 2013; Chap. Volcanic and Deformation History of the Campi Flegrei Volcanic Field, Italy).

Down to 4–5 km depth beneath the CFc, no evidence is found in the current 3D P-wave velocity images for magma bodies with volumes greater than 1 km^3. This would imply that the main magma reservoir feeding the Phlegraean magmatic system has to be located deeper, well within the carbonate basement. On the other hand, the results of the attenuation tomography study by De Siena et al. (2010) using local earthquake data have suggested the presence of small batches of molten rocks in the same depth range, based on the interpretation of low-velocity, high attenuation anomalies. However, since the resolution of the tomography method degrades with the size of such an anomaly, it is likely that 1 km is the lower limit of its spatial resolution to reconstruct subsurface seismic images in this area with the current data (e.g., Battaglia et al. 2008). Future higher-resolution observing systems (i.e., dense 2D or 3D arrays of seismic antennas) likely will be able to clarify this important but still unresolved issue.

Re-interpretation and modelling of local earthquake data of the 1982–1984 bradyseismic crisis, recorded by the University of Wisconsin mobile network, also provided seismic images of the shallow caldera structure (Vanorio et al. 2005; Chiarabba and Moretti 2006; Battaglia et al. 2008). With the aim of exploring the structure of the caldera and the role of hydrothermal fluids on velocity changes, Vanorio et al. (2005) adopted a multidisciplinary approach based on three-dimensional delay time tomography and rock physics characterisation. Selected arrival time data were modelled using a three-dimensional tomography method, based on an accurate finite difference travel-time

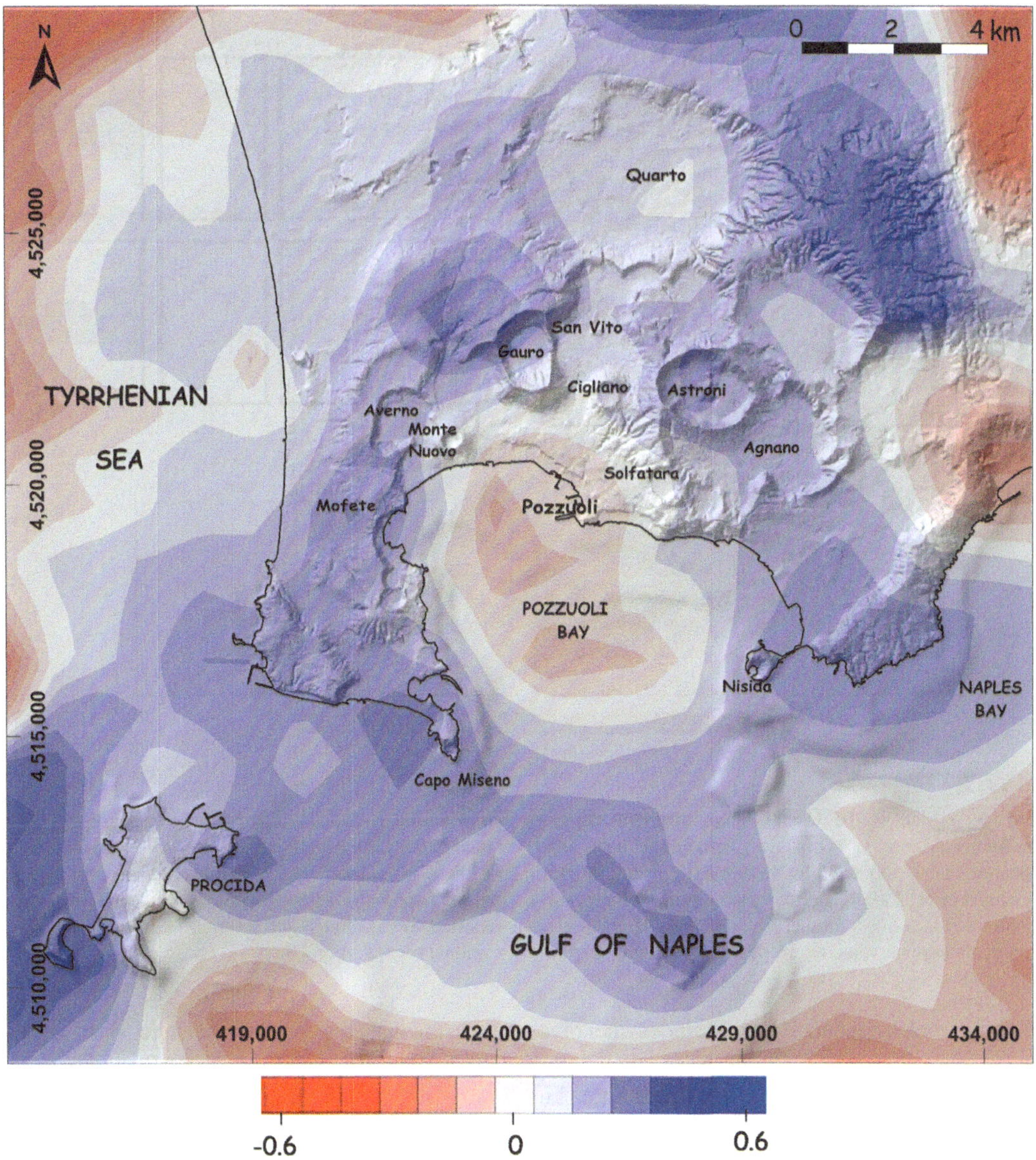

Fig. 8 Density horizontal slice between 2,200 and 2,600 m b.s.l. Colour scale represents density contrasts (g/cm^3). Figure modified after Capuano et al. (2013)

computation and a simultaneous inversion of P-wave and S-wave first-arrival times for both velocity model parameters and hypocentre locations (Latorre et al. 2004). Vanorio et al. (2005) interpreted the retrieved P-wave and S-wave velocity images as well as the deduced V_P/V_S images using experimental measurements of rock physical properties on Phlegraean samples to take into account steam-water phase transition mechanisms affecting the P-wave and S-wave velocities. Moreover, they constrained the role of over-pressurised fluids on velocity through modelling the petrophysical properties of site-relevant rocks. The authors located a flat and low V_P/V_S anomaly at 4 km depth under the town of Pozzuoli, below most of the events (Fig. 9). They suggested that this anomaly implies the occurrence of fractured, over-pressured gas-

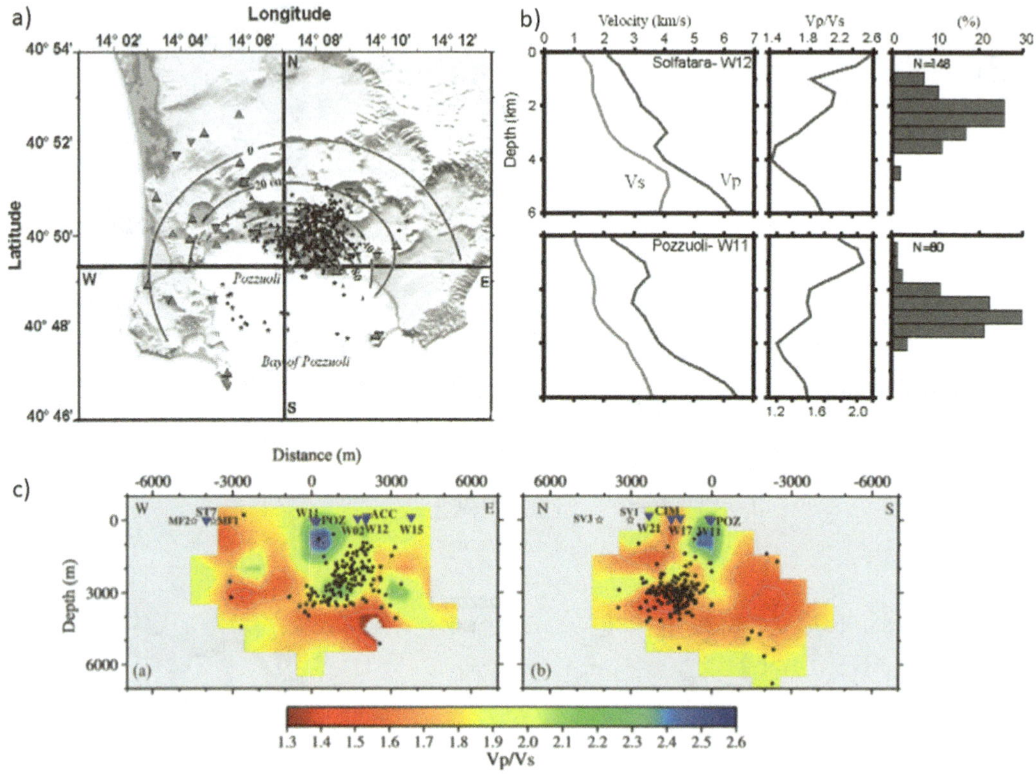

Fig. 9 **a** Map of the Campi Flegrei caldera showing seismometer stations (triangles) and final earthquake locations (stars) obtained by tomography inversion (from Vanorio et al. 2005). The map also reports the elevation contours (black lines) for the 1982–1984 uplift. **b** Reversal trends on vertical velocity profiles, V_P/V_S and histograms showing the earthquake distribution along profiles (radius equal to 0.5 km). **c** Vertical cross-sections of the variation of the V_P/V_S ratio and earthquake distribution (black points) along the east–west and north–south directions reported in **a**). Stars and triangles indicate wells and stations, respectively

bearing rock sequences and excludes the presence of molten rocks. Furthermore, they interpreted a high V_P/V_S anomaly found at 1 km depth as an evidence of the presence of rocks containing fluids in the liquid phase.

Battaglia et al. (2008) proposed a strategy for merging both active and passive datasets in a linearised tomography inversion and applied it to the whole active and passive Campi Flegrei datasets. This study provides the most comprehensive, integrated and best-constrained image reconstruction of the P-wave and S-wave velocity distribution beneath the caldera down to an approximate depth of 4–5 km. The P- and S-velocity models of the caldera structure were obtained by a tomography inversion based on travel-times recorded during two distinct experiments, the passive seismic data of the 1982–1984 crisis and the SERAPIS active seismic survey. Furthermore, their tomography inversion utilised an improved method based on an accurate finite-difference travel-time computation and a simultaneous inversion of both velocity models and earthquake locations.

The obtained images confirm the presence of a ring-like high P velocity in the southern part of the Pozzuoli Bay and extends its trace further inland as compared to previous results (Fig. 10). The V_P/V_S model confirms the presence of two characteristic features. A very high V_P/V_S anomaly at about 1 km depth below the town of Pozzuoli and interpreted as due to the presence of rocks containing fluids in the liquid phase. A low V_P/V_S body at about 3–4 km depth below a large

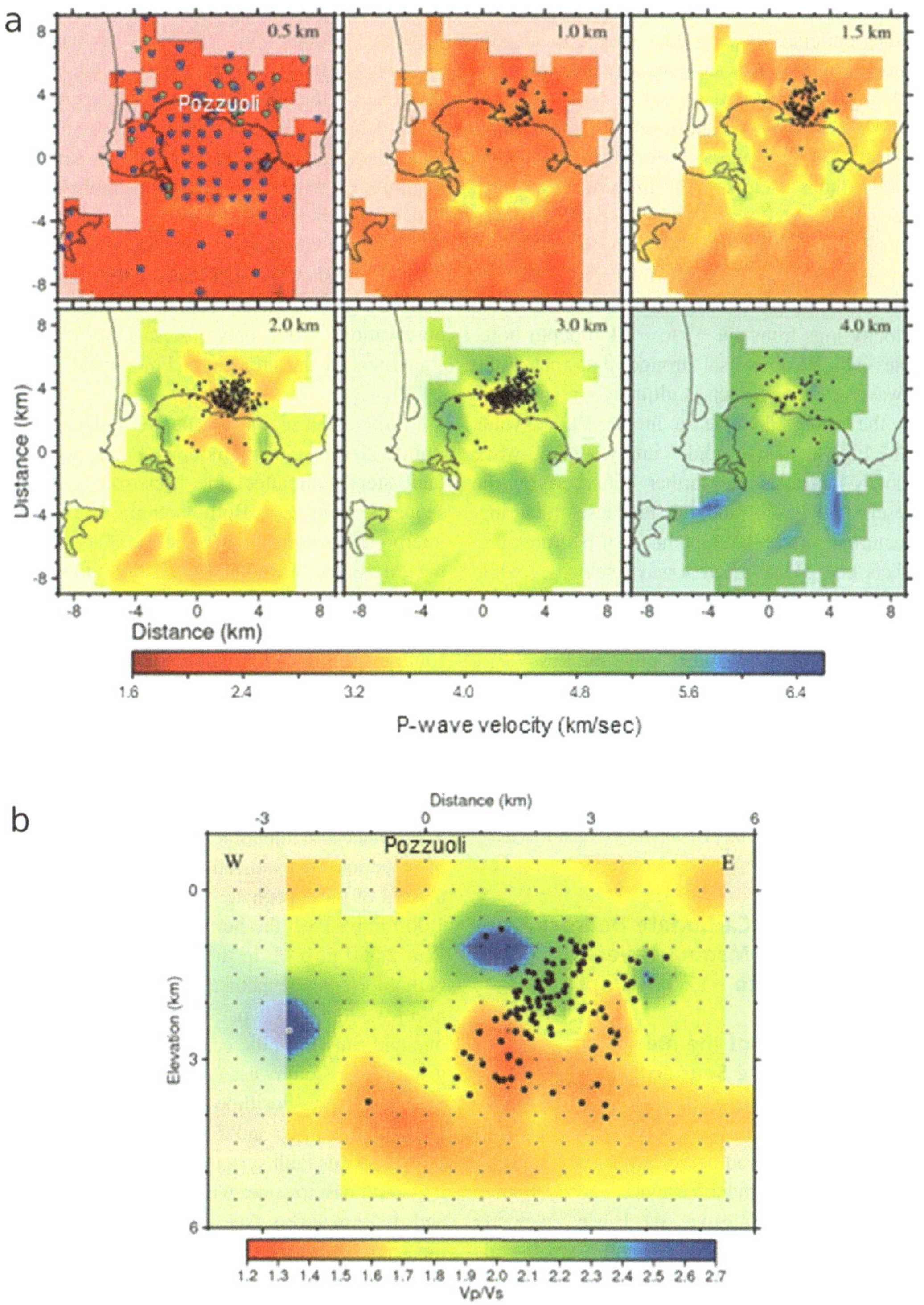

Fig. 10 Tomography results for Campi Flegrei obtained by inversion of active and passive data sets. **a** Horizontal sections of the P-wave velocity model at different depths (showed in each panel). **b** East–west cross-section at y = 2.5 km of the V_P/V_S ratio. Figure modified after Battaglia et al. (2008)

part of the caldera is interpreted as the top of rock bodies enriched in gas under supercritical conditions, as previously suggested by Vanorio et al. (2005).

As discussed by Vanorio et al. (2005), fluid overpressure that can result from fluid expansion, among other factors, can, in turn, cause the reversals in both P-wave and S-wave velocity logs, as it prevents the effective stress from increasing as fast as it would under normal hydrostatic or lithostatic pressure conditions. The P-wave velocity logs from the −1 to −3 km depth boreholes indicate a reversal stronger than that of the S-wave velocity, which is ultimately responsible for the observed decrease in the V_P/V_S value (cfr. Fig. 9). This would imply that P-wave velocity reversals are further enhanced by the presence of gas in addition to dense fracturing. Fracturing, under the hypothesis of overpressure, affects both P-wave and S-wave velocities while earthquakes mainly occur both at the top of reversal trends and within the low V_P/V_S zone. This evidence supports the hypothesis that the low V_P/V_S anomaly occurring at 4 km depth beneath the centre of the caldera and below the reversal velocity zone is due to the presence of over-pressured gas-bearing rocks at supercritical conditions (temperatures of about 400 °C were reported in the San Vito well at 3 km depth (Agip 1987).

3.3 Mesozoic Carbonate Bedrock, Including Magma Layers (from ~4 to ~12 km Depth)

3.3.1 The Top of the Mesozoic Carbonate Sequence

Judenherc and Zollo (2004) processed the Serapis P-wave travel-time collection, extended with a previously acquired dataset from the TomoVes experiment at Somma-Vesuvio, and computed the most comprehensive 3D P-wave velocity model for the Gulf of Naples and a small-scale high-resolution model for the CFc. The joint interpretation of the velocity distribution together with the available gravity data and the geothermal drilling information (Agip 1987) permitted to identify and map lithological units at depth. In particular, the tomography images and the apparent velocity ($V_P \sim 6$ km/s) of secondary head wave arrivals on seismograms indicate the existence of the Mesozoic carbonate sequence at about 4 km depth beneath the caldera (Fig. 11). This discontinuity is a continuous structural feature well detected in 3D tomography images and marks the sharp P-wave velocity increase from 3–4 to 5–6 km/s between 2 and 4–5 km depth in the shallow crust beneath the Naples and Pozzuoli bays. These results corroborate the observations from previous seismic reflection soundings in the area (e.g., Finetti and Morelli 1974).

At the wider regional scale, the Judenherc and Zollo (2004) velocity model reveals two main fault steps that affect the Mesozoic carbonate sequence (Fig. 11). Both tectonic features are interpreted as normal fault systems that deformed the Mesozoic carbonate sequence during the middle Pleistocene northwest-southeast extension. The south-eastern step corresponds to a southwest-northeast trending normal fault system (Finetti and Morelli 1974; Bonasia et al. 1985) that likely reaches Somma-Vesuvio toward the northeast (Bruno et al. 1998). The model also indicates a vertical offset of about 1,000 m, which affects the 1–2 km and 2–3 km layers. This offset is compatible with the interpretation of reflection data, indicating a rapid deepening of the top of the carbonate sequence from less than 1,000 m b.s.l. off the Sorrento Peninsula to more than 2,500 m b.s.l. in the central portion of the Gulf of Naples (Finetti and Morelli 1974). The north-western step is parallel to the south-eastern one and runs from the eastern rim of the caldera toward the southwest, along the 35-km-long segment of the Posillipo—Banco di Fuori fault system (Judenherc and Zollo 2004). The vertical slip along this fault system is similar to that along the south-eastern one, which affects the 3–4 km and 4–5 km layers (see Fig. 11). The height of the scarp is thus between 1,000 and 2,000 m.

The evidence for a geometric relation between the location of the CFc and the north-western normal fault affecting the carbonate sequence led Judenherc and Zollo (2004) to suggest that the deep magmatic feeding system of the Campi

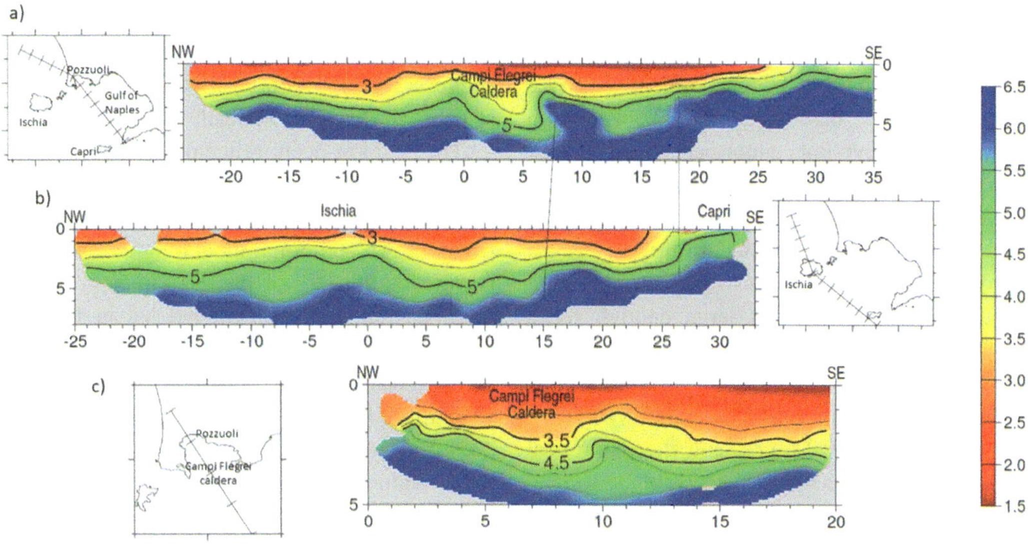

Fig. 11 Selected cross sections in the Gulf of Naples (top and centre) and Bay of Pozzuoli (bottom) tomography models (no vertical exaggeration). The inset maps show the traces of the cross sections, and the tick marks are at 5 km spacing. The two cross sections in the Gulf of Naples clearly show the normal faults affecting the carbonate sequence; their locations are indicated by the thick lines (from Judenherc and Zollo 2004)

Flegrei volcanoes could be related to this fault system, as the caldera structure lies just above it. Therefore, the normal fault system affecting the carbonate sequence could provide a preferential pathway to molten material located in a deeper magma reservoir, embedded within or beneath the carbonate sequence, for ascent. They also indicate that a similar geometric relation between Somma-Vesuvio and an analogous normal fault with the same orientation exists and suggest that this fault controls the magma displacement toward the surface in an equivalent way to the similarly-trending fault system beneath the CFc (Moretti et al. 2013 and references therein).

3.3.2 Main Crustal Discontinuities Beneath the Caldera from Seismic Reflection Analysis

The tomography velocity models from first-arrivals derived for the CFc reach a maximum depth of about 5–6 km. For a greater depth penetration of rays, either larger source-receiver distances or (earthquake) sources at greater depth would be required, or secondary waves such as reflections and conversions can be used. With the aim of characterising also the deeper structure beneath the CFc, Zollo et al. (2008) analysed reflected and converted wave amplitudes from main seismic horizons in a large subset of the SERAPIS data available within the caldera. The seismic reflection amplitudes depend on the impedance (product of velocity and density) contrast at the reflectors and the ray incidence angles or the source-receiver offset, thereby providing information on the velocities in a layer below the reflector. To include both P-wave and S-wave information and to remove unwanted source and receiver site effects, Zollo et al. (2008) analysed the ratio of the PP to the PS converted phase from the same reflector. Details of the method as well as its advantages and limitations are described in Maercklin and Zollo (2009).

About 75,000 three-component records acquired at 30 sea-bottom stations have been initially arranged in Common Midpoint gathers with 500 × 500 m cell size and more than 120 records per gather in the centre of the Pozzuoli Bay. Data pre-processing included a 5–15 Hz band-pass and additional trace normalisation for phase identification. PP reflection and PS conversion move-out analyses were applied using an average 1D velocity model derived from the 3D tomography models, and reflection events were identified based on lateral waveform coherency, move-out alignments and stack amplitudes of both vertical and horizontal component sections. The PP and PS amplitudes at the same interface were extracted from the seismograms at several source-receiver offsets, covering a distance range that was large enough for a significant PS-to-PP amplitude ratio variation with offset. The P-wave velocity and the V_P/V_S value above the reflecting interface were constrained by the travel-times, and the theoretical PS-to-PP amplitude ratios were computed using dynamic ray modelling, including the correction for geometrical divergence and assuming a constant rock density. The unconstrained model parameters V_P and V_P/V_S below the reflector were then varied to minimise the misfit between the observed and the theoretical ratios, and the best-fit model was found by inversion (Zollo et al. 2008; Maercklin and Zollo 2009). The final 1D average model was constructed in a layer-stripping procedure, starting with the shallowest reflector and moving subsequently downward.

Zollo et al. (2008) found three major reflectors around 0.6, 2.7, and 7.5 km depth, respectively (Fig. 12). The shallowest reflector corresponds to the base of the marine unconsolidated sediments discussed above. The 2.7-km-deep reflector is characterised by a well-constrained positive P-wave velocity contrast and a small value of V_P/V_S value below the reflector. Zollo et al. (2008), following a previous interpretation and rock physics modelling by Vanorio et al. (2005), attributed this reflector to the top of a supercritical fluid- or gas-bearing, thermo-metamorphic rock layer. The type of fluid or gas (i.e., CO_2 or vapour) could not be inferred from this seismic study because their physical properties at the high in-situ temperatures (up to 400 °C) are very similar and do not affect seismic velocities significantly. The thickness of the thermo-metamorphic layer also remains unclear, and the top of the Mesozoic carbonate sequence could not be identified as a contiguous reflector. However, the tomography P-velocities of 6 km/s (Judenherc and Zollo 2004) and some P-wave reflection energy imaged by Blacic et al. (2009) suggests the transition to the carbonate sequence at about 4 km depth beneath the CFc.

The 7.5-km-deep reflector shows a strong negative velocity contrast with a wide range of possible values for V_P and V_P/V_S below the reflector. An additional modelling of the relative P-wave amplitudes of the 2.7-km- and the 7.5-km-deep reflectors constrained V_P in the layer below the deepest reflector to about 2.8 km/s, which means a strong negative V_P contrast with an estimated V_P/V_S increase from 1.65 to 2.35 or even higher. According to Zollo et al. (2008) these values indicate that the 7.5-km-deep reflector corresponds to the top of a low velocity layer, whose seismic velocities are consistent with the values expected for a magma body set in a densely fractured volume of rock (Fig. 13). Under the simplifying assumption of a constant velocity within this layer, later reflections detected at some locations suggested that this layer is about 1.5 km thick. This led to the conclusion of an extended magma sill beneath CFc similar to the one found beneath the neighbouring Somma-Vesuvio (Auger et al. 2001).

Taking the P-wave reflection time picks of Zollo et al. (2008) for the three reflectors, Vassallo et al. (2010) made a first attempt to obtain a 3D model of the morphology of each reflector using a tomography inversion scheme (Hobro et al. 2003). The model is parameterised as a stack of layers separated by interfaces. The inversion process starts with a highly-smoothed model whose complexity increases at each step by adding smaller wavelengths until a satisfactory fit to the data is obtained. Using an optimised 1D P-velocity model as a starting model, the 3D morphology of each reflector beneath the caldera was seismically imaged separately, following a layer-stripping procedure beginning with the shallowest reflector. Figure 14 shows the obtained interface models along with their associated uncertainties. The 0.6-km-deep reflector indicate the approximate thickness of the very shallow and most recent layer composed of water-saturated volcanic deposits. The

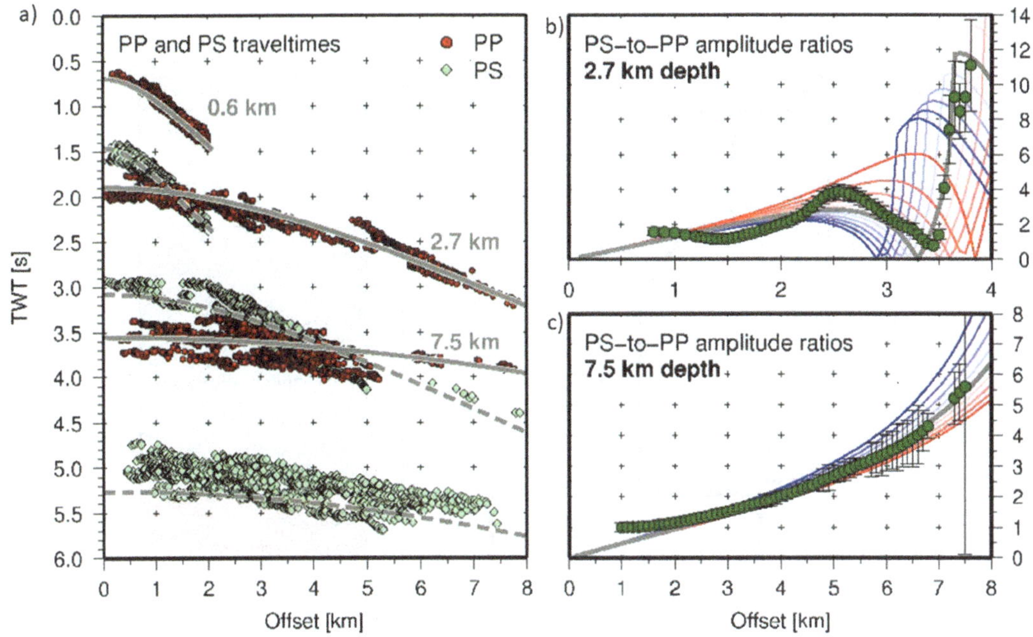

Fig. 12 Reflection travel-time curves. **a** P-to-P and P-to-S travel-time picks (dots) for three major reflectors overlaid with the theoretical times (grey lines) from a 1D average model. Average of the amplitude ratio variations with offset for reflectors at **b** 2700 m and **c** 7500 m depth. The figure shows the observed amplitude ratio (green dots), as compared with a set of theoretical curves computed for the V_P and V_P/V_S velocity contrast models. Given the best-fit model parameters at each interface, the blue and red curves indicate 10% positive and negative model variation with respect to V_P. Error bars for the estimated amplitude ratios are also shown. Density contrasts at both interfaces have not been included in the analysis, as this parameter has a minor effect on the amplitude ratio variation with offset (from Zollo et al. 2008)

2.7-km-deep reflector shows a well-resolved morphological high coinciding with the buried, southern caldera rim and a basin-like structure toward its central portion. The 7.5-km-deep reflector appears essentially flat, being characterised by variations smaller than the estimated depth uncertainty with a minor up-doming feature toward the southwest.

3.3.3 A Magma Sill Beneath the Caldera

The interpretation of the seismic low-velocity layer below the 7.5-km-deep reflector as an extended magma sill (Auger et al. 2001; Zollo et al. 2008) is supported also by petrological data (Mangiacapra et al. 2008; Pappalardo and Mastrolorenzo 2012a, b; Moretti et al. 2013). It should be noted that Ferrucci et al. (1992) previously found evidence for a magmatic body between 4 and 6 km depth beneath the caldera. These authors based their claim on direct P-wave and converted-transmitted PS-waves from a single 1-ton explosion at sea, recorded in a fan-layout onshore at 60–65 km distance. In particular, they analysed the delay time between direct and converted wave and modelled the amplitude ratio between the two phases, indicating a low-rigidity (low-velocity) body.

Considering extrapolations of the average geothermal gradients below the critical point of water (at ~3 km depth), where the thermal regime is due to pure conduction, the temperatures in the layer below 7.5 km should be in the range of 800–1,600 °C (e.g., Carlino and Somma 2009). This is in favour of the interpretation of seismic low velocities as due to the presence of melt instead of, for example, porous, brine-saturated rocks. The melt fraction in this layer

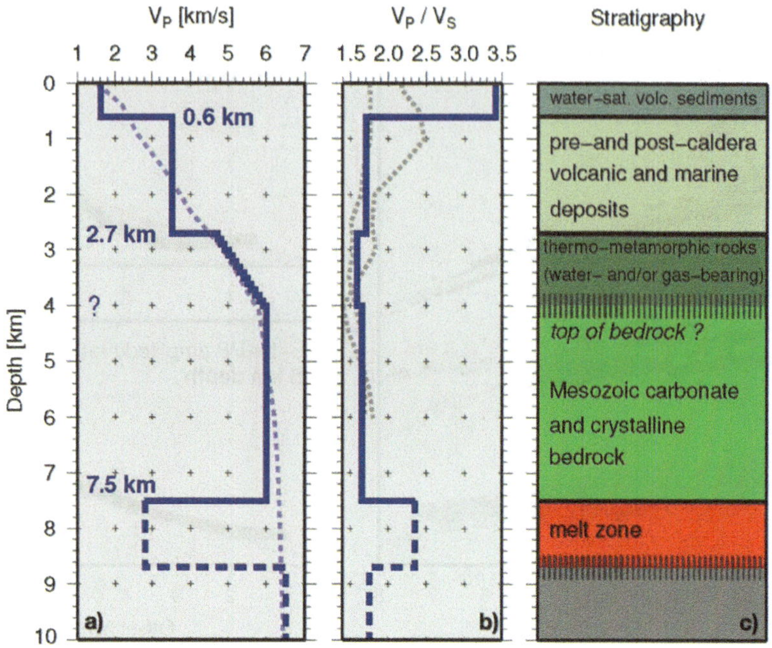

Fig. 13 Geophysical and structural model of the deep structure of the Campi Flegrei caldera. **a** Average 1D P-velocity model for the Campi Flegrei caldera, based on PP and PS travel-times, and on PS-to-PP amplitude ratios. The dashed line is the average of the 3D P-velocity model in the study area. **b** V_P/V_S ratio as a function of depth, as estimated in Zollo et al (2008). The dotted lines are two V_P/V_S depth profiles estimated from the local earthquake tomography (Vanorio et al. 2005). **c** Stratigraphic model. Figure modified after Zollo et al. (2008)

can be estimated using an appropriate rock physics model to relate the detected seismic velocities to the physical properties of a mixture of melt and solid. To get such an estimate, Zollo et al. (2008) computed Hashin–Shtrikman bounds (e.g., Mavko et al. 1998) on seismic P-wave and S-wave velocities for a two-phase medium consisting of solid rock and pure melt. The Hashin–Shtrikman bounds provide the narrowest possible range of the elastic moduli (or seismic velocities) by assuming an isotropic mixture of the two constituents without the need of specifying their geometry. Zollo et al. (2008), based on the results of laboratory measurements on similar rock types (Murase and McBirney 1973), used values of $V_P = 5.5$ km/s and $V_S = 3.2$ km/s for the solid phase, and of $V_P = 2.4$ km/s and $V_S = 0$ km/s for the pure melt, while the density was held constant at 2,600 kg/m^3. These boundary values indicate the presence of 65–75 vol% melt directly beneath the reflector, taking the obtained seismic velocities for the magma layer ($V_P = 2.8$ km/s and $V_P/V_S = 2.35$). However, it has been proposed that magmas at depth could crystallise in cooler regions generating non-zero Vs, which would lead to an even higher percentage of melt fraction (for details see discussion in Zollo et al. 2008). On the other hand, Chu et al. (2010) found a similar seismic velocity structure beneath the Yellowstone caldera from the analysis of teleseismic P-waves, but estimated a melt fraction of only 32 vol% (see also Fedi et al. 2018). Although initially developed for unconsolidated sediments, Chu et al. (2010) used a rock physics model based on the Gassmann relations for porous, fluid-saturated media and the concept of a critical melt fraction, at which the solid medium would break down and become suspended in the fluid (e.g., Mavko et al. 1998). Albeit these variations, seismic evidence shows clearly an extended sill containing a high melt fraction at mid-crustal depths beneath the CFc, and the presence of a similar feature at the same depth beneath Somma-Vesuvio could suggest a

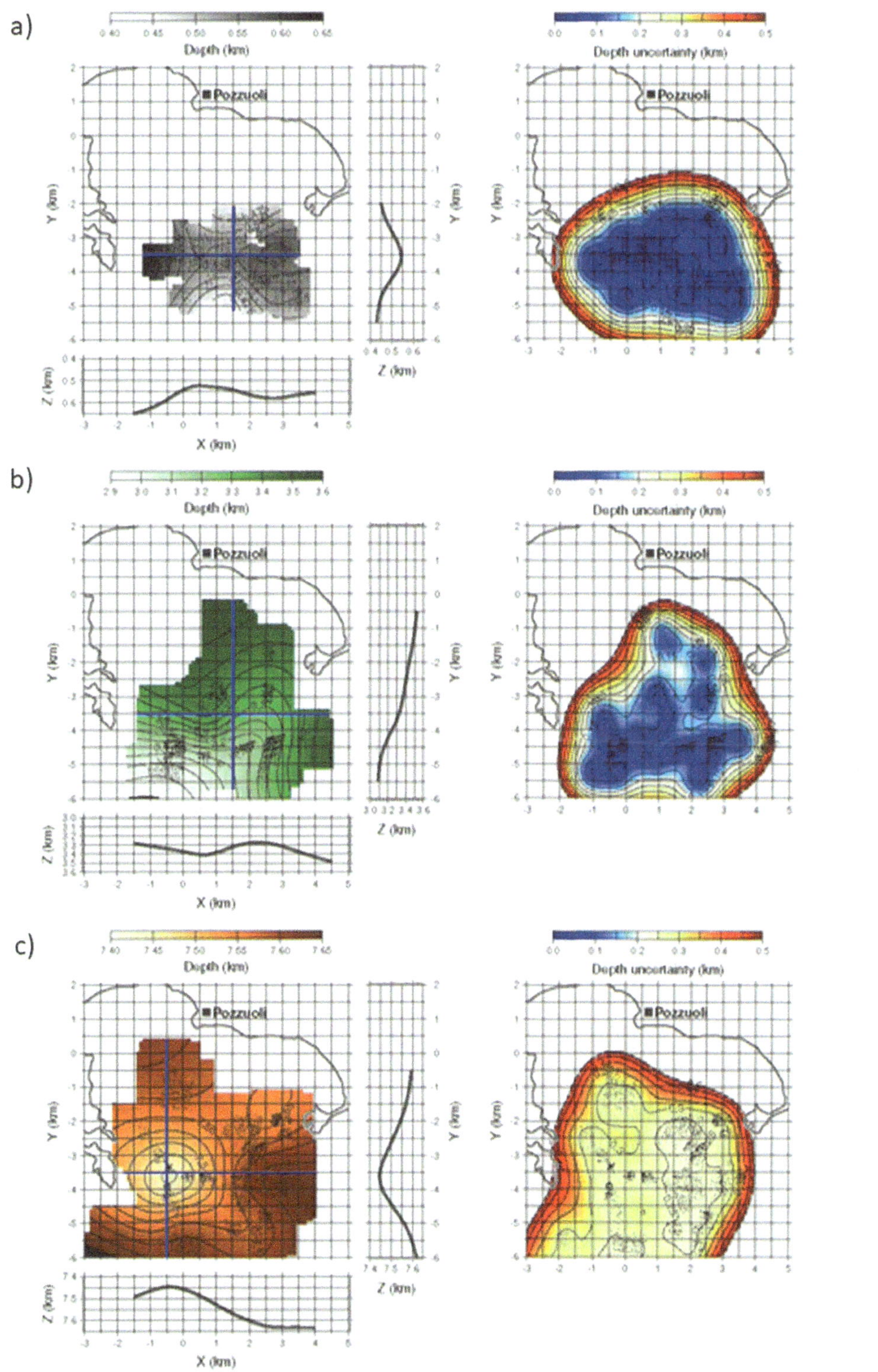

Fig. 14 Reflection interfaces obtained from the inversion of P-to-P travel-times associated with phases reflected from the first (**a**), second (**b**) and third (**c**) discontinuity. The left-side panels show the areal distribution of the reflectors and the vertical sections computed along the north–south and east–west oriented blue lines. The right-side panels show the computed uncertainties associated to the parameters of the models. The small black dots are the positions of the reflected impact points (from Vassallo et al. 2010)

common magma reservoir of these two volcanoes, as also proposed by Pappalardo and Mastrolorenzo (2012a, b). The existence of a shallow, extended and fractured rock layer saturated with supercritical fluids, above a mid-crustal melt zone, is also consistent with similar observations at other unrest calderas, such as Yellowstone, Socorro and Long Valley (DeNosaquo et al. 2009; Fialko et al. 2001).

3.4 Lower Crustal Structure and Moho (from ~12 to ~30 km Depth)

Rather little is known on the lower crust beneath the CFc from seismic data. In a regional crustal thickness model for Italy, Di Stefano et al. (2011) suggested that the Moho beneath the CFc and the Somma-Vesuvio is at 27 km using an interpolated value as in other regional Moho maps. Ferrucci et al. (1989) found a major Moho upheaval beneath the CFc, using data from two large-explosion-source locations at sea (3 $\times$ 1-ton and 2 $\times$ 0.5-ton) recorded inland in arc-shaped receiver spreads (fan-shooting geometry). The source-receiver distances ranged from $\sim$ 45 to 100 km, covering both the CFc and Somma-Vesuvio. Ferrucci et al. (1989) developed an essentially 1D P-velocity model using detected reflection travel-times and amplitudes, and Gaussian-beam ray-tracing modelling. In their model the lower-crustal P-velocities show typical values for a continental lower-crust in which the P-velocity increases from 6.2 to 6.6 km/s between 12 and 18 km depth. This layer is underlain by a 7-km-thick high-velocity layer with a P-velocity of 7.2–7.3 km/s. Ferrucci et al. (1989) identified the Moho depth at 25 km beneath the CFc and at 35 km beneath the nearby Somma-Vesuvio. To date these are the only published images of the lower crust below the magmatic layer at 7.5 km depth based on the results of active seismic surveys. However, the structure of lower crust has been recently hypothesised through an interpretation of the large gravity low centred beneath the Gulf of Naples in terms of density distribution (Fedi et al. 2018). The modelling suggests large volumes of partially molten rock ($\sim$30 vol%) distributed randomly throughout the entire lower crust sector, likely made up of crystalline rocks largely produced by accumulation of crystals left by magma risen in the past.

In a complementary work, Nunziata and Costanzo (2010) derived average crustal S-wave velocity profiles from surface wave dispersion curves for a three travel paths between local earthquakes around Somma-Vesuvio and permanent seismic stations located within the CFc. To extract the fundamental Rayleigh wave mode, the authors applied a Frequency-Time Analysis to instrument-corrected radial-component seismograms, and then used data in the period range 1–5 s to invert for a 1D S-wave velocity model along each observation path. The typical lower-crustal S-velocity is $\sim$3.6 km/s. Besides that, the models by Nunziata and Costanzo (2010) feature an S-wave velocity reduction at 7–15 km depth, again consistent with the presence of melt at these depths.

4 CFc Structure Deduced by Attenuation, Scattering and Ambient Noise Imaging

4.1 Direct-Wave Attenuation Images

Seismic attenuation tomography has been performed at the CFc using the waveform dataset from both the 1982–1984 unrest episode and the SERAPIS experiment. de Lorenzo et al. (2001) obtained the first 3D Q_P image of the CFc. Three-component waveform data from 87 micro-earthquakes recorded in March–April 1984 imaged Q_P in the shallowest 3 km of the central part of the caldera. The authors estimated the rise time and the total pulse width for direct P-waves, obtaining around 1,000 pulse data. These were inverted with a non-linear approach, retrieving attenuation images at a maximum resolution of $\sim$1 km. Q_P models were transformed into temperatures by calibration with surface and borehole temperature data.

De Siena et al. (2010) expanded the dataset in time (January–June 1984) to 246 earthquakes (location errors < 300 m). They used the coda-normalisation method (Aki 1980; Del Pezzo et al. 2006), removing source and site trade-offs by dividing direct energy. Coda energy is assumed constant in space. A ray-bending technique was carried out in the velocity model obtained by Battaglia et al. (2008). The work performed resolution tests, allowing for a theoretical maximum resolution of 0.5 km across the wider caldera, between 0 and 4 km depth. De Siena et al. (2010) applied their method jointly with the more standard spectral-decay method, which allowed a joint description of both Q_P and Q_S. The P- and S-waves attenuation features from De Siena et al. (2010) are compared with the V_P/V_S anomalies from Battaglia et al. (2008) in Fig. 15. Horizontal slices have been obtained at depths of 200, 1,200, 2,200 and 3,200 m b.s.l. and redrawn after space interpolation with the methodology described by Del Pezzo and Bianco (2013b). Due to the maximum resolution available, each image reflects the average attenuation structure in a layer with a thickness of 500 m, centred at the depth values indicated above. A dashed line borders the areas with the highest resolution in the attenuation tomography, and symbols (X1 through X7 in Fig. 16) mark the main volcanological features of the area under study, listed in the caption of Fig. 16. Serlenga et al. (2016) obtained the only attenuation tomography model for the offshore portion of the caldera in the Pozzuoli Bay and independent of the 1982–1984 unrest using data from the SERAPIS experiment. They use the spectral decay method in analogy to De Siena et al. (2010), detecting structures at a resolution higher than 1 km across the offshore portion of the caldera down to a maximum depth of 2 km.

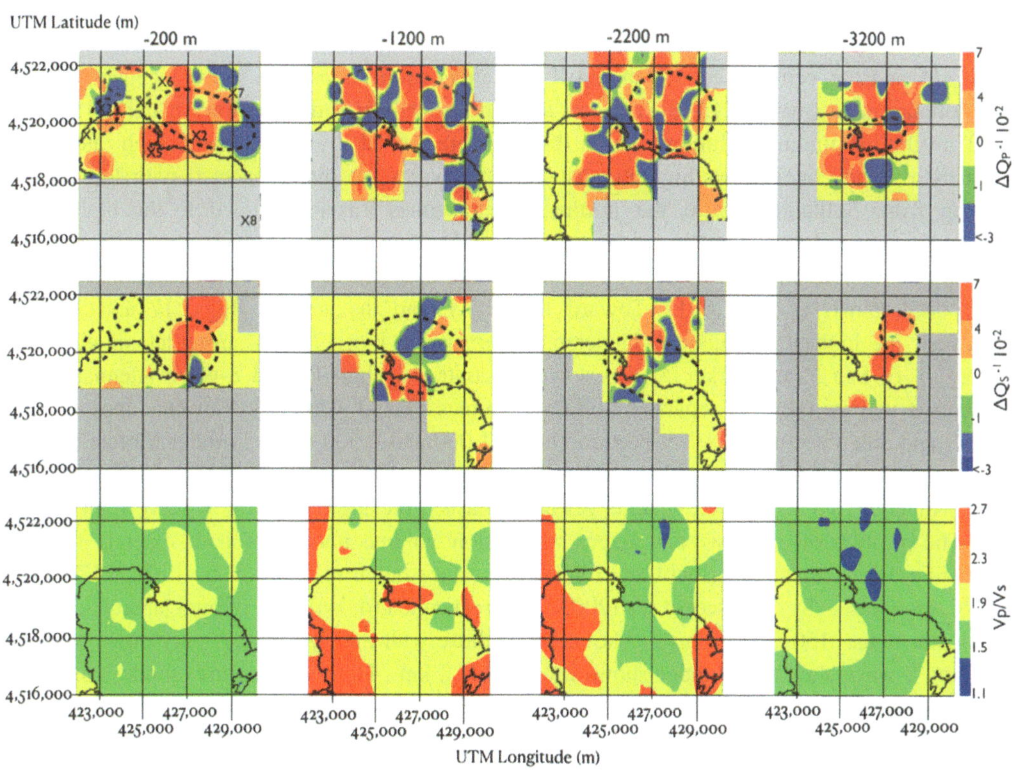

Fig. 15 Horizontal slices intersecting CFc at different depths showing variations of P-wave (top row) and S-wave (middle row) attenuation relative to the average quality factors. The V_P/V_S model (bottom row), redrawn after Battaglia et al. (2008), is compared with the attenuation models. The maximum resolution areas (500 m) are contoured with bold black dashed lines

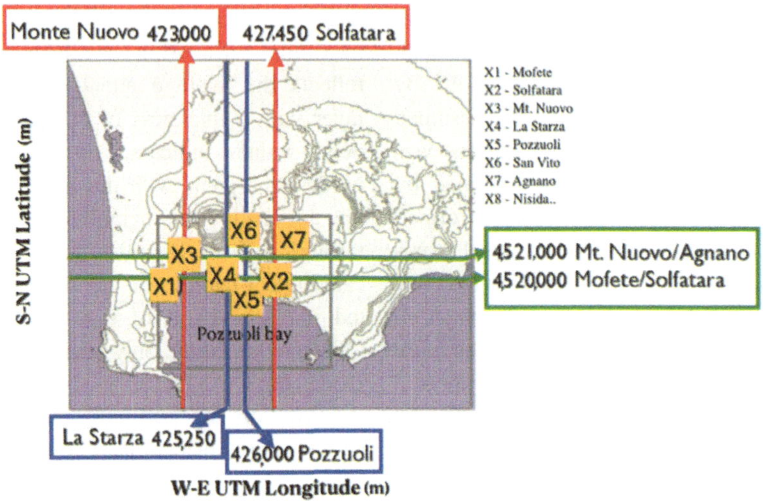

Fig. 16 Map of Campi Flegrei; the rectangle highlights the area investigated in De Siena et al. (2010). The lines define the spatial locations of the vertical sections associated with the tomography images of Figs. 17 and 18. The locations of the area of major volcanological interest are indicated as follows: X1 = Mofete, X2 = Solfatara, X3 = Monte Nuovo, X4 = La Starza, X5 = Pozzuoli, X6 = San Vito, X7 = Agnano

The upper 500 m at the CFc show small P-wave attenuation contrasts (see slice at -200 m in Fig. 15 and the vertical sections in Fig. 17). The Solfatara area (marker X2, Fig. 15 top left panel) in the maximum resolution volume (bordered by a dashed line) shows high attenuation. Moving from Solfatara toward the northeast, high-attenuation values extend to the Astroni - Agnano area (marker X7, northeast of Solfatara) as in de Lorenzo et al. (2001). Toward the northwest, the highest attenuation characterises the region of San Vito (marker X6). A lower-attenuation volume (yellow) separates Solfatara, Astroni and San Vito (markers X5–Pozzuoli and X4–La Starza zone) from high-attenuation rocks under Monte Nuovo and Mofete (marker X1 and X3). The position of this last zone agrees with the aquifer with temperatures between 100 and 130 °C recognised by Todesco et al. (2003) in the depth range 150–300 m b.s.l. The corresponding S-wave attenuation image shows lower attenuation in this depth range. At Solfatara, S-wave attenuation is high, similar to what found for the P-waves. Unlike P-waves, this zone is bordered east by a low attenuation anomaly, at the location of the high-velocity caldera rim (Battaglia et al. 2008).

The images in the three rightmost panels of Fig. 15 representing the space distribution of the attenuation anomalies, show a general laterally heterogeneous distribution pattern. Volumes characterised by rocks with high attenuation anomalies are interconnected with low attenuation zones. It is also noteworthy that high S-wave attenuation zones depict a vertically extended high attenuation pattern between about 3,500 m b.s.l. and the surface (Figs. 17 and 18). The images in Fig. 17 show strong lateral attenuation contrasts. High P- and S-wave attenuation volumes extend vertically between approximately 1,000 and 2,500 m b.s.l. under Mofete (Fig. 17, dashed rectangle #2) and Monte Nuovo (Fig. 17, dashed rectangle #1). The P-wave high attenuation volume under Solfatara extends from the surface down to 3,500 m b.s.l. (Fig. 17, dashed rectangle #6). Volumes of high attenuation match those with the V_P/V_S ratio only down to depths of 2 km. High-temperature fluids comprising carbon dioxide (Caliro et al. 2007) or supercritical fluids (Vanorio et al. 2005) developing under the CFc "caprock" could equally be responsible for these high attenuation volumes (Fig. 17, dashed rectangle #1). The high Q_S^{-1} volumes characterised under Mofete (X1) and Solfatara

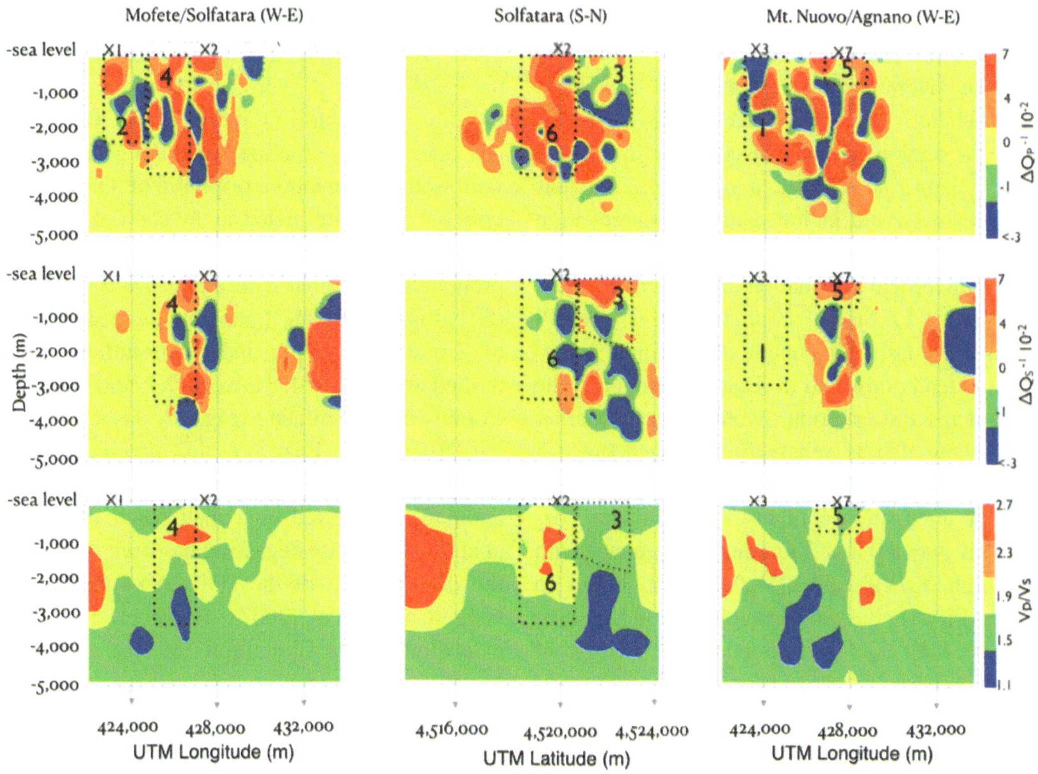

Fig. 17 Attenuation tomography results showing P-wave (top row) and S-wave (middle row) attenuation variations along three vertical sections passing through Mofete/Solfatara (east–west), Solfatara (north–south) and Monte Nuovo/Agnano (east–west). The last line shows V_P/V_S variations from Battaglia et al. (2008, redrawn). The columns are labelled with the name of the area crossed by the section and with the marker highlighting the area of volcanological interest (Fig. 16). The colour scale represents the variations from the average S- and P-wave inverse quality factor. The dashed numbered polygons contour the volumes discussed in the text

(X2) in Fig. 17 (dashed rectangle #4) coincide with the position of deep aquifers (1,250–1,600 m b.s.l.) recognised by Todesco et al. (2003). Under Agnano (Fig. 17, marker X7), high $Q_P^{-1}$and Q_S^{-1} in the shallowest kilometres of the crust coincide with a zone of moderate ($\geq$ 1.9) V_P/V_S anomaly (Fig. 17, dashed rectangle #5), where temperatures are on the order of 250 °C (de Lorenzo et al. 2001; Todesco et al. 2003). North of the Solfatara (X2), a very-low Q_P^{-1} anomaly characterises volumes down to the depth of 2,500 m b.s.l. (Fig. 17, dashed polygon #3). V_P/V_S shows a relative minimum in this area, while Q_S^{-1} is generally high (Fig. 17, dashed polygon #3). As high-temperature rocks (de Lorenzo et al. 2001), conductive heat transfer (Todesco et al. 2003), and low V_P (Battaglia et al. 2008) characterise this region at shallow depths (Fig. 17, dashed rectangle #5), De Siena et al. (2010) inferred the presence of a small reservoir of gas (mainly CO_2) feeding fumaroles (Caliro et al. 2007). The results confirm that fumaroles are the consequence of fluid migration from a magmatic to a geothermal reservoir and from there to the surface, favouring a magmatic interpretation of the 1982–1984 unrest (Todesco et al. 2003; Caliro et al. 2007; Amoruso et al. 2008; Berrino et al. 2008; Moretti et al. 2013, 2020). The observation of very-low Q_S^{-1} contrasts with this interpretation (Fig. 17, dashed rectangle #6); however, the assumption used by the coda-normalisation method suffers in the presence of sharp horizontal interfaces, where coda waves get trapped or reverberate at low frequency. This feature was interpreted by De Siena et al. (2010) as due to alternating heterogeneous layers that affected the coda wavefield, destabilising coda-normalised measurements. In the next section, we will discuss how this interface (seismic horizon) at a depth of 2 km has been resolved and included in direct-attenuation forward modelling (De Siena et al. 2017b).

A subvertical attenuation contrast images the high-angle "La Starza" fault (Di Vito et al. 1999—Fig. 18, dashed rectangle #7), which apparently extends down to at least 2 km depth. The area of the last eruption (Monte Nuovo, marker X3 in Fig. 15) is relatively homogeneous for both S-wave and V_P/V_S (Fig. 17). The high P-wave attenuation anomaly located under Pozzuoli, between 0 and 2 km depth, correlates with the high V_P/V_S value ($V_P/V_S = 2.6$) down to 2 km (Fig. 18, dashed rectangle #8). In the same volume, high Q_S^{-1} corresponds to the location of high V_P/V_S. The anomaly appears as an inclusion inside very-low attenuation volumes, supporting the existence of a hydrothermal reservoir under the city of Pozzuoli (Vanorio et al. 2005). The low-attenuation structure underlying the reservoir is the attenuation response of the seismic horizon, later interpreted as a "caprock" rich in hot hydrothermal fluids. One of the most prominent and debated results of De Siena et al. (2010) is the high Q_S^{-1} anomaly located between of 2 and 3 km depth under Pozzuoli (Fig. 17, dashed rectangle #4, between X1 and X2) that extends under Solfatara (Fig. 17, dashed rectangle #6). Under Pozzuoli, this anomaly corresponds precisely to the location of a high-gravity anomaly interpreted as a melt batch (Amoruso et al. 2008), whose degassing has been central for the development of the 1982–1984 unrest episode (Caliro et al. 2007). However, its lateral extension is at the limit of the method's

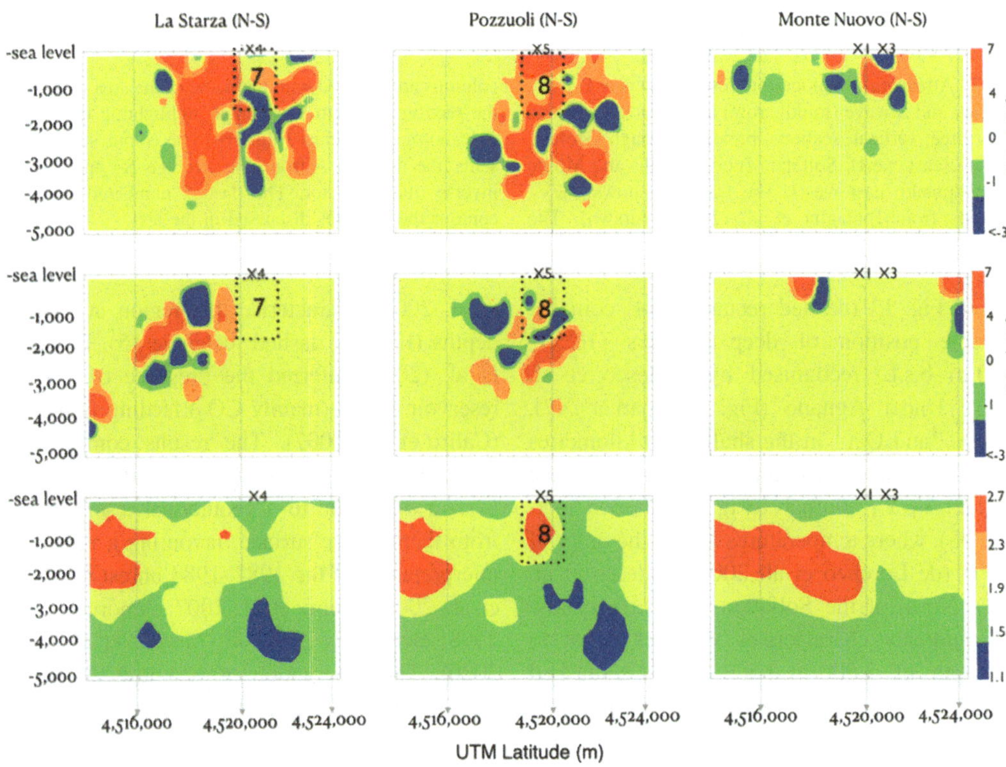

Fig. 18 Same as Fig. 17 for three vertical sections crossing La Starza (north–south), Pozzuoli (north–south) and Monte Nuovo (north–south). See Fig. 16 for section locations and labels

resolution at these depths. It also contrasts with the interpretations based only on the velocity images, as V_P/V_S value remains low under Pozzuoli and Solfatara (Vanorio et al. 2005). This trade-off has been tackled by following direct and coda-dependent methods applied to the same dataset, which remove faulty approximations from scattering dependent techniques (De Siena et al. 2017a, b; Akande et al. 2019) and use novel seismic tomography results at the caldera (Calò and Tramelli 2018).

The V_P/V_S model of Battaglia et al. (2008) shows two characteristic features across the offshore caldera. At about 1 km depth, a very high V_P/V_S anomaly occurs offshore the town of Pozzuoli and is interpreted as due to the presence of rocks that contain fluids in the liquid phase. A low V_P/V_S body extending at about 3–4 km depth below a large part of the caldera is interpreted as the top of rock bodies enriched in gas under supercritical conditions, as previously suggested by Vanorio et al. (2005). These fluids also produce high attenuation when they percolate in the shallower systems, either saturating sediments or remaining in a vapour phase. The strongest attenuation characterises the offshore caldera rim, which appears fluid-saturated in the uppermost 2 km of the crust (Serlenga et al. 2016).

4.2 Coda and Ambient Noise Imaging

Direct-wave attenuation combines both intrinsic and scattering effects: by separating these attenuation mechanisms and mapping them in space, we improve our ability to interpret structures. For example, low-velocity or high V_P/V_S anomalies corresponding to high attenuation can be interpreted as being due to the presence of: (1) a fluid or magma body, which would increase the shear-wave intrinsic attenuation with increasing temperature; (2) a partially saturated fracture network, which would produce an increase in scattering attenuation.

The joint interpretation of total attenuation and body-wave velocity tomography images suffers from this trade-off when locating and defining the extent of fluid reservoirs, magma batches, or the characteristics of a buried or submerged caldera rim. Scattering tomography detects sharp discontinuities in the elastic parameters (the position of the "strong scatterers"). It helps discriminate between zones with a high density of fractures (characteristic, e.g., of the caldera borders) and zones with large fluid reservoirs and magma intrusions (which can be associated with the active caldera centres). Scattering imaging has been applied to the CFc using different approaches (Tramelli et al. 2006; Maercklin 2008; De Siena et al. 2011, 2017a; Akande et al. 2019).

Tramelli et al. (2006) used a method developed by Nishigami (1991) that inverts the S-wave coda envelope fluctuations observed at a given lapse time in the coda of local earthquakes in terms of position and strength of the "strong" scatterers present in the Earth medium. The method computes seismic intensities at late lapse times in the coda and searches for the location and scattering coefficients that best fit it, with a search grid algorithm. The technique depends on the frequency-dependent transport mean free path and the assumption of diffusion. The equations of this method were originally written by Nishigami (1991) in the S-to-S single scattering assumption. Tramelli et al. (2009) demonstrated that this method is independent from the scattering model used to mathematically fit the average coda envelope, assuming that the coda envelope fluctuations are generated by S-to-S single scattering phenomena occurring at the "strong" scatterers. In Fig. 19, the stacks (over the components) of the values obtained from the inversion are showed separately for each frequency band. The inversion process returns the value of the scattering coefficient, $g = g_0\alpha_i$, in the i-th block. α takes into account the fluctuations of the scattering coefficient averaged over space (g_0). Figure 19 shows the space distributions of $\log(\alpha)$. The crosses and circles represent the scattering coefficients higher and lower than g_0, respectively. The results obtained in three layers spanning the shallowest 4 km of the upper crust show that the most heterogeneous materials correspond inland to the central portion of the

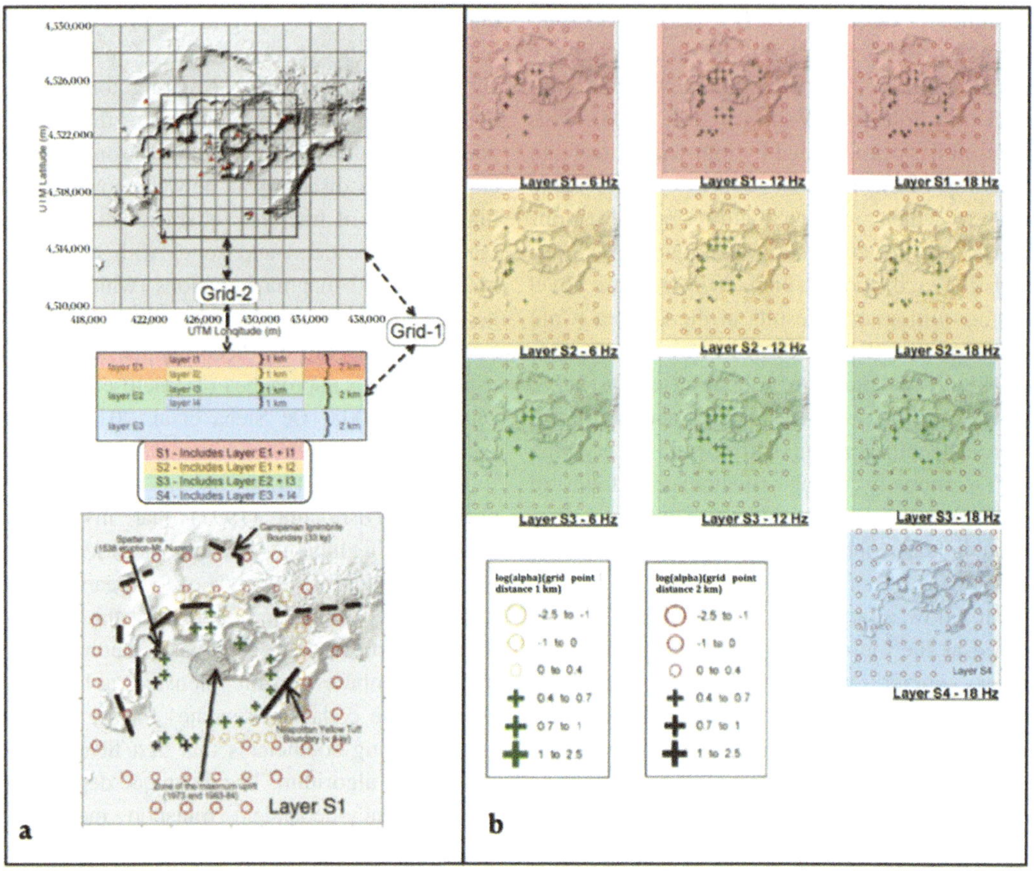

Fig. 19 Scattering image of the CFc: **a** at the top the scheme of the subdivisions with depth for the inversion, while the bottom part shows some geological features superimposed on the S1 layer at 18 Hz of panel b; **b** results of the inversions of the stacked (over the components) signals for several frequency. The crosses and the circles represent the areas with high and low scattering strengths, respectively as indicated in legend

caldera as mapped by velocity tomography. The authors deduced that similar anomalies in the submerged part of the caldera coincide with the submerged part of this structure. The assumption was confirmed by the offshore velocity tomography of Battaglia et al. (2008). The only high-scattering anomaly not directly connected to the rim appears in the eastern part of the caldera and corresponds to the high-temperature fractured body previously deduced by de Lorenzo et al. (2001) and later recognised by De Siena et al. (2010) from the contrasting P- and S-wave attenuation characteristics (Fig. 17, dashed rectangle #6).

Maercklin (2008) adopted an imaging method based on array beam-forming and coherence analysis of the P to P scattered phase, using the SERAPIS dataset. After grouping into arrays the signal sources, the author calculated the travel-times connecting any potential scatterer on a regular grid in the image volume to the source and receiver, using a 1D background velocity model. A high coherence measured on aligned traces indicates the presence of a scatterer at the corresponding location. A stack of the individual images from several pairs of a shot array and a single receiver provides the final image. Even in this approach, as in the previously described one, the prominent region of seismic scattering roughly coincides with the high-gravity (Capuano et al. 2013) eastern buried caldera rim (Zollo et al. 2003). This work was the first to detect this volume in the offshore caldera.

De Siena et al. (2011) proposed a new quantitative approach for the joint interpretation of velocity and attenuation tomography images, which allows the separation of scattering and intrinsic attenuation. These authors used the P-wave tomography structure of Battaglia et al. (2008) to calculate the autocorrelation functions (ACF) of P-wave vertical velocity fluctuations from which the P-wave quality factor is estimated (see Eq. 6 in De Siena et al. 2011). A map showing the lateral variations of the ACF values (see Fig. 2 in De Siena et al. 2009) is then compared with the total-QP map (De Siena et al. 2010) using cluster analysis. The result is finally interpreted in terms of separation of intrinsic and scattering attenuation on a 2D plane. The resulting map (Fig. 20) shows:

1. a north–south scattering attenuation region located below the zone of the 1982–1984 maximum uplift, offshore the Pozzuoli Bay but also extending below Monte Nuovo. De Siena et al. (2011) based on the previous images in attenuation (De Siena et al. 2010), velocity (Battaglia et al. 2008) and S-to-S scattering (Tramelli et al. 2006), favour the interpretation in terms of a solid but fractured body, contoured by strong S-wave scatterers. They speculate that the high-scattering area corresponds to a section of the residual magma body associated with the AD 1538 Monte Nuovo eruption;
2. that the high-temperature volume originally detected by de Lorenzo et al. (2001) is the region where P-wave intrinsic dissipation dominates. The anomaly suggests a connection, valid for the uppermost 4 km of the crust, between the region offshore Pozzuoli and the Solfatara crater. The results agree with the existence of a paired Pozzuoli-Solfatara deformation source, active throughout the last 40 years (Amoruso et al. 2014).

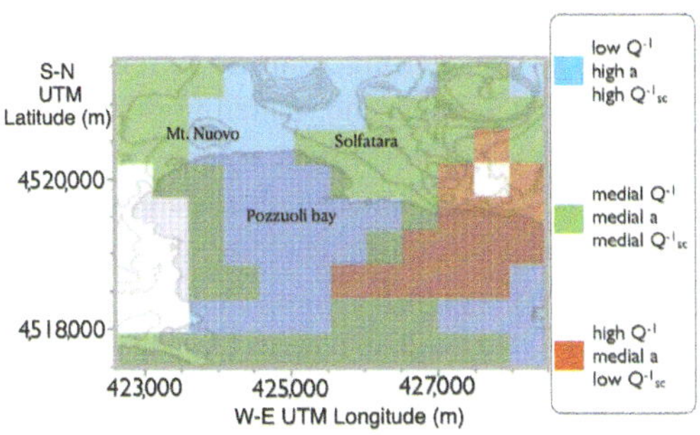

Fig. 20 Results obtained applying the cluster analysis to data from: total-$Q^{-1}{}_{P}$ structure (De Siena et al. 2010), $Q^{-1}{}_{Sc}$ measured by velocity fluctuations and correlation length, a (De Siena et al. 2011). This analysis shows the existence of three clusters (blue, green and orange). The physical characteristics of each cluster are indicated in the legend at the right side the of the map

During the last decade, researchers have developed a single open-access code that inverts for total, scattering and intrinsic attenuation from earthquakes and active seismic data simultaneously (MuRAT; De Siena et al. 2014a, b; Prudencio et al. 2015; Sketsiou et al. 2020). De Siena et al. (2017a, b) and Akande et al. (2019) applied this code to the 1982–1984 dataset. De Siena et al. (2017a) used 2D diffusive sensitivity kernels (Del Pezzo et al. 2016) to develop an inversion scheme for coda-wave attenuation (Qc) at multiple frequencies, allowing to model in space a quantity equivalent to absorption in the diffusive regime (Sketsiou et al. 2020). Akande et al. (2019) used 3D multiple-scattering kernels and the previous inversion scheme to obtain the 3D distribution of Qc in space. De Siena et al. (2017b) employed a revised coda-normalisation method that measures direct-wave attenuation, dropping the assumption of homogeneity of the coda wavefield. The authors interpreted the velocity and attenuation maps after re-localising the entire 1982–1984 earthquake dataset with a non-linear localisation approach (NonLinLoc; Lomax 2001). By pre-calculating source-station coda attenuation, they improved inversion residual by 80% relative to De Siena et al. (2010), removing ghosts in the outer caldera and obtaining maximum resolutions of 500 m in the eastern portion of the caldera. This work includes a time-dependent analysis of the 1982–1984 unrest through the joint description of seismic locations, attenuation and V_P/V_S anomalies.

The 3D P-wave velocity images described in previous sections find no evidence of magma bodies with volumes larger than 1 km^3 beneath the caldera. This would imply that the main magma reservoir feeding the Phlegraean magmatic system is deeper, within the carbonate sequence (Vanorio et al. 2005). The attenuation tomography study by De Siena et al. (2010) suggested the presence of low-velocities and high-attenuation batches of molten rocks (<1 km in size), correlated to high gravity anomalies at depths of ~3 km (Amoruso et al. 2008). However, the resolution of their tomography method degrades with the size of such an anomaly. One kilometre was the lower limit of spatial resolution to reconstruct subsurface seismic images in this area with the previously-mentioned techniques (e.g., Battaglia et al. 2008). More recently, Calò and Tramelli (2018) have reanalysed the 1982–1984 dataset using enhanced seismic tomography techniques, allowing theoretical resolutions of a few hundred metres. They found a high V_P/V_S anomaly between 2.3 and 4 km depth under the centre-eastern sector or the caldera, interpreting it as a plumbing system with a diameter of ~1 km. This is the first tomographic study to show prominent high V_P/V_S anomalies in this depth range and onshore. The model supports the presence of magmatic fluids or magma (Berrino et al. 2008; Amoruso et al. 2008) in the first 4 km of the crust in 1982–1984.

Calò and Tramelli (2018) also reconstructed a low V_P/V_S and high scattering anomaly (~500 m-thick) above the high V_P/V_S plumbing system. This is the same region where De Siena et al. (2010) had hypothesised the existence of a sequence of highly-scattering layers trapping coda intensities. De Siena et al. (2017b) imaged a low-attenuation seismic horizon under the entire caldera and particularly in this zone, compatible with the existence of the "caprock" recovered by rock-physics measurements at ~1–2 km depth under the caldera (Vanorio and Kanitpanyacharoen 2015). Under Pozzuoli and below this "caprock", De Siena et al. (2017b) imaged a low-velocity and high-attenuation anomaly, likely the response to deep fluids and magma. This anomaly, a reservoir of supercritical fluids (Vanorio et al. 2005) possibly intruded by small batches of magma (Amoruso et al. 2008) was the location of repeated seismic injections that opened the upper hydrothermal system in 1982–1984 (De Siena et al. 2017b). Once crossed with geochemical information, seismic tomography and source data confirm that this was also the primary location for processes causing geochemical and seismic unrest at the caldera in 1982–1984 (Chiodini et al. 2016, 2021). The seismic signature does not allow to discriminate dike intrusions from magmatic fluid injections but supports the existence of a magmatic component during the unrest (Amoruso et al. 2008; De Siena et al. 2010, 2017b; Moretti et al. 2020).

The deepest high-attenuation anomaly in De Siena et al. (2017b) plateaus at ~4 km depth and correlates laterally with the stationary deformation source active during the last four decades (Amoruso et al. 2014; Pepe et al. 2019). De Siena et al. (2017a) used the enhanced lateral resolution provided by the coda kernels to demonstrate that this high-attenuation structure produces a bulge in the low-attenuation caprock. This bulge corresponds to the source of deformation in the offshore central caldera, producing a single aseismic low coda attenuation anomaly (De Siena et al. 2017a). The ~4 km deep high-attenuation anomaly extends toward the eastern high-gravity sector of the caldera, connecting to the high V_P/V_S plumbing system (Calò and Tramelli 2018). Starting from de Lorenzo et al. (2001), all attenuation and scattering studies have found evidence of a fractured volume, producing high temperatures between 2 and 3 km depth across the eastern onshore sector of the caldera (see previous section).

Akande et al. (2019) used the 3D diffusive kernels to constrain the geometry of the connection between the deep source of the deformation and the Solfatara fumarole. They built a single interpretational model of the caldera using their coda-attenuation results, existing interdisciplinary knowledge, vents distribution (Orsi et al. 2004; Vilardo et al. 2010; Vitale and Isaia 2014; Chaps. Volcanic and Deformation History of the Campi Flegrei Volcanic Field, Italy; Volcanic Hazard Assessment at the Campi Flegrei Caldera, Italy) and the novel V_P/V_S anomalies (Calò and Tramelli 2018). They confirmed previous inferences (de Lorenzo et al. 2001; Maercklin et al. 2008) by imaging a unique high-Qc^{-1} anomalous volume associated with the high-gravity region where intense fracturing occurs during caldera collapse and resurgence (Capuano et al. 2013). They additionally detected its connection to the primary deformation source in the central caldera. Figure 21 is a sketch that adds to their finding the results of velocity, attenuation, scattering and gravimetric analyses performed during the last four decades, showing them across an east–west profile passing through Pozzuoli. Akande et al. (2019) concluded that the high V_P/V_S, high Q_P^{-1} and Q_S^{-1}, high temperature, high-scattering and high-absorption capped pathway from deformation source to the eastern portion of the caldera represents preferential paths for magmas to intrude the crust in their ascent toward the surface and eventually feed eruptions. Recent seismic, gravity and electromagnetic investigations have focussed on the eastern sector of the caldera to resolve this issue (De Landro et al. 2017; Siniscalchi et al. 2019; Young et al. 2020), confirming the existence of a resistive plume structure in the uppermost 2 km of the crust (Siniscalchi et al. 2019).

The CFc has been mostly aseismic after its 1982–1984 unrest. The new unrest phase started at the end of 2004 and has been characterised by very low levels of seismicity until 2015 (Bellucci Sessa et al. 2021). Seismicity has recently increased, but ambient noise imaging has become the primary way to seismologically assess the caldera structure. De Siena et al. (2018) used phase-weighted stacking (Shimmel and Paulssen 1997) and the FMTOMO inversion framework (Rawlinson et al. 2004) to image surface group velocities between the surface and about 2 km depth. Pepe et al. (2019) used the ambient noise map to interpret the results of deformation gradiometry. These works suggest that the high V_P/V_S reservoir first detected by Vanorio et al. (2005) at ~ 1 km depth (Fig. 21) has expanded north and west and become aseismic under the city of Pozzuoli. A seismic high-velocity structure north–south striking under Solfatara appears as a structural barrier for eastward deformation, in correspondence to the highly resistive plume (Siniscalchi et al. 2019). The analyses confirm the likelihood of an eastward-directed feeding system for fumaroles degassing at Solfatara and the new (2009) Pisciarelli vent (Chiodini et al. 2021).

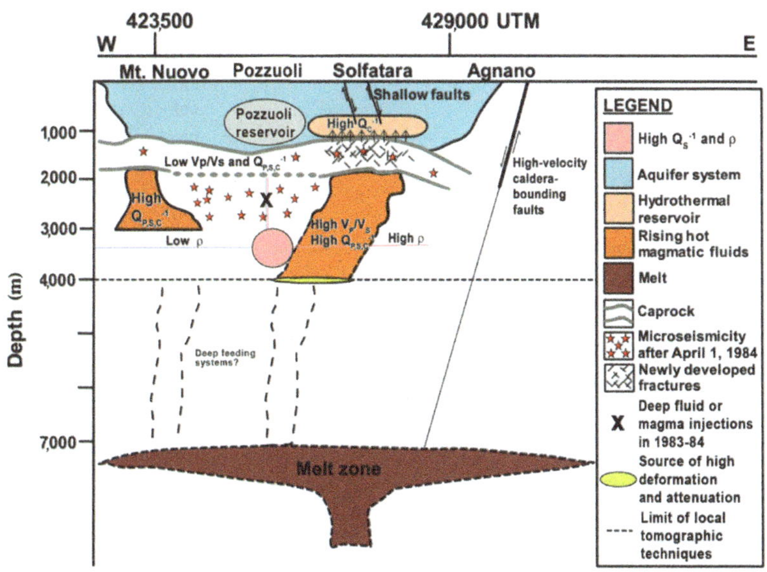

Fig. 21 Schematic illustration of the uppermost crust beneath the Campi Flegrei along an east–west vertical section, based on a joint interpretation of velocity, attenuation, scattering and gravity inversions. Figure modified after Akande et al. (2019)

5 Conclusive Remarks

A review of seismic and gravity data, and tomographic models produced during the past four decades has been here presented in order to give a picture of the structure of the crust beneath the CFc.

The defined lithological and geological discontinuities have been widely depicted, the constraints they give to the definition of magmatic and volcanic processes have been discussed.

The elastic properties of the shallowest 4 km of the upper crust have been reconstructed using velocity tomography elaborated on active and passive seismic data. Information complementary to velocity tomography, improved by seismic attenuation tomography, permit to define lateral lithological changes and the geometry of onshore and offshore fluid and/or small magma bodies (possibly placed during the 1982–1984 bradyseismic crisis) down to 4 km under Pozzuoli town.

Active experiments have permitted to recognise a partially molten body at a depth of $\sim$7.5 km interpreted as an extended magma sill under the caldera.

Acknowledgements This paper has been prepared in the frameworks of the Projects UNREST and V2 (Precursori) funded by the Department of the Civil Protection of Italy and by I.N.G.V, and partially funded by the Spanish research project CGL2011-29499-C02-01, EPHESTOS Istituto Andaluz de Geofisica - Universidad de Granada, Spain.

References

Achauer U, Evans JR, Stauber DA (1988) High-resolution seismic tomography of compressional wave velocity structure at Newberry Volcano, Oregon cascade range. J Geophys Res 93(B9):10135–10147. https://doi.org/10.1029/JB093iB09p10135

AGIP (1987) Modello geotermico del sistema flegreo. Sintesi. Servizi Centrali per l'Esplorazione, SERG-MESG, San Donato Milanese, Italy, 23 pp

Akande WG, De Siena L, Gan Q (2019) Three-dimensional kernel-based coda attenuation imaging of caldera structures controlling the 1982–1984 Campi Flegrei unrest. J Volcanol Geotherm Res 381:273–283

Aki K (1980) Attenuation of shear-waves in the lithosphere for frequencies from 0.05 to 25 Hz. Phys Earth Planet Int 21:50–60

Amoruso A, Crescentini L, Berrino G (2008) Simultaneous inversion of deformation and gravity changes in a horizontally layered half-space: evidences for magma intrusion during the 1982–1984 unrest at Campi Flegrei caldera (Italy). Earth Planet Sci Lett 272(1–2):181–188

Amoruso A, Crescentini L, Sabbetta I (2014) Paired deformation sources of the Campi Flegrei caldera (Italy) required by recent (1980–2010) deformation history. J Geophys Res Solid Earth 119(2):858–879

Aster RC, Meyer RP (1988) Three-dimensional velocity structure and hypocenter distribution in the Campi Flegrei caldera, Italy. Tectonophysics 149:195–218. https://doi.org/10.1016/00401951(88)90173-4

Auger E, Gasparini P, Virieux J, Zollo A (2001) Seismic evidence for an extended magmatic sill under Mt. Vesuvius. Science 294:1510–1512. https://doi.org/10.1126/science.1064893

Auger E, Virieux J, Zollo A (2003) Locating and quantifying the seismic discontinuities in a complex medium through the migration and AVA analysis of reflected and converted waves: an application to the Mt Vesuvius volcano. Geophys J Int 152(2):486–496. https://doi.org/10.1046/j.1365-246X.2003.01864.x

Barberi F, Cassano E, La Torre P, Sbrana A (1991) Structural evolution of Campi Flegrei caldera in light of volcanological and geophysical data. J Volcanol Geotherm Res 48:33–49

Battaglia J, Zollo A, Virieux J, Dello Iacono D (2008) Merging active and passive data sets in travel-time tomography: the case study of Campi Flegrei Caldera (Southern Italy). Geophys Prospect 56:555–573. https://doi.org/10.1111/j.1365-2478.2007.00687.x

Bellucci Sessa E, Castellano M, Ricciolino P (2021) GIS applications in volcano monitoring: the study of seismic swarms at the Campi Flegrei volcanic complex, Italy. Adv Geosci 52:131–144

Benz HM, Chouet BA, Dawson PB, Lahr JC, Page RA, Hole JA (1996) Three-dimensional P and S wave velocity structure of Redoubt Volcano, Alaska. J Geophys Res 101:8111–8128. https://doi.org/10.1029/95JB03046

Berrino G, Corrado G, Riccardi U (1998) Sea gravity data in the Gulf of Naples: a contribution to delineating the structural pattern of the Vesuvian area. J Geophys Res 82:139–150

Berrino G, Corrado G, Riccardi U (2008) Sea gravity data in the Gulf of Naples. A contribution to delineating the structural pattern of the Phlegrean volcanic district. J Volcanol Geotherm Res 175:241–252. https://doi.org/10.1016/j.jvolgeores.2008.03.007

Blacic TM, Latorre D, Virieux J, Vassallo M (2009) Converted phases analysis of the Campi Flegrei caldera using active seismic data. Tectonophysics 470(3–4):243–256. https://doi.org/10.1016/j.tecto.2008.12.006

Bonafede M (1991) Hot fluid migration: An efficient source of round deformation: application to the 1982–1985 crisis at Campi Flegrei-Italy. J Volcanol Geotherm Res 48:187–198

Bonasia V, Del Pezzo E, Pingue F, Scandone R, Scarpa R (1985) Eruptive history, seismic activity and ground deformation at Mt. Vesuvius, Italy. Ann Geofis 3:395–406

Brenguier F, Shapiro NM, Campillo M, Nercessian A, Ferrazzini V (2007) 3-D surface wave tomography of the Piton de la Fournaise volcano using seismic noise correlations. Geophys Res Lett 34:L02305. https://doi.org/10.1029/2006GL028586

Bruno PPG, Cippitelli G, Rapolla A (1998) Seismic study of the Mesozoic carbonate basement around Mt. Somma-Vesuvius, Italy. J Geophys Res 84:311–322. https://doi.org/10.1016/S0377-0273(98)00023-7

Bruno PPG, Rapolla A, Di Fiore V (2003) Structural setting of the Bay of Naples (Italy) seismic reflection data: implications for Campanian volcanism. Tectonophysics 372(3–4):193–213. https://doi.org/10.1016/S0040-1951(03)00327-5

Caliro S, Chiodini G, Moretti R, Avino R, Granieri D, Russo M, Fiebig J (2007) The origin of the fumaroles of La Solfatara (Campi Flegrei, South Italy). Geochim Cosmochim Acta 71:3040–3055. https://doi.org/10.1016/j.gca.2007.04.007

Calò M, Tramelli A (2018) Anatomy of the campi flegrei caldera using enhanced seismic tomography models. Sci Rep 8(1):1–12

Campos-Enriquez JO, Arredondo-Fragoso JJ (1992) Gravity study of Los Humeros caldera complex, Mexico: structure and associated geothermal system. J Volcanol Geotherm Res 49(1–2):69–90

Capuano P, Achauer U (2003) Gravity field modeling in the Vesuvius and Campanian area. In: Zollo A et al. (eds) The TomoVes seismic project: looking inside Mt. Vesuvius. Cuen, Naples, Italy

Capuano P, Russo G, Vanorio T, Prevete R, Auger E, Bonagura M, Caielli G, Convertito V, Damiano N, D'Auria L, Emolo A, Lovisa L, Moretti M (2006) 1984 Campi Flegrei seismic waveforms compilation. In: Zollo A et al. (eds) Geophysical exploration of the Campi Flegrei (Southern Italy) Caldera' interiors: data, methods and results. Doppiavoce, Naples, Italy, pp 15–24. ISBN: 88-89972-04-1

Capuano P, Russo G, Civetta L, Orsi G, D'Antonio M, Moretti R (2013) The active portion of the Campi Flegrei caldera structure imaged by 3D inversion of gravity data. Geochem Geophys Geosyst 14:4681–4697

Carlino S, Somma R (2009) Eruptive versus non-eruptive behaviour of large calderas: the example of Campi Flegrei caldera (Southern Italy). Bull Volcanol 72 (7):871–886. https://doi.org/10.1007/s00445-010-0370-y

Casertano L, Oliveri A, Quagliariello MT (1976) Hydrodynamics and geodynamics in the Phlegraean Fields area of Italy. Nature 264:161–164

Cassano E, La Torre P (1987) Geophysics. In: Rosi M, Sbrana A (eds) Phlegrean Fields. Quaderni de "La Ricerca Scientifica", vol 114(9). CNR, Rome, Italy, pp 94–103

Chiarabba C, Moretti M (2006) An insight into the unrest phenomena at the Campi Flegrei caldera from V_P and V_P/V_S tomography. Terra Nova 18:373–379. https://doi.org/10.1111/j.1365-3121.2006.00701.x

Chiodini G, Todesco M, Caliro S, Del Gaudio C, Macedonio G, Russo M (2003) Magma degassing as a trigger of bradyseismic events: the case of Phlegrean Fields (Italy). Geophys Res Lett 30(8):1434. https://doi.org/10.1029/2002GL016790

Chiodini G, Paonita A, Aiuppa A, Costa A, Caliro S, De Martino P, Acocella V, Vandemeulebrouck J (2016) Magmas near the critical degassing pressure drive volcanic unrest towards a critical state. Nat Commun 7 (1):1–9

Chiodini G, Caliro S, Avino R, Bini G, Giudicepietro F, De Cesare W, Ricciolino P, Aiuppa A, Cardellini C, Petrillo Z, Selva J, Siniscalchi A, Tripaldi S (2021) Hydrothermal pressure-temperature control on CO_2 emissions and seismicity at Campi Flegrei (Italy). J Volcanol Geotherm Res 414:107245

Chouet B (1988) Resonance of a fluid-driven crack: radiation properties and implications for the source of long-period events and harmonic tremor. J Geophys Res 93(B5):4375–4400. https://doi.org/10.1029/JB093iB05p04375

Chu R, Helmberger DV, Sun D, Jackson JM, Zhu L (2010) Mushy magma beneath Yellowstone. Geophys Res Lett 37:L01306. https://doi.org/10.1029/2009GL041656

Cinque A, Rolandi G, Zamparelli V (1985) L'estensione dei depositi marini olocenici nei Campi Flegrei in relazione alla vulcano-tettonica. Boll Soc Geol It 104 (2):327–348

Cinque A, Patacca E, Scandone P, Tozzi M (1993) Quaternary kinematic evolution of the Southern Apennines. Relationships between surface geological features and deep lithospheric structures. Ann Geophys 36:249–260

Davy BW, Caldwell TG (1998) Gravity, magnetic and seismic surveys of the caldera complex, Lake Taupo, North Island, New Zealand. J Volcanol Geotherm Res 81:69–89

De Bonitatibus A, Latmiral G, Mirabile L, Palumbo A, Sarpi E, Scalera A (1970) Rilievi sismici per riflessione: strutturali, ecografici (fumarole) e batimetrici del Golfo di Pozzuoli. Boll Soc Natur Napoli Italy 79:97–115

De Landro G, Serlenga V, Russo G, Amoroso O, Festa G, Bruno PP, Gresse M, Vandemeulebrouck J, Zollo A (2017) 3D ultra-high resolution seismic imaging of shallow Solfatara crater in Campi Flegrei (Italy): new Insights on deep hydrothermal fluid circulation processes. Sci Reports 7(1). https://doi.org/10.1038/s41598-017-03604-0

de Lorenzo S, Zollo A, Monelli F (2001) Source parameters and three-dimensional attenuation structure from the inversion of microearthquake pulse width data: Qp imaging and inferences on the thermal state

of Campi Flegrei caldera (Southern Italy). J Geophys Res 106. https://doi.org/10.1029/2000JB900462
De Siena L, Del Pezzo E, Bianco F, Tramelli A (2009) Multiple resolution seismic attenuation imaging at Mt Vesuvius. Phys Earth Planet Inter 173:17–32
De Siena L, Del Pezzo E, Bianco F (2010) Seismic attenuation imaging of Campi Flegrei: evidence of gas reservoirs, hydrothermal basins, and feeding systems. J Geophys Res 115:B09312. https://doi.org/10.1029/2009JB006938
De Siena L, Del Pezzo E, Bianco F (2011) A scattering image of Campi Flegrei from the auto correlation functions of velocity tomograms. Geophys J Int 184 (3):1304–1310. https://doi.org/10.1111/j.1365-246X.2010.04911.x
De Siena L, Thomas C, Aster R (2014a) Multi-scale reasonable attenuation tomography analysis (MuRAT): an imaging algorithm designed for volcanic regions. J Volcanol Geotherm Res 277:22–35
De Siena L, Thomas C, Waite GP, Moran SC, Klemme S (2014b) Attenuation and scattering tomography of the deep plumbing system of Mount St. Helens. J Geophys Res Solid Earth 119(11):8223–8238
De Siena L, Amoruso A, Del Pezzo E, Wakeford Z, Castellano M, Crescentini L (2017a) Space-weighted seismic attenuation mapping of the aseismic source of Campi Flegrei 1983–1984 unrest. Geophys Res Lett 44(4):1740–1748
De Siena L, Chiodini G, Vilardo G, Del Pezzo E, Castellano M, Colombelli S, Tisato N, Ventura G (2017b) Source and dynamics of a volcanic caldera unrest: Campi Flegrei, 1983–1984. Sci Rep 7(1):1–13
De Siena L, Sammarco C, Cornwell DG, La Rocca M, Bianco F, Zaccarelli L, Nakahara H (2018) Ambient seismic noise image of the structurally controlled heat and fluid feeder pathway at Campi Flegrei caldera. Geophys Res Lett 45(13):6428–6436
Del Pezzo E, Bianco F, De Siena L, Zollo A (2006) Small scale shallow attenuation structure at Mt. Vesuvius, Italy. Phys Earth Planet Inter 157:257–268
Del Pezzo E, Bianco F (2013a) A reappraisal of seismic Q evaluated in Campi Flegrei caldera. Receipt for the application to risk analysis. J Seismol 17:829–837. https://doi.org/10.1007/s10950-012-9349-9
Del Pezzo E, Bianco F (2013b) Inside Mt. Vesuvius, a new method to look at the seismic tomographic imaging. Ann Geophys 56(4):S0443. ISSN: 2037-416X. https://doi.org/10.4401/ag-6449
Del Pezzo E, Ibanez J, Prudencio J, Bianco F, De Siena L (2016) Absorption and scattering 2-D volcano images from numerically calculated space-weighting functions. Geophys J Int 206(2):742–756
Dello Iacono D, Zollo A, Vassallo M, Vanorio T, Judenherc S (2009) Seismic images and rock properties of the very shallow structure of Campi Flegrei caldera (Southern Italy). Bull Volcanol 71(3):275–284. https://doi.org/10.1007/s00445-008-0222-1
DeNosaquo KR, Smith RB, Lowry AR (2009) Density and lithospheric strength models of the Yellowstone-Snake River Plain volcanic system from gravity and heat flow data. J Volcanol Geotherm Res 188:108–127. https://doi.org/10.1016/j.jvolgeores.2009.08.006
Di Stefano R, Bianchi I, Ciaccio MG, Carrara G, Kissling E (2011) Three-dimensional Moho topography in Italy: new constraints from receiver functions and controlled source seismology. Geochem Geophys Geosys 12:Q09006. https://doi.org/10.1029/2011GC003649
Di Vito MA, Isaia R, Orsi G, Southon J, de Vita S, D'Antonio M, Pappalardo L, Piochi M (1999) Volcanism and deformation since 12,000 years at the Campi Flegrei caldera (Italy). J Volcanol Geotherm Res 91:221–246
Di Vito MA, Acocella V, Aiello G, Barra D, Battaglia M, Carandente A, Del Gaudio C, de Vita S, Ricciardi GP, Ricco C, Scandone R, Terrasi F (2016) Magma transfer at Campi Flegrei caldera (Italy) before the 1538 AD eruption. Sci Rep 6:32245. https://doi.org/10.1038/srep32245
Dvorkin J, Prasad M, Sakai A, Lavoie D (1999) Elasticity of marine sediments: rock physics modeling. Geophys Res Lett 26(12):1781–1784. https://doi.org/10.1029/1999GL900332
Eberhart-Phillips D, Reyners M, Chadwick M, Chiu JM (2005) Crustal heterogeneity and subduction processes: 3-D Vp, Vp/Vs and Q in the southern North Island, New Zealand. Geophys J Int 162(1):270–288
Fedi M, Nunziata C, Rapolla A (1991) The Campania-Campi Flegrei area: a contribution to discern the best structural model from gravity interpretation. J Volcanol Geotherm Res 48:51–59
Fedi M, Cella F, D'Antonio M, Florio G, Paoletti V, Morra V (2018) Gravity modeling finds a large magma body in the deep crust below the Gulf of Naples, Italy. Sci Rep 8:8229. https://doi.org/10.1038/s41598-018-26346-z
Ferrazzini V, Aki K (1987) Slow waves trapped in a fluid-filled infinite crack: implication for volcanic tremor. J Geophys Res 92(B9):9215–9223. https://doi.org/10.1029/JB092iB09p09215
Ferrucci F, Gaudiosi G, Pino NA, Luongo G, Hirn A, Mirabile L (1989) Seismic detection of a major Moho upheaval beneath the Campania volcanic area (Naples, Southern Italy). Geophys Res Lett 16(11):1317–1320. https://doi.org/10.1029/GL016i011p01317
Ferrucci F, Hirn A, De Natale G, Virieux J, Mirabile L (1992) P-SV conversions at a shallow boundary beneath Campi Flegrei caldera (Italy): evidence for the magma chamber. J Geophys Res 97(B11):15351–15359. https://doi.org/10.1029/92JB00888
Fialko Y, Simons M, Khazan Y (2001) Finite source modelling of magmatic unrest in Socorro, New Mexico, and Long Valley, California. Geophys J Int 146:191–200. https://doi.org/10.1046/j.1365-246X.2001.00453.x
Finetti I, Morelli C (1974) Esplorazione sismica a riflessione nei Golfi di Napoli e Pozzuoli. Boll Geofis Teor Appl 16:175–222
Florio G, Fedi M, Cella F, Rapolla A (1999) The Campanian Plain and Phlegrean Fields: structural

setting from potential field data. J Volcanol Geotherm Res 91:361–379
Fournier N, Rymer H, Williams-Jones G, Brenes J (2004) High-resolution gravity survey: investigation of subsurface structures at Poas volcano, Costa Rica. Geophys Res Lett 31:L15602. https://doi.org/10.1029/2004GL020563
Gaeta FS, De Natale G, Peluso F, Castagnolo D, Troise C, Pingue F, Mita DG, Rossano S (1998) Genesis and evolution of unrest episodes at Campi Flegrei caldera: the role of the thermal fluid dynamical processes in the geothermal system. J Geophys Res 103:20921–20933
Gailler LS, Lénat JF, Lambert M, Levieux G, Villeneuve N, Froger JL (2009) Gravity structure of Piton de la Fournaise volcano and inferred mass transfer during the 2007 crisis. J Volcanol Geotherm Res 184(1–2):31–48. https://doi.org/10.1016/j.jvolgeores.2009.01.024
Gasparini P and TomoVes Working Group (1998) Looking inside Mount Vesuvius. EOS Trans Am Geophys Un 79(19):229–232
De Gori P, Chiarabba C, Patanè D (2005) Qp structure of Mount Etna: constraints for the physics of the plumbing system. J Geophys Res Solid Earth 110(B5)
Hansen S, Thurber C, Mandernach M, Haslinger F, Doran C (2004) Seismic velocity and attenuation structure of the east rift zone and south flank of Kilauea Volcano, Hawaii. Bull Seismol Soc Am 94 (4):1430–1440
Hashin Z, Shtrikman S (1963) A variational approach to the elastic behavior of multiphase minerals. J Mech Phys Solids 11(2):127–140. https://doi.org/10.1016/0022-5096(63)90060-7
Hellweg M (2000) Physical models for the source of Lascar's harmonic tremor. J Volcan Geotherm Res 101(1–2):83–198
Hobro JW, Singh SC, Minshull TA (2003) Three-dimensional tomographic inversion of combined reflection and refraction seismic traveltime data. Geophys J Int 152(1):79–93. https://doi.org/10.1046/j.1365-246X.2003.01822.x
Hole JA (1992) Nonlinear high-resolution three-dimensional seismic travel time tomography. J Geophys Res 97(B5):6553–6562. https://doi.org/10.1029/92JB00235
Huffman A, Castagna J (2001) The petrophysical basis for shallow water flow prediction using multicomponent seismic data. Lead Edge 20(9):1030–1052
Imbò G, Bonasia V, Gasparini P (1964) Rilievo gravimetrico dell'isola di Procida. Ann Oss Ves, Naples, Italy 6:117–138
Jolivet L, Faccenna C, Piromallo C (2009) From mantle to crust: Stretching the Mediterranean. Earth and Planetary Science Letters 285(1–2):198–209. https://doi.org/10.1016/j.epsl.2009.06.017
Jousset P, Neuberg J, Sturton S (2003) Modelling the time-dependent frequency content of low-frequency volcanic earthquakes. J Volcanol Geotherm Res 128:201–223
Judenherc S, Zollo A (2004) The Bay of Naples (southern Italy): constraints on the volcanic structures inferred from a dense seismic survey. J Geophys Res 109 (B10):B10312. https://doi.org/10.1029/2003JB002876
Latorre D, Virieux J, Monfret T, Monteiller V, Vanorio T, Got JL, Lyon-Caen H (2004) A new seismic tomography of Aigion area (Gulf of Corinth-Greece) from a 1991 dataset. Geophys J Int 159:1013–1031
Locke CA, Cassidy J, MacDonald A (1993) Three-dimensional structure of relict stratovolcanoes in Taranaki, New Zealand: evidence from gravity data. J Volcanol Geotherm Res 59(1–2):121–130. https://doi.org/10.1016/0377-0273(93)90081-2
Lomax A, Zollo A, Capuano P, Virieux J (2001) Precise, absolute earthquake location under Somma Vesuvius volcano using a new three dimensional velocity model. Geophys J Int 146:313–331
Maercklin N (2008) Seismic scatterer imaging using shot array beamforming: method and application to the Campi Flegrei caldera. In: Marzocchi W, Zollo A (eds) Conception, verification, and application of innovative techniques to study active volcanoes. Doppiavoce, Naples, Italy, pp 261–268. ISBN: 978-88-89972-09-0
Maercklin N, Zollo A (2009) Estimation of elastic contrasts in a layered model from seismic PS-to-PP amplitude ratios. Geophys J Int 179(3):1617–1626. https://doi.org/10.1111/j.1365-246X.2009.04350.x
Maino A, Tribalto G (1971) Rilevamento gravimetrico di dettaglio dell'isola d'Ischia, Napoli. Boll Serv Geol Ital 92:109–122
Malone SD, Boyko C, Weaver CS (1983) Seismic precursors to the Mount St. Helens eruptions in 1981 and 1982. Science 221(4618):1376–1378
Mangiacapra A, Moretti R, Rutherford M, Civetta L, Orsi G, Papale P (2008) The deep magmatic system of the Campi Flegrei caldera (Italy). Geophys Res Lett 35:L21304. https://doi.org/10.1029/2008GL035550
Masturyono, McCaffrey R, Wark DA, Roecker SW, Fauzi, Ibrahim G, Sukhyar (2001) Distribution of magma beneath Toba caldera complex, north Sumatra, Indonesia, constrained by three-dimensional P wave velocities, seismicity, and gravity. Geochem Geophys Geosyst 2. https://doi.org/10.1029/2000GC000096
Mavko G, Mukerji T, Dvorkin J (1998) The rock physics handbook: tools for seismic analysis in porous media. Cambridge University Press, Cambridge, UK
Milia A, Torrente M (2003) Late-quaternary volcanism and transtensional tectonics in the Bay of Naples, Campanian continental margin, Italy. Miner Petrol 79:49–65
Monteiller V, Got JL, Virieux J, Okubo P (2005) An efficient algorithm for double-difference tomography and location in heterogeneous media, with an application to the Kilauea volcano. J Geophys Res 110: B12306. https://doi.org/10.1029/2004JB003466
Mora MM, Lesage P, Valette B, Alvarado GE, Leandro C, Metaxian JP, Dorel J (2006) Shallow velocity structure and seismic site effects at Arenal volcano, Costa

Rica. J Volcanol Geotherm Res 152(1–2):121–139. https://doi.org/10.1016/j.jvolgeores.2005.09.013
Moretti R, Arienzo I, Civetta L, Orsi G, Papale P (2013) Multiple magma degassing sources at an explosive volcano. Earth Planet Sci Lett 367:95–104. https://doi.org/10.1016/j.epsl.2013.02.013
Moretti R, De Natale G, Troise C (2020) Hydrothermal versus magmatic: geochemical views and clues into the unrest dilemma at Campi Flegrei. In: Vesuvius, Campi Flegrei, and Campanian volcanism. Elsevier, pp 371–406
Murase T, McBirney AR (1973) Properties of some common igneous rocks and their melt at high temperatures. Geol Soc Am Bull 84(11):3563–3592
Nakajima J, Hasegawa A (2003) Tomographic imaging of seismic velocity structure in and around the Onikobe volcanic area, northeastern Japan: implications for fluid distribution. J Volcanol Geother Res 127(1–2):1–18
Nakamichi H, Tanaka S, Hamaguchi H (2002) Fine S wave velocity structure beneath Iwate volcano, northeastern Japan, as derived from receiver functions and travel times. J Volcanol Geotherm Res 116(3–4):235–255. https://doi.org/10.1016/S0377-0273(02)00218-4
Nercessian A, Hirn A, Tarantola A (1984) Three dimensional seismic transmission prospecting of the Mont Dore volcano, France. Geophys J R Astr Soc 76:307–315
Neuberg J, Luckett, Ripepe M, Braun T (1994) Highlight from a seismic broadband array on Stromboli volcano. Geophys Res Lett 21:749–752
Neuberg JW, Tuffen H, Collier L, Green D, Powell T, Dingwell D (2006) The trigger mechanism of low-frequency earthquakes on Montserrat. J Volcanol Geotherm Res 153(1–2):37–50. https://doi.org/10.1016/j.jvolgeores.2005.08.008
Nishigami KY (1991) A new inversion method of coda waveforms to determine spatial distribution of coda scatterers in the crust and uppermost mantle. Geophys Res Lett 18(12):2225–2228
Nunziata C, Costanzo M (2010) Low Vs crustal zones in the Campanian plain (Southern Italy). Miner Petrol 100(3):215–225. https://doi.org/10.1007/s00710-010-0129-3
Nur A, Marion D, Yin H (1991) Wave velocities in sediments. In: Hovem J, Richardson MD, Stoll RD (eds) Shear waves in marine sediments. Kluwer Academic, pp 131–140
Nur A, Mavko G, Dvorkin J, Galmundi D (1995) Critical porosity: the key to relating physical properties to porosity in rocks. In: Proceedings of 65th annual international meeting social expl and geophysics, p 878
Oliveri del Castillo A, Quagliariello MT (1969) Sulla genesi del bradisismo flegreo. Atti Assoc Geofis Ital 4:1–4
Operto S, Ravaut C, Improta L, Virieux J, Herrero A, Dell'Aversana P (2004) Quantitative imaging of complex structures from dense wide-aperture seismic data by multiscale traveltime and waveform inversions: a case study. Geophys Prosp 52:625–651. https://doi.org/10.1111/j.1365-2478.2004.00452.x
Orsi G, D'Antonio M, de Vita S, Gallo G (1992) The Neapolitan Yellow Tuff, a large-magnitude trachytic phreatoplinian eruption: eruptive dynamics, magma withdrawal and caldera collapse. J Volcanol Geotherm Res 53:275–287. https://doi.org/10.1016/0377-0273(92)90086-S
Orsi G, de Vita S, Di Vito M (1996) The restless, resurgent Campi Flegrei nested caldera (Italy): constraints on its evolution and configuration. J Volcanol Geotherm Res 74:179–214
Orsi G, Civetta L, Del Gaudio C, de Vita S, Di Vito MA, Isaia R, Petrazzuoli SM, Ricciardi G, Ricco C (1999) Short-term ground deformations and seismicity in the resurgent Campi Flegrei caldera (Italy): an example of active block-resurgence in a densely populated area. J Volcanol Geotherm Res 91(2–4):415–451
Orsi G, De Vita S, Di Vito M, Isaia R, Nave R, Heiken, G (2003) Facing volcanic and related hazards in the neapolitan area. In: Heiken G, Fakuntiny R, Sutter J (eds) Earth science in the city: a reader, vol 56. Wiley
Orsi G, Di Vito MA, Isaia R (2004) Volcanic hazard assessment at the restless Campi Flegrei caldera. Bull Volcanol 66:514–530. https://doi.org/10.1007/s00445-003-0336-4
Pappalardo L, Matrolorenzo G (2012a) Rapid differentiation in a sill-like magma reservoir: a case study from the Campi Flegrei caldera. Sci Rep 2(712). https://doi.org/10.1038/srep00712
Pappalardo L, Mastrolorenzo G (2012b) Rapid differentiation in a sill-like magma reservoir: a case study from the Campi Flegrei caldera. Sci Rep 2:712. https://doi.org/10.1038/srep00712
Patacca E, Sartori R, Scandone P (1990) Tyrrhenian basin and apenninic arcs: kinematic relations since late Tortonian times. Mem Soc Geol It 45:425–451
Patanè D, Barberi G, Cocina O, De Gori P, Chiarabba C (2006) Time-resolved seismic tomography detects magma intrusions at Mount Etna. Science 313 (5788):821–823. https://doi.org/10.1126/science.1127724
Pepe S, De Siena L, Barone A, Castaldo R, D'Auria L, Manzo M, Casu F, Fedi M, Lanari R, Bianco F, Tizzani P (2019) Volcanic structures investigation through SAR and seismic interferometric methods: the 2011–2013 Campi Flegrei unrest episode. Remote Sens Environ 234(111440)
Pescatore T, Diplomatico G, Senatore MR, Tramutoli M, Mirabile L (1984) Contributi allo studio del Golfo di Pozzuoli: aspetti stratigrafici e strutturali. Mem Soc Geol Ital 27:133–149
Prudencio J, De Siena L, Ibáñez JM, Del Pezzo E, Garcia-Yeguas A, Diaz-Moreno A (2015) The 3D attenuation structure of deception island (Antarctica). Surv Geophys 36(3):371–390
Rawlinson N, Sambridge M (2004) Wave front evolution in strongly heterogeneous layered media using the fast marching method. Geophys J Int 156(3):631–647

Rosi M, Sbrana A (eds) (1987) Phlegrean Fields. Quad Ric Sci 114(9):1–175. ISSN: 0556-9664

Rosi M, Sbrana A, Principe C (1983) The phlegrean fields: structural evolution, volcanic history and eruptive mechanisms. J Volcanol Geoth Res 17(1–4):273–288

Rowe CA, Aster RC, Kyle PR, Schlue JW (1998) RR dibble broadband recording of strombolian explosions and associated very-longperiod seismic signals on Mount Erebus Volcano, Ross Island, Antarctica. Geophys Res Lett 25(13):2297–2300

Rowe CA, Aster RC, Kyle PR, RR Dibble, Schlue JW (2000) Seismic and acoustic observations at Mount Erebus volcano, Ross island, Antarctica, 1994–1998. J Volcanol Geotherm Res 101(1–2):105–128

Rymer H (1994) Microgravity change as a precursor to volcanic activity. J Volcanol Geotherm Res 61:311–328

Sacchi M, Alessio G, Aquino I, Esposito E, Molisso F, Nappi R, Porfido S, Violante C (2009) Risultati preliminari della campagna oceanografica CAFE_07—Leg 3 nei Golfi di Napoli e Pozzuoli, Mar Tirreno Orientale. Quad Geofis 64:1–28. ISSN: 1590-2595

Sanders CO, Ponko SC, Nixon LD, Schwartz FA (1995) Seismological evidence for magmatic and hydrothermal structure in Long Valley caldera from local earthquake attenuation and velocity tomography. J Geophys Res 100:8311–8326

Sato H, Fehler MC, Maeda T (2012) Seismic wave propagation and scattering in the heterogeneous earth, vol 496. Springer, Berlin, Germany

Satriano C, Zollo A, Rowe C (2008) Iterative tomographic analysis based on automatic refined picking. Geophys Pros 56:467–475. https://doi.org/10.1111/j.1365-2478.2008.00700.x

Schimmel M, Paulssen H (1997) Noise reduction and detection of weak, coherent signals through phase-weighted stacks. Geophys J Int 130(2):497–505

Schmeling H (1985) Numerical models on the influence of partial melt on elastic, anelastic and electric properties of rocks. Part I: elasticity and an elasticity. Phys Earth Planet Int 41(1):34–57

Schurr B, Asch G, Rietbrock A, Trumbull R, Haberland C (2003) Complex patterns of fluid and melt transport in the central Andean subduction zone revealed by attenuation tomography. Earth Planet Sci Lett 215 (1–2):105–119

Serlenga V, de Lorenzo S, Russo G, Amoroso O, Garambois S, Virieux J, Zollo A (2016) A three-dimensional QP imaging of the shallowest subsurface of Campi Flegrei offshore caldera, Southern Italy. Geophys Res Lett 43(21):11209–11218

Siniscalchi A, Tripaldi S, Romano G, Chiodini G, Improta L, Petrillo Z, D'Auria L, Caliro S, Avino R (2019) Reservoir structure and hydraulic properties of the Campi Flegrei geothermal system inferred by audiomagnetotelluric, geochemical, and seismicity study. J Geophys Res Solid Earth 124(6):5336–5356

Sketsiou P, Napolitano F, Zenonos A, De Siena L (2020) New insights into seismic absorption imaging. Phys Earth Planet Int 298:106337

Stabile TA, Zollo A, Vassallo M, Iannaccone G (2007) Underwater acoustic channel properties in the Gulf of Naples and their effects on digital data transmission. Ann Geophys 50:313–328

Takei Y (2002) Effect of pore geometry on VP/VS: from equilibrium geometry to crack. J Geophys Res Solid Earth 107(B2):ECV-6

Thorpe S, Locke CA, Brown GC, Francis PW, Randal M (1981) Magma chamber below Poàs volcano, Costa Rica. J Geol Soc 138:367–373. https://doi.org/10.1144/gsjgs.138.3.0367

Thurber CH (1984) Seismic detection of the summit magma complex of Kilauea volcano, Hawaii. Science 223:165–167

Todesco M, Chiodini G, Macedonio G (2003) Monitoring and modeling hydrothermal fluid emission at La Solfatara (Phlegrean Fields, Italy). An interdisciplinary approach to the study of diffuse degassing. J Volcanol Geotherm Res 125:57–79. https://doi.org/10.1016/S0377-0273(03)00089-1

Tramelli A, Del Pezzo E, Bianco F, Boschi E (2006) 3-D scattering image of the Campi Flegrei caldera (Southern Italy). New hints on the position of the old caldera rim. Phys Earth Planet Int 155:269–280

Tramelli A, Del Pezzo E, Fehler MC (2009) 3-D scattering image of Mt. Vesuvius. Bull Seismol Soc Am 99(3):1962–1972. https://doi.org/10.1785/0120080273

Vanorio T, Kanitpanyacharoen W (2015) Rock physics of fibrous rocks akin to Roman concrete explains uplifts at Campi Flegrei caldera. Science 349(6248):617–621. https://doi.org/10.1126/science.aab1292

Vanorio T, Virieux J, Capuano P, Russo G (2005) Three-dimensional seismic tomography from P wave and S wave microearthquake travel times and rock physics characterization of the Campi Flegrei caldera. J Geophys Res 110(B3):B03201. https://doi.org/10.1029/2004JB003102

Vassallo M, Zollo A, Dello Iacono D, Maercklin N, Virieux J (2008) Converted phase identification and retrieval of Vp/Vs ratios from move-out reflection analysis: application to the Campi Flegrei caldera. In: Marzocchi W, Zollo A (eds) Conception, verification, and application of innovative techniques to study active volcanoes. Doppiavoce, Naples, Italy, pp 349–360

Vassallo M, Zollo A, Festa G, Maercklin N (2010) Campi Flegrei (Southern Italy) calderic model from the joint inversion of reflected (PP, PS) and first arrival seismic traveltimes. 7th EGU General assembly geophysics research abstract 12:13157

Vilardo G, Isaia R, Ventura G, De Martino P, Terranova C (2010) InSAR permanent scatterer analysis reveals fault re-activation during inflation and deflation episodes at Campi Flegrei caldera. Remote Sens Environ 114(10):2373–2383

Vitale S, Isaia R (2014) Fractures and faults in volcanic rocks (Campi Flegrei, Southern Italy): insight into volcano-tectonic processes. Int J Earth Sci 103 (3):801–819

Yokoyama I (1989) Microgravity and height changes caused by volcanic activity: four Japanese examples. Bull Volcanol 51:333–345

Yokoyama S, Mena M (1991) Structure of La Primavera caldera, Jalisco, Mexico. J Volcanol Geotherm Res 47:183–194

Young N, Isaia R, Gottsmann J (2020) Gravimetric constraints on the hydrothermal system of the Campi Flegrei caldera. J Geophys Res Solid Earth 125(7): e2019JB019231

Zhdanov MS (2002) Geophysical inverse theory and regularization problems. Elsevier, Amsterdam, The Netherlands, p 605

Zimmer M (2003) Controls on the seismic velocities of unconsolidated sands: measurements of pressure, porosity and compaction effects. Ph.D. thesis, Stanford University, Stanford, CA

Zollo A, Judenherc S, Auger E, D'Auria L, Virieux J, Capuano P, Chiarabba C, De Franco R, Makris J, Michelini A, Musacchio G (2003) Evidence for the buried rim of Campi Flegrei caldera from 3-D active seismic imaging. Geophys Res Lett 30(19). https://doi.org/10.1029/2003GL018173

Zollo A, Maercklin N, Vassallo M, Dello Iacono D, Virieux J, Gasparini P (2008) Seismic reflections reveal a massive melt layer feeding Campi Flegrei caldera. Geophys Res Lett 35(12):L12306. https://doi.org/10.1029/2008GL034242

An Evolutionary Model for the Magmatic System of the Campi Flegrei Volcanic Field (Italy) Constrained by Petrochemical Data

Lorenzo Fedele

Abstract

The petrochemical characteristics of the Campi Flegrei juvenile products are here reviewed in order to give a comprehensive portrait of the main processes governing magma evolution. Particular emphasis is devoted to the definition of the shallow-level differentiation processes (fractional crystallisation ± open-system assimilation/mixing/mingling/magma chamber refilling) which modify the composition of primitive magmas during their ascent to the surface. Constraints on the intensive parameters characterising the Campi Flegrei magmas as retrieved from mineral geothermobarometry, melt inclusions studies, experimental petrology and phase equilibria modelling from the existing literature are also presented and critically evaluated. On such basis, a model for the Campi Flegrei magmatic system is proposed, in an attempt to construct a reference framework from which reasonable speculations of its possible future behaviour could be obtained.

L. Fedele (✉)
Dipartimento di Scienze della Terra dell'Ambiente e delle Risorse, Università degli Studi di Napoli Federico II, Napoli, Italy
e-mail: lofedele@unina.it

1 Introduction

As reported in Chap. Volcanic and Deformation History of the Campi Flegrei Volcanic Field, Italy, the volcanic history of the Campi Flegrei volcanic field (CFvf) was characterised by the two main caldera-forming eruptions of the Campanian Ignimbrite (CI) and Neapolitan Yellow Tuff (NYT). These landmark events can be thus used to make a broad distinction into five main periods of activity: (1) pre-CI; (2) CI; (3) pre-NYT; (4) NYT and (5) post-NYT. This subdivision will be employed here in reviewing the extensive dataset of published whole-rock and mineral phase compositions for Campi Flegrei rocks (Di Girolamo 1970; Di Girolamo et al. 1973, 1984; Ghiara et al. 1977, 1979; Barberi et al. 1978; Di Girolamo and Rolandi 1979; Armienti et al. 1983; Lirer et al. 1987; Rosi and Sbrana 1987; Civetta et al. 1988, 1991a, 1997; Ghiara 1989–90; Beccaluva et al. 1990, 1991; Scarpati et al. 1993; Melluso et al. 1995, 2012; Orsi et al. 1995; D'Antonio et al. 1999a, b, 2007; de Vita et al. 1999; Pappalardo et al. 1999, 2002a, b, 2008; Signorelli et al. 1999; Webster et al. 2003; Fulignati et al. 2004; Tonarini et al. 2004, 2009; D'Oriano et al. 2005; Piochi et al. 2005; Insinga et al. 2006; Marianelli et al. 2006; Cannatelli et al. 2007; Fowler et al. 2007; Fedele et al. 2008, 2009, 2011, 2015, 2016; Mangiacapra et al. 2008; Pabst et al. 2008; Arienzo et al. 2009, 2010, 2015, 2016; Isaia et al. 2009; Di

© Springer-Verlag GmbH Germany, part of Springer Nature 2022
G. Orsi et al. (eds.), *Campi Flegrei*, Active Volcanoes of the World,
https://doi.org/10.1007/978-3-642-37060-1_3

Vito et al. 2011; Cannatelli 2012; Fourmentraux et al. 2012; Forni et al. 2016, 2018a, b; Stock et al. 2016; Iovine et al. 2017; Astbury et al. 2018; Voloschina et al. 2018; Albert et al. 2019).

As a part of the Phlegraean Volcanic District, the CFvf shares many geological, volcanological and petrological similarities with the neighbouring volcanic fields of the Procida and Ischia islands (e.g., Di Girolamo et al. 1984, 1995; De Astis et al. 2004). Notwithstanding this, numerous are the lines of evidence suggesting a substantial independence of the three, including clear differences on both mineral chemistry (Melluso et al. 2012, 2014) and isotope ratios grounds (e.g., Peccerillo 2005, 2017 and references therein; Conticelli et al. 2010 and references therein; D'Antonio et al. 2013; Iovine et al. 2018). For the sake of completeness, in this contribution the petrochemical features of the juvenile products from Procida and Ischia are employed for comparative purposes only (data from Di Girolamo and Stanzione 1973; Albini et al. 1977; Ghiara et al. 1977, 1979; Di Girolamo and Rolandi 1979; Poli et al. 1987; Vezzoli 1988; Crisci et al. 1989; Beccaluva et al. 1991; Civetta et al. 1991b; Orsi et al. 1992; D'Antonio and Di Girolamo 1994; Di Girolamo et al. 1995; D'Antonio et al. 1999a, b, 2007, 2013; Piochi et al. 1999; De Astis et al. 2004; Tonarini et al. 2004; D'Oriano et al. 2005; Fedele et al. 2006; Brown et al. 2008, 2014; Sbrana et al. 2009; Melluso et al. 2014; Casalini et al. 2017; Iovine et al. 2018).

2 Petrography, Mineralogy and Geochemistry of the Juvenile Products

Given its mainly explosive volcanic style, the CFvf is predominantly characterised by pyroclastic products made of highly vesiculated juvenile components (pumice and scoria clasts), and subordinate lavas, both typically showing aphyric to slightly-moderately porphyritic textures (e.g., Morra et al. 2010), i.e., with ~10–30 vol% of phenocrysts. Rock compositions (filtered to exclude low-confidence data with loss on ignition, LOI, > 4 wt%) range from very rare shoshonites to abundant trachytes and phonolites, defining a typical slightly silica-undersaturated shoshonitic series (e.g., Di Girolamo et al. 1984; Rosi and Sbrana 1987; Peccerillo 2005, 2017; Conticelli et al. 2010; Morra et al. 2010; Fig. 1a, b). A similar compositional spectrum can be observed at Ischia volcanic field (Ivf), whose highly differentiated products reach slightly higher SiO_2 contents (up to ~66 wt%; Fig. 1c). On the other hand, rocks from the Procida volcanic field (Pvf) are mainly of basic-intermediate composition (shoshonitic basalts, K-trachybasalts and shoshonites), with much fewer intermediate and evolved lithotypes.

When periods of activity are taken into account, it appears evident that Campi Flegrei

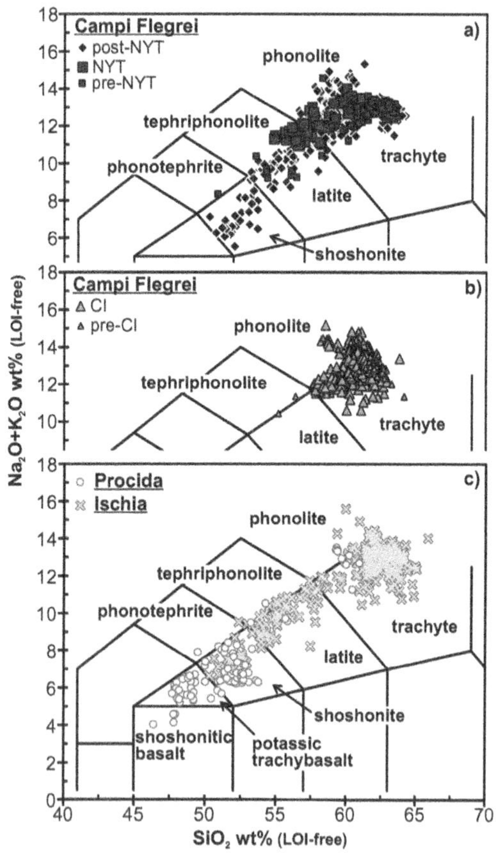

Fig. 1 Total Alkali versus Silica classification diagram (TAS; Le Maitre 2002) for the juvenile products from the existing literature for Campi Flegrei **a** post-NYT, NYT and pre-NYT, and **b** CI and pre-CI, and for **c** Procida and Ischia volcanic fields (see text for data sources)

rocks define three well-recognizable compositional groups. The rocks of the pre-CI activity and those of the CI eruption generally show a highly evolved trachyte-phonolite composition, whereas pre-NYT rocks and NYT ones range from latite to trachyte-phonolite (plus some NYT tephriphonolites). Finally, post-NYT rocks of the so-called "recent" activity span the entire compositional range of Phlegraean volcanism, from the very few shoshonites to trachyte-phonolite compositions.

The most abundant mineral phases making up the parageneses of CFvf juvenile products are unquestionably clinopyroxene and feldspars, both as phenocrysts and as groundmass phases (see Morra et al. 2010; Fig. 2). Clinopyroxene is the main phenocryst of the least differentiated products (up to ~50 vol% of the total phenocryst load) and is progressively less abundant, Mg-poorer and Fe-richer as the degree of differentiation increases. It displays a relatively constant composition in the least evolved and intermediate rocks, basically spanning the range $Ca_{45-49}Mg_{37-46}Fe_{5-16}$ (Fig. 3a). Compositions are on the whole similar to those of clinopyroxene crystals from Pvf and Ivf of comparable degrees of evolution. In the most evolved CFvf rocks, clinopyroxene occurs both as Mg-rich crystals, similar in composition to those from the least-evolved and intermediate rocks, and as Fe-richer crystals (mostly $Ca_{45-48}Mg_{17-35}Fe_{18-37}$ but even up to $Ca_{45}Mg_{11}Fe_{44}$). Exceptionally Na-rich crystals (i.e., aegirine, with Na_2O ~9–13 wt %), similar to those more commonly found in some trachytic and phonolitic products from Ivf, have been occasionally analysed in the groundmass of the Cuma, Punta Marmolite and Accademia trachytic lava domes (Melluso et al. 2012). Plagioclase feldspar crystals are also very abundant in the less-evolved rocks (30–50 vol% of the phenocrysts) and then are progressively replaced by alkali feldspar in the most-evolved ones. In CFvf plagioclase crystals, a general CaO decrease and Na_2O and K_2O increase is recorded with increasing degree of rock differentiation. In the least-evolved shoshonites, plagioclase has a quite constant bytownite composition (An_{75-89}), then it reaches Ca-poorer labradorite composition in latitic rock (An_{55-93}). Plagioclase from the least evolved (shoshonitic basalts) and intermediate (shoshonites) rocks from Pvf is also Ca-rich, whereas shoshonites and latites from Ivf include also plagioclase crystals that are both Na- and K-richer (i.e., up to andesine-oligoclase compositions, with $An_{27-50}Ab_{36-60}Or_{5-1.3}$). As observed for clinopyroxene, plagioclase from the most evolved CFvf rocks (as well as their counterparts from Ivf) covers a very wide compositional range from bytownite to andesine-oligoclase.

Olivine is basically confined to the least-evolved lithotypes, where it generally features Mg-Cr spinel inclusions, shows a quite narrow compositional spectrum (Fo_{80-88}; Ghiara et al. 1979; Armienti et al. 1983; Beccaluva et al. 1990; Cannatelli et al. 2007; Mangiacapra et al. 2008) and is commonly totally or partially replaced by iddingsite. A larger variability is recorded for Ivf olivines, ranging from Fo_{89} to Fo_{72}, the latter being found in equilibrium with latitic-trachytic host rocks. Very Mg-rich olivine crystals (up to ~Fo_{89}) of xenocrystic origin, and Fe-rich phenocrysts and microcrysts ($Mg_{1-4}Fe_{86-96}Ca_{1-2}Mn_{2-8}$) and Fe- and Mn-rich groundmass crystals ($Mg_{7-16}Fe_{48-78}Ca_1Mn_{14-35}$), have been reported for the Accademia and for the Cuma lava domes, respectively (Melluso et al. 2012).

Sanidine alkali feldspar is quite rare in the least-evolved rocks, where it has been observed only as sparse groundmass crystals with K-rich composition (Or_{74-80} up to Or_{90}), then it becomes increasingly important as rock compositions pass to latites (~10–20 vol% of phenocrysts), where it includes also Na-richer varieties (Or_{62-87}). In trachytes and phonolites sanidine accounts for ~80–90% of the total phenocryst assemblage, with compositions up to ~Or_{87} in trachytes, then evolving to Na- and Ca-richer crystals (~Or_{63}, with CaO as high as 2.83 wt%) in the most-evolved rocks (Fig. 3b). In Ivf rocks, alkali feldspar is frequently found both as thin rims on plagioclases of all compositions or as single crystals, and is generally poorer in Ca (CaO < 1.62 wt%) and richer in

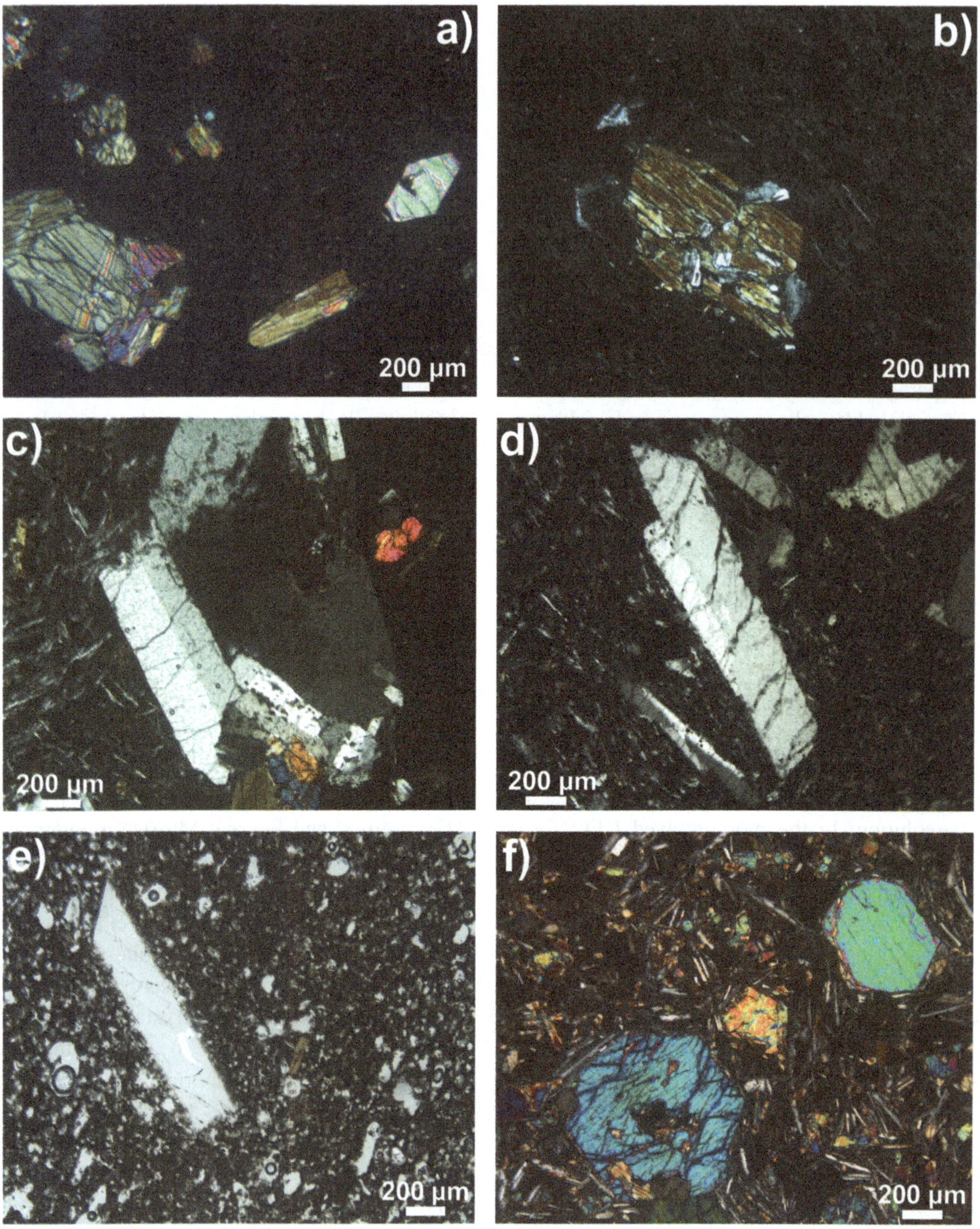

Fig. 2 Thin section photomicrographs (crossed polarisers, except 1**e**, plane polarised light) of some representative samples for the Campi Flegrei juvenile products. **a** shoshonite scoria clast from the Minopoli volcano, showing olivine (right end) and clinopyroxene phenocrysts set into a glassy groundmass; **b** shoshonitic-latitic scoria clast from the Montagna Spaccata volcano showing clinopyroxene and intergrowing plagioclase microphenocrysts in a hypocrystalline/cryptocrystalline groundmass; **c** trachytic lava of Accademia, showing sanidine, plagioclase (upper centre) and clinopyroxene (lower centre) phenocrysts and Mg-olivine xenocrysts (right end) in a trachytic groundmass mainly made of alkali feldspar microliths; **d** trachytic lava of Cuma, showing big sanidine phenocrysts in a trachytic groundmass almost entirely made of alkali feldspar microliths; **e** trachytic pumice clast from the post-NYT activity (Porto Miseno volcano), with phenocrysts (centre left) and microphenocrysts (lower right) of sanidine and a brown mica microphenocryst (centre) set in a glassy vesicular groundmass. In **f**, a shoshonitic basalt scoria clast from the Solchiaro volcano of Procida is shown, with olivine (centre and upper right) and clinopyroxene phenocrysts set into a hypocrystalline groundmass with abundant plagioclase microcrysts (white needles). Further details on the mentioned volcanoes and products can be found in Chap. Volcanic and Deformation History of the Campi Flegrei Volcanic Field, Italy

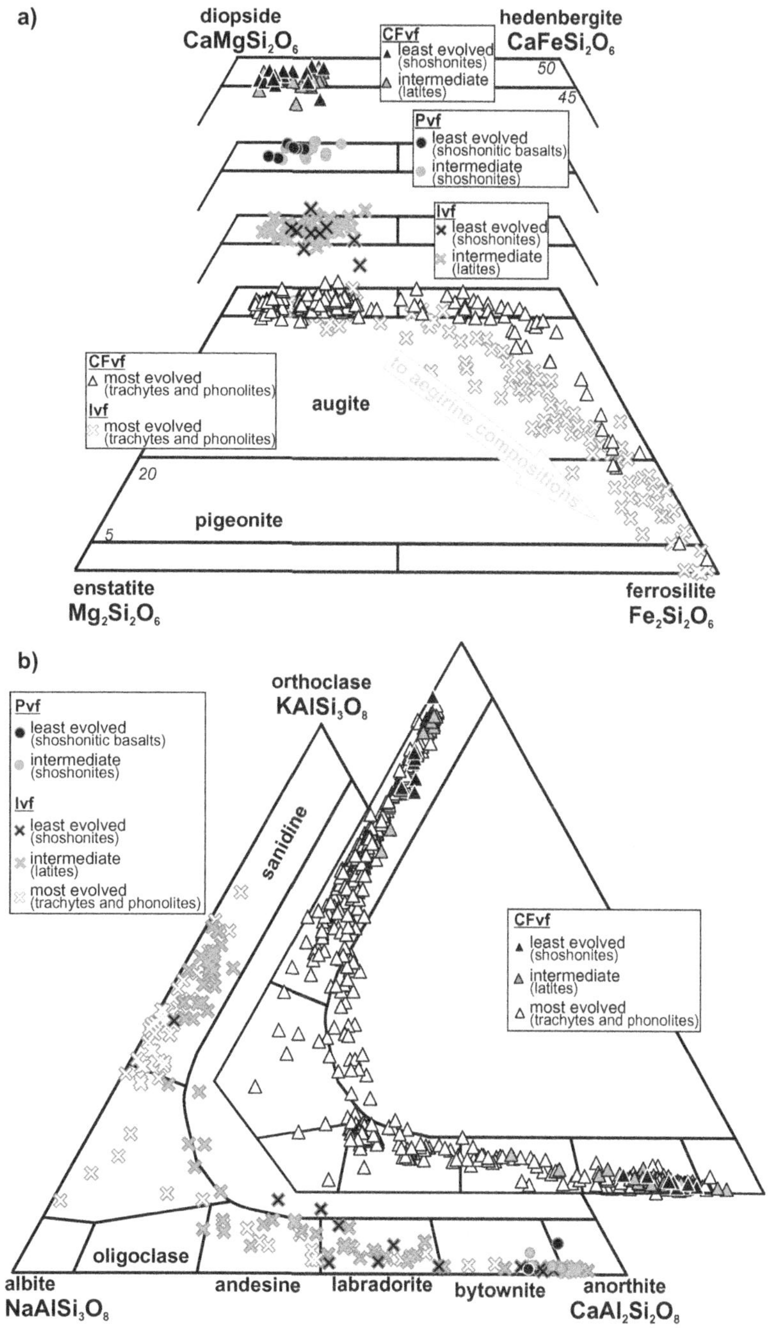

Fig. 3 Composition of **a** clinopyroxene (after Morimoto 1988, with grey arrow pointing to aegirine-rich compositions) and **b** feldspar crystals analysed in juvenile products of variably evolved compositions from the existing literature for Campi Flegrei (CFvf), Procida (Pvf) and Ischia (Ivf) volcanic fields (see text for data sources)

Na_2O, both in the least-evolved ($Or_{45\text{-}63}$) and in the most-evolved rocks (mostly $Or_{34\text{-}52}$), compared with CFvf counterparts. Unusually Na-rich sanidine crystals, similar to those reported for some Ivf trachytes, were found also in the CFvf Cuma lava dome (up to Or_{34}; Melluso et al. 2012). Sporadic anorthoclase crystals occur in the most-evolved lithotypes of the two volcanic districts, with Campi Flegrei rocks being characterised by slightly Ca-richer and K-poorer compositions ($Ab_{52\text{-}62}Or_{18\text{-}30}An_{9\text{-}23}$) with respect to Ischia ones ($Ab_{53\text{-}62}Or_{23\text{-}35}An_{3\text{-}19}$).

Almost ubiquitous in the CFvf products are also opaque minerals and brown mica ("biotite") accessory crystals, the latter limited to the evolved lithotypes (plus some sporadic occurrences in intermediate rocks). Opaque minerals are mainly Ti-magnetite crystals showing a slight increase in ulvöspinel contents with increasing degree of rock evolution (i.e., from ~20 mol% in shoshonites and latites to ~17–30 mol% in trachytes and phonolites; Melluso et al. 1995, 2012; Civetta et al. 1997; Fedele et al. 2008, 2016; Cannatelli 2012). Exceptionally ulvöspinel-rich (~62 mol%) and relatively Mn-rich crystals (MnO up to 4.5 wt%) have been also reported for the Cuma lava dome (Melluso et al. 2012), in some instances associated with inclusions of manganoan ghanite (MnO ~13 wt%, ZnO ~28 wt%, Al_2O_3 ~60 wt%). Chromiferous spinel has been analysed only in Mg-rich olivine xenocrysts within the Accademia lava dome (Cr_2O_3 ~38–46 wt%; Melluso et al. 2012). Brown mica is generally present as unzoned crystals with moderate TiO_2 contents (~4–6 wt%) and Mg# [Mg*100/(Mg + Fe^{2+})] mostly in the range of 62–70 (although values as low as 47 and as high as 82 were also obtained for crystals from the Monte Nuovo and Accademia lava dome deposits, respectively; Armienti et al. 1983; Melluso et al. 1995, 2012; Civetta et al. 1997; Fedele et al. 2008, 2016). Other less common accessory phases include apatite, zircon, brown amphibole, titanite and feldspathoids (nepheline and sodalite), typically observed in the most-evolved rock types and reported also for the most-evolved products from Ischia. Interestingly, amphibole compositions have been proven to be one of the most diagnostic features distinguishing trachytic rocks from CFvf and Ivf, as in the first amphibole is remarkably Ca-Al-richer and alkali-poorer with respect to the latter (Melluso et al. 2014). In addition, very peculiar peralkaline (i.e., aenigmatite and the aforementioned aegirine and anorthoclase), Cl-F-rich (fluorite) and incompatible elements-rich (britholite, pyrochlore, baddeleyite, zirconolite, monazite, thorite, låvenite) phases are reported for the groundmass of the Accademia, Cuma and Punta Marmolite lava domes (Melluso et al. 2012).

Chondrite- and primitive-mantle normalised multielemental diagrams for the least-evolved CFvf rocks typically show marked LREE/HREE (Light/Heavy Rare Earth Elements) and LILE/HFSE (Large Ion Lithophile Elements/High Field Strength Elements) enrichments (Fig. 4). Small Eu troughs [$Eu/Eu^* = Eu_N/(Sm_N{*}Gd_N)^{1/2}$ = 0.92–0.95, N for chondrite-normalised abundances], marked HFSE troughs (Ta, Nb, Hf and Ti) and a peak at Pb are also evident. No substantial differences can be observed between the least-evolved rocks from the three Phlegraean volcanic fields, except for those from the Pvf showing generally lower abundances, likely due to their lower degree of differentiation. Intermediate and evolved rock compositions, plotted for illustrative purposes only, are characterised by an overall linear increment of all the elements, accompanied by the appearance of evident troughs at Ba, Eu, Sr, P and Ti. The observed multielemental patterns for the CFvf least-evolved products, typical of magmas emplaced in an orogenic context, are similar to those shown by the most primitive rocks from the other districts of the Pleistocene Roman Magmatic Province (including the Latian Vulsini, Vico, Sabatini, Alban Hills, Middle Latin Valley and Roccamonfina, the Umbrian San Venanzo, Cupaello and Polino, and the Campanian Campi Flegrei, Procida, Ischia and Somma Vesuvio; Conticelli et al. 2010). However, with respect to the latter the CFvf rocks display less pronounced LILE/HFSE enrichments and Nb–Ta troughs, lower U and Th abundances and no Ba trough (see Conticelli et al. 2010 and references therein).

As for isotope systematics, for which a more detailed treatment is provided in Chap. Origin and Differentiation History of the Magmatic System Feeding the Campi Flegrei Volcanic Field (Italy) Constrained by Radiogenic and Stable Isotope Data, a general observation is that $^{87}Sr/^{86}Sr$ is notably higher for the Campi Flegrei with respect to both Procida and Ischia volcanic fields. Moreover, while for the two islands a crude positive correlation with the degree of differentiation can be observed (although the most evolved Ivf rocks span a large range of $^{87}Sr/^{86}Sr$), a clear negative correlation

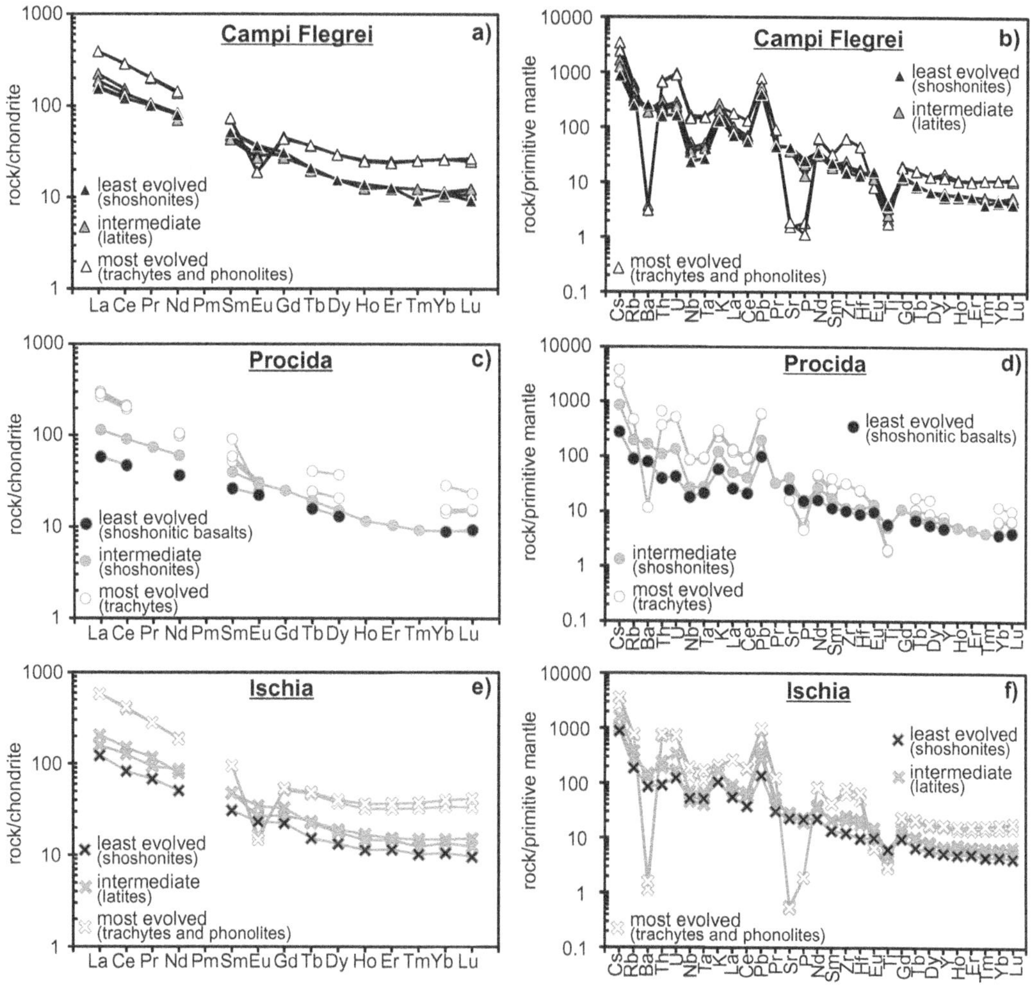

Fig. 4 Chondrite and primitive mantle-normalised (after Sun and McDonough 1989) multielemental diagrams for representative **a**, **b** Campi Flegrei, **c**, **d** Procida and **e**, **f** Ischia juvenile products of variably evolved compositions from the existing literature (see text for data sources)

characterises the CFvf juvenile products, which also appear to depict a general increase of radiogenic Sr with time (Fig. 5a). Neodymium isotope values seem uncorrelated with both age and degree of differentiation for CFvf rocks, whereas both Pvf and Ivf ones appear to be characterised by a rough decrease of $^{143}Nd/^{144}Nd$ with increasing differentiation (Fig. 5b). The fewer Pb isotope ratios available for the CFvf show a large variability that does not allow to define any clear compositional trend, although a general decrease of $^{206}Pb/^{204}Pb$, $^{207}Pb/^{204}Pb$ and $^{208}Pb/^{204}Pb$ (not shown; see Chap. Origin and Differentiation History of the Magmatic System Feeding the Campi Flegrei Volcanic Field (Italy) Constrained by Radiogenic and Stable Isotope Data) is observed moving to both the youngest and the least evolved rocks (Fig. 5c, d). On the other hand, Pvf rocks display a large scatter of Pb isotope values, whereas data for Ivf rocks quite regularly increase with increasing degree of differentiation.

In the wider framework of the Pleistocene Italian magmatism, CFvf rocks plot on the radiogenic Sr-depleted and radiogenic Nd- and Pb-enriched end of the global arrays (see

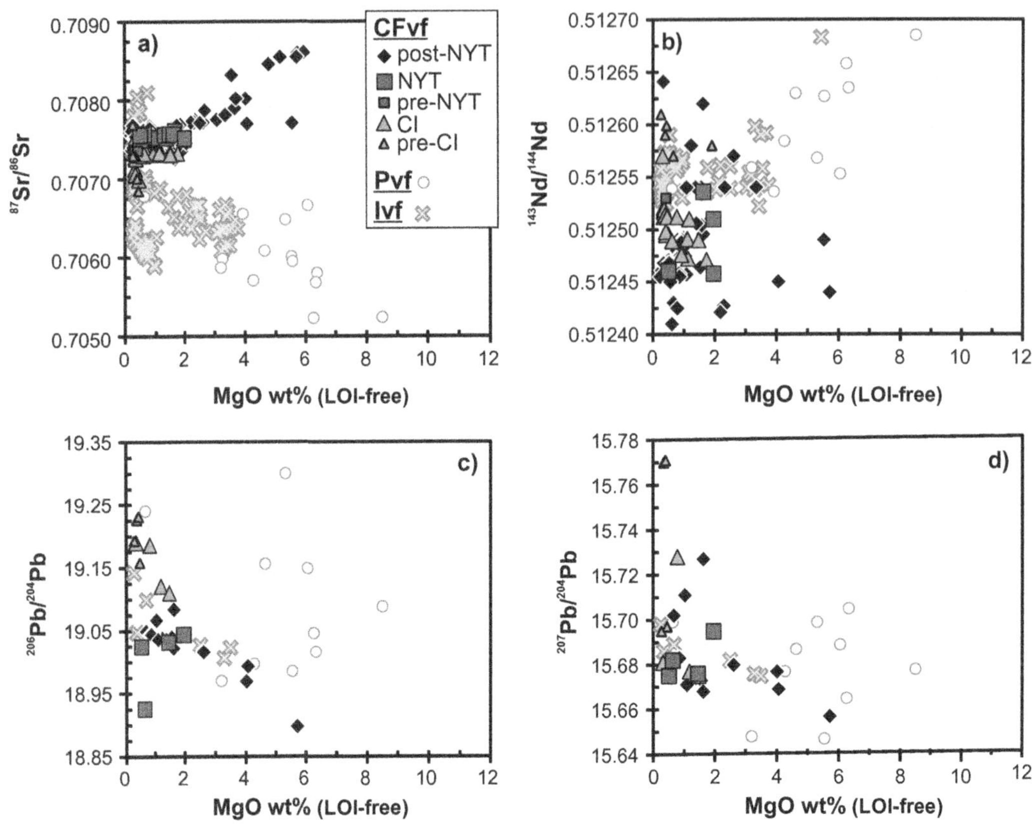

Fig. 5 **a** $^{87}Sr/^{86}Sr$, **b** $^{143}Nd/^{144}Nd$, **c** $^{206}Pb/^{204}Pb$ and **d** $^{207}Pb/^{204}Pb$ vs. MgO diagrams for the juvenile products from the Campi Flegrei (CFvf), Procida (Pvf) and Ischia (Ivf) volcanic fields from the existing literature (see text for data sources)

Conticelli et al. 2010 and references therein). Such alignment defines a clear east/southeast-directed trend (following the progressive migration of the Apenninic subduction system hinge) marked by an evident decrease of a crustal component in the mantle sources moving from the Tuscan Province to the Latian and Neapolitan districts (Roman Province) and finally to the Lucanian Province. These geochemical features are interpreted as reflecting a complex mantle source (phlogopite-veined lherzolite/wehrlite), including a subducted sedimentary component (as also indicated by B isotope ratios; Tonarini et al. 2004; D'Antonio et al. 2007) and an asthenospheric intraplate-like component (Avanzinelli et al. 2009; Conticelli et al. 2010; Mazzeo et al. 2014).

3 The Differentiation of Campi Flegrei Magmas

The unquestionable dominance of evolved compositions over the much rarer least-evolved and intermediate ones in the Campi Flegrei juvenile products, represents a self-evident proof for the paramount role played by magma differentiation processes (as well as for the efficiency of crustal filtering). During their ascent to the surface, magmas must have found the conditions to achieve neutral buoyancy and accumulate at various crustal levels, where magma chambers developed and differentiation processes have acted, ultimately producing highly evolved volatile-rich magmas and thus determining the peculiar, mainly explosive, eruptive style of the volcanic field. In order to shed light on the dynamics of the CFvf, it is thus crucial to define the processes that modified the composition of the magmas.

Differentiation trends of the CFvf juvenile products are characterised by a clear decrease of TiO_2, Fe_2O_3tot, MgO, CaO, P_2O_5 (not shown), V, Sc, Cr and Ni, whereas SiO_2, Na_2O, Rb, Y and Nb (not shown) and Zr markedly increase (Figs. 6 and 7). Potassium and Ba (the latter less linearly) show a more complex differentiation behaviour, characterised by an initial slight increase (from shoshonites to latites), followed by a sudden decrease as magmas approach trachyte compositions (i.e., MgO ~1.0–1.5 wt%), with the highest concentration values ($K_2O > 9$ wt%, Ba > 2,500 ppm) likely reflecting sanidine accumulation. A similar behaviour is observed also for Sr, which is initially constant at ~900–950 ppm and then abruptly falls (except for some samples with > 1,000 ppm probably indicative of plagioclase accumulation).

Rocks from Procida also follow these general trends, which extend to less-evolved shoshonitic basalt and potassic trachybasalt compositions, and are apparently characterised by TiO_2, Fe_2O_3tot and Sr being initially constant or even slightly increasing up to ~4–5 wt% MgO. On the other hand, Ischia juveniles show some small differences with respect to CFvf ones, i.e., slightly higher TiO_2 and Na_2O, lower CaO and K_2O and a different Sr and Ba differentiation behaviour (both almost continuously decreasing and with a gentler slope).

The quite regular major and trace element variation trends suggest an overall comagmatic character for the CFvf suite. Chemical evolution can be mainly ascribed to the fractionation of clinopyroxene (CaO, MgO, Fe_2O_3, V and Sc decrease), plagioclase (CaO and Sr, the latter being initially constant because of the counter-balancing effect of clinopyroxene) and Fe-Ti oxides (TiO_2, Fe_2O_3, V and Sc), consistent with their ubiquitous presence in the observed parageneses. Additional fractionating phases are represented by olivine, in the earliest differentiation stages (MgO, Cr and Ni decrease), and, more significantly, alkali feldspar in the latest, as K_2O and Ba "bell-shaped" differentiation trends suggest that after an initial stage of substantially no fractionation, abundant alkali feldspar segregation characterises trachytic-phonolitic magmas. Procida products might be taken as representative of the least-evolved, near-primitive end of such trends, more strongly influenced by olivine segregation (see the near-horizontal TiO_2, Fe_2O_3 and CaO trends and the initial increase of Sr). On the other hand, the peculiarities reported for Ischia products suggest small differences in magma composition and/or in the fractionating assemblage, the latter likely characterised by a higher feldspars/clinopyroxene ratio in the early/intermediate stages, possibly involving also an alkali feldspar phase (as observed by Crisci et al. 1989). This is also consistent with petrographic evidence (see Sect. 2) and with the slightly lower slope of the CaO/Al_2O_3 differentiation trend (Fig. 2).

Diagrams of Fig. 8a, b show even more evidently the above inferences regarding the evolutionary processes experienced by the Campi Flegrei magmas. The CaO and SiO_2 versus MgO plots highlight the paramount role of plagioclase and clinopyroxene fractionation (plus opaque minerals; grey arrow), as well as the stronger olivine control (coupled with a smaller plagioclase segregation; black arrow) in the very first stages of evolution, represented by the basic lithotypes from Procida. The Petrogeny's residua diagram of Fig. 8c, on the other hand, highlights the predominant alkali feldspar fractionation (and its progressive Na-enrichment) that dominates the latest evolutionary stages. Interestingly, Ivf differentiates generally show a different crystallization path, slightly displaced to the left portion of the diagram, characterised by alkali feldspar compositions that are relatively richer in Na. Such distinctive trait, consistent with mineral chemistry (see Sect. 2), again strongly argues in favour of a difference in terms of magma compositions and/or evolutionary processes/conditions (e.g., fractionating phases, pressure, volatile contents, volatile compositions) for CFvf and Ivf rock suites. What is also to note is that, as previously observed by Melluso et al. (2012), though falling in the leucite stability field, the CFvf rocks lack

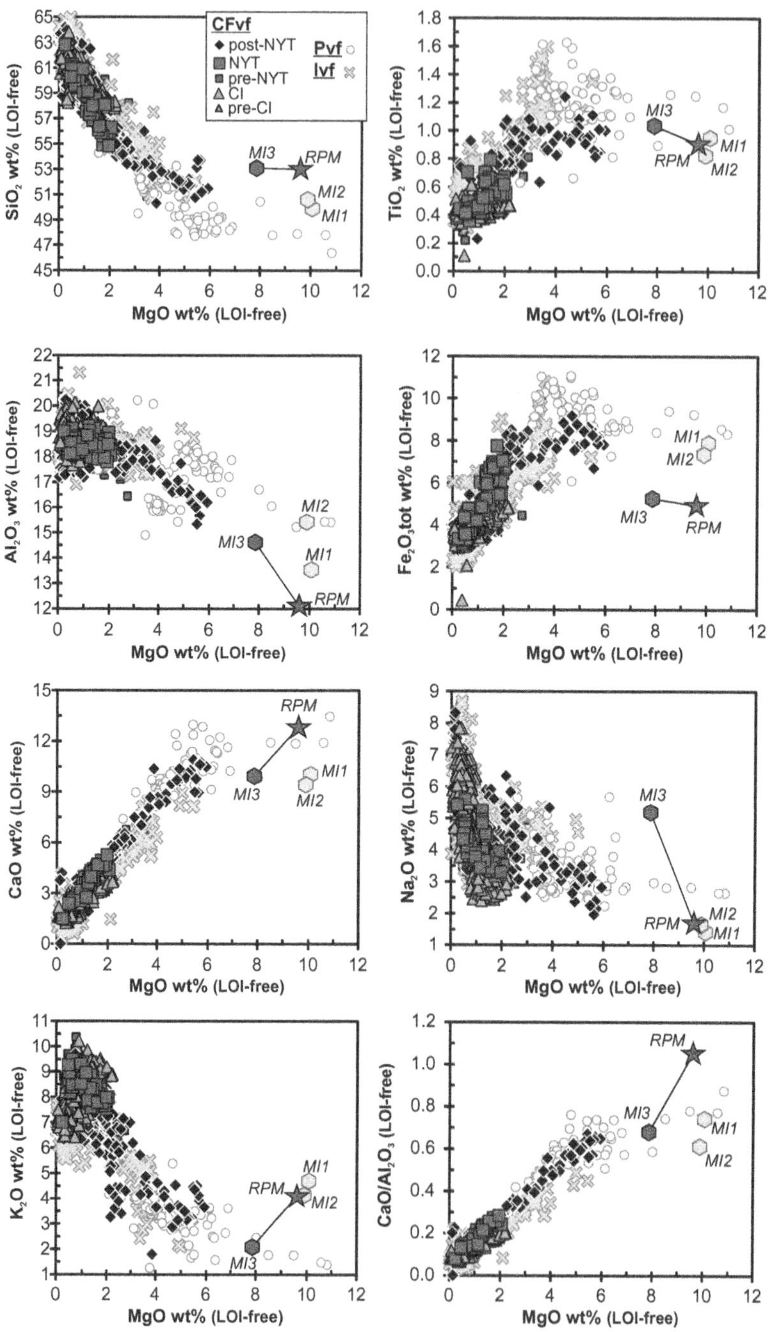

Fig. 6 Selected major elements and CaO/Al_2O_3 versus MgO diagrams for juvenile products from the Campi Flegrei (CFvf), Procida (Pvf) and Ischia (Ivf) volcanic fields from the existing literature (see text for data sources). The light grey hexagons MI1 and MI2 are for the melt inclusions used by Cannatelli (2012) for thermodynamic modelling. The dark grey hexagon and the connected star are for the melt inclusion (MI3) and the reconstructed parental melt (RPM) used by Bohrson et al. (2006) and Fowler et al. (2007) for thermodynamic modelling. See Sect. 4 for details

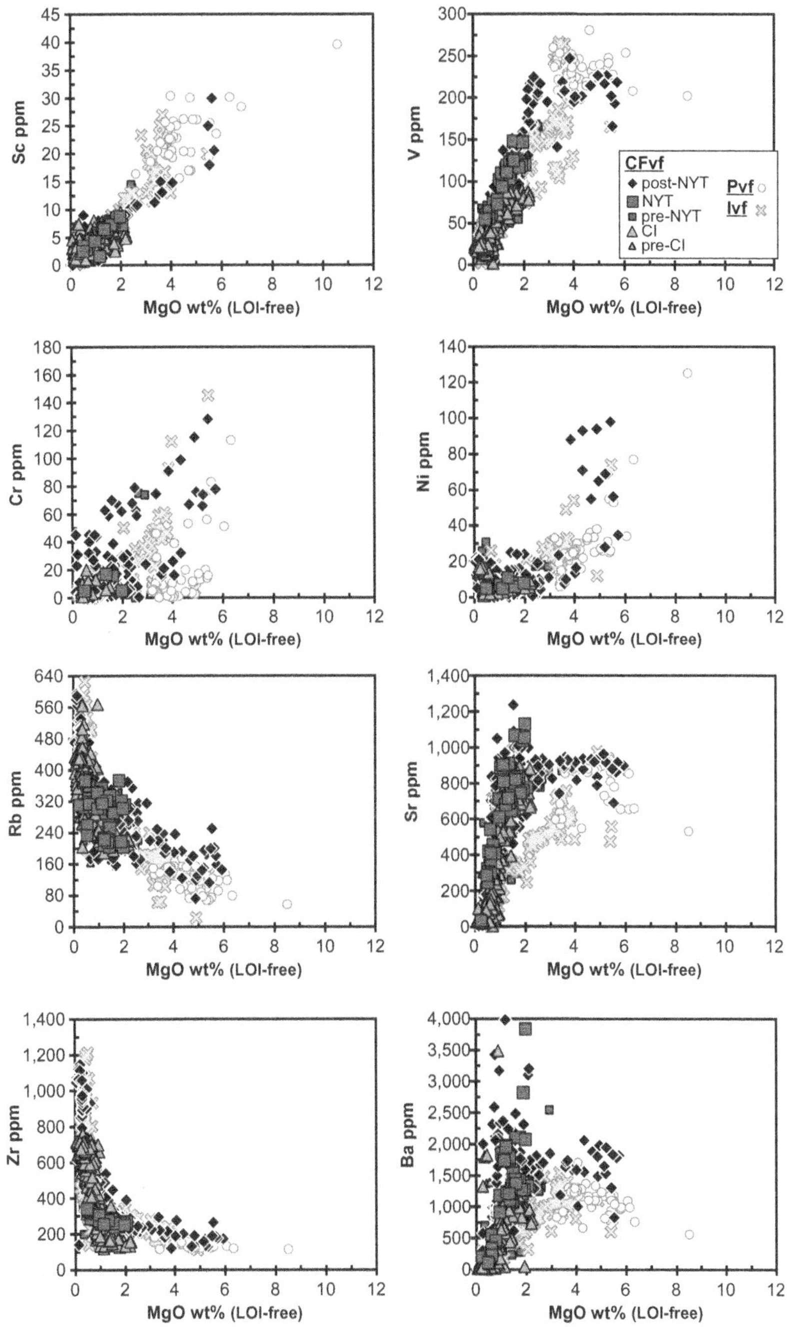

Fig. 7 Selected trace elements versus MgO diagrams for the juvenile products from the Campi Flegrei (CFvf), Procida (Pvf) and Ischia (Ivf) volcanic fields from the existing literature (see text for data sources)

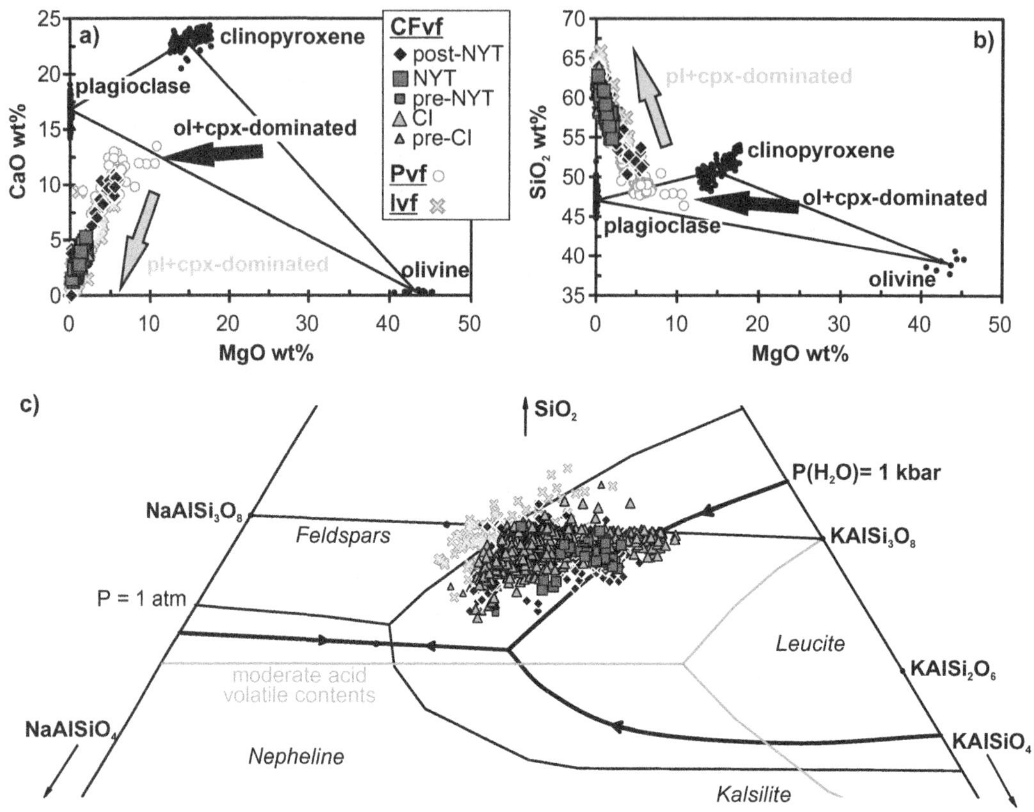

Fig. 8 **a** CaO versus MgO, **b** SiO_2 versus MgO and **c** Petrogeneny's residua (Hamilton and MacKenzie 1965) diagrams for the juvenile products from the Campi Flegrei (CFvf), Procida (Pvf) and Ischia (Ivf) volcanic fields from the existing literature (see text for data sources). In **a** and **b** the composition of CFvf Ca-rich plagioclase and Mg-rich clinopyroxene (data sources as in Fig. 1), and of CFvf and Pvf Mg-rich olivine (data from Armienti et al. 1983; Beccaluva et al. 1990; D'Antonio and Di Girolamo 1994; D'Antonio et al. 2013; Esposito et al. 2011; Ghiara et al. 1979; Melluso et al. 2012, 2014) are reported as small black dots in order to emphasise the fractionating assemblages driving magma differentiation (black and grey arrows; see text for further explanations). In **c** only trachyte and phonolite compositions are plotted and equilibrium curves are drawn for P = 1 atm (solid line), $P(H_2O)$ = 1 kbar (solid bold line) and for moderate acid volatile contents (solid grey line, following the hypothetical proposal of Kogarko et al. 1974)

leucite. The reduction of leucite stability field, which can be achieved both by H_2O or acid volatiles enrichment of the melts (see Fig. 8c), is highly indicative of Cl- and F-rich evolved melts, supported by the recognition of halogen-rich phases in the groundmass of some CFvf lavas (and, on the other hand, by the relative paucity of hydrous phases; Melluso et al. 2012).

Similar conclusions can be drawn also observing the multielemental normalised patterns of Fig. 4 for CFvf variously evolved juvenile compositions. The observed appearance of "fictitious" troughs at Ba, Eu, Sr, P and Ti with increasing rock differentiation, can be indeed ascribed to crystal fractionation processes, mainly involving alkali feldspar, plagioclase, apatite and Fe-Ti oxides, respectively. Interestingly, the small Eu troughs displayed by the least-evolved CFvf and Ivf rocks are absent from the Pvf counterparts. This is likely an effect of a different degree of differentiation, suggesting that in Procida samples (MgO ~8.5 wt%) plagioclase fractionation was not yet significant (see the initial Sr increase, as discussed above), differently from the Campi Flegrei and Ischia cases (MgO ~5.5 wt%). The deepening of Eu troughs with increasing differentiation is first subtle up to intermediate compositions (due to continued fractionation of plagioclase, plus lesser alkali feldspar), then extremely evident when significant fractionation of alkali feldspar (plus lesser plagioclase) marks the transition to the most evolved trachyte-phonolite magmas.

As a summary, the above observations, consistent with the described petrography of the juvenile products, argue for a substantially closed-system evolution of the CFvf magmas, which differentiated following a "two-steps" process basically consisting of (1) clinopyroxene and plagioclase and (2) alkali feldspar and plagioclase fractionation.

A quantitative approach to the differentiation processes operating in the CFvf can be obtained from major and trace element mass balance calculations, whose results are summarised in Tables 1 and 2. Major element models were performed according to the method of Stormer and Nicholls (1978), using some representative juvenile and mineral compositions. Trace element models are based on Rayleigh's fractional crystallisation equation $C_l/C_0 = F^{(D-1)}$. C_l and C_0 represent the composition of the final and initial magmas and F is the residual melt fraction as obtained from major element models. Bulk partition coefficients D were calculated ($D = Kd_i * x_i$) using the fraction of crystallised phase (x_i) obtained from major element mass balance calculations and mineral/liquid partition coefficients (Kd_i) from the existing literature for the CFvf rocks (Villemant et al. 1988; Fedele et al. 2009, 2015). Additional Kd_i values from Skluski et al. (1994), McKenzie and O'Nions (1991), Nielsen et al. (1994) and Lemarchand et al. (1987) were used where no data from CFvf rocks are available (mostly for least-evolved and intermediate melt compositions).

Major element models indicate that the entire differentiation processes occurred in the CFvf magmatic system can be approximated to a basically closed-system evolution involving ~93% fractionation of an assemblage made of clinopyroxene (37%), plagioclase (28%),

Table 1 Results of major element mass balance calculations for selected juvenile products from the existing literature covering the entire compositional spectrum for the Campi Flegrei volcanic field

	X_{ol}	X_{cpx}	X_{pl}	X_{kfs}	X_{op}	X_{ap}	X_{bt}	ΣR^2	F
SHO1[(1)]-SHO2[(2)]	0.01 Fo_{86}	0.67 $Mg_{45}Ca_{47}Fe_8$	0.24 An_{83}		0.06 Usp 15%	0.01		0.194	0.73
SHO2[(2)]-LAT[(3)]	0.05 Fo_{81}	0.56 $Mg_{39}Ca_{47}Fe_{14}$	0.32 An_{73}		0.05 Usp 15%	0.02		0.252	0.69
	0.03	*0.62*	*0.28*		*0.06*	*0.01*			*0.50*
LAT[(3)]-TRA[(4)]		0.13 $Mg_{37}Ca_{46}Fe_{17}$	0.33 An_{50}	0.38 Or_{85}	0.09 Usp 30%	0.02	0.05 Mg# 70	0.157	0.57
	0.02	*0.47*	*0.29*	*0.12*	*0.07*	*0.02*	*0.01*		*0.29*
TRA[(4)]-PHO[(5)]		0.03 $Mg_{35}Ca_{47}Fe_{18}$	0.22 An_{42}	0.71 Or_{73}	0.03 Usp 30%		0.01 Mg# 62	0.151	0.25
	0.02	*0.37*	*0.28*	*0.25*	*0.06*	*0.01*	*0.01*		*0.07*

SHO = shoshonite (1with MgO 5.47 wt%, 2 with MgO 3.58 wt%); LAT = latite; TRA = trachyte; PHO = phonolite. X_{ol}, X_{cpx}, X_{pl}, X_{kfs}, X_{op}, X_{ap} and X_{bt} refer to the fraction of removed olivine (Fo = forsterite content), clinopyroxene (as a function of Mg, Ca and Fe cations), plagioclase (An = anorthite content), alkali feldspar (Or = orthoclase content), opaques (Usp = ulvöspinel content), apatite and biotite [Mg# = Mg*100/(Mg+Fe^{2+})], respectively. F = residual melt fraction (values in italics refer to the mass of starting least-evolved magma); ΣR^2 = sum of squared residuals. Superscripts in brackets are for data sources: 1 = Ghiara et al. (1977); 2 = D'Antonio et al. (1999b); 3 = Orsi et al. (1995); 4 = Civetta et al. (1997); 5 = Pappalardo et al. (2002a)

Table 2 Results of trace element mass balance calculations for selected juvenile products from the existing literature covering the entire compositional spectrum for the Campi Flegrei volcanic field (see Table 1 for data sources)

		Sc	V	Cr	Co	Ni	Rb	Sr	Y	Zr	Nb	Ba	La	Ce	Eu	F(inc)
SHO1-SHO2	C_0			128	25	98	112	913				1,310				0.54^{Rb}
	C_l(obs)	13	205	17	18	10	207	900	27	207	26	1,639	54	109	2.3	
	C_l(calc)			0	24	13	152	1,033				1,787				
SHO2-LAT	C_0	13	205	17	18	10	207	900	27	207	26	1,639	54	109	2.3	0.71^{Rb}
	C_l(obs)	8	127				293	800	28	268	36	1,359	67	133	2.3	
	C_l(calc)	4	72				268	980	nc	288	38	1,918	67	152	2.3	
LAT-TRA	C_0	8	127				293	800	28	268	36	1,359	67	133	2.3	0.64^{Nb}
	C_l(obs)	4	40				324	203	34	359	56	178	85	148	1.7	
	C_l(calc)	5	100				336	763	nc	403	nc	959	104	nc	2.2	
TRA-PHO	C_0	4	40				324	203	34	359	56	178	85	148	1.7	0.57^{Nb}
	C_l(obs)		19		11	2	420	33	54	60	98	30	122	227	1.5	
	C_l(calc)		20				566	4	119	1,371	212	36	281	524	1.2	

SHO = shoshonite (1 with MgO 5.47 wt%, 2 with MgO 3.58 wt%); LAT = latite; TRA = trachyte; PHO = phonolite. F(inc) = residual melt fraction as obtained assuming D = 0 for the most incompatible trace element (indicated as superscript)

alkali feldspar (25%), opaque oxides (6%), olivine (2%) and accessory biotite and apatite. Each of the investigated differentiation steps was satisfactory modelled ($\Sigma R^2 < 0.26$) and the obtained fractionation assemblages resulted consistent with petrographic observations, magma differentiation trends (the change in the crystallization assemblages, from cpx + pl- to kfs + pl-dominated) and variations in mineral compositions (decreasing MgO in clinopyroxene and olivine and CaO in plagioclase, K_2O firstly increasing and then decreasing in the alkali feldspar). Trace element models reveal a good correspondence between observed and calculated final magma compositions (notwithstanding the cited paucity of appropriate Kd_i values). As F and x_i values were taken from major element models, the consistency of trace element models can be taken as a further confirmation of the method viability. In addition, it should be also observed that if for each of the modelled transitions F values are calculated, a broad correspondence with values obtained from major element models can be envisaged [compare values of F and F(inc) in Table 2].

The results of the fractional crystallisation model are overall in line with those presented in the existing CFvf literature (e.g., Melluso et al. 1995, 2012; Civetta et al. 1997; Marianelli et al. 2006; Fedele et al. 2008; Pappalardo et al. 2008; Arienzo et al. 2009; D'Antonio 2011; Fedi et al. 2018; Forni et al. 2018a). The differences between these are mostly related to the choice of rocks and minerals compositions, but also reflect the intrinsic simplification of such models assuming a single magma lineage through time, whereas numerous liquid lines of descent are more likely to have been involved (see Sect. 5). Bearing this in mind, the results of the performed model can be used for retrieving rough, semi-quantitative estimates for the involved volumes of magmas and crystal cumulates, using the available estimates for the emplaced products. The latter can be approximated to the sum of the volumes of the largest NYT and CI eruptions, for which 50 km^3 of DRE (Dense Rock Equivalent) and between 49–72 and 300 km^3 of DRE have been respectively proposed (according to the various models employed; see discussion in Scarpati et al. 2014). Such total volume of 99–122 to 350 km^3 DRE (ρ = 2,500 kg/m^3) correspond to 105–130 to 372 km^3 of trachytic-phonolitic magma (ρ = 2,350 kg/m^3; density values from Polacci et al. 2004; Chap. Volcanic and Deformation History of the Campi Flegrei Volcanic Field, Italy). Since this represents ~30% of residue from a parental shoshonitic magma (Table 1), indicative volumes of 351–433 to 1,241 km^3 of starting magma and 246–303 to 869 km^3 of crystallised cumulates can be inferred.

Notwithstanding the overall reliability of the above closed-system evolution models, numerous lines of evidence suggest that open-system processes such as crustal assimilation, magma mixing and/or mingling, must have played some role (though limited) in the differentiation of CFvf magmas (e.g., Di Girolamo et al. 1984; Rosi and Sbrana 1987; Villemant 1988; Ghiara 1989; Beccaluva et al. 1990; Melluso et al. 1995, 2012; Civetta et al. 1997; D'Antonio et al. 1999b, 2007; Pappalardo et al. 1999, 2002b, 2008; Bohrson et al. 2006; Fowler et al. 2007; Pabst et al. 2008; Fedele et al. 2008, 2009, 2016; Arienzo et al. 2009, 2010, 2011; Tonarini et al. 2009; Di Renzo et al. 2011; Di Vito et al. 2011; Pappalardo and Mastrolorenzo 2012; Gebauer et al. 2014; Iovine et al. 2017, 2018). Some of the most significant of these include: (1) petrographic disequilibrium textures (complex zoning patterns, corroded crystal margins), (2) xenocrystic Mg-olivine with Cr-spinel inclusions, Mg-clinopyroxene and Ca-plagioclase within some highly evolved products (see Sect. 2); (3) mafic lithic clasts/xenoliths within some highly evolved products; (4) occurrence of andesitic xenoliths and quartz xenocrysts; (5) occurrence of hornfels, marls and argillaceous-arenaceous xenoliths; (6) isotopic disequilibrium between phenocryst phases and host rocks; (7) recognition of isotopically different magma batches feeding the various CFvf eruptions (see also Chap. Origin and Differentiation History of the Magmatic System Feeding the Campi Flegrei Volcanic Field (Italy) Constrained by Radiogenic and Stable Isotope Data for the latter two). In addition, in the recent literature some growing evidence is being brought about the role of processes such as crystal accumulation, melt extraction, magma recharge and cumulate melting and remobilization (still subordinate with respect to fractional crystallisation; Forni et al. 2016, 2018a, b; Voloschina et al. 2018).

In this light, the differentiation of CFvf magmas should be more properly ascribed to a complex interplay of (mainly) closed- and (subordinate) open-system processes, and tentatively quantified by means of open-system AFC model (Assimilation and Fractional Crystallisation). In order to develop such a model, the assimilated material must be necessarily inferred. Data coming from both direct (i.e., borehole samples and lithic fragments hosted within the pyroclastic deposits) and indirect approaches (i.e., geophysical investigations, knowledge of the general geology of the area and comparison with similar volcanic areas), do not give unequivocal indications regarding the composition of basement rocks (e.g., Di Girolamo et al. 1984; Rosi and Sbrana 1987; Pappalardo et al. 2002b; Zollo et al. 2008; D'Antonio 2011; Mazzeo et al. 2014). Candidates more commonly taken into account include: (1) Meso-Cenozoic platform carbonates and (2) their Lower Miocene arenaceous sedimentary cover; (3) crystalline igneous rocks representing the product of cumulate processes (mainly monzonitic-syenitic subvolcanic/intrusive rocks, possibly accompanied by olivine-clinopyroxenites in the deeper portions); (4) metamorphic rocks of the Calabrian Hercynian crust; (5) metamorphic rocks of the African-type continental crust. In this framework, interesting general considerations and suggestions were made by D'Antonio (2011), who argued that the basement at > 4 km depth mainly consists of crystalline igneous rocks originated by the prolonged fractional crystallisation processes producing the huge volumes of evolved melts that fed the CFvf eruptions. Such hypothesis, based on simple mass balance calculations and corroborated by evidence for analogues worldwide, appears consistent with geophysical data, with the geochemical and isotopic features of the magmas feeding the recent volcanism, with the composition of fumarole emissions and with the remarkable scarcity of carbonate lithic clasts in the CFvf pyroclastic products (which, on the other hand, predominantly feature igneous lithic material).

On these grounds, the proposal of a likely scenario for the CFvf basement is a very hard task and, as a consequence, the correct definition of the assimilation process(es) appears highly speculative and elusive. This is particularly true if it is also considered that such processes might have involved the overprinting of various types of interactions (i.e., with different basement lithologies) at various stages of magma differentiation, some of which are thus possibly obliterated.

This huge uncertainty has brought to AFC models requiring largely variable amounts of assimilation, depending on the chosen contaminant. Models assuming Hercynian/African-type continental crust generally require low degrees of assimilation (from traces to 1–2%; D'Antonio et al. 2007; Gebauer et al. 2014; Mazzeo et al. 2014), although values up to 10% have also been proposed (Pappalardo et al. 2002b, 2008). Similar low values have been obtained using syenitic shallow intrusives (1%; Gebauer et al. 2014). For carbonatic assimilants, both comparable (0.7–3%; Arienzo et al. 2009; Pappalardo and Mastrolorenzo 2012) and much higher degrees of assimilation (~15–30%; Pappalardo et al. 1999; Voloschina et al. 2018) have been retrieved. These latter are commonly retained unrealistic, as they would notably increase CaO and decrease K_2O contents, thus driving magmas to strongly silica-undersaturated compositions. An alternative crustal contaminant is that proposed by Fowler et al. (2007), who argued for the involvement of 5–10% of a mixed assimilant consisting of skarns and foid-syenites. Recently, based on oxygen isotope evidence, Iovine et al. (2018) have discarded contaminants from both the continental crust and the carbonate pile, and invoked a maximum of 9% assimilation of hydrothermally-altered intra-caldera pyroclastic deposits.

4 Constraints from Geothermobarometry, Experimental Petrology, Melt Inclusions and Thermodynamic Modelling

Numerous attempts have been made in order to estimating the values of the intensive variables that characterised the magmas feeding the CFvf volcanism. Temperature values as low as 800–840 and 836–878 °C were initially proposed based on two-feldspar geothermometry on CI trachytes-phonolites by Melluso et al. (1995) and Civetta et al. (1997). Later estimates on similar magma compositions based on updated versions of the same method generally yielded higher temperatures, in the range of 850–991 (CI; Fulignati et al. 2004), 850–1,100 (Monte Nuovo phonolites; D'Oriano et al. 2005) and 882–973 °C (Agnano-Monte Spina trachytes-phonolites; Iovine et al. 2017). Geothermometers based on the clinopyroxene-liquid equilibria also yielded quite variable T ranges, i.e., 1,071–1,156 (±51, Agnano-Monte Spina; Iovine et al. 2017), 945–1,070 and 879–972 (CI trachytes and trachytes-phonolites, respectively; Forni et al. 2016), 930–1,095 and 910–1,065 °C (NYT latites-trachytes and trachytes-phonolites, respectively; Forni et al. 2018a).

Temperature estimates that are broadly comparable with those reported above have been proposed also based on alternative methods, such as models for trace elements distribution in clinopyroxene (875–900 °C, CI; Fedele et al. 2009), models for Ba diffusion in sanidine (930 °C, Agnano-Monte Spina; Iovine et al. 2017), and in a study on the probability distribution of two-feldspar (870 ± 30 °C) and clinopyroxene-melt geothermometers (955–970 °C) for the Astroni trachytes-phonolites (Astbury et al. 2018). These are also in line with homogenisation temperatures obtained for clinopyroxene-hosted melt inclusions from CI juvenile clasts (Fulignati et al. 2004; Marianelli et al. 2006), with values of 870–980 (for Fe-rich "salite" compositions) and 950–1,080 °C (in zoned crystals at the salite-diopside interface). The authors also obtained T > 1,100 °C for diopside crystals, which however are likely too high, since the investigated melt inclusions were totally recrystallised and were thus re-heated, which might have altered their original composition (see discussion in Esposito et al. 2018). The same might also apply to the clinopyroxene- and olivine-hosted melt inclusions from the less evolved shoshonites-latites of the Minopoli 1 and Fondo Riccio eruptions, for which Cannatelli et al. (2007) obtained T values averaging at ~1,132–1,159 ± 3 °C.

The large uncertainties in the above T estimates are likely caused by a number of factors, including (1) progressive recalibration of the employed geothermometric equations, (2) mineral phases not crystallising at the same time (e.g., clinopyroxene preceding plagioclase and sanidine), (3) presence of phases that are not in equilibrium with each other and/or with the host melt (see Sects. 2 and 3), (4) influence of other intensive variables. As for the latter, pressure is known to generally play a smaller role in modifying crystal-liquid equilibria and, in any case, it is generally well-constrained to a tight interval of values by several melt inclusion studies. Based on the saturation of the fluid phase, pressures in the range of 100–200 MPa (with values as low as 40 and as high as 211 MPa) have been indeed obtained from melt inclusions hosted in clinopyroxene (plus few olivine) crystals from CFvf trachytes-phonolites (CI, Agnano-Monte Spina and Averno 2; Signorelli et al. 2001; Marianelli et al. 2006; Arienzo et al. 2010; Fourmentraux et al. 2012) and shoshonites-latites (Mangiacapra et al. 2008).

As a result, the most obvious factor possibly influencing the large variability of T estimates for the CFvf magmas is represented by the composition and the concentrations of the fluid phase. This is well evident from the results of several experimental petrology investigations performed on CFvf magmas (mainly on trachytes-phonolites from the CI, Agnano-Monte Spina and Monte Nuovo eruptions). With the only exception of two studies conducted at very high (1 GPa; Misiti et al. 2006) or very low P (1 atm; Iezzi et al. 2008), these were undertaken at pressures consistent with those obtained from the above melt inclusion investigations (i.e., 20–200 MPa; Di Matteo et al. 2004; Roach 2005; Fabbrizio and Carroll 2008; Fabbrizio et al. 2009; Calzolaio et al. 2010; Arzilli and Carroll 2013; Arzilli et al. 2015b, 2016). Oxygen fugacity conditions were between quartz-fayalite-magnetite (QFM) and nickel-nickel oxide + 1.3 buffers, broadly consistent with the QFM + 1 estimate obtained by Forni et al. (2016) using a spinel-melt oxybarometer.

One of the most relevant outcomes of these experiments, all performed in water-saturated conditions, is that only for the least evolved trachyte compositions (MgO ~1 wt%, K_2O ~9 wt%; Fabbrizio and Carroll 2008) a close match between natural and synthetic samples is obtained at 725–825 °C. On the other hand, more evolved trachyte-phonolite compositions typically fail to reproduce natural sample mineralogy as they lack plagioclase. This might suggest that the fluid was not pure H_2O, but more likely a H_2O-CO_2 mixture, which would also explain why in the matching experimental runs (i) T is well below estimates from geothermometric and melt inclusions investigations (see above), (ii) plagioclase is the latest phase to appear on the liquidus (and is very close to the solidus, especially at low P ~50 MPa) and (iii) plagioclase has high An (mostly > 77, up to 86). A possible confirmation comes from Roach (2005), who performed crystallisation experiments with variable CO_2 in the fluid phase, and replicated natural sample mineralogy and crystallinity with 50:50 H_2O-CO_2 fluid at 900–950 °C. It should be however noted that studies on melt inclusions and matrix glasses from the CI products generally report very low CO_2 concentrations (typically below detection limits; Signorelli et al. 1999, 2001; Fulignati et al. 2004; Marianelli et al. 2006). On the other hand, glasses of the post-NYT eruptions seem relatively CO_2-rich (up to 488, 901 and 1,070 ppm, respectively for the Agnano-Monte Spina trachytes, Fondo Riccio latites and Minopoli 1 shoshonites, and in the ranges of 240–530 and 21–423 ppm for Nisida latites and Astroni 1 trachytes-phonolites; Mangiacapra et al. 2008; Arienzo et al. 2010, 2016; Stock et al. 2016). Whether this represents a clear time-related trend, and whether other volatile species (e.g., Cl, F, S, all variably present in CFvf melt inclusions and matrix glasses) might have played a similar role on mineral phase stability, are some potential key-aspects that still need to be fully ascertained. This is particularly true also considering that volatile concentrations in melt inclusions usually underestimate the original volatile content of the magma (see discussion in Esposito et al. 2018), which thus further adds uncertainties to the general portrait.

In order to retrieve some additional constraints on the physico-chemical conditions characterising the CFvf magmas during their evolution, some authors have performed thermodynamic phase equilibria models using the MELTS software (Ghiorso and Sack 1995). The simulations presented by Bohrson et al. (2006), Fowler et al. (2007) and Cannatelli (2012) suggest that magma evolution, driven mainly by isobaric fractional crystallisation at low depth (~6–7 km, i.e., P ~0.15 GPa), was characterised by an abrupt change in physical parameters (density, viscosity, water contents) and degree of crystallisation at a pseudo-invariant temperature of 870/883 °C. Such point, marked by the simultaneous appearance on the liquidus of alkali feldspar, plagioclase and biotite, was interpreted as driving the magma to a condition of dynamic instability that triggered eruptive events. The late crystallisation of plagioclase assumed by these models is remarkably in contrast with petrography (plagioclase being present in the least evolved CFvf lithotypes), mineral chemistry (decrease in An with increasing degree of evolution), whole-rock geochemistry (major- and trace element trends, Eu troughs in the least-evolved magmas) and mass balance calculations (see Sect. 3), and other lines of evidence argue against it.

First, plagioclase is a common phase not only in the least evolved and intermediate rocks from the CFvf (as well as of the Pvf and Ivf), but also for other basic-intermediate shoshonitic rocks of all ages and from all over the world (e.g., Rogers and Setterfield 1994; Duchesne et al. 1998; Müller et al. 2001; Jiang et al. 2002; Carr et al. 2003; Cvetković et al. 2004; Eklund and Shebanov 2005; Ujike et al. 2007; Conticelli et al. 2010; Gao et al. 2010; Tupinambá et al. 2012; Campbell et al. 2014; Plechov et al. 2015; Rezeau et al. 2018; Çoban et al. 2019).

Second, crystallisation experiments largely document that plagioclase is actually a liquidus phase for shoshonitic systems of basaltic to latitic composition (e.g., Conte et al. 2006; Iezzi et al. 2008; Mollo et al. 2011; Agostini et al. 2013; Arzilli et al. 2015a; Perinelli et al. 2019) at temperatures between 900 and 1,235 °C.

A third and possibly major line of evidence is based on the observation that the chosen starting compositions for the above thermodynamic models are those of melt inclusions that likely are an inadequate representative for the CFvf magmas. Indeed, as observed in the review from Esposito et al. (2018), many CFvf and Pvf melt inclusions have anomalous compositions due to a number of chemical modifications that make them remarkably different from whole-rock samples. The models by Cannatelli (2012) are built starting from two melt inclusions significantly deviating from the composition of CFvf and Pvf rocks (the latter being the closest analogues for the CFvf parental melts; see Sect. 3), especially for SiO_2, Al_2O_3, Fe_2O_3tot, CaO, Na_2O and K_2O (MI1 and MI2 light grey hexagons of Fig. 6). This likely had the effect to restrict the stability field for plagioclase, as also indicated by the fact that plagioclase appears on the liquidus earlier (i.e., at 970 °C versus 870 °C) in the simulation starting from a melt inclusion that is more in line with whole-rock Al_2O_3 (MI2 in Fig. 6). As for the models by Bohrson et al. (2006) and Fowler et al. (2007), these not only started from a melt inclusion that is again quite different from CFvf + Pvf whole-rock samples (especially for SiO_2, Al_2O_3, Fe_2O_3tot and Na_2O; MI3 dark grey hexagon in Fig. 6), but the authors also arbitrarily corrected its composition adding 20% of host clinopyroxene, in order to account for the effects of post-entrapment modifications. This supposed "reconstructed parental melt" is even more deviating from the reference CFvf + Pvf whole-rock field for its notably lower Al_2O_3 (~12.1 vs ~15 wt%, at a comparable degrees of evolution), Fe_2O_3tot (4.93 vs ~8–8.5 wt%) and Na_2O (1.70 vs ~2–3 wt%) and higher SiO_2 (53.0 vs ~48 wt%) and CaO/Al_2O_3 (1.06 vs. ~0.8 wt%; RPM dark grey star in Fig. 6).

Finally, it should be remarked that more recent phase equilibria models for the CFvf and Pvf magmas actually report plagioclase as a much earlier liquidus phase (i.e., ~1,100 °C for Pvf shoshonitic basalts, 1,020 and 980 °C for Baia-Fondi di Baia latites and trachytes; Esposito et al. 2018; Voloschina et al. 2018). This could be due to both (i) choice of appropriate starting compositions (i.e., melt inclusion and matrix glasses consistent with whole-rock compositions, as also acknowledged by the authors themselves), and (ii) employment of the updated rhyolite-MELTS algorithm (Ghiorso and Gualda 2015), specifically calibrated for hydrous silicic systems.

5 The Campi Flegrei Magmatic System

The above inferences made on the petrochemical characteristics of the juvenile products, the intensive variables and the processes governing magma differentiation can be useful in an attempt to propose a synthetic, coherent model for the magmatic system feeding the Campi Flegrei volcanism. The primary role of fractional crystallization has been acknowledged since the earliest works of Armienti et al. (1983) and Di Girolamo et al. (1984), which argued for a closed-system evolution within a single, shallow level (~4–5 km deep) magma chamber, whose emptying would have caused the eruptions to progressively shift towards the central portions of the district. Successive works (e.g., Rosi and Sbrana 1987; Villemant 1988; Civetta et al. 1991a) added some complexity by invoking the occasional refilling of the magma chamber with newly produced primitive magmas (plus mixing of the different magma batches) and some interaction between magmas and hydrothermal fluids at ~11–10 ka.

The growing evidence testifying for magma evolution also involving open-system processes (see Sect. 3) made more and more clear that the CFvf magmatic system is likely to be much more complex. While it was still evident that the huge, highly explosive CI and NYT eruptions were basically fed by a single magma chamber, whose geologically instantaneous emptying determined the development of the nested structure of the Campi Flegrei caldera (CFc; see Chap. Volcanic and Deformation History of the Campi Flegrei Volcanic Field, Italy), the several lower-energy eruptions that punctuated the eruptive history of the CFvf should be more adequately collocated within a magmatic framework made of several small-sized independent magma chambers (Beccaluva et al. 1990). Magmas filling such "autonomous" reservoirs could have evolved separately in different physico-chemical conditions (possibly also within different types of host rocks), episodically interacting with each other through fracture zones, and being occasionally refilled by arrivals of new magma batches (possibly acting as an eruption trigger; e.g., Cañón-Tapia 2014 and references therein), as it has been also testified by extensive isotopic geochemistry evidence (see Chap. Origin and Differentiation History of the Magmatic System Feeding the Campi Flegrei Volcanic Field (Italy) Constrained by Radiogenic and Stable Isotope Data). This is possibly further confirmed by the observation that the dynamics of magma storage and evolution for the numerous eruptions that have occurred in the CFvf are highly variable, and did not follow a single, linear evolution through time. As indeed observed by Melluso et al. (2012), both extreme closed-system fractional crystallisation (producing peralkaline evolved melts rich in incompatible elements, Cl and F) and open-system magmatic processes (as testified by evidence for petrographic and geochemical disequilibrium) were active throughout the entire volcanological history of the volcanic field. In addition, while some eruptions were clearly fed by a vertically-zoned magma chamber (e.g., CI, NYT, Agnano-Monte Spina, Averno, Fondi di Baia; Civetta et al. 1988, 1997; Scarpati et al. 1993; Melluso et al. 1995; Wohletz et al. 1995; de Vita et al. 1999; Pappalardo et al. 2002a; Fedele et al. 2008, 2011, 2016; Arienzo et al. 2010; Di Vito et al. 2011), reflected in the chemostratigraphical zoning of their juvenile products, others appear to have tapped a substantially homogeneous magma reservoir, as their juvenile products show a markedly constant chemical composition (e.g., Pomici Principali, Porto Miseno, Capo Miseno, Bacoli, Nisida, Monte Nuovo; Di Girolamo et al. 1984; Lirer et al. 1987; Rosi and Sbrana 1987; D'Oriano et al. 2005; Insinga et al. 2006; Fedele et al. 2011). This is again consistent with a model assuming that each volcanic vent was fed by an independent magma chamber, possibly branched off from a common reservoir.

According to some recent models, the development of chemical gradients might be related also to the existence of a stratification of the magma reservoirs in terms of degree of crystallinity. More in detail, in the CI case Forni et al. (2016) proposed the existence of a high-crystallinity crystal mush region where low-density, volatile-rich, crystal-poor highly evolved melts concentrated at the topmost regions, and were consequently extracted during the earliest, pre-collapse stages of the eruption. After the collapse event, the eruption mainly tapped the lowermost crystal-rich portions of the reservoir, representing remobilised and partially remelted cumulates, rejuvenated after the arrival of more mafic magmas. On the other hand, in the NYT case, only crystal-poor magmas were withdrawn (Forni et al. 2018a). The first magma batches (reflected in the lowermost emplaced units) were interpreted as evolved melts that formed an easily eruptible cap of the reservoir (i.e., as in the CI event). The complex chemical gradients characterising the deposits emplaced during the successive stages were ascribed to significant interaction of residing melts with more mafic recharge magmas (much more pronounced with respect to the CI, for which the higher volumes of residing magmas probably had a dilution effect), accompanied by melting of sanidine and lesser biotite from the cumulate mush.

In summary, the presented observations strongly argue in favour of a model of a very complex magmatic system for the CFvf, well-represented by that outlined in Chap. Origin and Differentiation History of the Magmatic System Feeding the Campi Flegrei Volcanic Field (Italy) Constrained by Radiogenic and Stable Isotope Data. The deepest portions are likely to have remained substantially unchanged with time, as demonstrated by the overall homogeneity of data coming from the least-evolved products (see Sect. 2). The mantle-derived primitive melts from which these originated were likely underplated beneath the crust, where these differentiated leaving a load of ultramafic and mafic cumulates (e.g., Fedi et al. 2018). On the other hand, the shallower-level evolutionary processes experienced by the CFvf magmas can be shown to have acted in many different ways (possibly resulting in several, slightly different liquid lines of descent), and thus more probably reflect the existence of numerous branches connected with the main reservoir structure. Based on the composition of the emitted magmas in the recognised 5 major phases of activity, the shallow magmatic system seems to have followed a somewhat cyclic behaviour. Pre-CI and CI activities were characterised mainly by the emplacement of highly evolved magmas, passing to slightly less evolved compositions in the latest stages of the caldera-forming CI eruption. The successive pre-NYT stage saw some relatively less-evolved latitic eruption, but was again mostly fed by evolved trachytic magmas. The latter compositions were also emitted during the earliest phases of the caldera-forming NYT eruption, followed by some less evolved compositions as the eruption progressed. Finally, post-NYT volcanism was again characterised by eruptions of less evolved shoshonite magmas, but again progressed to trachytic melts and eventually reached the highly evolved compositions of the latest historical eruption of Monte Nuovo. A strong control was also that played by local stress distribution, allowing the processes of magma upraise, eruption, storage (in some cases producing the huge reservoirs feeding the major CI and NYT eruptions) and interaction, as also demonstrated by the concentration of recent, post-NYT eruptive vents along the margin of either the CFc or the resurgent block (e.g., Orsi et al. 1996, 2004; Di Vito et al. 1999; Insinga et al. 2006; Fedele et al. 2011; Selva et al. 2012; see also Chap. Volcanic and Deformation History of the Campi Flegrei Volcanic Field, Italy). The very few occurrences of poorly-evolved juvenile products have again to be ascribed to the development of particularly favourable structural conditions, allowing a relatively rapid magma ascent and substantially negligible storage + differentiation processes.

The existence of a linkage between tectonics and volcanic phenomena is possibly also reflected by the alignment of CFvf, Pvf and Ivf roughly along a southwest-northeast direction, likely related to the presence of regional scale

Antiapenninic (northeast-southwest) structural elements (e.g., Acocella and Funiciello 2006). This is even more evident from the chronology of volcanic activity, which is generally younger moving from southwest to northeast, from Ischia (where volcanism mainly concentrated from 150 to 55 ka; Vezzoli 1988; Brown et al. 2008; Sbrana et al. 2009) to Procida and Campi Flegrei (where the oldest products date back to ~75–80 ka; e.g., Morra et al. 2010; Fedele et al. 2012b; Scarpati et al. 2013). In addition, the somewhat "peculiar" volcanism of Procida (i.e., only five eruptions of low explosivity emplacing volcanic products of mainly basic-intermediate composition) is located exactly in between these two ends, thus still arguing in favour of tectonically-driven volcanism. In this perspective, Moretti et al. (2013) suggested that the similarities in both volcanic style and rock compositions between the CFvf and Ivf are related to the fact that both districts are located at the intersection of northwest-southeast and southwest-northeast structures, which favoured the development of large magma chambers where strongly-evolved melts could collect and drive explosive volcanism. The completely different volcanic style and magma compositions observed at the Pvf would be instead related to the presence of northwest-southeast fault systems only, which allowed rapid magma transfer *en route* to the surface, preventing significant crustal storage. In any case, it should be observed that structural linkages between the three Phlegraean volcanic fields do not imply that these are also linked on a magma-genetic point of view. On the contrary, the mentioned geochemical distinctive traits of Ischia rocks (see Sect. 3), the differences in isotope systematics (see Sect. 2), as well as the differences between Cr-spinel inclusions in high-Mg olivine crystals (Melluso et al. 2012, 2014) suggest that a model for a uniform magmatic system can be hardly held. The cited evidence seems to more reasonably point to a model assuming (i) the existence of (slightly) different mantle sources or (ii) a basically common mantle source from which three different shallower reservoirs have developed and then fed the activity of Campi Flegrei, Procida and Ischia independently from each other (i.e., in a somewhat similar fashion to what, at a smaller scale, was proposed for the numerous branches of the CFvf magmatic system).

In the light of the definition of the general architecture of the CFvf magmatic system, a couple of words are also worth to be spent on its relationships with the nearby Somma Vesuvio Volcanic System (SVvs), the most iconic and worldwide known of the active centres of the Neapolitan Volcanic Area. Notwithstanding their obvious connection to the same global magmatic processes responsible for the Pleistocene Italian magmatism, the similar mantle source characteristics and the overall coincident period of activity (see Peccerillo 2005, 2017; Conticelli et al. 2010), the existence of a common magmatic system for the CFvf and SVvs appears at least highly questionable. Seismic models based on the recognition of low-velocity layers of similar depth (e.g., Nunziata et al. 2006; Piana Agostinetti and Chiarabba 2008; Zollo et al. 2008; Nunziata 2010; Costanzo and Nunziata 2017) apparently support such hypothesis, together with the comparable estimates for the depth and timescales of magma storage obtained for the last 50 kyrs of their activity (Pappalardo and Mastrolorenzo 2012). However, simple petrological observations on the very different nature of the respective magmas strongly argue against this. In fact, even the slightly silica-undersaturated products of the SVvs (which also features leucite-bearing mildly and strongly silica-undersaturated rocks; see Peccerillo 2005, 2017; Conticelli et al. 2010; Morra et al. 2010 and references therein) are remarkably different from the CFvf rocks due to the ubiquitous presence of abundant leucite crystals. In this perspective, the recognition of peculiar groundmass phases in the very final stages of the Campi Flegrei magmatic evolution (see Sect. 2; Melluso et al. 2012), testifying for a relatively Cl- and F-rich composition of highly differentiated melts, can give very interesting clues. Indeed, as observed in Sect. 3, in alkaline systems halogens are known to strongly limit the stability of leucite, whose massive occurrence in SVvs rocks can instead point to CO_2-richer, Cl- and F-poorer magmas (see discussions in Edgar 1987; Veksler et al. 1998; Melzer and Foley

2000; Conticelli et al. 2010; Dallai et al. 2011; Melluso et al. 2012, 2014) with respect to the leucite-free CFvf ones. A substantial independence of the shallow plumbing systems feeding the activity at CFvf and SVvs, as well as those of Ivf and Pvf, has been recently supported also by modelling of the gravity data for the Neapolitan Volcanic Area by Fedi et al. (2018). Based on the recognition of a wide gravity low, the authors proposed a low-density source represented by a mush zone at intermediate/lower crustal levels (8–24 km depth), consisting of large volumes of melts, cumulates and country rocks, with either (i) a layered distribution of density (i.e., percentages of melts, cumulates and country rocks vary with depth), or (ii) a fractal distribution of density (melts distributed in sill-like pockets within solid rocks). In this model, small, shallow ephemeral reservoirs (2–4 km depth) develop upon differentiation and upraise of the magmas from this zone feeds individual eruptions.

A final key-aspect that needs to be addressed in order to define the characteristics of the Campi Flegrei magmatic system, regards its evolution with time. As observed above, composition of the emitted products seems to have followed a somewhat cyclic behaviour, and the progression of volcanic phenomena is similarly suggestive of a repetitive pattern. Volcanism indeed basically shifted from an initial period characterised by a small number of eruptions, two of which (CI and NYT) of cataclysmic impact, to a phase punctuated by a great number of events of much smaller explosivity and volumes of emitted magmas (see Chap. Volcanic and Deformation History of the Campi Flegrei Volcanic Field, Italy). Such a general trend, accompanied by a progressive migration of the eruptive vents towards the inner caldera rim (Armienti et al. 1983; Rosi and Sbrana 1987; Orsi et al. 1996, 2004, 2009; Di Vito et al. 1999), could suggest that the magmatic system is currently shrinking and emptying, as originally proposed by Armienti et al. (1983). On the contrary, Forni et al. (2018b) observed that the variations in the composition of emplaced magmas with time was accompanied by similar oscillations in both T and H_2O contents (respectively decreasing and increasing with increasing rock evolution). The authors produced a thermomechanical model for the recent activity that they used to make inferences on the present state of the CFvf magmatic system. The similar physical conditions (i.e., low T, H_2O saturation, high crystallinity of the cumulate mush) characterising the Monte Nuovo eruption and both the pre-CI and pre-NYT phases were taken as indicative of the magmatic system having potentially entered a new caldera cycle.

6 Conclusive Remarks

The large number of the data presently available for the juvenile products of the CFvf has provided a solid and robust base for the current knowledges about the present state and the temporal evolution of its magmatic system. However, different interpretations and even opposing models are still being proposed, clearly speaking for the need of additional studies shedding new light on such controversies. In addition, and even more importantly, it is also worth of note that all such speculations may be undermined by the overall current knowledges on the Campi Flegrei volcanism (see Chap. Volcanic and Deformation History of the Campi Flegrei Volcanic Field, Italy). In fact, most of the current data is focused on the two main eruptions of the CI and NYT and on the most recent phase of activity (whose complete understanding, however, still necessitates improvements; see Insinga et al. 2006; Fedele et al. 2011, 2012a; Smith et al. 2011; Isaia et al. 2012). On the other hand, the pre-NYT and the ancient pre-CI activity are very poorly characterised, basically due to their unfavourable outcrop conditions related to both the caldera collapses and the burial by deposits of the subsequent activity. This can result in a substantially incorrect evaluation of these two (likely underestimated) periods of activity, and many of the hypotheses regarding the chronological evolution of the magmatic system could need to be variably corrected. Future research aimed to fill such gaps will probably reveal completely new data and add notable complications to the general model, and will surely bring to a more detailed and comprehensive understanding of the global dynamics driving the Campi Flegrei magmatism.

Acknowledgements Financial support was provided by Ricerca Dipartimentale and FFABR2017 CUP E61I18001620005 grants. The comments from the official Reviewers Massimo Coltorti and Dan Morgan and from the Editors provided very constructive criticism that greatly contributed to increasing the quality and the consistency of the paper.

References

Acocella V, Funiciello R (2006) Transverse systems along the extensional Tyrrhenian margin of central Italy and their influence on volcanism. Tectonics 25: TC2003. https://doi.org/10.1029/2005TC001845

Agostini C, Fortunati A, Arzilli F, Landi P, Carroll MR (2013) Kinetics of crystal evolution as a probe to magmatism at Stromboli (Aeolian Archipelago, Italy). Geochim Cosmochim Acta 110:135–151. https://doi.org/10.1016/j.gca.2013.02.027

Albert PG, Giaccio B, Isaia R, Costa A, Niespolo EM, Nomade S, Pereira A, Renne PR, Hinchliffe A, Mark DF, Brown RJ, Smith VC (2019) Evidence for a large-magnitude eruption from Campi Flegrei caldera (Italy) at 29 ka. Geology 47(7):595–599. https://doi.org/10.1130/G45805.1

Albini A, Cristofolini R, Di Girolamo P, Nardi G, Rolandi G, Stanzione D (1977) Rare earth element and thorium distribution in volcanic rocks of the potassic kindred from Procida Island and the Phlegraen Fields, Southern Italy. Accad Naz Lincei 63:416–429

Arienzo I, Civetta L, Heumann A, Wörner G, Orsi G (2009) Isotopic evidence for opens system processes within the Campania Ignimbrite (Campi Flegrei—Italy) magma chamber. Bull Volcanol 71:285–300

Arienzo I, Moretti R, Civetta L, Orsi G, Papale P (2010) The feeding system of Agnano-Monte Spina eruption (Campi Flegrei, Italy): dragging the past into present activity and future scenarios. Chem Geol 270:135–147

Arienzo I, Heumann A, Wörner G, Civetta L, Orsi G (2011) Processes and timescales of magma evolution prior to the Campanian Ignimbrite eruption (Campi Flegrei, Italy). Earth Planet Sci Lett 306:217–228

Arienzo I, D'Antonio M, Di Renzo V, Tonarini S, Minolfi G, Orsi G, Carandente A, Belviso P, Civetta L (2015) Isotopic microanalysis sheds light on the magmatic endmembers feeding volcanic eruptions: the Astroni 6 case study (Campi Flegrei, Italy). J Volcanol Geotherm Res 304:24–37. https://doi.org/10.1016/j.jvolgeores.2015.08.003

Arienzo I, Mazzeo FC, Moretti R, Cavallo A, D'Antonio M (2016) Open-system magma evolution and fluid transfer at Campi Flegrei caldera (Southern Italy) during the past 5 ka as revealed by geochemical and isotopic data: the example of the Nisida eruption. Chem Geol 427:109–124. https://doi.org/10.1016/j.chemgeo.2016.02.007

Armienti P, Barberi F, Bizouard H, Clocchiatti R, Innocenti F, Metrich N, Rosi M, Sbrana A (1983) The Phlegrean Fields: magma evolution within a shallow chamber. J Volcanol Geotherm Res 17:289–311

Arzilli F, Carroll MR (2013) Crystallization kinetics of alkali feldspars in cooling and decompression-induced crystallization experiments in trachytic melt. Contrib Mineral Petrol 166:1011–1027. https://doi.org/10.1007/s00410-013-0906-1

Arzilli F, Agostini C, Landi P, Fortunati A, Mancini L, Carroll MR (2015a) Plagioclase nucleation and growth kinetics in a hydrous basaltic melt by decompression experiments. Contrib Mineral Petrol 170:55. https://doi.org/10.1007/s00410-015-1205-9

Arzilli F, Mancini L, Voltolini M, Cicconi MR, Mohammadi S, Giuli G, Mainprice D, Paris E, Barou F, Carroll MR (2015b) Near-liquidus growth of feldspar spherulites in trachytic melts: 3D morphologies and implications in crystallization mechanisms. Lithos 216–217:93–105. https://doi.org/10.1016/j.lithos.2014.12.003

Arzilli F, Piochi M, Mormone A, Agostini C, Carroll MR (2016) Constraining pre-eruptive magma conditions and unrest timescales during the Monte Nuovo eruption (1538 AD; Campi Flegrei, Southern Italy): integrating textural and CSD results from experimental and natural trachy-phonolites. Bull Volcanol 78:72. https://doi.org/10.1007/s00445-016-1062-z

Astbury RL, Petrelli M, Ubide T, Stock MJ, Arienzo I, D'Antonio M, Perugini D (2018) Tracking plumbing system dynamics at the Campi Flegrei caldera, Italy: high-resolution trace element mapping of the Astroni crystal cargo. Lithos 318–319:464–477. https://doi.org/10.1016/j.lithos.2018.08.033

Avanzinelli R, Lustrino M, Mattei M, Melluso L, Conticelli S (2009) Potassic and ultrapotassic magmatism in the circum-Tyrrhenian region: the role of carbonated pelitic vs. pelitic sediment recycling at destructive plate margin. Lithos 113:213–227

Barberi F, Innocenti F, Lirer L, Munno R, Pescatore TS, Scandone R (1978) The Campanian Ignimbrite: a major prehistoric eruption in the Neapolitan area (Italy). Bull Volcanol 41:10–22

Beccaluva L, Di Girolamo P, Morra V, Siena F (1990) Phlegraean Fields volcanism revisited: a critical re-examination of deep eruptive systems and magma evolutionary processes. N Jb Geol Paläont Mh 5:257–271

Beccaluva L, Di Girolamo P, Serri G (1991) Petrogenesis and tectonic setting of the Roman volcanic province, Italy. Lithos 26:191–221

Bohrson WA, Spera FJ, Fowler SJ, Belkin HE, De Vivo B, Rolandi G (2006) Petrogenesis of the Campanian Ignimbrite: implications for crystal-melt separation and open-system processes from major and trace elements and Th isotopic data. In: De Vivo B (ed) Volcanism in the Campania Plain: Vesuvius, Campi Flegrei and Ignimbrites. Developments in volcanology, vol 9. Elsevier, Amsterdam, pp 249–288

Brown RJ, Orsi G, de Vita S (2008) New insights into late pleistocene explosive volcanic activity and caldera formation on Ischia (southern Italy). Bull Volcanol 70:583–603. https://doi.org/10.1007/s00445-007-0155-0

Brown RJ, Civetta L, Arienzo I, D'Antonio M, Moretti R, Orsi G, Tomlinson EL, Albert PG, Menzies MA (2014) Geochemical and isotopic insights into the assembly, evolution and disruption of a magmatic plumbing system before and after a cataclysmic caldera-collapse eruption at Ischia volcano (Italy). Contrib Mineral Petrol 168:1035. https://doi.org/10.1007/s00410-014-1035-1

Calzolaio M, Arzilli F, Carroll MR (2010) Growth rate of alkali feldspars in decompression-induced crystallization experiments in a trachytic melt of the Phlegraean Fields (Napoli, Italy). Eur J Mineral 22:485–493. https://doi.org/10.1127/0935-1221/2010/0022-2012

Campbell IH, Stepanov AS, Liang H-Y, Allen CM, Norman MD, Zhang Y-Q, Xie Y-W (2014) The origin of shoshonites: new insights from the tertiary high-potassium intrusions of eastern Tibet. Contrib Mineral Petrol 167:983. https://doi.org/10.1007/s00410-014-0983-9

Cannatelli C (2012) Understanding magma evolution at Campi Flegrei (Campania, Italy) volcanic complex using melt inclusions and phase equilibria. Mineral Petrol 104:29–42

Cannatelli C, Lima A, Bodnar RJ, De Vivo B, Webster JD, Fedele L (2007) Geochemistry of melt inclusions from the Fondo Riccio and Minopoli 1 eruptions at Campi Flegrei (Italy). Chem Geol 237:418–432. https://doi.org/10.1016/j.chemgeo.2006.07.012

Cañón-Tapia E (2014) Volcanic eruption triggers: a hierarchical classification. Earth-Sci Rev 129:100–119. https://doi.org/10.1016/j.earscirev.2013.11.011

Carr PF, Ferguson CL, Pemberton JW, Colquhoun GP, Murray SI, Watkins J (2003) Late Ordovician island-arc volcanic rocks, northern Capertee Zone, Lachlan Fold Belt, New South Wales. Aust J Earth Sci 50:319–330

Casalini M, Avanzinelli R, Heumann A, de Vita S, Sansivero F, Conticelli S, Tommasini S (2017) Geochemical and radiogenic isotope probes of Ischia volcano, Southern Italy: constraints on magma chamber dynamics and residence time. Am Mineral 102:262–274. https://doi.org/10.2138/am-2017-5724

Civetta L, Innocenti F, Sbrana A, Taddeucci G (1988) Variazioni petrografiche e geochimiche nei prodotti di Averno: implicazioni sulla zonatura del sistema di alimentazione. Boll GNV 4:201–217

Civetta L, Carluccio E, Innocenti F, Sbrana A, Taddeucci G (1991a) Magma chamber evolution under the Phlegrean Fields during the last 10 ka: trace element and isotope data. Eur J Mineral 3:415–428

Civetta L, Gallo G, Orsi G (1991b) Sr- and Nd-isotopes and trace-element constraints on the chemical evolution of the magmatic system of Ischia (Italy) in the last 55 ka. J Volcanol Geotherm Res 46:213–230

Civetta L, Orsi G, Pappalardo L, Fisher RV, Heiken G, Ort M (1997) Geochemical zoning, mingling, eruptive dynamics and depositional processes—the Campanian Ignimbrite, Campi Flegrei caldera, Italy. J Volcanol Geotherm Res 75:183–219

Çoban H, Karsli O, Caran S, Yilmaz K (2019) What processes control the genesis of absarokite to shoshonite-banakite series in an intracontinental setting, as revealed by geochemical and Sr-Nd-Pb isotope data of Karadağ Stratovolcano in Central Anatolia, Turkey. Lithos 324–325:609–625. https://doi.org/10.1016/j.lithos.2018.11.034

Conte AM, Perinelli C, Trigila R (2006) Cooling kinetics experiments on different Stromboli lavas: effects on crystal morphologies and phases composition. J Volcanol Geotherm Res 155:179–200

Conticelli S, Laurenzi MA, Giordano G, Mattei M, Avanzinelli R, Melluso L, Tommasini S, Boari E, Cifelli F, Perini G (2010) Leucite-bearing (kamafugitic/leucititic) and -free (lamproitic) ultrapotassic rocks and associated shoshonites from Italy: constraints on petrogenesis and geodynamics. In: Beltrando M, Peccerillo A, Mattei M, Conticelli S, Doglioni C (eds) Geology of Italy. J Virtual Expl 36, paper 20. https://doi.org/10.3809/jvirtex.2009.00251

Costanzo MR, Nunziata C (2017) Inferences on the lithospheric structure of Campi Flegrei District (Southern Italy) from seismic noise cross-correlation. Phys Earth Planet Int 265:92–105. https://doi.org/10.1016/j.pepi.2017.02.010

Crisci GM, De Francesco AM, Mazzuoli M, Poli G, Stanzione D (1989) Geochemestry of recent volcanics of Ischia Island, Italy: evidences for fractional crystallization and magma mixing. Chem Geol 78:15–33

Cvetković V, Prelević D, Downes H, Jovanović M, Vaselli O, Pécskay Z (2004) Origin and geodynamic significance of tertiary postcollisional basaltic magmatism in Serbia (Central Balkan Peninsula). Lithos 73:161–186

D'Antonio M (2011) Lithology of the basement underlying the Campi Flegrei caldera: volcanological and petrological constraints. J Volcanol Geotherm Res 200:91–98

D'Antonio M, Di Girolamo P (1994) Petrological and geochemical study of mafic shoshonitic volcanics from Procida-Vivara and Ventotene Islands. Acta Vulcanol 5:69–80

D'Antonio M, Civetta L, Di Girolamo P (1999a) Mantle source heterogeneity in the Campanian region (South Italy) as inferred from geochemical and isotopic features of mafic volcanic rocks with shoshonitic affinity. Mineral Petrol 67:163–192

D'Antonio M, Civetta L, Orsi G, Pappalardo L, Piochi M, Carandente A, de Vita S, Di Vito MA, Isaia R (1999b) The present state of the magmatic system of the Campi

Flegrei Caldera based on a reconstruction of its behaviour in the past 12 ka. J Volcanol Geotherm Res 91:247–268
D'Antonio M, Tonarini S, Arienzo I, Civetta L, Di Renzo V (2007) Components and processes in the magma genesis of the Phlegrean volcanic district, Southern Italy. In: Beccaluva L, Bianchini G, Wilson M (eds) Cenozoic volcanism in the Mediterranean area. Geol Soc Am Spec Pap 418:203–220
D'Antonio M, Tonarini S, Arienzo I, Civetta L, Dallai L, Moretti R, Orsi G, Andria M, Trecalli A (2013) Mantle and crustal processes in the magmatism of the Campania region: inferences from mineralogy, geochemistry, and Sr-Nd-O isotopes of young hybrid volcanics of the Ischia island (South Italy). Contrib Mineral Petrol 165:1173–1194. https://doi.org/10.1007/s00410-013-0853-x
D'Oriano C, Poggianti E, Bertagnini A, Cioni R, Landi P, Polacci M, Rosi M (2005) Changes in eruptive style during the A.D. 1538 Monte Nuovo eruption (Phlegrean Fields, Italy): the role of syn-eruptive crystallization. Bull Volcanol 67:601–621. https://doi.org/10.1007/s00445-004-0397-z
Dallai L, Cioni R, Boschi C, D'Oriano C (2011) Carbonate-derived CO_2 purging magma at depth: influence on the eruptive activity of Somma-Vesuvius, Italy. Earth Planet Sci Lett 310:84–95. https://doi.org/10.1016/j.epsl.2011.07.013
De Astis G, Pappalardo L, Piochi M (2004) Procida volcanic history: new insights into the evolution of the Phlegraean volcanic district (Campanian region, Italy). Bull Volcanol 66:622–641
de Vita S, Orsi G, Civetta L, Carandente A, D'Antonio M, Deino A, di Cesare T, Di Vito MA, Fisher RV, Isaia R, Marotta E, Necco A, Ort M, Pappalardo L, Piochi M, Southon J (1999) The Agnano-Monte Spina eruption (4100 years BP) in the restless Campi Flegrei caldera (Italy). J Volcanol Geotherm Res 91:269–301
Di Girolamo P (1970) Differenziazione gravitativa e curve isochimiche nella "Ignimbrite Campana." Rend Soc It Mineral Petrol 26:3–44
Di Girolamo P, Rolandi G (1979) Vulcani shoshonitici Nell'area Flegrea. Per Mineral 48:93–114
Di Girolamo P, Stanzione D (1973) Lineamenti geologici e petrologici dell'Isola di Procida. Rend Soc It Mineral Petrol 29:81–126
Di Girolamo P, Ghiara MR, Lirer L, Munno R, Rolandi G, Stanzione D (1984) Vulcanologia e petrologia dei Campi Flegrei. Boll Soc Geol It 103:349–413
Di Girolamo P, Melluso L, Morra V, Secchi FGA (1995) Evidence of interaction between mafic and differentiated magmas in the youngest phase of activity at Ischia Island (Italy). Per Mineral 64:393–411
Di Matteo V, Carroll MR, Behrens H, Vetere F, Brooker RA (2004) Water solubility in trachytic melts. Chem Geol 213:187–196. https://doi.org/10.1016/j.chemgeo.2004.08.042
Di Renzo V, Arienzo I, Civetta L, D'Antonio M, Tonarini S, Di Vito MA, Orsi G (2011) The magmatic feeding system of the Campi Flegrei caldera: architecture and temporal evolution. Chem Geol 281:227–241
Di Vito MA, Isaia R, Orsi G, Southon J, de Vita S, D'Antonio M, Pappalardo L, Piochi M (1999) Volcanism and deformation since 12,000 years at the Campi Flegrei caldera (Italy). J Volcanol Geotherm Res 91:221–246
Di Vito MA, Arienzo I, Braia G, Civetta L, D'Antonio M, Di Renzo V, Orsi G (2011) The Averno 2 fissure eruption: a recent small-size explosive event at the Campi Flegrei Caldera (Italy). Bull Volcanol 73:295–320
Duchesne J, Berza T, Liégeois J-P, Vander Auwera J (1998) Shoshonitic liquid line of descent from diorite to granite: the late Precambrian post-collisional Tismana pluton (South Carpathians, Romania). Lithos 45:281–303
Edgar AD (1987) The genesis of alkaline magmas with emphasis on their source regions: inferences from experimental studies. In: Fitton JG, Upton BGJ (eds) Alkaline igneous rocks. Geol Soc Sp Publ 30:29–52
Eklund O, Shebanov A (2005) Post-collisional shoshonitic magmatism in the southern Svecofennian domain —a case study of the Åva granite-lamprophyre ring complex. Lithos 80:229–247
Esposito R, Bodnar RJ, Danyushevsky L, De Vivo B, Fedele L, Hunter J, Lima A, Shimizu N (2011) Volatile evolution of magmaassociated with the Solchiaro eruption in the Phlegrean volcanic district (Italy). J Petrol 52(12):2431–2460
Esposito R, Badescu K, MacInnis MS, Cannatelli C, De Vivo B, Lima A, Bodnar RJ, Manning CE (2018) Magmatic evolution of the Campi Flegrei and Procida volcanic fields, Italy, based on interpretation of data from well-constrained melt inclusions. Earth-Sci Rev 185:325–356. https://doi.org/10.1016/j.earscirev.2018.06.003
Fabbrizio A, Carroll MR (2008) Experimental constraints on the differentiation process and pre-eruptive conditions in the magmatic system of Phlegraean Fields (Naples, Italy). J Volcanol Geotherm Res 171:88–102. https://doi.org/10.1016/j.jvolgeores.2007.11.002
Fabbrizio A, Scaillet B, Carroll MR (2009) Estimation of pre-eruptive magmatic water fugacity in the Phlegrean Fields, Naples, Italy. Eur J Mineral 21:107–116. https://doi.org/10.1127/0935-1221/2009/0021-1873
Fedele L, Morra V, Perrotta A, Scarpati C (2006) Volcanological and geochemical features of the products of the Fiumicello eruption, Procida island, Campi Flegrei (southern Italy). Per Mineral 75:43–72
Fedele L, Scarpati C, Lanphere M, Melluso L, Morra V, Perrotta A, Ricci G (2008) The Breccia Museo formation, Campi Flegrei, southern Italy: geochronology, chemostratigraphy and relationship with the Campanian Ignimbrite eruption. Bull Volcanol 70:1189–1219
Fedele L, Zanetti A, Morra V, Lustrino M, Melluso L, Vannucci R (2009) Clinopyroxene/liquid trace

element partitioning in natural trachyte-trachyphonolite systems: insights from Campi Flegrei (southern Italy). Contrib Mineral Petrol 158:337–356

Fedele L, Insinga DD, Calvert AT, Morra V, Perrotta A, Scarpati C (2011) $^{40}Ar/^{39}Ar$ dating of tuff vents in the Campi Flegrei caldera (Southern Italy): toward a new chronostratigraphic reconstruction of the Holocene volcanic activity. Bull Volcanol 73:1323–1336

Fedele L, Insinga DD, Calvert AT, Morra V, Perrotta A, Scarpati C (2012a) Reply to the comment on the article "$^{40}Ar/^{39}Ar$ dating of tuff vents in the Campi Flegrei caldera (southern Italy): toward a new chronostratigraphic reconstruction of the Holocene volcanic activity" by Isaia et al. Bull Volcanol 74:297–299

Fedele L, Morra V, Perrotta A, Scarpati C, Sbrana A, Putignano ML, Orrù, PE, Schiattarella M, Aiello G, Budillon F, Conforti B, D'Argenio B, Calcaterra D, Sgrosso A, Vecchio E (2012b). Note illustrative della Carta Geologica alla scala 1:10.000. Fogli 465 e 464 Isola di Procida. Carta Geologica della Regione Campania, SystemCart, Rome, Italy

Fedele L, Lustrino M, Melluso L, Morra V, Zanetti A, Vannucci R (2015) Trace-element partitioning between plagioclase, alkali feldspar, Ti-magnetite, biotite, apatite, and evolved potassic liquids from Campi Flegrei (Southern Italy). Am Mineral 100:233–249. https://doi.org/10.2138/am-2015-4995

Fedele L, Scarpati C, Sparice D, Perrotta A, Laiena F (2016) A chemostratigraphic study of the Campanian Ignimbrite eruption (Campi Flegrei, Italy): insights on magma chamber withdrawal and deposit accumulation as revealed by compositionally zoned stratigraphic and facies framework. J Volcanol Geotherm Res 324:105–117. https://doi.org/10.1016/j.jvolgeores.2016.05.019

Fedi M, Cella F, D'Antonio M, Florio G, Paoletti V, Morra V (2018) Gravity modeling finds a large magma body in the deep crust below the Gulf of Naples, Italy. Sci Rep 8:8229. https://doi.org/10.1038/s41598-018-26346-z

Forni F, Bachmann O, Mollo S, De Astis G, Gelman SE, Ellis BS (2016) The origin of a zoned ignimbrite: insights into the Campanian Ignimbrite magma chamber (Campi Flegrei, Italy). Earth Planet Sci Lett 449:259–271. https://doi.org/10.1016/j.epsl.2016.06.003

Forni F, Petricca E, Bachmann O, Mollo S, De Astis G, Piochi M (2018a) The role of magma mixing/mingling and cumulate melting in the Neapolitan Yellow Tuff caldera-forming eruption (Campi Flegrei, Southern Italy). Contrib Mineral Petrol 173:45. https://doi.org/10.1007/s00410-018-1471-4

Forni F, Degruyter W, Bachmann O, De Astis G, Mollo S (2018b) Long-term magmatic evolution reveals the beginning of a new caldera cycle at Campi Flegrei. Sci Adv 4(11):eaat9401. https://doi.org/10.1126/sciadv.aat9401

Fourmentraux C, Metrich N, Bertagnini A, Rosi M (2012) Crystal fractionation, magma step ascent, and syn-eruptive mingling; the Averno 2 eruption (Phlegraean Fields, Italy). Contrib Mineral Petrol 163:1121–1137. https://doi.org/10.1007/s00410-012-0720-1

Fowler SJ, Spera FJ, Bohrson WA, Belkin HE, De Vivo B (2007) Phase equilibria constraints on the chemical and physical evolution of the Campanian Ignimbrite. J Petrol 48:459–493

Fulignati P, Marianelli M, Proto M, Sbrana A (2004) Evidences for disruption of a crystallizing front in a magma chamber during caldera collapse: an example from the Breccia Museo unit (Campanian Ignimbrite eruption, Italy). J Volcanol Geotherm Res 133:141–155

Gao Y, Yang Z, Hou Z, Wei R, Meng X, Tian S (2010) Eocene potassic and ultrapotassic volcanism in south Tibet: new constraints on mantle source characteristics and geodynamic processes. Lithos 117:20–32

Gebauer SK, Schmitt AK, Pappalardo L, Stockli DF, Lovera OM (2014) Crystallization and eruption ages of Breccia Museo (Campi Flegrei caldera, Italy) plutonic clasts and their relation to the Campanian ignimbrite. Contrib Mineral Petrol 167:953. https://doi.org/10.1007/s00410-013-0953-7

Ghiara MR (1989–1990) Studio evolutivo del sistema magmatico flegreo negli ultimi 10 Ka. Boll Soc Natu Napoli 98–99:41–70

Ghiara MR, Rolandi G, Stanzione D (1977) Evoluzione delle vulcaniti ad affinità shoshonitica dei Campi Flegrei s.l.: studio petrologico e geochimico. Per Mineral 46:99–124

Ghiara MR, Lirer L, Munno R (1979) Mineralogy and geochemistry of the "low-potassium series" of the Campania volcanics (South Italy). Chem Geol 26:29–49

Ghiorso MS, Gualda GAR (2015) An H_2O-CO_2 mixed fluid saturation model compatible with rhyolite-MELTS. Contrib Mineral Petrol 169:53. https://doi.org/10.1007/s00410-015-1141-8

Ghiorso MS, Sack RO (1995) Chemical mass transfer in magmatic processes IV. A revised internally consistent thermodynamic model for the interpolation and extrapolation of liquid-solid equilibria in magmatic systems at elevated temperatures and pressures. Contrib Mineral Petrol 119:197–212

Hamilton DL, MacKenzie WS (1965) Phase equilibrium studies in the system $NaAlSiO_4$ (nepheline)—$KAlSiO_4$ (kalsilite)—SiO_2—H_2O. Mineral Mag 34:214–231

Iezzi G, Mollo S, Ventura G, Cavallo A, Romano C (2008) Experimental solidification of anhydrous latitic and trachytic melts at different cooling rates: the role of nucleation kinetics. Chem Geol 253:91–101

Insinga D, Calvert AT, Lanphere M, Morra V, Perrotta A, Sacchi M, Scarpati C, Saburomaru J, Fedele L (2006) The Late-Holocene evolution of the south-western sector of Campi Flegrei as inferred by stratigraphy, petrochemistry and $^{40}Ar/^{39}Ar$ dating. In: De Vivo B (ed) Volcanism in the Campania Plain: Vesuvius, Campi Flegrei and Ignimbrites. Developments in volcanology, vol 9. Elsevier, Amsterdam, pp 97–125

Iovine RS, Fedele L, Mazzeo FC, Arienzo I, Cavallo A, Wörner G, Orsi G, Civetta L, D'Antonio M (2017) Timescales of magmatic processes prior to the ~4.7 ka Agnano-Monte Spina eruption (Campi Flegrei caldera, Southern Italy) based on diffusion chronometry from sanidine phenocrysts. Bull Volcanol 79:18. https://doi.org/10.1007/s00445-017-1101-4

Iovine RS, Mazzeo FC, Wörner G, Pelullo C, Cirillo G, Arienzo I, Pack A, D'Antonio M (2018) Coupled $\delta^{18}O$-$\delta^{17}O$ and $^{87}Sr/^{86}Sr$ isotope compositions suggest a radiogenic and ^{18}O-enriched magma source for Neapolitan volcanoes (Southern Italy). Lithos 316–317:199–211. https://doi.org/10.1016/j.lithos.2018.07.009

Isaia R, Marianelli P, Sbrana A (2009) Caldera unrest prior to intense volcanism in Campi Flegrei (Italy) at 4.0 ka B.P.: implications for caldera dynamics and future eruptive scenarios. Geophys Res Lett 36: L21303. https://doi.org/10.1029/2009GL040513

Isaia R, Di Vito MA, de Vita S, Rosi M, Sbrana A (2012) Comment to "$^{40}Ar/^{39}Ar$ dating of tuff vents in the Campi Flegrei caldera (southern Italy): toward a new chronostratigraphic reconstruction of the Holocene volcanic activity" by Fedele et al. [Bull Volcanol 73:1323–1336]. Bull Volcanol 74:293–296

Jiang Y, Jiang S, Ling H, Zhou X, Rui X, Yang W (2002) Petrology and geochemistry of shoshonitic plutons from the western Kunlun orogenic belt, Xinjiang, northwestern China: implications for granitoid geneses. Lithos 63:165–187

Kogarko LN, Ryabchikov ID, Sørensen H (1974) Liquid fractionation. In: Sørensen H (ed) Alkaline rocks. Wiley, New York, pp 488–500

Lemarchand F, Benoit V, Calais G (1987) Trace element distribution coefficients in alkaline series. Geochim Cosmochim Acta 51:1071–1081

Le Maitre RW (2002) Igneous rocks. A classification and glossary of terms. Recommendations of the International Union of Geological Sciences Subcommission on the Systematics of Igneous Rocks. Cambridge University Press, Cambridge, 256 pp

Lirer L, Mastrolorenzo G, Rolandi G (1987) Un evento pliniano nell'attività recente dei Campi Flegrei. Boll Soc Geol It 106:461–473

Mangiacapra A, Moretti R, Rutherford M, Civetta L, Orsi G, Papale P (2008) The deep magmatic system of the Campi Flegrei caldera (Italy). Geophys Res Lett 35:L21304. https://doi.org/10.1029/2008GL035550

Marianelli P, Sbrana A, Proto M (2006) Magma chamber of the Campi Flegrei supervolcano at the time of the eruption of the Campanian Ignimbrite. Geology 34:937–940. https://doi.org/10.1130/G22807A.1

Mazzeo FC, D'Antonio M, Arienzo I, Aulinas M, Di Renzo V, Gimeno D (2014) Subduction-related enrichment of the Neapolitan volcanoes (Southern Italy) mantle source: new constraints on the characteristics of the slab-derived components. Chem Geol 386:165–183. https://doi.org/10.1016/j.chemgeo.2014.08.014

McKenzie D, O'Nions RK (1991) Partial melt distributions from inversion of rare earth element concentrations. J Petrol 32:1021–1091

Melluso L, Morra V, Perrotta A, Scarpati C, Adabbo M (1995) The eruption of the Breccia Museo (Campi Flegrei, Italy): fractional crystallization processes in a shallow, zoned magma chamber and implications for the eruptive dynamics. J Volcanol Geotherm Res 68:325–339

Melluso L, de' Gennaro R, Fedele L, Franciosi L, Morra V (2012) Evidence of crystallization in residual, Cl-F-rich, agpaitic, trachyphonolitic magmas and primitive Mg-rich basalt-trachyphonolite interaction in the lava domes of the Phlegrean Fields (Italy). Geol Mag 149:532–550

Melluso L, Morra V, Guarino V, de'Gennaro R, Franciosi L, Grifa C (2014) The crystallization of shoshonitic to peralkaline trachyphonolitic magmas in a H_2O–Cl–F-rich environment at Ischia (Italy), with implications for the feeder system of the Campania Plain volcanoes. Lithos 210–211:242–259. doi:https://doi.org/10.1016/j.lithos.2014.10.002

Melzer S, Foley SF (2000) Phase relations and fractionation sequences in potassic magma series modelled in the system $CaMgSi_2O_6$-$KAlSiO_4$-Mg_2SiO_4-SiO_2-F_2O_{-1} at 1 bar to 18 kbar. Contrib Mineral Petrol 138:186–197

Misiti V, Freda C, Taddeucci J, Romano C, Scarlato P, Longo A, Papale P, Poe BT (2006) The effect of H_2O on the viscosity of K-trachytic melts at magmatic temperatures. Chem Geol 235:124–137. https://doi.org/10.1016/j.chemgeo.2008.04.008

Mollo S, Putirka K, Iezzi G, Del Gaudio P, Scarlato P (2011) Plagioclase-melt (dis)equilibrium due to cooling dynamics: implications for thermometry, barometry and hygrometry. Lithos 125:221–235

Moretti R, Arienzo I, Orsi G, Civetta L, D'Antonio M (2013) The deep plumbing system of Ischia: a physico-chemical window on the fluid-saturated and CO_2-sustained Neapolitan Volcanism (Southern Italy). J Petrol 54:951–984. https://doi.org/10.1093/petrology/egt002

Morimoto N (1988) Nomenclature of Pyroxenes. Am Mineral 73:1123–1133

Morra V, Calcaterra D, Cappelletti P, Colella A, Fedele L, de' Gennaro R, Langella A, Mercurio M, de' Gennaro M (2010) Urban geology: relationships between geological setting and architectural heritage of the Neapolitan area. In: Beltrando M, Peccerillo A, Mattei M, Conticelli S, Doglioni C (eds) Geology of Italy. J Virtual Expl 36, paper 26. https://doi.org/10.3809/jvirtex.2010.00261

Müller D, Franz L, Herzig PM, Hunt S (2001) Potassic igneous rocks from the vicinity of epithermal gold mineralization, Lihir Island, Papua New Guinea. Lithos 57:163–186

Nielsen RL, Forsythe LM, Gallahan WE, Fisk MR (1994) Major- and trace-element magnetite-melt equilibria. Chem Geol 117:167–191

Nunziata C (2010) Low shear-velocity zone in the Neapolitan-area crust between the Campi Flegrei and the Vesuvio volcanic areas. Terra Nova 22:208–217
Nunziata C, Natale M, Luongo G, Panza GF (2006) Magma reservoir at Mt. Vesuvius: size of the hot molten, crust material detected deeper than 8 km. Earth Planet Sci Lett 42:51–57
Orsi G, Gallo G, Heiken G, Wohletz K, Yu E, Bonani G (1992) A comprehensive study of pumice formation and dispersal: the Cretaio Tephra of Ischia (Italy). J Volcanol Geotherm Res 53:329–354
Orsi G, Civetta L, D'Antonio M, Di Girolamo P, Piochi M (1995) Step-filling and development of a three-layer magma chamber: the NYT case history. J Volcanol Geotherm Res 67:291–312
Orsi G, de Vita S, Di Vito M (1996) The restless, resurgent Campi Flegrei nested caldera (Italy): constraints on its evolution and configuration. J Volcanol Geotherm Res 74:179–214
Orsi G, Di Vito MA, Isaia R (2004) Volcanic hazard assessment at the restless Campi Flegrei caldera. Bull Volcanol 66:514–530. https://doi.org/10.1007/s00445-003-0336-4
Orsi G, Di Vito MA, Selva J, Marzocchi W (2009) Long-term forecast of eruption style and size at Campi Flegrei caldera (Italy). Earth Planet Sci Lett 287:265–276. https://doi.org/10.1016/j.epsl.2009.08.013
Pabst S, Wörner G, Civetta L, Tesoro R (2008) Magma chamber evolution prior to the Campanian Ignimbrite and Neapolitan Yellow Tuff eruptions (Campi Flegrei, Italy). Bull Volcanol 70:961–976
Pappalardo L, Mastrolorenzo G (2012) Rapid differentiation in a Sill-like Magma Reservoir: a case study from the Campi Flegrei Caldera. Sci Rep 2:712. https://doi.org/10.1038/srep00712
Pappalardo L, Civetta L, D'Antonio M, Deino A, Di Vito M, Orsi G, Carandente A, de Vita S, Isaia R, Piochi M (1999) Chemical and Sr-isotopic evolution of the Phlegrean magmatic system before the Campanian Ignimbrite and the Neapolitan Yellow Tuff eruptions. J Volcanol Geotherm Res 91:141–166
Pappalardo L, Civetta L, de Vita S, Di Vito M, Orsi G, Carandente A, Fisher RV (2002a) Timing of magma extraction during the Campanian Ignimbrite eruption (Campi Flegrei Caldera). J Volcanol Geotherm Res 114:479–497
Pappalardo L, Piochi M, D'Antonio M, Civetta L, Petrini R (2002b) Evidence for multi-stage magmatic evolution during the past 60 kyr at Campi Flegrei (Italy) deduced from Sr, Nd and Pb isotope data. J Petrol 43:1415–1434
Pappalardo L, Ottolini L, Mastrolorenzo G (2008) The Campanian Ignimbrite (southern Italy) geochemical zoning: insight on the generation of a super-eruption from catastrophic differentiation and fast withdrawal. Contrib Mineral Petrol 156:1–26
Peccerillo A (2005) Plio-Quaternary volcanism in Italy. Petrology, geochemistry, geodynamics. Springer, Heidelberg, Germany, p 365
Peccerillo A (2017) Cenozoic volcanism in the Tyrrhenian Sea region. Advances in Volcanology, vol 399. Springer, Heidelberg, Germany, p 399
Perinelli C, Gaeta M, Bonechi B, Granati SF, Freda C, D'Antonio M, Stagno V, Sicola S, Romano C (2019) Effect of water on the phase relations of primitive K-basalts: implications for high-pressure differentiation in the Phlegraean volcanic district magmatic system. Lithos 342–343:530–541. https://doi.org/10.1016/j.lithos.2019.05.032
Piana Agostinetti N, Chiarabba C (2008) Seismic structure beneath Mt Vesuvius from receiver function analysis and local earthquake tomography: evidence for location and geometry of the magma chamber. Geophys J Int 175:1298–1308
Piochi M, Civetta L, Orsi G (1999) Mingling in the magmatic system of Ischia (Italy) in the past 5 ka. Mineral Petrol 66:227–258
Piochi M, Mastrolorenzo G, Pappalardo L (2005) Magma ascent and eruptive processes from textural and compositional features of Monte Nuovo pyroclastic products, Campi Flegrei, Italy. Bull Volcanol 67:663–678. https://doi.org/10.1007/s00445-005-0410-1
Plechov P, Blundy J, Nekrylov N, Melekhova E, Shcherbakov V, Tikhonova MS (2015) Petrology and volatile content of magmas erupted from Tolbachik Volcano, Kamchatka, 2012–13. J Volcanol Geotherm Res 307:182–199. https://doi.org/10.1016/j.volgeores.2015.08.011
Polacci M, Papale P, Del Seppia D, Giordano D, Romano C (2004) Dynamics of magma ascent and fragmentation in trachytic versus rhyolitic eruptions. J Volcanol Geotherm Res 131:93–108. https://doi.org/10.1016/S0377-0273(03)00319-6
Poli S, Chiesa S, Gillot PY, Gregnanin A, Guichard F (1987) Chemistry versus time in the volcanic complex of Ischia (Gulf of Naples, Italy): evidence of successive magmatic cycles. Contrib Mineral Petrol 95:322–335
Rezeau H, Leuthold J, Tayan R, Hovakimyan S, Ulianov A, Kouzmanov K, Moritz R (2018) Incremental growth of mid- to upper-crustal magma bodies during Arabia-Eurasia convergence and collision: a petrological study of the calc-alkaline to shoshonitic Meghri-Ordubad Pluton (Southern Armenia and Nakhitchevan, Lesser Caucasus). J Petrol 59:931–966. https://doi.org/10.1093/petrology/egy050
Roach AL (2005) The evolution of silicic magmatism in the post-caldera volcanism of the Phlegrean Fields, Italy. Unpublished PhD Thesis, Department of Geological Sciences, Brown University. Providence, Rhode Island (USA), 171 pp
Rogers NW, Setterfield TN (1994) Potassium and incompatible-element enrichment in shoshonitic lavas from the Tavua volcano, Fiji. Chem Geol 118:43–62
Rosi M, Sbrana A (1987) The Phlegrean Fields. C.N.R. Quaderni de La ricerca scientifica 114, vol 9, 175 pp
Sbrana A, Fulignati P, Marianelli P, Boyce AJ, Cecchetti A (2009) Exhumation of an active magmatic-

hydrothermal system in a resurgent caldera environment: the example of Ischia (Italy). J Geol Soc Lond 166:1061–1073. https://doi.org/10.1144/0016-76492009-030

Scarpati C, Cole P, Perrotta A (1993) The Neapolitan Yellow Tuff—a large volume multiphase eruption from Campi Flegrei, Southern Italy. Bull Volcanol 55:343–356

Scarpati C, Perrotta A, Lepore S, Calvert A (2013) Eruptive history of Neapolitan volcanoes: constraints from ^{40}Ar-^{39}Ar dating. Geol Mag 150:412–425. https://doi.org/10.1017/S0016756812000854

Scarpati C, Sparice D, Perrotta A (2014) A crystal concentration method for calculating ignimbrite volume from distal ash-fall deposits and a reappraisal of the magnitude of the Campanian Ignimbrite. J Volcanol Geotherm Res 280:67–75. https://doi.org/10.1016/j.jvolgeores.2014.05.009

Selva J, Orsi G, Di Vito MA, Marzocchi W, Sandri L (2012) Probability hazard map for future vent opening at the Campi Flegrei caldera, Italy. Bull Volcanol 74:497–510. https://doi.org/10.1007/s00445-011-0528-2

Signorelli S, Vaggelli G, Francalanci L, Rosi M (1999) Origin of magmas feeding the Plinian phase of the Campanian Ignimbrite eruption, Phlegrean Fields (Italy): constraints based on matrix-glass and glass-inclusion compositions. J Volcanol Geotherm Res 91:199–220

Signorelli S, Vaggelli G, Romano C, Carroll MR (2001) Volatile element zonation in Campanian Ignimbrite magmas (Phlegrean Fields, Italy): evidence from the study of glass inclusions and matrix glasses. Contrib Mineral Petrol 140:543–553. https://doi.org/10.1007/s004100000213

Skulski T, Minarik W, Watson EB (1994) High-pressure experimental trace-element partitioning between clinopyroxene and basaltic melts. Chem Geol 117:127–147. https://doi.org/10.1016/0009-2541(94)90125-2

Smith VC, Isaia R, Pearce NJG (2011) Tephrostratigraphy and glass compositions of post-15 kyr Campi Flegrei eruptions: implications for eruption history and chronostratigraphic markers. Quat Sci Rev 30:3638–3660. https://doi.org/10.1016/j.quascirev.2011.07.012

Stock MJ, Humphreys MCS, Smith VC, Isaia R, Pyle DM (2016) Late-stage volatile saturation as a potential trigger for explosive volcanic eruptions. Nat Geosci 9:249–255. https://doi.org/10.1038/NGEO2639

Stormer JC, Nicholls J (1978) XLFrac: a program for interactive testing of magmatic differentiation models. Comput Geosci 4:143–159

Sun S-S, McDonough WF (1989) Chemical and isotopic systematics of oceanic basalts: implications for mantle compositions and processes. In: Saunders AD, Norry MJ (eds) Magmatism in the Ocean Basins, Geol Soc Lond Spec Publ 42:313–345

Tonarini S, Leeman WP, Civetta L, D'Antonio M, Ferrara G, Necco A (2004) B/Nb and $\delta^{11}B$ systematics in the Phlegrean volcanic district, Italy. J Volcanol Geotherm Res 133:123–139

Tonarini S, D'Antonio M, Di Vito MA, Orsi G, Carandente A (2009) Geochemical and B-Sr-Nd isotopic evidence for mingling and mixing processes in the magmatic system that fed the Astroni volcano (4.1-3.8 ka) within the Campi Flegrei caldera (Southern Italy). Lithos 107:135–151

Tupinambá M, Heilbron M, Valeriano C, Porto Júnior R, Blanco de Dios F, Machado N, do Eirado Silva LG, Horta de Almeida JC (2012) Juvenile contribution of the Neoproterozoic Rio Negro Magmatic Arc (Ribeira Belt, Brazile): implications for Western Gondwana amalgamation. Gondwana Res 21:422–438. https://doi.org/10.1016/j.gr.2011.05.012

Ujike O, Goodwin AM, Shibata T (2007) Geochemistry and origin of Archean volcanic rocks from the upper Keewatin assemblage (ca 2.7 Ga), lake of the woods Greenstone Belt, Western Wabigoon Subprovince, Superior Province, Canada. Island Arc 16:191–208

Veksler IV, Fedorchuk YM, Nielsen TFD (1998) Phase equilibria in the silica-undersaturated part of the $KAlSiO_4$-Mg_2SiO_4-Ca_2SiO_4-SiO_2-F system at 1 atm and the larnite-normative trend of melt evolution. Contrib Mineral Petrol 131:347–363

Vezzoli L (ed) (1988) Island of Ischia. C.N.R. Quaderni de La ricerca scientifica 114, vol 10, 134 pp

Villemant B (1988) Trace element evolution in the Phlegrean Fields (Central Italy): fractional crystallization enrichment. Contrib Mineral Petrol 98:169–183

Voloschina M, Pistolesi M, Bertagnini A, Métrich N, Pompilio M, Di Roberto A, Di Salvo S, Francalanci L, Isaia R, Cioni R, Romano C (2018) Magmatic reactivation of the Campi Flegrei volcanic system: insights from the Baia–Fondi di Baia eruption. Bull Volcanol 80:75. https://doi.org/10.1007/s00445-018-1247-8

Webster JD, Raia F, Tappen C, De Vivo B (2003) Pre-eruptive geochemistry of the ignimbrite-forming magmas of the Campanian volcanic zone, southern Italy, determined from silicate melt inclusions. Mineral Petrol 79:99–125. https://doi.org/10.1007/s00710-003-0004-6

Wohletz K, Orsi G, de Vita S (1995) Eruptive mechanisms of the Neapolitan Yellow Tuff interpreted from stratigraphic, chemical and granulometric data. J Volcanol Geotherm Res 67:263–290

Zollo A, Maercklin N, Vassallo M, Dello Iacono D, Virieux J, Gasparini P (2008) Seismic reflections reveal a massive melt layer feeding Campi Flegrei caldera. Geophys Res Lett 35:L12306. https://doi.org/10.1029/2008GL034242

Origin and Differentiation History of the Magmatic System Feeding the Campi Flegrei Volcanic Field (Italy) Constrained by Radiogenic and Stable Isotope Data

Massimo D'Antonio, Ilenia Arienzo,
Valeria Di Renzo, Lucia Civetta, Antonio Carandente,
and Sonia Tonarini

Abstract

A review of the recent geochemical investigations carried out on products of the Campi Flegrei volcanic field is presented, with particular attention on radiogenic (Sr, Nd, Pb) and stable (B, O) isotope data. The studied volcanic rocks span the entire period of best known Campi Flegrei volcanic field activity, since ~58 ka to the last event, the AD 1538 Monte Nuovo eruption. The isotope data, combined with mineralogical, geochemical, geochronological and volcanological data have allowed reconstructing a complex behaviour of the Campi Flegrei volcanic field magmatic feeding system. They suggest that several, isotopically-distinct magmatic components fed the activity over various time periods. In particular, over the past 15 kyrs at least three isotopically-distinct magmatic components, located either in a deep and large reservoir or in shallower and smaller chambers, have variably interacted, eventually feeding the volcanic activity. These magmatic components originated through a complex interplay of magma supply from a variably-enriched mantle source, and crustal contamination processes occurring at depth. Mingling/mixing among these isotopically-distinct magmatic components, as well as closed-system differentiation of variable hybrid magmas and interaction with cumulate mushes, were the dominant processes acting in the Campi Flegrei caldera feeding system over the past 15 kyrs. Three examples including the Agnano-Monte Spina, Astroni and Averno 2 eruptions of this period are described to illustrate the interplay among the above-mentioned different processes.

M. D'Antonio (✉) · V. Di Renzo
Dipartimento di Scienze della Terra, dell'Ambiente e delle Risorse, Università degli Studi di Napoli Federico II, Napoli, Italy
e-mail: masdanto@unina.it

L. Civetta
Emeritus Professor of Geochemistry and Volcanology, Università degli Studi di Napoli Federico II, Napoli, Italy

I. Arienzo · A. Carandente
Istituto Nazionale di Geofisica e Vulcanologia, Sezione Osservatorio Vesuviano, Napoli, Italy

V. Di Renzo
Dipartimento Provinciale di Napoli, Agenzia Regionale Protezione Ambientale Campania, Napoli, Italy

S. Tonarini
Via Chiaro 22, 23823 Colico, Lucca, Italy

1 Introduction

The products of the Campi Flegrei volcanic field (CFvf) activity have attracted the attention of scientists since the early work of Rittmann (1950), who first extensively studied their petrographic and geochemical characteristics. Since

© Springer-Verlag GmbH Germany, part of Springer Nature 2022

G. Orsi et al. (eds.), *Campi Flegrei*, Active Volcanoes of the World,
https://doi.org/10.1007/978-3-642-37060-1_4

then, much petrological work has been carried out on these rocks highlighting their peculiar petrographic, geochemical and isotopic features. In particular, isotopic compositions of key elements in a few CFvf rocks revealed strong variability since the pioneering work of Hurley et al. (1966) and Hoefs and Wedepohl (1968). This variability was later confirmed by studies carried out on more representative sets of samples, using modern mass spectrometry facilities (e.g., Civetta et al. 1991, 1997; D'Antonio et al. 1999b; de Vita et al. 1999; Pappalardo et al. 1999, 2002a, b; Tonarini et al. 2004, 2009; Arienzo et al. 2010, 2014, 2015, 2016; Di Renzo et al. 2011; Iovine et al. 2018; Di Salvo et al. 2020). Details on the petrographical and petrochemical features of the CFvf rocks, and their implications for the shallow-level differentiation processes acting in the CFvf magmatic system are provided in Chap. An Evolutionary Model for the Magmatic System of the Campi Flegrei Volcanic Field (Italy) Constrained by Petrochemical Data. Here, recently published radiogenic (Sr, Nd, and Pb) and stable (B, O) isotope data on CFvf rocks representative of the activity of the past ~58 kyrs are reviewed in order to highlight inferences on both the source region of this magmatism and the complex evolution processes undergone by magmas during their way up to the surface. Particular emphasis will be given to the variable, isotopically-identified magmatic components that have fed the CFvf magmatic system through time, and to the open-system evolution processes such as crustal contamination and mingling/mixing that CFvf magmas have experienced, especially over the past 15 kyrs (Di Renzo et al. 2011 and references therein).

2 Geochronological and Petrological Outlines

Campi Flegrei, together with Ischia and Procida constitute a unique volcanic district, named the Phlegraean Volcanic District (Orsi et al. 1996). The geological history of the CFvf is reported in Chap. Volcanic and Deformation History of the Campi Flegrei Volcanic Field, Italy. Therefore, only the most significant volcanic events are summarised here. The time of onset of volcanism in the CF area is unknown. Several pyroclastic units occur below the products of the first cataclysmic, caldera-forming Campanian Ignimbrite eruption (CI; ~40 ka; Giaccio et al. 2017 and references therein). The oldest dated volcanic unit exposed inside the CF has an age of ~58 ka (Pappalardo et al. 1999), although volcanic rocks older than 78 ka have been found in the city of Napoli (Scarpati et al. 2013). These, and older units identified in distal outcrops (Di Vito et al. 2008; Giaccio et al. 2012, 2017; Wulf et al. 2012; Albert et al. 2019; Petrosino et al. 2019; Chap. Tephrochronology and Geochemistry of Tephra from the Campi Flegrei Volcanic Field, Italy) are related to volcanism extending outside the limits of the Campi Flegrei caldera (CFc; Orsi et al. 1996; Chap. Volcanic and Deformation History of the Campi Flegrei Volcanic Field, Italy). Another caldera forming event, the Neapolitan Yellow Tuff eruption (NYT) dated at ~15 ka (Deino et al. 2004a, b) occurred inside the CI caldera. According to recent studies (Orsi et al. 1996; Scarpati et al. 2013; Vitale and Isaia 2014), three cycles of activity can be identified in the CF volcanic history since 78 ka, separated by the two major eruptions: the first cycle is from 78 to 40 ka, the second one spans from 40 to 15 ka and the last cycle is from 15 ka until today. At least 69 eruptions have occurred in the past 15 kyrs, grouped in three epochs of activity separated by quiescence periods: epoch 1, ~15 to ~10.6 ka; epoch 2, ~9.6 to ~9.1 ka, epoch 3, ~5.5 to ~3.5 ka (Di Vito et al. 1999; Orsi et al. 2004, 2013; Smith et al. 2011; Chaps. Volcanic and Deformation History of the Campi Flegrei Volcanic Field, Italy; Volcanic Hazard Assessment at the Campi Flegrei Caldera, Italy). The eruptions originated from vents mostly located either along the structural boundary of the NYT caldera or along some of the boundaries of the resurgent block developed inside the NYT caldera (Chap. Volcanic and Deformation History of the Campi Flegrei Volcanic Field, Italy). The last eruption occurred in AD 1538, after 3 kyrs of quiescence, and produced the

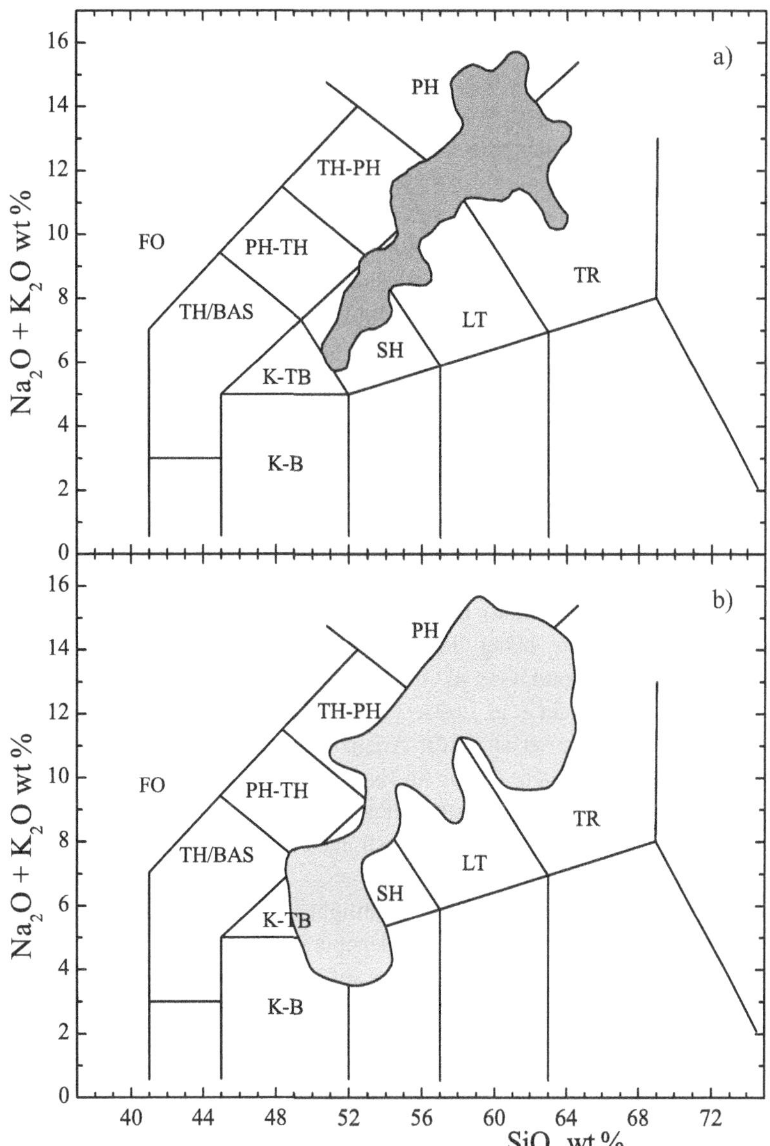

Fig. 1 **a** Total Alkali versus Silica classification grid (Le Maitre 2002) for representative Campi Flegrei volcanic field rocks from ~58 ka to the last eruption (dark grey field) (data from D'Antonio et al. 1999b; Pappalardo et al. 1999; Tonarini et al. 2009; Di Vito et al. 2011; Melluso et al. 2012; Forni et al. 2018b; Voloschina et al. 2018; Di Salvo et al. 2020). **b** Total alkali versus silica classification grid for Campi Flegrei volcanic field melt inclusions (light grey field) (data from Cecchetti et al. 2002–2003; Marianelli et al. 2006; Cannatelli et al. 2007; Cipriani et al. 2008; Mangiacapra et al. 2008; Arienzo et al. 2010, 2016; Mormone et al. 2011; Fourmentraux et al. 2012; Voloschina et al. 2018; Moretti et al. 2019)

monogenetic Monte Nuovo tuff cone (Di Vito et al. 2016; Liedl et al. 2019 and references therein). The CF volcanism is still active as demonstrated by fumarolic and seismic activities and by recurrent episodes of unrest in the past 70 years (Chaps. The Hydrothermal System of the Campi Flegrei Caldera, Italy; Historic Unrest of the Campi Flegrei Caldera, Italy and references therein).

The volcanic rocks older than CI are mostly phonolitic in composition (Fig. 1a) (Pappalardo et al. 1999; Wulf et al. 2004; Di Renzo et al. 2007; Pabst et al. 2008; Tomlinson et al. 2012; Belkin et al. 2016; Forni et al. 2018a; Petrosino et al. 2019). During the CI eruption, about 300 km^3 (Dense Rock Equivalent, DRE) (Chap. Volcanic and Deformation History of the Campi Flegrei Volcanic Field, Italy and references therein) of trachytic-to-phonolitic magma (Civetta et al. 1997; Pappalardo et al. 2002a; Bohrson et al. 2006; Arienzo et al. 2009, 2011; Gebauer et al. 2014; Forni et al. 2018a).

Magmas varying in composition from trachyte to (peralkaline) phonolite (Fig. 1a) fed the later volcanism until the NYT eruption (e.g., Pappalardo et al. 1999; Pabst et al. 2008; Tomlinson et al. 2012). The latter, the largest known trachytic phreatoplinian event, extruded at least 40 km^3 (DRE) of latitic-to-trachytic magma (Orsi et al 1992, 1995; Scarpati et al. 1993; Forni et al. 2018b). After the NYT eruption, volcanic activity continued within the younger caldera, and an ongoing resurgence has affected the caldera floor (Orsi et al. 1996; Chap. Volcanic and Deformation History of the Campi Flegrei Volcanic Field, Italy). The volcanic rocks of the post-NYT activity, as well as the entire compositional spectrum of CFvf rocks emplaced since more than 78 ka until the AD 1538 Monte Nuovo last eruption, range in composition from shoshonite to (peralkaline) phonolite, with trachyte and phonolite being the most abundant products (e.g., D'Antonio et al. 1999b, 2007; de Vita et al. 1999; Isaia et al. 2004, 2009; Di Vito et al. 2011; Melluso et al. 2012; Arienzo et al. 2016; Mollo et al. 2016; Chap. An Evolutionary Model for the Magmatic System of the Campi Flegrei Volcanic Field (Italy) Constrained by Petrochemical Data).

All the previously cited articles highlight that the CFvf products define a silica-undersaturated, potassic alkaline series, with degree of silica undersaturation increasing with Differentiation Index (DI = normative Or + Ab + Ne). The large amount of erupted differentiated magmas imply that large volumes of cumulates must have been left at depth (D'Antonio 2011 and references therein; Forni et al. 2018a, b; Di Salvo et al. 2020). The products are enriched in K and related trace elements (e.g., Cs, Rb, Ba, Sr) with respect to high-field strength elements (Ta, Nb, Hf, Zr), and in light with respect to heavy rare earth elements (Chap. An Evolutionary Model for the Magmatic System of the Campi Flegrei Volcanic Field (Italy) Constrained by Petrochemical Data and references therein). The least-differentiated products (from Minopoli 2 and Pigna San Nicola eruptions), emplaced during epoch 1 of the post-NYT activity along portions of northeast-southwest oriented regional fault systems bordering the NYT caldera (Capuano et al. 2013; Chap. Volcanic and Deformation History of the Campi Flegrei Volcanic Field, Italy), contain crustal xenoliths classified as andesite (Pappalardo et al. 2002b). No rocks representative of very primitive magmas have ever been found among the CFvf products; however, mafic rocks possibly representative of mantle-derived magmas that might have generated all more evolved CFvf magmas through fractional crystallisation processes occur on the nearby Procida island (K-basalts and K-trachybasalts of Solchiaro eruption; D'Antonio et al. 1999a, 2007; De Astis et al. 2004; Fedele et al. 2006; Avanzinelli et al. 2008).

3 Radiogenic and Stable Isotope Data

Campi Flegrei volcanic field magmas are characterised by relatively variable $^{87}Sr/^{86}Sr$ (0.7068–0.7086) and $\delta^{11}B$ (-6.8 to -10.6 ‰), and rather limited ranges of $^{143}Nd/^{144}Nd$ (0.51236–0.51252), $^{206}Pb/^{204}Pb$ (18.85–19.25) and $\delta^{18}O$ (+7 to +9 ‰) (Civetta et al. 1997; D'Antonio et al. 1999b, 2007; Pappalardo et al. 1999, 2002a, 2002b; Tonarini et al. 2004, 2009; Di Renzo et al. 2007, 2011; Pabst et al. 2008; Arienzo et al. 2009, 2010, 2015, 2016; Di Vito et al. 2011; Iovine et al. 2018; Forni et al. 2018b; Di Salvo et al. 2020). These variations are highlighted by isotope-isotope diagrams (Fig. 2). $^{87}Sr/^{86}Sr$ and $^{143}Nd/^{144}Nd$ ratios of CF samples of various ages exhibit an evident anti-correlation, with the exception of some pre-CI samples (Fig. 2a). Interestingly, the highest $^{87}Sr/^{86}Sr$ (>0.7080) and lowest $^{143}Nd/^{144}Nd$ (<0.5124) are shown by the least evolved volcanic rocks, i.e., the shoshonites of both Minopoli 1 and 2 eruptions ($\sim$12 ka) occurred during epoch 1 of the post-NYT activity.

A good anti-correlation occurs also between $^{87}Sr/^{86}Sr$ and $^{206}Pb/^{204}Pb$ ratios, although a limited set of data is available (Fig. 2b). More complex is the anti-correlation between $^{87}Sr/^{86}Sr$ and $\delta^{11}B$. In order to better clarify its meaning, Sr and B isotope data for representative volcanic rocks from the volcanic fields of Procida and

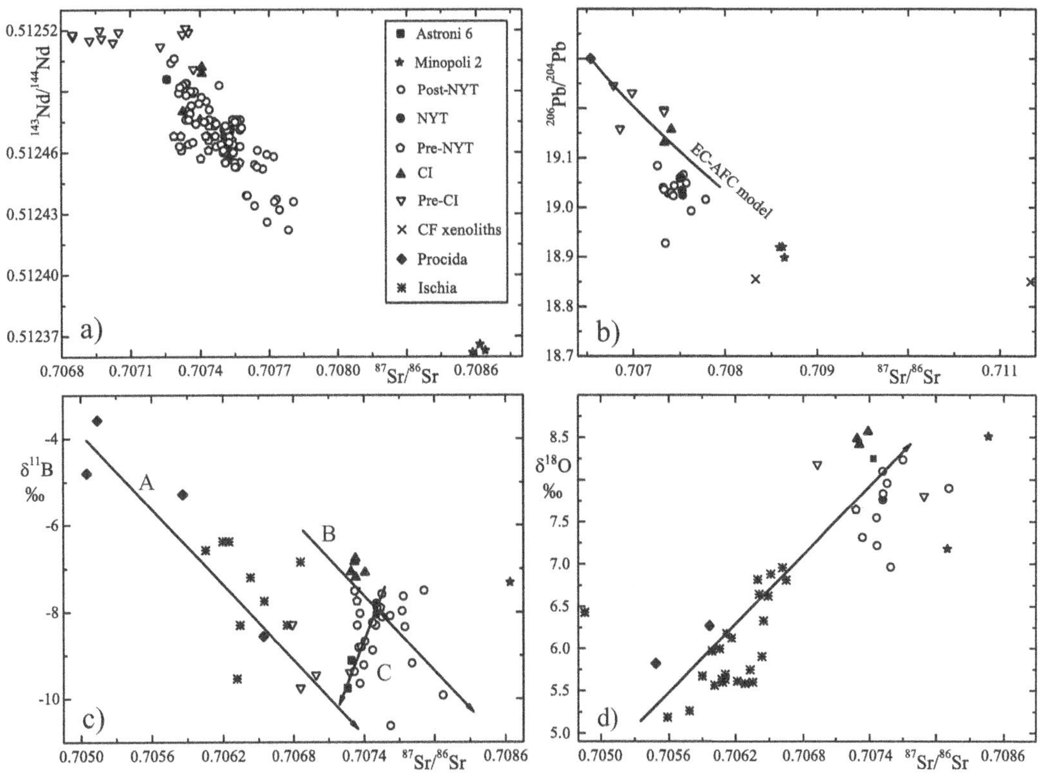

Fig. 2 **a** $^{87}Sr/^{86}Sr$ versus $^{143}Nd/^{144}Nd$; **b** $^{206}Pb/^{204}Pb$ versus $^{87}Sr/^{86}Sr$; **c** $\delta^{11}B$ ‰ versus $^{87}Sr/^{86}Sr$; **d** $\delta^{18}O$ ‰ versus $^{87}Sr/^{86}Sr$ diagrams for Campi Flegrei volcanic field rocks (data from Tonarini et al. 2004, 2009; D'Antonio et al. 2007; Pabst et al. 2008; Arienzo et al. 2009, 2010, 2016; Di Renzo et al. 2007, 2011; Di Vito et al. 2011; Iovine et al. 2018). The Sr, Nd, Pb and B isotope ratios reported in this review were measured with a ThermoFinnigan Triton TI mass spectrometer (see Avanzinelli et al. 2005; Arienzo et al. 2013 for details on the analytical conditions) and with a ThermoFinnigan MAT 262 mass spectrometer (Tonarini et al. 2004 and references therein). The O isotope data were determined on separate minerals with a ThermoFinnigan MAT 253 mass spectrometer (Iovine et al. 2018 and references therein). The EC-AFC curve, and trends A, B, C are described in the text. Figure modified after Di Renzo et al. (2011)

Ischia islands have been plotted together with those from CFvf (Fig. 2c; D'Antonio et al. 2007). Two distinct, roughly parallel, anti-correlation trends are evident in the Sr-B isotopic diagram: one trend includes Procida, Ischia and some pre-CI CFvf rocks (trend A), and another one includes all the other CFvf rocks (trend B). Moreover, trends A and B are crossed by a third trend, named C, that includes mostly volcanic rocks from the Astroni activity (epoch 3 of post-NYT activity; Isaia et al. 2004; Tonarini et al. 2009; Arienzo et al. 2015, 2016). The Sr-O isotopic diagram (Fig. 2d) is less clear, and only a positive correlation between $^{87}Sr/^{86}Sr$ and $\delta^{18}O$ is evident. It must be pointed out that this diagram has been constructed using Sr and O isotope data on minerals, and only one sample from the Astroni products is available. A possible interpretation of these trends is discussed in the next section.

4 Inferences from Isotope Data on the Origin and Evolution of the Campi Flegrei Volcanic Field Magmas

The large variability of both geochemical and isotopic tracers of the CFvf rocks is not easy to interpret. The difficulty arises from the high degree of differentiation of most of these rocks (trachyte to phonolite); only few are sufficiently poorly evolved, although not primitive (shoshonite to latite) (Fig. 1a), to provide more direct information on the source characteristics. Fractional crystallisation is the dominant evolution process responsible for the major element and most of the trace-element geochemical variations throughout the entire series (Chap. An Evolutionary Model for the Magmatic System of the Campi Flegrei Volcanic Field (Italy) Constrained by Petrochemical Data and references therein). However, the radiogenic (Sr, Nd, Pb) and stable (B, O) isotopic variations and the mineralogical disequilibria require open-system processes that must have been active in the mantle sources and/or in the CFvf magmatic system through time. According to the interpretation of petrological and isotope data of the CFvf products, together with those from Ischia and Procida islands, at least part of their geochemically enriched signature is inferred to be inherited from mantle sources. These sources have been variably enriched in K and related incompatible trace elements, radiogenic Sr and Pb, ^{11}B and ^{18}O by fluids and melts released by subduction processes (e.g., Rogers et al. 1985; Di Girolamo 1987; Serri 1990; Beccaluva et al. 1991; D'Antonio et al. 1996, 1999a, 2007, 2013; Tonarini et al. 2004, 2009; Peccerillo 2005, 2017; Avanzinelli et al. 2008; Mazzeo et al. 2014; Iovine et al. 2018). Moreover, crustal contamination and/or mingling/mixing processes among distinct magmas and crystal mushes have been recently invoked to explain part of the isotopic variability of the CFvf products (e.g., Civetta et al. 1991, 1997; D'Antonio et al. 1999b, 2007; de Vita et al. 1999; Pappalardo et al. 2002b; Tonarini et al. 2004; 2009; Piochi et al. 2005; De Campos et al. 2008; Arienzo et al. 2009, 2011, 2015, 2016; Perugini et al. 2010, 2015; Di Renzo et al. 2011; Di Vito et al. 2011; Forni et al. 2018a, b; Di Salvo et al. 2020). The latter open-system processes were superimposed on the mantle-inherited signature of magmas.

A careful investigation of trace element contents and Sr, Nd, Pb, O isotopic compositions of selected volcanic rocks allowed us to distinguish two major processes in the genesis and differentiation of the CFvf magmas (Tonarini et al. 2004; D'Antonio et al. 2007; Mazzeo et al. 2014; Iovine et al. 2018). The first process should have occurred in a transitional-MORB-type asthenospheric mantle portion and consisted of source contamination by subducting slab-derived fluids and sediments (or melts thereof). The observed enrichment in incompatible trace elements concurs with this hypothesis (Peccerillo 2017 and references therein; Chap. An Evolutionary Model for the Magmatic System of the Campi Flegrei Volcanic Field (Italy) Constrained by Petrochemical Data and references therein). The variable enrichment of this mantle portion can be inferred from trend A of Fig. 2c, d. As demonstrated by high-T-P experiments (Perinelli et al. 2019), partial melting of such a source originated K-basaltic primary magmas, found at CFc only as poorly evolved melt inclusions (MIs) trapped in Mg-rich olivines and clinopyroxenes (Fig. 1b; see caption for references). Such mafic MIs were also found in mafic minerals from the nearby Ischia and Procida islands (Moretti et al. 2013; Esposito et al. 2011, 2018). Primary K-basaltic magmas, according to the isotopic features of the least differentiated rocks of the Phlegraean Volcanic District, must have been characterised by Sr, Nd and Pb isotopic values of ~ 0.705, ~ 0.5127 and ~ 19.0, respectively (D'Antonio et al. 2007, 2013; Di Renzo et al. 2011; Mazzeo et al. 2014), with $\delta^{11}B$ and $\delta^{18}O$ values of -5 and +5, respectively (Tonarini et al. 2004; D'Antonio et al. 2007, 2013; Iovine 2017b, 2018).

The second process occurred at crustal depth, where K-basaltic magmas, during their ascent and storage, differentiated to K-trachybasalt and then to shoshonite through combined fractional crystallisation and $\sim 2\%$ assimilation (AFC) of continental crust (D'Antonio et al. 2007; Di Renzo et al. 2011; Iovine et al. 2018). The AFC process is responsible for further changes in Sr, Nd, Pb and B isotopic compositions of magmas, visible in trend B (Fig. 2c). In the Sr-O isotopic diagram (Fig. 2d), an AFC process could be envisaged in the near-vertical array of CFvf data points, as suggested by Iovine et al. (2018). This process accounts for the geochemical

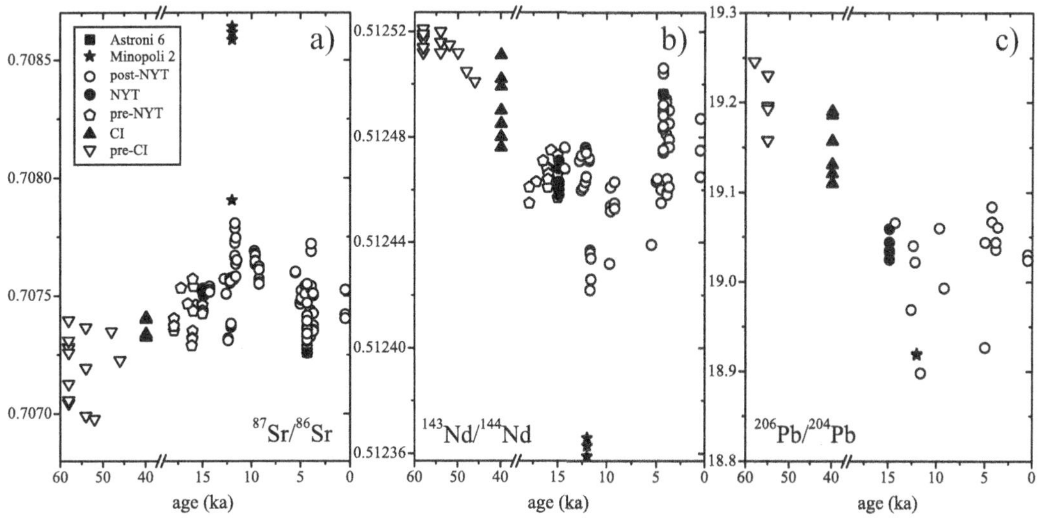

Fig. 3 **a** $^{87}Sr/^{86}Sr$, **b** $^{143}Nd/^{144}Nd$, and **c** $^{206}Pb/^{204}Pb$ ratios through time of rocks emplaced at the Campi Flegrei volcanic field over the past ~58 kyrs. Data sources as for Fig. 2. Figure modified after Di Renzo et al. (2011)

and most of the isotopic features of the CFc rocks younger than 40 ka. It is also supported by the occurrence of millimetre- to centimetre-sized crustal xenoliths (with $^{87}Sr/^{86}Sr$ ratio up to ~0.7113) in the Minopoli 2 and Pigna San Nicola shoshonites (Fig. 2a). It must be stressed that analyses of whole-rock and glass from Minopoli 2 products show highly radiogenic values: the whole-rock $^{87}Sr/^{86}Sr$ ranges from 0.7079 to 0.7086, whereas the glass clusters at 0.7086 (Di Renzo et al. 2011). These xenoliths, andesitic in composition and chemically similar to the crystalline rocks of the Calabrian continental crust, have been interpreted as products of partial melting of the metamorphic basement that likely occurs in the deep crust underlying CF (Pappalardo et al. 2002b and references therein; Fedi et al. 2018).

The interplay of the two hypothesised processes during the evolution of the CFvf magmatic system is highlighted by the $^{87}Sr/^{86}Sr$, $^{143}Nd/^{144}Nd$ and $^{206}Pb/^{204}Pb$ ratios vs. age of magma eruption (Fig. 3). Increasing Sr- (and $\delta^{18}O$, not shown) and decreasing Nd and Pb isotopic ratios in the ~58–10 ka time interval suggest that contamination with the crystalline metamorphic basement, characterised by high $^{87}Sr/^{86}Sr$ and $\delta^{18}O$, and low $^{143}Nd/^{144}Nd$ and $^{206}Pb/^{204}Pb$ ratios, could have become progressively more important with time. This process was probably coupled with a decreasing magma supply from the mantle (e.g., Hammersley and De Paolo 2006). Conversely, after ~12 ka (Minopoli 2 eruption) the isotopic trends with time appear inverted, as the Sr isotopic ratio generally decreases, whereas Nd and Pb isotopic ratios increase (Fig. 3). During the last epoch of the post-NYT activity, the erupted magmas were less enriched in radiogenic Sr, and more enriched in radiogenic Nd and Pb, suggesting a decreasing role of crustal contamination in their evolution, so that isotopic features closer to those of the mantle source are evident in the rocks. Part of the isotopic variations through time have been also attributed to mingling/mixing processes involving magmas with variable geochemical and isotopic features. Such processes occurred very often in the CF magmatic system (Di Renzo et al. 2011; Arienzo et al. 2016; see next section).

5 Magmatic Components

It is evident that, from ~58 to ~12 ka, the Nd and Pb isotopic compositions of the CF rocks were progressively decreasing ($^{143}Nd/^{144}Nd$ varying from 0.51252 to 0.51236, and

$^{206}Pb/^{204}Pb$ varying from 19.2 to 18.9), whereas the Sr isotopic composition became more radiogenic (from 0.7068 to 0.7086). Over the past 12 kyrs instead, the isotopic trends have changed (Fig. 3). This was interpreted as due to isotopically distinct magmatic components feeding the CF magmatic system in the past ~58 kyrs (Di Renzo et al. 2011).

5.1 Pre-CI and CI Components

The products emplaced during the 58–40 kyrs time span are characterised by variable $^{87}Sr/^{86}Sr$, displaying the lowest values ever detected at CF (0.7068), whereas the $^{143}Nd/^{144}Nd$ and $^{206}Pb/^{204}Pb$ values are the highest (~0.51252 and ~19.2, respectively) (Fig. 3; see caption for references). This activity was fed by a trachytic to phonolitic magmatic component that was then progressively replaced by another component, more enriched in radiogenic Sr and "unradiogenic" Nd and Pb, akin to the one that fed the CI eruption (D'Antonio et al. 2007, and references therein). Not any post-CI erupted magma has ever had Sr, Nd and Pb isotopic ratios similar to those of the CFvf products older than ~55 ka, meaning that the first identified, pre-CI component was exhausted after the CI eruption.

According to Pabst et al. (2008), Arienzo et al. (2009, 2011), Di Renzo et al. (2011), and Tomlinson et al. (2012), the CI system included two isotopically distinct components. The 1st component (CI 1st), compositionally similar to the pre-CI erupted magmas, was phonolitic in composition and characterised by values of $^{87}Sr/^{86}Sr$ ~0.7074, $^{143}Nd/^{144}Nd$ ~0.51250, $^{206}Pb/^{204}Pb$ 19.2, $\delta^{11}B$ -7.9‰ and $\delta^{18}O$ +8.6‰ (Iovine et al. 2018). This component was erupted for the last time during the CI eruption as the most-radiogenic, most-differentiated, earlier-erupted magma. It has been considered to be the resident magma stored at shallow depth, resulting from contamination of a pre-CI magma batch by shallow sediments (Arienzo et al. 2009). The 2nd component (CI 2nd), trachytic in composition, has slightly lower $^{87}Sr/^{86}Sr$ (~0.7073), $^{143}Nd/^{144}Nd$ (~0.51248) and $\delta^{18}O$ (+8.4‰) values, with respect to the CI 1st component. It represents a new magma that entered the CI reservoir during the course of the eruption (Civetta et al. 1997; De Campos et al. 2008; Pabst et al. 2008; Arienzo et al. 2009). Combining the results of the isotopic investigations on CI magmatic end-members with geophysical and geological data, the following scenario was proposed: (1) a parental, poorly differentiated magma rose into the middle crust, and evolved through combined fractional crystallisation and crustal assimilation processes; (2) the differentiated magma rose to shallower depth, fed the pre-CI activity and evolved by further open-system processes into the most-evolved and most-radiogenic CI 1st component; (3) new magma (CI 2nd component), isotopically distinct from both the CI 1st component and pre-CI magmas, recharged the most-evolved pre-CI magma chamber, and formed the large and stratified CI magmatic system. During the course of the eruption, the two layers were tapped separately and/or simultaneously and gave rise to the range of chemical and isotopic values displayed by the CI pumice, glass and minerals (Arienzo et al. 2009, 2011). Alternative models have been proposed by Borhson et al. (2006) and Fowler et al. (2007), who pointed out the important role of the crystal fractionation process in generating the geochemical variability of the CI magmas, using MELTS and EC-AFC modelling. More recently Forni et al. (2018a) and Di Salvo et al. (2020) suggested the occurrence of re-melting of a crystal mush to explain the geochemical variability of the CI magmas. However, isotope data allow discriminating among distinct magma batches.

5.2 Pre-NYT/NYT Component

The pre-NYT/NYT component (D'Antonio et al. 2007; Pabst et al. 2008; Tomlinson et al. 2012) arrived into the shallow Campi Flegrei caldera magmatic system a few thousand years before the NYT eruption, as testified by the isotopic composition of pre-NYT products younger than 16–18 ka (Pappalardo et al. 1999), changing toward that of NYT products with time (Fig. 3). The latter range in composition from earlier-erupted trachyte to later-erupted latite, with Nd, Pb, B and O isotopic compositions fairly homogeneous ($^{143}Nd/^{144}Nd = 0.51246$, $^{206}Pb/^{204}Pb = 19.04$, and $\delta^{11}B = -7.9$ ‰; $\delta^{18}O = +7.8$‰), and $^{87}Sr/^{86}Sr$ slightly variable from 0.70753 to 0.70750 during the course of the eruption (Di Renzo et al. 2011, and references therein; Iovine et al. 2018). The Sr and Nd isotopic values, and the geochemical data of the pre-NYT/NYT rocks have been interpreted by Pabst et al. (2008) as the result of a process of coalescence of pre-NYT magma batches, triggered by the arrival at shallow depths of a latitic magma slightly less enriched in radiogenic Sr ($^{87}Sr/^{86}Sr = 0.70750$) (Orsi et al. 1995). This coalescence process did not cause isotopic disequilibria in the NYT magma, as testified by the relative homogeneity of Sr isotope data of separated feldspars (Orsi et al. 1995), as well as dark and light glass fragments (Di Renzo et al. 2011). On the basis of geochemical and mineralogical data, Forni et al. (2018b) suggested that the main process was mingling/mixing among a new arrived magma and two crystal mushes at shallow depth. In this case the isotope data, being homogeneous, do not allow to choose between these two hypotheses.

5.3 Post-NYT Components

According to isotopic and geochemical data, and thermal modelling results, residuals of the NYT magma are hypothesised to have fed the Campi Flegrei caldera magmatic system over the past 15 kyrs (D'Antonio et al. 1999b, 2007; Di Renzo et al. 2011; Arienzo et al. 2016; Iovine et al. 2018). However, the post-NYT rocks display a large range of isotopic composition ($^{87}Sr/^{86}Sr$ varies from 0.70726 to 0.70864, $^{143}Nd/^{144}Nd$ from 0.51236 to 0.51250, $^{206}Pb/^{204}Pb$ from 18.92 to 19.07, $\delta^{11}B$ from -10.6 to -7.3‰ and $\delta^{18}O$ from +7.11 to +8.8‰) requiring the involvement of further magmatic components. Isotopic disequilibrium between glass and minerals, and isotopic variability through time are the dominant compositional features of most of the studied post-NYT volcanoes/eruptions (e.g., Tonarini et al. 2004, 2009; Morgan et al. 2004; D'Antonio et al. 2007; Di Renzo et al. 2011; Di Vito et al. 2011; Voloschina et al. 2018). In particular, the most evident mineralogical, geochemical and isotopic disequilibria occur in shoshonites and latites, i.e., the least evolved CF products, emplaced during epochs 1 and 2 of the post-NYT activity. This was taken as an evidence for arrival of a new magmatic component into the CFc system at ~12 ka, named Minopoli 2, shoshonitic in composition and isotopically well distinct from the other components (Tonarini et al. 2004, 2009; Morgan et al. 2004; D'Antonio et al. 2007; Di Renzo et al. 2011).

A further new component, named Astroni 6, trachytic in composition, less enriched in radiogenic Sr and unradiogenic Nd and Pb, with a strongly negative $\delta^{11}B$ and positive $\delta^{18}O$, arrived into the shallow feeding system at ~4 ka. These isotopic features, mostly the strongly negative $\delta^{11}B$, allow its distinction from the CI 2nd magmatic component. The Astroni 6 component was first erupted during the course of the Astroni activity (Tonarini et al. 2009), and then mixed with other magmatic components prior to or during some of the most recent eruptions (Di Renzo et al. 2011), as described below. This scenario is suggested also by the negative correlation between the Sr and Nd isotopic ratios of whole-rock and glass samples representative of the NYT and post-NYT magmas (Fig. 2a), that form a unique trend between Astroni 6 trachyte and Minopoli 2 shoshonite, as well as by the negative and positive correlation displayed by the Sr versus Pb and B isotopic ratios, respectively (Fig. 2b, c).

The identification of distinct magmatic end-members that have fed the CFc system over the past 15 kyrs has been improved by investigating the frequency distribution of the isotope data of the post-NYT magmas (Fig. 4) (Di Renzo et al. 2011). The results display wide ranges of Sr, Nd, Pb and B isotopic ratios of all the post-NYT products, highlighting the two previously described magmas with extreme isotopic compositions: Minopoli 2 ($^{87}Sr/^{86}Sr = 0.70860$, $^{143}Nd/^{144}Nd = 0.51236$, $^{206}Pb/^{204}Pb = 18.92$) and Astroni 6 ($^{87}Sr/^{86}Sr = 0.70726$, $^{143}Nd/^{144}Nd = 0.51250$,

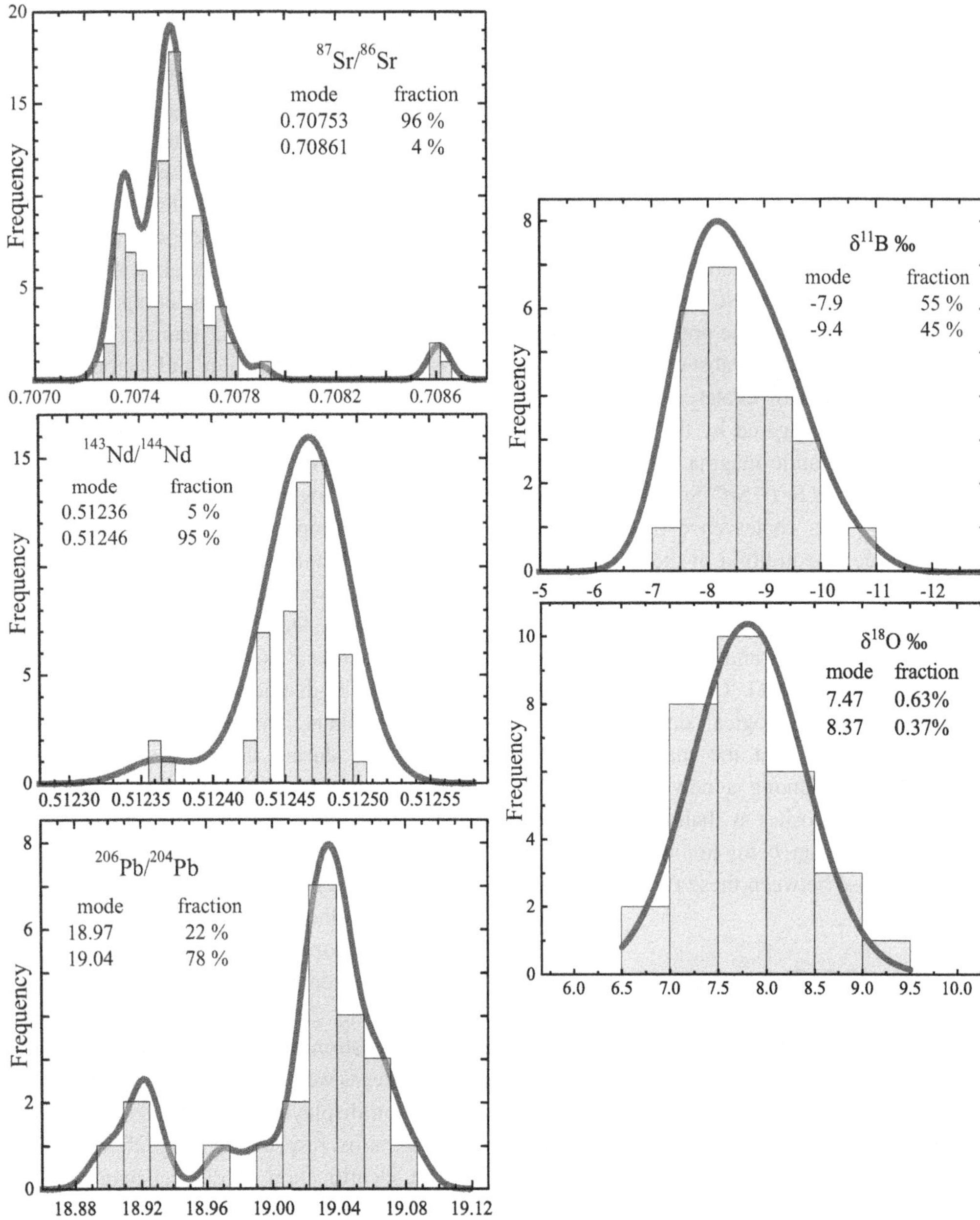

Fig. 4 Frequency histograms of $^{87}Sr/^{86}Sr$, $^{143}Nd/^{144}Nd$, $^{206}Pb/^{204}Pb$, $\delta^{11}B$ and $\delta^{18}O$ data for the post-NYT Campi Flegrei caldera volcanism. Gaussian curves, and the identified modes and their fractions are reported. Sr, Nd, Pb and B data are from Di Renzo et al. (2011), whereas O data are from Iovine et al. (2018). Figure modified after Di Renzo et al. (2011)

$^{206}Pb/^{204}Pb$ = 19.08). The main mode, recurrent in all isotope ratio histograms but that of $\delta^{18}O$ ($^{87}Sr/^{86}Sr$ = 0.70753, $^{143}Nd/^{144}Nd$ = 0.51246, $^{206}Pb/^{204}Pb$ = 19.04, $\delta^{11}B$ = -8‰), corresponds to the NYT magma isotopic values.

In summary, the isotope data highlight the involvement of at least three isotopically distinct magmatic components in the CFc reservoir feeding the volcanism of the past 15 kyrs. One component, similar to the NYT magma and trachytic in composition, was likely a residue left in the CFc system after the NYT eruption (D'Antonio et al. 1999b; Di Renzo et al. 2011). The second component, Minopoli 2, shoshonitic in composition, was the most enriched in radiogenic Sr, $\delta^{18}O$ and unradiogenic Nd of the whole CF volcanic field history, characterised by a $\delta^{11}B$ value similar to that of the NYT: thus, it was generated most likely through crustal contamination of mafic magmas prior to 12 ka. The third one, Astroni 6, trachytic in composition, was slightly less enriched in radiogenic Sr and unradiogenic Nd and characterised by a more negative $\delta^{11}B$ value than those of the NYT and Minopoli 2 components. Its isotopic features can be attributed to a less significant role of crustal contamination, and a greater role of the source in magma production than those required for the Minopoli 2 and, to a lesser extent, for the NYT magmatic components.

The $\delta^{11}B$ differences among Astroni 6, NYT and Minopoli 2 components should derive mostly from source heterogeneity, probably produced by addition of variable amounts of radiogenic Sr- and Pb-enriched, ^{11}B-depleted and ^{18}O-enriched slab materials (fluids and/or melts) to the mantle wedge, and subordinately by crustal contamination. We emphasise that the least evolved magmas, i.e., Minopoli 2, display a large variation in $\delta^{18}O$ (+7.2 to +9.2; Iovine et al. 2018) and $^{87}Sr/^{86}Sr$ (0.7067–0.7095; Morgan et al. 2004; D. Morgan, pers. comm. 2014) values. Various hypotheses have been proposed invoking source heterogeneity and/or crustal contamination during magma ascent. Indeed, interaction of CFc magmas with crustal material, particularly with the lower crust, characterised by very low B content ($\sim$5 ppm; Tonarini et al. 2004, 2009 and references therein), appears to be unable to modify significantly the boron concentration and its isotopic composition in the CFc erupted magmas, characterised by more than 30 ppm even in the least evolved samples.

Thus, a complex interplay of magma supply from a variably enriched mantle source and limited crustal contamination generated the identified, isotopically distinct magmatic components. Mingling/mixing among these components, as well as closed-system differentiation of various hybrid magmas and likely remelting of crystal mushes, were the dominant processes acting in the CFc feeding system over the past 15 kyrs. This is well displayed by Sr and Nd isotope versus incompatible trace element contents relationships (Fig. 5). The CFc volcanic activity after the NYT eruption until $\sim$9.1 ka, end of epoch 2, was fed by magma batches resulting from mixing between NYT and Minopoli 2 magmatic components (curves a and b in Fig. 5), and closed-system differentiation processes (horizontal dashed lines in Fig. 5) of hybrid magmas (Di Renzo et al. 2011). However, the number of identified magmatic components must be considered as a minimum estimate as it cannot be ruled out that further investigations will allow recognition of more components.

6 Current Knowledge on the Architecture of the Magmatic Feeding System of the Campi Flegrei Caldera

Volcanological, petrological, geochemical and geophysical data recently collected for the CF volcanic field have been used to elaborate a model for its magmatic feeding system. The CF area is filled down to 3 km depth by a sequence of volcanic and non-volcanic sediments; temperature profiles measured in geothermal wells located within the caldera floor give maximum values of about 350 °C at about 2.5 km depth (AGIP 1987). Seismic reflection studies have shown that between 3 and 4 km depth there are thermo-metamorphic rocks bearing water and/or gas that overlie the carbonate and/or crystalline basement (Judenherc and Zollo 2004; Zollo et al. 2008). The nature of the basement occurring in the $\sim$4–8 km depth range is still debated (see D'Antonio 2011 for a review). Velocity tomography studies provide clear evidence for the occurrence of a 1-km-thick, low-velocity layer, at about 8 km depth, interpreted as a partial melting zone (Zollo et al. 2008). Passive high-resolution attenuation tomography (De Siena et al. 2010)

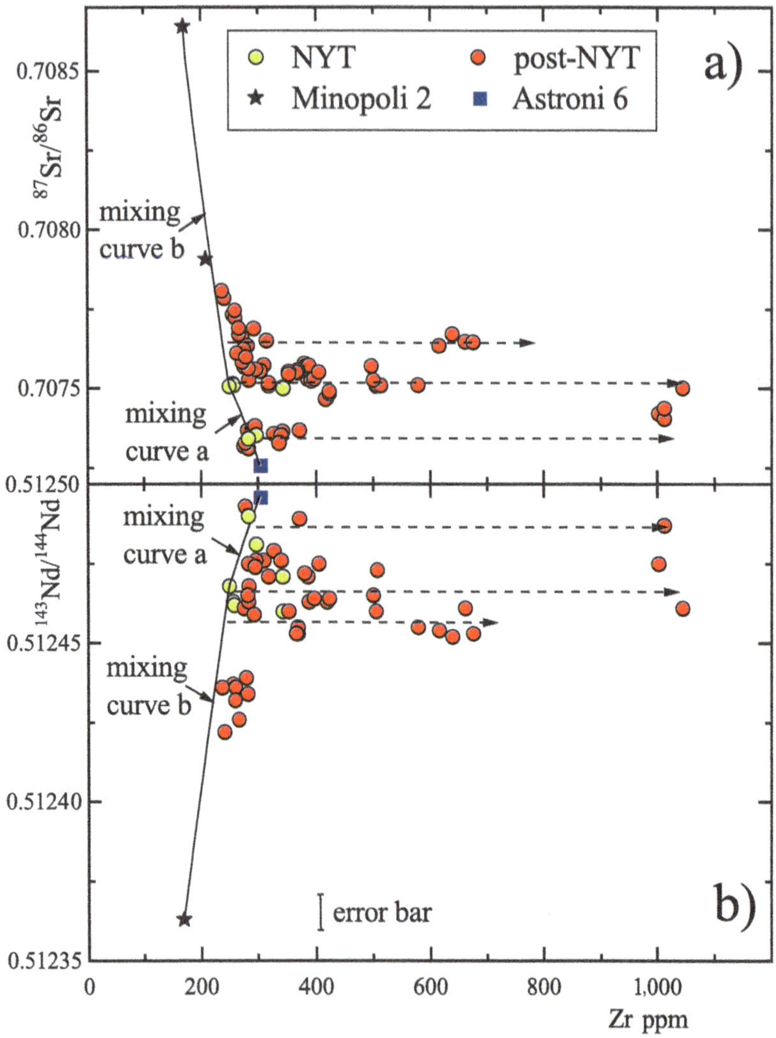

Fig. 5 **a** $^{87}Sr/^{86}Sr$ versus Zr and **b** $^{143}Nd/^{144}Nd$ versus Zr for NYT and post-NYT volcanic rocks. Mixing curve a was calculated assuming NYT and Astroni 6 components as end-members; mixing curve b was calculated assuming Minopoli 2 and NYT components as end-members; dashed lines = possible fractional crystallisation (FC) trends. Mixing curves are calculated according to Langmuir et al. (1978). $^{87}Sr/^{86}Sr$ uncertainty is within symbols. Figure modified after Di Renzo et al. (2011)

and coda-wave attenuation (Akande et al. 2019) imaging highlights two high-attenuation zones located at 2–4 km depth one beneath Solfatara and the other beneath Monte Nuovo. Akande et al. (2019) unveil the differences in the feeding systems of Solfatara, where hot fluids find pathways to the surface, and Monte Nuovo, where these fluids are stored below a thicker, possibly less fractured caprock. Moreover, De Siena et al. (2010) evidenced a possible small volume of magma at > 3 km depth beneath Pozzuoli during the 1982–1984 bradyseismic crisis.

Studies of melt inclusions (MIs) in minerals suggest stagnation and crystallisation depths between ~2 and 12 km for the NYT and post-NYT erupted magmas (Fig. 6). The largest depth values are in good agreement with the location of a partially molten zone hypothesised by means of seismic tomography (Zollo et al. 2008) and geobarometry (Astbury et al. 2018). However, more recent results of experimental petrology (Roach 2005; Fanara et al. 2015) and geophysical and thermal modelling (e.g., Bagagli et al. 2017; Amoruso et al. 2017; Chaps. Seismic and Gravity Structure of the Campi Flegrei Caldera, Italy; Magma Chamber Dynamics at the Campi Flegrei Caldera, Italy; Source Modelling from Ground Deformation and Gravity Changes at the Campi Flegrei Caldera, Italy), agree with the occurrence in the past of storage regions at variable depth and a feeding zone deeper than 8 km (Arienzo et al. 2010, 2016; Moretti et al. 2019).

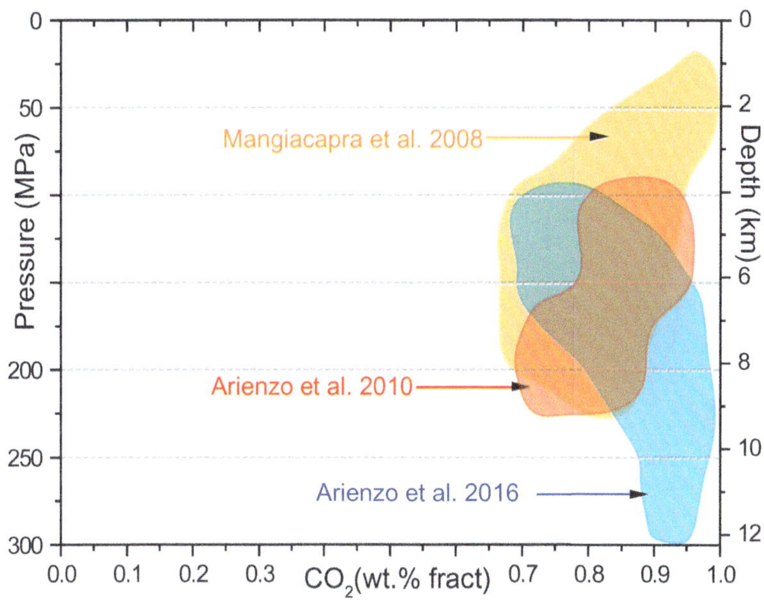

Fig. 6 Pressure estimates based on melt inclusion volatiles (CO_2 and H_2O) composition detected in post-NYT volcanic rocks (data from: Mangiacapra et al. 2008; Arienzo et al. 2010, 2016)

The identified magmatic components, hypothetically localised at variable depth in the CFc magmatic feeding system, likely arrived many times at shallow depth from the deeper reservoir. Minopoli 2 component fed the system mostly during epochs 1 and 2, whereas Astroni 6 during epoch 3. These magmatic components were erupted in different sectors of the CFc (Fig. 7). During epochs 1 and 2, magmas similar to the NYT component, and those resulting from mixing between Minopoli 2 and NYT components were erupted from vents located mostly on the marginal faults of the NYT caldera. As the latter are parts of regional tectonic structures (Chap. Volcanic and Deformation History of the Campi Flegrei Volcanic Field, Italy), it is likely that they acted as preferred path for mafic magmas rising from the deeper reservoir (~8 km depth). During epoch 3, magmas either similar to NYT, or resulting from mixing between Astroni 6 and NYT components were erupted from vents located along faults bordering the La Starza resurgent block and, subordinately, the NYT caldera. Moreover, magmas resulting from mixing between Minopoli 2 and NYT components were erupted from vents located along the northeast-southwest regional fault system activated during caldera resurgence (Di Renzo et al. 2011).

In summary, the architecture of the CFc magmatic system, during the three epochs of the post-NYT volcanic activity, consists of: (i) a deep magmatic reservoir, whose top is located at ~8 km depth, where mantle-derived magmas arrive, stagnate and differentiate to shoshonite-trachyte, and (ii) shallow magma chambers, formed at different depths (between 8 and 2 km depth) and in different sectors of the crust underlying the CFc. In these magma chambers, mingling/mixing among distinct components (volatile-rich magmas of deep provenance and crystal-rich mushes left by previous eruptions), differentiation, and crustal contamination processes occur.

7 Mingling/Mixing Processes in the Past 5 Kyrs: Agnano-Monte Spina, Astroni and Averno 2 Eruptions Case-Studies

In this section, three eruptions of the past 5 kyrs are presented as case-studies. For each of these eruptions a great amount of Sr and Nd isotopic

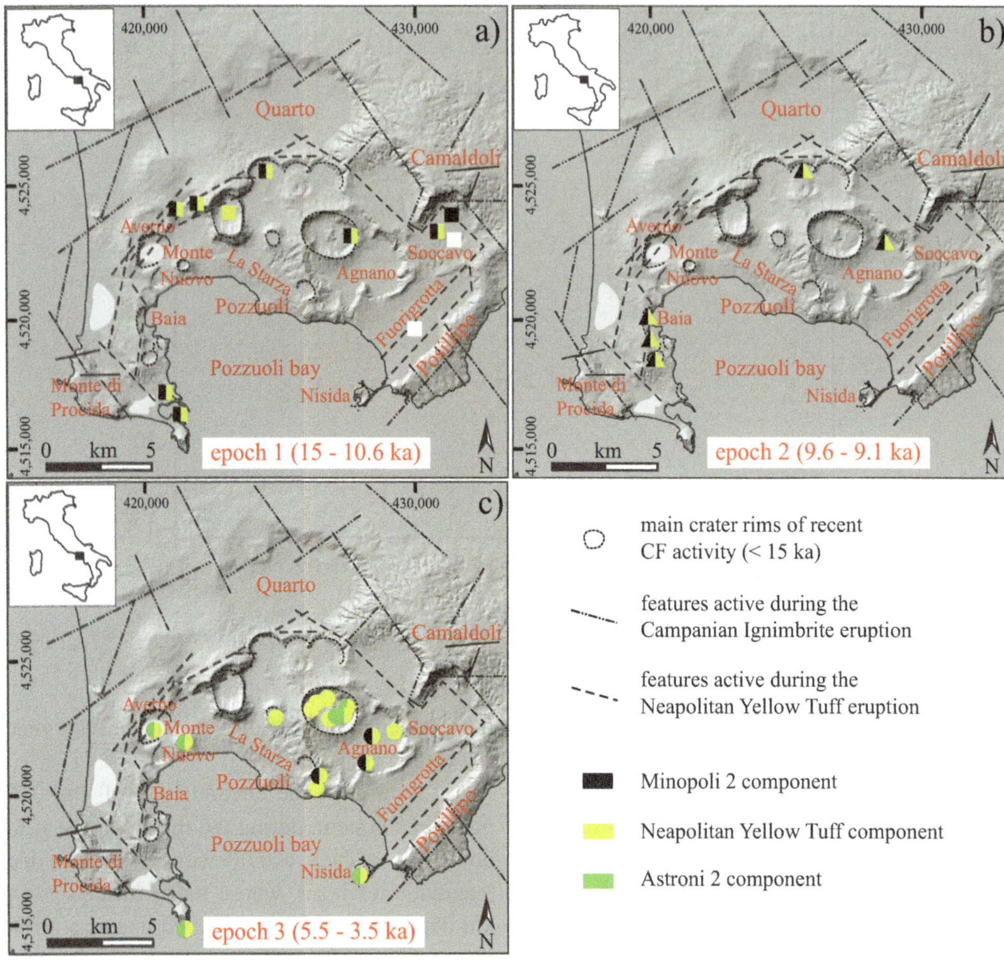

Fig. 7 Location of vent position of the post-NYT eruptions in the structural map of Campi Flegrei caldera. **a** epoch 1; **b** epoch 2; **c** epoch 3 and Monte Nuovo eruption. Different colours identify the dominant isotopically distinct magmatic components, or in case of mixing, the two components involved. The white squares refer to Santa Teresa and Soccavo 4 eruption vents, for which magmatic components were not identified. Figure modified after Di Renzo et al. (2011)

data along with chemical parameters for rock samples arranged in stratigraphic order, i.e., chemostratigraphy (Figs. 8, 9 and 10) can be found in the literature. The chemostratigraphy well illustrates the occurrence of mingling/mixing processes during the course of each volcanic event. For the majority of the presented sequences very few, in some cases only one, Pb, O and B isotope data are available; therefore, they are not either discussed or reported in the chemostratigraphy.

7.1 Agnano-Monte Spina

The Agnano-Monte Spina eruption (A-MS: 4,482–4,625 cal. years BP; Chap. Volcanic and Deformation History of the Campi Flegrei Volcanic Field, Italy and references therein) occurred in epoch 3 of activity in the eastern sector of the NYT caldera (Fig. 7c). The eruption was accompanied by a volcano-tectonic collapse and emplaced a complex pyroclastic sequence, subdivided into six Members (A through F), each

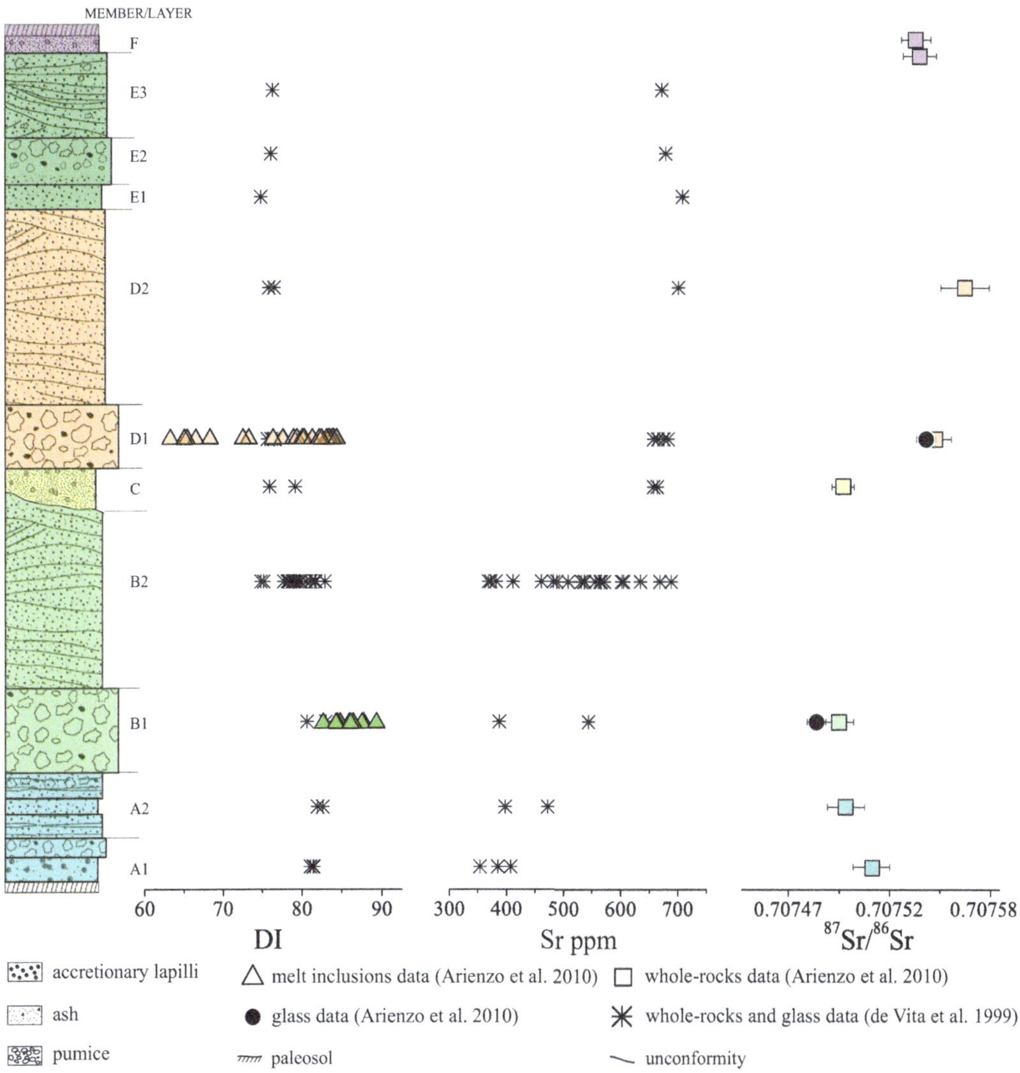

Fig. 8 Chemostratigraphy of the Agnano-Monte Spina pyroclastic sequence. Figure modified after de Vita et al. (1999) and Arienzo et al. (2010)

representing the deposit of one eruption phase. The sequence is dominated by surge beds generated by phreatomagmatic explosions, with only two major fallout deposits. The latter, named Members B and D, are the products of magmatic phases that generated Plinian columns reaching heights between 20 and 30 km. The rocks of the entire sequence range in composition from trachyte to phonolite. The whole-rock DI generally decreases from bottom to top of the sequence (Fig. 8). All the variation diagrams describe a single compositional trend; moreover, the total range of variation for several trace elements is large in spite of the limited variation of major oxide contents (de Vita et al. 1999). The $^{87}Sr/^{86}Sr$ ratios range from 0.70748 near the base of the sequence to as high as 0.70756 at the top, outside the analytical uncertainty. The $^{143}Nd/^{144}Nd$ ratio of whole-rocks and glass is homogenous within the errors (~0.51246; Arienzo et al. 2010). According to the quoted authors, two magmatic components that mixed prior to and/or during the eruption can be distinguished (Fig. 8). One is the earlier-erupted,

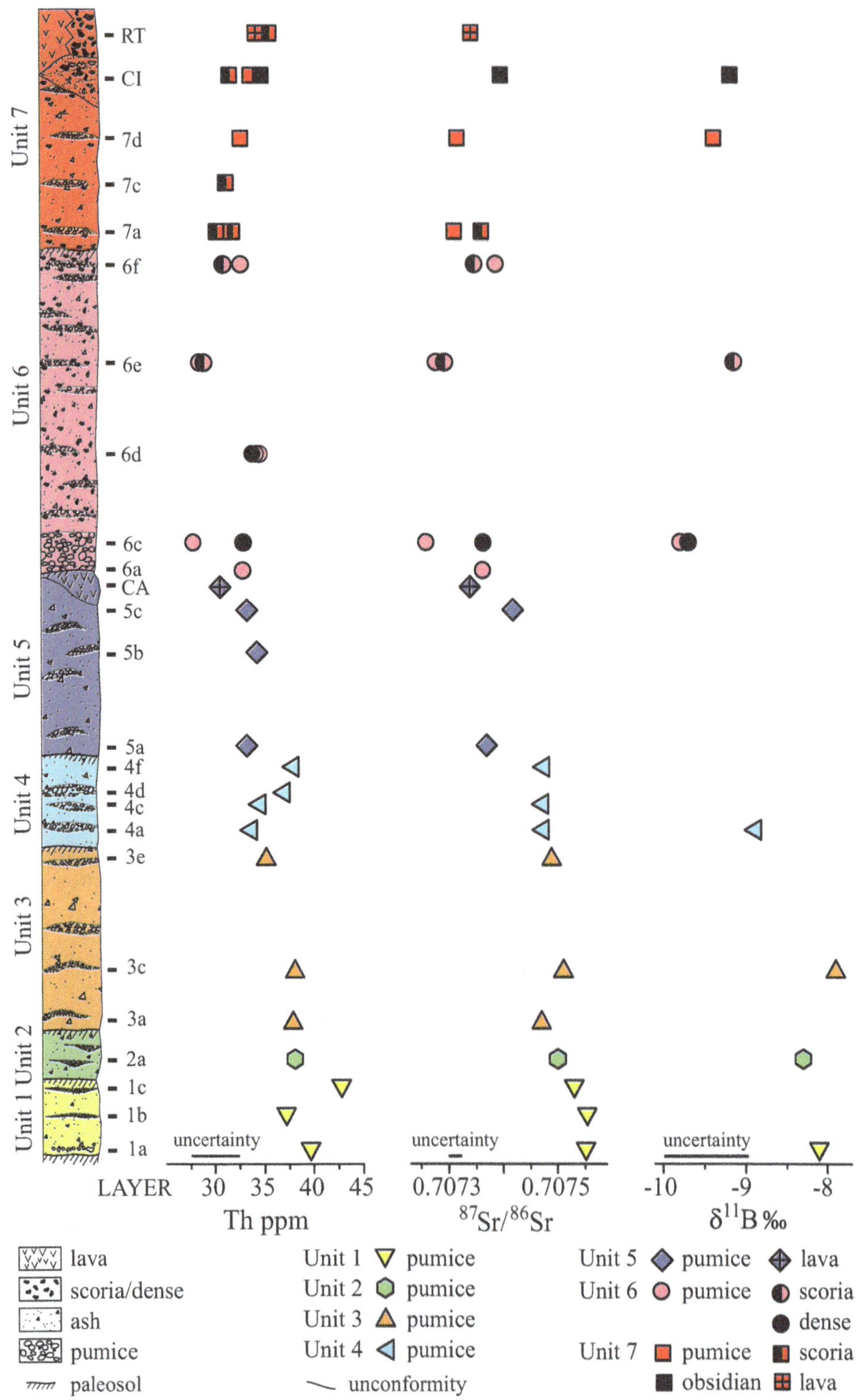

Fig. 9 Chemostratigraphy of the Astroni pyroclastic sequence. Symbols in legend refer to different types of clasts except those for lava, which refer to lava domes. Figure modified after Tonarini et al. (2009)

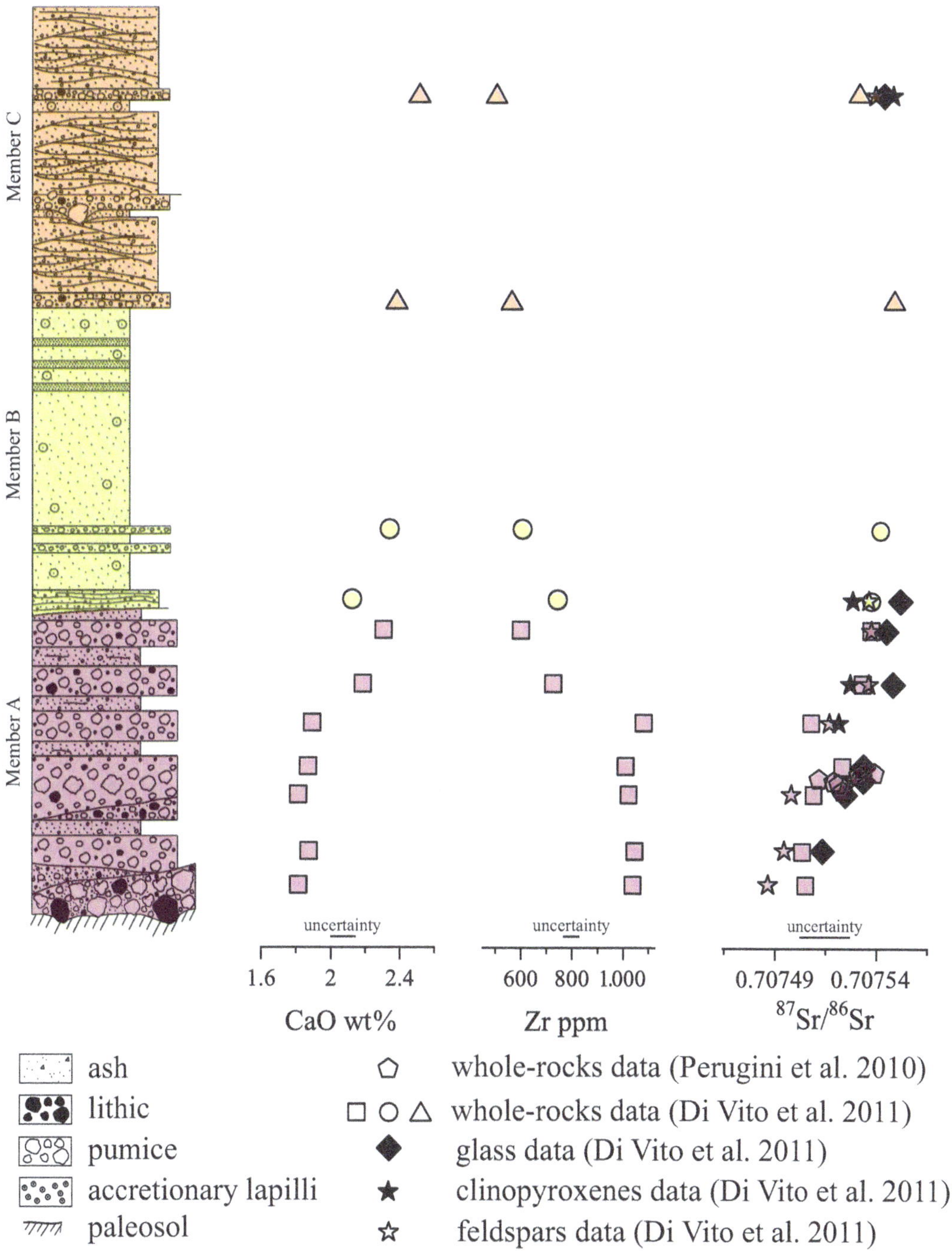

Fig. 10 Chemostratigraphy of the Averno 2 pyroclastic sequence. Figure modified after Di Vito et al. (2011)

less radiogenic and slightly more differentiated magma ($^{87}Sr/^{86}Sr$ ~0.70748), while the other is the later-erupted, more radiogenic and less differentiated magma ($^{87}Sr/^{86}Sr$ ~0.70756). The earlier component, largely dominant in the system, represents a residing phonolitic melt, isotopically and compositionally similar to the NYT magma, whereas the later component has some features similar to those of the Minopoli 2 shoshonite. These results are constrained by the strong mineral-melt isotopic disequilibria (de Vita et al. 1999; Arienzo et al. 2010; S. Tonarini, unpublished data). According to geobarometric estimates (Arienzo et al. 2010), the magma feeding the A-MS eruption was stored in a vertically extended plumbing system, which was continuously subject to addition of CO_2-dominated volatiles. This allowed the magmatic system to switch progressively from relatively H_2O-rich background conditions of the system to relatively dehydrated and CO_2 enriched ones, typical of a gas-buffered system that becomes nearly insensitive to gas–melt chemical exchanges. Under these gas-rich conditions, more primitive magmas rose from depths ≥ 8 km and mixed with the evolved phonolite that fed the B phase of the eruption (Fig. 8). Mingling/mixing among different components could explain the variability found in crystals and melt/groundmass in the course of the eruption (Pelullo et al. 2020, 2021). Diffusion times based on BaO contents and grey-scale swath profiles modelling on sanidine phenocrysts yielded timescales <60 years between mixing and eruption (Iovine et al. 2017a).

7.2 Astroni

The Astroni activity was built through at least seven eruptions named Astroni 1 through 7 (Astroni 1 dated at 4,153–4,345 cal. years BP; Astroni 6 dated at 4,098–4,297 cal. years BP, Smith et al. 2011; Chaps. Volcanic and Deformation History of the Campi Flegrei Volcanic Field, Italy; Volcanic Hazard Assessment at the Campi Flegrei Caldera, Italy and references therein) and following each other at short time intervals within the same vent area (Fig. 7c; Isaia et al. 2004; Orsi et al. 2004). The rock composition ranges from trachyte to phonolite. The Sr and B isotopic compositions of representative volcanic rocks from the seven Astroni eruptions display a good positive correlation (trend C in Fig. 2c), interpreted as a result of both mingling and mixing processes (Tonarini et al. 2009). In particular, the variation in $^{87}Sr/^{86}Sr$ ratio from 0.70728 to 0.70755 is significantly greater than the analytical uncertainty (Fig. 9), whereas the variation in $^{143}Nd/^{144}Nd$ ratio from 0.512506 to 0.512474 is only slightly greater than the error. The ranges of both Sr and Nd isotope ratios of the Astroni rocks partially overlap with those of the CFc volcanic rocks younger than 40 ka ($^{87}Sr/^{86}Sr$ from 0.70729 to 0.70864; $^{143}Nd/^{144}Nd$ from 0.51240 to 0.51254). Also, the Astroni $\delta^{11}B$ measured values from -7.9 to -9.8‰ overlap with those of the CFc volcanic rocks younger than 40 ka (from -6.8 to -10.6‰). Interestingly, such variations in the isotopic compositions are well correlated with the degree of chemical evolution of the Astroni rocks (Fig. 9), allowing the reconstruction of the open-system magmatic evolution superimposed over fractional crystallisation processes. Mineralogical disequilibria, whole-rock B, Nd, Sr isotopic variations versus the degree of chemical evolution and through time (Fig. 9), as well as the significant Sr isotopic disequilibria between phenocrysts and glass, can be accounted for by mingling/mixing processes. The geochemical and isotopic evidence suggests that mingling/mixing involved at least two main magmatic end-members: one more evolved (phonolite), and with higher radiogenic Sr and $\delta^{11}B$ contents, the other less evolved (tephriphonolite) and with lower radiogenic Sr and more negative $\delta^{11}B$.

The earlier-erupted more evolved end-member (phonolite) is similar to the NYT, and shows strong geochemical and isotopic similarities with the magmas that fed the final phases of the A-MS eruption, occurred a few hundred years earlier in the same vent area. The less evolved end-member is likely a mantle-derived magma that evolved into a tephriphonolite through dominant fractional crystallisation. This end-member represents the previously described Astroni 6 new magmatic component, that arrived at shallow depth at $\sim$4.2 ka, invaded the plumbing system and mingled/mixed with the resident magma (Tonarini et al. 2009). Timescale of the mixing process has been estimated on the order of few hours to days before the eruption by using high-resolution maps of compatible trace

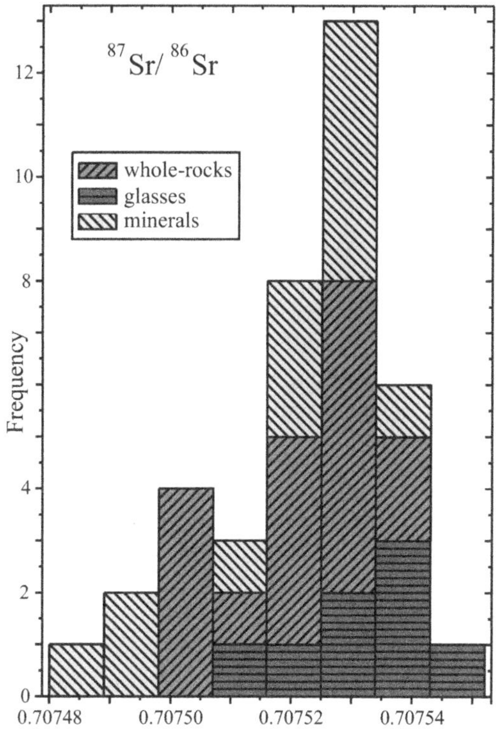

Fig. 11 Frequency histogram of $^{87}Sr/^{86}Sr$ measured in whole-rocks, glasses and minerals (feldspar and clinopyroxene) of representative Averno 2 tephra samples. Figure modified after Di Vito et al. (2011)

elements in K-feldspar (Ba, Sr) and clinopyroxenes (Cr, Ni) (Astbury et al. 2018).

7.3 Averno 2

The Averno 2 eruption (3,895–4,205 cal. years BP; calibration carried out using OxCal 4.4, IntCal09 curve on the ^{14}C date originally determined by Alessio et al. 1971) occurred in the western sector of NYT caldera (Fig. 7c). The eruption was a complex explosive event, including mostly magmatic explosions in the earliest part of the event, followed by phreatomagmatic explosions (Di Vito et al. 2011). Therefore, it generated a pyroclastic sequence dominated by fallout beds in its lower portion, surmounted by surge deposits. It was fed by a small volume of magma (~0.07 km^3 DRE) extruded through a vent that migrated along a 2-km-long fissure, part of a northeast-southwest fault system frequently re-activated in the last 15 kyrs (Di Vito et al. 2011). The erupted magma included two compositionally distinct, alkali-trachytic magma batches. $^{87}Sr/^{86}Sr$ of whole-rock and separated glass, feldspars and clinopyroxenes vary from ~0.70749 to ~0.70754 from base upward of the pyroclastic sequence, with the Sr isotopic composition of the feldspars slightly less radiogenic than the glass in the lower portion of the sequence (Fig. 10). A frequency histogram of $^{87}Sr/^{86}Sr$ isotope ratios (Fig. 11) shows two modes. One mode ($^{87}Sr/^{86}Sr$ = 0.707507) mostly includes the earlier-erupted, more differentiated and less radiogenic magma from the lower portion of the stratigraphic sequence. The second mode ($^{87}Sr/^{86}Sr$ = 0.707531) is defined by the later-erupted, less differentiated and more radiogenic magma from the upper portion of the stratigraphic sequence. Conversely, $^{143}Nd/^{144}Nd$ ratios are similar within the analytical error, around an average value of ~0.51246. The frequency histograms of selected trace element contents also well evidence two extreme modes that represent the earlier- and the later-erupted, isotopically distinct magma compositions (Di Vito et al. 2011). The isotopic similarity between the earlier-erupted Averno 2 magma and the later-erupted NYT magma has provided new evidence in favour of the hypothesis that residues of the magma feeding the NYT eruption, the last catastrophic event of the CFc, may still take part in the dynamics of the magmatic system, as they can be intercepted either by new magma batches of deeper provenance or as re-melting of crystal mush. Recently, Di Renzo et al. (2016) have performed a 3D conductive thermal modelling of an 8 km^3 residual NYT magma resident at 4 km depth, showing that today, after 15 kyrs, its temperature would be 700 °C. Therefore, it is still eruptible, especially if it mixes with fluids and/or hotter magma of deeper provenience, e.g., Minopoli 2 or Astroni 6. This result can be used as a proof that also the younger components Minopoli 2 and Astroni 6 can erupt, especially if intercepted by deeper, hot magmas. Indeed, the last CFc event, Monte Nuovo, has been probably fed by two magmas that mingled/mixed during the eruption (Fig. 7c): one is a residue of the

NYT component and the other is probably the Astroni 6 component (Di Renzo et al. 2011; Di Vito et al. 2011).

8 Conclusive Remarks

The peculiar isotopic features of the volcanic rocks emplaced over the past ~58 kyrs bear witness for the complex behaviour through time of the Campi Flegrei volcanic field magmatic feeding system. Many distinct magmatic components have fed the volcanism in different time spans. In the past 15 kyrs, the best studied period of activity, at least three distinct magmatic components have fed the eruptions and are still eruptible: a residue of the NYT (well recognisable in the products of Averno 2, Nisida and Monte Nuovo eruptions), Minopoli 2 (well recognisable in Agnano-Monte Spina and Averno 2 products) and Astroni 6. The latter appeared not before ~4.2 ka. Even though no CFc eruptions without any evidence of mingling/mixing have been found to date, the possibility that a fresh magmatic component might rise from depth to the surface and feed an eruption without stagnation and interaction with residues of previous eruptions cannot be ruled out at present.

The results of the isotopic studies carried out on the volcanic rocks, integrated with volcanological, geochronological e petrological data illustrated in this chapter and others (Chaps. Volcanic and Deformation History of the Campi Flegrei Volcanic Field, Italy; An Evolutionary Model for the Magmatic System of the Campi Flegrei Volcanic Field (Italy) Constrained by Petrochemical Data; Tephrochronology and Geochemistry of Tephra from the Campi Flegrei Volcanic Field, Italy; Rheological Properties of the Magmas Feeding the Campi Flegrei Caldera (Italy) and Their Influence on Mixing Processes; Magma Chamber Dynamics at the Campi Flegrei Caldera, Italy) provide important constraints to the definition of the volcanic hazard at the CFc (Chap. Volcanic Hazard Assessment at the Campi Flegrei Caldera, Italy). Either a fresh magma batch coming from the deep reservoir or a magma produced by reactivation of a cumulate mush flushed by deep volatiles, should be preceded and accompanied by significant volatile discharge upwards when approaching the surface. This, and possible magma movement, might be recognised by the efficient monitoring system operating at Campi Flegrei (see Chap. The Permanent Monitoring System of the Campi Flegrei Caldera, Italy). The short timescales estimated for the pre-eruptive magmatic processes of the past eruption contribute to the definition of the present-day volcanic hazard at the Campi Flegrei caldera as significantly high.

Acknowledgments The authors wish to thank M. Coltorti and D. Morgan for their fruitful comments and suggestions.

References

AGIP (1987) Modello geotermico del sistema flegreo - Sintesi. Servizi Centrali per l'Esplorazione, SERG-MESG, San Donato Milanese, Italy, 23 pp

Akande WG, De Siena L, Gan Q (2019) Three-dimensional kernel-based coda attenuation imaging of caldera structures controlling the 1982–84 Campi Flegrei unrest. J Volcanol Geotherm Res 381:273–283

Albert PG, Giaccio B, Isaia R, Costa A, Niespolo EM, Nomade S, Pereira A, Renne PR, Hinchliffe A, Mark DF, Brown RJ, Smith VC (2019) Evidence for a large-magnitude eruption from Campi Flegrei caldera (Italy) at 29 ka. Geology 47:595–599. https://doi.org/10.1130/G45805.1

Alessio M, Bella F, Improta S, Belluomini G, Cortesi C, Turi B (1971) University of Rome carbon-14 dates IX. Radiocarbon 13(2):395–411

Amoruso A, Crescentini L, D'Antonio M, Acocella V (2017) Thermally-assisted magma emplacement explains restless calderas. Sci Rep 7:7948. https://doi.org/10.1038/s41598-017-08638-y

Arienzo I, Civetta L, Heumann A, Wörner G, Orsi G (2009) Isotopic evidence for open system processes within the Campanian Ignimbrite magma chamber. Bull Volcanol 71:285–300

Arienzo I, Moretti R, Civetta L, Orsi G, Papale P (2010) The feeding system of Agnano-Monte Spina eruption (Campi Flegrei, Italy): dragging the past into the present activity and future scenarios. Chem Geol 270 (1–4):135–147

Arienzo I, Heumann A, Wörner G, Civetta L, Orsi G (2011) Time scales of magma evolution prior to the Campanian Ignimbrite eruption (Campi Flegrei, Italy). Earth Planet Sci Lett 306:217–228. https://doi.org/10.1016/j.epsl.2011.04.002

Arienzo I, Carandente A, Di Renzo V, Belviso P, Civetta L, D'Antonio M, Orsi G (2013) Sr and Nd isotope analysis at the Radiogenic Isotope Laboratory of the Istituto Nazionale di Geofisica e Vulcanologia, Sezione di Napoli - Osservatorio Vesuviano. Rapporti Tecnici INGV 260:1–18. ISSN 2039-7941

Arienzo I, Di Renzo V, Minolfi G, Mazzeo F, Carandente A, Belviso P, D'Antonio M, Orsi G, Civetta L (2014) Sr isotope microanalysis at the Radiogenic Isotope Laboratory of the Istituto Nazionale di Geofisica e Vulcanologia, Sezione di Napoli—Osservatorio Vesuviano. Rapporti Tecnici INGV 271:1–17. ISSN 2039-7941

Arienzo I, D'Antonio M, Di Renzo V, Minolfi G, Tonarini S, Orsi G, Carandente A, Belviso P, Civetta L (2015) Isotopic microanalysis sheds light on the magmatic endmembers feeding volcanic eruptions: the Astroni 6 case study (Campi Flegrei, Italy). J Volcanol Geotherm Res 304:24–37. https://doi.org/10.1016/j.jvolgeores.2015.08.003

Arienzo I, Mazzeo FC, Moretti R, Cavallo A, D'Antonio M (2016) Open-system magma evolution and fluid transfer at Campi Flegrei caldera (Southern Italy) during the past 5 ka as revealed by geochemical and isotopic data: the example of the Nisida eruption. Chem Geol 427:109–124. https://doi.org/10.1016/j.chemgeo.2016.02.007

Astbury RL, Petrelli M, Ubide T, Stock MJ, Arienzo I, D'Antonio M, Perugini D (2018) Tracking plumbing system dynamics at the Campi Flegrei caldera, Italy: high-resolution trace element mapping of the Astroni crystal cargo. Lithos 318–319:464–477. https://doi.org/10.1016/j.lithos.2018.08.033

Avanzinelli R, Boari E, Conticelli S, Francalanci L, Guarnieri L, Perini G, Petrone CM, Tommasini S, Ulivi M (2005) Highprecision Sr, Nd, and Pb isotopic analyses using the new generation thermal ionisation mass spectrometer ThermoFinniganTriton-Ti®. Per Mineral 74(3):147–166

Avanzinelli R, Elliott T, Tommasini S, Conticelli S (2008) Constraints on the genesis of potassium-rich Italian volcanic rocks from U/Th disequilibrium. J Petrol 49(2):195–223

Bagagli M, Montagna CP, Papale P, Longo A (2017) Signature of magmatic processes in strainmeter records at Campi Flegrei (Italy). Geophys Res Lett 44:718–725. https://doi.org/10.1002/2016GL071875

Beccaluva L, Di Girolamo P, Serri G (1991) Petrogenesis and tectonic setting of the Roman Volcanic Province, Italy. Lithos 26:191–221

Belkin HE, Rolandi G, Jackson JC, Cannatelli C, Doherty AL, Petrosino P, De Vivo B (2016) Mineralogy and geochemistry of the older (>40 ka) ignimbrites on the Campanian Plain, Southern Italy. J Volcanol Geotherm Res 323:1–18

Bohrson WA, Spera FJ, Fowler SJ, Belkin HE, De Vivo B, Rolandi G (2006) Petrogenesis of the Campanian Ignimbrite: implications for crystal-melt separation and open-system processes from major and trace elements and Th isotopic data. In: De Vivo B (ed) Volcanism in the Campania plain: Vesuvius, Campi Flegrei and Ignimbrites. Developments in volcanology 9. Elsevier, Amsterdam, The Netherlands, pp 249–288

Cannatelli C, Lima A, Bodnar RJ, De Vivo B, Webster JD, Fedele L (2007) Geochemistry of melt inclusions from the Fondo Riccio and Minopoli1 eruptions at Campi Flegrei (Italy). Chem Geol 237(3–4):418–432

Capuano P, Russo G, Civetta L, Orsi G, D'Antonio M, Moretti R (2013) The Campi Flegrei caldera structure imaged by 3-D inversion of gravity data. Geochem Geophys Geosyst 14:4681–4697. https://doi.org/10.1002/ggge.20276

Cecchetti A, Marianelli P, Sbrana A (2002–2003) L'eruzione di Astroni (caldera dei Campi Flegrei): dati preliminari dallo studio di inclusioni silicatiche. Atti S Tosc Sci Nat, Mem, Ser A 108:59–63

Cipriani F, Marianelli P, Sbrana A (2008) Studio di una sequenza piroclastica del vulcano della Solfatara (Campi Flegrei). Considerazioni vulcanologiche e sul sistema di alimentazione. Atti S Tosc Sci Nat, Mem, Ser A 113:39–48

Civetta L, Carluccio E, Innocenti F, Sbrana A, Taddeucci G (1991) Magma chamber evolution under the Phlegraean Field during the last 10 ka: trace element and isotope data. Eur J Mineral 3:415–428

Civetta L, Orsi G, Pappalardo L, Fisher RV, Heiken G, Ort M (1997) Geochemical zoning, mingling, eruptive dynamics and depositional processes—the Campanian ignimbrite, Campi Flegrei Caldera, Italy. J Volcanol Geotherm Res 75:183–219

D'Antonio M (2011) Lithology of the basement underlying the Campi Flegrei caldera: volcanological and petrological constraints. J Volcanol Geotherm Res 200:91–98. https://doi.org/10.1016/j.jvolgeores.2010.12.006

D'Antonio M, Tilton GR, Civetta L (1996) Petrogenesis of Italian alkaline lavas deduced from Pb–Sr–Nd isotope relationships. In: Basu A, Hart S (eds) Earth processes: reading the isotopic code. AGU Monograph Series 95. Washington, DC, USA, pp 253–267

D'Antonio M, Civetta L, Di Girolamo P (1999a) Mantle source heterogeneity in the Campanian region (South Italy) as inferred from geochemical and isotopic features of mafic volcanic rocks with shoshonitic affinity. Mineral Petrol 67:163–192

D'Antonio M, Civetta L, Orsi G, Pappalardo L, Piochi M, Carandente A, de Vita S, Di Vito MA, Isaia R (1999b) The present state of the magmatic system of the Campi Flegrei caldera based on a reconstruction of its behavior in the past 12 ka. J Volcanol Geotherm Res 91:247–268

D'Antonio M, Tonarini S, Arienzo I, Civetta L, Di Renzo V (2007) Components and processes in the magma genesis of the Phlegrean Volcanic District, (Southern Italy). In: Beccaluva L, Bianchini G, Wilson M (eds), Cenozoic volcanism in the Mediterranean Area. Boulder, CO, USA, Geol Soc Am Sp Pap 418, pp 203–220

D'Antonio M, Tonarini S, Arienzo I, Civetta L, Orsi G, Andria M, Dallai L, Trecalli A (2013) Mantle and crustal processes in the magmatism of the Campania region: inferences from mineralogy, geochemistry, and Sr-Nd-O isotopes of young hybrid volcanics of the Ischia island (south Italy). Contrib Mineral Petrol 165(6):1173–1194. https://doi.org/10.1007/s00410-013-0853-x

De Astis G, Pappalardo L, Piochi M (2004) Procida volcanic history: new insights into the evolution of the Phlegraean volcanic district (Campania region, Italy). Bull Volcanol 66:622–641

De Campos CP, Dingwell DB, Perugini D, Civetta L, Fehr TK (2008) Heterogeneities in magma chambers: insights from the behavior of major and minor elements during mixing experiments with natural alkaline melts. Chem Geol 256:131–145

De Siena L, Del Pezzo E, Bianco F (2010) Seismic attenuation imaging of Campi Flegrei: evidence of gas reservoirs, hydrothermal basins, and feeding systems. J Geophys Res B09312. https://doi.org/10.1029/2009JB006938

de Vita S, Orsi G, Civetta L, Carandente A, D'Antonio M, Deino A, di Cesare T, Di Vito MA, Fisher RV, Isaia R, Marotta E, Necco A, Ort M, Pappalardo L, Piochi M, Southon J (1999) The Agnano-Monte Spina eruption (4100 years BP) in the restless Campi Flegrei caldera (Italy). J Volcanol Geotherm Res 91:269–301

Deino AL, Orsi G, de Vita S, Piochi M (2004a) The age of the Neapolitan Yellow Tuff caldera–forming eruption (Campi Flegrei caldera – Italy) assessed by $^{40}Ar/^{39}Ar$ dating method. J Volcanol Geotherm Res 133:157–170

Deino AL, Orsi G, Piochi M, de Vita S (2004b) The age of the Neapolitan Yellow Tuff caldera-forming eruption (Campi Flegrei caldera - Italy) assessed by 40Ar/39Ar dating method. J Volcanol Geotherm Res 133:157–170. https://doi.org/10.1016/S0377-0273(03)00396-2

Di Girolamo P (1987) Orogenic and anorogenic mantle-source components in the "anomalous" post-collisional peri-tyrrhenian volcanics (Italy). Boll Soc Geol Ital 106:757–766

Di Salvo S, Avanzinelli R, Isaia R, Zanetti A, Druitt T, Francalanci L (2020) Crystal-mush reactivation by magma recharge: evidence from the Campanian Ignimbrite activity, Campi Flegrei volcanic field, Italy. Lithos 376–377:105780

Di Renzo V, Di Vito MA, Arienzo I, Carandente A, Civetta L, D'Antonio M, Giordano F, Orsi G, Tonarini S (2007) Magmatic history of Somma-Vesuvius on the basis of new geochemical and isotopic data from a deep borehole (Camaldoli della Torre). J Petrol 48 (4):753–784. https://doi.org/10.1093/petrology/egl081

Di Renzo V, Arienzo I, Civetta L, D'Antonio M, Tonarini S, Di Vito MA, Orsi G (2011) The magmatic feeding system of the Campi Flegrei caldera: architecture and temporal evolution. Chem Geol 281:227–241

Di Renzo V, Wohletz K, Civetta L, Moretti R, Orsi G, Gasparini P (2016) The thermal regime of the Campi Flegrei magmatic system reconstructed through 3D numerical simulations. J Volcanol Geotherm Res 328:210–221

Di Vito MA, Isaia R, Orsi G, Southon J, de Vita S, D'Antonio M, Pappalardo L, Piochi M (1999) Volcanism and deformation since 12,000 years at the Campi Flegrei caldera (Italy). J Volcanol Geotherm Res 91:221–246

Di Vito MA, Sulpizio R, Zanchetta G, D'Orazio M (2008) The late Pleistocene pyroclastic deposits of the Campanian Plain: new insights into the explosive activity of Neapolitan volcanoes. J Volcanol Geotherm Res 177:19–48

Di Vito MA, Arienzo I, Braia G, Civetta L, D'Antonio M, Orsi G (2011) The Averno 2 fissure eruption: a recent small size explosive event at the Campi Flegrei caldera. Bull Volcanol 73:295–320. https://doi.org/10.1007/s00445-010-0417-0

Di Vito MA, Acocella V, Aiello G, Barra D, Battaglia M, Carandente A, Del Gaudio C, de Vita S, Ricciardi GP, Ricco C, Scandone R, Terrasi F (2016) Magma transfer at Campi Flegrei caldera (Italy) before the 1538 AD eruption. Sci Rep 6:32245. https://doi.org/10.1038/srep32245

Esposito R, Bodnar RJ, Danyushevsky LV, De Vivo B, Fedele L, Hunter J, Lima A, Shimizu N (2011) Volatile evolution of magma associated with the Solchiaro eruption in the Phlegrean volcanic district (Italy). J Petrol 52(12):2431–2460

Esposito R, Badescu K, Steele-MacInnis M, Cannatelli C, De Vivo B, Lima A, Bodnar RJ, Manning CE (2018) Magmatic evolution of the Campi Flegrei and Procida volcanic fields, Italy, based on interpretation of data from well-constrained melt inclusions. Eart-Sci Rev 185:325–356

Fanara S, Botcharnikov RE, Palladino DM, Adams F, Buddensieck J, Mulch A, Behrens H (2015) Volatiles in magmas related to the Campanian Ignimbrite eruption: experiments vs. natural findings. Am Mineral 100:2284–2297

Fedele L, Morra V, Perrotta A, Scarpati C (2006) Volcanological and geochemical features of the products of the Fiumicello eruption, Procida island, Campi Flegrei (Southern Italy). Per Mineral 75:43–72

Fedi M, Cella F, D'Antonio M, Florio G, Paoletti V, Morra V (2018) Gravity modeling finds a large magma body in the deep crust below the Gulf of Naples, Italy. Sci Rep 8:8229. https://doi.org/10.1038/s41598-018-26346-z

Forni F, Degruyter W, Bachmann O, De Astis G, Mollo S (2018a) Long-term magmatic evolution reveals the beginning of a new caldera cycle at Campi Flegrei Sci Adv 4:eaat9401

Forni F, Petricca E, Bachmann O, Mollo S, De Astis G, Piochi M (2018b) The role of magma mixing/mingling and cumulate melting in the Neapolitan Yellow Tuff caldera-forming eruption (Campi Flegrei, Southern

Italy). Contrib Mineral Petrol 173:45. https://doi.org/10.1007/s00410-018-1471-4

Fourmentraux C, Métrich N, Bertagnini A, Rosi M (2012) Crystal fractionation, magma step ascent, and syn-eruptive mingling: the Averno 2 eruption (Phlegraean Fields, Italy). Contrib Mineral Petrol 163:1121–1137

Fowler SJ, Spera FJ, Bohrson WA, Belkin HE, De Vivo B (2007) Phase equilibria constraints on the chemical and physical evolution of the campanian ignimbrite. J Petrol 48:459–493

Gebauer SK, Schmitt AK, Pappalardo L, Stockli DF, Lovera OM (2014) Crystallization and eruption ages of Breccia Museo (Campi Flegrei caldera, Italy) plutonic clasts and their relation to the Campanian Ignimbrite. Contrib Mineral Petrol 167:953. https://doi.org/10.1007/s00410-013-0953-7

Giaccio B, Nomade S, Wulf S, Isaia R, Sottili G, Cavuoto G, Galli P, Messina P, Sposato A, Sulpizio R, Zanchetta G (2012) The late MIS 5 Mediterranean tephra markers: a reappraisal from peninsular Italy terrestrial records. Quat Sci Rev 56:31–45

Giaccio B, Hajdas I, Isaia R, Deino A, Nomade S (2017) High-precision ^{14}C and $^{40}Ar/^{39}Ar$ dating of the campanian ignimbrite (Y-5) reconciles the time-scales of climatic-cultural processes at 40 ka. Sci Rep 7:45940. https://doi.org/10.1038/srep45940

Hammersley L, DePaolo DJ (2006) Isotopic and geophysical constraints on the structure and evolution of the Clear Lake volcanic system. J Volcanol Geotherm Res 153:331–356

Hoefs J, Wedepohl KH (1968) Strontium isotope studies on young volcanic rocks from Germany and Italy. Contrib Mineral Petrol 19:328–338

Hurley PM, Fairbairn HW, Pinson WH Jr (1966) Rb-Sr isotopic evidence in the origin of potash-rich lavas of western Italy. Earth Planet Sci Lett 5:301–306

Iovine RS, Fedele L, Mazzeo FC, Arienzo I, Cavallo A, Wörner G, Orsi G, Civetta L, D'Antonio M (2017a) Timescales of magmatic processes occurred prior to the ~4.7 ka Agnano-Monte Spina eruption (Campi Flegrei caldera, Southern Italy) based on diffusion chronometry on sanidine phenocrysts. Bull Volcanol 79:18. https://doi.org/10.1007/s00445-017-1101-4

Iovine RS, Mazzeo FC, Arienzo I, D'Antonio M, Wörner G, Civetta L, Pastore Z, Orsi G (2017b) Source and magmatic evolution inferred from geochemical and Sr-O-isotope data on hybrid lavas of Arso, the last eruption at Ischia island (Italy; 1302 AD). J Volcanol Geotherm Res 331:1–15. https://doi.org/10.1016/j.jvolgeores.2016.08.008

Iovine RS, Mazzeo FC, Wörner G, Pelullo C, Cirillo G, Arienzo I, Pack A, D'Antonio M (2018) Coupled $\delta^{18}O$-$\delta^{17}O$ and $^{87}Sr/^{86}Sr$ isotope compositions suggest a radiogenic and ^{18}O-enriched magma source for Neapolitan volcanoes (Southern Italy). Lithos 316–317:199–211. https://doi.org/10.1016/j.lithos.2018.07.009

Isaia R, D'Antonio M, Dell'Erba F, Di Vito MA, Orsi G (2004) The Astroni volcano: the only example of close eruptions within the same vent area in the recent history of the Campi Flegrei caldera (Italy). J Volcanol Geotherm Res 133:171–192

Isaia R, Marianelli P, Sbrana A (2009) Caldera unrest prior to intense volcanism in Campi Flegrei (Italy) at 4.0 ka B.P.: implications for caldera dynamics and future eruptive scenarios. Geophys Res Lett 36: L21303. https://doi.org/10.1029/2009GL040513

Judenherc S, Zollo A (2004) The Bay of Naples (southern Italy): constraints on the volcanic structures inferred from a dense seismic survey. J Geophys Res 109: B10312. https://doi.org/10.1029/2003JB002876

Langmuir CH, Vocke RD Jr, Hanson GN, Hart SR (1978) A general mixing equation with applications to Icelandic basalts. Earth Planet Sci Lett 37:380–392

Le Maitre RW (2002) Igneous rocks: a classification and glossary of terms: recommendations of the international union of geological sciences, subcommission on the systematics of igneous rocks. Cambridge University Press, Cambridge, UK, p 256

Liedl A, Buono G, Lanzafame G, Dabagov SB, Della Ventura G, Hampai D, Mancini L, Marcelli A, Pappalardo L (2019) A 3D imaging textural characterization of pyroclastic products from the 1538 AD Monte Nuovo eruption (Campi Flegrei, Italy). Lithos 340–341:316–331

Mangiacapra A, Moretti R, Rutherford M, Civetta L, Orsi G, Papale P (2008) The deep magmatic system of the Campi Flegrei caldera (Italy). Geophys Res Lett 35:L21304. https://doi.org/10.1029/2008GL035550

Marianelli P, Sbrana A, Proto M (2006) Magma chamber of the Campi Flegrei supervolcano at the time of the eruption of the campanian ignimbrite. Geology 34:937–940. https://doi.org/10.1130/G22807A.1

Mazzeo FC, D'Antonio M, Arienzo I, Aulinas M, Di Renzo V, Gimeno D (2014) Subduction-related enrichment of the Phlegrean volcanic district (Southern Italy) mantle source: new constraints on the characteristics of the sedimentary components. Chem Geol 386:165–183

Melluso L, de' Gennaro R, Fedele L, Franciosi L, Morra V (2012) Evidence of crystallization in residual, Cl-F-rich, agpaitic, trachyphonolitic magmas and primitive Mg-rich basalt-trachyphonolite interaction in the lava domes of the Phlegrean Fields (Italy). Geol Mag 149:532-550

Mollo S, Forni F, Bachmann O, Blundy JD, De Astis G, Scarlato P (2016) Trace element partitioning between clinopyroxene and trachy-phonolitic melts: a case study from the Campanian Ignimbrite (Campi Flegrei, Italy). Lithos 252–253:160–172

Moretti R, Arienzo I, Orsi G, Civetta L, D'Antonio M (2013) The deep plumbing system of the Ischia island (southern Italy): a physico-chemical and geodynamic window on the fluid-sustained and CO_2-dominated magmatic source of Campanian volcanoes. J Petrol 54 (5):951–984. https://doi.org/10.1093/petrology/egt002

Moretti R, Arienzo I, Di Renzo V, Orsi G, Arzilli F, Brun F, D'Antonio M, Mancini L, Deloule E (2019) Volatile segregation and generation of highly vesiculated explosive magmas by volatile-melt fining processes: the case of the Campanian ignimbrite eruption. Chem Geol 503:1–14. https://doi.org/10.1016/j.chemgeo.2018.10.001

Morgan DJ, Davidson JP, Pearson DG, Nowell GM, Civetta L (2004) Strontium isotopic heterogeneity among phenocrysts of the Minopoli 2 eruption of the Campi Phlegrei caldera complex (9500 BP). Geophys Res Abstr 6:04003 SRef-ID:1607-7962/gra/EGU04-A-04003

Mormone A, Piochi M, Bellatreccia F, De Astis G, Moretti R, Della Ventura G, Cavallo A, Mangiacapra A (2011) A CO_2-rich magma source beneath the Phlegrean volcanic district (Southern Italy): evidence from a melt inclusion study. Chem Geol 287:66–80

Orsi G, D'Antonio M, de Vita S, Gallo G (1992) The Neapolitan Yellow Tuff, a large-magnitude trachytic phreatoplinian eruption: eruptive dynamics, magma withdrawal and caldera collapse. J Volcanol Geotherm Res 53:275–287

Orsi G, Civetta L, D'Antonio M, Di Girolamo P, Piochi M (1995) Step–filling and development of three layer magma chamber: the Neapolitan Yellow Tuff case history. J Volcanol Geotherm Res 67:291–312

Orsi G, de Vita S, Di Vito MA (1996) The restless, resurgent Campi Flegrei nested caldera (Italy): constraints on its evolution and configuration. J Volcanol Geotherm Res 74:179–214

Orsi G, Di Vito MA, Isaia R (2004) Volcanic hazard assessment at the restless Campi Flegrei caldera. Bull Volcanol 66:514–530

Orsi G, Cioni R, Di Renzo V (2013) The Campanian plain during the Bronze age: development of volcanism and impact of the Vesuvius Avellino eruption in a densely populated area. In: Meller H et al (eds) Proceedings of the 4th archaeological conference of Central Germany "1600—cultural change in the shadow of the thera-eruption?", pp 117–134

Pabst S, Wörner G, Civetta L, Tesoro R (2008) Magma chamber evolution prior to the Campanian Ignimbrite and Neapolitan Yellow Tuff eruptions (Campi Flegrei, Italy). Bull Volcanol 70:961–976

Pappalardo L, Civetta L, D'Antonio M, Deino A, Di Vito MA, Orsi G, Carandente A, de Vita S, Isaia R, Piochi M (1999) Chemical and Sr—isotopic evolution of the Phlegrean magmatic system before the Campanian Ignimbrite and the Neapolitan Yellow Tuff eruptions. J Volcanol Geotherm Res 91:141–166

Pappalardo L, Civetta L, de Vita S, Di Vito M, Orsi G, Carandente A, Fisher RV (2002a) Timing of magma extraction during the Campanian Ignimbrite eruption (Campi Flegrei Caldera). J Volcanol Geotherm Res 114:479–497

Pappalardo L, Piochi M, D'Antonio M, Civetta L, Petrini R (2002b) Evidence for multi–stage magmatic evolution during the past 60 kyr at Campi Flegrei (Italy) deduced from Sr, Nd and Pb isotope data. J Petrol 43(8):1415–1434

Peccerillo A (2005) Plio-quaternary volcanism in Italy. Springer Ed Berlin Heidelberg, New York, pp 365

Peccerillo A (2017) Cenozoic volcanism in the Tyrrhenian sea region 2nd edn. Advances in volcanology series. Springer International Publishing AG, pp 399

Pelullo C, Arienzo I, Cambeses A, Chakraborty S, D'Antonio M, Dohmen R, Nazzari M, Pappalardo L, Petrosino P (2020) Compositionally zoned clinopyroxene crystals record fluctuations in the Agnano Monte Spina (Campi Flegrei, Italy) magma plumbing system. In: AIV-INGV–4th Rittmann conference for young researchers. Catania, Italy, 12–14 February 2020, Miscellanea INGV 52:228 ISSN: 1590-2595

Pelullo C, Chakraborty S, Cambeses A, Dohmen R, Arienzo I, D'Antonio M, Pappalardo L, Petrosino P (2021) Zoned clinopyroxene crystals record decadal scale evolution of the magma plumbing system of the Agnano-Monte Spina Plinian eruption (Campi Flegrei, Italy). Geochim Cosmochim Acta (submitted)

Perinelli C, Gaeta M, Bonechi B, Granati SF, Freda C, D'Antonio M, Stagno V, Sicola S, Romano C (2019) Effect of water on the phase relations of primitive K-basalts: implications for high-pressure differentiation in the Phlegraean volcanic district magmatic system. Lithos 342–343:530–541. https://doi.org/10.1016/j.lithos.2019.05.032

Perugini D, Poli G, Petrelli M, De Campos CP, Dingwell DB (2010) Time-scales of recent Phlegrean Fields eruptions inferred from the application of a 'diffusive fractionation' model of trace elements. Bull Volcanol 72(4):431–447

Perugini D, De Campos CP, Petrelli M, Dingwell DB (2015) Concentration variance decay during magma mixing: a volcanic chronometer. Sci Rep 5:14225. https://doi.org/10.1038/srep14225

Petrosino P, Arienzo I, Mazzeo FC, Natale J, Petrelli M, Milia A, Perugini D, D'Antonio M (2019) The San Gregorio magno lacustrine basin (Campania, Southern Italy): improved characterization of the tephrostratigraphic markers based on trace elements and isotopic data. J Quat Sci 34(6):393–404. https://doi.org/10.1002/jqs.3107

Piochi M, Bruno PP, De Astis G (2005) Relative roles of rifting tectonics and magma ascent processes: Inferences from geophysical, structural, volcanological, and geochemical data for the Neapolitan volcanic region (Southern Italy). Geochem Geophy Geosys 6: Q07005. https://doi.org/10.1029/2004GC000885

Rittmann A (1950) Sintesi geologica dei Campi Flegrei. Boll Soc Geol It 69:117–128

Roach AL (2005) The evolution of silicic magmatism in the post-caldera volcanism of the Phlegrean Fields, Italy. PhD thesis, Brown University, Providence, USA, 151 pp

Rogers NW, Hawkesworth CJ, Parker RJ, Marsh JS (1985) The geochemistry of potassic lavas from Vulsini, central Italy and implications for mantle enrichment processes beneath the Roman region. Contrib Mineral Petrol 90:244–257

Scarpati C, Cole P, Perrotta A (1993) The Neapolitan Yellow Tuff—a large volume multiphase eruption

from Campi Flegrei, Southern Italy. Bull Volcanol 55:343–356
Scarpati C, Perrotta A, Lepore S, Calvert A (2013) Eruptive history of Neapolitan volcanoes: constraints from 40Ar-39Ar dating. Geol Mag 150:412–425
Serri G (1990) Neogene-quaternary magmatism of the Tyrrhenian region: characterization of the magma sources and geodynamic implications. Mem Soc Geol Ital 41:219–242
Smith VC, Isaia R, Pearce NJG (2011) Tephrostratigraphy and glass compositions of post-15 kyr Campi Flegrei eruptions: implications for eruption history and chronostratigraphic markers. Quat Sci Rev 30:3638–3660. https://doi.org/10.1016/j.quascirev.2011.07.012
Tomlinson EL, Arienzo I, Civetta L, Wulf S, Smith VC, Hardiman MS, Lane CS, Carandente A, Orsi G, Rosi M, Muller W, Menzies MA (2012) Geochemistry of the Phlegraean Fields (Italy) proximal sources for major Mediterranean tephras: Implications for the dispersal of Plinian and co-ignimbritic components of explosive eruptions. Geochim Cosmochim Acta 93:102–128
Tonarini S, Leeman WP, Civetta L, D'Antonio M, Ferrara G, Necco A (2004) B/Nb and $\delta^{11}B$ systematics in the Phlegrean volcanic district (PVD). J Volcanol Geotherm Res 113:123–139
Tonarini S, D'Antonio M, Di Vito MA, Orsi G, Carandente A (2009) Geochemical and B-Sr-Nd isotopic evidence for mingling and mixing processes in the magmatic system feeding the Astroni volcano (4.1-3.8 ka) within the Campi Flegrei caldera (South Italy). Lithos 107:135–151
Vitale S, Isaia R (2014) Fractures and faults in volcanic rocks (Campi Flegrei, southern Italy): insight into volcano-tectonic processes. Int J Earth Sci (geol Rundsch) 103:801–819. https://doi.org/10.1007/s00531-013-0979-0
Voloschina M, Pistolesi M, Bertagnini A, Métrich N, Pompilio M, Di Roberto A, Di Salvo S, Francalanci L, Isaia R, Cioni R, Romano C (2018) Magmatic reactivation of the Campi Flegrei volcanic system: insights from the Baia–Fondi di Baia eruption. Bull Volcanol 80:75. https://doi.org/10.1007/s00445-018-1247-8
Wulf S, Kraml M, Brauer A, Keller J, Negendank JFW (2004) Tephrochronology of the 100 ka lacustrine sediment record of Lago Grande di Monticchio (southern Italy). Quat Int 122:7–30. https://doi.org/10.1016/j.quaint.2004.01.028
Wulf S, Keller J, Paterne M, Mingram J, Lauterbach S, Opitz S, Sottili G, Giaccio B, Albert PG, Satow C, Tomlinson EL, Viccaro M, Brauer A (2012) The 100–133 ka record of Italian explosive volcanism and revised tephrochronology of Lago Grande di Monticchio. Quat Sci Rev 58:104–123
Zollo A, Maercklin N, Vassallo M, Dello Iacono D, Virieux J, Gasparini P (2008) Seismic reflections reveal a massive melt layer feeding Campi Flegrei caldera. Geophys Res Lett 35:L12306. https://doi.org/10.1029/2008GL034242

Tephrochronology and Geochemistry of Tephra from the Campi Flegrei Volcanic Field, Italy

Emma L. Tomlinson, Paul G. Albert, and Martin A. Menzies

Abstract

Layers of volcanic ash (tephra) can be used as isochronous markers linking different sedimentary archives and allowing detailed analysis of the relative timing of climatic or evolutionary events. Conversely, high-resolution sedimentary tephra archives can shed light on the geochemical evolution of magmatic systems. The last ~110 kyrs of activity at the Campi Flegrei volcanic field is punctuated by several large eruptions, including the caldera forming Campanian Ignimbrite (~40 ka) and Neapolitan Yellow Tuff (~15 ka) eruptions and the more recent Pomici Principali (~12.1 ka) and Agnano-Monte Spina (~4.5 ka) all of which are preserved in distal settings several hundred kilometres away. In addition, there are a number of important distal Campi Flegrei caldera tephra layers for which the source eruption is not known (X-5 at ~105 ka and X-6 at ~109 ka) or only recently established (Y-3; ~29 ka), indicating the occurrence of large eruptions that are unknown or poorly constrained from proximal deposits. In this contribution we discuss the dispersal and age significance of these 7 key Campi Flegrei caldera eruptions. The value of distal tephra layers is dependent on the ability to identify a distinctive geochemical fingerprint for each layer, for example, the Campanian Ignimbrite and Neapolitan Yellow Tuff events tapped zoned magma chambers and thus produced unique and compositionally variable tephra layers. Micron-beam major and trace element analyses of distal and proximal volcanic glasses allow direct distal-distal and proximal–distal tephra comparisons. In addition, good stratigraphic and chronological control is important in order to identify and correlate tephra with confidence. Here, we discuss the distinctive geochemical compositions of each of the 7 highlighted Campi Flegrei caldera marker tephra layers.

E. L. Tomlinson (✉)
Department of Geology, Trinity College, Dublin, Ireland
e-mail: TOMLINSE@tcd.ie

P. G. Albert
Research Laboratory for Archaeology and the History of Art, Oxford University, Oxford, UK

P. G. Albert
Department of Geography, Swansea University, Swansea, UK

M. A. Menzies
Department of Earth Sciences, Royal Holloway University of London, London, UK

1 Tephrostratigraphy and Tephrochronology

Tephra (volcanic ash < 2 mm) is erupted over very short time intervals: explosive volcanic eruptions typically last for a few hours or days

© Springer-Verlag GmbH Germany, part of Springer Nature 2022
G. Orsi et al. (eds.), *Campi Flegrei*, Active Volcanoes of the World,
https://doi.org/10.1007/978-3-642-37060-1_5

with some persisting for a few weeks or months. The time taken for the tephra particles to fall through the atmosphere is of the order of a few minutes to days (Rose and Durant 2009) and the time taken for tephra particles to sink to the seabed or lake bottom is also rapid (Carey 1997). Therefore, tephra layers provide isochronous markers in the stratigraphic record, allowing archaeological, terrestrial (lacustrine, peat), marine and ice records to be correlated (tephrostratigraphy). In addition, if the precise age of the tephra is known from an independent absolute dating method, then tephra horizons provide age markers within the stratigraphy (tephrochronology). Few, if any, geochronological methods can offer the temporal and spatial precision offered by tephrochronology (Lowe 2011), with the exception of single grain dating of distal phenocrysts from fall deposits.

Traditional tephrochronology focuses on tephra layers that are visible to the naked eye, however, these only occur relatively close to the source volcano. In addition, non-visible ash deposits, or "cryptotephra" can be identified (e.g., Dugmore et al. 1995). Cryptotephra can be detected in sediment sequences far beyond the range of visible tephra records, greatly expanding the geographic footprint of a given tephra. This increases the likelihood of the superposition of tephra from different source regions at a single location, allowing regional tephrochronologies to be integrated (e.g., Lane et al. 2011). In addition, new tephra have been discovered that presently are known in cryptotephra form only, because their visible equivalents have not yet been reported, this is often because they relate to the mid- to low-intensity explosive eruptions that are frequently lost from the near-source eruption record.

1.1 Geochemical Fingerprinting

Tephrochronology requires tephra layers to be characterised and identified, or 'fingerprinted'. The most effective tool is geochemical analysis of the volcanic glass. Grain specific analyses are preferred to bulk tephra analysis for both distal and proximal tephra, both because of the small size (typically 20–100 μm) and paucity (few tens to hundreds of shards) of volcanic material recovered in distal settings. In addition, this avoids problems of (a) variable phenocryst and lithic contents with distance from the source, and (b) averaging of chemical variability when using bulk rock techniques. Furthermore, the range in observed compositions may itself be diagnostic. Major element data is widely reported, however many evolved magmas especially those erupted from the same volcano, are compositionally similar and cannot always be confidently separated using major elements alone (Allan et al. 2008), in such cases, trace elements are an important diagnostic tool. Trace elements show a wider variability than major elements because they are more strongly affected by differences in source composition and by sub-volcanic magmatic processes such as fractional crystallisation and assimilation. In the last decade, it has become routine to report trace element data for distal tephra layers. Unfortunately, much trace element data for proximal volcanic deposits is whole rock rather than glass data, complicating detailed proximal–distal comparison. Two datasets present glass data for a range of proximal CFc deposits spanning 60–0 kyrs, these are Smith et al. (2011) and Tomlinson et al. (2012).

Additional correlative information may be derived from mineral phase assemblages and the compositions of loose mineral grains within tephra layers (Giordano et al. 2014). However, the morphology of volcanic shards cannot be used to fingerprint individual tephra layers as tephra typically shows a range of shard morphologies and microlite abundances within the deposits of a single eruption (Smith et al. 2011). These variations do not correspond with compositional differences but instead reflect rheological differences associated with degassing, decompression and shear and viscous heating within the conduit (Costa et al. 2007).

1.2 Controls on Tephra Dispersal

Eruption style and magnitude is a major influence on tephra dispersal, controlling both the volume and granularity of the ash produced and the altitude that it is propelled to. The most explosive volcanic eruptions vary from phreatomagmatic (mafic to felsic magmas) to Plinian/ultra-Plinian (mostly felsic but also foiditic magmas) eruptions. Tephra deposits originate from within the convective buoyant fountain/column and from pyroclastic density currents (PDCs) produced during fountain/column collapse and the associated co-PDC ash cloud. Both mechanisms can produce tephra which is then deposited several thousand kilometres from source, for example tephra from the 74 ka Younger Toba Tuff form a visible layer > 200 km from the source (Matthews et al. 2012).

Wind is a major control on tephra dispersal. High-level stratospheric winds have a major role in dispersing Plinian fall tephra, while tropospheric winds control the dispersal of sub-Plinian column tephra and co-PDC ash. The present day average wind direction in southern Italy is from the west-west northwest and is independent of height in the troposphere and stratosphere, however there is some seasonal variability and in summer the dominant wind is from the east at altitudes greater than 14 km (Cornell et al. 1983; Mastrolorenzo et al. 2008). A comparison of the relative frequency of recent wind directions with the dispersal axes of past Campanian volcanic eruptions suggests that atmospheric conditions prevailing during the last 100 kyrs are similar to those occurring over Southern Italy today (Fig. 1) (Mastrolorenzo et al. 2008). Below 2 km, there is considerable seasonal variability, low-level winds are dominantly from the west-southwest and southwest for most of the year, but from the northwest during summer months (Cornell et al. 1983).

1.3 Applications of Tephrochronology

Tephrochronology is routinely utilised for paleoclimatic reconstructions. Ice-core records have demonstrated that pronounced climatic shifts with severe environmental consequences are possible within twenty years or less (Steffensen et al. 2008), this is beyond the centennial precision that can be achieved by radiocarbon dating in terrestrial climate archives. Furthermore, volcanic marker layers also offer a vital tool for directly synchronising sedimentary records with high-precision polar ice-core climatic records. Therefore, tephrochronology is critical for reconstructing the relative timing, rate and duration of abrupt environmental transitions and hence for testing theories about the causes and impacts of those changes. Indeed, tephrochronology has been used to demonstrate that the Younger Dryas cold period (equivalent to Greenland Stadial 1 [GS-1]) was time transgressive across Europe (Lane et al. 2013). Tephrochronology has also been applied as a chronological tool in archaeological studies, for example Italy derived tephra layers were critical in dating occupancy period of the Haua Fteah cave in Libya (Douka et al. 2013).

Medial and distal tephra archives provide valuable information about the longevity, eruptive frequencies and repose periods of their source volcanoes. This is particularly useful in cases where unravelling eruptive history from proximal deposits is hampered by burial, resurgent uplift, erosion and weathering. Analysis of tephra from the Auckland Volcanic Field allowed the order of activity at 48 volcanic centres to be deciphered and indicated that about half of these volcanic centres have repose periods shorter than 1,000 years (Hopkins et al. 2017). The distal annually laminated and intensely ^{14}C dated sediment of Lake Suigestu, Honshu Island, Japan, have been used to reconstruct a high-

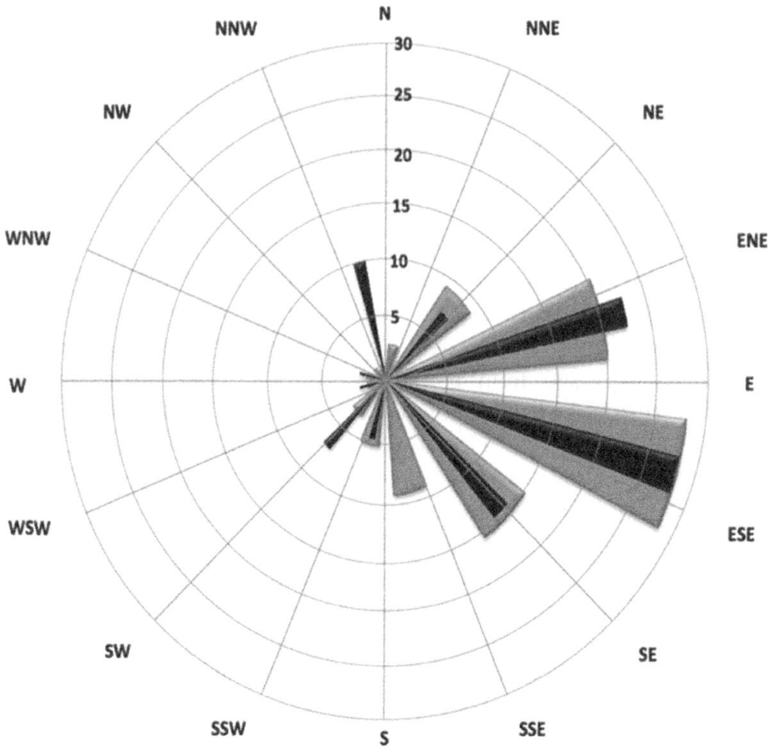

Fig. 1 Frequency of wind directions collected at Brindisi (Southern Italy) meteorological station in the period 1953–2000 in the height interval 0–34 km (grey) compared to the observed tephra dispersions of Campanian eruptions in the last 100 kyrs (black). Figure modified after Mastrolorenzo et al. (2008)

precision eruption chronology for two nearby stratovolcanoes, Daisen and Sambe, with closely spaced explosive eruptions at the former separated by as little as six years within the sedimentary record (Albert et al. 2018). Medial and distal tephra layers may be used to investigate the long-term geochemical evolution of the volcanic-magmatic system, for example the compositions of tephra preserved in marine cores from the Izu-Bonin forearc record the temporal evolution in magma composition during the transition from extensional back arc to arc volcanism (Bryant et al. 2003). Finally, tephrochronology has been used to investigate the link between climate and volcanism, revealing an increase in volcanic activity following glacial unloading in places such as Chile (Rawson et al. 2015) and Alaska (Praetorius et al. 2016). Further examples of tephrochronology applications can be found in Lowe et al. (2015) and Lane et al. (2017).

An important archive of Phlegraean eruptions is Lago Grande di Monticchio (LGdM), a varved lake in the Monte Vulture volcanic complex, 125 km east of Campi Flegrei and on the dominant tephra dispersal axis of most eruptions. The LGdM record spans 133 kyrs and preserves > 340 distinct tephra layers, of which ~215 are thought to originate from Campi Flegrei (Wulf et al. 2004, 2008, 2012). The laminated sediments of the lake preserve an independent varve chronology that can provide ages for distal Phlegraean tephra, although it should be noted that there are varying offset between the ages of directly dated tephra and their varve calendar ages (Wulf et al. 2012).

2 Geochemistry, Dispersal and Age of Key Tephra from the Campi Flegrei Volcanic Field

Volcanism at Campi Flegrei volcanic field (CFvf; see Chap. Volcanic and Deformation History of the Campi Flegrei Volcanic Field, Italy) began more than 78 kyrs ago (Scarpati et al. 2013). The current Campi Flegrei caldera (CFc) is a nested structure formed within the CFvf as a result of the highly explosive eruptions associated with the ~40 ka Campanian Ignimbrite (CI) and the ~15 ka Neapolitan Yellow Tuff (NYT) (Orsi et al. 1996; see Chap. Volcanic and Deformation History of the Campi Flegrei Volcanic Field, Italy). Volcanism before the CI eruption was characterised by high-energy explosive activity from vents outside the present Campi Flegrei depression (Orsi et al. 1996; Pappalardo et al 1999; Scarpati et al. 2013). Between the CI and NYT eruptions, numerous minor fall deposits were generated from intra-caldera phreatomagmatic eruptions (Orsi et al. 1996; Pappalardo et al. 1999). Within this interval, Albert et al. (2019) provide evidence for at least one large magnitude eruption, Masseria del Monte Tuff, which produced PDC that travelled beyond CFc. After the NYT up to 70 eruptions occurred during three epochs of volcanic activity (Orsi et al. 2004; Smith et al. 2011). The most recent eruption formed the Monte Nuovo tuff cone in AD 1538 (Guidoboni and Ciuccarelli 2011; Di Vito et al. 2016 and references therein). Here, we focus on key CFc eruptions with > 0.5 km^3 DRE that are widely reported in distal tephra archives (Figs. 2 and 3). These eruptions include those that generated the terrestrial deposits of CI, NYT, Pomici Principali (PP) and Agnano-Monte Spina (A-MS), and the distal tephra layers of X-5, X-6 and Y-3.

2.1 Campanian Ignimbrite

The caldera-forming CI was the most explosive eruption of the CFvf and the largest known volcanic event in the Mediterranean region in the last 200 kyrs. The CI eruption is dated at 39.85 ± 0.14 ka by $^{40}Ar/^{39}Ar$ (Giaccio et al. 2017). The eruption dispersed ~250–300 km^3 of ash (Rolandi et al. 2003; Pyle et al. 2006; Fedele et al 2007; Costa et al. 2012; Marti et al. 2016) during two eruptive phases. The first phase was Plinian and produced a column that reached a height of ~44 km and deposited pumice and ash fall predominantly to the east (Fig. 2) (Rosi et al. 1999). The second phase accompanied caldera collapse and was characterised by highly inflated PDCs that flowed over the sea and crossed mountain ridges more than 70 km from the vent (Fisher et al. 1993; Ort et al. 2003). Numerical modelling indicates that the mass eruption rate was an order of magnitude higher during the PDC phase (~10^{10} kg^*s^{-1}) relative to the fall phase (10^9 kg^*s^{-1}) and therefore most of the tephra dispersal was likely to be associated with the dilute pyroclastic density current phase of the CI eruption (Costa et al. 2012).

Chemical zoning in proximal fall and flow deposits (Fig. 4) is interpreted to reflect the

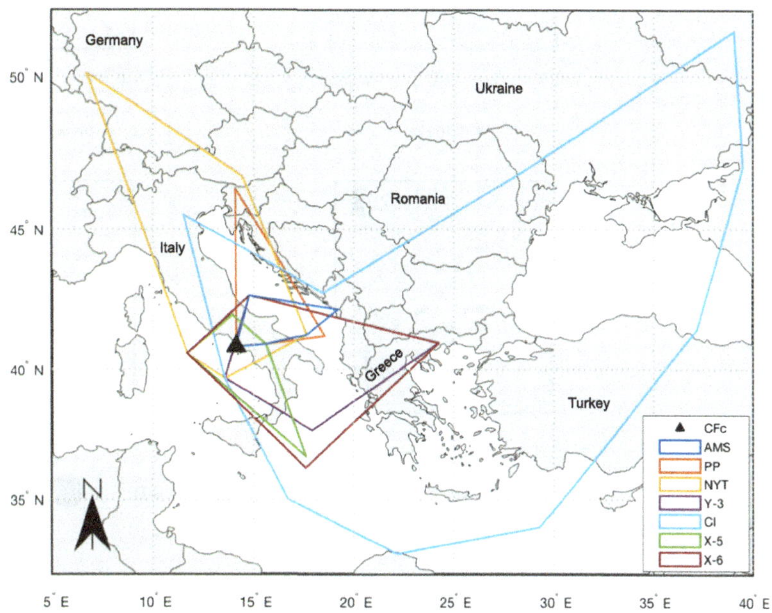

Fig. 2 Known dispersal of 7 key CFc tephra based on sites and references in Fig. 3

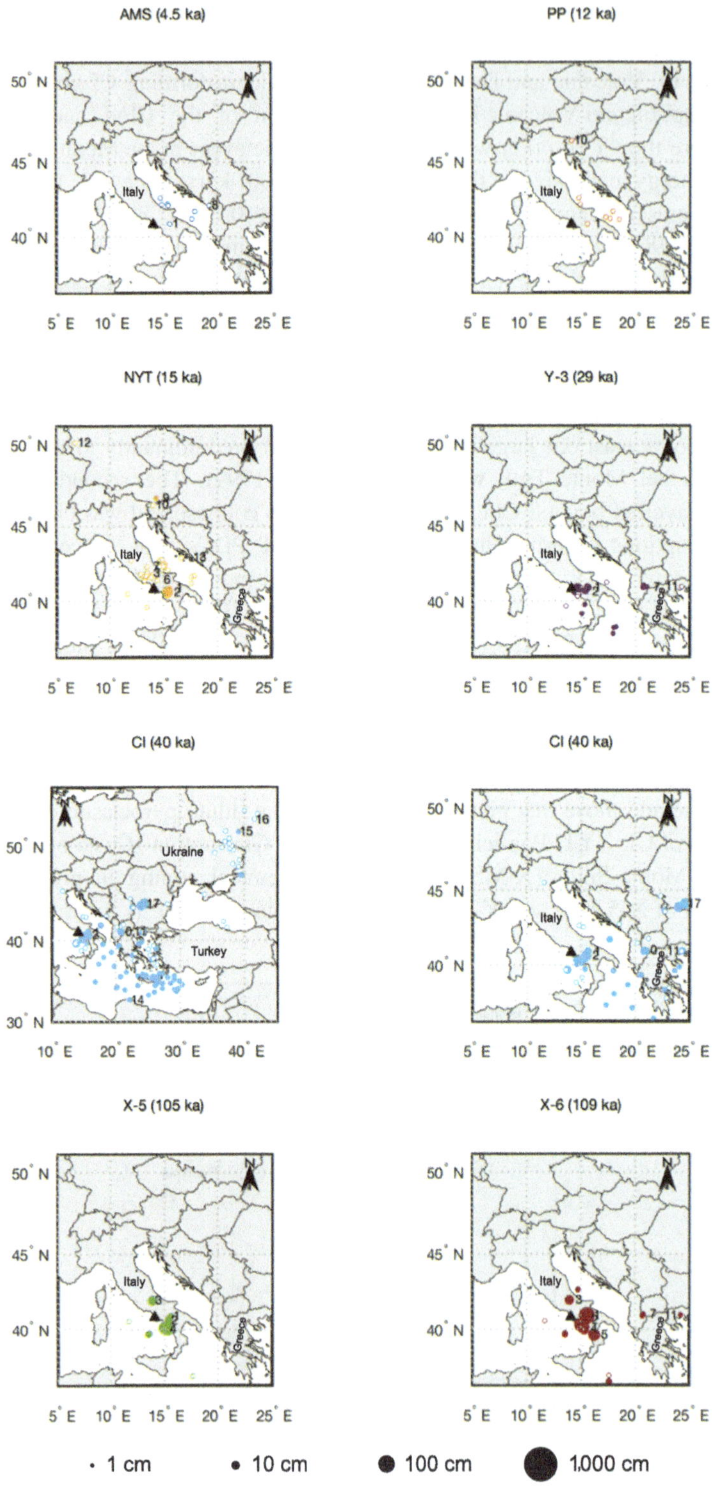
AMS (4.5 ka)
PP (12 ka)
NYT (15 ka)
Y-3 (29 ka)
CI (40 ka)
CI (40 ka)
X-5 (105 ka)
X-6 (109 ka)
Italy
Greece
Ukraine
Turkey
1 cm
10 cm
100 cm
1,000 cm

◀ **Fig. 3** Distal occurrences of tephra: **a** A-MS; **b** PP/C-1; **c** NYT/C-2; **d** Y-3; **e** CI/Y-5 full dispersal; **f** CI/Y-5 Mediterranean dispersal for comparison; **g** X-5; **h** X-6 (Keller et al. 1978; Thunell et al. 1979; Cornell et al. 1983; Vinci 1985; Paterne et al. 1988, 2008; Cramp et al. 1989; Vezzoli 1991; St. Seymour and Christanis 1995; Frezzotti and Narcisi 1996; Calanchi et al. 1998; Schmidt et al. 2002; Sulpizio et al. 2003, 2010; Munno and Petrosino 2004, 2007; St Seymour et al. 2004; Wulf et al. 2004, 2018; Siani et al. 2004; Pyle et al. 2006; Margari et al. 2007; Radić et al. 2007; Di Vito et al. 2008; Giaccio et al. 2008, 2012; Marciano et al. 2008; Wagner et al. 2008; Zanchetta et al. 2008; Bourne et al. 2010, 2015; Caron et al. 2010; Vogel et al. 2010; Golovanova et al. 2010; Lane et al. 2011, 2015; Morley and Woodward 2011; Lowe et al. 2012; Damaschke et al. 2013; Veres et al. 2013; Fitzsimmons et al. 2013; Amato et al. 2014; Douka et al. 2014; Insinga et al. 2014; Iorio et al. 2014; Engwell et al. 2014; Matthews et al. 2015; Regattieri et al. 2015; Donato et al. 2016; Ivanova et al. 2016; Leicher et al. 2016; Di Roberto et al. 2018; Mannella et al. 2019). Tephra layer thickness indicated by bubble size (closed symbol) unless unknown or cryptotephra (open symbol). Colours as in Fig. 2. Position of the Campi Flegrei caldera is indicated (black triangle). Key sites and locations: (1) LGdM; (2) San Gregorio Magno; (3) Fucino basin; (4) Cilento coast; (5) Tarsia; (6) Boiano basin (locations 1–6: Italy); (7) Lakes Ohrid and Prespa (Macedonia/Albania); (8) Lake Shkodra (Montenegro); (9) Langsee (Austria); (10) Lake Bled (Slovenia); (11) Tenaghi Philipon (Greece); (12) Meerfelder Maar (Germany); (13) Vela Spila (Croatia); (14) Haua Fteah (Libya); (15) Kostenski (Russia); (16) Tambov (Russia); (17) Daneasa (Romania)

emptying of a compositionally zoned magma chamber comprising two distinct magmas (trachytic and phonolitic), which mixed during the eruption (Civetta et al. 1997). The layered CI magma chamber underwent late recharge by an isotopically distinct and less evolved trachytic magma (Arienzo et al. 2011). The involvement of two isotopically distinct magmas is reflected in the chemical heterogeneity of CI tephra (Fig. 4). Glasses from the CI fall and the lower and intermediate flows straddle the trachyte-phonolite boundary and have a narrow range of CaO, MgO, K_2O and Na_2O contents (1.75 ± 0.11, 0.33 ± 0.05, 7.31 ± 0.27 and 6.24 ± 0.27 wt%, respectively) and are intermediate to evolved with $Zr/Sr = 5 - 31$ with negative anomalies in Sr ($Sr/Pr_N = 0.01 - 0.08$) and Eu ($Eu/Eu^*_N = 0.24 - 0.36$) consistent with feldspar fractionation (Tomlinson et al. 2012).

The upper PDC flows sampled the less evolved late recharge magma and extend to poorly evolved compositions with higher CaO, MgO, K_2O and lower Na_2O and $Zr/Sr = 0.4$ with negligible Eu and Sr anomalies ($Eu/Eu^*_N = 1.1$; $Sr/Pr_N = 0.7$) (Tomlinson et al. 2012).

2.1.1 Distal Occurrences of Campanian Ignimbrite Tephra

The CI is correlated with the Y-5 ash layer, which is recognised in marine cores from across the Eastern Mediterranean (e.g., Keller et al. 1978) and the C-13 layer in the Tyrrhenian Sea (Paterne et al. 1988). In the varved LGdM core, the CI is linked with tephra layer TM-18 (Wulf et al. 2004). The CI is found in terrestrial sites as far as Russia ~2,500 km to the northeast (Pyle et al. 2006) and Libya ~1,120 km to the southeast (Lowe et al. 2012), in total an area of > 3.7 million km^2 was covered by > 5 mm of ash (Costa et al. 2012). Distal occurrences of CI tephra are shown in Figs. 2 and 3e, f.

In many medial to distal locations, the fall unit is overlain by a fine ash deposited from the co-PDC cloud (Smith et al. 2016). Indeed, both fall and flow (co-PDC) components are recorded in sites up to 850 km from the source, indicated by the occurrence of poorly evolved glasses (Tomlinson et al. 2012) and by bi-modal size distributions (Sparks and Huang 1980; Cornell et al. 1983; Engwell et al. 2014; Smith et al. 2016), for example at LGdM TM-18 comprises a 17-cm-thick pumice fall layer overlain by a 9 cm vitric ash layer (Wulf et al. 2004). Engwell et al. (2014) estimate that, most of the tephra dispersal ($60 \pm 5\%$ by volume) is fallout from the co-PDC cloud, while the rest ($40 \pm 5\%$) is associated with the Plinian phase of the eruption.

In some medial settings, the CI tephra layer is immediately underlain by chemically similar tephra produced by minor eruptions at CFc. At LGdM four CFc tephra occur within 600 varve years of the CI tephra, these have similar major

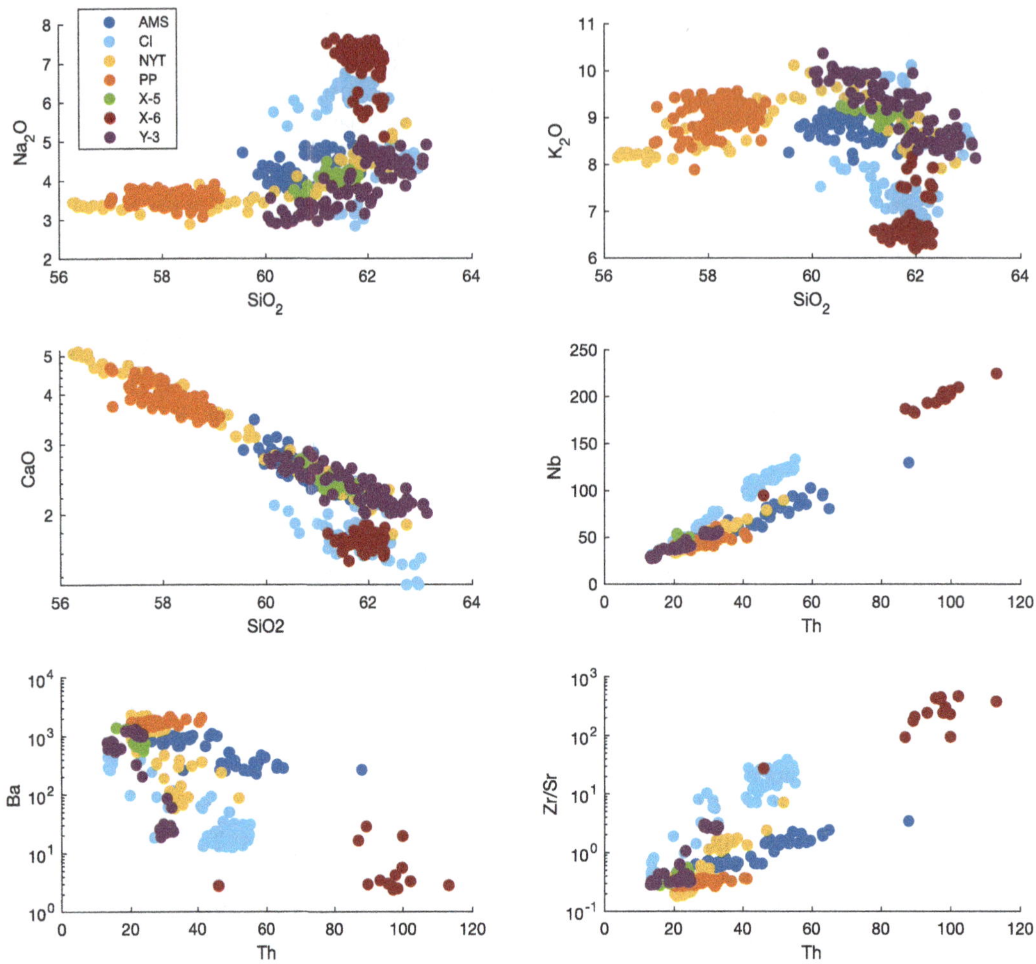

Fig. 4 Major and trace element glass compositions of key CFc tephra layers, proximal data for A-MS, PP, NYT and CI from Smith et al. (2011) and Tomlinson et al. (2012); distal data for Y-3 from Albert et al. (2015) and for X-5 and X-6 from Donato et al. (2016)

element compositions but can be distinguished using trace elements (Wutke et al. 2015). CI-like tephra are also found just below the CI tephra in cores from the Adriatic Sea (PRAD-1752; Bourne et al. 2010) and the Balkans (PT0915-8; Damaschke et al. 2013), however they are found in the presence of the CI and can be distinguished with careful geochemical analyses.

2.1.2 Climatic Significance of the Campanian Ignimbrite

The CI eruption injected ~200 Tg sulphur dioxide into the stratosphere, numerical modelling suggests that this is enough to cause hemispheric cooling of ~2 °C lasting ~2 years (Costa et al. 2012; Black et al. 2015), however there is no evidence of a volcanic winter associated with the CI. Deposition of volcanic ash over a large continental area likely resulted in increased seasonal climate variability on a decadal scale (Jones et al. 2007). The region that was affected by significant ash deposition (the eastern Mediterranean, Balkans and part of the Caucasus region) was heavily vegetated at that time (e.g., Huntley and Allen 2003) and may have been severely affected by (1) acid rain as a result of sulphur, fluorine and chlorine aerosols being injected into the troposphere; and (2) soil and water contamination as volcanic aerosols are leached from volcanic ash deposits. This may have led to partial or total destruction of forests and grasslands (Delmelle 2003; Kockum et al. 2006). The CI eruption also affected lacustrine environments, as indicated by the rapid change in diatom communities in two lakes on the Balkan Peninsula (Jovanovska et al. 2016).

The CI eruption occurred during the Last Glacial, shortly after the onset of Heinrich Event 4 (HE4) (Ton-That et al. 2001), a northern hemisphere-wide cold and arid period related to the discharge of Arctic ice into the North Atlantic. In the eastern Mediterranean, the onset of HE4 is marked by a change in pollen species and an increase in atmospheric dust, which occurs below the CI marker tephra in lacustrine and marine cores indicating that the CI eruption did not cause the HE4 event (Lowe et al. 2012). However, various climatic proxies indicate that the cold and arid conditions of HE4 were the most pronounced of all the Heinrich events to have occurred during the Last Glacial. This has led Fedele et al. (2003, 2007) to suggest that the CI eruption may have interacted with the onset of HE4, leading to a more amplified and prolonged climatic deterioration across the whole Mediterranean basin.

2.1.3 Timing of Dispersal and Development of Humans in Europe

The CI occurs within the timeframe of two key changes in the dispersal and development of humans in Europe. The first is the replacement of indigenous Neanderthals by anatomically modern humans (AMH) between 45 and 30 ka ago. The second is the European Middle to Upper Palaeolithic transition ($\sim$40 ka; Fedele et al. 2008), one of several intervals of faster-paced changes in human cognition and innovation in Palaeolithic societies. These changes have been attributed to severe climatic deterioration (Tzedakis et al. 2007; Muller et al. 2011) and/or the environmental stresses caused by the CI eruption itself (Fedele et al. 2008). In either case, AMH are assumed to have had competitive advantages that allowed them to adapt, recolonise and survive a climatically unstable Europe more effectively than Neanderthals. Crucial to testing these hypotheses is the establishment of accurate chronologies for archaeological and climatic events in each region and their synchronisation on a continental scale.

CI tephra is found stratigraphically above lithic assemblages of the earliest Upper Palaeolithic associated with the spread of modern humans in several European sites from Italy to Greece to Libya (Giaccio et al. 2008; Lowe et al. 2012). This applies in particular to some forms of the Aurignacian, a lithic industry with sophisticated art objects and musical instruments (Higham et al. 2012). This finding demonstrates that neither the CI eruption nor the HE4 cold/arid conditions could have triggered the cultural changes or population replacements witnessed around this time. Rather, it suggests that AMH proved a greater competitive threat to indigenous Neanderthals than major environmental change or explosive eruptions (Lowe et al. 2012). However, this remains a contentious area, with other workers arguing that there is no significant time gap between the latest early Palaeolithic occupation and the CI tephra layer (Giaccio et al. 2017) and that the coincidence of the CI eruption and cultural change support an evolutionary role for the combined CI-HE4 event (Fedele et al. 2008; Giaccio et al. 2017) especially where CI deposits are thick (Fitzsimmons et al. 2013).

2.2 Neapolitan Yellow Tuff

The NYT was the second caldera-forming eruption from the CFvf, it has been dated by $^{40}Ar/^{39}Ar$ both proximally at 14.9 $\pm$ 0.4 ka (Deino et al. 2004), and distally at 14.5 $\pm$ 0.4 ka (Galli et al. 2017) and by radiocarbon to 14.32–13.90 ka cal. BP (Blockley et al. 2008) and so occurred $\sim$25 kyrs after the CI. The NYT erupted at least 40 km^3 dense rock equivalent

(DRE) of magma (Orsi et al. 1995; Wohletz et al. 1995; Deino et al. 2004) during two main phases. During the first phase, eruption occurred from a central vent and consisted of an initial dilute PDC followed by an alternating sequence of dilute PDCs and Plinian fall episodes that were dispersed to the northeast (Fig. 2). The second phase was associated with caldera collapse, during which pyroclastic density currents were erupted from multiple vents (Orsi et al. 1992; Wohletz et al. 1995). Deposits from phase 1 and phase 2 are termed the lower member (LM) and upper member (UM), respectively.

Deposits of the NYT are crystal poor and chemically zoned and were produced from three distinct magma batches (Orsi et al. 1992), as a result the NYT is the most chemically variable of the CFvf tephra (Fig. 4) spanning 56–63 wt% SiO_2 and 1.8–5.2 wt% CaO. The LM tephra form two populations separated by a compositional gap. The initial dilute PDC deposit is characterised by trachytic glass with < 2.6 wt% CaO and > 61 wt% SiO_2 and poorly evolved trace element compositions (Sr/Zr = 0.4–1.8) with weak anomalies relating to feldspar fractionation (Eu/Eu^*_N = 0.6–0.9; Sr/Pr_N = 0.8–1.2) (Tomlinson et al. 2012). Subsequent LM glasses are phono-trachytic with > 3.8 wt% CaO and < 59 wt% SiO_2 and also only weakly evolved (Sr/Zr = 0.3–0.4) but with higher incompatible element concentrations and larger Eu and Sr anomalies (Eu/Eu^*_N = 0.6–0.8; Sr/Pr_N = 0.7–0.9) than LM1 (Tomlinson et al. 2012). These two LM glass compositions represent magma 1 and magma 2 described by Orsi et al. (1995). The UM glasses straddle the phonolite-trachyte boundary, spanning a wide compositional range between the two LM clusters and extending to less evolved compositions with 56.0–62.1 wt% SiO_2 and 2.0–5.1 wt% CaO and Zr/Sr = 0.2–7.3 (Tomlinson et al. 2012).

Compositional variability in NYT deposits do not reflect the inverted contents of a compositionally zoned magma body, rather it is the result of complex interactions between different magmatic components stored in a heterogeneous reservoir (Orsi et al. 1992, 1995; Forni et al. 2018). These include an evolved magma, a more mafic magma and an intermediate magma, interpreted as fractionated liquid, mafic recharge and remelted cumulate crystal mush melt, respectively (Forni et al. 2018).

2.2.1 Distal Occurrences of Neapolitan Yellow Tuff Tephra

The NYT has been correlated to marine tephra C-2 in distal settings up to 250 km from Campi Flegrei in the Tyrrhenian and Adriatic Seas (Paterne et al. 1988; Siani et al. 2004) and with tephra TM-8 in the varved LGdM core (Wulf et al. 2004). The most easterly recorded occurrence of NYT tephra is at Vela Spila, Croatia (Radic 2007). The NYT is found in a number of terrestrial sites up to 1,200 km north of Campi Flegrei including Lake Fucino (Italy; Giaccio et al. 2017), Lake Bled (Slovenia; Lane et al. 2011), Langsee (Austria; Schmidt et al. 2002) and Meerfelder Maar (Lane et al. 2015). At Meerfelder Maar, the NYT tephra is found with the Icelandic Vedde Ash and German Laacher See Tephra and so contributes to a Europe-wide tephrostratigraphic framework (Lane et al. 2015). The distribution of sites containing NYT tephra implies dominant dispersal to the north or northeast (Figs. 2 and 3c).

2.2.2 Paleoclimate Reconstruction

The NYT is important for paleoclimate reconstruction because it resides just above the onset of the Bølling-Allerød interstadial, or Greenland interstadial (GI-1) in key climate archives (Siani et al. 2004; Lane et al. 2015). The discovery of NYT tephra in a varved lake at Längsee (Austria) allowed the discontinuous annually laminated Late Glacial timescale to be linked to calendar years (Schmidt et al. 2002), using the NYT age from LGdM (Wulf et al. 2004). Thus, tephra can be used to anchor floating chronologies in paleoenvironmental archives. The refined chronology for Längsee enabled comparison with paleoclimate records from the Swiss Plateau and north-western Germany and indicated that climatic oscillations at 14.35 and 13.94 ka were synchronous across central Europe (Schmidt et al. 2002).

2.3 Pomici Principali

The Pomici Principali (PP; also known as Agnano Pomici Principali) eruption has been dated at 11.92–12.16 ka cal. BP (Smith et al. 2011). PP was a sub-Plinian to Plinian eruption, which ejected ~1.8 km^3 ash (Sulpizio 2005). The PP eruption started with a phreatomagmatic vent-opening phase followed by a mainly magmatic phase with a pulsating Plinian column and sporadic phreatomagmatic explosions (Di Vito et al. 1999). This deposited an alternating sequence of pumice/ash fall units and pyroclastic density current deposits (Smith et al. 2011) mainly towards the east (Fig. 2).

PP glasses are phonolitic and extend into the tephri-phonolite field with increasing stratigraphic height. The glasses show a continuous but limited compositional range (Fig. 4) with the most differentiated glass at the base of the sequence (SiO_2 58.5 ± 0.4 wt%; CaO 3.80 ± 0.24 wt%; FeO 4.46 ± 0.11 wt%; MgO 1.09 ± 0.11 wt%; Eu/Eu^*_N = 0.91 ± 0.05) and least differentiated at the top (SiO_2 57.7 ± 0.4 wt %; CaO 4.27 ± 0.25 wt%; FeO 4.72 ± 0.21 wt %; MgO 1.39 ± 0.14 wt%; Zr/Sr = 1.8 ± 0.6; Eu/Eu^*_N = 0.74 ± 0.07). Trace element variation in PP glasses is limited (Zr 215–306 ppm, Th 20–29 ppm) and indicate low degrees of evolution, with high Sr contents (Zr/Sr = 0.27–0.36) and weak Eu anomalies (Eu/Eu^*_N = 0.92 ± 0.06) (Smith et al. 2011; Tomlinson et al. 2012).

2.3.1 Distal Occurrences of Pomici Principali Tephra

PP is correlated with tephra C-1 in marine cores from the South Adriatic (Paterne et al. 1988; Siani et al. 2004; Calanchi et al. 2008) and with TM-7b in LGdM and is found as a crypto-tephra in Lake Bled, Slovenia (Lane et al. 2011). However, PP has not been found in marine cores from the Central Adriatic or Tyrrhenian Seas as previously reported. NYT/C-1 tephra in the Central Adriatic Sea cores do not match the chemistry of the proximal PP; similarly, reported occurrences of PP tephra at Lake Shkodra and Lake Accesa are not consistent with the proximal PP glass compositions (Tomlinson et al. 2012). Therefore, distal occurrences of PP tephra are of limited areal extent and are dominantly to the north and east (Figs. 2 and 3b).

2.3.2 Constraining the Younger Dryas

The PP is a useful marker for the Younger Dryas, a cold and arid stadial that occurred at ~12.9 to 11.6 ka during the transition from cold glacial to a warmer interglacial state. In the North Atlantic and northern Europe, the timing of the Younger Dryas is constrained by the widespread Icelandic Vedde Ash dated at ~12.1 ± 0.1 ka (Rasmussen et al. 2006). In Southern Europe the age of the PP eruption is within error of the onset of the Younger Dryas. Numerical age estimates for the Vedde Ash and PP tephra are indistinguishable within errors. In Lake Bled, Slovenia (Lane et al. 2011), the PP tephra lies just above the Vedde Ash, proving that the Icelandic Vedde eruption pre-dated the PP eruption and allowing the age of the PP eruption to be better constrained. The discovery of Italian and Icelandic tephra layers at Lake Bled provides a pivotal link between tephra archives across a wide region that encompasses the North Atlantic, Europe and the Mediterranean, paving the way for a better understanding of the relative timing of environmental response to the Younger Dryas event across large distances (Lane et al. 2011).

2.4 Agnano-Monte Spina

Agnano-Monte Spina was a moderate sized sub-Plinian to Plinian eruption and is the highest magnitude event to have occurred in the Campi Flegrei caldera in the last 5 kyrs (Arienzo et al. 2010). The A-MS eruption is dated at 4,482–4,625 cal. years BP (Smith et al. 2011). The eruption ejected 2.0 km^3 ash (Sulpizio 2005) in a series of phreatic and magmatic explosions. de Vita et al. (1999) divided the sequence of A-MS tephra into six members A to F. Two major fall deposits (members B and D) were generated from Plinian columns of between 20 and 30 km height. Fall deposits (tens of cm) can be found up to 50 km from the eruption vent and an area of ~1,000 km^2 was covered by a pyroclastic fall deposit more than 10-cm-thick dominantly towards the east-northeast (de Vita et al. 1999). The PDC of member E overcame the topographic high of the caldera wall and flowed ~15 km from the vent (Smith et al. 2011). Pyroclastic density currents covered an area of ~200 km^2 (de Vita et al. 1999) dominantly towards the north-northeast (Fig. 2).

A-MS glasses span a narrow, continuous compositional range (Fig. 4), straddling the phonolite-trachyte boundary with 59.3–62.0 wt% SiO_2 and approximately constant Na_2O and K_2O (4.3 ± 0.3 wt% and 8.9 ± 0.3 wt%, respectively). The most evolved compositions are found at the base of the sequence (SiO_2 61.26 ± 0.31 wt%; CaO 2.33 ± 0.11 wt%; FeO 3.55 ± 0.14 wt%; MgO 0.59 ± 0.07 wt%; Zr/Sr = 1.8 ± 0.6; Eu/Eu^*_N = 0.74 ± 0.07) and there is a decrease in the degree of evolution towards the top of the stratigraphy (SiO_2 60.25 ± 0.40 wt%; CaO 2.69 ± 0.20 wt%; FeO 3.71 ± 0.19 wt%; MgO 0.75 ± 0.13 wt%; Zr/Sr = 0.7 ± 0.1; Eu/Eu^*_N = 0.83 ± 0.04) (Smith et al. 2011). Total trace element variation is large given the limited variation in major oxides (Zr 300–840 ppm, Th 26–88 ppm). Compositional variability likely results from mixing between two geochemically and isotopically distinct magmas (de Vita et al. 1999; Arienzo et al. 2010).

2.4.1 Distal Occurrences of Agnano-Monte Spina Tephra

A-MS tephra is found in the central and south Adriatic Sea (Calanchi et al. 1998; Siani et al. 2004; Lowe et al. 2007) and in terrestrial settings at Lake Shkodra, Albania and Montenegro (Sulpizio et al. 2010), Lake Accesa (Magny et al. 2006) and LGdM (Wulf et al. 2008). A-MS is one of the most widespread tephra deposits in the Adriatic marine record (Figs. 2 and 3a) and marks an important and complex phase of environmental changes during the mid- to late-Holocene climatic transition.

2.4.2 Constraining Short Time Intervals

In proximal stratigraphies, the two fall deposits are cut by an erosional unconformity, suggesting the occurrence of a time break between the deposition of members B and D (de Vita et al 1999; Chap. Volcanic and Deformation History of the Campi Flegrei Volcanic Field, Italy). Two tephra layers linked to A-MS are recorded in a deep-sea sediment core from the South Adriatic with overlapping ages of 4.60 ± 0.7 ka and 4.35 ± 0.7 ka (Siani et al. 2004). In LGdM, A-MS tephra occurs as two discrete layers, TM-5b and TM-5a, separated by 40 varve years (Wulf et al. 2008). Layers TM-5b and TM-5a are correlated with the lower and upper A-MS units, indicating a hiatus of ~40 years between the two phases of the A-MS eruption (Smith et al. 2011). Thus, medial-distal tephra archives can be used to constrain the relative ages of eruptive events, even those separated by short time intervals.

3 Distal Tephra Archives

The previous section discussed the four most widely used CFc tephra, those produced by known, moderate to high magnitude eruptions for which the age and dispersal are reasonably well constrained. However, CFc is the most active caldera volcano in Europe with more than 70 eruptions within the last 15 kyrs (Orsi et al. 2004) and a number of other Campi Flegrei tephra are recorded in medial and distal sedimentary archives. Campi Flegrei tephra are recognised by their high-K-trachytic to phonolitic compositions. CF tephra can be distinguished from those produced at other volcanoes in the Neapolitan Volcanic Area on the basis of major element composition. CF tephra are characterised by $K_2O > Na_2O$ and often $K_2O >> Na_2O$, which is inconsistent with Ischia type glasses, which typically show higher Na_2O, often with $Na_2O \geq K_2O$, in addition, CFc glasses have CaO contents (for a given MgO) that are intermediate between those of Ischia and Somma-Vesuvio (Tomlinson et al. 2015).

Sedimentary archives recording CFc tephra indicate that the proximal volcano-stratigraphy is often incomplete. Proximal deposits produced during the older eruptions are recorded sporadically at intra-caldera localities. The scarcity of proximal preservation of older eruptions is largely the result of burial by the products of more recent activities and caldera collapse. Indeed, some widespread, first order tephra markers recognised in multiple distal archives still remain poorly constrained to volcanic source at CFc. This has driven investigations into medial tephra on the Campanian Plain (i.e., Di Vito et al. 2008).

Sedimentary archives of CFc tephra also indicate that tephra from smaller eruptions may occur in medial locations, for example, tephra from the Soccavo 1 eruption is a visible layer in the Tyrrhenian Sea (Albert et al. 2012). Tephra from small eruptions may also be preserved as cryptotephra (non-visible tephra) in distal settings. Recognition of a larger number of CFc tephra from a number of smaller eruptions offers the potential for increasing the number of tie points for tephra correlations in medial locations. However, this requires that each tephra can be uniquely distinguished from geochemically similar tephra produced during other eruptions closely spaced in time. Conversely, if a tephra cannot be uniquely identified then there is potential for miss-correlating tephra layers and inferring erroneous relative and absolute ages.

Details of the major CFc tephra reported in a number of extended tephra archives are given in Table 1. In addition, each of these archives contains a number of other tephra for which proximal/medial and distal equivalents are unknown or poorly constrained. In the following sections we discuss the discrepancy between the number of CFc eruptions known proximally and distally for the time periods bracketed by the CI and NYT eruptions. We also summarise the geochemistry and distribution of key CFc tephra found in distal archives for which the source eruption is not known.

3.1 Eruptions Before the Campanian Ignimbrite (> 39 ka)

Twelve stratigraphically constrained eruptions older than the CI are preserved at Trefola quarry along the northern margin of the CFc, with the oldest $^{40}Ar/^{39}Ar$ dated at 58.9 ± 1.8 ka (Pappalardo et al. 1999). Pyroclastic deposits older than the CI have been found in cores taken at various locations north and east of Naples, the highest number of pre-CI Phlegraean eruptions (eleven) were recognised at Ponti Rossi (Pappalardo et al. 1999), four of which are also found in the Camaldoli della Torre borehole on the southern flank of Somma-Vesuvio (Di Renzo et al. 2007). In addition, several pyroclastic deposits attributed to the activity of the CFvf are described in the Campanian Plain, including Santa Lucia (51 ± 3 ka) and CA1-a (55–105 ka) (Di Vito et al. 2008).

This time period was characterised by the eruption of moderately evolved phonolitic magmas showing limited variation in major element composition (SiO_2 58.7–60.2 wt%; CaO 1.6–2.3 wt%; MgO 0.28–0.43 wt%, FeO 3.2–3.7 wt%). Pre-CI tephra are clearly distinguished from CI products using major element composition. Pabst et al. (2008) recognised three compositional clusters within the Pre-CI section at Trefola, where differences between these groups are most clearly seen in Na_2O, K_2O, Zr and Nb contents (pre-CI 1: Zr 680–805 ppm; pre-CI 2: Zr 470–620 ppm; and pre-CI 3: Zr ~690 ppm) (Di Renzo et al. 2007; Pabst et al. 2008; Tomlinson et al. 2012). However, identification of tephra from individual eruptions within each group is difficult as many eruptions share similar major and trace element compositions and because the age and chemistry of many tephra layers produced during this time remain poorly constrained.

Rolandi et al. (2003) and Belkin et al. (2016) identified a number of older (> 100 ka) trachytic PDC deposits, these include the Durazzano (116.1 ka), Taurano 14 (157.4 ka), Taurano 9 (183.8 ka), Moschiano (184.7 ka), Taurano 7 (205.6 and 210.4 ka), Seiano Valley A (245.9 ka) and Seiano Valley B (289.6 ka).

Table 1 Age and distal occurrences of key CFc tephra layers

Unit	Age	Region								
		Marine			Terrestrial Italy				Terrestrial Balkans	
		Location								
		Ionian Sea	Tyrrhenian Sea	Adriatic Sea	Lago Grande di Monticchio	San Gregorio Magno	Cilento coast	Fucino Lake	Ohrid - Macedonia	Tenaghi Philipon - Greece
		Reference								
		Keller et al. (1978)	Paterne et al. (1988, 2008)	Bourne et al. (2010, 2012)	Wulf et al. (2004, 2008)	Munno and Petrosino (2007)	Marciano et al. (2008)	Mannella et al. (2019)	Vogel et al. (2010)	Wulf et al. (2018)
		Cores								
		RC9-191	KET 8004	PRAD 1–2	LGM-B/D/E/J	-	outcrop	F1-F3	OT0702	TP-2005
		Layers								
Agnano-Monte Spina	4.48 - 4.63 ka cal. BP[a]	–	–	PRAD-055	TM-5a, b	–	–	–	–	–
Pomici Principali	11.92 - 12.16 ka cal. BP[a]	–	C-1	PRAD-203	TM-7b	–	–	–	–	–
Neapolitan Yellow Tuff	14.9 ± 0.4 ka[b]	–	C-2	PRAD-218	TM-8	S21	–	TF-2	–	–
Masseria del Monte Tuff	28.68 - 29.42 ka cal. BP[c]	Y-3	C-7	–	TM-15	S19	CPL1/CPL2	–	OT0702-4	TP05-9.70
Campanian Ignimbrite	39.85 ± 0.14 ka[d]	Y-5	C-13	PRAD-1752	TM-18	S17	PL1/PL2/PL3/AS	TF-5a	OT0702-6	TP05-12.87
unknown	105 ± 2 ka[e]	X-5	C-27	–	TM-25	S11	–	TF-12	OT0702-8	–
unknown	108.9 ± 1.8 ka[f]	X-6	C-31	PRAD-2812	TM-27	S10	SM1/SM2/SA	TF-13	OT0702-9	TP05-31.255

Ages from a Smith et al. (2011), b Deino et al. (2004), c Albert et al. (2015), d Giaccio et al. (2017), e Giaccio et al. (2012), f Iorio et al. (2014)

These medial deposits are highly altered, with loss on ignition from 8 to 17 wt% and depletion of Na_2O and K_2O, ratios of immobile elements are similar to those measured in the CI (Belkin et al. 2016).

Distal tephra archives suggest that activity at the CFvf may have extended back to at least ~732 ka, the age of CFc-like Parmenide/V5 tephra found at several locations in southern and central Italy (Ciaranfi et al. 2010; Capraro et al. 2011; Giaccio et al. 2013, 2014). Other > 200 ka, CFc-like tephra layers found in southern Italy include the ~687 ka V7 tephra from Montalbano Jonico (Petrosino et al. 2015), the ~514 ka A11/A12 and SC2 tephra from the Acerno and Mercure basins (Giaccio et al. 2014; Petrosino et al. 2014) and the ~374 ka Morphi ash (Pyle et al. 1998). This highlights the potential of medial tephra archives for identifying previously unknown eruptions and constraining the longevity of volcanic fields.

3.1.1 The X-5 Tephra

The distal X-5 tephra is a homogenous trachy-phonolite with 61.3 ± 0.3 wt% SiO_2 and relatively high CaO (2.45 ± 0.15 wt%), MgO (0.62 ± 0.77 wt%) and K_2O contents (9.10 ± 0.13 wt%) and $K_2O/Na_2O > 2$ (Donato et al. 2016) (Fig. 4). X-5 tephra lack an Eu anomaly and can be distinguished from younger tephra, such as TM-24, by their lower Ba and Sr for a given Th (Donato et al. 2016).

The X-5 tephra represents a significant eruption that is not known in the proximal stratigraphic record. The X-5 is a K-trachytic tephra, first recognised as a visible layer in the sediments of the Ionian Sea (Keller et al. 1978) and directly $^{40}Ar/^{39}Ar$ dated to 105 ± 2 ka (Kraml 1997) and resides in interstadial conditions associated with MIS5.4, equivalent to Greenland Interstadial 24 (Giaccio et al. 2012). Distal occurrences of the X-5 tephra are shown in Figs. 2 and 3g. The X-5 has been correlated with the K-trachytic C-27 tephra from the Tyrrhenian Sea (Paterne et al. 2008). It has also been recognised in terrestrial sequences from the central Mediterranean, including on Salina Island in the Tyrrhenian Sea (Lucchi et al. 2008) and on the Italian mainland along the Cilento coastline (CIL1 and LeS1; Giaccio et al. 2012; Donato et al. 2016) and in the Sulmona (POP-3; Giaccio et al. 2012), Fucino (TF-12; Giaccio et al. 2017; Mannella et al. 2019) and San Gregorio Magno (S11; Munno and Petrosino 2007) basins. In the LGdM archive, layer TM-25 is correlated with the X-5 (Wulf et al. 2012). Previous correlations to LGdM TM-24a and TM-24b (Wulf et al. 2004) and TAU-1 in the Campanian Plain (Di Vito et al. 2008) have since been shown to be erroneous (Wulf et al. 2012). Identification of the proximal equivalent of the X-5 tephra is hampered both by limited preservation of pre-CI eruptive deposits and by the altered nature of those eruptive deposits that are preserved, thus the proximal counterpart of the X-5 remains elusive.

3.1.2 The X-6 Tephra

The distal X-6 tephra has a trachy-phonolite composition with 61.9 ± 0.3 wt% SiO_2 and 1.72 ± 0.07 wt% CaO and a bimodal distribution of MgO and alkalis (Fig. 4). Cluster 1 has higher K_2O (7.68 ± 0.29 wt%), MgO (0.46 ± 0.07 wt%) and lower Na_2O (5.79 ± 0.42 wt%), while cluster 2 has lower K_2O (6.41 ± 0.68 wt%) and MgO (0.28 ± 0.05 wt%) and higher Na_2O (7.07 ± 0.97 wt%) (Donato et al. 2016). X-6 glasses have elevated concentrations of Rb, Nb, Zr, Ta, Th, U and REE (e.g., Th 43–114 ppm) and are depleted in Ba and Sr with a negative Eu anomaly.

The X-6 tephra represents another significant CFc eruption that is not known in the proximal stratigraphy. The X-6 is a relatively homogenous, amphibole-bearing trachytic tephra, that forms a visible layer in several Ionian Sea cores (Keller et al. 1978). X-6 has been correlated with the C-31 tephra from the Tyrrhenian Sea (Paterne et al. 2008; Insinga et al. 2014; Iorio et al. 2014) but is mainly found in the eastern Mediterranean: in the Adriatic Sea (PRAD 2812; Bourne et al. 2015) and at Tenaghi Philippon in Greece (TP05-915; Wulf et al. 2018). The X-6 is found on the Italian mainland at LGdM (TM-27; Wulf et al. 2012) and Fucino (TF-13; Giaccio et al. 2017; Mannella et al. 2019) and Sulmona (POP-4;

Giaccio et al. 2012) basins and on along the Cilento coastline (CIL2 and LeS2; Giaccio et al. 2012; Donato et al. 2016). Distal occurrences of the X-6 tephra are shown in Fig. 2. It is also recognised on Salina Island in the southern Tyrrhenian Sea (Albert 2012; Lucchi et al. 2013). The X-6 has been directly $^{40}Ar/^{39}Ar$ dated to 108.9 ± 1.8 ka (Iorio et al. 2014). The X-6 provides an excellent marker for the Marine isotope stage (MIS) 5.4 stadial (Bourne et al. 2015), equivalent to the terrestrial Melisey 1 stadial event recorded in LGdM (Brauer et al. 2007). As with the X-5, the proximal counterpart of the X-6 remains elusive. The X-6 tephra is more widely dispersed and forms thicker layers than the X-5 (Figs. 2 and 3h), suggesting that the X-6 may have been a larger event. Indeed, distal X-6 tephra layers rival the CI in thickness.

3.2 Eruptions Between the Campanian Ignimbrite and Neapolitan Yellow Tuff (40–15 ka)

Post-CI, pre-NYT (40–15 ka) units are collectively known as Tufi Biancastri, however they represent the products of several eruptions (see Chap. Volcanic and Deformation History of the Campi Flegrei Volcanic Field, Italy and references therein). At least 9 eruptions are recorded at Ponti Rossi (Orsi et al. 1996) and 12 tephra layers are preserved in LGdM (Wulf et al. 2004, 2008). The first erupted magmas are trachytic and are some of the most highly evolved compositions in the CFc stratigraphy, with SiO_2 ~64 wt% and Zr up to 840 ppm (pre-NYT 1 of Pabst et al. 2008). There is then a switch to less evolved compositions with lower Zr, Nb and SiO_2, these later erupted tephra may be subdivided into two populations: moderately enriched trachyte with CaO ~2.2 wt% and Zr ~350 ppm (pre-NYT 2 of Pabst et al. 2008), and weakly enriched phono-trachyte with CaO ~4.0 wt% and Zr ~450 ppm (pre-NYT 3 of Pabst et al. 2008). It is extremely difficult to distinguish between tephra from individual eruptions within each group because there is extensive overlap in both major and trace element composition. Some pre-NYT eruptions are bimodal containing from both pre-NYT 2 and 3 (Tomlinson et al. 2012). Pre-NYT 2 and 3 tephra overlap extensively with tephra from the NYT LM initial dilute PDC and phreatoplinian deposits, respectively, making identification of the NYT problematic in the absence of the NYT UM.

In addition to the proximal deposits exposed within the CFc, several pyroclastic deposits of similar age and attributed to activity at the caldera are described from the slopes of Somma-Vesuvio. GM1 stratigraphically overlies Lagno Amendolare, dated at 15.4 ± 0.2 ka cal. BP (Damaschke et al. 2013), both of which have been suggested as the proximal equivalents of tephra found in the Adriatic Sea (Siani et al. 2004; Bourne et al. 2010). GM1 and Lagno Amendolare have been correlated to LGdM tephra TM-9 and TM-10, respectively, largely on the basis of age (Wulf et al. 2004). A 2.6 m thick pyroclastic layer, CdT-k2, is found within the Camaldoli della Torre borehole from the southern slope of Somma-Vesuvio (Di Renzo et al. 2007). Further CFc pyroclastic deposits are described from colluvial and alluvial deposits at the foot of Apennine mountains that border the Campanian Plain to the east of Campi Flegrei, including TAU1-e (22–23 cal. ka BP), SMP1-e (30.7 ± 0.23 cal. ka BP) and SMP1-d (33–36 cal. ka BP) (Di Vito et al. 2008). In particular, SMP1-e has been suggested as the proximal equivalent of the Y-3 tephra (Di Vito et al. 2008).

3.2.1 The Y-3 Tephra

The distal Y-3 tephra is bi-modal in composition (Fig. 4), component 1 glasses are high SiO_2 (~61.6–62.4 wt%) trachytic glasses with ~8.3–8.7 wt% K_2O, and are enriched in incompatible trace elements (e.g., ~30–33 ppm Th). Component 2 glasses are phono-trachytic with lower SiO_2 (59.8–60.6 wt%) and higher K_2O (~9.7–10.3 wt%) content and are less enriched in incompatible trace elements (e.g., 14–15 ppm Th) (Albert et al. 2015, 2019).

The Y-3 tephra was first recognised in the Ionian Sea (Keller et al. 1978) and since reported in the Tyrrhenian (C-7, Paterne et al. 1988; Munno and Petrosino 2004) and Adriatic Seas (MD90-917; Zanchetta et al. 2008). Terrestrially, the Y-3 is reported in Italy, in lacustrine sequences at LGdM and San Gregorio Magno (TM-15 and S-19) (Wulf et al. 2004; Munno and Petrosino, 2007) and at Lakes Ohrid and Prespa in the Balkans (OT0701-4, JO-188, PT90115-05; Caron et al. 2010; Vogel et al. 2010; Damaschke et al. 2013) and at Tenaghi Philippon in Greece (TP 9.70; Albert et al. 2015). The ^{14}C age model for Tenaghi Philippon provides an age of 28,680–29,420 cal. years BP (IntCal13) for the Y-3 (Albert et al. 2015). The distribution of the Y-3 tephra is shown in Figs. 2 and 3d.

The 40–15 ka period in CFc's volcanic history is characterised by the eruption of a number of geochemically similar tephra of different ages. Furthermore, the age and composition of the Y-3 tephra was poorly constrained at its type locality in the Ionian Sea (Zanchetta et al. 2008). Consequently, numerous other distal tephra occurring in broadly similar stratigraphic positions and showing a similar K-trachyte chemistry have been correlated to the Y-3 without precise geochemical validation. This issue was addressed by detailed characterisation of the type locality Y-3 from the Ionian Sea, and subsequent validation of its other occurrences across the Mediterranean region (Albert et al. 2015). Crucially the distinctive bi-modal glass chemistry of the Y-3 tephra helped verify its widespread distribution and identified several ash layers that had been incorrectly correlated to the Y-3 event (e.g., Adriatic tephra layer, PRAD-1332).

Proposed near-source equivalents of the Y-3 tephra include a unit in the Ponti Rossi sequence and a nearby borehole (S19) both from Naples, north-east of the caldera, this unit has been ^{40}Ar/^{39}Ar dated to 29.3 $\pm$ 0.7 ka and crucially spans the diagnostic compositional range of Y-3 tephra (Albert et al. 2019). This eruption unit comprises a lower sub-unit of fall containing moderately sorted pumices, and a thicker upper sub-unit of poorly sorted, ash dominated, PDC deposits. Albert et al. (2019) suggest that this eruption was a large magnitude (VEI 6) event that dispersed $\sim$40 km^3 ash (16 km^3 DRE) over an area > 150,000 km^2. The eruption was named the Masseria del Monte Tuff by Albert et al. (2019) owing to its probable chrono-stratigraphic association with the most proximal Tufi Biancastri pyroclastic succession along the western slope of the Camaldoli Hills first reported by Calcaterra et al. (2007), and more recently by Isaia et al. (2016), which in part comprises the tuff unit of Masseria del Monte with a bracketing age of between 39 and 18 ka (Calcaterra et al. 2007).

Previously the Y-3 had been linked to the near-source VRa eruption deposit exposed in Verdolino valley (Orsi et al. 1996; Di Vito et al. 2008; Zanchetta et al. 2008) and dated at 30.3 $\pm$ 0.2 ka using ^{40}Ar/^{39}Ar (Pappalardo et al. 1999), whilst also the SMP1-e unit from a medial location on the Sorrento Peninsula (Zanchetta et al. 2008), ^{14}C dated at 30.31–32.11 ka cal. BP (Di Vito et al. 2008, recalibrated using Reimer et al. 2009). A correlation with the VRa deposit was subsequently excluded on the basis of detailed trace element glass chemistry (Albert et al. 2015), whilst the SMP1-e units thickness at distance from CFc, its age and stratigraphic position make a correlation with the Y-3 tephra plausible, the deposit did not satisfy the full compositional range (trachytic end-member only) diagnostic of the distal tephra prohibiting an unequivocal correlation (Albert et al. 2015).

The widespread Y-3 tephra was recognised as offering a key maker for the MIS 2/3 transition in Mediterranean marine archives, and thus the onset of the North Atlantic Stadial, Heinrich Event 3 (HE3), or Greenland Stadial 5 (GS-5) conditions which are recognised by reduce total tree pollen percentages in Mediterranean archives, including LGdM (Zanchetta et al. 2008; Albert et al. 2015). However, assessment of the chrono-stratigraphic position of the Y-3 tephra in the higher resolution pollen record of Tenaghi Philippon illustrates that the ash fall in Greece post-dates the onset of this cooling event by as much as 2,300 years (Albert et al. 2015).

4 Post Neapolitan Yellow Tuff Eruptions (<15 ka)

Seventy-one eruptions younger than the NYT are recognised in the proximal deposits (Orsi et al. 2004; Smith 2011), while the LGdM core records at least 20 tephra layers in this time frame (Wulf et al. 2004). This period has been subdivided into epochs 1 ($\sim$15 to $\sim$10.6 ka), 2 ($\sim$9.6 to $\sim$9.1 ka) and 3 ($\sim$5.5 to $\sim$3.5 ka), which are separated by periods of quiescence (Di Vito et al. 1999; Chap. Volcanic and Deformation History of the Campi Flegrei Volcanic Field, Italy). Epoch 1, which includes PP, saw the eruption of magmas with the least evolved compositions and lowest incompatible trace element contents in the post-NYT stratigraphy (D'Antonio et al. 1999; Smith et al. 2011). Sometimes, these SiO_2-poor melts were erupted with more differentiated magmas producing deposits that vary continuously or contain discrete populations. The deposits of epoch 1 eruptions fall into two broad compositional groups: (1) Soccavo 1, PP and Soccavo 4 have higher SiO_2 and Th and lower CaO, FeO than (2) Paleo-pisani 2, Pisani 1, and Pisani 2, and these eruptions can be identified using precise major element chemistry (D'Antonio et al. 1999; Smith et al. 2011). Epoch 2 sees a step to more evolved compositions. Tephra from epochs 2 and 3 span a narrow compositional range, the larger eruptions (Pigna San Nicola, Agnano 3, Monte Sant'Angelo, Paleo-Astroni 2, Astroni 3 and Astroni 4) cannot be separated, while A-MS and Astroni 6 can only be distinguished using minor trace element differences (Smith et al. 2011). During epoch 3, simultaneous eruption of evolved and less evolved magmas took place at Averno 2 and Solfatara at 4.3 ka (Pistolesi et al. 2016).

5 Conclusive Remarks

1. Tephra from CFvf eruptions was widely dispersed, tephra from the Plinian CI has been found 2,500 km from the source, while tephra from the NYT, PP, A-MS, Y-3, X-5 and X-6 eruptions are found in locations throughout the central and eastern Mediterranean.
2. Glass geochemistry has a pivotal role to play in "fingerprinting" juvenile magmatic compositions allowing for proximal–distal tephra correlations. Major-minor elements constrain petrogenetic lineages, such as the pre-NYT magma groups (Pabst et al. 2008) or the post-NYT epochs (Smith et al. 2011). Trace element concentrations and ratios provide an additional level of discrimination allowing, in some cases, discrimination between similar magmas produced over short time intervals.
3. Distal tephra layers reveal the occurrence CFvf eruptions that are not known from proximal deposits, such as the X-6 eruption, which may rival the magnitude of the NYT on the basis of the thickness and dispersal of its tephra.
4. Distal tephra archives may be used to constrain (a) the relative timing of eruptions from the stratigraphic superposition of tephra layers in that they retain crucial stratigraphic context, and (b) absolute timing from varve and ^{14}C ages and direct $^{40}Ar/^{39}Ar$ dating of phenocryst phases in distal tephra layers, for example revealing that the two phases of the CFc A-MS eruption were separated by $\sim$40 years (Smith et al. 2011).
5. CFc derived tephra are pivotal for constraining the timing and synchronicity in key environmental transitions such as the Younger Dryas (PP), Bølling-Allerød interstadial (NYT) and Heinrich Event 3 (Y-3), Heinrich Event 4 (CI) in Europe.

Acknowledgements The authors wish to thank the NERC RESET consortium members and associates for their ongoing interest and collaboration, in particular Victoria Smith, Christine Lane, Sabine Wulf, Lucia Civetta and Giovanni Orsi. We also thank David Pyle and Paola Petrosino for their valuable reviews and Giovanni Orsi, Lucia Civetta, Massimo D'Antonio for their insight and editorial handling.

References

Albert PG (2012) Volcanic glass chemistry of Italian proximal deposits link to distal archives in the Central Mediterranean region. PhD Thesis, Royal Holloway University of London

Albert PG, Smith VC, Suzuki T, Tomlinson EL, Nakagawa T, McLean D, Yamada M, Staff RA, Schlolaut G, Takemura K, Nagahashi Y, Kimura J-I, Suigetsu 2006 Project Members (2018) Constraints on the frequency and dispersal of explosive eruptions at Sambe and Daisen volcanoes (South-West Japan Arc) from the distal Lake Suigetsu record (SG06 core). Earth-Sci Rev 185:1004–1028

Albert PG, Tomlinson EL, Smith VC, Di Roberto A, Todman A, Rosi M, Marani M, Muller M, Menzies MA (2012) Marine-continental tephra correlations: volcanic glass geochemistry from the Marsili Basin and the Aeolian islands, southern Tyrrhenian Sea, Italy. J Volcanol Geotherm Res 229–230:74–94

Albert PG, Hardiman M, Keller J, Tomlinson EL, Smith VC, Bourne AJ, Wulf S, Zanchetta G, Sulpizio R, Müller UC, Pross J, Ottolini L, Matthews IP, Blockley SPE, Menzies MA (2015) Revisiting the Y-3 tephrostratigraphic marker: a new diagnostic glass geochemistry, age estimate, and details on its climatostratigraphical context. Quat Sci Rev 118:105–121

Albert PG, Giaccio B, Isaia R, Costa A, Niespolo EM, Nomade S, Pereira A, Renne PR, Hinchliffe A, Mark DF, Brown RJ, Smith VC (2019) Evidence for a large-magnitude eruption from Campi Flegrei caldera (Italy) at 29 ka. Geology 47:595–599

Allan ASR, Baker JA, Carter L, Wysoczanksi RJ (2008) Reconstructing the quaternary evolution of the world's most active silicic volcanic system: insights from a similar to 1.65 Ma deep ocean tephra record sourced from Taupo volcanic zone, New Zealand. Quat Sci Rev 27:2341–2360

Amato V, Aucelli PPC, Cesarano M, Jicha B, Lebreton V, Orain R, Pappone G, Petrosino P, Russo Ermolli E (2014) Quaternary evolution of the largest intermontane basin of the Molise Apennine (Central-Southern Italy). Rend Lincei 25:197–216

Arienzo I, Moretti R, Civetta L, Orsi G, Papale P (2010) The feeding system of Agnano Monte Spina eruption (Campi Flegrei, Italy): dragging the past into present activity and future scenarios. Chem Geol 270:135–147

Arienzo I, Heumann A, Worner G, Civetta L, Orsi G (2011) Processes and timescales of magma evolution prior to the Campanian Ignimbrite eruption (Campi Flegrei, Italy). Earth Planet Sci Lett 306:217–228

Belkin HE, Rolandi G, Jackson JC, Cannatelli C, Doherty AL, Petrosino P, De Vivo B (2016) Mineralogy and geochemistry of the older (>40 ka) ignimbrites on the Campanian Plain, Southern Italy. J Volcanol Geotherm Res 323:1–18

Black BA, Neely RR, Manga M (2015) Campanian Ignimbrite volcanism, climate and the final decline of the Neanderthals. Geology 43:411–414

Blockley SPE, Bronk Ramsey C, Pyle DM (2008) Improved age modeling and high-precision age estimates of late quaternary tephras, for accurate palaeoclimate reconstruction. J Volcanol Geotherm Res 177:251–262

Bourne AJ, Lowe JJ, Trincardi F, Asioli A, Blockley SPE, Wulf S, Matthews IP, Piva A, Vigliotti L (2010) Distal tephra record for the last ca 105,000 years from core PRAD 1–2 in the central Adriatic Sea: implications for marine tephrostratigraphy. Quat Sci Rev 29:3079–3094

Bourne AJ, Albert PG, Matthews IP, Wulf S, Lowe JJ, Asioli A, Blockley SPE, Trincardi F (2015) Tephrochronology of core PRAD1-2 from the Adriatic Sea: insights into Italian explosive volcanism for the period 200–80 ka. Quat Sci Rev 116:28–43

Brauer A, Allen JRM, Mingram J, Dulski P, Wulf S, Huntley B (2007) Evidence for the last interglacial chronology and environmental change from Southern Europe. P Natl Acad Sci 104:450–455

Bryant CJ, Arculus RJ, Eggins SM (2003) The geochemical evolution of the Izu-Bonin arc system: a perspective from tephras recovered by deep-sea drilling. Geochem Geophys Geosys 1094

Calanchi N, Cattaneo A, Dinelli E, Gasparotto G, Lucchini F (1998) Tephra layers in late quaternary sediments of the Central Adriatic Sea. Mar Geol 149:191–209

Calanchi N, Dinelli E (2008) Tephrostratigraphy of the last 170 ka in sedimentary successions from the Adriatic Sea. J Volcanol Geotherm Res 177:81–95

Calcaterra D, Coppin D, de Vita S, Di Vito MA, Orsi G, Palma B, Parise M (2007) Slope processes in weathered volcaniclastic deposits within the city of Naples: the Camaldoli Hills case. Geomorphology 87(3):132–157

Capraro L, Massari F, Rio D, Fornaciari E, Backman J, Channell JET, Macri P, Prosser G, Speranza F (2011) Chronology of the lower-middle Pleistocene succession of the south western part of the Crotone Basin (Calabria, Southern Italy). Quat Sci Rev 30:1185–1200

Carey S (1997) Influence of convective sedimentation on the formation of widespread tephra fall layers in the deep sea. Geology 25:839–842

Caron B, Sulpizio R, Zanchetta G, Siani G, Santacroce R (2010) The late Holocene to Pleistocene tephrostratigraphic record of Lake Ohrid (Albania). Comptes Rendus Geosci 342:453–466

Ciaranfi N, Lirer F, Lirer L, Lourens LJ, Maiorano P, Marino M, Petrosino P, Sprovieri M, Stefanelli S, Brilli M, Girone A, Joannin S, Pelosi N, Vallefuoco M (2010) Integrated stratigraphy and astronomical tuning of lower middle Pleistocene Montalbano Jonico section (Southern Italy). Quat Int 219:109–120

Civetta L, Orsi G, Pappalardo L, Fisher RV, Heiken G, Ort M (1997) Geochemical zoning, mingling, eruptive dynamics and depositional processes; the Campanian Ignimbrite, Phlegrean Fields caldera, Italy. J Volcanol Geotherm Res 75:183–219

Cornell W, Carey S, Sigurdsson H (1983) Computer simulation of transport and deposition of the Campanian Y-5 ash. J Volcanol Geotherm Res 17:89–109
Costa A, Melnik O, Vedeneeva E (2007) Thermal effects during magma ascent in conduits. J Geophys Res 112: B12205
Costa A, Folch A, Macedonio G, Giaccio B, Isaia R, Smith VC (2012) Quantifying volcanic ash dispersal and impact of the Campanian Ignimbrite super-eruption. Geophys Res Lett. https://doi.org/10.1029/2012GL051605
Cramp A, Vitaliano CJ, Collins MB (1989) Identification and dispersion of the campanian ash layer (Y-5) in the sediments of the Eastern Mediterranean. Geo-Marine Lett 9:19–25
Damaschke M, Sulpizio R, Zanchetta G, Wagner B, Böhm A, Nowacyk N, Rethemeyer J, Hilgers A (2013) Tephrostratigraphic studies on a sediment core from Lake Prespa in the Balkans. Clim past 9:267–287
D'Antonio M, Civetta L, Orsi G, Pappalardo L, Piochi M, Carandente A, de Vita S, Di Vito MA, Isaia R (1999) The present state of the magmatic system of the Campi Flegrei caldera based on a reconstruction of its behavior in the past 12 ka. J Volcanol Geotherm Res 91:247–268
de Vita S, Orsi G, Civetta L, Carandente A, D'Antonio M, Deino A, Di Cesare T, Di Vito MA, Fisher RV, Isaia R, Marotta E, Necco A, Ort M, Pappalardo L, Piochi M, Southon J (1999) The Agnano-Monte Spina eruption (4100 years BP) in the restless Campi Flegrei caldera (Italy). J Volcanol Geotherm Res 91:269–301
De Vivo B, Rolandi G, Gans PB, Calvert A, Bohrson WA, Spera FJ, Belkin HE (2001) New constraints on the pyroclastic eruptive history of the Campanian volcanic Plain (Italy). Miner Petrol 73:47–65
Deino AL, Orsi G, de Vita S, Piochi M (2004) The age of the Neapolitan Yellow Tuff caldera-forming eruption (Campi Flegrei caldera Italy) assessed by Ar-40/Ar-39 dating method. J Volcanol Geotherm Res 133:157–170
Delmelle P (2003) Environmental impacts of tropospheric volcanic gas plumes. Geol Soc London Sp Publ 213:381–399
Di Renzo V, Di Vito MA, Arienzo I, Carandente A, Civetta L, D'Antonio M, Giordano F, Orsi G, Tonarini S (2007) Magmatic history of Somma-Vesuvius on the basis of new geochemical and isotopic data from a deep borehole (Camaldoli della Torre). J Petrol 48:753–784
Di Roberto A, Smedile A, Del Carlo P, De Martini PM, Iorio M, Petrelli M, Pantosti D, Pinzi S, Todrani A (2018) Tephra and cryptotephra in a ~60,000-year-old lacustrine sequence from the Fucino Basin: new insights into the major explosive events in Italy. Bull Volcanol 80:20
Di Vito M, Isaia R, Orsi G, Southon J, di Vita S, D'Antonio M, Pappalardo L, Piochi M (1999) Volcanism and deformation since 12,000 years at the Phlegrean Fields caldera (Italy). J Volcanol Geotherm Res 91:221–246
Di Vito MA, Sulpizio R, Zanchetta G, D'Orazio M (2008) The late Pleistocene pyroclastic deposits of the Campanian Plain: new insights into the explosive activity of Neapolitan volcanoes. J Volcanol Geotherm Res 177:19–48
Di Vito MA, Acocella V, Aiello G, Barra D, Battaglia M, Carandente A, Del Gaudio C, de Vita S, Ricciardi GP, Ricco C, Scandone R, Terrasi F (2016) Magma transfer at Campi Flegrei caldera (Italy) before the 1538 AD eruption. Sci Rep 6:32245
Donato P, Albert PG, Crocitti M, De Rosa R, Menzies MA (2016) Tephra layers along the Southern Tyrrhenian coast of Italy: links to the X-5 & X-6 using volcanic glass geochemistry. J Volcanol Geotherm Res 317:30–41
Douka K, Jacobs Z, Lane C, Grün R, Farr L, Hunt C, Inglis RH, Reynolds T, Albert P, Aubert M, Cullen V, Hill E, Kinsley L, Roberts RG, Tomlinson EL, Wulf S, Barker G (2013) The chronostratigraphy of the Haua Fteah cave (Cyrenaica, Northeast Libya). J Hum Evol 66:39–63
Douka K, Jacobs Z, Lane C, Gruen R, Farr L, Hunt C, Inglis RH, Reynolds T, Albert P, Aubert M, Cullen V, Hill E, Kinsley L, Roberts RG, Tomlinson EL, Wulf S, Barker G (2014) The chronostratigraphy of the Haua Fteah cave (Cyrenaica, Northeast Libya). J Hum Evol 66:39–63
Dugmore AJ, Larsen G, Newton AJ (1995) Seven tephra isochrones in Scotland. The Holocene 5:257–266
Engwell SL, Sparks RSJ, Carey S (2014) Physical characteristics of tephra layers in the deep sea realm: the Campanian Ignimbrite eruption. In: Austin WEN, Abbott PM, Davies SM, Pearce NJG, Wastegard S (eds) Marine tephrochronology, vol 398. Geol Soc London Sp Publ, UK, pp 47–64
Fedele FG, Giaccio B, Isaia R, Orsi G (2003) The Campanian Ignimbrite eruption, heinrich event 4, and the palaeolithic change in Europe: a high-resolution investigation, in Volcanism and the earth's atmosphere. Geophys Monogr Ser 139:301–325
Fedele FG, Giaccio B, Isaia R, Orsi G, Carroll M, Scaillet B (2007) The Campanian Ignimbrite factor: towards a reappraisal of the middle to upper palaeolithic "transition". In: Grattan J, Torrence R (eds) Living under the shadow: the cultural impacts of volcanic eruptions. One world archaeology series, vol 53. Left Coast Press, Walnutt Creek, CA, pp 19–41
Fedele FG, Giaccio B, Hajdas I (2008) Timescales and cultural process at 40,000 BP in the light of the Campanian Ignimbrite eruption, Western Eurasia. J Hum Evol 55:834–857
Fisher RV, Orsi G, Ort M, Heiken G (1993) Mobility of a large-volume pyroclastic flow emplacement of the Campanian Ignimbrite, Italy. J Volcanol Geotherm Res 56:205–220
Fitzsimmons KE, Hambach U, Veres D, Iovita R (2013) The Campanian Ignimbrite eruption: new data on

volcanic ash dispersal and its potential impact on human evolution. PLoS ONE 0065839
Forni F, Petricca E, Bachmann O, Mollo S, De Astis G, Piochi M (2018) The role of magma mixing/mingling and cumulate melting in the Neapolitan Yellow Tuff caldera-forming eruption (Campi Flegrei, Southern Italy). Contrib Mineral Petrol 173:45
Frezzotti M, Narcisi B (1996) Late quaternary tephra-derived paleosols in central Italy's carbonate Apennine range: stratigraphical and paleoclimatological implications. Quat Int 34–36:147–153
Galli P, Giaccio B, Messina P, Peronace E, Amato V, Naso G, Nomade S, Pereira A, Piscitelli S, Bellanova J, Billi A, Blamart D, Galderisi A, Giocoli A, Stabile T, Thil F (2017) Middle to Late Pleistocene activity of the northern Matese fault system (Southern Apennines, Italy). Tectonophysics 699:61–81
Giaccio B, Isaia R, Fedele FG, Di Canzio E, Hoffecker J, Ronchitelli A, Sinitsyn AA, Aniko-vich M, Lisitsyn SN, Popov VV (2008) The Campanian Ignimbrite and Codola tephra layers: two temporal/stratigraphic markers for the early upper palaeolithic in Southern Italy and Eastern Europe. J Volcanol Geotherm Res 177:208–226
Giaccio B, Nomade S, Wulf S, Isaia R, Sottili G, Cavuoto G, Galli P, Messina P, Sposato A, Sulpizio R, Zanchetta G (2012) The late MIS 5 Mediterranean tephra markers: a reappraisal from peninsular Italy terrestrial records. Quat Sci Rev 56:31–45
Giaccio B, Castorina F, Nomade S, Scardia G, Voltaggio M, Sagnotti L (2013) Revised chronology of the Sulmona lacustrine succession, Central Italy. J Quat Sci 28:545–551
Giaccio B, Galli P, Peronace E, Arienzo I, Nomade S, Cavinato GP, Mancini M, Messina P, Sottili G (2014) A 560–440 ka tephra record from the Mercure Basin, Southern Italy: volcanological and tephrostratigraphic implications. J Quat Sci 29:232–248
Giaccio B, Hajdas I, Isaia R, Deino A, Nomade S (2017) High-precision ^{14}C and ^{40}Ar/^{39}Ar dating of the Campanian Ignimbrite (Y-5) reconciles the time-scales of climatic-cultural processes at 40 ka. Sci Rep 7:45940
Giordano R, Sulpizio R, Caggianelli A (2014) Mineral phases as a tool for robust correlation of proximal–distal tephra in the central mediterranean area BT - STRATI 2013. In: Rocha R, Pais J, Kullberg JC, Finney S (eds). Springer International Publishing, Cham, pp 1227–1232
Golovanova LV, Doronichev VB, Cleghorn NE, Koulkova MA, Sapelko TV, Shackley MS (2010) Significance of ecological factors in the middle to upper paleolithic transition. Curr Anthropol 51:655–691
Guidoboni E, Ciuccarelli C (2011) The Campi Flegrei caldera: historical revision and new data on seismic crises bradyseisms the Monte Nuovo eruption and ensuing earthquakes (twelfth century 1582 AD). Bull Volcanol 73:655–677
Higham T, Basell L, Jacobi R, Wood R, Ramsey CB, Conard NJ (2012) Testing models for the beginnings of the aurignacian and the advent of figurative art and music: the radiocarbon chronology of Geißenklösterle. J Hum Evol 62:664–676
Hopkins JL, Wilson CJN, Millet M-A, Leonard GS, Timm C, McGee LE, Smith IEM, Smith EGC (2017) Multi-criteria correlation of tephra deposits to source centres applied in the Auckland volcanic field, New Zealand. Bull Volcanol 79:55
Huntley B, Alfano MJO, Allen JRM, Pollard D, Tzedakis PC, de Beaulieu J-L, Gruger E, Watts B (2003) European vegetation during marine oxygen isotope stage-3. Quat Res 59:195–212
Insinga DD, Tamburrino S, Lirer F, Vezzoli L, Barra M, De Lange GJ, Tiepolo M, Vallefuoco M, Mazzola S, Sprovieri M (2014) Tephrochronology of the astronomically-tuned KC01B deep-sea core, Ionian Sea: insights into the explosive activity of the Central Mediterranean area during the last 200 ka. Quat Sci Rev 85:63–84
Iorio M, Liddicoat J, Budillon F, Incoronato A, Insinga DD, Cassata WS, Lubritto C, Angelino A, Coe RS, Tamburrino S (2014) Combined palaeomagnetic secular variation and petro-physical records to time-constrain geological and hazardous events: an example from the Eastern Tyrrhenian sea over the last 120 ka. Glob Planet Change 113:91–109
Isaia R, Iannuzzi E, Sbrana A, Marianelli P, Donadio C, Conforti A, D'Argenio B (2016) Note Illustrative della Carta Geologica d'Italia alla scala 1: 50.000, Foglio 446–447 Napoli (aree emerse). Istituto Superiore per la Protezione e la Ricerca Ambientale (ISPRA). www.isprambiente.gov.it/Media/carg/note_illustrative/446_447_Napoli.pdf
Ivanova S, Gurova M, Spassov N, Hristova L, Tzankov N, Popov V, Marinova E, Makedonska J, Smith V, Ottoni C, Lewis M (2016) Magura Cave, Bulgaria: a multidisciplinary study of late pleistocene human palaeoenvironment in the Balkans. Quat Int 415:86–108
Jones M, Sparks RSJ, Valdes PJ (2007) The climatic impact of supervolcanic ash blankets. Clim Dyn 29:553–564
Jovanovska E, Cvetkoska A, Hauffe T, Levkov Z, Wagner B, Sulpizio R, Francke A, Albrecht C, Wilke T (2016) Differential resilience of ancient sister lakes Ohrid and Prespa to environmental disturbances during the late pleistocene. BiogeoSci 13:1149–1161
Keller J, Ryan WBF, Ninkovich D, Altherr R (1978) Explosive volcanic activity in the Mediterranean over the past 200,000 yr as recorded in deep-sea sediments. Geol Soc Am Bull 89:591–604
Kockum PCF, Herbert RB, Gislason SR (2006) A diverse ecosystem response to volcanic aerosols. Chem Geol 231:57–66
Kraml M (1997) Laser ^{40}Ar/^{39}Ar-Datierimgen an distalen marinen Tephren desjung-quartaren Mediterranean Vulkanismus (Ionisches Meer, METEOR-Fahrt

25/4). PhD Thesis, Albert Ludwings-Universitat Freiburg, Germany, p 216

Lane CS, Andric M, Cullen VL, Blockley SPE (2011) The occurrence of distal Icelandic and Italian tephra in the lateglacial of Lake Bled, Slovenia. Quat Sci Rev 30:1013–1018

Lane CS, Brauer A, Blockley SPE, Dulski P (2013) Volcanic ash reveals time-transgressive abrupt climate change during the Younger Dryas. Geology 41:1251–1254

Lane CS, Brauer B, Martín-Puertas C, Blockley SPE, Smith VC, Tomlinson EL (2015) The late quaternary tephrostratigraphy of annually laminated sediments from Meerfelder Maar, Germany. Quat Sci Rev 122:192–206

Lane CS, Lowe DJ, Blockley SPE, Suzuki T, Smith VC (2017) Advancing tephrochronology as a global dating tool: applications in volcanology, archaeology and paleoclimate research. Quat Geochronol 40:1–7

Leicher N, Zanchetta G, Sulpizio R, Giaccio B, Wagner B, Nomade S, Francke A, Del Carlo P (2016) First tephrostratigraphic results of the DEEP site record from Lake Ohrid (Macedonia and Albania). Biogeosciences 13:2151–2178

Lowe DJ (2011) Tephrochronology and its application: a review. Quat Geochronol 6:107–153

Lowe JJ, Blockley S, Trincardi F, Asioli A, Cattaneo A, Matthews IP, Pollard M, Wulf S (2007) Age modelling of late quaternary marine sequences in the adriatic: towards improved precision and accuracy using volcanic event stratigraphy. Cont Shelf Res 27:560–582

Lowe J, Barton N, Blockley S, Bronk Ramsey C, Cullen VL, Davies W, Gamble C, Grant K, Hardiman M, Housley R, Lane CS, Lee S, Lewis M, MacLeod A, Menzies M, Müller W, Pollard M, Price C, Roberts AP, Rohling EJ, Satow C, Smith V, Stringer C, Tomlinson EL, White D, Albert P, Arienzo I, Barker G, Carandente A, Civetta L, Farrand W, Ferrier C, Gaudelli JL, Karkanas P, Koumouzelis M, Muller UC, Orsi G, Pross J, Rosi M, Shalamanov-Korobar L, Sirakov N, Tzedakis PC, Boriæ D (2012) Volcanic ash layers illuminate the resilience of Neanderthals and early modern humans to natural hazards. P Natl Acad Sci 109(34):13532–13537

Lowe JJ, Housley RA, Tomlinson EL (2015) Synchronising environmental and archaeological records using volcanic isochrons. Quat Sci Rev 118:1–242

Lucchi F, Tranne CA, De Astis G, Keller J, Losito R, Morche W (2008) Stratigraphy and significance of Brown tuffs on the Aeolian islands (Southern Italy). J Volcanol Geotherm Res 177:49–70

Lucchi F, Gertisser R, Keller J, Forni F, De Astis G, Tranne CA (2013) Eruptive history and magmatic evolution of the island of Salina (central Aeolian archipelago). In: Lucchi,F, Peccerillo A, Keller J, Tranne CA, Rossi PL (eds) The Aeolian Islands Volcanoes. Geol Soc London Memoirs 37:155–211

Magny M, de Beaulieu J-L, Drescher-Schneider R, Vanniere B, Walter-Simonnet A-V, Millet L, Bossuet G, Peyron O (2006) Climatic oscillations in central Italy during the last glacial-Holocene transition: the record from Lake Accesa. J Quat Sci 21:311–320

Mannella G, Giaccio B, Zanchetta G, Regattieri E, Niespolo EM, Pereira A, Renne PR, Nomade S, Leicher N, Perchiazzi N, Wagner B (2019) Palaeoenvironmental and palaeohydrological variability of mountain areas in the central Mediterranean region: a 190 ka-long chronicle from the independently dated Fucino palaeolake record (Central Italy). Quat Sci Rev 210:190–210

Marciano R, Munno R, Petrosino P, Santangelo N, Santo A, Villa I (2008) Late quaternary tephra layers along the Cilento coastline (southern Italy). J Volcanol Geotherm Res 177:227–243

Margari V, Pyle DM, Bryant C, Gibbard PL (2007) Mediterranean tephra stratigraphy revisited: Results from a long terrestrial sequence on Lesvos island, Greece. J Volcanol Geotherm Res 163:34–54

Marti A, Folch A, Costa A, Engwell S (2016) Reconstructing the plinian and co-ignimbrite sources of large volcanic eruptions: a novel approach for the Campanian Ignimbrite. Sci Rep 6:21220

Mastrolorenzo G, Pappalardo L, Troise C, Panizza A, De Natale G (2008) Probabilistic tephra hazard maps for the Neapolitan area: quantitative volcanological study of Campi Flegrei eruptions. J Geophys Res 113: B07203

Matthews IP, Trincardi F, Lowe JJ, Bourne AJ, MacLeod A, Abbott PM, Andersen N, Asioli A, Blockley SPE, Lane CS, Oh YA, Satow CS, Staff RA, Wulf S (2015) Developing a robust tephrochronological framework for late quaternary marine records in the Southern Adriatic Sea: new data from core station SA03-11. Quat Sci Rev 118:84–104

Matthews NE, Smith VC, Costa A, Durrant AJ, Pyle DM, Pearce NJG (2012) Ultra-distal tephra deposits from super-eruptions: examples from Toba, Indonesia and Taupo Volcanic Zone, New Zealand. Quat Int 258:54–79

Morley MW, Woodward JC (2011) The Campanian Ignimbrite (Y5) tephra at Crvena Stijena Rockshelter, Montenegro. Quat Res 75:683–696

Müller UC, Pross J, Tzedakis PC, Gamble C, Kotthoff U, Schmiedl G, Wulf S, Christanis K (2011) The role of climate in the spread of modern humans into Europe. Quat Sci Rev 30:273–279

Munno R, Petrosino P (2004) New constraints on the occurrence of the Y-3 upper Pliestocene tephra marker layer in the Tyrrhenian Sea. Il Quaternario 17:11–20

Munno R, Petrosino P (2007) Tephra layers in the S. Gregorio Magno lacustrine succession. J Quat Sci 22:247–266

Orsi G, D'Antonio M, de Vita S, Gallo G (1992) The Neapolitan Yellow Tuff, a large-magnitude Trachytic phreatoPlinian eruption—eruptive dynamics, magma withdrawal and caldera collapse. J Volcanol Geotherm Res 53:275–287

Orsi G, Civetta L, D'Antonio M, Di Girolamo P, Piochi M (1995) Step-filling and development of a three-layer magma chamber: the neapolitan yellow tuff case history. J Volcanol Geotherm Res 67:291–312

Orsi G, de Vita S, Di Vito M (1996) The restless, resurgent Campi Flegrei nested caldera (Italy): constraints on its evolution and configuration. J Volcanol Geotherm Res 74:179–214

Orsi G, Di Vito MA, Isaia R (2004) Volcanic hazard assessment at the restless Campi Flegrei caldera. Bull Volcanol 66:514–530

Ort M, Orsi G, Pappalardo L, Fisher RV (2003) Anisotropy of magnetic susceptibility studies of depositional processes in the Campanian Ignimbrite, Italy. Bull Volcanol 65:55–72

Pabst S, Worner G, Civetta L, Tesoro R (2008) Magma chamber evolution prior to the Campanian Ignimbrite and Neapolitan Yellow Tuff eruptions (Campi Flegrei, Italy). Bull Volcanol 70:961–976

Pappalardo L, Civetta L, D'Antonio M, Deino A, Di Vito M, Orsi G, Carandente A, de Vita S, Isaia R, Piochi M (1999) Chemical and Sr-isotopical evolution of the Phlegrean magmatic system before the Campanian Ignimbrite and the Neapolitan Yellow Tuff eruptions. J Volcanol Geotherm Res 91:141–166

Paterne M, Guichard F, Labeyrie J (1988) Explosive activity of the south Italian volcanoes during the past 80,000 years as determined by marine tephrochronology. J Volcanol Geotherm Res 34:153–172

Paterne M, Guichard F, Duplessy JC, Siani G, Sulpizio R, Labeyrie J (2008) A 90,000–200,000 years marine tephra record of Italian volcanic activity in the Central Mediterranean Sea. J Volcanol Geotherm Res 177:187–196

Petrosino P, Jicha BR, Mazzeo FC, Russo Ermolli E (2014) A high resolution tephrochronological record of MIS 14–12 in the Southern Apennines (Acerno Basin, Italy). J Volcanol Geotherm Res 274:34–50

Petrosino P, Jicha BR, Mazzeo FC, Ciaranfi N, Girone A, Maiorano P, Marino M (2015) The Montalbano Jonico marine succession: an archive for distal tephra layers at the Early-Middle Pleistocene boundary in Southern Italy. Quat Int 383:89–103

Pistolesi M, Isaia R, Marianelli P, Bertagnini A, Fourmentraux C, Albert P, Tomlinson E, Menzies M, Rosi M, Sbrana A (2016) Simultaneous eruptions from multiple vents at Campi Flegrei (Italy) highlight new eruption processes at calderas. Geology 44 (6):487–490

Praetorius S, Mix A, Jensen B, Froese D, Milne G, Wolhowe M, Addison J, Prahl F (2016) Interaction between climate, volcanism, and isostatic rebound in Southeast Alaska during the last deglaciation. Earth Planet Sci Lett 452:79–89

Pyle DM, van Andel TH, Paschos P, van den Bogaard P (1998) An exceptionally thick Middle Pleistocene tephra layer from Epirus, Greece. Quat Res 49:280–286

Pyle DM, Ricketts GD, Margari V, van Andel TH, Sinitsyn AA, Praslov ND, Lisitsyn S (2006) Wide dispersal and deposition of distal tephra during the Pleistocene 'Campanian Ignimbrite/Y5' eruption, Italy. Quat Sci Rev 25:2713–2728

Radić D, Lugović B, Marjanac L (2007) Neapolitan Yellow Tuff (NYT) from the Pleistocene sediments in Vela Spila on the island of Korčula: a valuable chronostratigraphic marker of the transition from the Palaeolithic to the Mesolithic. Op Arch 31:7–26

Rasmussen SO, Andersen KK, Svensson AM, Steffensen JP, Vinther BM, Clausen HB, Siggaard-Andersen M-L, Johnsen SJ, Larsen LB, Dahl-Jensen D, Bigler M, Röthlisberger R, Fischer H, Goto-Azuma K, Hansson ME, Ruth U (2006) A new Greenland ice core chronology for the last glacial termination. J Geophys Res 111:D06102

Rawson H, Naranjo JA, Smith V, Fontijn K, Pyle DM, Mather TA, Moreno H (2015) The frequency and magnitude of post-glacial explosive eruptions at Volcán Mocho-Choshuenco, southern Chile. J Volcanol Geotherm Res 299:103–129

Regattieri E, Giaccio B, Zanchetta G, Drysdale R, Galli P, Nomade S, Peronace E, Wulf S (2015) Hydrological variability over the Apennines during the early last glacial precession minimum, as revealed by a stable isotope record from Sulmona basin, Central Italy. J Quat Sci 30:19–31

Reimer PJ, Baillie MGL, Bard E, Bayliss A, Beck JW, Blackwell PG, Ramsey BC, Buck CE, Burr GS, Edwards RL, Friedrich M, Grootes PM, Guiderson TP, Hajdas I, Heaton TJ, Hogg AG, Hughen KA, Kaiser KF, Kromer B, McCormac FG, Manning SW, Richards DA, Southon JR, Talamo S, Turney CSM, van der Plicht J, Weyhenmeyer CE (2009) IntCal09 and Marine09 radiocarbon age calibration curves, 0-50,000 years cal BP. Radiocarbon 51:1111–1150

Rolandi G, Bellucci F, Heizler MT, Belkin HE, De Vivo B (2003) Tectonic controls on the genesis of ignimbrites from the Campanian volcanic zone, Southern Italy. Miner Petrol 79:3–31

Rose WI, Durant AJ (2009) Fine ash content of explosive eruptions. J Volcanol Geotherm Res 186:32–39

Rosi M, Vezzoli L, Castelmenzano A, Grieco G (1999) Plinian pumice fall deposit of the Campanian Ignimbrite eruption (Phlegraean Fields, Italy). J Volcanol Geotherm Res 91:179–198

Scarpati C, Perrotta A, Lepore S, Calvert A (2013) Eruptive history of Neapolitan volcanoes: constraints from ^{40}Ar-^{39}Ar dating. Geol Mag 150:412–425

Schmidt R, van den Bogaard C, Merkt J, Muller J (2002) A new lateglacial chronostratigraphic tephra marker for the south-eastern Alps: The Neapolitan Yellow Tuff (NYT) in Langsee (Austria) in the context of a regional biostratigraphy and palaeoclimate. Quat Int 88:45–56

Siani G, Sulpizio R, Paterne M, Sbrana A (2004) Tephrostratigraphy study for the last 18,000 C-14 years in a deep-sea sediment sequence for the South Adriatic. Quat Sci Rev 23:2485–2500

Smith VC, Isaia R, Pearce NJG (2011) Tephrostratigraphy and glass compositions of post-15 kyr Campi Flegrei eruptions: implications for eruption history and chronostratigraphic markers. Quat Sci Rev 30:3638–3660

Smith VC, Isaia R, Engwell S, Albert PG (2016) Tephra dispersal during the Campanian Ignimbrite (Italy) eruption: implications for ultra-distal ash transport during the large caldera-forming eruption. Bull Volcanol 78:45

Sparks RSJ, Huang TC (1980) The volcanological significance of deep-sea ash layers associated with ignimbrites. Geol Mag 117:425–436

St Seymour K, Christanis K (1995) Correlation of a tephra layer in western Greece with a late pleistocene eruption in the Campanian Province of Italy. Quat Res 43:46–54

St Seymour K, Christanis K, Bouzinos A, Papazisimou S, Papatheodorou G, Moran E, Dénès G (2004) Tephrostratigraphy and tephrochronology in the Philippi peat basin, Macedonia, Northern Hellas (Greece). Quat Int 121:53–65

Steffensen et al (2008) High-resolution Greenland ice core data show abrupt climate change happens in few years. Science 321:680–684

Sulpizio R (2005) Three empirical methods for the calculation of distal volume of tephra-fall deposits. J Volcanol Geotherm Res 145:315–336

Sulpizio R, Zanchetta G, Paterne M, Siani G (2003) A review of tephrostratigraphy in central and Southern Italy during the last 65 ka. Quat 16:91–108

Sulpizio R, van Welden A, Caron B, Zanchetta G (2010) The Holocene tephrostratigraphic record of Lake Shkodra (Albania and Montenegro). J Quat Sci 25:633–650

Thunell R, Federman A, Sparks S, Williams D (1979) The age, origin, and volcanological significance of the Y-5 ash layer in the Mediterranean. Quat Res 12:241–253

Tomlinson EL, Arienzo I, Wulf S, Smith VC, Carandente A, Civetta L, Hardiman M, Lane CS, Orsi G, Rosi M, Muller W, Menzies MA (2012) Geochemistry of the Phlegrean fields (Italy) proximal sources for major Mediterranean tephras: Implications for the dispersal of Plinian and co-ignimbritic components of explosive eruptions. Geochim Cosmochim Acta 93:102–128

Tomlinson EL, Smith VC, Albert PG, Aydar E, Civetta L, Cioni R, Çubukçu HE, Gertisser R, Isaia R, Menzies MA, Orsi G, Rosi M, Zanchetta G (2015) The major and trace element glass compositions of the productive Mediterranean volcanic sources: tools for correlating distal tephra layers in and around Europe. Quat Sci Rev 118:48–66

Ton-That T, Singer B, Paterne M (2001) $^{40}Ar/^{39}Ar$ dating of latest pleistocene (41 ka) marine tephra in the Mediterranean Sea: implications for global climate records. Earth Planet Sci Lett 184:645–658

Tzedakis PC, Hughen KA, Cacho I, Harvati K (2007) Placing late Neanderthals in a climatic context. Nature 449:206–208

Veres D, Lane CS, Timar-Gabor A, Hambach U, Constantin D, Szakács A, Fülling A, Onac BP (2013) The Campanian Ignimbrite/Y5 tephra layer—a regional stratigraphic marker for isotope stage 3 deposits in the lower Danube region, Romania. Quat Int 293:22–33

Vezzoli L (1991) Tephra layers in Bannock Basin (Eastern Mediterranean). Mar Geol 100:21–34

Vinci A (1985) Distribution and chemical composition of tephra layers from Eastern Mediterranean abyssal sediments. Mar Geol 64:143–155

Vogel H, Zanchetta G, Sulpizio R, Wagner B, Nowaczyk N (2010) A tephrostratigraphic record for the last glacial-interglacial cycle from Lake Ohrid, Albania and Macedonia. J Quat Sci 25:320–338

Wagner B, Sulpizio R, Zanchetta G, Wulf S, Wessels M, Daut G, Nowaczyk N (2008) The last 40 ka tephrostratigraphic record of Lake Ohrid, Albania and Macedonia: a very distal archive for ash dispersal from Italian volcanoes. J Volcanol Geotherm Res 177:71–80

Wohletz K, Orsi G, de Vita S (1995) Eruptive mechanisms of the Neapolitan Yellow Tuff interpreted from stratigraphic, chemical, and granulometric data. J Volcanol Geotherm Res 67:263–290

Wulf S, Kraml M, Brauer A, Keller J, Negendank JFW (2004) Tephrochronology of the 100 ka lacustrine sediment record of Lago Grande di Monticchio (southern Italy). Quat Int 122:7–30

Wulf S, Kraml M, Keller J (2008) Towards a detailed distal tephrostratigraphy in the Central Mediterranean: the last 20,000 yrs record of Lago Grande di Monticchio. J Volcanol Geotherm Res 177:118–132

Wulf S, Keller J, Paterne M, Mingram J, Lauterbach S, Opitz S, Sottili G, Giaccio B, Albert PG, Satow C, Tomlinson EL, Viccaro M, Brauer A (2012) The 100–133 ka record of Italian explosive volcanism and revised tephrochronology of Lago Grande di Monticchio. Quat Sci Rev 58:104–123

Wulf S, Hardiman MJ, Staff RA, Koutsodendris A, Appelt O, Blockley SPE, Lowe JJ, Manning CJ, Ottolini L, Schmitt AK, Smith VC, Tomlinson EL, Vakhrameeva P, Knipping M, Kotthoff U, Milner AM, Müller UC, Christanis K, Kalaitzidis S, Tzedakis PC, Schmiedl G, Pross J (2018) The marine isotope stage 1–5 cryptotephra record of Tenaghi Philippon, Greece: towards a detailed tephrostratigraphic framework for the Eastern Mediterranean region. Quat Sci Rev 186:236–262

Wutke K, Wulf S, Tomlinson EL, Hardiman M, Dulski P, Luterbacher J, Brauer A (2015) Geochemical properties and environmental impacts of seven Campanian tephra layers deposited between 40 and 38 ka BP in the varved lake sediments of Lago Grande di Monticchio, Southern Italy. Quat Sci Rev 118:67–83

Zanchetta G, Sulpizio R, Giaccio B, Siani G, Paterne M, Wulf S, D'Orazio M (2008) The Y-3 tephra: a last Glacial stratigraphic marker for the central Mediterranean basin. J Volcanol Geotherm Res 177:145–154

Rheological Properties of the Magmas Feeding the Campi Flegrei Caldera (Italy) and Their Influence on Mixing Processes

Cristina P. De Campos, Kai-Uwe Hess, Diego Perugini, and Donald B. Dingwell

Abstract

This chapter reviews and discusses the main rheological and physical properties and models (viscosity and density) of the melts feeding the Campi Flegrei caldera volcanism. Viscosity and density control flow and diffusion, and thus multicomponent convection and chemical mobility in the magma chamber. These, in turn, are thought to constrain magma mixing processes in the system. Our main goal is to summarise and analyse mixing experiments using natural volcanic products of the caldera as starting material. The mixing experiments have been performed using different devices (Taylor-Couette and centrifuge). Results show how easily Campi Flegrei caldera melts may mix. They confirm that different chemical elements homogenise in the melt at differing rates, providing an innovative quantitative approach, the estimation of a new parameter for measuring multi-component diffusion in magmas: Relaxation of Concentration Variance or Concentration Variance Decay. This enables the measurement of mobility for all elements present in the melt simultaneously. Comparing experimental and natural data clearly confirm the unavoidability of mixing during the replenishment history of the caldera reservoir(s).

1 Introduction

The most active magmatic system in the Mediterranean region is certainly the Neapolitan Volcanic Area. For more than 300 ka (Rolandi et al. 2003), the predominantly explosive character of the Phlegraean volcanism has greatly affected the surroundings, also well beyond this area. The Campi Flegrei caldera (CFc) accounts for the most devastating eruptions. Detailed information on the geology and volcanology of CFc will be found in Chap. Volcanic and Deformation History of the Campi Flegrei Volcanic Field, Italy. For the magmatic system, Chaps. An Evolutionary Model for the Magmatic System of the Campi Flegrei Volcanic Field (Italy) Constrained by Petrochemical Data; Origin and Differentiation History of the Magmatic System Feeding the Campi Flegrei Volcanic Field (Italy) Constrained by Radiogenic and Stable Isotope Data; Tephrochronology and Geochemistry of Tephra from the Campi Flegrei Volcanic Field, Italy will provide a thorough review of the state of the art, and this chapter will illustrate physical modelling and numerical simulation of its dynamics.

C. P. De Campos (✉) · K.-U. Hess · D. B. Dingwell
Department of Earth and Environmental Sciences, Ludwig Maximilians University, Munich, Germany
e-mail: decampos@lmu.de

D. Perugini
Dipartimento di Fisica e Geologia, Università degli Studi di Perugia, Perugia, Italy

© Springer-Verlag GmbH Germany, part of Springer Nature 2022
G. Orsi et al. (eds.), *Campi Flegrei*, Active Volcanoes of the World,
https://doi.org/10.1007/978-3-642-37060-1_6

The eruptive behaviour and compositional variations of volcanic systems is defined by the physical and chemical properties of their feeding magmas (e.g., Dingwell et al. 1993; Mungall et al. 1999). From all properties governing magmatic processes, viscosity is mostly considered to be the single most important property (e.g., Sparks 2004; Dingwell 2006). Varying by over 15 orders of magnitude within the range of natural magma compositions, it controls the formation, transport and crystallisation of magmas in all volcanic activities, from explosive to effusive flows (e.g., Dingwell 1996; Giordano et al. 2008). Viscosity has also been recognised as one of the main parameters controlling convection and diffusion and, therefore, the rheology of mixing. Several laboratory experiments with analogue fluids at room temperature have been used to quantify the circumstances under which fluids can be mixed by natural convection at high flux (Rayleigh numbers, see De Campos et al. 2011 and references therein). The injection of one buoyant fluid into another at a fixed rate was the most common procedure. In this way the nature of the resulting compositional convection has been found to depend on two key dimensionless parameters: the Reynolds number (Re) and the ratio between the ambient fluid viscosity and the input fluid viscosity (U) (e.g., Jellinek et al. 1999). With time, convection slows down and will be controlled by the progressive smaller buoyancy of the input and mixed component. Thus, the more the two end-members have mingled, the less intense will be the convection, as shown in a recent numerical approach (e.g., Montagna et al. 2015; Morgavi et al. 2017 and references therein). As a result, the viscosity of the initial magmas involved, and its time as the process evolves, is the most important of all physical properties and is thought to control the efficiency of mixing. Therefore, the accurate determination of viscosity, for the volcanic products of the CFc, is discussed in this chapter. An additional parameter leading to compositional convection is density. Accordingly, we review high temperature density measurements for the CFc melts. A short focus on rheology starts up this review. Following the summary on viscosity and density, different mixing experiments using natural CFc volcanic products as starting materials are reviewed and evaluated in the same way. A list of the CFc eruptions and related samples, whose data are reviewed here, is presented in Table 1. The location of the samples is reported in Fig. 1. Samples have been collected from the pyroclastic sequences of the following 8 eruptions, listed from the youngest to the oldest: Monte Nuovo (MN); Astroni 6 (AST6), Fondo Riccio (FR), Averno 2 (AVE2), Agnano-Monte Spina (A-MS), Minopoli 2 (MIN2), Neapolitan Yellow Tuff (NYT), Campanian Ignimbrite (CI). A discussion on time-scales of hybridisation sums up the main implications of the experimental simulation of mixing processes for magmatic differentiation and, consequently, volcanic hazards.

2 Rheological Properties of Magmas

One aspect, which makes volcanic processes so unique in their place in Earth Sciences, is their tremendously variable ability for mass transport by viscous flow. Rheology concerns itself with the properties of materials, which influence their flow and deformation (e.g., Dingwell 2006). The products of volcanic eruptions owe a great deal of their complexity and variety to the variable manner in which feeding magmas flow, evolve and deform prior to and during eruptions. The most evolved products of the CFc magmatic system are predominantly phonolitic and trachytic. The magmatic system, however, is thought to be fed by deeper sources ranging towards least evolved trachybasalts, mostly only to be found as melt inclusions (e.g., Moretti et al. 2013, 2019; Arienzo et al. 2016 and references therein; Chap. Origin and Differentiation History of the Magmatic System Feeding the Campi Flegrei Volcanic Field (Italy) Constrained by Radiogenic and Stable Isotope Data). They may lead to peralkaline compositions, whose kinetics, transport properties and rheology are likely to be particularly path-dependent, immediately preceding and during eruption, due to the combined enhanced efficiency of crystallisation and degassing (e.g., Giordano et al. 2008; Andújar and Scaillet 2012;

Table 1 Campi Flegrei caldera eruptions and related samples reviewed in this chapter

Eruption	Magma composition	Measurement or experiment temperature range	Sample	References
Monte Nuovo	Trachytic	deformation** 600–800 °C	MN1	1, 2, 3
Monte Nuovo	Trachytic	deformation** 600–800 °C	MN2	1, 2, 3
Monte Nuovo	Trachytic	viscosity * CC/MP	MN	3, 18, 19
Averno 2	Trachytic	timescales of mixing 1,300 °C	AVE2	4, 5, 6
Astroni 6	Trachytic	timescales of mixing 1,300 °C	AST6	6, 7
Agnano-Monte Spina	Trachytic	viscosity* 700–1,450 °C	A-MS_B1	3, 17, 18, 19
Agnano-Monte Spina	Trachytic	viscosity* 700–1,450 °C	A-MS_D1	9, 17
Agnano-Monte Spina	Phonolitic	viscosity*, density mixing** 1,200 °C	A-MS_B_1 + B_2	8, 10, 11
Fondo Riccio	Trachytic	viscosity* 700–1,450 °C	FR + FRd1	3, 13
Fondo Riccio	Latitic	viscosity	FR	18
Minopoli 2	Shoshonitic	viscosity, density mixing** 1,200 °C	MIN 2a + 2b + ad1	3, 9, 10, 11, 13, 17
CI	Trachytic	viscosity*	CI-IGC, CI-OF	9, 18, 19
CI	Trachytic	viscosity*, mixing 1,300 °C	CI-OF 152b2	3, 6, 12, 14, 15, 16
CI	Phonolitic	mixing** 1,300 °C	CI-OF 104f	6, 12, 14, 15, 16
Neapolitan Yellow Tuff	Trachytic	viscosity* 700–1,450 °C	NYT	9

*viscosity measurements: for dry melts high temperature, concentric cylinder (CC); for hydrous melts high temperature, piston cylinder; for low temperature, micropenetration (MP), **deformation experiments at magmatic temperature.
References: (1) D'Oriano et al. 2005; (2) Vona et al. 2013; (3) Giordano et al. 2004; (4) Di Vito et al. 2011; (5) D'Antonio et al. 2007; (6) Perugini et al. 2010; (7) Isaia et al. 2004; (8) de Vita et al. 1999; (9) Romano et al. 2003; (10) Di Renzo et al. 2011; (11) Perugini et al. 2013; (12) Civetta et al. 1997; (13) Giordano et al. 2008; (14) De Campos et al. 2004; (15) De Campos et al. 2008; (16) Perugini et al. 2008; (17) Misiti et al. 2006; (18) Giordano et al. 2006; (19) Giordano et al. 2009

Moretti et al. 2019). The effect of melt composition on viscosity and thus magma rheology, also includes the effects of temperature and volatile content. Here, a compilation of effective viscosities from melts obtained from CFc volcanic products is provided. For the most abundant compositions, we present a comparative review of CFc data and viscosity models. In the Appendix 1 to this chapter, additional viscosity data for less abundant compositions is also presented and discussed. Density is a further physical parameter controlling the efficiency and extent of magmatic emplacement and differentiation. The attainment of liquid density data at high temperatures is of critical importance to petrogenetic models focused on crystal-melt and melt-melt density relations (e.g. Lange and Carmichael 1987; Lange 1994; Knoche et al. 1995). For a better understanding flow processes, an accurate study of density of natural CFc melts can only be obtained from models resting on reliable experimental density determinations.

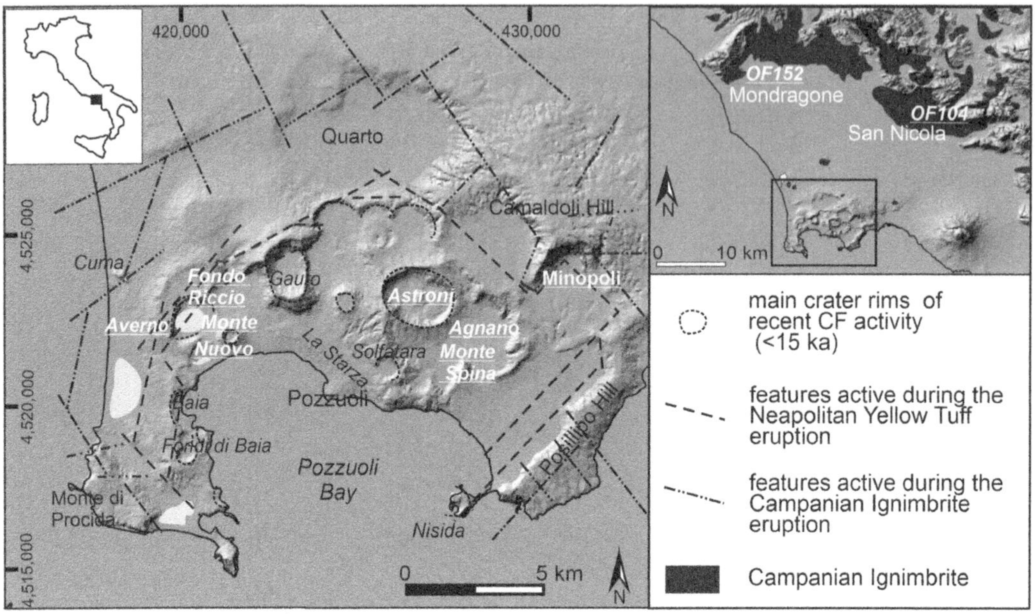

Fig. 1 **a** Schematic structural map of the Campi Flegrei caldera with indication of the volcanoes of the past 15 kyrs activity (in white) from which pyroclastic sequences, the samples reviewed in this chapter have been collected. **b** Label and location of the Campanian Ignimbrite reviewed samples. Maps in **a** and **b** are modified after Chap. Volcanic and Deformation History of the Campi Flegrei Volcanic Field, Italy

2.1 Viscosity—Evaluation of Viscosity Models for Campi Flegrei Caldera Melts: Trachytic Compositions

Viscosity is sensitive to melt chemistry and structure (e.g., Mysen 1988). The main factors controlling the viscosity of magmas for most of their differentiation history are temperature and chemical composition. Composition can be seen as controlling the relationship between "network-modifying" cations and "network-forming/stabilising" cations (e.g., Dingwell 2006). The main challenge for modelling viscosity in natural systems is deriving rational means for distributing the effects of melt composition across the non-Arrhenian model parameters (e.g., Persikov 1991; Richet and Bottinga 1995; Hess et al. 1996; Rössler et al. 1998; Toplis 1998).

The primary volatiles important for melt viscosity are H_2O and F, and to a lesser extent CO_2. The magma rheology is additionally impacted by the presence of crystals and bubbles (e.g., Dingwell et al. 1993; Giordano and Dingwell 2003; Giordano et al. 2008; Vona et al. 2013). Amongst these controlling parameters, the role of H_2O (e.g., Fischer 2008 and references therein), the most abundant volatile phase in magmas is known to produce dramatic changes in viscosity and, hence, in rheological properties of magmas (e.g., Hess and Dingwell 1996; Hess et al. 1996; Dingwell et al. 1998; Dingwell 2006; Giordano et al. 2008). Dissolved H_2O in the CFc magmatic system is of paramount importance for the whole volcanic system (e.g., Papale 2005; Chiodini et al 2012; Moretti et al. 2013, 2019 and references therein). It is a decisive control parameter of its viscosity and, therefore, of its rheology. Since CFc is still active (see Chaps. The Permanent Monitoring System of the Campi Flegrei Caldera, Italy; The Hydrothermal System of the Campi Flegrei Caldera, Italy; Historic Unrest of the Campi Flegrei Caldera, Italy), a first attempt to estimate the multiphase rheology of the youngest MN trachytic eruption (AD 1538) was performed by Vona et al. (2013) and previously by Caricchi et al. (2008). As expected, with increasing complexity, the viscosity, and therefore, the rheology of the system, changes from Newtonian to non-Newtonian. The use of high-temperature deformation to investigate the multiphase (liquid + crystals + bubbles) rheology of natural samples from MN trachyte does not permit an appropriate comparison with the most common viscometric methods: the high-temperature concentric cylinder viscometry and the low-temperature micropenetration (see Dingwell et al. 1993 and Giordano et al. 2008, for experimental details). A new attempt to measure viscosity at low and high temperature for dry, hydrous and CO_2-bearing samples from FR eruption sequence (see Table 1 for compositions) has been published by Di Genova et al. (2014).

The calculation of melt viscosity as a function of temperature and composition has been proposed first for trachytic (e.g., Romano et al. 2003; Giordano et al. 2004; Misiti et al. 2006) and then for latitic and shoshonitic melts (Misiti et al. 2011), for this system. Two general numerical viscosity models are available in the literature: (a) by Hui and Zhang (2007) and (b) Giordano et al. (2008). For these general models, a large dataset of dry trachytic compositions, including most of the compositions referred to here, as well as hydrous trachytic melts, has been taken into account for calibration. In the following we briefly compare the available models. Because trachytic compositions are the most abundant, they are the most studied CFc melts. Romano et al. (2003) and Misiti et al. (2006) focus on the 4,482–4,625 cal. years BP A-MS eruption (e.g., Smith et al. 2011) and Giordano et al. (2004) on the 39.85 ± 0.14 ka CI (Giaccio et al. 2017) and AD 1538 MN eruptions (e.g., Di Vito et al. 2016 and references therein). For a more detailed evaluation of viscosity models of latitic (from the FR rocks) and of shoshonitic (from the MIN2 eruption sequence) compositions, see Appendix 1.

2.1.1 Viscosity Model for Agnano-Monte Spina Eruption Melt

Additional references and information on the petrology and geochemistry of the different studied samples are in Table 1, as well as in Chaps. An Evolutionary Model for the Magmatic System of the Campi Flegrei Volcanic Field (Italy) Constrained by Petrochemical Data; Origin and Differentiation History of the Magmatic System Feeding the Campi Flegrei Volcanic Field (Italy) Constrained by Radiogenic and Stable Isotope Data; Tephrochronology and Geochemistry of Tephra from the Campi Flegrei Volcanic Field, Italy. To study the dry and hydrous viscosities of trachytic melt from CFc, Romano et al. (2003) used a combination of micropenetration and concentric cylinder techniques to cover the high temperature (superliquidus) to low temperature (near the glass transition temperature) range. Viscosity data varied between 10^2 to 10^{12} Pa*s and the water content ranged from (0.02 wt%) dry to 3.8 wt%. Only glassy parts from crushed pumice samples were chosen, then remelted above the liquidus temperature in air and quenched to a glass. The resultant glass was used for the viscosity determinations. Anhydrous glass compositions were chemically analysed by electron microprobe (Table 2, A-MS_D1). Parts of the dry glass from the concentric cylinder experiments were drilled and used for hydration experiments in a piston cylinder apparatus. The homogeneity and stability of water content were checked by Fourier Transform Infrared spectroscopy before and after each viscosity experiment. Viscosity determinations for the A-MS_D1 representative sample are presented in Table 3. In order to fit the temperature dependence of the viscosity data, the non-Arrhenian Vogel-Fulcher-Tammann (VFT) model was applied:

$$\log \eta = \mathrm{a} + \mathrm{b}/(\mathrm{T} - \mathrm{c}) \quad (1)$$

where η(eta) is the viscosity in Pa*s, T is the absolute temperature, a, b and c being fit parameters with values depending on the dissolved water content. The best functional forms of parameters a, b and c in Eq. 1 were found to be:

$$a = a_1 + a_2 ln w_{H_2O} \quad (2)$$

Table 2 Chemical composition of the analysed dry glasses from CFc eruptions

Sample	A-MS_D1 *+	A-MS_B1 #+	MN #+	MN ++	CI *+	CI ++	CI-OF #
Type	trachyte	trachyte	trachyte	trachyte	trachyte	trachyte	trachyte
SiO_2	60.86	62.09	64.58	63.88	62.09	60.74	68.80
TiO_2	0.39	0.28	0.31	0,31	0.28	0.27	0.23
Al_2O_3	18.27	19.65	17.29	17.10	19.65	19.22	12.58
FeOt	3.88	3.44	2.93	2.90	3.44	3.44	3.37
MnO	0.12	0.18	0.13	0.13	0.18	0.18	0.14
MgO	0.90	0.29	0.24	0.24	0.29	0.28	1.24
CaO	2.96	2.16	1.84	1.82	2.16	2.11	3.43
Na_2O	4.12	5.40	5.73	5.67	5.40	5.28	4.01
K_2O	8.50	6.46	6.89	6.82	6.46	6.32	6.18
P_2O_5	NR	0.06	0.05	0.05	0.06	0.06	0.03
Sample	MIN2a*	MIN2b*	MINad1*	FRd1*	FRa#	FR**	NYT*
Type	shosho	shosho	shosho	latite	latite	latite	trachyte
SiO_2	64.58	53.10	53.10	55.87	55.41	56.63	58.77
TiO_2	0.31	0.84	0.84	0.89	0.72	0.82	0.50
Al_2O_3	17.29	16.35	16.35	18.76	18.38	18.00	18.39
FeOt	2.93	7.03	7.03	6.55	7.31	6.70	4.96
MnO	0.13	0.13	0.13	0.13	0.16	0.17	0.06
MgO	0.24	5.69	5.69	2.47	2.39	2.41	1.43
CaO	1.84	10.34	10.34	5.85	5.76	5.60	4.03
Na_2O	5.73	2.29	2.29	4.19	4.23	4.61	3.38
K_2O	6.89	3.81	3.81	4.65	4.58	4.56	7.67
P_2O_5	0.05	0.42	NR	0.64	NR	0.48	0.00

Total Fe as FeO. All data have been normalised to 100%, see text for reference. All oxides are expressed in wt% (NR not reported). Viscosity measurements from *, ++, #, + : for dry melts at high temperature, concentric cylinder; for hydrous melts high temperature, piston cylinder; for low temperature, micro-penetration; **: deformation experiment at magmatic temperature (Di Genova et al. 2014). Data source: *Romano et al. (2003); ++Giordano et al. (2004); #Giordano et al. (2006); +Giordano et al. (2009); **Di Genova et al. (2014). For specific information on the eruptions see Chaps. Volcanic and Deformation History of the Campi Flegrei Volcanic Field, Italy; An Evolutionary Model for the Magmatic System of the Campi Flegrei Volcanic Field (Italy) Constrained by Petrochemical Data; Origin and Differentiation History of the Magmatic System Feeding the Campi Flegrei Volcanic Field (Italy) Constrained by Radiogenic and Stable Isotope Data; Tephrochronology and Geochemistry of Tephra from the Campi Flegrei Volcanic Field, Italy.

$$b = b_1 + b_2 ln w_{H_2O} \quad (3)$$

$$c = c_1 + c_2 ln w_{H_2O} \quad (4)$$

where w_{H2O} is the dissolved water content in wt%. The results of the fit are shown in Table 3, the fit parameters are listed in Table 4 and presented in Fig. 2. All experimental data from this study are well fitted by the model with a deviation of about ± 0.15 or less. Because of the data amount and quality, it can be safely used to predict the viscosity at lower temperatures (glass transition range) in the range of water contents between 0.02 and 4 wt%. However, the high temperature regime has only been constrained by the measurement of the anhydrous melt.

Table 3 Selected measured and calculated viscosity data on dry and hydrous trachytic samples from the pyroclastic sequence of the Agnano-Monte Spina eruption

Experimental data				Model fits and deviation							
H_2O	T	log η			Δ		Δ		Δ		Δ
wt%	°C	Pa*s	Ref.	A	log η	B	log η	C	log η	D	log η
0.02	1,496	2.49	1	2.57	−0.08	2.51	−0.02	2.49	0.00	2.54	−0.05
0.02	1,446	2.74	1	2.81	−0.07	2.76	−0.02	2.77	−0.03	2.79	−0.05
0.02	1,397	3.01	1	3.07	−0.06	3.03	−0.02	3.06	−0.05	3.06	−0.05
0.02	1,348	3.30	1	3.34	−0.04	3.31	−0.01	3.36	−0.06	3.34	−0.04
0.02	1,299	3.62	1	3.63	−0.01	3.61	0.01	3.69	−0.07	3.65	−0.03
0.02	1,249	3.96	1	3.96	0.00	3.95	0.01	4.04	−0.08	3.98	−0.02
0.02	1,200	4.33	1	4.30	0.03	4.30	0.03	4.42	−0.09	4.34	−0.01
0.02	1,151	4.73	1	4.68	0.05	4.69	0.04	4.82	−0.09	4.73	0.00
0.02	814.1	8.45	1	8.59	−0.14	8.61	−0.16	8.65	−0.20	8.74	−0.29
0.02	765.3	9.32	1	9.47	−0.15	9.46	−0.14	9.43	−0.11	9.62	−0.30
0.02	736.5	9.77	1	10.05	−0.28	10.02	−0.25	9.93	−0.16	10.20	−0.43
0.02	712.0	10.56	1	10.58	−0.02	10.53	0.03	10.38	0.18	10.73	−0.17
0.02	700.2	10.75	1	10.85	−0.10	10.79	−0.04	10.61	0.14	11.00	−0.25
0.02	683.8	11.29	1	11.24	0.05	11.17	0.12	10.93	0.36	11.40	−0.11
3.75	450.5	9.90	1	10.04	−0.14	9.93	−0.03	10.03	−0.13	10.20	−0.30
3.75	436.2	10.31	1	10.38	−0.07	10.30	0.01	10.44	−0.13	10.52	−0.21
3.75	415.7	11.05	1	10.89	0.16	10.86	0.19	11.06	−0.01	11.01	0.04
0.18*	1,400	2.66	2	2.92	−0.26	2.63	0.03	2.88	−0.22	2.79	−0.13
0.18*	1,300	3.56	2	3.43	0.13	3.17	0.39	3.50	0.06	3.35	0.21
0.18*	1,200	3.66	2	4.02	−0.36	3.80	−0.14	4.21	−0.55	4.00	−0.34
5.81	1,400	0.11	2	1.07	−0.96	0.03	0.08	-0.04	0.15	1.25	−1.14
5.81	1,300	0.21	2	1.37	−1.16	0.36	−0.15	0.33	−0.12	1.63	−1.42
5.81	1,200	0.35	2	1.72	−1.37	0.74	−0.39	0.77	−0.42	2.05	−1.70

Including S of model fit
Without piston cylinder data: model A: 0.46; model B: 0.15; model C: 0.20; model D: 0.56
With piston cylinder data: model A: 0.11; model B: 0.10; model C: 0.14; model D: 0.20
Model A: Romano et al. (2003) (ref. 1); model B: Misiti et al. (2006) (ref. 2); model C: Misiti et al. (2011); model D: Giordano et al. (2009)
Note: * means that the estimated water content is uncertain
Standard deviation equation: $S = \sqrt{\frac{1}{n-1}\Sigma ai^2}$
where ai = measured value; n = number of measurements

Misiti et al. (2006) extended the experimental range by performing a "falling sphere" study on hydrated melts with the same base composition. Selected viscosity data are tabulated in Table 3 and the refitted parameter of Eq. 1 are tabulated in Table 4 (an alternative derived equation is omitted because of higher standard error of the global fit). As can be seen from the inspection of the deviations of single viscosity data and by the comparison of the standard deviation of the model fits, the new model fits the low temperature region as good as the previous one, but performs better at high temperatures and high water contents. One can note that there is a systematic mismatch in the fitting of the temperature dependence of viscosity for the highest water-

Table 4 Calibrated parameters of viscosity models for the Campi Flegrei caldera magmas

Model	Composition	a1/a	a2/b	b1/c	b2/d	c1/e	c2/g	equation
Romano et al. (2003)	Trachyte A-MS	−3.54	0.145	9,618.90	−498.79	191.78	−35.52	1, 2, 3, 4
Misiti et al. (2006)	Trachyte A-MS	−4.73	−0.004	10,788.93	−587.33	173.56	−26.78	1, 2, 3, 4
Giordano et al. (2009)	Trachyte A-MS	−4.55		10,382.00	−44.69	313.20	−227.70	1, 3, 4
Misiti et al. (2011)	Trachyte A-MS	−6.64	8,464.730	186.36	7,220.89	−129.20	−429.34	5
Giordano et al. (2004)	Trachyte CI	−4.42	0.226	9,243.00	−428.10	255.30	−127.00	1, 2, 3, 4
Giordano et al. (2009)	Trachyte CI	−4.55		8,889.00	82.10	492.90	−385.50	1, 3, 4
Giordano et al. (2004)	Trachyte MN	−5.86	−0.117	12,747.00	−673.50	103.40	−59.85	2
Giordano et al. (2009)	Trachyte MN	−4.55		10,449.00	−101.63	303.70	−193.80	1, 3, 4
Misiti et al. (2011)	Latite FR	−4.99	5,412.988	552.69	2,799.22	303.41	−3,956.44	6
Di Genova et al. (2014)	Latite FR	−4.55		7,744.97	−2,192.14	482.24	−255.88	7
Misiti et al. (2011)	Shoshonite MIN2	−5.57	7,812.046	321.73	874.68	770.04	−2,289.73	6

Values correspond to used wt% H_2O and absolute temperature in the equation, and yielded viscosity in Pa*s

containing melts. Therefore, to calculate the viscosities at high and low temperatures, in the range of water contents between 0.02 and 4 wt%, we would recommend the use of this model.

The same data set of Misiti et al. (2006) has been refitted in Misiti et al. (2011) (see Tables 3, 4 and Fig. 2, with a different functional form for the VFT equation. The fit quality has not improved. Especially due to the change in the functional form a completely different behaviour in the low temperature regime is anticipated, which is not in agreement with all low temperature data of Romano et al. (2003) and Giordano et al. (2004, 2006, 2008). The differences in standard deviation values may be due to the available dataset. Variations in the standard deviation have been calculated including and excluding the piston cylinder values from Misiti et al. (2006) (see Table 3). Although there are discrepancies in all models, for comparative purposes, we will subsequently use the model of Giordano et al. (2009).

2.1.2 Viscosity Model for Campanian Ignimbrite and Monte Nuovo Eruptions Melts

The viscosity of natural liquids representative of the glassy portion of pumice collected from deposits of the CI and MN eruptions has been measured from superliquidus temperatures down to the glass transition range by Giordano et al. (2004). The experimental strategy was similar to that of Romano et al. (2003). The high temperature range was only constrained by viscosity data of anhydrous compositions. When compared to A-MS (see Table 3), the variation of the dry base composition of the glasses is slightly enriched in silica with a minor loss in total iron and alkaline earth oxides. In Fig. 3 different models for trachytic compositions are compared to each other, for water contents covered by the source studies, i.e., ranging from (0.02) dry, 1, 2, to 4 wt% of water and using the approved model of Misiti et al. (2006) for A-MS. For the low-temperature regime a striking similarity in

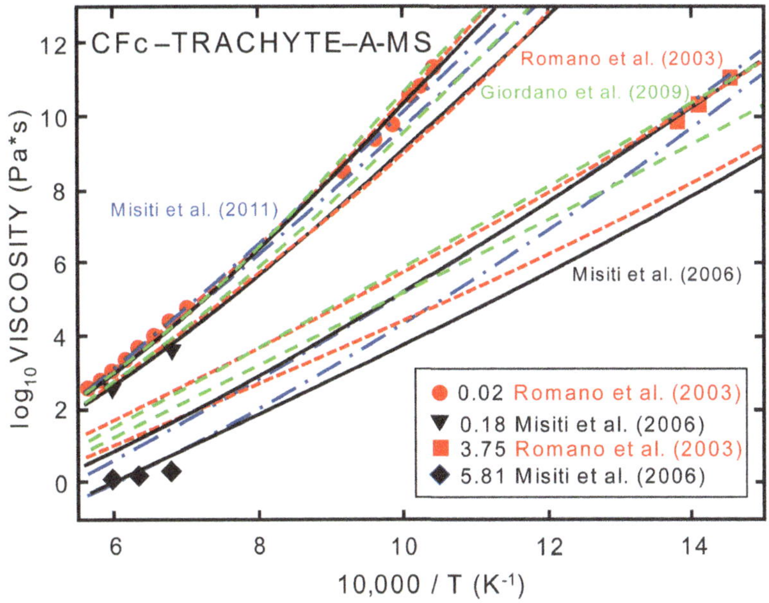

Fig. 2 Comparison of different viscosity models with selected viscosity data for the trachytic A-MS composition. Numbers in the label field indicate the water content of the experimental data. The fitted lines are calculated using 0.02, 0.18, 3.75 and 5.81 wt% H_2O; the viscosity curves decrease with increasing water content. The following models have been used for the calculations: Romano et al. (2003) (red dotted lines); Giordano et al. (2009) (green dashed lines); Misiti et al. (2006) (black lines)

Fig. 3 Comparison of different viscosity data recalculated after the model of Giordano et al. (2009) for trachytes: A-MS, CI and MN composition. The numbers indicate the water content (wt %) used for the fits

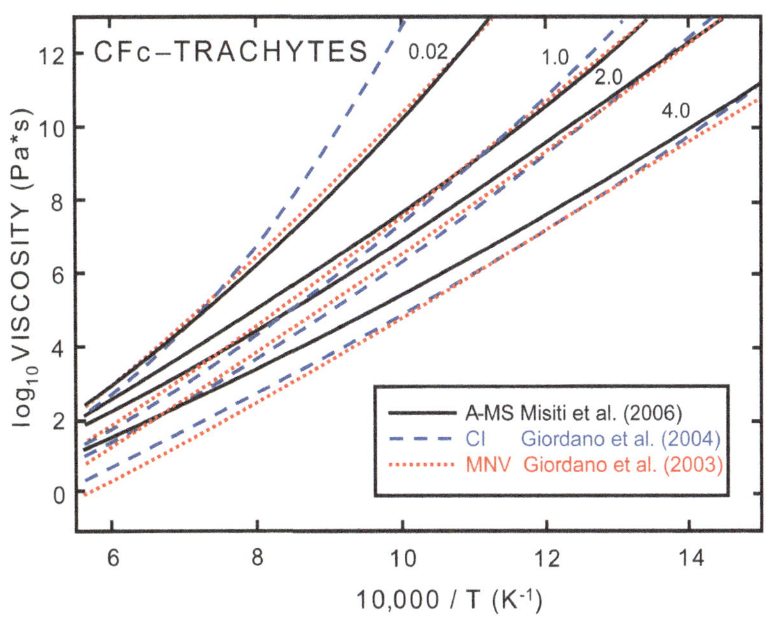

the absolute values and the temperature dependence of viscosity emerges. Only the dry data for CI, which will be discussed later, seem to depart. For the high-temperature regime, a similar pattern as in the Romano et al. (2003) study comes out. The viscosities of water-rich compositions predicted by unconstrained models (e.g., not covered by measurements) are systematically up to about 1.5 orders of magnitude lower. The model of Misiti et al. (2006) fulfils the goal for all natural trachytic (at least for the compositional range covered in Table 2) liquids of A-MS, CI and MN to calculate the viscosities at high and low temperatures in the range of water contents between 0.02 and 4 wt%.

2.1.3 Testing General Viscosity Models for Campi Flegrei Caldera Trachytic Melts

After the publication of the measured viscosity data for A-MS, CI and MN from 2003 to 2006 two general viscosity models by Hui and Zhang (2007), here HZ, and Giordano et al. (2008), here GRD, were available. In Fig. 4 a comparison among the GRD and the HZ models for the trachytic A-MS composition is made with water contents ranging from 0.02 (dry), 1, 2–4 wt%. For the low-temperature regime, a striking similarity in the absolute values and the temperature dependence of viscosity is once more seen. For the high-temperature regime, a similar pattern as in the Romano et al. (2003) and Giordano et al. (2004) studies arise. An explanation for this behaviour might be due to a bias of high temperature viscosity data. In contrast, the standard deviations of both the GRD and HZ general models are similar, especially concerning the prediction of viscosities of hydrous trachytic melts, in the high temperature range, ranging from 0.35 to 0.57. All calibrated parameters for the different viscosity models from CFc are shown in Table 4.

2.2 Density—Thermodynamic Calculations and Laboratory High-Temperature Measurements on Campi Flegrei Caldera Volcanic Products

The further focus on density variations along the reservoir depth, at high temperatures, may provide an additional insight into understanding the influence of the combined effect of all elements in the convection dynamics. This will be important for the mixing experiments to be commented further in this text. The big challenge for density measurements using trachytic melt

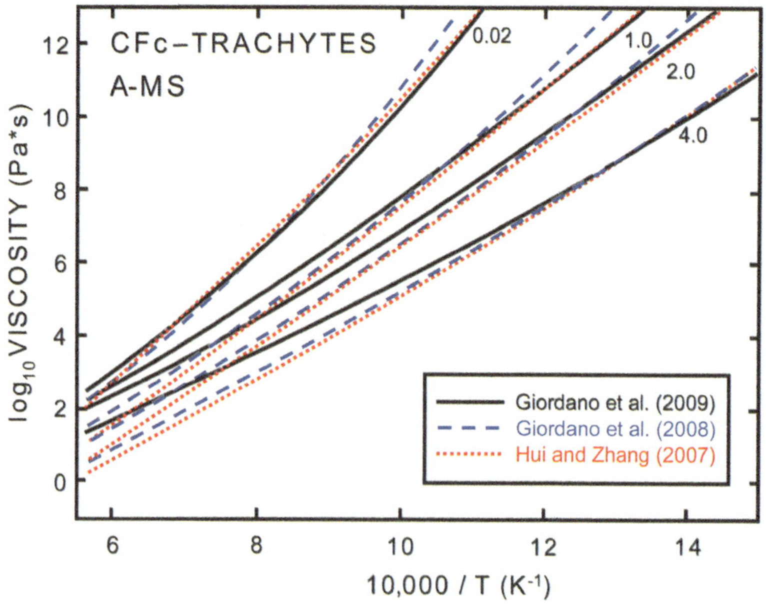

Fig. 4 Comparison between different multi-component viscosity models (e.g., Hui and Zhang 2007; Giordano et al. 2008, 2009) for A-MS trachytic composition with different water contents (wt %). The numbers indicate the water content used for the fittings

compositions, like those so common in the CF, especially the CI, is the high viscosity. This is the main reason why, up to now, only shoshonitic compositions yielded good experimental data.

The measurements discussed in this section were conducted using the Pt-based double-bob Archimedean method (see Courtial and Dingwell 2005, for experimental details). The Archimedean method is based on two platinum bobs with different volumes, which are immersed in the high temperature melts. Several buoyancy determinations of the bobs are performed, including correction for air, measurements in the melt and are recorded at regular time intervals for different temperatures. The submerged volume of the bobs is then computed from their masses, using a volume-temperature relationship for Pt, geometric considerations of the immersed length of leader wire, and the melt height displacement caused by the submerged bob. The buoyancy data and the volumes of the bobs are entered into Eq. 5 to solve for density (ρ):

$$\rho = (\mathrm{B1} - \mathrm{B2})/(\mathrm{V1} - \mathrm{V2}) \qquad (5)$$

MIN2 is the least evolved potential end-member erupted from the CF magma reservoir (e.g., Di Renzo et al. 2011). This volcanic product should give the upper density limit (highest density values) in the CF. Density measurements for this MIN2 shoshonite at high temperatures are depicted on Fig. 5.

The Campanian Ignimbrite (CI) magma chamber is thought to include at least two cogenetic magma layers, separated by compositional gaps (e.g., Civetta et al. 1997; Signorelli et al. 1999; Pappalardo et al. 2002; Di Renzo et al. 2011). For the CI, detailed isotopic and chronostratigraphic studies point towards the existence of, at least, two large relatively homogeneous layers of similar compositions (a more evolved phonotrachytic and a least evolved trachytic layer) (e.g., Civetta et al. 1997; Pappalardo et al. 2002). In this case, no significant temperature and density gradient in the magma chamber is the most probable scenario. Convection may then be driven by small compositional gradients, which may become an important motor for the flow process.

To test the hypothesis of a two-layer magma chamber for the CI, De Campos et al. (2004, 2005) compared micro-analytical data from mixing experiments and glassy melt inclusions in pyroxene phenocrysts (data from Signorelli et al. 1999) in order to calculate a density gradient along the hypothetical magma chamber. For this purpose, molar volumes of the different phases present in the micro-volumes of experimental

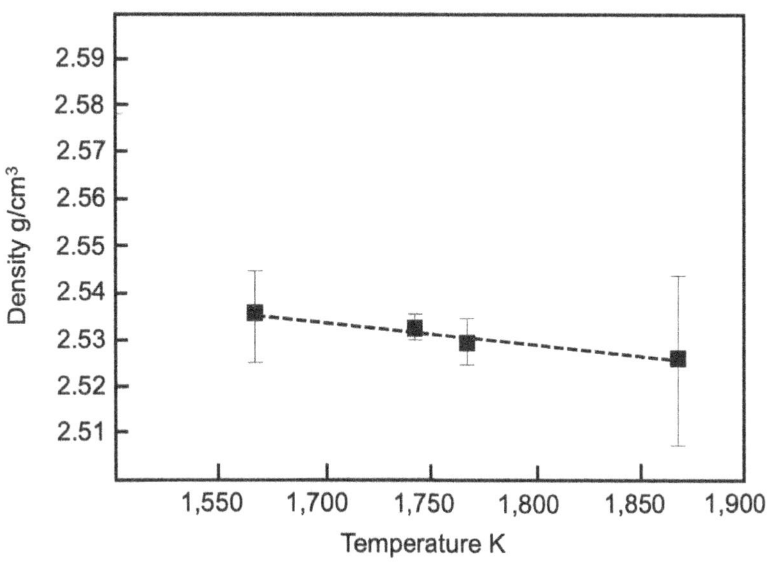

Fig. 5 Density measurement data from MIN2 shoshonite with the double bob method at high temperature (see Courtial and Dingwell 2005 for experimental methods)

melts were calculated at 1,300 °C, using the model of Lange and Carmichael (1987). This way, the density distribution data displayed on Fig. 6 has been obtained. Calculated densities for glasses in natural samples were than plotted against SiO_2, as a depth related parameter. In the uppermost layer, density values show a small variation around 1%. They have been interpreted as well-mixed. Below the well-mixed layer, density values grade downward into a diffusive layer, varying from 2.32 up to 2.39 g/cm^3, a variation range close to 4%. This approximates in more detail the stratified magma chamber depicted by Signorelli et al. (1999).

The existence of a big homogeneous and overheated magma chamber of single trachytic composition relies on restricted geochemical data from melt inclusions (Marianelli et al. 2006) and do not depict the whole isotopic and melt inclusions dataset variation from CI samples from the literature (e.g., Arienzo et al. 2009, 2010, 2016; Di Renzo et al. 2011; Moretti et al. 2019 and references therein). To our best knowledge, this hypothesis has not yet been tested by means of numerical models or experiments.

3 Magma Rheology and Magma Mixing—Principles and Background *(What is Mixing? Importance of an Adequate Numerical Modelling)*

Following an emphasis on the two main physical parameters (viscosity and density) for CF melts, we will summarise and discuss different mixing experiments from the literature, performed by using CFc volcanic products as starting materials. Numerical modelling and low-temperature analogue models, using low-viscosity fluids, provided the first picture of the basic dynamics associated with fluid transport (convection/ advection) and its potential in modulating geochemical variabilities in any igneous systems (e.g., Snyder and Tait 1996; Jellinek et al. 1999; Petrelli et al. 2011; Spera et al. 2016). Due to the decisive constrains on mixing efficiency, geophysicists and petrologists generally assume that mingling, rather than chemical mixing, dominates the interaction between melts with high viscosity contrast (e.g., Sparks and Marshall 1986; Bateman 1995).

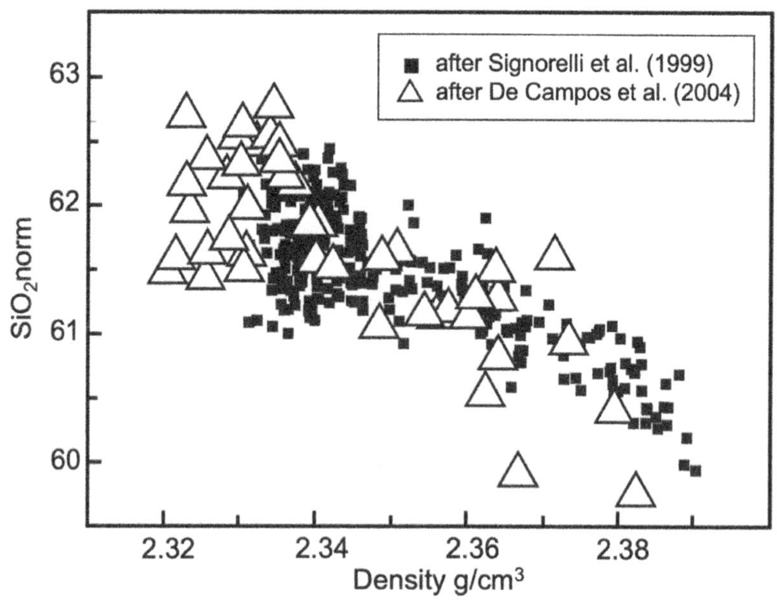

Fig. 6 Calculated densities for glasses from the CI (geochemical data on melt inclusions in pyroxene crystals from Signorelli et al. 1999) are compared to calculated densities of experimental melts along a two-layers magma chamber model. Experimental data from De Campos et al. (2004). Density values calculated after Lange and Carmichael (1987)

Mixing and mingling have been widely reported in several petrogenetic models for the CF reservoir (e.g., Civetta et al. 1997; Pappalardo et al. 2002; Tonarini et al. 2009; Arienzo et al. 2009, 2010, 2016; Di Renzo et al. 2011). For simulating the natural mixing process in this system, several experimental time series, at high temperature, using different techniques and geometries, from experiments with natural CFc volcanic products, will be resumed and compared with each other. The last review point to be addressed in this chapter will be, if and how results of the mixing experiments (using natural CFc compositions as end-members) can be merged with information from numerical simulations of magma mixing, in order to derive mixing timescales and, further, a new chronometer for unrest episodes.

As already discussed in previous sections, it is generally accepted that magma chambers may be density- and compositionally-stratified. This will necessarily lead to free convection and, therefore mixing. Where do magmas mix: in a chamber or in the conduit? If magmas mix in a chamber, what are the main control parameters? How efficient is this process? And how can mixing evolve in time? Important steps towards unscrambling these questions were taken during the1980s with a series of laboratory analogue experiments, where aqueous solutions were used to depict fluid dynamics in a convecting magma chamber (e.g., Blake and Ivey 1986; Sparks and Marshall 1986; Martin et al. 1987), during replenishment (Huppert et al. 1982) or along the flow in a conduit. Numerical modelling of the fluid dynamics in conduits has pointed towards the importance of viscosity variations during a single eruption (e.g., Papale et al. 1998; Longo et al. 2006, 2012; Peruzzo et al. 2010 and references therein). The viscosity ratio between two fluids has been shown to have a significant effect upon the rate and different styles of convection, leading to turbulence and mixing (e.g., Campbell and Turner 1989; Jellinek et al. 1999). For a review in magma mixing processes, especially for field and isotopic evidence, see Morgavi et al. (2017).

In a magma chamber, convection may be generated by thermal variations within the system (Furbish 1997), whereas an outside agent, such as the replenishment of the chamber or assimilation of wall rocks, may cause forced convection (e.g., De Campos et al. 2008 and references therein). In general, if compositional heterogeneities remain, after thermal equilibration, the mobility of the different components of silicate melts will be driven by their chemical potentials (e.g., Dingwell 2006). This later process may have implications not only on the diffusive but also on the convective process. As a consequence, at any point during the life span of a magmatic system, compositional gradients may develop and lead to subsequent processes of magma hybridisation, or "magma mixing" (e.g., Perugini et al. 2013 and references therein).

In summary, the two main parameters controlling mixing are: (1) convection and (2) diffusion (e.g., Ottino 1989; Liang et al. 1996). Numerical modelling of magma dynamics have showed how the combined process of convection and diffusion in a magma chamber may be described by chaotic dynamics and fractal patterns (e.g., Perugini et al. 2004). This means that, despite their complexity, flow patterns and processes, irrespective of the length scale, are statistically similar and repetitive. This conclusion opened new perspectives on small-scale experiments with natural magmas finalised to study the dynamics of magma chambers (e.g., De Campos et al. 2004, 2008; Perugini et al. 2008, 2013). Given the competing roles of diffusion and convection, this insight is particularly significant as a source of information on chemical behaviour and mobility of major and trace elements in a dynamic environment (e.g., De Campos et al. 2004, 2008; Perugini et al. 2006, 2008, 2013). Since stretching and folding has been recognised as the fundamental physical process producing mechanical mixing, leading to chaotic behaviour (e.g., Ottino 1989) in the past years there has been increasing attention paid to the application of chaos theory to the study of magma mixing (e.g., Perugini et al. 2004, 2006). It follows that a major implication for experimenting with natural volcanic products was to develop a device to simulate chaotic dynamics at magmatic temperatures (e.g., De Campos et al. 2011; Perugini et al. 2012, 2013; Morgavi et al. 2013a, b, c, 2017 and references therein).

Recently, the injection of a basalt into a tonalite has been studied by means of high-pressure mixing experiments. These have been performed through torsion and at lower temperature (from 1,200 °C for the basalt, down to 650 °C, for the tonalite), to mimic the generation of andesites in the crust (Laumonier et al. 2014). In these experiments, changes in viscosity have been tracked for the whole range of the experiment and could motivate similar experiments with CFc natural products as well.

3.1 Motivation for Experimenting with Mixing in the Campi Flegrei Caldera Magmatic System

Plenty of geochemical and isotopic evidence of mixing in the CFc magmatic system, especially for the CI reservoir (e.g., Civetta et al. 1997; Signorelli et al. 1999; Pappalardo et al. 2002; Di Renzo et al. 2011; Arienzo et al. 2016; Chaps. An Evolutionary Model for the Magmatic System of the Campi Flegrei Volcanic Field (Italy) Constrained by Petrochemical Data; Origin and Differentiation History of the Magmatic System Feeding the Campi Flegrei Volcanic Field (Italy) Constrained by Radiogenic and Stable Isotope Data), has been the motivation for performing mixing experiments with CF natural volcanic products. Among the plethora of evidence, we call attention to: (1) abrupt changes in oxides and isotopic contents along the chemostratigraphy (e.g., Civetta et al. 1997); and, (2) structural and textural patterns, such as filament-like structures, enclaves, and mineral phases with clear physicochemical disequilibrium, so often described in the literature (e.g., Civetta et al. 1997; Pappalardo et al. 2002; Tonarini et al. 2009; Chap. An Evolutionary Model for the Magmatic System of the Campi Flegrei Volcanic Field (Italy) Constrained by Petrochemical Data). In addition to the petrographic evidence, some geochemical trends for major, minor and trace elements may plot along straight lines or not and may change slopes drastically, especially for K_2O and Na_2O and SiO_2 against CaO, both for whole rock and melt inclusions data. In the last case this is also true for Al_2O_3, K_2O and FeO (e.g., Civetta et al. 1997; Signorelli et al. 1999; Pappalardo et al. 2002; Cannatelli 2012; Arienzo et al. 2016; Chap. An Evolutionary Model for the Magmatic System of the Campi Flegrei Volcanic Field (Italy) Constrained by Petrochemical Data). During the CI event, the erupted magmas ranged in composition from trachyte to alkali-trachyte or phonolitic trachyte. A new input of more radiogenic magma is thought to have mixed with a resident less radiogenic magma (e.g., Civetta et al. 1997; Arienzo et al. 2009, 2016). Although claimed by some authors as very homogeneous (Marianelli et al. 2006), isotope ratios from latitic compositions in the CI episode (e.g., Civetta et al. 1997; Pappalardo et al. 2002; Tonarini et al. 2009; Di Renzo et al. 2011; Arienzo et al. 2016) point towards clear mixing in the magma reservoir.

Additional evidence of mixing in the CFc reservoir comes from the study of melt inclusions. Least evolved compositions are trapped as micro volumes of primitive magma in phenocrysts from more evolved matrix of the CI (Signorelli et al. 1999). Evidence of a deep source component in the more evolved magmas has been confirmed by several studies (Cannatelli et al. 2007; Mangiacapra et al. 2008; Arienzo et al. 2016; Morgavi et al. 2017; Moretti et al. 2019). While magmatic compositions, ranging from latite to trachyte, differentiate from a more primitive magma (trachyandesite according to Cannatelli 2012, shoshonite according to Mangiacapra et al. 2008 and Morgavi et al. 2017, or trachybasalt to basalt according to Arienzo et al. 2016 and Moretti et al. 2019), convection and diffusion must have proceeded in the magma reservoir. An additional explanation for contrasting compositions, between melt inclusions and matrix, could be replenishment of the differentiating more evolved magma batch by a least evolved magma, triggering the crystallisation of olivine and clinopyroxene. This process would implicitly originate more expanded geochemical trends in the system.

3.2 Mixing Experiments with Natural Campi Flegrei Caldera Melts

To the best of our knowledge, two different Taylor-Couette and Centrifuge mixing experiments have been performed using CFc natural volcanic products as starting compositions. The first mixing experiments with CFc products (Taylor-Couette geometry) were performed with starting compositions thought to be the best candidates as end-members for the mixing process in the CI reservoir: samples OF152b and OF104f (starting compositions in Table 5; sample location in Fig. 1). For further information on the end-members see Civetta et al. (1997), and on the experiments see De Campos et al. (2004, 2008) and Perugini et al. (2008). In this case convection was driven by the combined effect of the applied forced convection and local

Table 5 Starting compositions for Taylor-Couette experiments

Oxide (wt%)	EM A OF152b	EM B OF140f	Element (ppm)	EM A OF152b	EM B OF140f
SiO_2	62.11	59.96	Rb	273	474
TiO_2	0.43	0.51	Sr	433	31
Al_2O_3	18.24	17.87	Y	15	46
FeO_t	3.22	4.23	Zr	205	721
MnO	0.27	0.17	Nb	30	115
MgO	0.38	1.07	Ba	621	62
CaO	1.98	3.57	La	39	123
Na_2O	5.74	3.91	Ce	86	257
K_2O	7.20	8.31	Pr	8.71	22
P_2O_5	0.05	0.22	Nd	34	91
Cl^-	0.39	0.18	Sm	6.01	15
ρ (g/cm^3)	2.38	2.33	Eu	2.00	1.14
η (Pa*s)	5,754	4,571	Gd	4.19	11
			Tb	0.57	1.51
			Dy	3.00	8.14
			Ho	0.60	1.72
			Er	1.59	4.66
			Tm	0.23	0.75
			Yb	1.63	5.40
			Lu	0.20	0.77
			Hf	3.05	12
			Ta	1.23	5.83
			Th	11	52
			U	5.06	21

Major and minor elements mean concentration values from normalised microprobe analysis from 1,300 °C starting glasses. For details in trace elements ICP-MS Laser Ablation analyses see Perugini et al. (2008). Calculated densities after Lange and Carmichael (1987); measured viscosities after Potuzak and Mangiacapra (pers. comm.) (De Campos et al. 2008 for details).

Table 6 Starting compositions for Centrifuge Experiments

Oxide (wt%)	Felsic EM1 AMS $B_1 + D_1$	Mafic EM2 MIN2	Element (ppm)	Felsic EM1 AMS $B_1 + D_1$	Mafic EM2 MIN2
SiO_2	60.60	52.56	Rb	372	180
TiO_2	0.45	0.93	Sr	511	1,007
Al_2O_3	18.32	15.60	Y	39.29	24.05
FeO_t	3.46	7.68	Zr	506	151
MnO	0.16	0.15	Nb	62	29
MgO	1.15	5.59	Cs	20.53	10.70
CaO	3.74	10.70	Ba	620	2,018
Na_2O	4.05	2.16	La	97	46
K_2O	7.96	4.05	Ce	178	88
P_2O_5	0.15	0.63	Pr	19.5	10.8
Cl^-	0.41	0.20	Nd	72	42
ρ (g/cm^3)	2.50	2.82	Sm	13	9
η (Pa*s)	~ 20.10^3	231	Eu	3.20	1.68
			Gd	10	6.3
			Tb	1.41	0.9
			Dy	7.50	5.11
			Ho	1.57	0.87
			Er	4.14	2.20
			Tm	0.68	0.32
			Yb	4.50	1.91
			Lu	0.68	0.26
			Hf	10.93	3.90
			Ta	2.80	0.75
			Pb	60.08	30.27
			Th	45	13.07
			U	14	3

For measured densities see Sect. 6.1.2; EM = end-member. Measured viscosities at 1,200 °C (D. Giordano, pers. comm.). Calculated densities at 1,200 °C after Lange and Carmichael (1987). See Perugini et al. (2013, 2015) for details

compositional gradients enhancing diffusion along the sample.

For the second set of experiments with the high temperature centrifuge (Perugini et al. 2013, 2015), more contrasting compositions have been used: A-MS (see Table 6 and Fig. 1 for sample composition and location) and an olivine-free shoshonite from the Minopoli 2 event (see Table 6 and Fig. 1 for sample composition and location; for details on the experiment Perugini et al. 2013). For the preparation of the least evolved end-member, most of the olivine phenocrysts have been removed from the matrix (Perugini et al. 2013). The reason for this removal was due to detected strongly contrasting isotope ratios ($^{87}Sr/^{86}Sr$ and $^{143}Nd/^{144}Nd$) between phenocrysts and glass-matrix (Civetta et al. 1997). This observation suggests that these phenocrysts could not have crystallised directly from the matrix but have been carried upwards by the flow from deeper layers in the magma reservoir.

3.2.1 Taylor-Couette Experiments—Review of Results

From the first mixing experiments with natural CI products as starting materials (De Campos et al. 2004), the hypothesis of pre-eruptive mixing for the CI has been tested for the first time by stirring the two end-members during a time series ranging from 1 h up to one week at constant 1,300 °C and room pressure conditions.

The two starting compositions were loaded as cylinders and convection was forced via rotation of a spindle at a velocity of 0.5 rpm. After 16 and 25 h, under constant stirring, Al-Ti-Fe spinel crystals were still present, so that flow directions could be traced. Separate convection cells were then observed. Under the microprobe a complex layering bounded by clear compositional gaps in oxide ratios was further detected. The very low Reynolds numbers (10^{-7} to 10^{-9}) in these experiments simulated mixing under absolutely laminar flow conditions. The applied forced convection combined with local diffusion effect along the sample, lead to a layered compositional and density distribution (De Campos et al. 2004). By comparing experimental data and natural glass compositions, the authors support the evidence that mixing might be an important process in the CI reservoir, mainly before the onset of fractional crystallisation. The combined effect of convection and diffusion may originate a vertically and laterally zoned chamber, whose different models are discussed in the literature (e.g., Civetta et al. 1997; Signorelli et al. 1999; Pappalardo et al. 2002). In a later work using the same starting materials from the first experiments, De Campos et al. (2008) confirmed homogenisation of the starting melts after 16 h stirring on the basis of isotopic analyses. A separation in layers with different compositions and therefore densities after 25 h was reinforced.

Oxide contents along the experimental glasses and calculated density distribution patterns highlighted three main stages in the mixing process:

1. after 16 h: homogenisation with development of a well-mixed layer of intermediary hybrid composition;
2. after 25 h: growth of a well-mixed layer on the top of the crucible and development of a diffusive layer below (see Fig. 3 in De Campos et al. 2004) with unmixing (segregation) of a spreading horizon ranging in density from 2.34 to 2.37 g/cm^3 at the interface of the two layers;
3. after 1 week: reduction of density range and reversal between lower- and higher-density layers with change in the slope directions of oxides content in a complex zigzag pattern along the layers.

This work additionally shows that major and minor elements do not behave as expected and do not always produce linear trends in interelemental diagrams. Observations confirm what was previously perceived and modelled by Perugini et al. (2006) in natural samples. Therefore, it complements previous data and confirms the development of a diffusive fractionation in silicate melts strengthening the role of chaotic dynamics in generating heterogeneities in a magma chamber.

The CI is fairly aphyric and phenocrysts abundance is less than 5%. Despite the narrow range of reproduced compositions, comparison with the existing models for the CI magma chamber points towards a good to very good correlation between natural and artificial products (De Campos et al. 2004, 2008; Perugini et al. 2008). Results from the experiments performed by De Campos et al. (2004, 2008) and Perugini et al. (2008) indicate that convection can be driven by the combined effect of the applied forced convection (simulating small replenishment) and local compositional gradients (situation after replenishment), leading to diffusion along the sample. Density differences have been exclusively originated by the combination between diffusion and convection in the liquidus phase. Such mixing conditions could have prevailed in the magma chamber, short before the eruption, when turbulence took over. This hypothesis would therefore imply syn-eruptive mixing as suggested by Civetta et al. (1997). However, despite the relative compositional

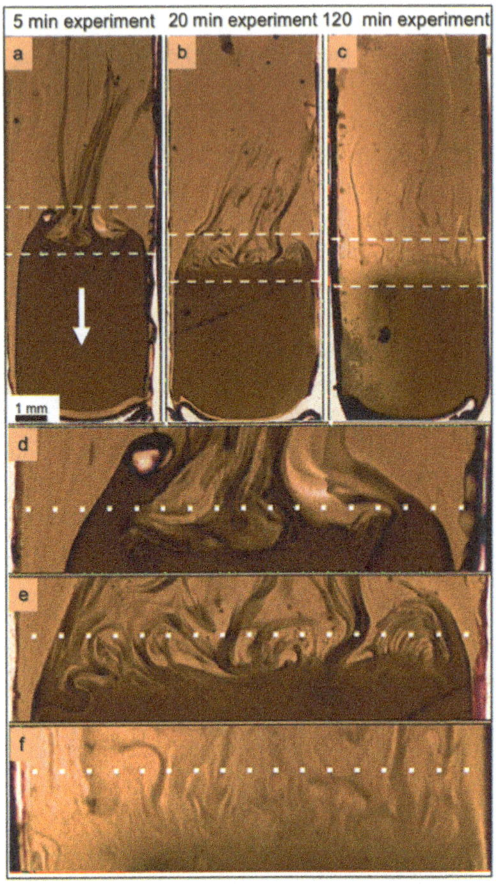

Fig. 7 Resultant glasses from mixing experiments with A-MS and MIN2 products as end-members (high-T centrifuge at 1200 °C). See text for details. Figure modified after Perugini et al. (2013)

similarity between these end-members, as highlighted by Marianelli et al. (2006), the interaction between melts with very distinct isotopic initial ratios and distinct geochemical trends (Civetta et al. 1997; Signorelli et al. 1999; Pappalardo et al. 2002; D'Antonio et al. 2007; Arienzo et al. 2009) are clear evidence of an open magmatic system under complex convection–diffusion processes.

3.2.2 Centrifuge Experiments—Review of Results

A set of mixing experiments with melts from natural CFc rocks has been published by Perugini et al. (2013, 2015). Experiments have been performed by mixing phonolitic (A-MS, B1 + B2-A-MS) and alkali-basaltic/shoshonitic (MIN 2, without the olivine phenocrysts) melts at 1,200 ± 1 °C. The mixing process was induced by using a high-temperature centrifuge apparatus. Experiments were performed in a time series of 5, 20 and 120 min. Samples were arranged in a buoyantly unstable geometry with the denser material (alkali-basaltic/shoshonitic) placed at the inner side of the rotating circle. The combination of high rotating speed (1,850 rpm) and acceleration (1,000 g) during all experiments thus resulted in the injection of the mafic into the felsic melt during rotation. The experiments were performed at a temperature lower than that necessary for the previous set of experiments (Taylor-Couette). The obtained glasses highlighted a clear increase of the resultant hybridised melt volume (Fig. 7; Perugini et al. 2013, 2015). Vortex-like structures at the interface between the two initial end-member melts (Fig. 7d, e and f) are generated by stretching and folding dynamics.

Microchemical analyses of major, minor and trace element concentrations (Fig. 7) along the dotted lines in the resultant glasses, from the time series a, b and c, showed that concentrations variance for major to trace elements decays exponentially with time. Once more it is reassured that different chemical elements homogenise in the melt at differing rates, confirming results from previous experiments with a different geometry and under a higher temperature (e.g., De Campos et al. 2004, 2008; Perugini et al. 2008). The calculation of the differential mobility of each oxide/element has been quantitatively obtained by means of a new parameter, which mimics both the different initial compositions and the rheological properties of the contrasting melts coexisting in the same system (Fig. 8). Concentration variance takes into account the compositional and rheological dependence of diffusion, which the diffusion coefficient does not.

For the A-MS-MIN2 system, Na is the fastest amongst the major elements, followed by Al, Mg, Ca, K and Si. Trace elements such as: Ba,

Rb, Sr, Nb and Zr evidence similar mobilities. The REE, with the exception of Eu, display the lowest mobilities and show a systematic decrease from light towards heavy elements (see Fig. 8; Perugini et al. 2013, 2015).

The proposed parameter, the Relaxation of Concentration Variance (RCV; Perugini et al. 2013, 2015), is shown as an effective tool for quantifying the homogenisation of chemical elements during magma mixing and, therefore, can be an alternative to the diffusion coefficient. Results from the centrifuge experiments have reassured that different chemical elements homogenise in the melt at differing rates, confirming results from previous experiments with a different rheology (e.g., De Campos et al. 2004, 2008; Perugini et al. 2008), furthermore, they have enabled a more quantitative approach, resulting in the estimation of RCV for all elements simultaneously present in the melts (Fig. 8).

3.2.3 Chaotic Mixing—Preparing First Experiments with Campi Flegrei Caldera Volcanic Products

After recognition of the stretching and folding as the fundamental physical process producing mechanical mixing, leading to chaotic behaviour (e.g., Ottino 1989), the application of the chaos theory to the study of magma mixing (e.g., Perugini et al. 2004), the development of a new device to simulate chaotic dynamics at magmatic temperatures was the next natural step. This goal has been recently achieved (De Campos et al. 2011; Perugini et al. 2012 and references therein) and since then, is being used, especially with very contrasting end-members, such as rhyolite and basalt (Morgavi et al. 2013a, b, c). For the first chaotic mixing experiments, analogue silicate melts with a high viscosity contrast have been used, namely a peralkaline haplogranite and a haplobasalt, because the extreme viscosity ratio

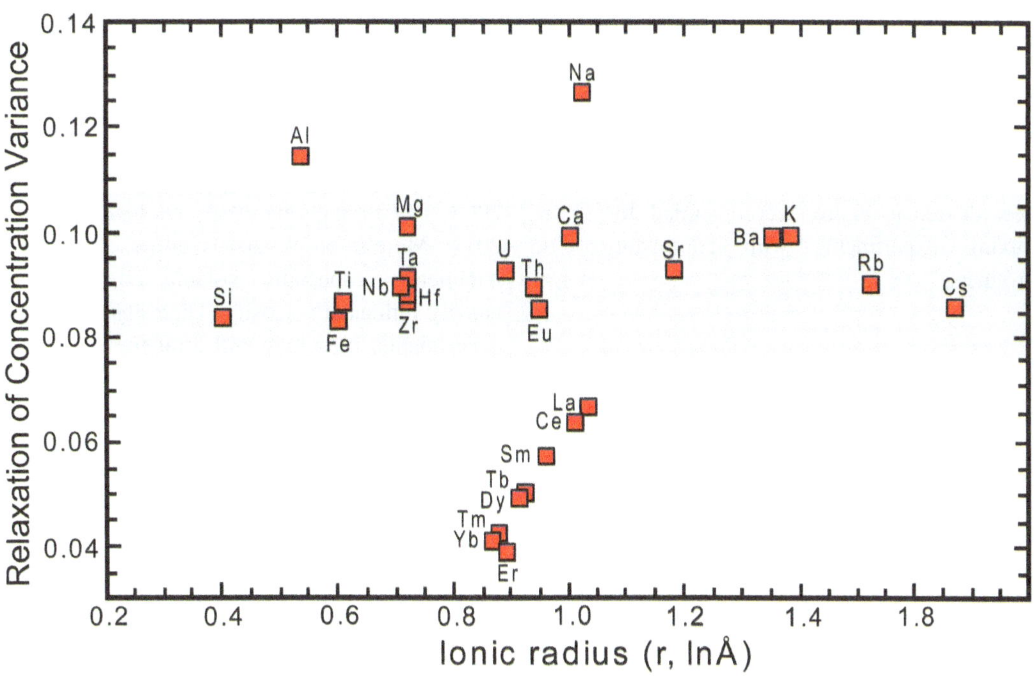

Fig. 8 Estimated Relaxation of Concentration Variance values for all analysed chemical elements in the system A-MS-MIN2 plotted against ionic radius. Figure modified after Perugini et al. (2013)

between these two melts is in the order of 10^3 (De Campos et al. 2011). With this device, the original end-member compositions nearly entirely disappeared from the filaments in very short mixing times. Notably, due to diffusive fractionation highly heterogeneous volumes of melt are generated, in which depletion or enrichment of chemical elements occur, depending on their potential to spread via chemical diffusion within the magma mixing system (e.g., De Campos et al. 2011; Perugini et al. 2012). Additional insights from experiments with the new device: (1) small volumes of least evolved magma can be hybridised with contrasting high viscosity melts, thus extending significantly the spectrum of geological conditions under which magma mixing processes can occur efficiently; (2) the mixing process is not linear; and (3) the chemical compositions on short-length scales represent snapshots within the process of mixing and therefore may not reflect the final composition of the magmatic system. A major implication from these results is that microanalysis on short-length scales, such as melt inclusions, may provide misleading information on the parental composition of magmas.

Since viscosity ratios in the CFc system have not yet been detected to be higher than 10^2 (e.g., A-MS vs MIN2, after Perugini et al. 2013), the new device is highly recommended for future mixing experiments with contrasting CFc products.

3.3 Reproducing Magma Dynamics at Laboratory Conditions – Challenges and Limitations

It is well known that physical and chemical properties of magmas determine both style of an eruption and compositional variations of the erupted rocks. In this review we called attention to the fact that, among all properties governing magmatic processes, viscosity is probably the most important one.

Taylor-Couette experiments are not as efficient as those using the chaotic mixing set. However, as shown by De Campos et al. (2004, 2008), the homogenisation during mixing experiments with CFc products can be easily attained. Complex unmixing and new layering may emerge after hybridisation, as previously discussed in Sect. 3.2.1. Therefore, we still encourage further Taylor-Couette experiments for highly alkaline systems such as the CFc system.

As previously discussed, both convection and diffusion are nowadays recognised as important processes in the magma chamber. Thus, any appropriate set for magma mixing should necessarily mimic the stretching and folding process at high temperature. The use of analogue materials instead of natural melts should be avoided. Clear differences between experimental results from simple analogue systems and complex natural systems have been studied in detail (e.g., De Campos et al. 2011; Perugini et al. 2013; Morgavi et al. 2013a, b, c, 2017) and support this conclusion.

When comparing the previously discussed experiments with the natural scenario, at least two major constrains emerge: (1) it is not yet possible to simulate adequate pressure under dynamic conditions and, (2) it is not yet possible to simulate adequate volatile contents, as those expected to be present in the natural magma chamber (e.g., Papale 2005; Mangiacapra et al. 2008; Chiodini et al. 2012; Di Genova et al. 2014; Moretti et al. 2019) either. These two parameters, especially volatile content, may heavily influence viscosity. Their absence in the experimental runs may still limit the implementation of obtained results. Nevertheless, preliminary attempts of comparing geochemical data from experimental and natural samples ended up in good agreement (e.g., De Campos et al. 2004, 2008; Perugini et al. 2010). This may be due to the fact that, under higher viscosity ratios, the process will only be less efficient but the behaviour of all elements in the mixing system will be very similar to the real world. In addition, the high temperature of some experiments (e.g., Taylor-Couette) may compensate for the role of dissolved water at high pressure bringing viscosity values towards more realistic values.

4 Conclusions and Implications for the Campi Flegrei Caldera System

Modelling viscosity and density in natural systems implies the accomplishment of thorough experimental measurements of a wide compositional range, and the use of different techniques. In the last decade there has been an enormous effort to decrease uncertainties. Results from both mixing experiments and numerical modelling suggest that mixing processes in igneous systems may strongly influence the mobility of trace elements inducing a 'diffusive fractionation' phenomenon (e.g., Perugini et al. 2006, 2008, 2013, 2015; De Campos et al. 2008, 2011), whose extent depends on the mixing timescale.

4.1 A New Chronometer for Unrest Episodes

A first effort to directly apply results from mixing experiments with natural CF products to compare them to natural volcanic sequences has been performed by Perugini et al. (2010). The pyroclastic sequences of the AST6 and AVE2 eruptions of the CFc were studied. The variation of chemical element compositions of both sequences indicated a compositionally zoned magma chamber prior to eruption. A clear dichotomy between the behaviour of major vs. trace elements was observed in both sequences, with major elements displaying nearly linear inter-elemental trends and trace elements showing a variable scattered behaviour. Perugini et al. (2010) merged the results of numerical simulations of magma mixing with those of magma mixing experiments (using natural compositions as end-members) to derive a relationship relating the degree of 'diffusive fractionation' to the mixing timescales.

The application of the 'diffusive fractionation' model to the two studied pyroclastic sequences enabled the authors to make an estimation of the mixing timescale. Results indicated that mixing processes in AST6 and AVE2 systems lasted for approximately 2 and 9 days, respectively, prior to eruption. A second direct application of the results from the centrifuge experiments compares additional natural pyroclastic sequences from the CFc with the experimental results, by means of a new reworked numerical parameter Concentration Variance Decay (Perugini et al. 2015). Concentration Variance Decay is the same parameter as the Relaxation of Concentration Variance or the variation of the normalised Concentration Variance (e.g., Perugini et al. 2013; Morgavi et al. 2013b) but for the refinement of the fitting coefficients (C0, R, C1) used to fit the decay of concentration variance for the chemical elements measured on experimental samples to estimate the mixing-to-eruption time for the three pyroclastic sequences of AVE2, AST6 and A-MS. The variance calculated for the same elements on the three pyroclastic sequences and the corresponding timescale (in minutes) are reported. The timescale of each eruption is the average value of the timescales measured for each element: i.e., 18 ± 5 (s.d.), 13 ± 4 (s.d.) and 15 ± 4 (s.d.) minutes for AVE2, AST6 and A-MS, respectively. This method allows using the variation of representative major and trace elements across the pyroclastic sequences to estimate the mixing-to-eruption time.

4.2 Conclusive Remarks

Rheological properties of magmas reflect the inherent structures of molten silicates and depend on the basic nature of silicate liquids. Mixing experiments with natural silicate melts have systematically shown how easily these melts may mix, so that, in a short time span, the initial compositions disappear from the system. Small volumes of least evolved magma can be hybridised with contrasting high viscosity melts, thus extending significantly the spectrum of geological conditions under which magma mixing processes can occur efficiently. Another major conclusion from this review is that the mixing process has been shown not to be linear, meaning that complex non-linear two-element plots may still be a sign of mixing in the system, in our case, in the CFc system. A third implication is that, on short-length scales, different chemical compositions simply represent snapshots within the process of mixing and therefore may not reflect the final composition of the magmatic system. A major implication from these results is that microanalysis on melt inclusions may provide misleading information on the parental character of magmas.

This review summarizes a present state of knowledge. Some discrepancies in the available viscosity data and modelling of the CFc products, which have been discussed in this review chapter, point towards the need of additional experimental measurements, especially for hydrous and least evolved components of the CFc system, in order to improve the mathematical parametrisation. Along with recent rheological studies, it may additionally help on interpretations of the CFc magmatic system and associated differentiation processes. This could be the case for the replenishment history in the shallow magmatic reservoir(s), the unavoidable mixing and eruptive dynamics. Results may shed new light on the differentiation processes and timescales of unrest in the CFc reservoir.

Acknowledgements This work was supported by the DFG-Project DI 431/31 and INGV Campi Flegrei UNREST project (Italy). Perugini (DP) and Dingwell (DBD) acknowledge research grants from the University of Perugia (DP), the MIUR-PRIN2010TT22SC_004 (DP) Humboldt Foundation (Germany) that awarded DP a Humboldt Fellowship at the Ludwig Maximilians University (Munich, Germany), and a Research Professorship (LMUexcellent) of the Bundesexellenzinitiative as well as ERC Advanced Grant 247076 "EVOKES" (DBD). Constructive comments by the editors, Daniele Giordano and anonymous reviewers are gratefully acknowledged.

Appendix 1 Viscosity Model for Latitic and Shoshonitic Compositions

Dry and hydrous viscosity for latitic and shoshonitic compositions from the CFc have been studied by Giordano et al. (2006, 2009), Misiti et al. (2011) and Di Genova et al. (2014). Misiti et al. (2011) selected two natural scoria samples from the pyroclastic sequences of the FR (latitic) and MIN2 (shoshonitic) eruptions. In contrast to other studies (Romano et al. 2003; Giordano et al. 2004; Misiti et al. 2006), instead of selecting the natural glassy phase, the bulk rock was remelted in air and quenched to glass (see Table 2). Parts of these then hydrated for measurements with the "falling sphere" and the micropenetration-techniques. The correspondent dry melts were further measured by the concentric cylinder and the micropenetration methods (see Table 4). A modified VFT-equation (Eq. 6) has been used in the description of the viscosity model:

$$\log \eta = a + \frac{b}{(T - c)} + \frac{d}{(T - e)} \cdot \exp\left(g \cdot \frac{w}{T}\right) \tag{6}$$

where η(eta) is the viscosity in Pa*s, T is the absolute temperature, w is the amount of water in wt% and a, b, c, d, e and g being fit parameters (there is no "f" fit parameter). Note: Confusingly the fitted parameter reported in the Misiti et al. (2011) paper have been fitted with "-a" instead of the "a" parameter. In Table 4 the "a" parameter is therefore the negative "a" parameter of Misiti et al. (2011). The viscosity data set for latitic melts of the FR eruption is rather limited in compositional space. At high temperatures there are only two data points for hydrous melts, with very similar water contents of 2.84 and 3.28 wt %, respectively. At lower temperatures the data covers restricted water contents up to a maximum of 1.2 wt%. The standard deviation of the model fit (0.35) is the worst for a specific model. Thus, the fit model could be further improved. A more recent viscosity model for water-and CO_2-bearing CFc latite has been obtained by Di Genova et al. (2014). These authors parameterised the measured viscosities of the FR magma with different amounts of CO_2 and H_2O using the equation reported in Di Genova et al. (2014), who modelled the viscosity by using the VFT equation log η = A + B / (T − C)] (Eq. 1) (Vogel 1921; Fulcher 1925; Tammann and Hesse 1926) with 4 fit parameters as follows:

$$\log\eta = A_{VFT} + \frac{b1 + b2.\log(1 + H_2O)}{T(K) - C1 + C2.\log(1 + H_2O)} \quad (7)$$

where η is the viscosity in Pa*s, T is the absolute temperature, H_2O is the water concentration in wt%, AVFT is the pre-exponential factor, b1 and b2 are related to the pseudo-activation energy, c1 and c2 are related to the Vogel temperature.

The viscosity data has been fitted by assuming that AVFT is a constant and independent of composition (e.g., Richet and Bottinga 1995; Whittington et al. 2001). The value of the preexponential parameter AVFT is taken as −4.55 Pa*s (Giordano et al. 2009). Fitted values and standard deviations of the b and c parameters are on Table 4. Equation 7 reproduces the experimental data from Di Genova et al. (2014) with a standard error of estimation of 0.57 log units.

The viscosity dataset for a shoshonitic melt feeding the MIN2 eruption is quantitatively not sufficient. Only 12 data points cover the low temperature region with maximum water content of 2.43 wt%. In the high temperature region two data points, for relatively similar water contents (2.35 and 3.30 wt%), constrained the dataset.

Both models predict the unlikely behaviour of a distinct difference of a dry curve and a curve where all hydrous melts collapse. It remains unclear if this result is due to the sparse compositional range (water content) covered, or to the modified VFT-equation, or to both. Also, the GRD and HZ general viscosity models cannot solve this problem. Up to this point it seems that for hydrous melts of shoshonitic composition there is no sufficient dataset range and a pertinent model available yet. Thus, both dataset and fit models still need improvements.

The numerical consequences of fitting viscosity-temperature datasets using natural samples from CF to non-Arrhenian rheological models have been explored by several authors (e.g., Giordano and Dingwell 2003; Russell and Giordano 2005; Giordano et al. 2006, 2008, 2009; Misiti et al. 2011; Di Genova et al. 2014). This kind of analysis shows that strong correlations of model parameters (e.g., ATVF, BTVF, CTVF) are inherent to non-Arrhenian models (Giordano et al. 2009). Uncertainties on model parameters and covariances between parameters are strongly affected by the quality and distribution of the experimental data, as well as the degree of non-Arrhenian behaviour as pointed out by Giordano et al. (2009).

The task of modelling viscosity in natural systems is therefore the product of a joint effort of decreasing uncertainties, through thorough experimental measurements of a wide compositional range, with different techniques, and improving mathematical parameterisation. This way, step by step, we are getting closer to the natural system, towards a best-fitting model.

References

Andújar J, Scaillet B (2012) Relationships between pre-eruptive conditions and eruptive styles of phonolite-trachyte magmas. Lithos 152:122–131

Arienzo I, Civetta L, Heumann A, Wörner G, Orsi G (2009) Isotopic evidence for open system processes within the Campanian Ignimbrite magma chamber. Bull Volcanol 71:285–300

Arienzo I, Moretti R, Civetta L, Orsi G, Papale P (2010) The feeding system of Agnano-Monte Spina eruption (Campi Flegrei, Italy): dragging the past into the present activity and future scenarios. Chem Geol 270 (1–4):135–147

Arienzo I, Mazzeo FC, Moretti R, Cavallo A, D'Antonio M (2016) Open-system magma evolution and fluid transfer at Campi Flegrei caldera (Southern Italy) during the past 5 ka as revealed by geochemical and isotopic data: the example of the Nisida eruption. Chem Geol 427:109–124

Bateman R (1995) The interplay between crystallization, replenishment and hybridisation in large felsic magma chambers. Earth-Sci Rev 39:91–106

Blake S, Ivey GN (1986) Density and viscosity gradients in zoned magma chambers, and their influence in withdraw dynamics. J Volcanol Geotherm Res 30:201–230

Campbell IH, Turner JS (1989) Fountains in magma chambers. J Petrol 30(4):885–923

Cannatelli C (2012) Understanding magma evolution at Campi Flegrei (Campania, Italy) volcanic complex using melt inclusions and phase equilibria. Miner Petrol 104:29–42. https://doi.org/10.1007/s00710-011-0182-6

Cannatelli C, Lima A, Bodnar RJ, De Vivo B, Webster JD, Fedele L (2007) Geochemistry of melt inclusions from the Fondo Riccio and Minopoli 1 eruptions at Campi Flegrei (Italy). Chem Geol 237:418–432

Caricchi L, Giordano D, Burlini L, Ulmer P, Romano C (2008) Rheological properties of magma from the 1538 eruption of Monte Nuovo (Phlegrean Fields, Italy): an experimental study. Chem Geol 256:158–171

Chiodini G, Caliro S, De Martino P, Avino R, Gerardi F (2012) Early signals of new volcanic unrest at Campi Flegrei caldera? Insights from geochemical data and physical simulations. Geology 10:943–946

Civetta L, Orsi G, Pappalardo L, Fisher RV, Heiken G, Ort M (1997) Geochemical zoning, mingling, eruptive dynamics and depositional processes, the Campanian Ignimbrite, Campi Flegrei Caldera, Italy. J Volcanol Geotherm Res 75:183–219

Courtial P, Dingwell DB (2005) Non-linear composition dependence of melt volume in the $CaO\text{-}Al_2O_3\text{-}SiO_2$ system. Geochim Cosmochim Acta 59:3685–3695

D'Antonio M, Tonarini S, Arienzo I, Civetta L, Di Renzo, V (2007) Components and processes in the magma genesis of the Phlegrean volcanic district (southern Italy). In: Beccaluva L, Bianchini G, Wilson M (eds) Cenozoic volcanism in the Mediterranean area. Geol Soc Am Sp Paper 418:203–220

D'Oriano C, Poggianti E, Bertagnini A, Cioni R, Landi P, Polacci M, Rosi M (2005) Changes in eruptive style during the A.D. 1538 Monte Nuovo eruption (Phlegrean Fields, Italy): the role of syn-eruptive crystallization. Bull Volcanol 67:601–621. https://doi.org/10.1007/s00445-004-0397-z

De Campos CP, Dingwell DB, Fehr KT (2004) Decoupled convection cells from mixing experiments with alkaline melts from Campi Flegrei. Chem Geol 213:227–251

De Campos CP, Dingwell DB, Fehr KT (2005) Double diffusive convection in alkaline silicate melts: first experimental results. Phys Chem Glasses-B 46 (4):330–333

De Campos CP, Dingwell DB, Perugini D, Civetta L, Fehr TK (2008) Heterogeneities in magma chambers: insights from the behaviour of major and minor elements during mixing experiments with natural alkaline melts. Chem Geol 256:131–145

De Campos CP, Perugini D, Ertel-Ingrisch W, Dingwell DB, Poli G (2011) Enhancement of magma mixing efficiency by chaotic dynamics: an experimental study. Contrib Mineral Petrol 161:863–881

de Vita S, Orsi G, Civetta L, Carandente A, D'Antonio M, Deino A, di Cesare T, Di Vito A, Fisher V, Isaia R, Marotta E, Necco A, Ort M, Pappalardo L, Piochi M, Southon J (1999) The Agnano-Monte Spina eruption (4100 BP) in the restless Campi Flegrei caldera (Italy). J Volcanol Geotherm Res 91:269–301. https://doi.org/10.1016/S0377-0273(99)00039-6

Di Genova D, Romano C, Alletti M, Misiti V, Scarlato P (2014) The effect of CO_2 and H_2O on Etna and Fondo Riccio (Phlegrean Fields) liquid viscosity, glass transition temperature and heat capacity. Chem Geol 377:72–86. https://doi.org/10.1016/j.chemgeo.2014.04.001

Di Renzo V, Arienzo I, Civetta L, D'Antonio M, Tonarini S, Di Vito MA, Orsi G (2011) The magmatic feeding system of the Campi Flegrei caldera: architecture and temporal evolution. Chem Geol 281:227–241

Di Vito MA, Isaia R, Orsi G, Southon J, de Vita S, D'Antonio M, Pappalardo L, Piochi M (1999) Volcanic and deformational history of the Campi Flegrei caldera in the past 12 ka. J Volcanol Geotherm Res 91:221–246

Di Vito MA, Arienzo I, Braia G, Civetta L, D'Antonio M, Di Renzo V, Orsi G (2011) The Averno 2 fissure eruption: a recente small-size explosive event at Campi Flegrei Caldera (Italy). Bull Volcanol 73:295–320

Di Vito MA, Acocella V, Aiello G, Barra D, Battaglia M, Carandente A, Del Gaudio C, de Vita S, Ricciardi G, Ricco C, Scandone R, Terrasi F (2016) Magma transfer at Campi Flegrei caldera (Italy) before the 1538 AD eruption. Sci Rep 6:32245. https://doi.org/10.1038/srep32245

Dingwell DB (1996) Volcanic dilemma: flow or blow? Science 273:1054–1055

Dingwell DB (2006) Transport properties of magmas: diffusion and rheology. Elements 2:281–286

Dingwell DB, Bagdassarov NS, Bussod GY, Webb SL (1993) Magma rheology. In: Luth RW (ed) Short course handbook on experiments at high pressure and applications to the earth's mantle. Mineral Assoc Canada 21, pp 131–196

Dingwell DB, Hess KU, Romano C (1998) Extremely fluid behavior of hydrous peralkaline rhyolites. Earth Planet Sci Lett 158:31–38

Fischer TP (2008) Fluxes of volatiles (H_2O, CO_2, N_2, Cl, F) from arc volcanoes: Geochem J 42:21–38. https://doi.org/10.2343/geochemj.42.21

Fulcher GS (1925) Analysis of recent measurements of the viscosity of glasses. J Am Ceramic Soc 8(6):339–355. https://doi.org/10.1111/j.1151-2916.1925.tb16731.x

Furbish DJ (1997) Fluid physics in geology: an introduction to fluid motions on Earth's surface and within its crust. Oxford Univ Press Inc., USA, p 476

Giaccio B, Haidas I, Isaia R, Deino A, Nomade S (2017) High-precision ^{14}C and $^{40}Ar/^{39}Ar$ dating of the Campanian Ignimbrite (Y-5) reconciles the time-scales of climatic-cultural processes at 40 ka. Sci Rep 7:45940

Giordano D, Dingwell DB (2003) Non-Arrhenian multicomponent melt viscosity: a model. Earth Planet Sci Lett 208(3):337–349

Giordano D, Romano C, Papale P, Dingwell DB (2004) The viscosity of trachytes, and comparison with basalts, phonolites, and rhyolites. Chem Geol 213:49–61

Giordano D, Mangiacapra A, Potuzak M, Russell JK, Romano C, Dingwell DB, Di Muro A (2006) An expanded non-Arrhenian model for silicate melt viscosity: a treatment for metaluminous, peraluminous and peralkaline liquids. Chem Geol 229:42–56. https://doi.org/10.1016/j.chemgeo.2006.01.007

Giordano D, Russell JK, Dingwell DB (2008) Viscosity of magmatic liquids: a model. Earth Planet Sci Lett 271:123–134

Giordano D, Ardia P, Romano C, Dingwell DB, Di Muro A, Schmidt MW, Mangiacapra A, Hess KU (2009) The rheological evolution of alkaline Vesuvius magmas and comparison with alkaline series from the Phlegrean Fields, Etna, Stromboli and Teide. Geochim

Cosmochim Acta 73:6613–6630. https://doi.org/10.1016/j.gca.2009.07.033

Hess KU, Dingwell DB (1996) Viscosities of hydrous leucogranitic melts: a non-Arrhenian model. Am Mineral 81:1297–1300

Hess KU, Dingwell DB, Rössler E (1996) Parameterization of viscosity temperature relationships of aluminosilicate melts. Chem Geol 128:155–163

Hui H, Zhang Y (2007) Toward a general viscosity equation for natural anhydrous and hydrous silicate melts. Geochim Cosmochim Acta 71:403–416

Huppert HE, Turner JS, Sparks RSJ (1982) Replenished magma chambers: effects of compositional zonation and input rates. Earth Planet Sc Let 57(2):345–357

Isaia R, D'Antonio M, Dell'Erba F, Di Vito M, Orsi G (2004) The Astroni volcano: the only example of closely spaced eruptions in the same vent area during the recent history of the Campi Flegrei caldera (Italy). J Volcanol Geotherm Res 133:171–192. https://doi.org/10.1016/S0377-0273(03)00397-4

Jellinek AM, Kerr RC, Griffiths RW (1999) Mixing and compositional stratification produced by natural convection: 1. Experiments and their applications to Earth's core and mantle. J Geophys Res 104 (B4):7183–7201

Knoche R, Dingwell DB, Webb SL (1995) Leucogranitic and pegmatitic melt densities: partial molar volumes for SiO_2, Al_2O_3, Na_2O, K_2O, Rb_2O, CsO, Li_2O, BaO, SrO, CaO, MgO, TiO_2, B_2O_3, P_2O_5, F_2O^{-1}, Ta_2O_5, Nb_2O_5, and WO_3. Geochim Cosmochim Acta 59:4645–4652

Lange RA (1994) The effect of H_2O, CO_2 and F on the density and viscosity of silicate melts. In: Carroll M, Holloway JR (eds) Volatiles in magmas. Rev Mineral vol 30, Mineral Soc Am, pp 331–169

Lange RA, Carmichael ISE (1987) Densities of Na_2O–K_2O–CaO–MgO–FeO–Fe_2O_3–Al_2O_3–TiO_2–SiO_2 liquids: new measurements and derived partial molar properties. Geochim Cosmochim Acta 51:2931–2946

Laumonier M, Scaillet B, Pichavant M, Champallier R, Andujar J, Arbaret L (2014) On the conditions of magma mixing and its bearing on andesite production in the crust. Nat Commun 5. https://doi.org/10.1038/ncomms6607

Liang Y, Richter FM, Watson EB (1996) Diffusion in silicate melts: II. Multicomponent chemical diffusion in CaO–Al_2O_3–SiO_2 at 1500 °C and 1 GPa. Geochim Cosmochim Acta 60:5021–5036

Longo A, Vassalli M, Papale P, Barsanti M (2006) Numerical simulation of convection and mixing in magma chambers replenished with CO_2-rich magma. Geophys Res Lett 33(21). https://doi.org/10.1029/2006GL027760

Longo A, Papale P, Vassalli M, Saccorotti G, Montagna CP, Cassioli A, Giudice S, Boschi E (2012) Magma convection and mixing dynamics as a source of ultra-long-period oscillations. Bull Volcanol 74 (4):873–880. https://doi.org/10.1007/s00445-011

Mangiacapra A, Moretti R, Rutherford M, Civetta L, Orsi G, Papale P (2008) The deep magmatic system of the Campi Flegrei caldera (Italy). Geophys Res Lett 35:L21304. https://doi.org/10.1029/2008GL035550

Marianelli P, Sbrana A, Proto M (2006) Magma chamber of the Campi Flegrei supervolcano at the time of the eruption of the Campanian Ignimbrite. Geology 34 (11):937–940

Martin D, Griffiths RW, Campbell IH (1987) Compositional and thermal convection in magma chambers. Contrib Mineral Petrol 96:465–475

Misiti V, Freda C, Taddeucci J, Romano C, Scarlato P, Longo A, Papale P, Poe BT (2006) The effect of H_2O on the viscosity of K-trachytic melts at magmatic temperatures. Chem Geol 235:124–137

Misiti V, Vetere F, Freda C, Scarlato P, Behrens H, Mangiacapra A, Dingwell DB (2011) A general viscosity model of Campi Flegrei (Italy) melts. Chem Geol 290(1–2):50–59

Montagna CP, Papale P, Longo A (2015) Timescales of mingling in shallow magmatic reservoir. In: Caricchi L, Blundy JD (eds) Chemical, physical and temporal evolution of volcanic systems. Geol Soc London, UK, Spec Publ 422, pp 131–140

Moretti R, Arienzo I, Civetta L, Orsi G, Papale P (2013) Multiple magma degassing sources at an explosive volcano. Earth Planet Sci Lett 367:95–104

Moretti R, Arienzo I, Di Renzo V, Orsi G, Arzilli F, Brun F, D'Antonio M, Mancini L, Deloule E (2019) Volatile segregation and generation of highly vesiculated explosive magmas by volatile-melt fining processes: the case of the Campanian Ignimbrite eruption. Chem Geol 503:1–14

Morgavi D, Perugini D, De Campos CP, Ertl-Ingrisch W, Dingwell DB (2013a) Time evolution of chemical exchanges during mixing of rhyolitic and basaltic melts. Contrib Mineral Petrol 166(2):615–638

Morgavi D, Perugini D, De Campos CP, Ertl-Ingrisch W, Dingwell DB (2013b) Morphochemistry of patterns produced by mixing of rhyolitic and basaltic melts. J Volcanol Geotherm Res 253:87–96

Morgavi D, Perugini D, De Campos CP, Ertl-Ingrisch W, Lavallee Y, Morgan L, Dingwell DB (2013c) Interactions between rhyolitic and basaltic melts unraveled by chaotic magma mixing experiments. Chem Geol 346:119–212

Morgavi D, Arienzo I, Montagna C, Perugini D, Dingwell DB (2017) Magma mixing: history and dynamics of an eruption trigger. In: Gottsmann J, Neuberg J, Scheu B (eds) Volcanic unrest, from science to society. Advances in volcanology, pp 123–137. ISSN: 2364-3277, ISSN: 2364-3285 (electronic). ISBN: 978-3-319-58411-9 ISBN: 978-3-319-58412-6 (eBook). https://doi.org/10.1007/978-3-319-58412-6

Mungall JE, Dingwell DB, Chaussidon M (1999) Chemical diffusivities of 18 trace elements in granitoid melts. Geochim Cosmochim Acta 63:2599–2610

Mysen BO (1988) Structure and properties of silicate melts. Elsevier, Amsterdam, The Netherlands, pp 354

Ottino JM (1989) The kinematics of mixing: stretching. Cambridge University Press, Chaos and Transport

Papale P (2005) Determination of total H_2O and CO_2 budgets in evolving magmas from melt inclusion data. J Geophys Res 110:B03208. https://doi.org/10.1029/2004JB003033

Papale P, Neri A, Macedonio G (1998) The role of magma composition and water content in explosive eruption: I. Conduit ascent dynamics. J Volcanol Geotherm Res 87:75–93
Pappalardo L, Piochi M, D'Antonio M, Civetta L, Petrini R (2002) Evidence for multi-stage magmatic evolution during the past 60 kyr at Campi Flegrei (Italy) deduced from Sr, Nd and Pb isotope data. J Petrol 43(8):1415–1434
Persikov ES (1991) Viscosity of magmatic liquids: experiment, generalized patterns. A model for the calculation and prediction. In: Perchuk L, Kushiro I (eds) Adv Physic Geochem 9. Springer-Verlag, NY, pp 1–40
Perugini D, Ventura G, Petrelli M, Poli G (2004) Kinematic significance of morphological structures generated by mixing of magmas: a case study from Salina Island (Southern Italy). Earth Planet Sci Lett 222:1051–1066
Perugini D, Petrelli M, Poli G (2006) Diffusive fractionation of trace elements by chaotic mixing of magmas. Earth Planet Sci Lett 24:669–680
Perugini D, De Campos CP, Dingwell DB, Petrelli M, Poli G (2008) Trace element mobility during magma mixing: preliminary experimental results. Chem Geol 256:145–156. https://doi.org/10.1016/j.chemgeo.2008.06.032
Perugini D, Petrelli M, Poli G, De Campos CP, Dingwell DB (2010) Time-scales of recent Phlegrean Fields eruptions inferred by the application of the 'diffusive fractionation' model of trace elements. Bull Volcanol 72:431–447. https://doi.org/10.1007/s00445-009-0329-z
Perugini D, De Campos CP, Ertel-Ingrisch W, Dingwell DB (2012) The space and time complexity of chaotic mixing of silicate melts: Implications for igneous petrology. Lithos 155:326–340
Perugini D, De Campos CP, Dingwell DB, Dorfman A (2013) Relaxation of concentration variance: A new tool to measure chemical element mobility during mixing of magmas. Chem Geol 335:8–23
Perugini D, De Campos CP, Petrelli M, Dingwell DB (2015) Concentration variance decay during magma mixing: a volcanic chronometer. Sci Rep 5:14225. https://doi.org/10.1038/srep14225
Peruzzo E, Barsanti M, Flandoli F, Papale P (2010) The stochastic quantization method and its application to the numerical simulation of volcanic conduit dynamics under random conditions. Solid Earth 1:49–59. https://www.solid-earth.net/1/49/2010/doi:10.5194/se-1-49-2010
Petrelli M, Perugini D, Poli G (2011) Transition to chaos and implications for time-scales of magma hybridization during mixing processes in magma chambers. Lithos 125(1–2):211–220. https://doi.org/10.1016/j.lithos.2011.02.007
Richet P, Bottinga Y (1995) Rheology and configurational entropy of silicate liquids. In: Stebbins JF, McMillan PF, Dingwell DB (eds) Structure, dynamics and properties of Silicate melts. Min Soc Am Rev Mineral 32, pp 67–93
Rolandi G, Bellucci F, Heizler MT, Belkin HE, De Vivo B (2003) Tectonic controls on the genesis of ignimbrites from the Campanian volcanic zone, Southern Italy. Mineral Petrol 79:3–31
Romano C, Giordano D, Papale P, Mincione V, Dingwell DB, Rosi M (2003) The dry and hydrous viscosities of alkaline melts from Vesuvius and Phlegrean Fields. Chem Geol 202:23–38
Rössler E, Hess KU, Novikov VN (1998) Universal representation of viscosity in glass forming liquids. J Non-Cryst Solids 223:207–222
Russell JK, Giordano D (2005) A model for silicate melt viscosity in the system $CaMgSi_2O_6$–$CaAl_2Si_2O_8$–$NaAlSi_3O_8$. Geochim Cosmochim Acta 69:5333–5349
Signorelli S, Vaggelli G, Francalanci L, Rosi M (1999) Origin of magmas feeding the Plinian phase of the Campanian Ignimbrite eruption, Phlegrean Fields (Italy): constraints based on matrix-glass and glass-inclusion compositions. J Volcanol Geotherm Res 91:199–220
Smith VC, Isaia R, Pearce NJG (2011) Tephrostratigraphy and glass compositions of post-15 kyr Campi Flegrei eruptions: implications for eruption history and chronostratigraphic markers. Quat Sci Rev 30:3638–3660
Snyder D, Tait S (1996) Magma mixing by convective entrainment. Nature 379:529–531
Sparks RSJ (2004) Dynamics of magma degassing. Volcanic degassing. Geol Soc Lond Sp Publ 213:5–22
Sparks RSJ, Marshall IA (1986) Thermal and mechanical constrains on mixing between mafic and silicic magmas. J Volcanol Geotherm Res 29:99–124
Spera FJ, Schmidt JS, Bohrson BGA (2016) Dynamics and thermodynamics of magma mixing: Insights from a simple exploratory model. Am Mineral, Sp Col: Perspect Origins Evol Crustal Magmas 101:627–643
Tammann G, Hesse W (1926) Die Abhängigkeit der viscosität von der temperatur bie unterkühlten Flüssigkeiten. Zeitsch Anorgan Allg Chem 156(1):245–257. https://doi.org/10.1002/zaac.19261560121
Tonarini S, D'Antonio M, Di Vito MA, Orsi G, Carandente A (2009) Geochemical and B-Sr-Nd isotopic evidence for mingling and mixing processes in the magmatic system that fed the Astroni volcano (4.1-3.8 ka) within the Campi Flegrei caldera (Southern Italy). Lithos 107:135–151
Toplis MJ (1998) Energy barriers to viscous flow and the prediction of glass transition temperatures of molten silicates. Am Mineral 83:480–490
Vogel DH (1921) Das Temperaturabhaengigkeitsgesetz der Viskositaet von Fluessigkeiten. Physik Zeitsch 22:645
Vona A, Romano C, Giordano D, Russell JK (2013) The multiphase rheology of magmas from Monte Nuovo (Campi Flegrei, Italy). Chem Geol 346:213–227
Whittington A, Richet P, Linard Y, Holtz F (2001) The viscosity of hydrous phonolites and trachytes. Chem Geol 174:209–223

Magma Chamber Dynamics at the Campi Flegrei Caldera, Italy

Chiara P. Montagna, Paolo Papale, and Antonella Longo

Abstract

The Campi Flegrei caldera volcanic system is certainly a remarkable case study of magma chamber dynamics. Its magmatic and volcanic history appears to have been largely driven by magma chamber processes like fractional crystallisation, magma mixing, and volatile degassing. These processes have been intensely investigated with a variety of approaches that are described in many chapters of this book, and more specifically, in Chaps. An Evolutionary Model for the Magmatic System of the Campi Flegrei Volcanic Field (Italy) Constrained by Petrochemical Data; Rheological Properties of the Magmas Feeding the Campi Flegrei Caldera (Italy) and Their Influence on Mixing Processes. In this chapter, physical modelling and numerical simulations are employed in order to study the dynamics of magma convection and mixing in a vertically extended, geometrically complex, compositionally heterogeneous magmatic system representing a schematic simplification of an overall picture emerging from previous studies at Campi Flegrei caldera. Although clearly an idealisation, a number of first order characteristics of possible real magmatic systems at Campi Flegrei caldera are accounted for. They include the more chemically evolved, partially degassed nature of magmas emplaced at shallow depths, and the likely occurrence of multiple reservoirs with different depth, size and shape which can be connected at certain stages during system evolution. If that happens, deeper, CO_2-rich magmas may rise and rejuvenate the shallow magmas.

1 The Magmatic System of the Campi Flegrei Caldera

The Campi Flegrei caldera (CFc) is the dominant feature of the Campi Flegrei volcanic field (Chap. Volcanic and Deformation History of the Campi Flegrei Volcanic Field, Italy). It is a complex, nested structure formed by the Campanian Ignimbrite (CI; ~40 ka) and the Neapolitan Yellow Tuff (NYT; ~15 ka) eruptions (Orsi et al. 1992, 1996; Chap. Volcanic and Deformation History of the Campi Flegrei Volcanic Field, Italy). The NYT caldera has been the site of an intense volcanic and deformation activity (Orsi et al. 1996, 1999, 2004; Di Vito et al. 1999; Smith et al. 2011; Chap. Volcanic and Deformation History of the Campi Flegrei Volcanic Field, Italy). Volcanism has generated no less than 70 eruptions, among which Agnano-Monte Spina (4,482–4,625 cal. years BP; de Vita et al. 1999; Arienzo et al. 2010; Smith et al 2011), Minopoli 2

C. P. Montagna (✉) · P. Papale · A. Longo
Istituto Nazionale di Geofisica e Vulcanologia, Sezione di Pisa, Pisa, Italy
e-mail: chiara.montagna@ingv.it

© Springer-Verlag GmbH Germany, part of Springer Nature 2022
G. Orsi et al. (eds.), *Campi Flegrei*, Active Volcanoes of the World,
https://doi.org/10.1007/978-3-642-37060-1_7

(~11 ka; Di Vito et al. 1999; Di Renzo et al. 2011; Tomlinson et al. 2012) and Fondo Riccio (~11 ka; Di Vito et al. 1999; Di Renzo et al. 2011; Tomlinson et al. 2012).

The magmatic system of the CFc can be reconstructed based on information from various geophysical surveys as well as analyses of eruptive products of past eruptions, as detailed in Chaps. Seismic and Gravity Structure of the Campi Flegrei Caldera, Italy; An Evolutionary Model for the Magmatic System of the Campi Flegrei Volcanic Field (Italy) Constrained by Petrochemical Data; Origin and Differentiation History of the Magmatic System Feeding the Campi Flegrei Volcanic Field (Italy) Constrained by Radiogenic and Stable Isotope Data; Tephrochronology and Geochemistry of Tephra from the Campi Flegrei Volcanic Field, Italy; Rheological Properties of the Magmas Feeding the Campi Flegrei Caldera (Italy) and Their Influence on Mixing Processes. Analyses of seismic records allow us to infer the position and location of fluids within the host rock (Judenherc and Zollo 2004; Zollo et al. 2008; De Siena et al. 2010; Masterlark et al. 2010; Chouet and Matoza 2013). A seismic reflection horizon at around 8 km depth possibly identifies the top of a partially molten region, probably as wide as some tens of kilometres in the horizontal direction, most likely representing a magmatic sill (Judenherc and Zollo 2004; Zollo et al. 2008; Pappalardo and Mastrolorenzo 2012).

A variety of observations suggests that small batches of magma have been forming throughout the caldera history (De Siena et al. 2010; Di Renzo et al. 2011). Crystallised magma bodies, sized less than 1 km^3, have been recognised by processing offshore seismic lines, at depths ranging between less than 1 and 6 km (Piochi et al. 2014). Fluid-rich batches interpreted as partially molten magma have been identified by seismic attenuation tomography in the same depth range (De Siena et al. 2010). Reconstructed phase equilibria for the Agnano-Monte Spina eruption, as well as H_2O and CO_2 contents in melt inclusions from Minopoli 2, Fondo Riccio, and Agnano-Monte Spina confirm that magmas had been residing at all depths less than about 9 km (Roach 2005; Mangiacapra et al. 2008; Arienzo et al. 2010; Piochi et al. 2014; Astbury et al. 2018; Chaps. Volcanic and Deformation History of the Campi Flegrei Volcanic Field, Italy; Seismic and Gravity Structure of the Campi Flegrei Caldera, Italy; An Evolutionary Model for the Magmatic System of the Campi Flegrei Volcanic Field (Italy) Constrained by Petrochemical Data; Origin and Differentiation History of the Magmatic System Feeding the Campi Flegrei Volcanic Field (Italy) Constrained by Radiogenic and Stable Isotope Data). Small, shallow magma bodies have been identified as actively involved in past eruptions, which at least in some cases shortly followed the arrival of volatile-rich, less differentiated magmas from greater depth (Arienzo et al. 2009, 2010; Fourmentraux et al. 2012; Astbury et al. 2018; Chaps. An Evolutionary Model for the Magmatic System of the Campi Flegrei Volcanic Field (Italy) Constrained by Petrochemical Data; Origin and Differentiation History of the Magmatic System Feeding the Campi Flegrei Volcanic Field (Italy) Constrained by Radiogenic and Stable Isotope Data). Preferential pathways for magma ascent have often been related to caldera structures (Akande et al. 2019).

The composition of the erupted products ranges from shoshonitic to trachytic to phonolitic; geochemical analyses on melt inclusions suggest a variety of processes during inter-eruptive periods including magma recharge, intra-chamber mixing, and entrapment of phenocrysts left from previous eruptions (Arienzo et al. 2010; Forni et al. 2018; Chaps. An Evolutionary Model for the Magmatic System of the Campi Flegrei Volcanic Field (Italy) Constrained by Petrochemical Data; Origin and Differentiation History of the Magmatic System Feeding the Campi Flegrei Volcanic Field (Italy) Constrained by Radiogenic and Stable Isotope Data).

From all the above, a schematic picture of the CFc magmatic system can be derived (Fig. 1). It includes the deep magmatic sill containing the poorly evolved, CO_2-rich shoshonitic magma; and more or less persistent features, such as

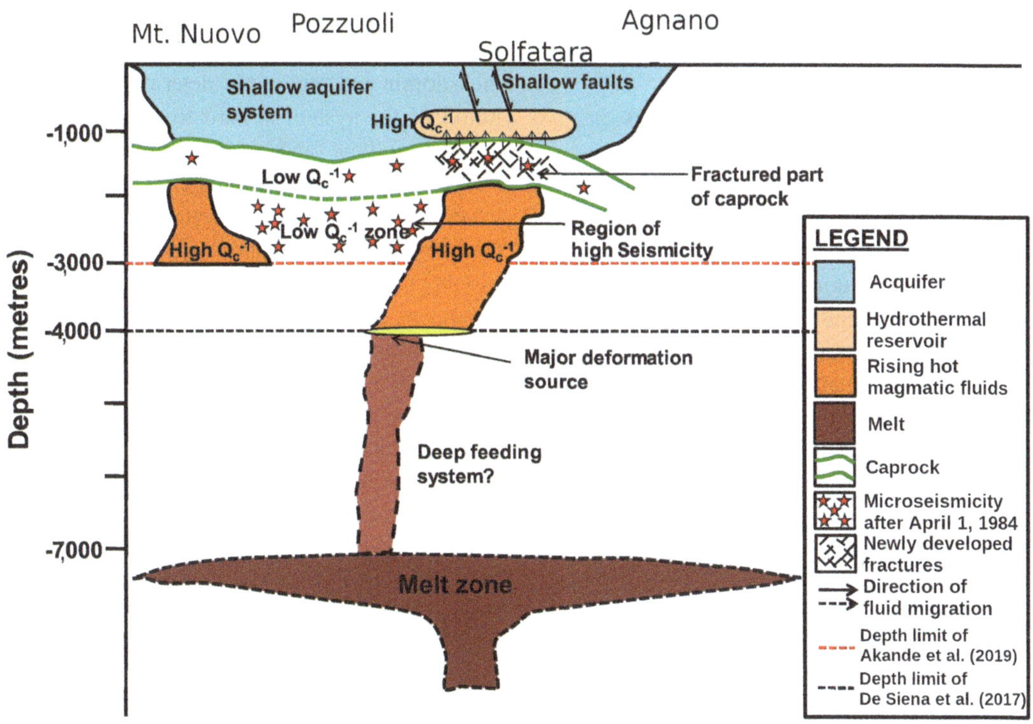

Fig. 1 Scheme of the shallow magmatic system beneath the Campi Flegrei caldera. Figure modified after Akande et al. (2019)

dikes, fractures or conduits, possibly related to caldera structures, linking the deep reservoir to shallower, degassed, more crystal-rich batches of magma. Shallower reservoirs contain magmas with compositions from trachyte to phonolite. Mingling and mixing of shallow magma with others coming from depth is thought to be a recurrent process in the CFc volcanism, often invoked as the main trigger of past eruptions (Arienzo et al. 2011; Fourmentraux et al. 2012; Astbury et al. 2018; Chaps. An Evolutionary Model for the Magmatic System of the Campi Flegrei Volcanic Field (Italy) Constrained by Petrochemical Data; Origin and Differentiation History of the Magmatic System Feeding the Campi Flegrei Volcanic Field (Italy) Constrained by Radiogenic and Stable Isotope Data). Eruptive products preserve the signature of such processes, and provide clues about the time scales of mixing, that in most cases appear to last only hours to days before the eruption (Perugini et al. 2010, 2015a, b; Morgavi et al. 2017; Chap. Rheological Properties of the Magmas Feeding the Campi Flegrei Caldera (Italy) and Their Influence on Mixing Processes). The regional and/or local stress regimes allow the deep magmas to rise to shallow levels, either along established pathways or cracking the host rocs and creating new ones (Orsi et al. 1996, 1999). When the rising magma encounters a molten or partially molten region, the residing magma gets rejuvenated and mixes with the incoming one (Bachmann and Bergantz 2003). At CFc, magmas rising from depth are typically rich in gas, especially CO_2 (Mangiacapra et al. 2008; Arienzo et al. 2010; Moretti et al. 2013). When such magmas reach an already established reservoir where more dense and viscous magma resides, they tend to rise under the action of buoyancy forces. The compositionally different magmas mix first physically, and, on longer time scales, chemically, producing intermediate compositions (Arienzo et al. 2009; Tonarini et al. 2009; Perugini et al. 2010; Di Renzo et al. 2011).

2 Physical Model for the Campi Flegrei Caldera Magmatic System

2.1 Magma Chamber Processes

Magma chamber dynamics is governed by a variety of microscopic and macroscopic physical and chemical processes, which range from periodic magma injections, to mixing and mingling of different magmas, to magma differentiation due to crystal fractionation and melt segregation, to heat and mass transfer between the phases and with the surrounding rocks, to interaction with the host rock and its tectonic stresses (Sigurdsson et al. 2015). The quantitative study of magma dynamics is complicated by the uncertainties about the characteristics of the magmatic system, which is out of direct observation. Nevertheless, all of the aforementioned processes bear consequences on the state of the reservoir, which can be inferred by analysing either the petrological records (e.g., Perugini and Poli 2012) or the geophysical signals indicating movement at depth (e.g., Bagagli et al. 2017).

A precise understanding of the physics of magmatic reservoirs is still lacking (Cashman et al. 2017). Magma is a multi-phase, multi-component fluid and as it rises and stalls in the Earth's crust it encounters very different pressure and temperature conditions, as well as stress and strain, that determine the phase assemblage and its evolution in space and time. The thermodynamics and fluid dynamics of such a mixture are quite complex as they involve phase changes between melt, gas and crystals, which in turn contribute to determine the physical and rheological properties governing its motion (Bachmann and Bergantz 2006; Degruyter and Huber 2014; Parmigiani et al. 2014; Bergantz et al. 2015; Montagna et al. 2015; Papale et al. 2017).

It is more and more commonly being assumed that the typical magmatic reservoir comprises a highly crystalline mushy region, and possibly a smaller, melt-rich portion at its top (Schleicher and Bergantz 2017; Cashman et al. 2017). Long-lived magma chambers can slowly solidify and originate mush zones that are dominated by crystals (Pistone et al. 2017). The relative importance of the mushy with respect to the melt-dominated regions in determining the evolution of the system is hard to assess, as is the relative volume and mass ratios of the two. Models that describe magma chamber dynamics have either focussed on one or the other of the two regimes (Bachmann and Bergantz 2006; Montagna et al. 2015; Papale et al. 2017); moreover, to model the mushy region different approaches have been used, from fluid-like to porous flow, to brittle fracturing (Parmigiani et al. 2014; Carrara et al. 2019).

Convection, mingling and mixing are among the most common processes in melt-dominated as well as mushy reservoirs (Schleicher and Bergantz 2017; Morgavi et al. 2017), but their features can significantly differ among the two (Parmigiani et al. 2014; Montagna et al. 2015). Mingling refers to mechanical mixing of two or more batches of magma, without chemical exchanges between them; on the contrary, mixing involves chemical exchanges whereby elements diffuse according to compositional gradients (e.g., Flinders and Clemens 1996).

One of the first numerical models for mixing of incompressible homogeneous magmas due to diffusion and thermal-compositional convection was developed by Oldenburg et al. (1989, and references therein). Diffusive convection can be originated by both thermal and compositional gradients, that give rise to a boundary layer instability developing on time scales that, for typical magmatic systems, are on the order of years to thousands of years, depending mostly on the viscosities of the magmas involved (Spera et al. 1982).

On the other hand, mingling in melt-dominated reservoirs (e.g., Andersen et al. 2019; Gualda et al. 2019) can be originated by Rayleigh–Taylor instability as a volatile-rich magma reaches a degassed reservoir (Ruprecht et al. 2008). This process can happen on much shorter time scales, on the order of hours to days (Montagna et al. 2015). Over such short time spans, the interacting magmas are effectively immiscible, and chemical diffusion is negligible, albeit a limited amount of numerical diffusion is inevitably present (Longo et al. 2012a).

2.2 Models of Magma Dynamics

The physical model in Longo et al. (2012a) describes the time-dependent dynamics of a multicomponent mixture consisting of liquid (or crystal suspension) in thermodynamic equilibrium with an $H_2O + CO_2$ gas phase at the local conditions of pressure, temperature and composition. The model can handle both compressible and incompressible flows, therefore it is particularly apt to solve for magmatic mixtures characterised by a very low compressibility at depth, where most volatiles are dissolved in the melt, that increases at shallower depths as volatiles begin to exsolve. The numerical algorithm used in the solution of the conservation equations is based on an extension of the finite element formulation by Hauke and Hughes (1998) to include multicomponent fluids (Longo et al. 2006), allowing the investigation of processes involving mixing of fluids, chemical changes, and phase transitions. The algorithm consists of a space–time discretisation with Galerkin least-squares and discontinuity capturing terms, with third order accuracy in time and space. This method allows the simulation of both compressible and incompressible flow dynamics (Shakib et al. 1991; Chalot and Hughes 1994; Hauke and Hughes 1998), and it is effective in stabilising the numerical solution without introducing excessive numerical diffusion. A large number of problems can be solved, such as natural and forced convection, shock waves and their interaction with contact discontinuities, evolution of internal interfaces in incompressible or compressible flows, and bubbly flows with evaporation or gas dissolution. All of these processes are relevant for magmatic systems, and the model has been successfully applied to obtain significant results on magma reservoir dynamics in different magmatic settings including basaltic, andesitic and dacitic systems (Longo et al. 2012b; Montagna et al. 2015; Papale et al. 2017; Garg et al. 2019). The conservation equations for the mass of single components and momentum of the whole mixture, together with the gas–liquid thermodynamic equilibrium model and the constitutive equations for mixture properties (density and viscosity), are discretised and solved for the primitive variables pressure, velocity, temperature, and concentration of components. Magmas are described as ideal mixtures with components that may be either in the liquid or gaseous state, with instantaneous phase change according to the non-ideal multicomponent $H_2O + CO_2$ saturation model of Papale et al. (2006). Gas bubbles are assumed to be non-deformable, a good approximation if the bubble size is smaller than ~ 10 m (Marchetti et al. 2004). This corresponds to a gas volume of 5–50% for bubble number densities in the range 10^{14} to 10^{15} m^{-3}. Such small bubble sizes justify the gas–melt coupled flow assumption, as well as the equilibrium thermodynamics (Burgisser et al. 2005; Yoshimura and Nakamura 2011). The role on the relevant properties (density and viscosity) of dissolved water and of the dispersed gas and solid phases, and the mutual roles of H_2O and CO_2 in affecting their saturation contents, are accounted for. Mixture density is calculated using the Lange (1994) equation of state for the liquid phase, real gas properties and standard mixture laws for multiphase fluids. Mixture viscosity (under the assumption of Newtonian rheology) is computed through standard rules of mixing (Reid et al. 1977) for one-phase mixtures and with a semi-empirical relation (Ishii and Zuber 1979) in order to account for the effect of non-deformable gas bubbles. Liquid viscosity is modelled as in Giordano et al. (2008), and it depends on liquid composition and dissolved water content.

The solid phase (suspended crystals) can be taken into account in the computation of mixture properties (density and viscosity). Nevertheless, in the models shown here the magmas are assumed crystal-free as most erupted products with the compositions used have been found to have very little phenocrysts (Mangiacapra et al. 2008). The model is meaningful for melt-dominated, crystal-poor magma reservoirs, which is expected to be the case for many eruptions at Campi Flegrei (Di Renzo et al. 2011;

Forni et al. 2018; Chaps. An Evolutionary Model for the Magmatic System of the Campi Flegrei Volcanic Field (Italy) Constrained by Petrochemical Data; Origin and Differentiation History of the Magmatic System Feeding the Campi Flegrei Volcanic Field (Italy) Constrained by Radiogenic and Stable Isotope Data). A crystal-rich, mushy region is also hypothesised to have been involved in the CI eruption (Forni et al. 2018); this work focuses on smaller eruptions that have not been energetic enough to tap the crystal-rich regions of the reservoirs.

The magmatic system of CFc is approximated as isothermal: temperature differences inferred from melt inclusion data are negligible (Mangiacapra et al. 2008), and solving the dynamics for an isothermal system greatly reduces computational needs as the energy conservation equation does not need to be solved. The effects of wall cooling can be neglected on the very short time scales analysed here (Sect. 4).

2.3 Modelling the Magmatic System of the Campi Flegrei Caldera

In order to understand the dynamics of magmas beneath the CFc, the magmatic system is simplified to retain its most essential features. The specific aim is to explore the mixing process, and the dependence of its efficiency and time scales on the simulation conditions. Other details on the time scales, the evolution of pressure in the different portions of the magmatic system as well as the ground deformation signals related to the dynamics described here are analysed in separate studies (Montagna et al. 2015, 2017; Bagagli et al. 2017; Morgavi et al. 2017; Papale et al. 2017).

The model describes the injection of CO_2-rich, shoshonitic magma coming from a deep reservoir into a shallower, much smaller chamber, containing more evolved and partially degassed phonolitic magma. The deep reservoir is schematised as a sill; the geometry of the shallow chamber has been varied from oblate to prolate in a set of simulations, as shown in Fig. 2. Initial conditions for the simulated cases are reported in Table 1.

The top of the deep reservoir is at 8 km depth, and its horizontal and vertical semi-axes measure 4 km and 0.5 km, respectively. The shallow reservoir has its top at 3 km depth below the surface; its surface area is kept to 0.25×10^6 m^2, representing a volume of magma in the order of a few tenths of km^3 (Chaps. Volcanic and Deformation History of the Campi Flegrei Volcanic Field, Italy; Seismic and Gravity Structure of the Campi Flegrei Caldera, Italy; An Evolutionary Model for the Magmatic System of the Campi Flegrei Volcanic Field (Italy) Constrained by Petrochemical Data; Origin and Differentiation History of the Magmatic System Feeding the Campi Flegrei Volcanic Field (Italy) Constrained by Radiogenic and Stable Isotope Data; Source Modelling from Ground Deformation and Gravity Changes at the Campi Flegrei Caldera, Italy). The oxides composition of the two magmas is detailed in Table 2.

At the initial time, the two magmas are placed in contact at the bottom of the shallow chamber (top of the feeding dyke; Fig. 2). Because of the higher volatile content, thus lower overall density of the deeper magma, the system is gravitationally unstable. Therefore, the deep magma tends to rise into the shallow chamber, pushing the resident magma down through the feeding dike.

The initial condition aims at reproducing the effects of chemical processes such as compositional evolution and degassing, which contribute to a density increase in the shallow regions of a spatially extended, interconnected magmatic system (e.g., Kazahaya et al. 1994), possibly through permanently open pathways (e.g., Di Renzo et al. 2011). As such a system is expected to be in continuous evolution, any choice of initial conditions would be equally arbitrary, therefore the extreme case is explored here, where buoyancy forces are most effective and acting at shallow chamber inlet. Analysis of results focuses on the long-term dynamics, neglecting the transient dictated by the choice of initial conditions.

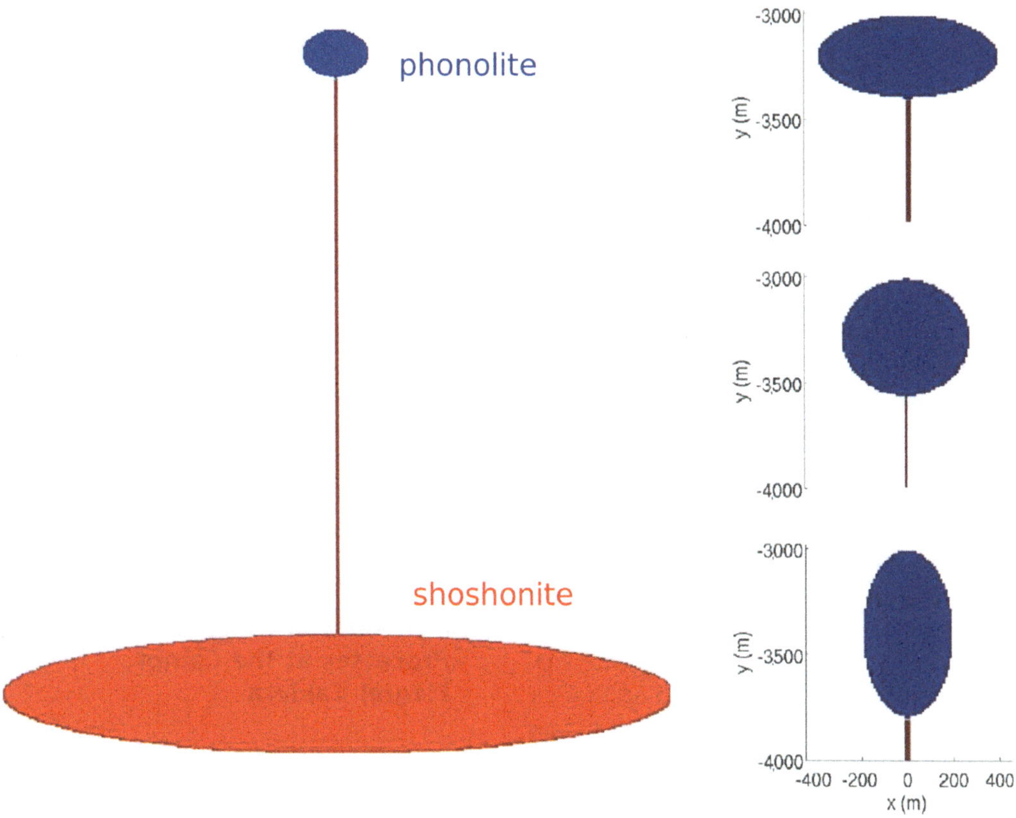

Fig. 2 The whole simulated system alongside the three different geometrical settings for the shallower magma chamber at the Campi Flegrei caldera. Top: horizontally elongated, sill-like chamber. Centre: Circular chamber. Bottom: vertically elongated, dike-like chamber. Figure modified after Montagna et al. (2017)

Table 1 List of simulations and their initial conditions

Simulation No.		1	2	3	4	5
Assumed initial conditions	CO_2 shoshonite (wt%)	1	1	1	1	1
	CO_2 phonolite (wt%)	0.3	0.3	0.3	0.3	0.3
	H_2O shoshonite (wt%)	2	2	2	2	2
	H_2O phonolite (wt%)	2.5	2.5	2.5	1	1
	Geometry of shallow chamber	Oblate	Prolate	Circular	Oblate	Prolate
Calculated initial conditions	Density contrast at initial interface (kg/m^3)	30	20	20	20	140
	Gas volume at initial interface, shoshonite (vol%)	10	8	9	9	8
	Gas volume at initial interface, phonolite (vol%)	5	4	5	2	2
	Gas volume at system top (vol%)	7	7	7	2	2
	Gas volume at system bottom (vol %)	3	3	3	3	3

Data from Montagna et al. (2015), Bagagli et al. (2017), Papale et al. (2017)

Table 2 Oxides contents (unit fraction) for the two end-member magmas, shoshonite (Sho) and phonolite (Pho)

	SiO_2	TiO_2	Al_2O_3	Fe_2O_3	FeO	MnO	MgO	CaO	Na_2O	K_2O
Sho	0.48	0.012	0.16	0.021	0.063	0.0014	0.10	0.12	0.028	0.015
Pho	0.54	0.0060	0.20	0.016	0.032	0.0014	0.017	0.068	0.047	0.079

Data from Montagna et al. (2015)

Different setups have been explored in order to highlight the roles played by the geometry of the magmatic system and by the extent of previous degassing of the shallow magma. Keeping the upper chamber top at a depth of 3 km and its 2D cross-sectional area constant, its geometry was varied from a sill to a circle and to a vertically elongated, dike-like ellipsoid. For the two ellipsoidal chambers, the volatile content of the shallow phonolitic magma was also reduced to 1 wt% H_2O and 0.1 wt% CO_2, compared to 2.5 wt % H_2O and 0.3 wt% CO_2 of the previous set of simulations. The deep shoshonitic magma contains in all cases 2 wt% H_2O and 1 wt% CO_2 (Table 1; Chaps. An Evolutionary Model for the Magmatic System of the Campi Flegrei Volcanic Field (Italy) Constrained by Petrochemical Data; Origin and Differentiation History of the Magmatic System Feeding the Campi Flegrei Volcanic Field (Italy) Constrained by Radiogenic and Stable Isotope Data). It is worth noting that such quantities are total volatile amounts, and they are distributed between liquid and gas phases according to the local conditions and the saturation model of Papale et al. (2006). The simulated systems are closed, with fixed boundaries. The effect of wall-rock elasticity on the chamber dynamics is neglected, as it is assumed to cause small pressure changes with respect to those originating from the magma dynamics.

The system setup is two-dimensional mainly to keep computational efforts affordable. Extension to full 3D systems should not bring about any substantial difference in the convective patterns observed: the third dimension would give rise to different dominant modes of the Rayleigh–Taylor instability (Ribe 1998), thus possibly modifying (either increasing or decreasing, depending on the amplitude of the initial perturbation) the efficiency of transport, but not substantially (Kaus and Podladchikov 2001).

The setups shown here are based on petrological and geophysical evidence, as detailed in Sect. 1; a more rigorous exploration of parameters' space is at the moment limited by computational resources. Nevertheless, the results are meaningful for a class of processes that have been hypothesised to be relatively common at CFc (e.g., Troise et al. 2019).

3 Dynamics of Magma Reservoir Processes at the Campi Flegrei Caldera

The simulations reveal detailed patterns of magma convection and mixing taking place in the upper chamber + feeding dyke domains. In order to illustrate those patterns, a reference simulation (corresponding to simulation no. 4 in Table 1) is chosen, and the other cases are described through comparison with the reference one.

The reference simulation takes into account a situation where the shallow chamber had previously degassed efficiently. Consequently, the density contrast between the two magma types at the initial magma interface is relatively large, corresponding to about 120 kg/m^3. The shallow reservoir is horizontally elongated.

Figure 3 summarises the results in terms of the relevant quantities—composition, density and gas volume fraction—that are shown at four different times up to 5 h of real time. The dynamics is triggered by a Rayleigh–Taylor instability, as the resident phonolitic magma has a higher density that the shoshonite. The development of the instability and its time scales are independent of the volumes of intruding magmas

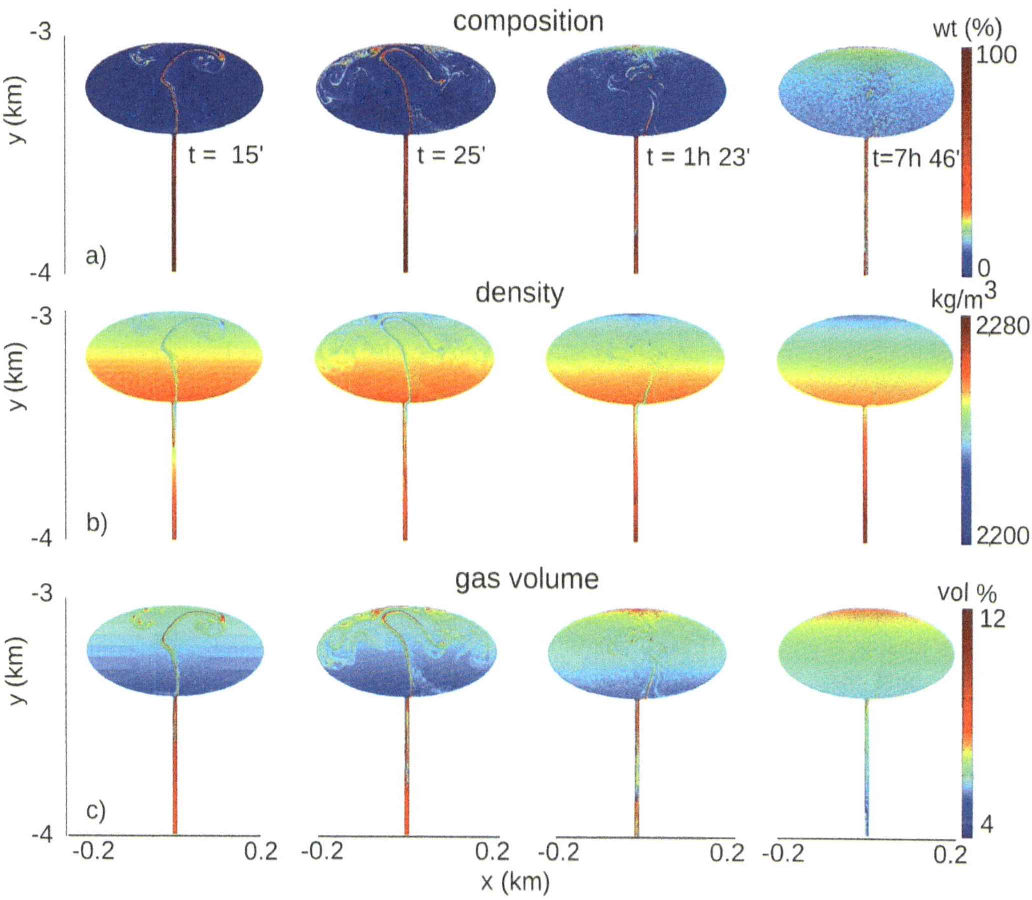

Fig. 3 Snapshots of composition, density and gas volume fractions at different simulated times in the shallower part of the system. Figure modified after Papale et al. (2017)

(e.g., Bellman and Pennington 1954; Seropian et al. 2018).

Soon after the start of the simulation, disruption of the initial interface by buoyancy forces produces a series of discrete plumes of light magma rising through the shallow chamber and reaching its top after having developed complex velocity fields. At the same time, part of the dense magma originally hosted in the shallow chamber sinks into the feeding dyke. Intense mixing originates at the dyke level, so that no pure shoshonitic end-member can be found in the shallow chamber; the rising plumes are rather made of a mixture with 30 to 50 wt% deep component. In this framework, only mechanical mixing is considered, as chemical reactions apart from volatile exsolution/dissolution are not implemented in the numerical scheme. Therefore, mixing at the scale of the simulation, with maximum resolution in the order of 1 m, refers to mechanical mingling rather than chemical mixing. The simulated dynamics suggests a complex pattern of convection and mixing (or mingling) whereby the original dense magma mixes with the volatile-rich, lighter magma coming from depth at chamber bottom or dyke level. The original phonolite carried down into the dyke mixes with the shoshonite; the mixed magma is overall lighter than the magma above it or in its immediate surroundings, therefore part of the initially sunk magma is carried up again into the shallow chamber, while other portions continue to sink down into the dyke. This complex process originates a compositional, density, and gas volume stratification inside the chamber, with maximum gas volumes at chamber top of about 6 vol%. In the meantime, the original density contrast at chamber bottom is disrupted and a smoother and smoother density profile is created, so that the convective process tends to slow down in time. Figure 4 shows the time distribution of the efficiency of convection η_C for all performed simulations, defined as the variation in time of phonolite mass in the shallow chamber, normalised to the initial amount so to allow direct comparison of the different cases:

$$\eta\eta_C = \frac{|m_R(t) - m_R(0)|}{m_R(0)} \quad (1)$$

In the above, m_R(t) is the mass of the initially resident phonolitic magma as a function of time.

A number of relevant aspects of the simulated dynamics must be highlighted. First of all, in all cases the total mass in the shallow chamber decreases with time as a consequence of average density decrease, due to the partial substitution of the initial dense phonolite with the lighter gas-rich shoshonite. In the most efficient case, corresponding to the reference case described above, the total mass in the shallow chamber has decreased by less than 2 wt% after about 5 h of convection and mixing. Second, the efficiency of convection increases substantially with the extent of degassing of the magma initially residing in the chamber (cases 1, 2 and 3 correspond to lower degassing extent, see Table 1; Fig. 4). Third, convection and mixing tend to slow down with time, as revealed by the progressive decrease of the slopes in Fig. 4. More than 80% of the dense, degassed magma initially residing in the chamber remains at chamber level after waning of the convective process (as in case 1, where an asymptotic trend clearly emerges). Thus, the complex patterns of convection and mixing described above are such that most of the degassed magma is not carried down to larger depths when a new stable profile has been created. While shallow chamber degassing creates unstable conditions by increasing the density of the magma at shallow levels, the ingression of small amounts of volatile-rich magma rapidly re-establishes a new "equilibrium". The new overall stability is a dynamic condition: even at the

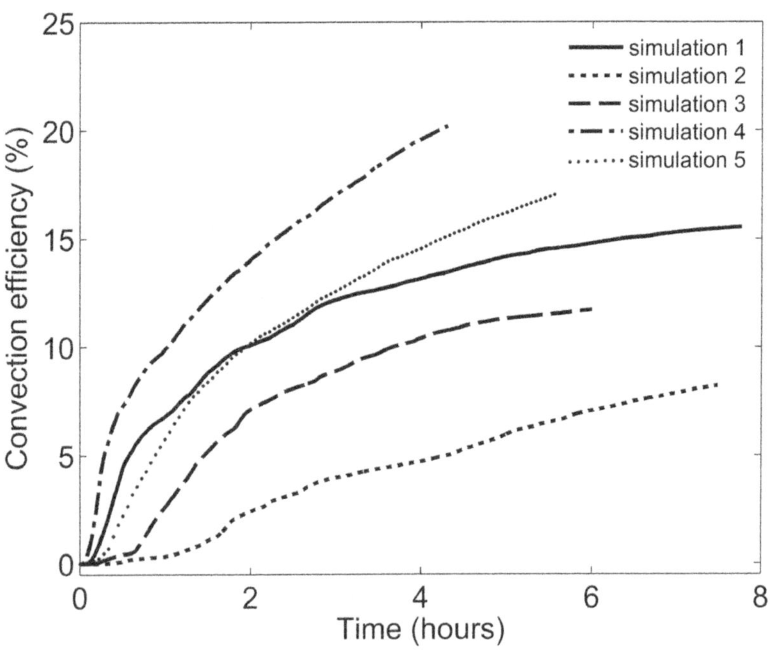

Fig. 4 Convection efficiency in the shallow reservoir for the different simulations as function of time. The simulation number is the same as in Table 1. Figure modified after Montagna et al. (2015)

longest simulated times, small batches of mixed magma still enter the chamber while other batches sink into the dyke, overall keeping a more or less constant balance of total mass and of single magmatic components in the chamber. The overall dynamics is not expected to drastically change after some hours, as in the simulated conditions the gravitational instability that triggers magma interaction is obliterated by the dynamics itself, as the density profile becomes more and more stable. On geological time scales, this time span seems irrelevant; nevertheless, petrological analyses on erupted products at CFc retrieve similar time scales for efficient mingling to take place (Perugini et al. 2010, 2015a, b; Chap. Rheological Properties of the Magmas Feeding the Campi Flegrei Caldera (Italy) and Their Influence on Mixing Processes). The results shown here provide a means for interpreting those data in relation to dynamical processes at depth.

Finally, the trends in Fig. 4 show that the efficiency of convection depends on the shallow chamber geometry in such a way that it decreases when its horizontal dimension also decreases. This is expected, as the efficiency of natural convection depends on the system dimension perpendicular to the driving force represented by gravity. That is also visible from the colour plots in Fig. 5, which show a progressive deeper blue (that is, lower proportion of deep magmatic component) when moving from oblate to circular to prolate chamber geometry. At the same time, however, the more prolate geometries result in more pronounced accumulation of volatile-rich deep component at the chamber top, thus more pronounced chamber stratification.

It is also noteworthy that the plumes of lighter magma tend to rise through the chamber following significantly different patterns. Rising plumes tend to follow the chamber walls for much longer distances in prolate geometries,

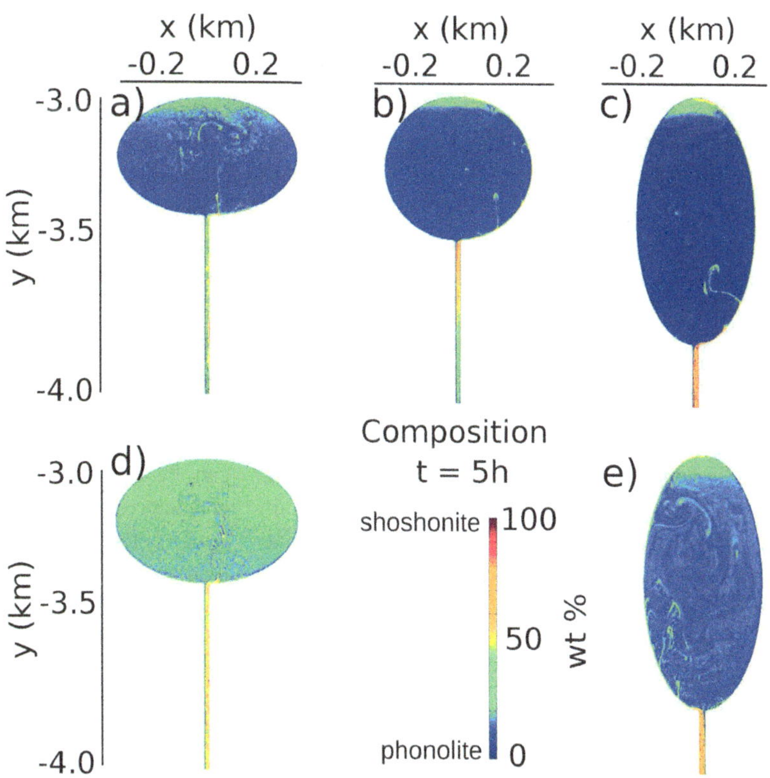

Fig. 5 Composition in the upper chamber after 5 h of simulated time. From left to right, top to bottom, order of simulations as in Table 1: **a** simulation 1, low density contrast, horizontal chamber; **b** simulation 2, low density contrast, circular chamber; **c** simulation 3, low density contrast, vertical chamber; **d** simulation 4, reference simulation, high density contrast, horizontal chamber; **e** simulation 5, high density contrast, vertical chamber. Figure modified after Montagna et al. (2015, 2017)

generating roughly one single convective cell in the chamber. This optimizes convection, thus partly balancing the effects of decreased horizontal chamber size diminishing the extent to which the convection efficiency is reduced when moving from oblate to prolate geometries.

4 Multi-disciplinary Interpretation of Modelling Results

This study focuses on the convection and mixing dynamics occurring in a shallow chamber when it is intersected by a dyke carrying more primitive, volatile-rich (especially CO_2) magma from a deeper reservoir. The system conditions refer to a general situation appropriate for the volcanism of the CFc, dominated by the emplacement of relatively small magma chambers at shallow levels, repeatedly recharged by deeper magma while undergoing shallow system degassing (Aiuppa et al. 2013; Di Renzo et al. 2016; Chaps. Volcanic and Deformation History of the Campi Flegrei Volcanic Field, Italy; Seismic and Gravity Structure of the Campi Flegrei Caldera, Italy; An Evolutionary Model for the Magmatic System of the Campi Flegrei Volcanic Field (Italy) Constrained by Petrochemical Data; Origin and Differentiation History of the Magmatic System Feeding the Campi Flegrei Volcanic Field (Italy) Constrained by Radiogenic and Stable Isotope Data). The volcanic products reveal magma mixing mostly from melt-crystal (Roach 2005) and isotopic (Arienzo et al. 2009, and references therein; Chap. Origin and Differentiation History of the Magmatic System Feeding the Campi Flegrei Volcanic Field (Italy) Constrained by Radiogenic and Stable Isotope Data) disequilibrium, as well as from melt inclusions within crystals (Arienzo et al. 2010). While mixing between different magmas is a widespread occurrence at the CFc, magmatic end-member components are usually not identified at the macroscopic scale in the volcanic products, especially for the deeper, less chemically evolved components the presence of which emerges only from detailed chemical and petrographic analyses (Astbury et al. 2018). On the other hand, time scales of magma mixing can be quite short, of the order of hours to days, as it is revealed by both time-scales of melt-crystal disequilibrium (Roach 2005) and by the application of a diffusive fractionation model of trace elements to a number of relevant eruptions in the CFc history (De Campos et al. 2008; Perugini et al. 2010, 2015a, b; Chap. Rheological Properties of the Magmas Feeding the Campi Flegrei Caldera (Italy) and Their Influence on Mixing Processes).

Although a full comparison is limited by the resolution of the computational grid, which is on the order of 1 m, the present numerical simulations provide a fully consistent picture: in all simulated cases, the deep, less evolved, volatile-rich end-member magma involved in the mixing process is rapidly lost as an individual component, while magma mixing results in a stratified but substantially laterally homogeneous (at the resolved spatial scale) composition on a short time-scale of a few to 10 h. In addition, the simulations suggest a relevant role of magma mixing at the feeding dyke level, and highlight complex patterns whereby the partially degassed magma originally hosted at shallow levels can sink down into the dyke, mix with the rising volatile-rich magma, then re-enter into the chamber as a mixture component masking the end-member characteristics of the deep rejuvenating magma. The evolution of pressure within the system is highly heterogeneous in space and time, as detailed in Papale et al. (2017). Calculated pressure variations are always smaller than 1 MPa, therefore large variations of the system volume, especially in the dyke, are not expected as a consequence of the exchange dynamics described above. Nevertheless, detectable ground deformation and seismicity can be generated by the convective dynamics, characterised by specific frequency patterns that could potentially be used to infer magma movement at depth from the monitoring records (Longo et al. 2012b; Bagagli et al. 2017; Chaps The Permanent Monitoring System of the Campi Flegrei Caldera, Italy; Source Modelling from Ground Deformation and Gravity Changes at the Campi Flegrei Caldera, Italy).

The numerical simulations shown here (see also Montagna et al. 2015, 2017; Papale et al.

2017) are limited by the huge computational time they require, therefore, only some hours of real time have been explored till now. A conceptual extrapolation to much longer times is allowed by the fact that in some of the simulated cases, and specifically in case 1 corresponding to the longest simulated time (about 8 h), a condition is reached where the initial gravitational instability is completely destroyed and new dynamically stable conditions are established. Thus, the simulation describes an entire mixing cycle in which the initial density contrast is totally removed, and a new dynamic equilibrium is reached. All other simulations show the same tendency, as it is suggested by the progressive flattening of the convection efficiency curves in Fig. 4, although over the simulated times a stable condition is in most cases approached but not reached. The trigger for the whole process is provided by the fact that the shallow magma is initially partially degassed, therefore its density is, at same pressure conditions, larger than the density of the magma rising from depth. This is the origin of the gravitational instability in the initial conditions.

Magmatic bodies emplaced at shallow level are known to degas efficiently, as a consequence of relatively large gas volumes at low pressure conditions and brittle rock mechanics allowing the generation of cracks providing pathways for gas escape towards the surface. In the specific case of the CFc, an abundant magmatic gas component is recognised in the fumaroles of La Solfatara crater (Caliro et al. 2007; Chiodini et al. 2015; Tamburello et al. 2019; Chaps. The Permanent Monitoring System of the Campi Flegrei Caldera, Italy; The Hydrothermal System of the Campi Flegrei Caldera, Italy), that globally discharge about 1,500 tons/day of CO_2 in the atmosphere (Caliro et al. 2007). The final condition in all simulations is the establishment of a gas-rich cap at the top of the shallow chamber (Figs. 3 and 5). It is expected that over a longer time this would feed a new efficient degassing phase, as gas-rich magma continues to slowly accumulate at the top of the system to then seep through the chamber roof and propagate through the porous host rock, over much longer time scales than those simulated here (Jasim et al. 2015). Large oscillations in the CO_2 content of the fumarolic gases at La Solfatara are well known (Chiodini et al. 1996, 2015), and more recently, similar oscillations have been proposed to occur in their magmatic component (Moretti et al. 2013). These oscillations follow the major seismic and deformation crisis at CFc in 1982–1984, and they are possibly related to the emplacement of a small magmatic body at 3–4 km depth as revealed by recent seismic attenuation tomography (De Siena et al. 2010; Akande et al. 2019; Chap. Seismic and Gravity Structure of the Campi Flegrei Caldera, Italy), although CO_2 flux can be independent of magmatic flux as it exsolves at large depths (Mangiacapra et al. 2008).

An overall picture emerges: after the emplacement of a small magma body at shallow depth, open system degassing feeding the fumarolic gases (and possibly, partial crystallisation of magma) increases magma density and triggers the arrival of new gas-rich, light magma batches from a deep reservoir, likely along the same pathways that brought the magma to a shallow level. This in turn gives rise to magma convection and mixing, that re-establishes dynamically stable conditions over a short time-scale of hours to a few days (increasing with the extent of shallow system degassing). New gas-rich, CO_2-rich magma is accumulated at the top of the shallow chamber, and over a longer time scale of months to years, open system degassing (and crystallisation) increases again the density in the upper system, triggering a new episode of magma injection and repeating the process. Such a sequence can be repeated several times, each time rejuvenating the shallow magmatic system through the injection of small batches of volatile-rich magma from larger depths, and providing new gas that will slowly escape from the system and feed the fumarolic system.

5 Conclusive Remarks

The present numerical simulations are consistent with a number of observations of the CFc volcanism (and to an extent, of several other volcanic products worldwide), although by no means they are able to describe in detail all of the processes within a deep, complex magmatic system. Such observations include (i) the chemical oscillations observed in the fumarolic gases, that appear to punctuate long-term degassing patterns (Chiodini et al. 2015; Chap. The Hydrothermal System of the Campi Flegrei Caldera, Italy); (ii) the chemical zoning in crystals, that marks oscillations in the P-T-X conditions at crystallisation levels (Astbury et al. 2018; Chaps. An Evolutionary Model for the Magmatic System of the Campi Flegrei Volcanic Field (Italy) Constrained by Petrochemical Data; Origin and Differentiation History of the Magmatic System Feeding the Campi Flegrei Volcanic Field (Italy) Constrained by Radiogenic and Stable Isotope Data); (iii) the presence of melt inclusions in crystals that are in chemical disequilibrium with their host (Tonarini et al. 2009; Chaps. An Evolutionary Model for the Magmatic System of the Campi Flegrei Volcanic Field (Italy) Constrained by Petrochemical Data; Origin and Differentiation History of the Magmatic System Feeding the Campi Flegrei Volcanic Field (Italy) Constrained by Radiogenic and Stable Isotope Data); (iv) the dispersion of volatile contents in melt inclusions that hardly mark simple closed or open system degassing patterns, rather they suggest a widespread occurrence of convection and mixing (a well-studied example is reported for Kīlauea in Barsanti et al. 2009; the melt inclusion data for the Agnano-Monte Spina eruption at the CFc by Arienzo et al. 2010 show similar patterns); (v) the tendency towards rapid homogenisation of magmas upon mixing, preventing macroscopic identification of end-member components (Astbury et al. 2018; Forni et al. 2018; Chaps. An Evolutionary Model for the Magmatic System of the Campi Flegrei Volcanic Field (Italy) Constrained by Petrochemical Data; Origin and Differentiation History of the Magmatic System Feeding the Campi Flegrei Volcanic Field (Italy) Constrained by Radiogenic and Stable Isotope Data; Rheological Properties of the Magmas Feeding the Campi Flegrei Caldera (Italy) and Their Influence on Mixing Processes); (vi) the time-scales of magma mixing as revealed by separate approaches like melt-crystal disequilibrium and trace element diffusive fractionation modelling (Tonarini et al. 2009; Perugini et al. 2015a, b; Chaps. An Evolutionary Model for the Magmatic System of the Campi Flegrei Volcanic Field (Italy) Constrained by Petrochemical Data; Origin and Differentiation History of the Magmatic System Feeding the Campi Flegrei Volcanic Field (Italy) Constrained by Radiogenic and Stable Isotope Data; Rheological Properties of the Magmas Feeding the Campi Flegrei Caldera (Italy) and Their Influence on Mixing Processes).

References

Aiuppa A, Tamburello G, Di Napoli R, Cardellini C, Chiodini G, Giudice G, Grassa F, Pedone M (2013) First observations of the fumarolic gas output from a restless caldera: implications for the current period of unrest (2005–2013) at Campi Flegrei. Geochem Geophys Geosys 14:4153–4169

Akande WG, De Siena L, Gan Q (2019) Three-dimensional kernel-based coda attenuation imaging of caldera structures controlling the 1982–84 Campi Flegrei unrest. J Volcanol Geotherm Res 38:273–283

Andersen NL, Singer BS, Coble MA (2019) Repeated rhyolite eruption from heterogeneous hot zones embedded within a cool, shallow magma reservoir. J Geophys Res-Sol Ea 124:2582–2600

Arienzo I, Civetta L, Heumann A, Horner GW, Orsi G (2009) Isotopic evidence for open system processes within the Campanian Ignimbrite (Campi Flegrei, Italy) magma chamber. Bull Volcanol 71:285–300. https://doi.org/10.1007/s00445-008-0223-0

Arienzo I, Moretti R, Civetta L, Orsi G, Papale P (2010) The feeding system of Agnano Monte Spina eruption (Campi Flegrei, Italy): dragging the past into present activity and future scenarios. Chem Geol 270:135–147

Arienzo I, Heumann A, Horner GW, Civetta L, Orsi G (2011) Processes and timescales of magma evolution prior to the Campanian Ignimbrite eruption (Campi Flegrei, Italy). Earth Planet Sci Lett 306:217–228

Astbury RL, Petrelli M, Ubide T, Stock MJ, Arienzo I, D'Antonio M, Perugini D (2018) Tracking plumbing system dynamics at the Campi Flegrei caldera, Italy: high-resolution trace element mapping of the Astroni crystal cargo. Lithos 318–319:464–477

Bachmann O, Bergantz GW (2003) Rejuvenation of the Fish Canyon magma body: a window into the evolution of large-volume silicic magma systems. Geology 31:789–792

Bachmann O, Bergantz GW (2006) Gas percolation in upper-crustal silicic crystal mushes as a mechanism for upward heat advection and rejuvenation of near-solidus magma bodies. J Volcanol Geotherm Res 149:85–102

Bagagli M, Montagna CP, Papale P, Longo A (2017) Signature of magmatic processes in strainmeter records at Campi Flegrei (Italy). Geophys Res Lett 44:718–725

Barsanti M, Papale P, Barbato D, Moretti R, Boschi E, Hauri EH, Longo A (2009) Heterogeneous large total CO_2 abundance in the shallow magmatic system of Kilauea volcano, Hawaii. J Geophys Res 144:B12201

Bellman R, Pennington RH (1954) Effects of surface tension and viscosity on Taylor instability. Q Appl Math 12:151–162

Bergantz GW, Schleicher JM, Burgisser A (2015) Open-system dynamics and mixing in magma mushes. Nat Geosci 8:793–796

Burgisser A, Bergantz GW, Breidenthal RE (2005) Addressing complexity in laboratory experiments: the scaling of dilute multiphase flows in magmatic systems. J Volcanol Geotherm Res 141:245–265

Caliro S, Chiodini G, Moretti R, Avino R, Granieri D, Russo M, Fiebig J (2007) The origin of the fumaroles of La Solfatara (Campi Flegrei, South Italy). Geochim Cosmochim Acta 71:3040–3055

Carrara A, Burgisser A, Bergantz GW (2019) Lubrication effects on magmatic mush dynamics. J Volcanol Geotherm Res 380:19–30

Cashman KV, Sparks RSJ, Blundy JD (2017) Vertically extensive and unstable magmatic systems: a unified view of igneous processes. Science 355:eaag3055

Chalot F, Hughes TJR (1994) A consistent equilibrium chemistry algorithm for hypersonic flows. Comput Methods Appl Mech Eng 112:25–40

Chiodini G, Cioni R, Magro G, Marini L, Panichi C, Raco B, Russo M (1996) Gas and water geochemistry chemical and isotopic variations of Bocca Grande fumarole (Solfatara volcano, Phlegrean Fields). Acta Vulcanol 8:228–232

Chiodini G, Pappalardo L, Aiuppa A, Caliro S (2015) The geological CO_2 degassing history of a long-lived caldera. Geology 43:767–770

Chouet BA, Matoza RS (2013) A multi-decadal view of seismic methods for detecting precursors of magma movement and eruption. J Volcanol Geotherm Res 252:108–175

De Campos CP, Dingwell DB, Perugini D, Civetta L, Fehr TK (2008) Heterogeneities in magma chambers: Insights from the behavior of major and minor elements during mixing experiments with natural alkaline melts. Chem Geol 256:131–145

De Siena L, Del Pezzo E, Bianco F (2010) Seismic attenuation imaging of Campi Flegrei: evidence of gas reservoirs, hydrothermal basins, and feeding systems. J Geophys Res 115:1–18

de Vita S, Orsi G, Civetta L, Carandente A, D'Antonio M, Deino A, Di Cesare T, Di Vito MA, Fisher RV, Isaia R, Marotta E, Necco A, Ort M, Pappalardo L, Piochi M, Southon J (1999) The Agnano-Monte Spina eruption (4100 years BP) in the restless Campi Flegrei caldera (Italy). J Volcanol Geotherm Res 91:269–301

Degruyter W, Huber C (2014) A model for eruption frequency of upper crustal silicic magma chambers. Earth Planet Sci Lett 403:117–130

Di Renzo V, Arienzo I, Civetta L, D'Antonio M, Tonarini S, Di Vito MA, Orsi G (2011) The magmatic feeding system of the Campi Flegrei caldera: architecture and temporal evolution. Chem Geol 281:227–241

Di Renzo V, Wohletz K, Civetta L, Moretti R, Orsi G, Gasparini P (2016) The thermal regime of the Campi Flegrei magmatic system reconstructed through 3D numerical simulations. J Volcanol Geotherm Res 328:210–221

Di Vito M, Isaia R, Orsi G, Southon J, de Vita S, D'Antonio M, Pappalardo L, Piochi M (1999) Volcanism and deformation since 12,000 years at the Campi Flegrei caldera (Italy). J Volcanol Geotherm Res 91:221–246

Flinders J, Clemens JD (1996) Non-linear dynamics, chaos, complexity and enclaves in granitoid magmas. Earth Environ Sci Trans R Soc Edinb 87:217–223

Forni F, Petricca E, Bachmann O, Mollo S, De Astis G, Piochi M (2018) The role of magma mixing/mingling and cumulate melting in the Neapolitan Yellow Tuff caldera-forming eruption (Campi Flegrei, Southern Italy). Contrib Mineral Petrol 173:45

Fourmentraux C, Metrich N, Bertagnini A, Rosi M (2012) Crystal fractionation, magma step ascent, and syn-eruptive mingling: the Averno 2 eruption (Phlegraean Fields, Italy). Contrib Mineral Petrol 163:1121–1137

Garg D, Papale P, Colucci S, Longo A (2019) Long-lived compositional heterogeneities in magma chambers, and implications for volcanic hazard. Sci Rep 9:1–13

Giordano D, Russell JK, Dingwell DB (2008) Viscosity of magmatic liquids: a model. Earth Planet Sci Lett 271:123–134

Gualda GAR, Gravley DM, Deering CD, Ghiorso MS (2019) Magma extraction pressures and the architecture of volcanic plumbing systems. Earth Planet Sci Lett 522:118–124

Hauke G, Hughes TJR (1998) A comparative study of different sets of variables for solving compressible and incompressible flows. Comput Methods Appl Mech Eng 153:1–44

Ishii M, Zuber N (1979) Drag coefficient and relative velocity in bubbly, droplet or particulate flows. AiChE J 25:843–855

Jasim A, Whitaker FF, Rust AC (2015) Impact of channelized flow on temperature distribution and fluid flow in restless calderas: insight from Campi Flegrei caldera, Italy. J Volcanol Geotherm Res 303:157–174

Judenherc S, Zollo A (2004) The Bay of Naples (Southern Italy): constraints on the volcanic structures inferred from a dense seismic survey. J Geophys Res 109: B10312

Kaus BJP, Podladchikov YY (2001) Forward and reverse modeling of the three-dimensional viscous Rayleigh-Taylor instability. Geophys Res Lett 28:1095–1098

Kazahaya K, Shinohara H, Saito G (1994) Excessive degassing of Izu-Oshima volcano: magma convection in a conduit. Bull Volcanol 56:207–216

Lange RA (1994) The effect of H_2O, CO_2 and F on the density and viscosity of silicate melts. Rev Mineral 30:331–369

Longo A, Vassalli M, Papale P, Barsanti M (2006) Numerical simulation of convection and mixing in magma chamber replenished with CO_2-rich magma. Geophys Res Lett 33:L21305

Longo A, Barsanti M, Cassioli A, Papale P (2012a) A finite element Galerkin/least-squares method for computation of multicomponent compressible/ incompressible flows. Comput Fluids 67:57–71

Longo A, Papale P, Vassalli M, Saccorotti G, Montagna CP, Cassioli A, Giudice S, Boschi E (2012b) Magma convection and mixing dynamics as a source of ultra-long-period oscillations. Bull Volcanol 74:873–880

Mangiacapra A, Moretti R, Rutherford M, Civetta L, Orsi G, Papale P (2008) The deep magmatic system of the Campi Flegrei caldera (Italy). Geophys Res Lett 35:L21304

Marchetti E, Ichihara M, Ripepe M (2004) Propagation of acoustic waves in a viscoelastic two-phase system: influence of gas bubble concentration. J Volcano Geotherm Res 137:93–108

Masterlark T, Haney M, Dickinson H, Fournier T, Searcy C (2010) Rheologic and structural controls on the deformation of Okmok volcano, Alaska: FEMs, InSAR, and ambient noise tomography. J Geophys Res 115:B02409

Montagna CP, Papale P, Longo A (2015) Timescales of mingling in shallow magmatic reservoirs. Geol Soc London Spec Publ 422:131–140

Montagna CP, Papale P, Longo A, Bagagli M (2017) Magma chamber rejuvenation: insights from numerical models. In Gottsmann J, Neuberg J, Scheu B (eds) Volcanic unrest. Springer, pp 111–122

Moretti R, Arienzo I, Civetta L, Orsi G, Papale P (2013) Multiple magma degassing sources at an explosive volcano. Earth Planet Sci Lett 367:95–104

Morgavi D, Arienzo I, Montagna CP, Perugini D, Dingwell DB (2017) Magma mixing: history and dynamics of an eruption trigger. In Gottsmann J, Neuberg J, Scheu B (eds) Volcanic unrest. Springer, pp 123–137

Oldenburg CM, Spera FJ, Yuen DA, Sewell G (1989) Dynamic mixing in magma bodies: theory, simulations and implications. J Geophys Res 94:9215–9236

Orsi G, D'Antonio M, De Vita S, Gallo G (1992) The Neapolitan Yellow Tuff, a large-magnitude trachytic phreatoplinian eruption: eruptive dynamics, magma withdrawal and caldera collapse. J Volcanol Geotherm Res 53:275–287

Orsi G, de Vita S, di Vito M (1996) The restless, resurgent Campi Flegrei nested caldera (Italy): constraints on its evolution and configuration. J Volcanol Geotherm Res 74:179–214

Orsi G, Petrazzuoli SM, Wohletz K (1999) Mechanical and thermo-fluid behaviour during unrest at the Campi Flegrei caldera (Italy). J Volcanol Geotherm Res 91:453–470

Orsi G, Di Vito MA, Isaia R (2004) Volcanic hazard assessment at the restless Campi Flegrei caldera. Bull Volcanol 66:514–530

Papale P, Moretti R, Barbato D (2006) The compositional dependence of the saturation surface of H_2O+CO_2 fluids in silicate melts. Chem Geol 229:78–95

Papale P, Montagna CP, Longo A (2017) Pressure evolution in shallow magma chambers upon buoyancy-driven replenishment. Geochem Geophys Geosys 18:1214–1224

Pappalardo L, Mastrolorenzo G (2012) Rapid differentiation in a sill-like magma reservoir: a case study from the campi flegrei caldera. Sci Rep 2:712

Parmigiani A, Huber C, Bachmann O (2014) Mush microphysics and the reactivation of crystal-rich magma reservoirs. J Geophys Res-Sol Ea 119:6308–6322

Perugini D, Poli G, Petrelli M, Campos CP, Dingwell DB (2010) Time-scales of recent Phlegrean Fields eruptions inferred from the application of a diffusive fractionation model of trace elements. Bull Volcanol 72:431–447

Perugini D, Poli G (2012) The mixing of magmas in plutonic and volcanic environments: analogies and differences. Lithos 153:61–277

Perugini D, De Campos CP, Petrelli M, Dingwell DB (2015a) Concentration variance decay during magma mixing: a volcanic chronometer. Sci Rep 5:1–10

Perugini D, De Campos CP, Petrelli M, Morgavi D, Vetere FP, Dingwell DB (2015b) Quantifying magma mixing with the Shannon entropy: application to simulations and experiments. Lithos 236–237:299–310

Piochi M, Kilburn CRJ, Di Vito MA, Mormone A, Tramelli A, Troise C, De Natale G (2014) The volcanic and geothermally active Campi Flegrei caldera: an integrated multidisciplinary image of its buried structure. Int J Earth Sci 103:401–421

Pistone M, Blundy J, Brooker RA (2017) Water transfer during magma mixing events: insights into crystal mush rejuvenation and melt extraction processes. Am Mineral 102:766–776

Reid RC, Prausnitz J, Sherwood T (1977) The properties of gases and liquids. McGraw Hill, New York

Ribe NM (1998) Spouting and planform selection in the Rayleigh-Taylor instability of miscible viscous fluids. J Fluid Mech 377:27–45

Roach AC (2005) The evolution of silicic magmatism in the post-caldera volcanism of the Phlegrean Fields, Italy. PhD thesis, Brown University

Ruprecht P, Bergantz GW, Dufek J (2008) Modeling of gas-driven magmatic overturn: tracking of phenocryst dispersal and gathering during magma mixing. Geochem Geophys Geosys 9:Q07017

Schleicher JM, Bergantz GW (2017) The mechanics and temporal evolution of an open-system magmatic intrusion into a crystal-rich magma. J Petrol 58:1059–1072

Seropian G, Rust AC, Sparks RSJ (2018) The gravitational stability of lenses in magma mushes: confined Rayleigh-Taylor instabilities. J Geophys Res-Sol Ea 123:3593–3607

Shakib F, Hughes TJR, Johan Z (1991) A new finite element formulation for computational fluid dynamics: X. The compressible Euler and Navier-Stokes equations. Comput Methods Appl Mech Eng 89:141–219

Sigurdsson H, Houghton BF, McNutt SR, Rymer H, Stix J (2015) The encyclopedia of volcanoes. Elsevier, London, UK, p 1456

Smith VC, Isaia R, Pearce NJG (2011) Tephrostratigraphy and glass compositions of post-15 kyr Campi Flegrei eruptions: implications for eruption history and chronostratigraphic markers. Quat Sci Rev 30:3638–3660

Spera FJ, Yuen DA, Kirschvink SJ (1982) Thermal boundary layer convection in silicic magma chambers: effects of temperature-dependent rheology and implications for thermogravitational chemical fractionation. J Geophy Res 87(B10):8755. https://doi.org/10.1029/JB087iB10p08755

Tamburello G, Caliro S, Chiodini G, De Martino P, Avino R, Minopoli C, Carandente A, Rouwet D, Aiuppa A, Costa A, Bitetto M, Giudice G, Francoforte V, Ricci T, Sciarra A, Bagnato E, Capecchiacci F (2019) Escalating CO_2 degassing at the Pisciarelli fumarolic system, and implications for the ongoing Campi Flegrei unrest. J Volcanol Geotherm Res 384:151–157

Tomlinson EL, Arienzo I, Civetta L, Wulf S, Smith VC, Hardiman M, Lane CS, Carandente A, Orsi G, Rosi M, Müller W, Menzies MA (2012) Geochemistry of the Phlegraean Fields (Italy) proximal sources for major Mediterranean tephras: implications for the dispersal of Plinian and co-ignimbritic components of explosive eruptions. Geochim Cosmochim Acta 93:102–128

Tonarini S, D'Antonio M, Di Vito MA, Orsi G, Carandente A (2009) Geochemical and B-Sr-Nd isotopic evidence for mingling and mixing processes in the magmatic system that fed the Astroni volcano (4.1-3.8 ka) within the Campi Flegrei caldera (Southern Italy). Lithos 107:135–151

Troise C, De Natale G, Schiavone R, Somma R, Moretti R (2019) The Campi Flegrei caldera unrest: discriminating magma intrusions from hydrothermal effects and implications for possible evolution. Earth-Sci Rev 188:108–122

Yoshimura S, Nakamura M (2011) Carbon dioxide transport in crustal magmatic systems. Earth Planet Sci Lett 307:470–478

Zollo A, Maercklin N, Vassallo M, Dello Iacono D, Virieux J, Gasparini P (2008) Seismic reflections reveal a massive melt layer feeding Campi Flegrei caldera. Geophys Res Lett 35(12):L12306. https://doi.org/10.1029/2008GL034242

The Permanent Monitoring System of the Campi Flegrei Caldera, Italy

Francesca Bianco, Stefano Caliro, Prospero De Martino, Massimo Orazi, Ciro Ricco, Giuseppe Vilardo, Ida Aquino, Vincenzo Augusti, Rosario Avino, Emanuela Bagnato, Giuseppe Brandi, Antonio Caputo, Antonio Carandente, Giovanni Chiodini, Emilio Cuoco, Andrea D'Alessandro, Mario Dolce, Sergio Guardato, Carmine Minopoli, Fabio Sansivero, Alessandro Santi, Giovanni Scarpato, Anna Tramelli, and Mario Castellano

Abstract

We present the main features of the permanent monitoring system managed by the Istituto Nazionale di Geofisica e Vulcanologia-Osservatorio Vesuviano in the Campi Flegrei caldera. Eruptive history of this active volcano shows that the majority of the eruptive events has been characterised by high explosivity and was accompanied by pyroclastic density currents. Its last eruption occurred in AD 1538 and in the next centuries the Campi Flegrei caldera has experienced several episodes of bradyseism and also the progressive increasing of the urbanisation in the area (west of Naples). Monitoring the dynamics of a mainly explosive volcano completely embedded in a very populated area is a challenging task. In order to detect any variation in the physical and chemical parameters of the Campi Flegrei caldera, the Istituto Nazionale di Geofisica e Vulcanologia-Osservatorio Vesuviano manages a permanent multi-parametric monitoring system. All the recorded h24 continuous data are transmitted to the Monitoring Room of the Osservatorio Vesuviano in Naples, where they are acquired, processed and evaluated to define changes in the dynamical state of the volcano. The caldera, since the end of 2004, is experiencing a bradyseismic episode characterised by a low velocity rate uplift, low energy earthquakes and increasing in the magmatic components of fumarolic fluids. The monitoring and surveillance activity of the Campi Flegrei caldera plays a crucial role in the volcanic emergency plan that includes evacuation of approximately 500,000 people before the beginning of the eruption.

F. Bianco (✉) · S. Caliro · P. De Martino · M. Orazi · C. Ricco · G. Vilardo · I. Aquino · V. Augusti · R. Avino · E. Bagnato · G. Brandi · A. Caputo · A. Carandente · E. Cuoco · A. D'Alessandro · M. Dolce · S. Guardato · C. Minopoli · F. Sansivero · A. Santi · G. Scarpato · A. Tramelli · M. Castellano
Istituto Nazionale di Geofisica e Vulcanologia, Sezione Osservatorio Vesuviano, Napoli, Italy
e-mail: francesca.bianco@ingv.it

G. Chiodini
Istituto Nazionale di Geofisica e Vulcanologia, Sezione di Bologna, Bologna, Italy

© Springer-Verlag GmbH Germany, part of Springer Nature 2022
G. Orsi et al. (eds.), *Campi Flegrei*, Active Volcanoes of the World,
https://doi.org/10.1007/978-3-642-37060-1_8

1 Introduction

The Campi Flegrei caldera (CFc) is a complex and nested structure resulting mostly from two major collapses accompanying the catastrophic Campanian Ignimbrite ($\sim$40 ka) and the Neapolitan Yellow Tuff ($\sim$15 ka) eruptions (Orsi et al. 1996). Reviews of the volcanic and deformation history, and of the structure of the caldera are presented in Chaps. Volcanic and Deformation History of the Campi Flegrei Volcanic Field, Italy; Seismic and Gravity Structure of the Campi Flegrei Caldera, Italy and references therein, respectively. Also, the magmatic feeding system has a complex architecture with melts stagnating, differentiating, mingling and mixing at various depths within at least the shallowest $\sim$10 km of crust (see Chaps. An Evolutionary Model for the Magmatic System of the Campi Flegrei Volcanic Field (Italy) Constrained by Petrochemical Data; Origin and Differentiation History of the Magmatic System Feeding the Campi Flegrei Volcanic Field (Italy) Constrained by Radiogenic and Stable Isotope Data; Tephrochronology and Geochemistry of Tephra from the Campi Flegrei Volcanic Field, Italy; Rheological Properties of the Magmas Feeding the Campi Flegrei Caldera (Italy) and Their Influence on Mixing Processes; Magma Chamber Dynamics at the Campi Flegrei Caldera, Italy and references therein). Above the magmatic system is an active hydrothermal system that generates diffuse soil degassing, as well as fumarole and hot-water spring activity (see Chap. The Hydrothermal System of the Campi Flegrei Caldera, Italy and references therein). The caldera related to the Neapolitan Yellow Tuff eruption has been the site of intense volcanism and deformation and is the portion still active of the entire CFc (Orsi et al. 1996). The last eruption occurred in AD 1538 and formed the Monte Nuovo volcano (Guidoboni and Ciuccarelli 2011; Di Vito et al. 2016). The long-term deformation of the Neapolitan Yellow Tuff caldera has mostly consisted in the resurgence of its floor that has generated a maximum net uplift of $\sim$90 m of the La Starza marine terrace. After the AD 1538 Monte Nuovo eruption, several episodes of short-term deformation, known also as bradyseismic events and interpreted as transient short-term deformation events within the long-term deformation (Orsi et al. 1999) have been recorded (e.g., Del Gaudio et al. 2010; Chaps. The Hydrothermal System of the Campi Flegrei Caldera, Italy; Historic Unrest of the Campi Flegrei Caldera, Italy; Source Modelling from Ground Deformation and Gravity Changes at the Campi Flegrei Caldera, Italy). The last and ongoing bradyseismic event began in late 2004. Details on this activity may be found at the Istituto Nazionale di Geofisica e Vulcanologia-Osservatorio Vesuviano (INGV-OV) website: www.ov.ingv.it. As of this writing, its major characteristics are: the maximum uplift is of 72 cm; approximately 4,000 low magnitude (M_{max} = 3.3) earthquakes, located in the shallowest 3 km of the upper crust, have been recorded mainly in the Pozzuoli-Agnano area; magmatic components of fumarolic effluents have been consistently increasing.

History and present state of activity make the volcanic hazard of the CFc very high and its location in the densely populated Neapolitan area determines a potential volcanic risk among the highest in the world (see Chap. Volcanic Hazard Assessment at the Campi Flegrei Caldera, Italy and references therein). To mitigate the risk of such a potentially dangerous volcano, the Italian Civil Protection has produced an emergency plan that relies on high-quality data constantly available for the scientific community to monitor the state of the volcano. Due to the level of volcanic unrest as shown by the monitoring parameters, since December 2012 the CFc has been designated to be at the "Attention" alert level, the second level of a 4-tiered warning system (http://www.protezionecivile.gov.it).

INGV-OV is in charge of the surveillance of the three active Neapolitan volcanoes, namely the Campi Flegrei and Ischia resurgent calderas and the Somma-Vesuvio volcanic complex. As regards the CFc, it is committed to managing and improving the multi-parametric Monitoring Infrastructure with the aim of detecting with the highest possible instrumental sensitivity, any variation in the physical and chemical parameters of the volcano. This Infrastructure is composed of 6 different networks whose major characteristics will be described in this chapter.

Data acquired from all the networks, as well as those of gas and temperature monitoring are used for civil protection purposes and the results are included in the periodic surveillance reports shared with both authorities and the general public.

2 The Seismic Network

The CFc seismic network has been developed to record and store the seismic data of the volcanic structure that in its active portion hosts several cities, among which Pozzuoli and the western part of Naples, inhabited by several hundred thousand people. Such an intense urbanisation, the proximity of the sea and the presence of highways and railways determine a high level of cultural noise. Hence, installing and managing a seismic network in such an urbanised and, at the same time, risky and scientifically interesting area is an unavoidable scientific and technological challenge. The seismic data collected by this network are displayed in the INGV-OV Monitoring Room. The data and their subsequent analysis contribute to the compilation of the seismic catalogue and to the improvement of the knowledge on the state and evolution of the volcano.

For all these reasons, the Campi Flegrei seismic network plays a key role for both scientific and social purposes and is constantly updated and improved, thus its present configuration can be thought of as a snapshot image of an evolving system. The recorded seismicity consists mainly of Volcano-Tectonic (VT) earthquakes, with some anthropogenic signals (e.g., illegal fishing explosions, subway activity and cheering from the football stadium) (D'Auria et al. 2011). Long-Period (LP) earthquakes and Tremor are very rare and generally related to the dynamics of the hydrothermal system (Saccorotti et al. 2007). Currently the volcanic seismicity is mainly located in the central part of the active portion of the CFc, in an area of 3×2 km^2 that includes the Solfatara and Pisciarelli fumarolic fields. However, the historical seismic catalogue highlights that during the 1982–1984 bradyseismic crisis the seismicity was scattered within a larger area (see Chap. Historic Unrest of the Campi Flegrei Caldera, Italy and references therein).

The current CFc seismic network consists of 21 inland station sites that span, with variable density, the whole caldera area. The highest density of stations occurs around Solfatara and Pisciarelli, sources of most of the historically recorded seismicity (Fig. 1). The southern portion of the caldera is submerged and has represented a break in the network topology until the deployment of 4 marine underwater multiparametric stations, which are part of the MEDUSA marine monitoring system (Iannaccone et al. 2009; see http://portale.ov.ingv.it/medusa). For redundancy purposes or multiparametric applications, several network sites host more than one kind of seismic station. This network that includes a small number of analog stations installed in the 1970s and 1980s (Castellano et al. 2002), is mainly composed of modern digital stations. All these stations transmit the seismic data continuously in real-time by means of several transmission systems (e.g., UHF, Hiperlan, UMTS) to a remote acquisition centre and then to the INGV-OV data acquisition centre. "Analog seismic stations" consist of analog sensors where seismic electric signals are FM modulated and transmitted via analog FM radio to a remote centre. Data are received, demodulated and digitised by a multichannel digitiser PC controlled board. The digitised data are finally included in the whole digital data flow of the various seismic networks managed by INGV-OV. Because of their age and technical limitations, these stations are being progressively updated with more modern instruments. In order to have a controlled transition from one instrument to the other, both analog and digital stations are operated simultaneously for a limited period of time.

The main network improvement started in 2007 when the INGV-OV developed a high-resolution, low-power digitiser named GILDA (Orazi et al. 2006). This low-cost digitiser resulted in high resolution and high-quality data.

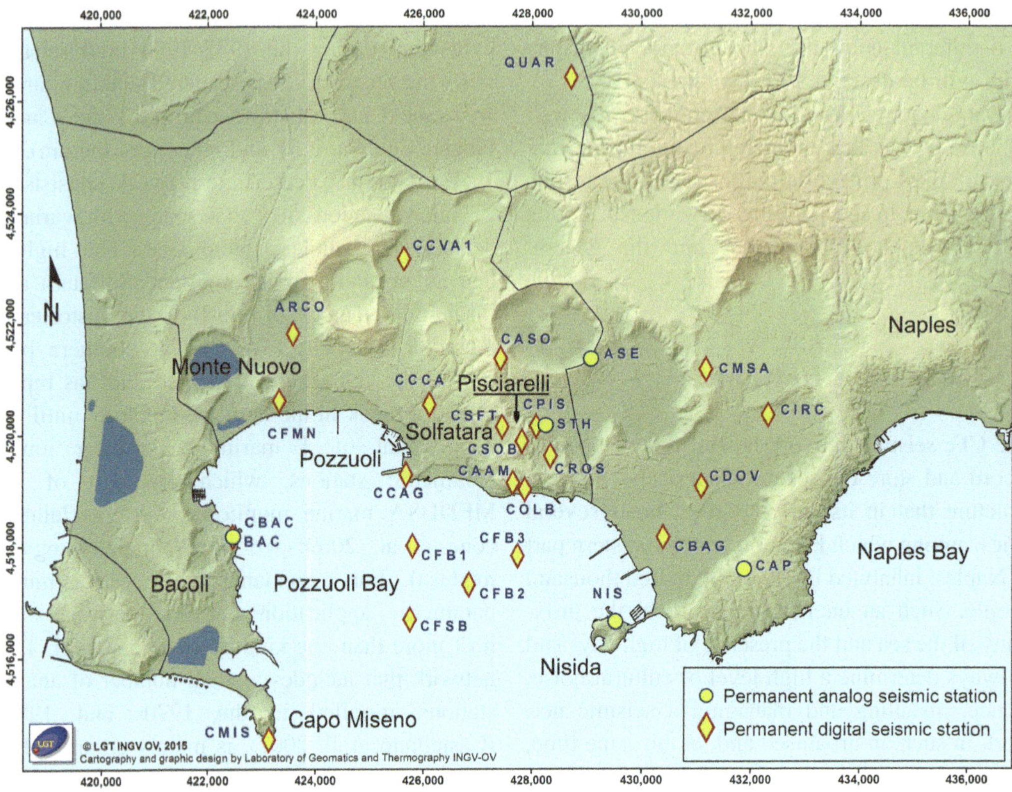

Fig. 1 Map of the Campi Flegrei seismic network

As reported in Table 1, five analog short-period inland seismic stations are currently operating within the Campi Flegrei seismic network (Fig. 1). Three are equipped with 1 Hz single-component velocimetric sensors (ASE, CAP and BAC) and two with triaxial 1 Hz sensors (STH and NIS). Two additional triaxial 1 Hz sensors are integrated with GILDA digitisers in two digital station sites (CDOV and CIRC). Twelve stations are equipped with broadband sensors (60s–50 Hz, 30s–50 Hz, 20s–40 Hz) and GILDA digitisers; two of these are also equipped with accelerometers. The inland seismic stations are topologically integrated across the Pozzuoli Bay by four MEDUSA sea bottom seismic stations equipped with very broadband velocimeters (120s–108 Hz) and MEMS accelerometers. Moreover, to improve seismic data quality in presence of the high levels of cultural noise, INGV-OV installed two borehole sensors equipped with very broadband velocimeters (120s–50 Hz) and accelerometers within the caldera (CBAG and CCCA). In addition, a borehole vertical array was installed close to the centre of the caldera (CCVA) until it was uninstalled in 2018 for instrument factory update. The seismic network is complemented by 8 infrasound sensors (CAAM, COLB, CPIS, CFSB and a four-sensors array at CDOV; Buonocunto et al. 2011). All these stations, including marine ones, transmit their data to the data acquisition centre through an INGV-OV proprietary data communication network mainly based on Wi-Fi and Hiperlan technology or, direct radio links, by using UHF analog or digital radio.

We estimated the theoretical seismic network sensitivity in order to illustrate the quality of the network. In Fig. 2 we plot the minimum magnitude of an earthquake located using the CFc network. Earthquake location is computed if at least three stations record the event with a Signal to Noise Ratio (SNR) greater than 2. The

Table 1 The Campi Flegrei seismic stations list

Station name	Coordinates and elevation above or below sea level	Installation date	Sensor	Digitiser	Sampling rate (sps)	Transmission method
ARCO	40.8438N 14.0933E 22 m	Feb 2006	Guralp 40T 60s–50Hz	GILDA 24 bit	100	ADSL
ASE	40.8402N 14.1587E 107 m	Jul 2000	Mark L4-C 1Hz	Analog - > 16 bit	100	Analogic FM radio
CAAM	40.8200N 14.1420E 100 m	Dec 2011	Guralp 40T 60s–50Hz Infracyrus infrasound	GILDA 24 bit	100	ADSL
CAP	40.8064N 14.1923E 166 m	Apr 2005	Mark L4-C 1Hz	Analog - > 16 bit	100	Direct
CASO	40.8401N 14.1390E 222 m	Jul 2015	Guralp 3TB 120s–50Hz vel Guralp 5TB accel	GILDA 24 bit	100 200	Radio Hiperlan
CBAC BAC	40.8110N 14.0807E 33 m	Sept 2009 May 1970	Lennartz 3D-Lite 1Hz Geotech S13	GILDA 24 bit Analog - > 16 bit	100 100	Radio Hiperlan Analogic UHF Radio
CBAG	40.8115N 14.1747E −169 m (borehole) 5 m (surface)	May 2004	Guralp 3TB 120s–50Hz vel Guralp 5TB accel Guralp 40T 30s–50Hz	GILDA 24 bit GILDA 24 bit GILDA 24 bit	100 200 100	Radio Wi-Fi
CCCA	40.8326N 14.1236E −26 m	May 2016	Guralp 3TB 120s–50Hz vel Guralp 5TB accel	GILDA 24 bit GILDA 24 bit	100 200	Radiomodem UMTS/LTE
CCVA	40.8561N 14.1177E −157 m	Mar 2015	6 × Guralp VSP Flute (60s–50Hz)	5 × GILDA 24 bit	100	*Temporary out of service*
CDOV CAID	40.8197N 14.1828E 18 m	May 2011	Lennartz 3D-Lite 1Hz 4 × Infracyrus infrasound	GILDA 24 bit GILDA 24 bit	100 100	Direct
CFB1	40.8103N 14.1212E −40 m	Jul 2016	Trillium Compact 120s Colibrys MEMS accelerometer	Quanterra Q330 Guralp DM24	100 200	Radio Wi Fi

(continued)

Table 1 (continued)

Station name	Coordinates and elevation above or below sea level	Installation date	Sensor	Digitiser	Sampling rate (sps)	Transmission method
CFB2	40.8042N 14.1338E −74 m	Jul 2016	Trillium Compact 120s Colibrys MEMS accelerometer	Quanterra Q330 Guralp DM24	100 200	Radio Wi Fi
CFB3	40.8090N 14.1442E −38 m	Jul 2016	Trillium Compact 120s Colibrys MEMS accelerometer	Quanterra Q330 Guralp DM24	100 200	Radio Wi Fi
CSFB	40.7978N 14.1198E −96 m	2008	Trillium Compact 120s Colibrys MEMS accelerometer	Quanterra Q330 Guralp DM24	100 200	Radio Wi Fi
CFMN	40.8329N 14.0904E 50 m	Sept 2014	Guralp 40T 60s–50Hz	GILDA 24 bit	100	Radio Hiperlan
CIRC	40.8314N 14.1972E 40 m	Feb 2016	Lennartz 3D-Lite 20s–40Hz	GILDA 24 bit	100	Radiomodem UMTS/LTE
CMIS	40.7782N 14.0890E 82 m	Oct 2015	Guralp 40T 60s–50Hz	GILDA 24 bit	100	Radio Hiperlan
CMSA	40.8386N 14.1836E 95 m	Nov 2008	Guralp 40T 60s–50Hz	GILDA 24 bit	100	ADSL
COLB	40.8190N 14.1446E 8 m	Jul 2011	Guralp 40T 60s–50Hz Infracyrus infrasound	GILDA 24 bit	100	Radio Hiperlan
CPIS	40.8292N 14.1470E 69 m	Jan 2010	Guralp 40 T 60s–50Hz Infracyrus infrasound	GILDA 24 bit	100	Radiomodem UHF
CPOZ	40.8212N 14.1186E 3 m	Jul 2011	Guralp 40T 60s–50Hz Episensor ES-T	GILDA 24 bit GILDA 24 bit	100 200	Radio Hiperlan
CSOB	40.8267N 14.1439E 177 m	2007	Guralp 40T 60s–50Hz Episensor ES-T	GILDA 24 bit GILDA 24 bit	100 200	Radio Hiperlan

(continued)

Table 1 (continued)

Station name	Coordinates and elevation above or below sea level	Installation date	Sensor	Digitiser	Sampling rate (sps)	Transmission method
CSFT	40.8290N 14.1395E 108 m	Jan 2013	Guralp 40T 60s–50Hz Infracyrus infrasound			Radio Wi Fi-Hiperlan
NIS	40.7978N 14.1646E 1 m	1983	Mark L4-3C	Analog - > 16 bit	100	CDA
QUAR	40.8856N 14.1538E 65 m	2006	Guralp 40T 60s–50Hz	GILDA 24 bit	100	*Temporary Out of service*
STH	40.8294N 14.1491E 65 m	1983	Mark L4-3C	Analog - > 16 bit	100	CDA

Stations location is shown in Fig. 1

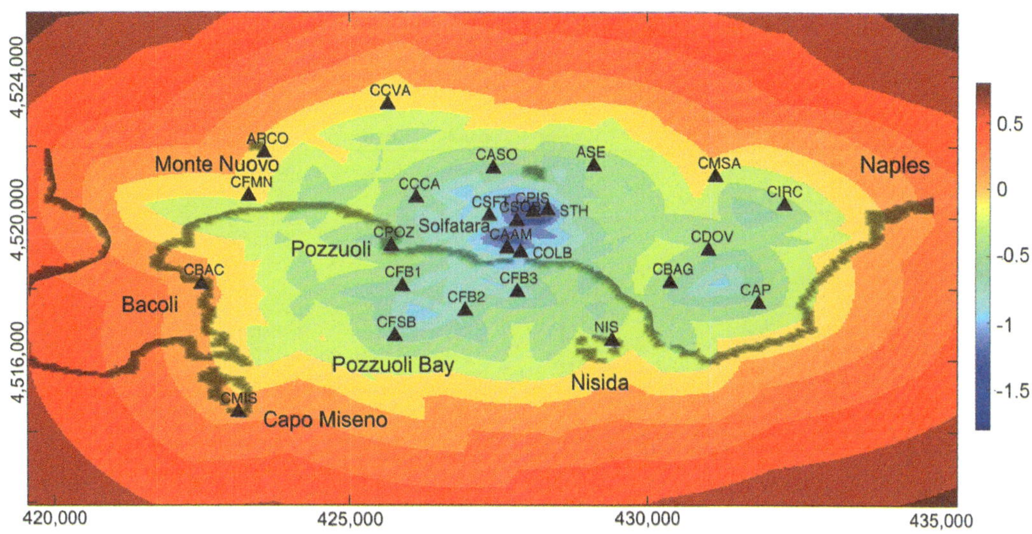

Fig. 2 Theoretical sensitivity of the Campi Flegrei seismic network. The picture shows the minimum magnitude of an earthquake able to be located with this network. It has been assumed for this simulation that located earthquakes occurred at 2 km b.s.l

earthquakes used for the simulation are located at depth of 2 km b.s.l., approximately equal to the mean depth of the earthquakes recorded in the last ten years in the CFc. The stress drop, S-wave velocity and attenuation coefficient used for the simulated earthquakes are the mean values found in the literature regarding the CFc (Tramelli et al. 2013). Figure 2 shows that in the central part of the caldera, where most of the current seismicity occurs, the network can also locate earthquakes with a magnitude close to -1. The current seismic network is able to locate earthquakes with magnitude greater than 0.5 and not deeper than 2 km beneath the caldera floor.

3 The Continuous Global Navigation Satellite System Network

The continuous Global Navigation Satellite System (cGNSS) network for monitoring both inland and seafloor ground deformation, currently consists of 25 stations (Fig. 3).

Some inland cGNSS stations have been in operation since 2000 (Table 2). In 2016, four stations were installed on the instrumented buoys (MEDUSA marine infrastructure) co-located with seismic stations in the Pozzuoli Bay (De Martino et al. 2014a; Iannaccone et al. 2018). All stations have been equipped with multi-frequencies GNSS receivers and antennas (Choke Ring type) to reduce multipath interference effects (Table 2). Due to the dense urbanisation of the CFc, inland stations are installed on the roof of stable buildings and the GNSS antennas are set up with three-dimensional forced centring devices. The stations are powered by commercial electricity with battery back-up or solar panels.

Every day, 24 h of observed multi-frequencies and multi-constellation GNSS data (30 s sampling rate and 10° cut-off angle) are downloaded from each station to the INGV-OV Monitoring Room by Hiperlan, UMTS/4G or HDSL transmission technologies. Data streaming at 1 Hz sampling rate is also available for all GNSS stations to real-time monitoring applications. The GNSS data quality has to be quickly and accurately tested before their processing and storage in a database. We use an in-house tool (SETA) for an automatic quality check. It uses the TEQC software (Estey and Meertens 1999) to translate from the binary receiver format to the standard RINEX format, perform a quality check of the data and transfer the data files to an

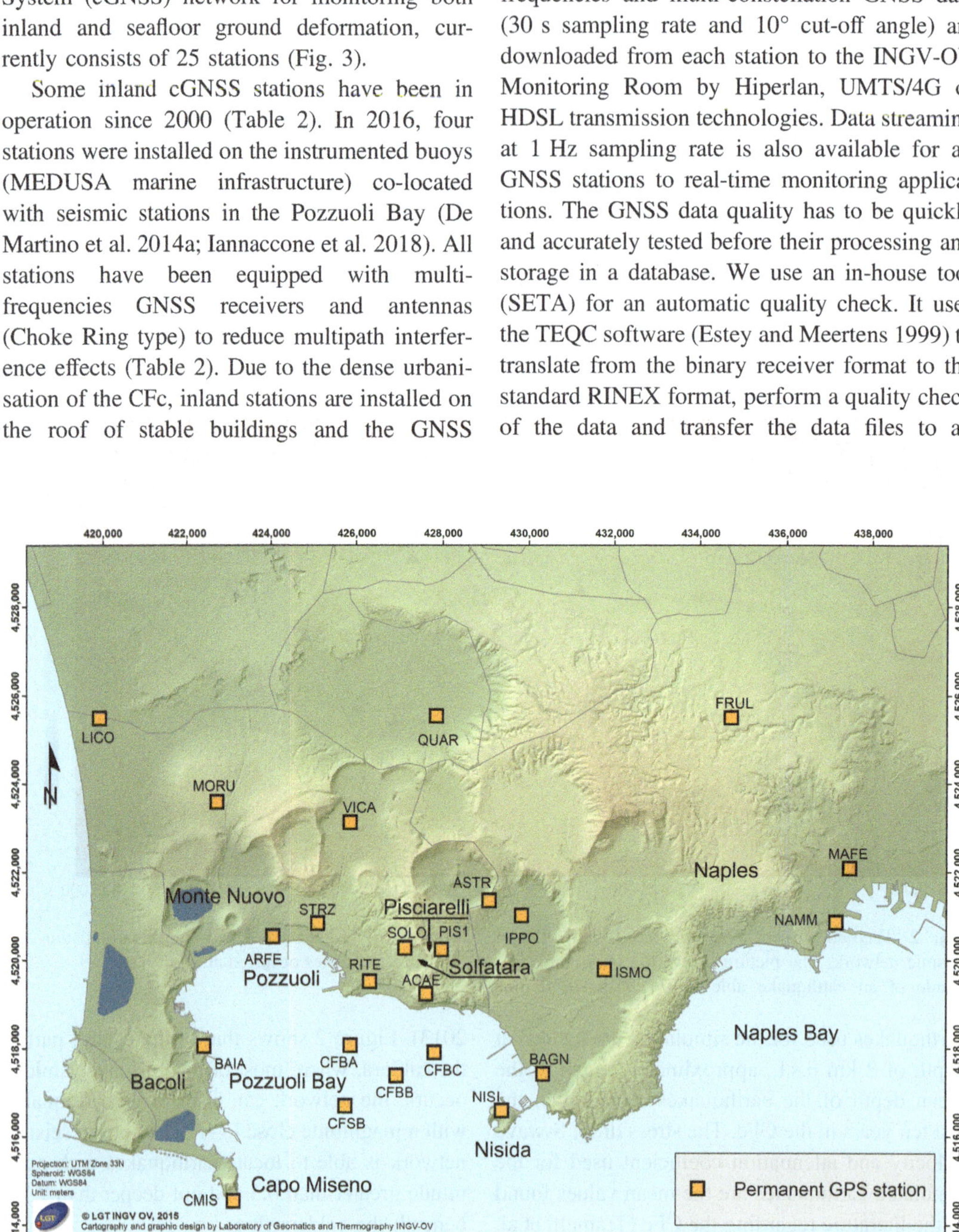

Fig. 3 Map of the Campi Flegrei cGNSS Network

Table 2 The Campi Flegrei cGNNS Network

Station name	Coordinates and elevation above sea level	First observation	Receiver antenna
ACAE	40.8209N 14.1417E 222 m	2000	Leica GR10 LEIAR25 LEIT
ARFE	40.8323N 14.0993E 69 m	2000	Leica GRX1200 LEIAT504 NONE
ASTR	40.8401N 14.1590E 159 m	2016	Leica GR10 LEIAR20 LEIM
BAGN	40.8048N 14.1741E 55 m	2012	Leica GR10 LEIAR25 LEIT
BAIA	40.8097N 14.0802E 108 m	1999	Leica GR10 LEIAR20 LEIM
CFBA	40.8097N 14.1202E 8 m	2016	Leica GR10 LEIAR20 LEIM
CFBB	40.8035N 14.1326E 8 m	2016	Leica GR10 LEIAR20 LEIM
CFBC	40.8057N 14.1432E 8 m	2016	Leica GR10 LEIAR20 LEIM
CFSB	40.7980N 14.1198E 8 m	2016	Leica GR10 LEIAR20 LEIM
CMIS	40.7781N 14.0889E 133 m	2015	Leica GR10 LEIAR20 LEIM
CUMA	40.8482N 14.0586E 61 m	2017	Leica GR10 LEIAR20 LEIM
FRUL	40.8779N 14.2252E 79 m	2003	Leica GR10 LEIAT504 LEIS
IPPO	40.8371N 14.1679E 66 m	1999	Leica GR10 LEIAR25 LEIT
ISMO	40.8262N 14.1911E 95 m	2016	Leica GR10 LEIAR20 LEIM
LICO	40.8764N 14.0496E 52 m	2003	Leica GR10 LEIAT504 LEIS
MAFE	40.8474N 14.2584E 95 m	2009	Leica GRX1200 LEIAT504 LEIS

(continued)

Table 2 (continued)

Station name	Coordinates and elevation above sea level	First observation	Receiver antenna
MORU	40.8596N 14.0830E 165 m	2000	Leica GRX1200 LEIAT504 LEIS
NAMM	40.8363N 14.2548E 50 m	2015	Leica GR10 LEIAR20 LEIM
NISI	40.7972N 14.1631E 153 m	2009	Leica GR10 LEIAT504 LEIS
PIS1	40.8300N 14.1460E 129 m	2015	Leica GR10 LEIAR20 LEIM
QUAR	40.8777N 14.1437E 107 m	1999	Leica GR10 LEIAR25 LEIT
RITE	40.8233N 14.1262E 104 m	2001	Leica GR10 LEIAR25 LEIT
SOLO	40.8303N 14.1357E 206 m	2006	Leica GR10 LEIAT504 LEIS
STRZ	40.8351N 14.1118E 120 m	2006	Leica GR10 LEIAT504 LEIS
VICA	40.8558N 14.1205E 112 m	2008	Leica GR10 LEIAT504 LEIS

internal archive. All the procedures for remote stations as configuration, raw data downloading, RINEX files generation and data quality control, work automatically and are described in Tammaro et al. (2013) and De Martino et al. (2014b).

Data processing is performed by the Bernese Processing Engine of the Bernese GNSS software v. 5.0 (Dach et al. 2007) in double-difference mode. The data are processed on a daily basis using the Ultra-rapid International GNSS Service (IGS) products in a fully automated processing chain. The data are reprocessed on a weekly basis once the IGS final orbits and Earth rotation parameters become available (De Martino et al. 2014b).

4 The Tiltmeter Network

Ground tilt monitoring is the continuous measurement of the rotation angle of the plane tangent to the geoid at a given measuring point (site station) that, if measured over time, can identify well-defined tilting directions. In the case of the CFc, the detected patterns are broadly consistent with those arising from cGNSS and Synthetic Aperture Radar (SAR) data. The greater detail offered by the signals recorded by the tiltmeters, however, allows detection of very slight changes of the volcano edifice caused by uplift or subsidence. In the following we present the CFc tiltmeter network and describe its development and

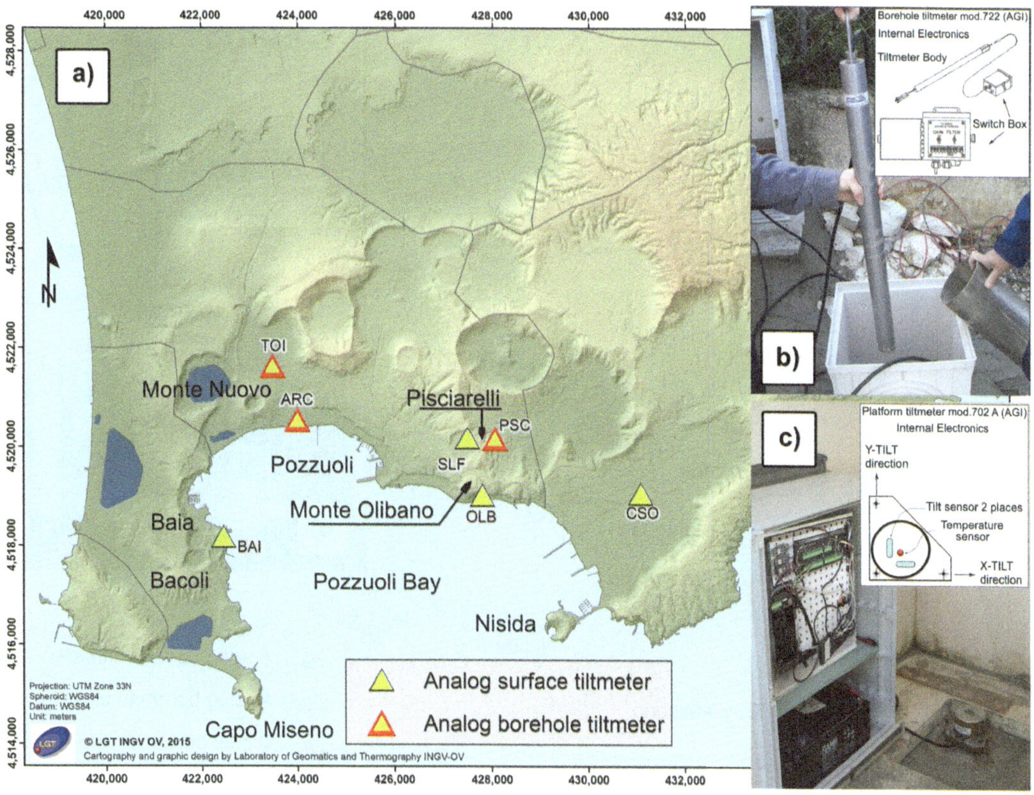

Fig. 4 **a** Map of the Campi Flegrei tiltmeter network improved from 1991 to 2013; **b** Borehole tiltmeter; **c** Platform tiltmeter

improvement over time to enable high-quality data collection.

At the end of 1985, the Osservatorio Vesuviano built the first CFc tiltmeter network to study the bradyseism events after those occurred in 1969–1972 and 1982–1984 (Orsi et al. 1999). This network, improved over time, continues to monitor the ongoing event that began in late 2004 (Ricco et al. 2007). Three stations were installed in 1985 at Pozzuoli, near the area of maximum ground uplift, and in a north–south aligned tunnel about 15 m deep, and were named DMB, DMA and DMC, respectively. A fourth station, named BAI, was installed in 1992 in a tunnel below the medieval Castle of Baia. All these instruments were installed in collaboration with the Institute de Physique du Globe de Paris (IPGP) (Ricco et al. 2003). Between 1991 and 1992, these four initial tilt stations were upgraded with surface sensors model AGI 702 (bi-axial, bubble type, manufactured by Applied Geomechanics) (AGI 1997). In order to reduce thermal excursions introducing further modulations on the tilt signal that also records the tidal deformation in the semi-diurnal, diurnal, fortnightly and monthly frequency bandwidth, two analog borehole tiltmeters were added in 2002–2003 (stations TOI and ARC in Fig. 4). These instruments included model AGI 722 (AGI 1999) and were installed at 7 m and 5 m depth, northeast and southeast of Monte Nuovo, respectively. In the same period, the DMA, DMB and DMC stations were disused.

From 2006 to 2011, the Phlegraean network was further extended with the installation of surface sensors at the CSO, OLB and SLF stations (Fig. 5). CSO is located in the basement of the INGV-OV building in Naples, whereas OLB is in a tunnel under Monte Olibano at about 70 m below ground level. Both stations are to the east of Pozzuoli. Finally, surface tilt station SLF is

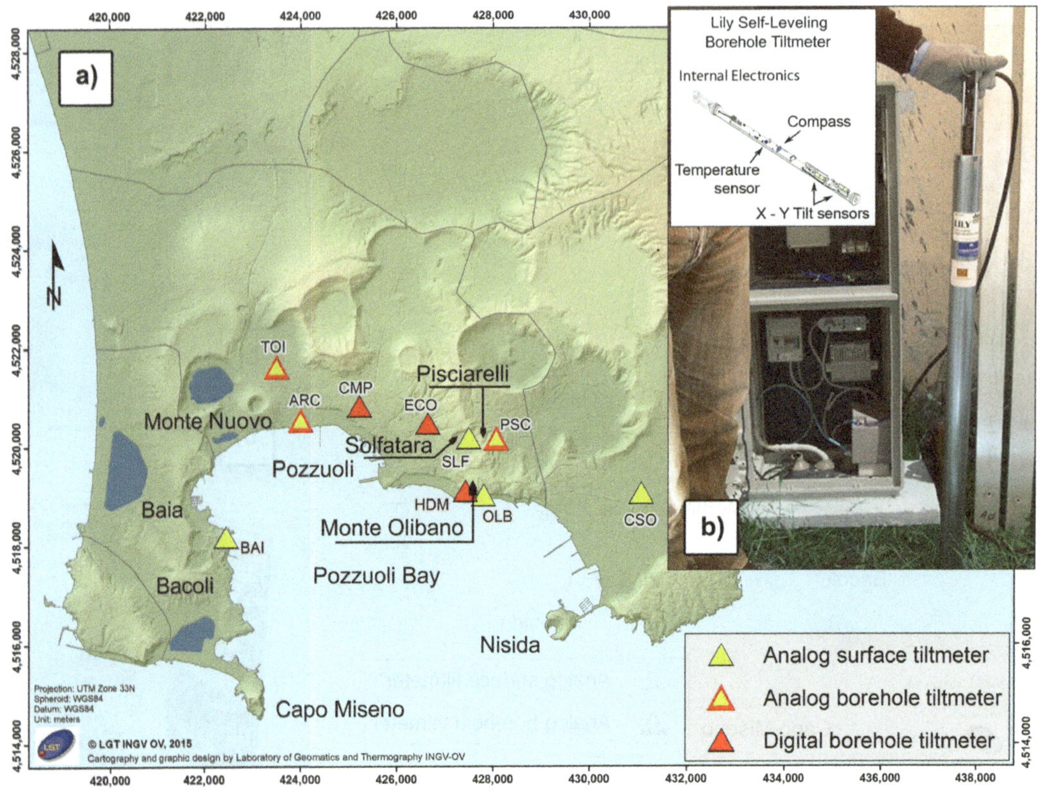

Fig. 5 **a** Map of the Campi Flegrei current real-time tiltmeter network; **b** Digital sensor Model Lily

located in the Solfatara Crater at Pozzuoli (Aquino et al. 2006; Ricco et al. 2018). In 2013 the last analog borehole tiltmeter (named PSC) was installed close to Pisciarelli fumarolic field at 1.5 m depth (Fig. 5).

The signals recorded by each surface and borehole sensor are digitised and automatically recorded at a sampling rate of 1 sample-per-minute. All sensors are connected to an electronic control unit containing a power supply system, a datalogger for data storage, and a data transmission system (AGI 2001). The transmission systems currently used are: direct connection to a computer, GSM modem, Wi-Fi Hiperlan and UMTS/4G connection to the INGV-OV Monitoring Room.

In early 2015, three borehole stations (CMP, ECO and HDM) equipped with digital sensors of the model "Lily Self-Leveling Borehole Tiltmeter" were installed in 25-m-deep wells (Fig. 5). These stations are close to the area of maximum ground uplift at Pozzuoli. The sensing device is a self-levelling sensor over a range of ±10°, with a dynamic range of ±330 μradians and a resolution less than 5 n radians (Aquino et al. 2016). The tilt signals are digitised and transmitted through RS485/RS422 or RS232 protocols (Jewell Instruments 2013). The data are also stored in an internal non-volatile flash memory card and subsequently transmitted to the surface, using an interface converter from serial-to-Ethernet (Lantronics uDS1100 model or similar). Currently the acquisition rate is 1 sample-per-minute and each acquisition is the mean value of 8,000 samples.

In summary, the current tiltmeter network operating at the CFc consists of 10 stations (Fig. 5), whose technical specifications are synthesised in Table 3. These stations use 3 different types of sensor: 4 surface-deployed analog, 3 borehole analog and 3 borehole digital sensors. Each surface station acquires 3 time-series

Table 3 Technical specifications of the Phlegraean tiltmeter stations

Station name	Coordinates and elevation above sea level	Installation year	Sensor	Depth	Sampling rate	Data transmission
BAI	40.8109N 14.0807E 35 m	1992	AGI 702	Ground level	1'	Hiperlan
SLF	40.8293N 14.1399E 104 m	2011	AGI 702	Ground level	1'	Hiperlan
OLB	40.8188N 14.1435E 30 m	2006	AGI 702	Ground level	1'	Hiperlan
CSO	40.8197N 14.1825E 19 m	2006	AGI 702	−5 m	1'	Telephone line
TOI	40.8421N 14.0924E 25 m	2002	AGI 702	−8 m	1'	GSM
ARC	40.8324N 14.0985E 3 m	2003	AGI 702	−6 m	1'	GSM
PSC	40.8300N 14.1456E 79 m	2013	AGI 702	−1 m	1'	GSM
CMP	40.8350N 14.1131E 61 m	2015	Jewell Lily	−25 m	1'	Hiperlan
ECO	40.8320N 14.1300E 112 m	2015	Jewell Lily	−25 m	1'	Router UMTS
HDM	40.8198N 14.1393E 115 m	2015	Jewell Lily	−25 m	1'	Router UMTS

including 2 orthogonal tilt values and 1 thermal. Borehole stations acquire 4 time-series, that is the previous ones plus the magnetic declination recorded at depth. All recorded signals, transmitted daily to the INGV-OV Monitoring Room, are processed through 3 sequential phases consisting of pre-processing, processing and data-analysis.

5 The Permanent Thermal Infrared Imagery Monitoring Network

The INGV-OV permanent Thermal InfraRed imagery Monitoring Network (TIRNet) continuously monitors selected thermal ground features of areas affected by diffuse degassing processes (Chiodini et al. 2007; Sansivero et al. 2013; Vilardo et al. 2015; Sansivero and Vilardo 2019). TIRNet consists of 5 stations acquiring daily images in the thermal infrared wavelengths range (LWIR) of target areas located inside and in the surroundings of Solfatara crater, and in the Monte Olibano-Pisciarelli area (Fig. 6). The investigated areas are: (1) part of a steep slope dominating the eastern side of the main Pisciarelli pool, targeted by station PS1; (2) Bocca Grande and Bocca Nuova fumarole areas, targeted by station SF1; (3) the base of northern inner slope of the Solfatara crater, targeted by station SF2, located in the same point of station

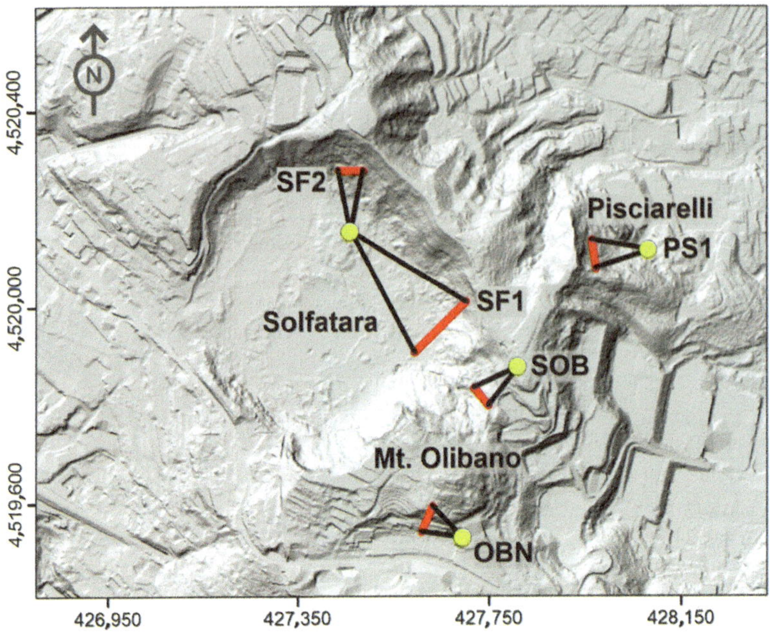

Fig. 6 Location of the TIRNet remote monitoring stations (yellow dots) at Solfatara crater and Monte Olibano-Pisciarelli area. The red lines indicate the target areas. Stations SF1 and SF2 are located in the same place and in order to avoid any confusion the station name is reported next to the target area

Table 4 The thermal infrared imagery monitoring network

Station name	Coordinates elevation above sea level	Installation date	Sensor	Data transmission
SF1	40.8292N 14.1397E 92 m	September 2004	FLIR A655SC (640 × 480) FoV: 25° × 19°	Wi-Fi
SF2	40.8292N 14.1397E 92 m	June 2013	FLIR A645SC (640 × 480) FoV: 15° × 11.9°	Wi-Fi
PS1	40.8289N 14.1471E 90 m	October 2006	FLIR A645SC (640 × 480) FoV: 15° × 11.9°	UMTS
OBN	40.8236N 14.1425E 150 m	March 2015	FLIR A645SC (640 × 480) FoV: 25° × 19°	Wi-Fi
IR-SOB	40.8268N 14.1439E 175 m	June 2016	FLIR A655SC (640 × 480) FoV: 25° × 19°	Wi-Fi

SF1; (4) the southern slope of Monte Olibano, targeted by the station OBN; (5) the top of the south-eastern rim of the Solfatara crater, targeted by station SOB.

The monitoring stations (Table 4) are equipped with FLIR A645 or A655SC IR cameras that use an uncooled focal plane array measuring system (microbolometer technology; 640 × 480 pixels) that operates in the 7.5 μm to 14 μm spectral band across a 15° (horizontal) × 11° (vertical) field of view (FLIR A645) and 25° (horizontal) × 19° (vertical) (FLIR A655sc).

The IR cameras are placed inside protective stainless-steel housing, with the lens protected by germanium glass and covered by a mechanical shutter. Three night-time images are acquired daily at 00:00, 02:00, and 04:00 UTC.

Image acquisition is directly managed by the station and an atmospheric correction is applied to the IR images by taking into account the detector-target distance, emissivity of the target, air temperature, and relative humidity. The acquisition server is located in the INGV-OV Monitoring Room and is connected to the remote station by the INGV-OV Wi-Fi-Hiperlan communication network. The acquired thermal images are uploaded to the server and processed according to a multi-step procedure (Sansivero and Vilardo 2019) performed by automated Matlab© software ASIRA (Automated System of IR Analysis). The main processing steps are: (a) data quality control; (b) accurate alignment of IR frames (co-registration by SIT flow algorithm); (c) IR frame seasonal temperature component removal by applying STL algorithm (Seasonal Trend decomposition based on Loess). Main post-processed products are: (a) maximum temperatures time-series; (b) radiative heat flux time-series; (c) maps of yearly rates of temperature change. The results are displayed in the INGV-OV Monitoring Room.

6 The Geochemical Network

The geochemical monitoring of a volcanic-hydrothermal system is based on the assumption that eruptions can be preceded and accompanied by an increase in magmatic degassing and the transfer of fluids and energy to the surface. Ascent of magma towards the surface causes the progressive exsolution of volatile constituents (H_2O, CO_2, SO_2, H_2S, HCl, He, etc.) that give rise to the formation of a gas phase. This gaseous phase, due to its high mobility, can reach the surface before the magma and during inter-eruptive periods, thus giving rise to surface hydrothermal activity, including fumarolic outflow and intense diffuse degassing. In pre-eruptive conditions, the increasing release of fluids from magmatic bodies rising toward the surface might cause variations in the amount and chemical-physical properties of both fumarolic gases and groundwater present in the system. Accordingly, current geochemical monitoring of CFc is based on the study of the chemical-isotopic compositions of the fumarolic fluids in order to determine their origin and the thermodynamic conditions of the hydrothermal system (Caliro et al. 2007, 2014; Chiodini et al. 2010, 2011, 2012, 2015a, b, 2016). These thermodynamic assessments combined with measurement of the diffuse degassing process (gas budget and energy released) provide crucial indications on the state of the hydrothermal-magmatic system (Chiodini et al. 2010, 2016; Cardellini et al. 2017).

Geochemical monitoring of the CFc consists of continuous and discrete measurements in the Solfatara crater, Pisciarelli and Agnano areas (Fig. 7). Continuous monitoring is performed through four stations (Table 5). Two are multi-parametric stations of the "GEochemical Multi Module Acquisition system" (GEMMA) type, developed and realised over the last five years as part of the technological research activities of the INGV-OV (Fig. 8).

The FLXOV5 multi-parametric station (Table 5) is installed in the Solfatara crater close to the main fumaroles. It measures soil CO_2 flux, temperature of the main fumarole, pressure and atmospheric temperature, CO_2, H_2S, SO_2 and H_2O concentrations in air (Multigas measurements). The other multi-parametric station (FLXOV8) is installed in the Pisciarelli area and is similarly equipped. Since 1997, station FLXOV1 monitors soil CO_2 flux in the Solfatara crater area; co-located GTS1 station measures the vertical thermal gradient of the soil. All the stations are connected via Wi-Fi to the INGV-OV Monitoring Room.

Additional campaign-style monitoring is carried out through periodic measurement of soil CO_2 flux and sampling of fumarolic gas. Geochemical monitoring activities carried out on a

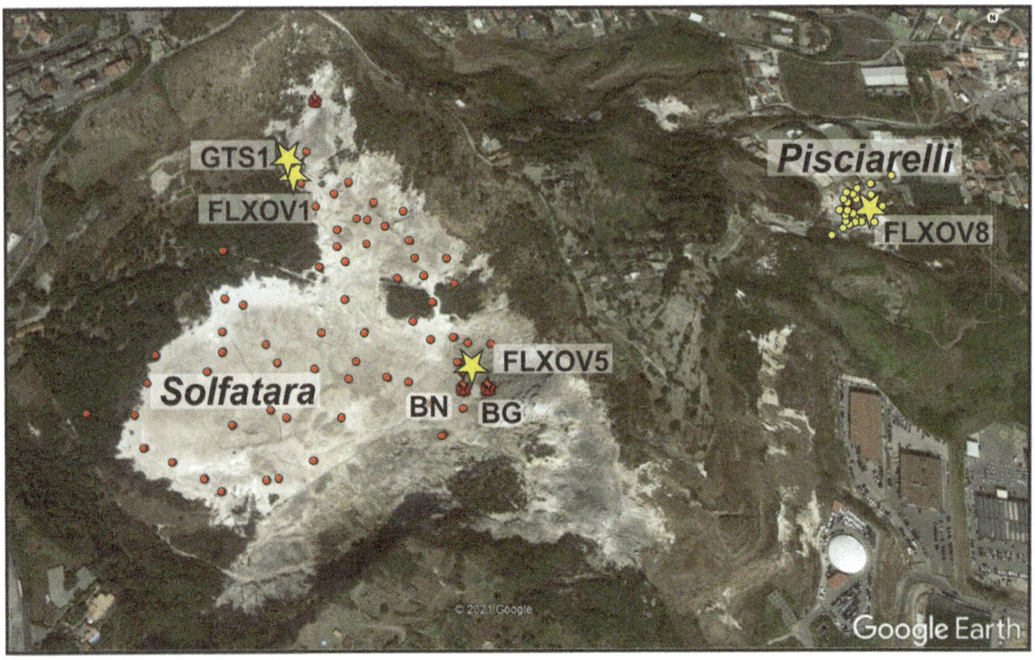

Fig. 7 Geochemical monitoring network at CFc made of the four geochemical stations located in the Solfatara-Pisciarelli area. Sites where geochemical monitoring activities are performed are also reported: the main fumaroles of the Solfatara (BG and BN) and Pisciarelli and the location of the monthly soil CO_2 flux measurements in Solfatara area (red dots) and in the Pisciarelli area (yellow dots)

Table 5 Description of stations forming the geochemical monitoring network of the INGV-OV

Station	Coordinates and elevation above sea level	Monitored parameters	Installation year
FLXOV1	40.8291N 14.1397E 103 m	Soil CO_2 flux;	1997
FLXOV8 (GEMMA)	40.8292N 14.1473E 74 m	Soil CO_2 flux; Fumarole Temperature, Multigas	2007 replace FLXOV3 Upgrade 2018
FLXOV5 (GEMMA)	40.8273N 14.1418E 105 m	Soil CO_2 flux; Fumarole Temperature, Multigas	2014
GTS1	40.8291N 14.1397E 103 m	Soil vertical thermal gradient	2016

monthly basis include (Fig. 8): (i) Sampling of fumarolic fluids discharged by the main fumaroles at Solfatara (BG and BN) and Pisciarelli, and chemical (H_2O, CO_2, H_2S, H_2, Ar, N_2, CH_4, He, CO) and isotopic ($^{40}Ar/^{36}Ar$, $\delta^{15}N_{N2}$, $\delta^{13}C_{CO2}$; $\delta^{18}O_{CO2}$, $\delta^{2}H_{H2O}$, $\delta^{18}O_{H2O}$) analyses of the collected fluids. (ii) Measurement of soil CO_2 flux and temperature at 10 cm depth, in about 60 fixed points in the Solfatara crater. (iii) Measurement of soil CO_2 flux and temperature at 10 cm depth, in about 28 fixed points in the Pisciarelli area.

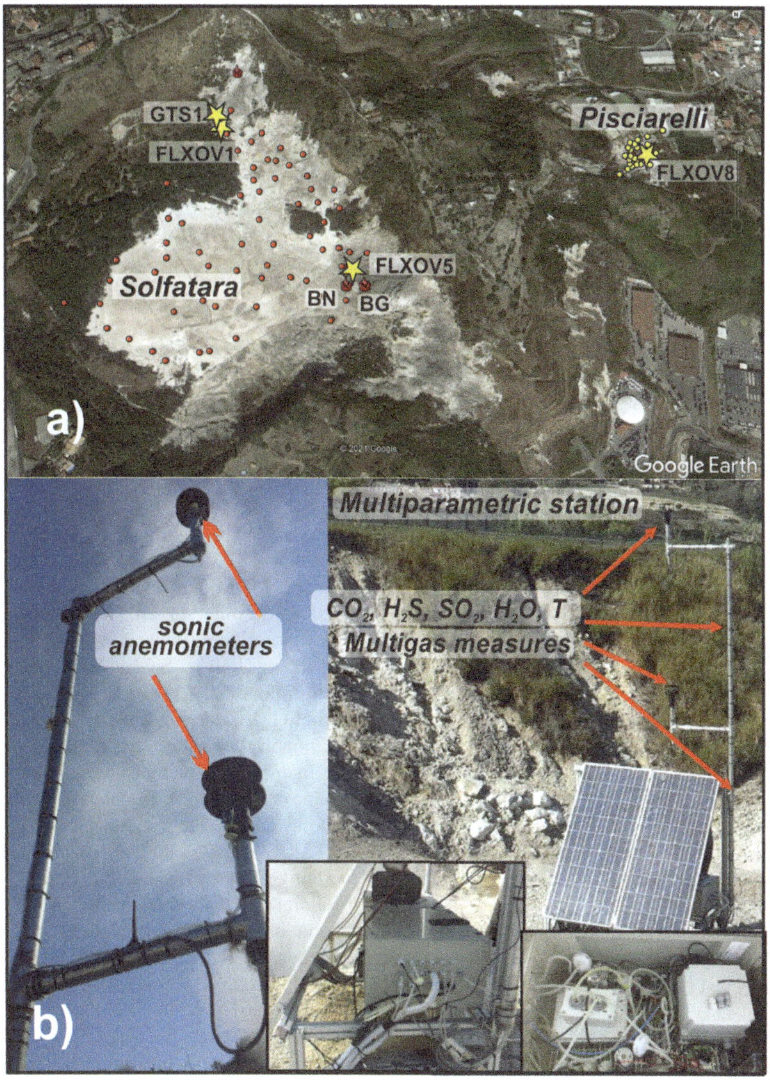

Fig. 8 Multi-parametric station "GEMMA" type installed at Pisciarelli degassing area

7 Conclusive Remarks

We have described the main features of the permanent monitoring system in the Campi Flegrei caldera managed by the INGV-OV. All the recorded h24 continuous data are transmitted to the Monitoring Room of the INGV-OV in Naples (Fig. 9) where they are processed and evaluated in order to define changes in the dynamical state of the volcano. Additional data from periodic measurements using levelling network, reoccupation of permanent gravity stations, discrete IR thermal stations (also periodic unmanned aerial vehicles–UAV overflights with FLIR VUEPRO camera) and mobile seismic stations, also contribute to defining levels of unrest or quiescence.

The monitoring system of the CFc is of sufficiently high spatial density and sensitivity to produce high quality data useful for monitoring this active and potentially extremely hazardous volcano. However, the quality of the monitoring increases following the growing of both the scientific and technological knowledge. Accordingly, the Campi Flegrei INGV-OV monitoring system is continuously upgraded in order to

Fig. 9 The Monitoring Room of the INGV-Osservatorio Vesuviano in Naples, operative 24h/7d. Photo by Giovanni Scarpato

assure the best performance of the multi-parametric system that plays a crucial role in the national emergency plan of the Campi Flegrei (Fig. 9).

References

AGI (1997) 700-Series platform and surface mount tiltmeters. User's Manual no B-88-1016 Rev E

AGI (1999) Model 722 Borehole Tiltmeter. User's Manual no B-91-1004 Rev B

AGI (2001) Model 798-A Handi-Logger user's manual telephone telemetry option. User's Manual no B-01-1008 Rev E

Aquino I, Ricco C, Del Gaudio C (2006) Rete tiltmetrica dell'area napoletana. Open file report INGV-OV n 4:23 pp

Aquino I, Ricco C, Del Gaudio C, Augusti V, Scarpato G (2016) Potenziamento delle reti tiltmetriche dell'area vulcanica campana: rapporto sull'attività svolta nell'ambito del Progetto VuLCAMED. Rapporti Tecnici INGV 348:51 pp, ISSN 2039–7941

Buonocunto C, D'Auria L, Caputo A, Martini M, Orazi M (2011) The InfraCyrus infrasound sensor. Rapporti Tecnici INGV 188, http://hdl.handle.net/2122/7306

Caliro S, Chiodini G, Moretti R, Avino R, Granieri D, Russo M, Fiebig J (2007) The origin of the fumaroles of La Solfatara (Campi Flegrei, South Italy). Geochim Cosmochim Acta 71:3040–3055. https://doi.org/10.1016/j.gca.2007.04.007

Caliro S, Chiodini G, Paonita A (2014) Geochemical evidences of magma dynamics at Campi Flegrei (Italy). Geochim Cosmochim Acta 132:1–15. https://doi.org/10.1016/j.gca.2014.01.021

Cardellini C, Chiodini G, Frondini F, Avino R, Bagnato E, Caliro S, Lelli M, Rosiello A (2017) Monitoring diffuse volcanic degassing during volcanic unrests: the case of Campi Flegrei (Italy). Sci Rep 7:6757. https://doi.org/10.1038/s41598-017-06941-2

Castellano M, Buonocunto C, Capello M, La Rocca M (2002) Seismic surveillance of active volcanoes: the Osservatorio Vesuviano Seismic Network (OVSN–Southern Italy). Seism Res Lett 73(2):177–184. https://doi.org/10.1785/gssrl.73.2.177

Chiodini G, Vilardo G, Augusti V, Granieri D, Caliro S, Minopoli C, Terranova C (2007) Thermal monitoring of hydrothermal activity by permanent infrared automatic stations: results obtained at Solfatara di Pozzuoli, Campi Flegrei (Italy). J Geophys Res 112: B12206. https://doi.org/10.1029/2007JB005140

Chiodini G, Caliro S, Cardellini C, Granieri V, Avino R, Baldini A, Donnini M, Minopoli C (2010) Long-term variations of the Campi Flegrei, Italy, volcanic system as revealed by the monitoring of hydrothermal activity. J Geophys Res 115:B03205. https://doi.org/10.1029/2008JB006258

Chiodini G, Avino R, Caliro S, Minopoli V (2011) Temperature and pressure gas geoindicators at the Solfatara fumaroles (Campi Flegrei). Ann Geophys 54:2. https://doi.org/10.4401/ag-5002

Chiodini G, Caliro S, De Martino P, Avino R, Gherardi F (2012) Early signals of new volcanic unrest at Campi Flegrei caldera? Insights from geochemical data and physical simulations. Geology 40(10):943–946. https://doi.org/10.1130/G33251.1

Chiodini G, Vandemeulebrouck J, Caliro S, D'Auria L, De Martino P, Mangiacapra A, Petrillo Z (2015a) Evidence of thermal driven processes triggering the 2005–2014 unrest at Campi Flegrei caldera. Earth Planet Sci Lett 414:58–67. https://doi.org/10.1016/j.epsl.2015.01.012

Chiodini G, Pappalardo L, Aiuppa A, Caliro S (2015b) The geological CO_2 degassing history of a long-lived caldera. Geology 43:767–770

Chiodini G, Paonita A, Aiuppa A, Costa A, Caliro S, De Martino P, Acocella V, Vandemeulebrouck J (2016) Magmas near the critical degassing pressure drive volcanic unrest towards a critical state. Nat Commun 7:13712. https://doi.org/10.1038/ncomms13712

Dach R, Hugentobler U, Fridez P, Meindl M (2007) Bernese GPS software version 5.0. Astronomical Institute, University of Bern

D'Auria L, Giudicepietro F, Aquino I, Borriello G, Del Gaudio C, Lo Bascio D, Martini M, Ricciardi GP, Ricciolino P, Ricco C (2011) Repeated fluid-transfer episodes as a mechanism for the recent dynamics of Campi Flegrei caldera (1989–2010). J Geophys Res Solid Earth 116(B4). https://doi.org/10.1029/2010JB007837

Del Gaudio C, Aquino I, Ricciardi GP, Ricco C, Scandone R (2010) Unrest episodes at Campi Flegrei: a reconstruction of vertical ground movements during 1905–2009, J Volcanol Geotherm Res 195:48–56. https://doi.org/10.1016/j.jvolgeores.2010.05.014

De Martino P, Guardato S, Tammaro U, Vassallo M, Iannaccone G (2014a) A first GPS measurement of vertical seafloor displacement in the Campi Flegrei caldera (Italy). J Volcanol Geotherm Res 276:145–151. https://doi.org/10.1016/j.jvolgeores.2014.03.003

De Martino P, Tammaro U, Obrizzo F (2014b) GPS time series at Campi Flegrei caldera (2000–2013). Ann Geophys 57(2):S0213. https://doi.org/10.4401/ag-6431

Di Vito MA, Acocella V, Aiello G, Barra D, Battaglia M, Carandente A, Del Gaudio C, de Vita S, Ricciardi GP, Ricco C, Scandone R, Terrasi F (2016) Magma transfer at Campi Flegrei caldera (Italy) before the 1538 AD eruption. Sci Rep 6(1):1–9. https://doi.org/10.1038/srep32245

Estey L, Meertens C (1999) TEQC: The multi-purpose toolkit for GPS/GLONASS data. GPS Solutions 3:42–49. https://doi.org/10.1007/PL00012778

Guidoboni E, Ciuccarelli C (2011) The Campi Flegrei caldera: historical revision and new data on seismic crises, bradyseisms, the Monte Nuovo eruption and ensuing earthquakes (twelfth century 1582 AD). Bull Volcanol 73:655–677. https://doi.org/10.1007/s00445-010-0430-3

Iannaccone G, Guardato S, Vassallo M, Elia L, Beranzoli L (2009) A new multidisciplinary marine monitoring system for the surveillance of volcanic and seismic areas. Seism Res Lett 80:203–213. https://doi.org/10.1785/gssrl.80.2.203

Iannaccone G, Guardato S, Donnarumma GP, De Martino P, Dolce M, Macedonio G, Chierici F, Beranzoli L (2018) Measurement of seafloor deformation in the marine sector of the Campi Flegrei caldera (Italy). J Geophys Res Solid Earth 123:66–83. https://doi.org/10.1002/2017JB014852

JEWELL INSTRUMENTS (2013) LILY Self-leveling borehole tiltmeter. User's manual no B-05-1003, Rev G

Orazi M, Martini M, Peluso R (2006) Data acquisition for volcano monitoring. EOS 87(38)

Orsi G, De Vita S, di Vito M (1996) The restless, resurgent Campi Flegrei nested caldera (Italy): constraints on its evolution and configuration. J Volcanol Geotherm Res 74:179–214. https://doi.org/10.1016/S0377-0273(96)00063-7

Orsi G, Civetta L, Del Gaudio C, De Vita S, Di Vito MA, Isaia R, Petrazzuoli SM, Ricciardi GP, Ricco C (1999) Short-term ground deformations and seismicity in the resurgent Campi Flegrei caldera (Italy): an example of active block-resurgence in a densely populated area. J Volcanol Geotherm Res 91:415–451

Ricco C, Aquino I, Del Gaudio C (2003) Ground tilt monitoring at Phlegraean fields (Italy): a methodological approach. Ann Geophys 46(6):1297–1314. https://doi.org/10.4401/ag-3474

Ricco C, Aquino I, Borgstrom SE, Del Gaudio C (2007) A study of tilt change recorded from July to October 2006 at the Phlegrean fields (Naples, Italy). Ann Geophys 50 (5):661–674. https://doi.org/10.4401/ag-3059

Ricco C, Aquino I, Augusti V, D'auria L, Del Gaudio C, Scarpato G (2018) Improvement and development of the tiltmetric monitoring networks of Neapolitan volcanoes. Ann Geophys 61(1). https://doi.org/10.4401/ag-7496

Saccorotti G, Petrosino S, Bianco F, Castellano M, Galluzzo D, La Rocca M, Del Pezzo E, Zaccarelli L, Cusano P (2007) Seismicity associated with the 2004–2006 renewed round uplift at Campi Flegrei caldera, Italy. Phys Earth Plan Int 165:14–24

Sansivero F, Scarpato G, Vilardo G (2013) The automated infrared thermal imaging system for the continuous long-term monitoring of the surface temperature of the Vesuvius crater. Ann Geophys 56:S0454. https://doi.org/10.4401/ag-6460

Sansivero F, Vilardo G (2019) Processing thermal infrared imagery time-series from volcano permanent ground-based monitoring network. Latest methodological improvements to characterize surface temperatures behavior of thermal anomaly areas. Remote Sens 11(5):553. https://doi.org/10.3390/rs11050553

Tammaro U, De Martino P, Obrizzo F, Brandi G, D'Alessandro A, Dolce M, Malaspina S, Serio C, Pingue F (2013) Somma Vesuvius volcano: ground deformations from CGPS observations (2001–2012). Ann Geophys 56(4):S0456. https://doi.org/10.4401/ag-6462

Tramelli A, Troise C, De Natale G, Orazi M (2013) A new method for optimization and testing of microseismic networks: an application to Campi Flegrei (Southern Italy). Bull Seismol Soc Am 103(3):1679–1691

Vilardo G, Sansivero F, Chiodini G (2015) Long-term TIR imagery processing for spatiotemporal monitoring of surface thermal features in volcanic environment: a case study in the Campi Flegrei (Southern Italy). J Geophys Res Solid Earth 120:812–826. https://doi.org/10.1002/2014JB011497

The Hydrothermal System of the Campi Flegrei Caldera, Italy

Giovanni Chiodini, Stefano Caliro, Rosario Avino, Emanuela Bagnato, Francesco Capecchiacci, Antonio Carandente, Carlo Cardellini, Carmine Minopoli, Giancarlo Tamburello, Simona Tripaldi, and Alessandro Aiuppa

Abstract

In this chapter, we review the state-of-the-art of the Campi Flegrei caldera (Naples) hydrothermal system, and its behaviour during the last decades. The Campi Flegrei caldera has been undergoing unrest since 1950, as evidenced by recurrent bradyseismic episodes accompanied by manifest changes in the degassing budget, degassing patterns and in the composition of the fumarolic fluids. In-depth analysis of geochemical and geophysical datasets acquired over decades has allowed identification of the mechanisms driving volcanic unrest at the Campi Flegrei caldera. We propose a conceptual model of the hydrothermal system feeding Solfatara fumaroles, where geochemical information is integrated with Audio Magneto Telluric measurements, which yields a realistic picture of the geometry of the system up to a depth of 2.5 km. The model identifies a ~2 km elongated vertical high resistivity structure in axis with the Solfatara fumaroles, which represents a relatively high permeability zone allowing hot fluid ascent from depth to the shallower portions of the hydrothermal system. Pulsed injections of hot magmatic fluids (CO_2-rich and CH_4-poor oxidised fluids) at the bottom of the hydrothermal system is thought to be one of the key processes that has controlled the evolution of the system during the last 40 years. The episodes of injection of magmatic fluids changed in frequency and intensity during time, ultimately causing an overall heating and pressurisation of the system since the early 2000s, as reflected by escalating degassing flux, increase in areal extension of the degassing areas, and in the composition of the fumaroles. In particular, the CO_2/CH_4 and He/CH_4 ratios of fumarolic fluids exhibited recurrent peaks, marking the episodes of injection of magmatic fluids. Moreover, the quasi-monotonic increasing trend of the fumarolic CO_2/H_2O ratio, from 0.15 to 0.18 in 2000 to ~0.4 in 2018–2019, has been interpreted as due to the combined action of partial steam condensation, and CO_2

G. Chiodini (✉) · C. Cardellini · G. Tamburello
Istituto Nazionale di Geofisica e Vulcanologia, Sezione di Bologna, Bologna, Italy
e-mail: giovanni.chiodini@ingv.it

S. Caliro · R. Avino · E. Bagnato · F. Capecchiacci · A. Carandente · C. Minopoli
Istituto Nazionale di Geofisica e Vulcanologia, Sezione Osservatorio Vesuviano, Napoli, Italy

C. Cardellini
Dipartimento di Fisica e Geologia, Università degli Studi di Perugia, Perugia, Italy

S. Tripaldi
Dipartimento di Scienze della Terra e Geoambientali, Università degli Studi di Bari Aldo Moro, Bari, Italy

A. Aiuppa
Dipartimento di Scienze della Terra e del Mare, Università degli Studi di Palermo, Palermo, Italy

© Springer-Verlag GmbH Germany, part of Springer Nature 2022

G. Orsi et al. (eds.), *Campi Flegrei*, Active Volcanoes of the World,
https://doi.org/10.1007/978-3-642-37060-1_9

addition from a magmatic source and possibly from de-carbonation of hydrothermal calcite favoured by the heating of the hydrothermal reservoir. These changes strongly suggest that the ongoing (since 2000) unrest is triggered by a degassing magma source, but also that the system's response is modulated by dynamics and structures of the overlying hydrothermal envelope. This evolution clearly requires careful scientific scrutiny and intensified monitoring in the years to come.

1 Introduction

The bradyseismic crises that started in the mid-1900s (Del Gaudio et al. 2010), and that in 1983–1984 caused the evacuation of the Pozzuoli town (Fig. 1) (Barberi et al. 1984), have prompted an increasing number of studies aiming at shading light onto the structure, chemical properties and temporal evolution of the hydrothermal system(s) hosted within the Campi Flegrei caldera (CFc). For a synthesis of the volcanic and deformation history and of the geophysical structure of the caldera see Chaps. Volcanic and Deformation History of the Campi Flegrei Volcanic Field, Italy; Seismic and Gravity Structure of the Campi Flegrei Caldera, Italy, respectively. Magmatism and volcanism of the structure were fed by a complex system whose architecture, behaviour and evolution are summarised in Chaps. An Evolutionary Model for the Magmatic System of the Campi Flegrei Volcanic Field (Italy) Constrained by Petrochemical Data; Origin and Differentiation History of the Magmatic System Feeding the Campi Flegrei Volcanic Field (Italy) Constrained by Radiogenic and Stable Isotope Data; Tephrochronology and Geochemistry of Tephra from the Campi Flegrei Volcanic Field, Italy; Rheological Properties of the Magmas Feeding the Campi Flegrei Caldera (Italy) and Their Influence on Mixing Processes; Magma Chamber Dynamics at the Campi Flegrei Caldera, Italy. For reviews of the results of the monitoring network and time variation of the physical parameters of the short-term deformation see Chaps. The Permanent Monitoring System of the Campi Flegrei Caldera, Italy; Historic Unrest of the Campi Flegrei Caldera, Italy; Source Modelling from Ground Deformation and Gravity Changes at the Campi Flegrei Caldera, Italy.

Knowledge of hydrothermal fluid circulation in the subsurface of Campi Flegrei was initially obtained from geothermal wells drilled in the 1950's (Penta 1954) and subsequently in the 1970's–early 1980's (Guglielminetti 1986). These drilling projects offered insights into the deep structure of CFc (Rosi and Sbrana 1987; De Vivo et al. 1989), and have revealed the existence of a saline multiple-reservoir field (sub-surface temperatures of 250–390 °C) composed of at least two distinct reservoirs fed by seawater (<2,000 m) and meteoric-magmatic fluids (>2,000 m) (Caprarelli et al. 1997). These deep reservoirs would supply magmatic volatiles and heat to the shallow groundwater system, whose surface manifestations (thermal wells and subaerial/submarine thermal springs; Martini et al. 1991; Celico et al. 1992; Valentino et al. 1999; Valentino and Stanzione 2003, 2004; Aiuppa et al. 2006; Di Napoli et al. 2016) generate a manifest thermal anomaly in the caldera's centre, in the surroundings of the Solfatara area (Fig. 1). The post-1980s clustering of degassing activity (Chiodini et al. 2001, 2005, 2010; Aiuppa et al. 2013), seismicity (Chiodini et al. 2017a; Giudicepietro et al. 2017; Di Luccio et al. 2015) and deformation (Del Gaudio et al. 2010) within (or nearby) the same Solfatara area, has motivated extensive research on the chemistry, isotope composition and flux of fluids discharged by its fumarolic field (Cioni et al. 1984; Chiodini et al. 2003, 2010, 2015a, b, 2016; Caliro et al. 2007, 2014; Cardellini et al. 2017). In this review, we concentrate on hydrothermal fluid circulation beneath Solfatara.

Solfatara is a tuff cone dated at 4,181–4,386 cal. years BP (Smith et al. 2011 and references therein; Chap. Volcanic and Deformation History of the Campi Flegrei Volcanic Field, Italy), whose volcanic products are strongly hydrothermally altered by pervasive circulation of hydrothermal fluid and intense CO_2 diffuse degassing from hot soils (Rosi and Sbrana 1987;

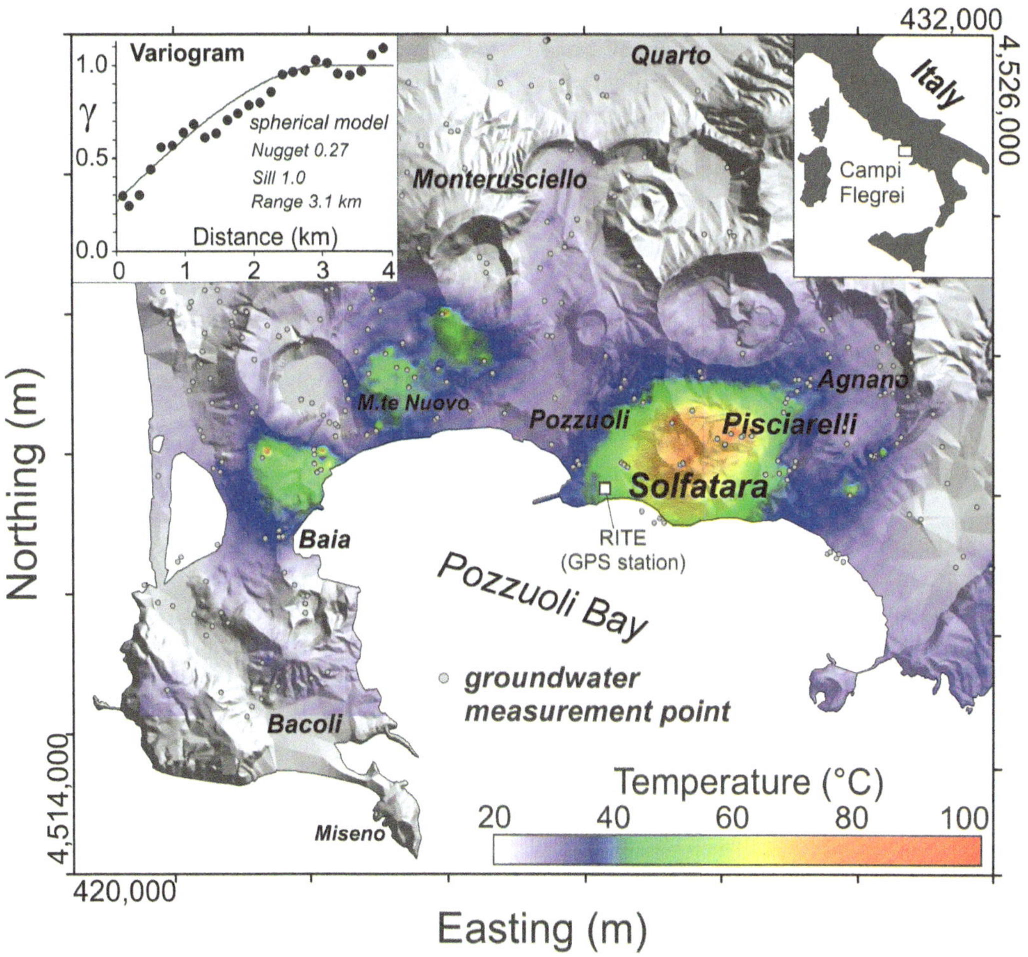

Fig. 1 Temperature distribution of the Campi Flegrei groundwaters. The map was produced using the Sequential Gaussian Simulation algorithm (Deutsch and Journel 1998) based on a dataset of 290 water temperatures, most of them refers to shallow wells with variable depths, generally lower than 100 m. Data from Petrillo et al. (2013), and data collected by Istituto Nazionale di Geofisica e Vulcanologia—Osservatorio Vesuviano in the period from 2013 to 2018. Coordinates expressed in metres refer to the UTM WGS84

Chiodini et al. 2001; Fig. 1). The crater and its surroundings host one of the better studied fumarolic fields of the world, with a fairly continuous and systematic record of geochemical data (acquired together with geophysical signals) dating back to the large 1983–1984 bradyseismic crises (e.g., Carapezza et al. 1984; Cioni et al. 1984; Tedesco et al. 1990; Allard et al. 1991; Chiodini and Marini 1998; Panichi and Volpi 1999; Tedesco and Scarsi 1999). In detail, an exceptionally long (~35 years) record of the chemical and isotopic compositions of fumaroles is available for the Solfatara (Chiodini et al. 2016; Istituto Nazionale di Geofisica e Vulcanologia—Osservatorio Vesuviano unpublished data), together with ~20 years of diffuse soil emission measurements of volcanic-hydrothermal CO_2 (Cardellini et al. 2017). In-depth analysis of this huge and unique dataset, acquired over decades of recurrent seismicity and ground deformation episodes, has been central to identifying the causal mechanisms driving volcanic unrest at CFc. The key conclusion that has arisen from these investigations is that observed indicators of volcanic unrest (escalating deformation, seismicity and degassing) are all caused by repeated pulses of hot magmatic fluid injection into the hydrothermal system feeding the Solfatara (e.g., Chiodini et al. 2003, 2012, 2016).

The aim of this chapter is to review our current understanding of the Solfatara hydrothermal system, putting the accent on both subjects where there is an agreement within the scientific community, and arguments that have remained intensely debated in the scientific community.

2 Mass and Energy Release at Solfatara

The Solfatara crater and its surroundings are home to numerous thermal manifestations, including fumarolic vents, boiling pools and hot soils diffusively degassing CO_2. In particular, the Solfatara hydrothermal system sustains one of the most actively degassing and well-characterised diffuse degassing structures (Chiodini et al. 2001) worldwide. Here, we refer to the work of Cardellini et al. (2017) that reports and discusses in detail the results of 30 surveys of soil diffuse CO_2 flux carried out at Solfatara from 1998 to 2016 using the accumulation chamber method (Chiodini et al. 1998). Each survey consisted of several hundred measurements (typically 400–500) and covered a $\sim 1.2 \times 1.2$ km^2 area. Cardellini et al. (2017), using the measurements of all the different campaigns (13,158 points), produced a very detailed image of the Solfatara diffuse degassing structure (DDS; Fig. 2). With the term Solfatara DDS we refer to the sector that, during the observation period, released volcanic-hydrothermal CO_2, either continuously or sporadically. The Solfatara DDS (yellow–red colours in Fig. 2) includes both the crater and sectors external to the cone, such as the Pisciarelli, Monte Olibano and via Antiniana areas (Fig. 2). It is worth noting that the geometry of the DDS correlates well with the main volcanic and extensional tectonic structures (northwest-southeast and northeast-southwest lineaments) channelling gas transfer from depth towards the surface.

The areal extent of the Solfatara DDS, and the total amount of CO_2 released by diffuse soil degassing, both varied considerably over time. The DDS surface increased from 0.45 km^2 in 1998 to more than 1 km^2 in many post-2012 surveys, and the total CO_2 flux consistently increased from $\sim$750 td^{-1} to $\sim$2,800 td^{-1} (Cardellini et al. 2017). The total CO_2 output includes both the deeply derived gas and that produced by biological activity in the soil. However, biogenic CO_2 output has been shown to account for only a few percent of the total release, using carbon isotopic composition of the CO_2 efflux (Chiodini et al. 2008; Cardellini et al. 2017). The diffuse CO_2 output at Solfatara is remarkably high, considering that its mean value (1,309 td^{-1}) is similar to the CO_2 flux to the atmosphere of a “medium-large” volcanic plume, while the maximum measured value (2,800 td^{-1}) would constitute the eighth highest CO_2 flux among volcanic plumes worldwide (see Table 2 in Burton et al. 2013).

The thermal energy release associated with the CO_2 emissions was estimated, in 1998, at $\sim$100 MW (Chiodini et al. 2001). Assuming this value as representative of the last 40 years, the energy associated with diffuse degassing would represent the main component of the CFc energy budget, as it is higher than i) the heat released by conduction over the entire caldera, ii) the energy associated to ground deformations, and iii) the elastic energy released by earthquakes (see discussion in Chiodini et al. 2001, 2017a).

3 Origin of the Solfatara Fluids Based on Fumarole Chemical and Isotopic Composition

The majority of the available information on the CF hydrothermal system stems from detailed analysis of the chemical and isotopic composition of the Solfatara fumaroles. Three of the numerous fumarolic vents located in the Solfatara DDS have temperatures above the water boiling point (Bocca Grande—BG, 150–165 °C; Bocca Nuova—BN, 140–145 °C; Pisciarelli—Pi, 95–120 °C; Fig. 2), and have been routinely

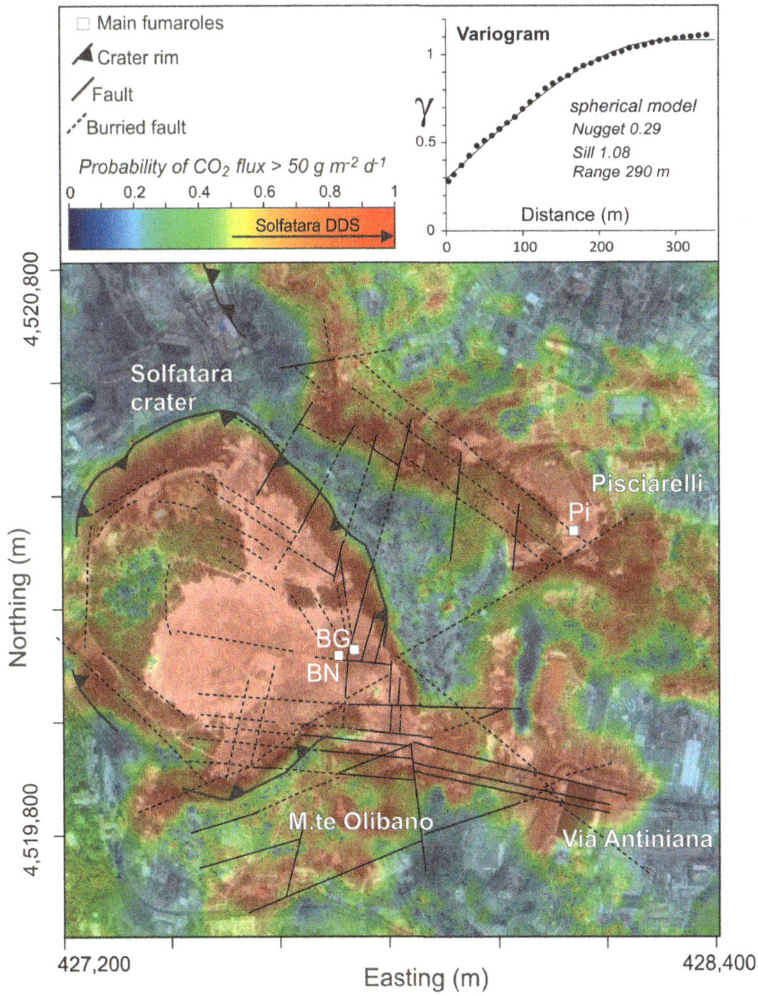

Fig. 2 Map of the Solfatara diffuse degassing structure (DDS). The map was realised by sequential Gaussian simulations (Deutsch and Journel 1998; Cardellini et al. 2003) based on the entire dataset of CO_2 fluxes from 1998 to 2016. The different colours represent the probability that the simulated CO_2 flux is greater than the biogenic threshold of 50 $gm^{-2}d^{-1}$ (Chiodini et al. 2008). Yellow to red colours indicate at each location probabilities higher than 0.5 of CO_2 flux > threshold, highlighting the areas where degassing of deeply derived CO_2 occurs (the Solfatara DDS). The Solfatara DDS well matches the volcano-tectonic lineaments of the area (Isaia et al. 2015). The locations of the main fumaroles, Bocca Nuova (BN), Bocca Grande (BG) and Pisciarelli (Pi) are reported. Coordinates are reported as meters projection UTM European Datum 50. The base map is from http://mapmaker.nationalgeographic.org/ (see Cardellini et al. 2017 for further details)

sampled and analysed in the frame of CFc volcanic surveillance. The BG highest temperature fumarole (Fig. 2), is thought to be the most representative of the deep fluids because it is less affected by shallow secondary processes (Gresse et al. 2018). Its composition is dominated by water vapour (X_{H_2O} from 0.73 to 0.87) and subordinately CO_2 (X_{CO_2} from 0.13 to 0.26). Other analysed species (including H_2S, N_2, H_2, CH_4, He, Ar and CO) exhibit concentrations ranging from less than 1 ppm (CO) to 1,000–2,000 ppm (H_2S).

Magmatic gas supply to the Solfatara hydrothermal system is supported by its fumaroles' stable water isotopes. Steam deuterium and ^{18}O compositions, once corrected for oxygen

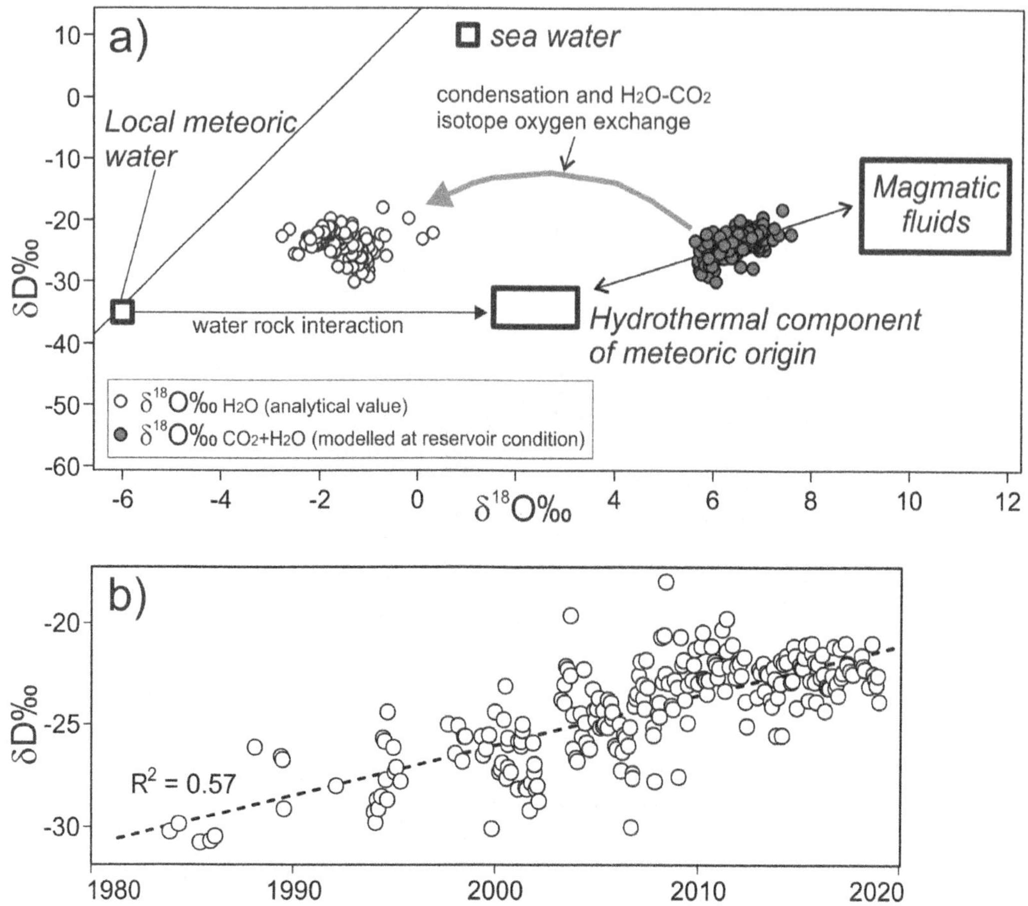

Fig. 3 **a** δD vs $\delta^{18}O$ diagram (modified after Chiodini et al. 2016). Starting from the analytical values (post-2000 samples of BG fumaroles), the equilibrium $\delta^{18}O - \delta D$ composition of hydrothermal vapour ($H_2O + CO_2$) was calculated considering: reservoir temperatures (Tr), reservoir CO_2 molar fractions (X_{CO_2}), and the fractions of steam (f) that condenses passing from reservoir to discharge. According to Chiodini et al. (2015a), Tr, X_{CO_2} and f were estimated applying gas equilibria in the H_2O-H_2-CO_2-CO gas system. Computations involved solving a set of isotope mass balance and fractionation equations. Fractionation during water condensation and H_2O-CO_2 isotope oxygen exchange (Chiodini et al. 2000) were taken into account. The re-computed $\delta^{18}O$ values refer to the whole $CO_2 + H_2O$ system. The 'Magmatic fluids' box refer to the Giggenbach (1992) andesitic water (δD from −10‰ to −20‰ and $\delta^{18}O$ from + 9‰ to +12‰) **b** Chronogram of the deuterium composition of BG condensates (data from Chiodini et al. 2016)

isotopic exchange between water vapor and CO_2 (Chiodini et al. 2000) and for partial H_2O removal by condensation, clearly depict a mixture between a classical magmatic component and hydrothermal fluids of meteoric origin (Fig. 3a).

The proportion of magmatic fluids in the mixture feeding the Solfatara fumaroles has increased in time, as suggested by δD values increasing from -30% in the early 1980's (close to the local meteoric waters values of δD ~ −35‰; Panichi and Volpi 1999) to more magmatic values (−25‰ to −20‰) in the post-2010 samples. This temporally increasing fraction of magmatic gas has paralleled with the escalating hydrothermal fluid output (see above) and the generally heating/pressurisation of the system, as derived from interpretation of gas equilibria (see Sect. 6).

A magmatic gas feeding into the hydrothermal system is also consistent with the isotopic signatures of fumarolic CO_2, He and N_2. The carbon isotopic composition of CO_2 ($\delta^{13}C$ = −1.3 ‰ ± 0.4‰; Caliro et al. 2014) supports a mantle source contaminated by crustal fluids delivered by a carbonate-rich subducted slab. The Solfatara CO_2 has, in fact, carbon isotope signature heavier than primary mantle carbon ($\delta^{13}C$ = -6 ‰ ± 2 ‰; Deines and Gold 1973), and lighter than those derived by crustal de-carbonation of marine limestone. It cannot be ruled out, however, that part of the emitted C is derived by de-carbonation of hydrothermal calcite, which is well present in the subsurface at CF, and which may have acted as a source and sink over the geological history of the caldera (Chiodini et al. 2015b). The $^3He/^4He$ ratio measured in the fumaroles (2.85 ± 0.12 R/Ra; Caliro et al. 2014) indicates a significant contribution of primordial (mantle-derived) ^3He, variably contaminated by crustal ^4He (Allard et al. 1997; Tedesco 1997). According to Martelli et al. (2004) these relatively low $^3He/^4He$ ratios, which are close to those measured in the olivine and pyroxenes of the volcanic products, do not result from shallow magma contamination, but rather reflect the composition of the mantle beneath the region. This is thought to have been metasomatised by the addition of crustal fluids from the subducted plate (Martelli et al. 2004; Peccerillo and Frezzotti 2015). The $\delta^{15}N$ of N_2, routinely analysed since 2007, points to a stable positive value of 6.3‰ ± 0.3‰ (Caliro et al. 2014), a value that is compatible with the nitrogen isotopic signature of the fluids emitted by volcanoes located in subduction zones (e.g., Guatemala volcanic arc; Fischer et al. 2002). Finally, the fumarolic He-Ar-N_2 relative compositions are similar to those discharged by the other volcanoes in the region (Vesuvio, Ischia; see Fig. 8 in Caliro et al. 2007) and point to magmatic fluid delivery from the metasomatised local mantle.

To summarise, consensus exists for a deep magmatic fluid feeding the hydrothermal system, followed by mixing with a shallower meteoric hydrothermal component. In this context, extensive scrubbing of hot magmatic gases within the hydrothermal environment would cause removal of original acidic species such as SO_2 (Cioni et al. 1984; Chiodini et al. 2001; Caliro et al. 2007). Sulphur gases are in fact present in detectable amount only as H_2S whose concentrations are controlled by reaction in the shallower part of the hydrothermal system (Chiodini et al. 2012). This hydrothermal control on sulphur species explains the very high C/S of the Solfatara fumaroles (100–200) with respect to ratios of 2 to 4 typical of magmatic fluids (Aiuppa et al. 2017).

4 Temperature–Pressure Estimation Based on Gas Equilibria

Since the 1980s, one of the major objectives of volcanic gas research at CF has been to extract information on temperature and pressure of the hydrothermal system from fumarolic composition (e.g., Cioni et al. 1984; Chiodini et al. 1992, 1996; Chiodini and Marini 1998). Early efforts were based upon resolving gas equilibria within the H_2O-H_2-CO_2-CO-CH_4 gas system, which has exhibited large variations during 1983–2016. A common conclusion reached by these initial studies, which is still valid today, is that the Solfatara fumaroles discharge fluids derived from (and equilibrating within) a vapour-dominated (liquid-free) zone. This distinctive vapour-dominated source area is in contrast with what found at the majority of the world's hydrothermal systems, whose surface gas manifestations are fed instead by boiling hot liquids (Chiodini and Marini 1998). Chiodini and Marini (1998) initially estimated an equilibrium temperature of 200–240 °C for the vapour-dominated gas equilibration zone underneath BG. Successively, Caliro et al. (2007) demonstrated, based on isotopic ($\delta^{13}C$ of coexisting CO_2 and CH_4) and chemical (CO_2-H_2O-CH_4) information, that CH_4-based geothermometers imply equilibrium temperatures (>360 °C) much higher than those returned by fast reacting species (H_2 and CO).

Ultimately, Caliro et al. (2007) questioned on the coupled use of species that, for kinetic reasons, equilibrate in separate zones of the hydrothermal system. Debate has also recently emerged on whether or not steam condensation (and consequent heating) takes place deep in the hydrothermal system (Moretti et al. 2017). Water removal from the gas phase, if occurring, would require reconsideration of the geothermometric-geobarometric approach (see discussion in Chiodini et al. 2015a).

In the attempt to clarify the potential role played by i) deep methane equilibration and ii) steam condensation, we here reconsider and update the log H_2/H_2O vs. log CO/CO_2 diagram (Fig. 4) originally used by Caliro et al. (2007). In the diagram, we distinguish the samples used by Caliro et al. (2007) (pre-2006 samples) from those acquired successively (post-2006 samples).

The pre-2006 Solfatara fumarole samples cluster along the theoretical composition expected from a typical hydrothermal redox buffer (D'Amore and Panichi 1980—DP redox buffer in Fig. 4, Campanian Volcanoes—CV redox buffer in Fig. 4; Chiodini and Marini 1998) at temperatures from 120 °C (Pisciarelli) to >200 °C (BG and BN). The same fumaroles also plot far from the distinctive field of high temperature "magmatic" fumaroles (SO_2-H_2S volcanic gas buffer; Giggenbach 1987). In Fig. 4, in order to illustrate the different behaviour of CH_4-based and H_2-CO-based geoindicators, we also report redox conditions (i.e., the log H_2/H_2O ratios) and equilibrium temperatures derived from CH_4-CO_2 isotopic equilibria (for those samples where carbon isotopic composition are available; Caliro et al. 2007; Bolognesi et al. 1986; Istituto Nazionale di Geofisica e Vulcanologia—Osservatorio Vesuviano unpublished data). The high temperatures estimated for these data-points ($T_{CH4\text{-}CO2}$ from 360 to 470 °C) are derived from the equilibrium carbon isotope exchange reaction between CH_4 and CO_2 (by using fractionation factors from Horita 2001). The log H_2/H_2O values are computed from the formation reaction of CH_4 from CO_2 that, expressed in terms of analytically detectable species H_2 and H_2O, is:

$$CO_2 + 4H_2 = CH_4 + 2H_2O \quad (1)$$

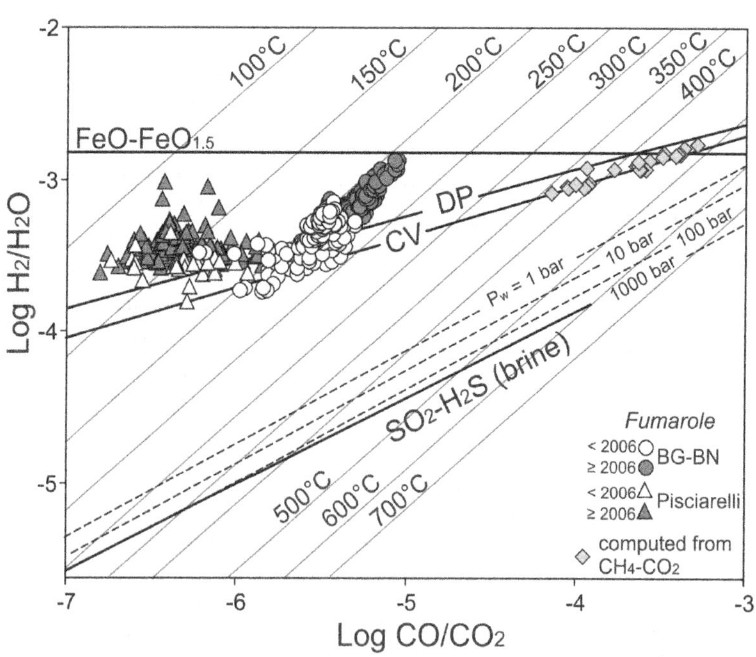

Fig. 4 Log H_2/H_2O vs Log CO/CO_2 diagram. The isotherms refer to equilibrium values for the H_2O-H_2-CO-CO_2 vapour phase. The theoretical values for redox condition typical of hydrothermal environments (FE, DP, CV) and volcanic gases (H_2S–SO_2, Giggenbach 1987) are also reported. Figure modified after Caliro et al. (2007)

Considering the dependence on temperature of the equilibrium constant of reaction 1 we can derive the following expression for the log H_2/H_2O:

$$\log H_2/H_2O = 0.25 \log CH_4/CO_2 - 0.5 \log f_{H_2O} + 2.416 - 2205.3/TK \quad (2)$$

Equation 2 is solved using (i) the analytical CH_4/CO_2 values, (ii) TK = $T_{CH4\text{-}CO2}$ + 273.15, and (iii) f_{H2O} fixed by the presence of a brine (log f_{H2O} = 4.9–1,820/TK; Giggenbach 1987). The estimated (CH_4-based) temperatures and log H_2/H_2O, plotted in Fig. 4, well illustrate the difference between the kinetically slow CH_4 and the fast-reactive species H_2 and CO: the first preserves indication on the deeper and hotter part of the system, while the latter re-equilibrate at shallower conditions upon gas ascent. It is worth noting that the CH_4-CO_2-derived temperatures and log H_2/H_2O align along the same CV and DP redox buffers that control redox conditions at near-surface (fumarolic vent) conditions (pre-2006 samples).

The post-2006 samples exhibit log H_2/H_2O ratio well above the pre-2006 samples and the theoretical ratios predicted from the DP and CV buffers. This is interpreted as due to steam condensation that has caused (since 2005–2006) the incondensable gases to increase relative to water (Chiodini et al. 2015a, b). This process, which is very efficient in heating the system, can also explain the increasing CO/CO_2 ratio (Fig. 4), the most suitable temperature geoindicator (Chiodini and Marini 1998). The recurrent episodes of condensate emission at Pisciarelli fumarole, observed since 2005–2006 (Chiodini et al. 2015a), are also direct evidence for condensation at depth.

Based on the considerations above, we estimate T-P of the shallowest part of the hydrothermal system from the functions listed in Table 1. These relationships do not consider deeply equilibrated methane and, being based on equimolar ratios between incondensable gases (Table 1; see also Chiodini et al. 2015a, 2016, 2017b), are unaffected by secondary processes involving H_2O (water condensation and/or addition).

The method again assumes the derivation of fumarolic fluids from a gas zone containing equilibrated vapour that, during its ascent, may be affected by either water addition or condensation. According to our calculations, the equilibrium temperatures and pressures of the gas equilibration zone (redox conditions are by the DP empirical function) oscillated in the 190–220 °C and 10–30 bar ranges during 1983–2010. More recently, they exhibited a gradual increase, reaching the maximum values (temperatures of ~240 °C and pressures of ~40 bar) in 2016, at the end of the observation period (note that these trends are still ongoing). We admit that choosing the DP redox buffer is somewhat arbitrary, but we stress that use of the CV buffer would only produce a shift of the computed trends to systematic higher estimated T (and P).

5 The Conceptual Geochemical Model and the Deep Resistivity Structure of the Hydrothermal System

The geochemical interpretation of the fumarolic composition in terms of P–T-redox conditions and liquid/gas phases, the mixing between meteoric and magmatic fluids, and the measured gas fluxes, all concur to define the conceptual model of the hydrothermal system feeding the Solfatara fumaroles. Geochemical information is sketched over a 2D resistivity model of Solfatara, derived by Audio Magneto Telluric measurements, which gives a realistic picture of the geometry of the system up to a depth of 2.5 km (Fig. 5). In particular, we here refer to the Solfatara portion of a longer, recently published, 2D resistivity model (Siniscalchi et al. 2019).

The image is dominated by the presence of a ~2 km elongated vertical structure of high resistivity in axis with the Solfatara fumaroles, and with top at 300–500 m depth. This is the core of the hydrothermal structure and, in our interpretation, represents a relatively high permeability zone that allows hot fluid ascent from the deeper to the shallower portions of the

Table 1 Assumptions and functions of the geobarometric-geothermometric approach (H_2O-H_2-CO_2-CO gas system)

Assumptions
Presence of an equilibrated gas phase in the 'gas equilibration zone'
Redox conditions fixed by the rock matrix (f_{O_2}-T function of D'Amore and Panichi 1980)
Saturated vapour (i.e., P_{H_2O} fixed by liquid–vapour coexistence)
Secondary processes can affect H_2O (i.e., water addition or vapour condensation)
Geobarometric and geothermometric functions
$T = 3{,}133.5/(0.933 - \mathrm{Log}\ XCO/X_{CO_2})$
$\mathrm{Log}\ P_{H_2O} = 5.51 - 2{,}048/TK$
$\mathrm{Log}\ P_{CO_2} = 3.025 + 201/TK - \mathrm{Log}\ X_{H_2}/X_{CO}$
$P_{tot} = P_{CO_2} + P_{H_2O}$

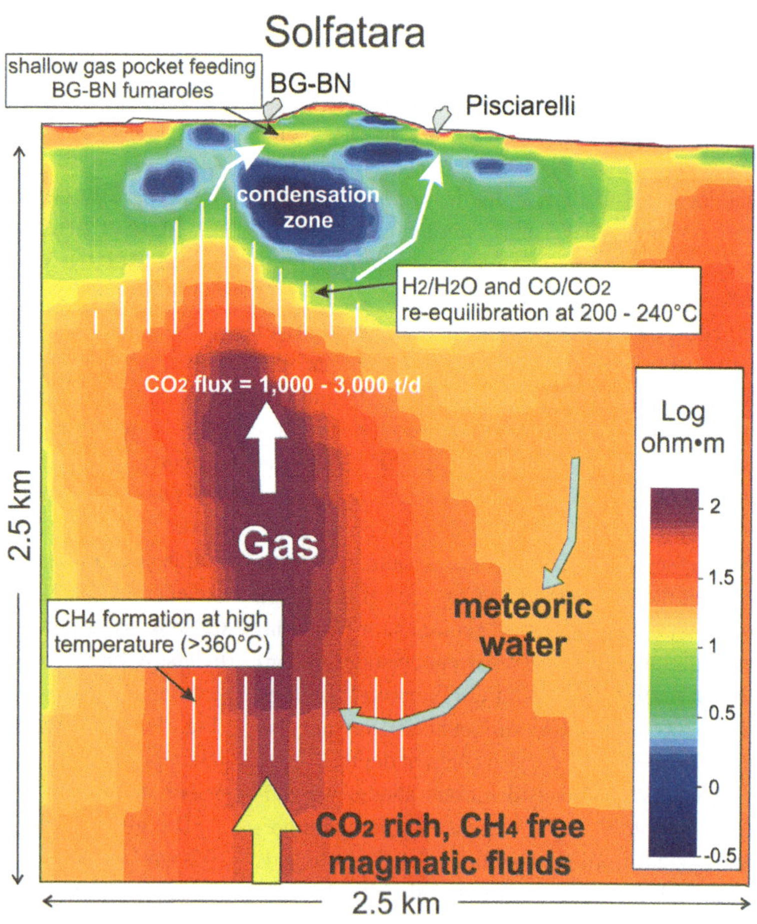

Fig. 5 Conceptual model of the hydrothermal system feeding the Solfatara fumaroles sketched over the 2D resistivity model of the Solfatara subsoil. Figure modified after Siniscalchi et al. (2019)

hydrothermal system. Relatively oxidised magmatic fluids (probably methane-free) enter the system at the bottom in a zone of very high temperature. For instance, temperatures as high as > 700 °C were recently modelled at depths of only 3–3.5 km (Amoruso et al. 2017). Here, the magmatic component mixes with and vaporises hydrothermal liquids of meteoric origin (Caliro et al. 2007), causing the more reducing redox conditions required for methane formation at the relatively high temperatures returned by the CH_4-based geoindicators (T > 360 °C). From that zone, the gas phase rises up to 300–500 m, where the resistive structure is interrupted by very conductive layers (cyan to blue colours), interpreted as reflecting hydrothermally altered zones (argillitic and phyllitic zones; Siniscalchi et al. 2019) and a liquid phase-dominated environment.

The upper part of the vertical permeable structure, covered by these impermeable layers, is likely the zone of H_2 and CO re-equilibration, i.e., the zone pertinent to our T-P estimations (gas equilibration zone). From that zone, the gas moves toward the surface through fractures and shallow gas pockets whose existence has been highlighted by detailed geoelectric surveys (Gresse et al. 2017, 2018). The main gas-ascent paths (white arrows in Fig. 5) connect the gas-equilibration zone directly to the gas pocket feeding the BG fumarole and, less directly, to the Pisciarelli fumaroles. It is during this transfer that condensation and/or addition of shallow waters may occur, modifying the composition of the original fluids.

6 Evolution of the Hydrothermal System During the Last 40 Years

The pulsed injection of magmatic fluids (IMF) into the hydrothermal system feeding the Solfatara fumaroles is thought to be one of the key processes that has controlled the evolution of the system during the last 40 years (see Chiodini et al. 2016 and references therein). During this time, the IMF episodes changed in frequency and intensity, ultimately causing the evident changes in degassing flux and geometry, and in the isotopic and chemical composition of the fumaroles (Fig. 6).

Chiodini et al. (2003) first demonstrated that both the large 1982–1984 bradyseismic crisis and the two mini-uplift episodes in 1989 and 1994 were followed, after some time, by evident peaks in the CO_2/H_2O ratio (Fig. 6a). Chiodini et al. (2003) were able to reproduce the shape and the timing of these peaks using a simple model that simulates the injection of a CO_2-rich gas phase (the magmatic component) into a porous media, ideally representing the hydrothermal system (see Fig. 6a and Fig. 5 in Chiodini et al. 2003).

In 2000, the pulsed behaviour of the fumarolic CO_2/H_2O ratio terminated and transitioned into a quasi-monotonic increasing trend (Fig. 6a). This led the fumarolic CO_2/H_2O ratio to increase from 0.15 to 0.18 in 2000 to ~0.4 in 2018–2019 (Fig. 6a). This large change has been interpreted as due to the combined action of (i) partial steam condensation, induced by the pressurisation of the system, (ii) CO_2 addition from decarbonation of hydrothermal calcite (Chiodini et al. 2015b), favoured by the heating of the system, and (iii) more frequent IMF events. That IMF events have frequently occurred in the last 20 years is testified by the recurrent peaks of the CO_2/CH_4 and He/CH_4 ratios (Fig. 6b; Chiodini 2009; Chiodini et al. 2012, 2015a, 2016).

Any magmatic fluid input at the base of the hydrothermal system causes, in fact, an increase of magmatic species such as CO_2 (and He) relative to CH_4, this latter being typically formed in hydrothermal environment (Chiodini 2009). These geochemically inferred IMF events are independently supported by seismicity clusters and pulsed deformations occurring few hundred days before the degassing anomaly is detected in the fumaroles (Fig. 7). Another key aspect to understand the 1990–2018 evolution of the CFc is the continuous decrease of the fumarolic N_2/He ratio (Fig. 6c). Considering the prevalent magmatic origin of both N_2 and He, and that the former (N_2) is the least soluble in primitive Campi Flegrei magmas, Caliro et al. (2014) interpreted the decreasing N_2/He ratios as caused by upward migration (lower pressure) of the gas–melt separation zone. Chiodini et al (2016) additionally argued that a lower magma degassing pressure would have resulted into a progressively more H_2O-rich magmatic gas phase, ultimately leading to the heating and

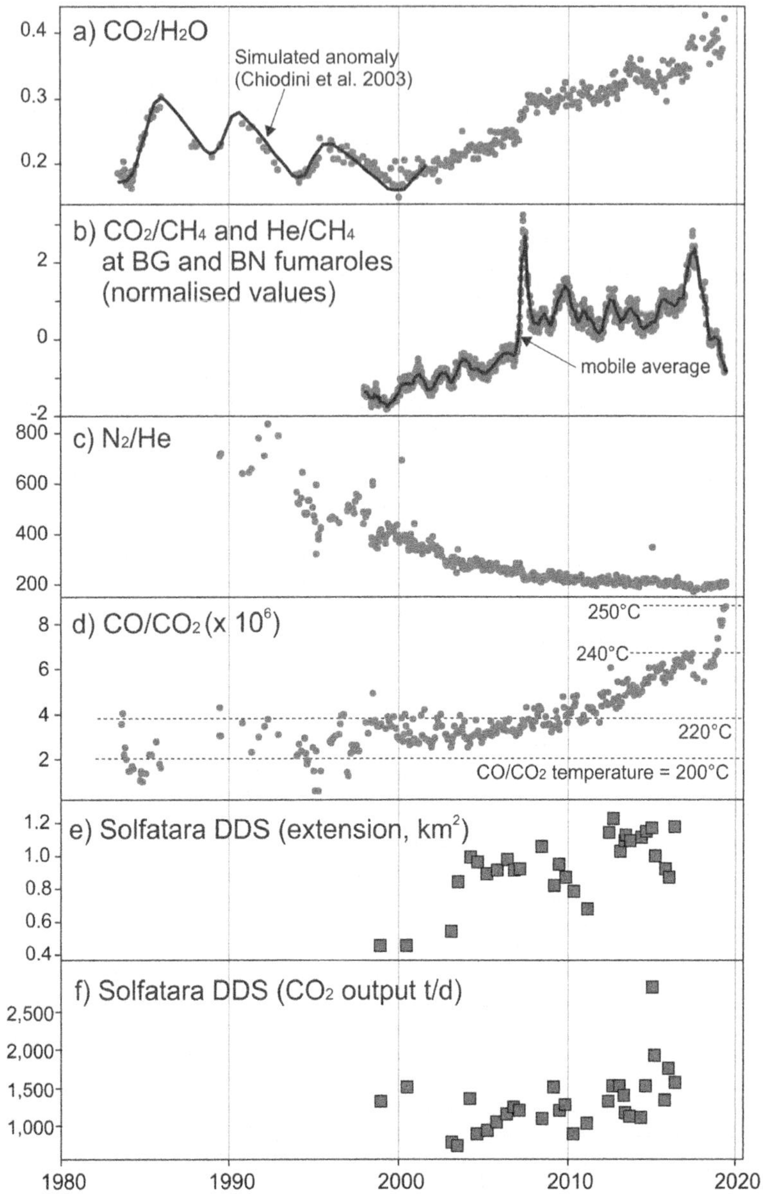

Fig. 6 Chronograms of different parameters in the 1983–2019 period. **a** CO_2/H_2O ratio. The dark grey line represents the values simulated by injecting CO_2 rich magmatic fluid at the base of the hydrothermal system (Chiodini et al. 2003). **b** CH_4 based geoindicators of magmatic fluids (Chiodini 2009). In order to compare the different signals, the measured data were normalised (standardised z-score) by removing the mean and dividing by their standard deviation (1998–2018 period). **c** N_2/He ratio (Caliro et al. 2014); **d** CO/CO_2 ratio. The isotherms refer to the $CO\text{-}CO_2$ temperatures (see Table 1); **e** extension of the Solfatara DDS; **f** CO_2 output from the Solfatara DDS (Cardellini et al. 2017). Figure modified after Chiodini (2009), Chiodini et al. (2003), Caliro et al. (2014), Cardellini et al. (2017)

pressurisation of the hydrothermal system. This is in line with results returned by fumarolic gas geoindicator for the last period of observation (e.g., CO/CO_2 ratio and temperature estimations in Fig. 6d, pressure estimations in Fig. 8d).

As noted above (cfr. Sect. 2), the post-2000 period has also been characterised by an enlargement of the Solfatara DDS, and by a remarkable increase of its CO_2 output. It is worth noting that, in addition to the DDS structure, CO_2 is also vigorously released by the fumarolic vents, especially at Pisciarelli (Aiuppa et al. 2013). At Pisciarelli, the main vent (Soffione) has shown a factor > 3 increase in CO_2 discharge since 2012, reaching in 2018–2019 levels (500–600 tons/day) that are similar to the CO_2 flux associated to medium-sized erupting arc volcanoes (Fig. 8d; Tamburello et al. 2019).

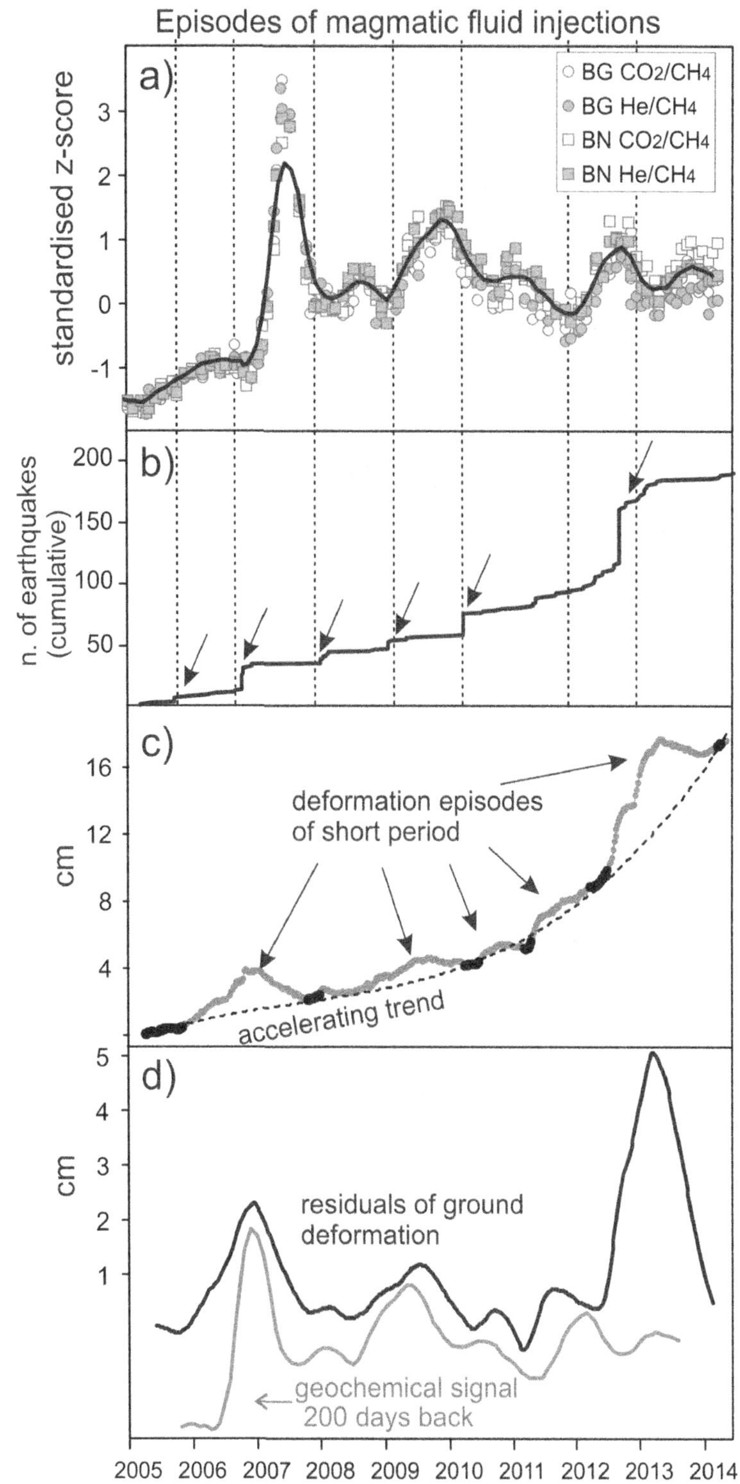

Fig. 7 **a** Geochemical signals; **b** cumulative number of Campi Flegrei earthquakes, and **c** deformation path (baseline length variation between the GPS stations ACAE and ARFE). **d** comparison of the residual of ground deformation path with respect to a polynomial fit with the geochemical signal. Figure modified after Chiodini et al. (2015a)

7 Conclusive Remarks

We have reviewed here the current state-of-the-art in our understanding of the CFc hydrothermal setting and behaviour. The emerging scenario is that of a very dynamic, rapidly evolving hydrothermal system. Results strongly suggest that the ongoing (since 2000) CFc unrest is triggered by a degassing magma source, but also that the system's response is strongly modulated by dynamics and structures of the overlying hydrothermal envelope (Fig. 8). The vertical displacement time-series (Fig. 8a, maximum values measured at RITE GPS station) clearly point to ongoing uplift since the first years of this millennium. Of particular relevance has been the 2012–2013 crises, during which ground deformation approached a power-law growth rate (red arrow in Fig. 8a), a fact that (with other concurrently observed signals) contributed to the

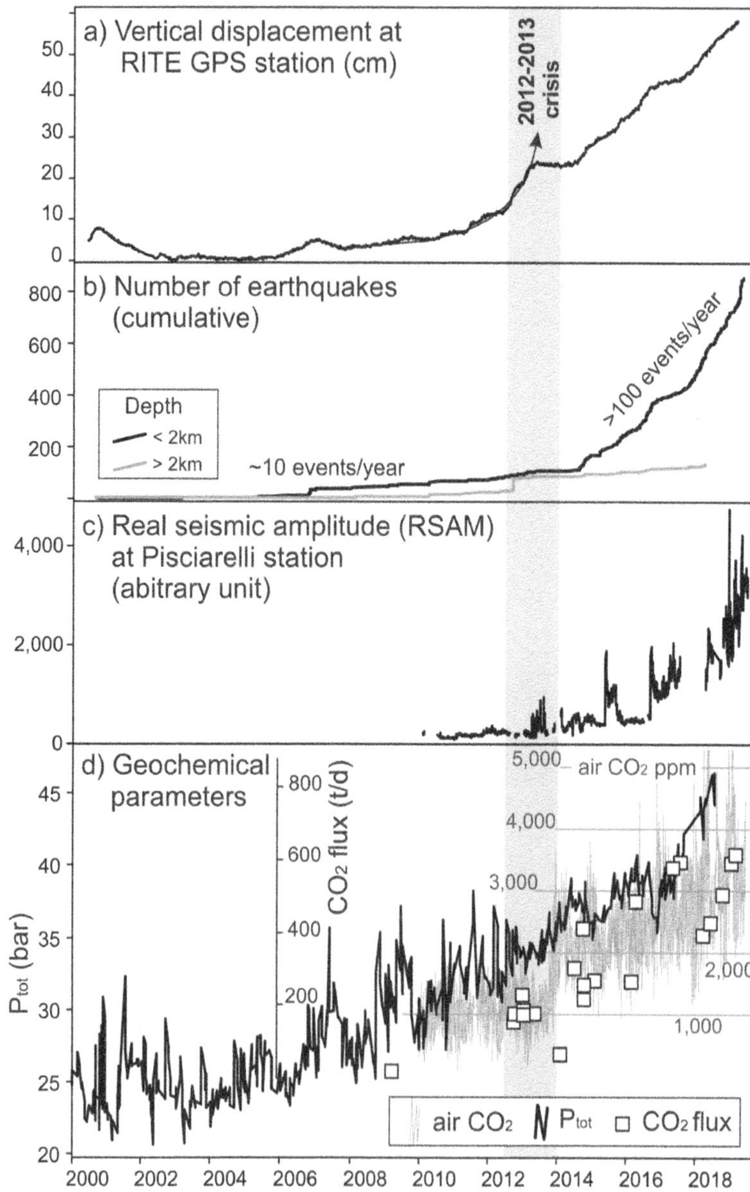

Fig. 8 Geophysical and geochemical signals at the Campi Flegrei caldera in the 2000–2019 period (modified after Giudicepietro et al. 2019). **a** Maximum vertical displacement at CFc (Tamburello et al. 2019); **b** Number of earthquakes (Giudicepietro et al. 2019). Shallower earthquakes (depth < 2 km) clustered in an area of few km^2 centred at Solfatara; **c** RSAM registered by the Pisciarelli seismic station (Chiodini et al. 2017a; Giudicepietro et al. 2019); **d** Geochemical parameters include: fluid pressure estimated at the Solfatara fumaroles (see Table 1), CO_2 fluxes measured at Pisciarelli vent (Tamburello et al. 2019; Aiuppa et al. 2015), air CO_2 concentration (grey line) automatically measured by a station sited at Pisciarelli (Tamburello et al. 2019)

decision taken by the Italian Dipartimento di Protezione Civile of raising the CFc alert level from green (quiet) to yellow (attention). The 2012–2013 crisis fortunately did not culminate into an eruption, yet may have represented a turning point in the unrest. In 2014, the number of shallow earthquakes (i.e., within the hydrothermal system feeding the Solfatara manifestation) has increased by a factor > 10 (Fig. 8b, Giudicepietro et al. 2019), concurrently with unequivocal signs of escalating hydrothermal activity (Fig. 6c, d). At Pisciarelli, the fumarolic CO_2 output has grown by a factor of 3 in only a few years (Fig. 8d). This increasing CO_2 degassing has been paralleled by strikingly similar temporal evolutions of (i) fluid pressure (Fig. 8d), (ii) air CO_2 concentrations in the surroundings of the vent (Tamburello et al. 2019; Chiodini et al. 2017b), and (iii) seismic tremor sourced within the vent (Fig. 8c, d; Chiodini et al. 2017b; Giudicepietro et al. 2019). This evolution clearly requires carefully scientific scrutiny and intensified monitoring in the years to come.

References

Aiuppa A, Avino R, Brusca L, Caliro S, Chiodini G, D'Alessandro W, Favara R, Federico C, Ginevra W, Inguaggiato S, Longo M, Pecoraino G, Valenza M (2006) Mineral control of arsenic content in thermal waters from volcano-hosted hydrothermal systems: insights from island of Ischia and Phlegrean fields (Campanian Volcanic Province, Italy). Chem Geol 229:313–330

Aiuppa A, Tamburello G, Di Napoli R, Cardellini C, Chiodini G, Giudice G, Grassa F, Pedone M (2013) First observations of the fumarolic gas output from a restless caldera: Implications for the current period of unrest (2005–2013) at Campi Flegrei. Geochem Geophys Geosys 14:4153–4169

Aiuppa A, Fiorani L, Santoro S, Parracino S, Nuvoli M, Chiodini G, Tamburello G (2015) New ground-based lidar enables volcanic CO_2 flux measurements. Sci Rep 5:13614

Aiuppa A, Fischer TP, Plank T, Robidoux P, Di Napoli R (2017) Along-arc, inter-arc and arc-to-arc variations in volcanic gas CO_2/S_T ratios reveal dual source of carbon in arc volcanism. Earth-Sci Rev 168:24–47

Allard P, Maiorani A, Tedesco D, Cortecci G, Turi B (1991) Isotopic study of the origin of sulfur and carbon in Solfatara fumaroles, Campi Flegrei caldera. J Volcanol Geotherm Res 48(1–2):139–159

Allard P, Jean-Baptiste P, D'Alessandro W, Parello F, Parisi B, Flehoc C (1997) Mantle-derived helium and carbon in groundwaters and gases of Mount Etna. Italy. Earth Planet Sci Lett 148(3–4):501–516

Amoruso A, Crescentini L, D'Antonio M, Acocella V (2017) Thermally-assisted magma emplacement explains restless calderas. Sci Rep 7:7948

Barberi F, Corrado G, Innocenti F, Luongo G (1984) Phlegrean Fields 1982–1984: brief chronicle of a volcano emergency in a densely populated area. Bull Volcanol 47:175–185

Burton MR, Sawyer GM, Granieri D (2013) Deep carbon emissions from volcanoes. Carbon in earth. Mineral Soc Am and Geochem Soc, Rev Mineral Geochem 75:323–354

Bolognesi L, Noto P, Nuti S (1986) Studio chimico ed isotopico della solfatara di Pozzuoli: ipotesi sull'origine e sulle temperature profonde dei fluidi. Rend Soc It Mineral Petrol 41(2):281–295

Caliro S, Chiodini G, Moretti R, Avino R, Granieri D, Russo M, Fiebig J (2007) The origin of the fumaroles of La Solfatara (Campi Flegrei, South Italy). Geochim Cosmochim Acta 71:3040–3055

Caliro S, Chiodini G, Paonita A (2014) Geochemical evidences of magma dynamics at Campi Flegrei (Italy). Geochim Cosmochim Acta 132:1–15

Caprarelli G, Tsutsumi M, Turi B (1997) Chemical and isotopic signatures of the basement rocks from the Campi Flegrei geothermal field, Naples, southern Italy: inferences about the origin and evolution of its hydrothermal fluids. J Volcanol Geotherm Res 76:63–82

Carapezza M, Nuccio PM, Valenza M (1984) Geochemical surveillance of the Solfatara of Pozzuoli (Phlegrean Fields) during 1983. Bull Volcanol 47:303–311

Cardellini C, Chiodini G, Frondini F (2003) Application of stochastic simulation to CO_2 flux from soil: Mapping and quantification of gas release. J Geophys Res 108:2425

Cardellini C, Chiodini G, Frondini F, Avino R, Bagnato E, Caliro S, Lelli M, Rosiello A (2017) Monitoring diffuse volcanic degassing during volcanic unrests: The case of Campi Flegrei (Italy). Sci Rep 7:6757

Celico P, Dall'Aglio M, Ghiara MR, Stanzione D, Brondi M, Prosperi M (1992) Geochemical monitoring of the thermal fluids in the Phlegrean Fields from 1970 to 1990. Boll Soc Geol Ital 111:409–422

Chiodini G (2009) CO_2/CH_4 ratio in fumaroles a powerful tool to detect magma degassing episodes at quiescent volcanoes. Geophys Res Lett 36:L02302

Chiodini G, Marini L (1998) Hydrothermal gas equilibria; the H_2O-H_2-CO_2-CO-CH_4 system. Geochim Cosmochim Acta 62(15):2673–2687

Chiodini G, Cioni R, Guidi M, Marini L, Raco B, Taddeucci G (1992) Gas geobarometry in boiling hydrothermal systems: A possible tool to evaluate the hazard of hydrothermal explosions. Acta Vulcanol 2:99–107

Chiodini G, Cioni R, Magro G, Marini L, Panichi C, Raco B, Russo M (1996) Chemical and isotopic variations of Bocca Grande fumarole (Solfatara volcano, Phlegrean Fields). Acta Vulcanol 8:129–138

Chiodini G, Cioni R, Guidi M, Raco B, Marini L (1998) Soil CO_2 flux measurements in volcanic and geothermal areas. Appl Geochem 13:543–552

Chiodini G, Allard P, Caliro S, Parello F (2000) ^{18}O exchange between steam and carbon dioxide in volcanic and hydrothermal gases; implications for the source of water. Geochim Cosmochim Acta 64 (14):2479–2488

Chiodini G, Frondini F, Cardellini C, Granieri D, Marini L, Ventura G (2001) CO_2 degassing and energy release at Solfatara Volcano, Campi Flegrei. Italy. J Geophys Res 106(8):16213–16221

Chiodini G, Todesco M, Caliro S, Del Gaudio C, Macedonio G, Russo M (2003) Magma degassing as a trigger of bradyseismic events; the case of Phlegrean Fields (Italy). Geophys Res Lett 30(8):1434

Chiodini G, Granieri D, Avino R, Caliro S, Costa A, Werner C (2005) Carbon dioxide diffuse degassing and estimation of heat release from volcanic and hydrothermal systems. J Geophys Res 110:B08204

Chiodini G, Caliro S, Cardellini C, Avino R, Granieri D, Schmidt A (2008) Carbon isotopic composition of soil CO_2 efflux, a powerful method to discriminate different sources feeding soil CO_2 degassing in volcanic-hydrothermal areas. Earth Planet Sci Lett 274:372–379

Chiodini G, Caliro S, Cardellini C, Granieri D, Avino R, Baldini A, Donnini M, Minopoli C (2010) Long-term variations of the Campi Flegrei, Italy, volcanic system as revealed by the monitoring of hydrothermal activity. J Geophys Res 115:B03205

Chiodini G, Caliro S, De Martino P, Avino R, Gherardi F (2012) Early signals of new volcanic unrest at Campi Flegrei caldera? Insights from geochemical data and physical simulations. Geology 40:943–946

Chiodini G, Vandemeulebrouck J, Caliro S, D'Auria L, De Martino P, Mangiacapra A, Petrillo Z (2015a) Evidence of thermal driven processes triggering the 2005–2014 unrest at Campi Flegrei caldera. Earth Planet Sci Lett 414:58–67

Chiodini G, Pappalardo L, Aiuppa A, Caliro S (2015b) The geological CO_2 degassing history of a long-lived caldera. Geology 43:767–770

Chiodini G, Paonita A, Aiuppa A, Costa A, Caliro S, De Martino P, Acocella V, Vandemeulebrouck J (2016) Magmas near the critical degassing pressure drive volcanic unrest towards a critical state. Nature Comm 7:13712

Chiodini G, Selva J, Del Pezzo E, Marsan D, De Siena L, D'Auria L, Bianco F, Caliro S, De Martino P, Ricciolino P, Petrillo Z (2017a) Clues on the origin of post-2000 earthquakes at Campi Flegrei caldera (Italy). Sci Rep 7:4472

Chiodini G, Giudicepietro F, Vandemeulebrouck J, Aiuppa A, Caliro S, De Cesare W, Tamburello G, Avino R, Orazi M, D'Auria L (2017b) Fumarolic tremor and geochemical signals during a volcanic unrest. Geology 45(12):1131–1134

Cioni R, Corazza E, Marini L (1984) The gas/steam ratio as indicator of heat transfer at the Solfatara fumaroles, Phlegraean Fields (Italy). Bull Volcanol 47:295–302

D'Amore F, Panichi C (1980) Evaluation of deep temperature of hydrothermal systems by a new gas-geothermometer. Geochim Cosmochim Acta 44:549–556

Deines P, Gold DP (1973) The isotopic composition of carbonatite and kimberlite carbonates and their bearing on the isotopic composition of deep-seated carbon. Geochim Cosmochim Acta 37:1709–1733

Del Gaudio C, Aquino I, Ricciardi GP, Ricco C, Scandone R (2010) Unrest episodes at Campi Flegrei: A reconstruction of vertical ground movements during 1905–2009. J Volcanol Geotherm Res 195:48–56

Deutsch CV, Journel AG (1998) GSLIB: Geostatistical Software Library and Users Guide, vol 136. Oxford University Press

De Vivo B, Belkin HE, Barbieri M, Chelini W, Lattanzi P, Lima A, Tolomeo L (1989) The Campi Flegrei (Italy) geothermal system: A fluid inclusion study of the Mofete and San Vito fields. J Volcanol Geotherm Res 36:303–326

Di Luccio F, Pino NA, Piscini A, Ventura G (2015) Significance of the 1982–2014 Campi Flegrei seismicity: Preexisting structures, hydrothermal processes, and hazard assessment. Geophys Res Lett 2:7498–7506

Di Napoli R, Aiuppa A, Sulli A, Caliro S, Chiodini G, Acocella V, Ciraolo G, Di Vito MA, Interbartolo F, Nasello C, Valenza M (2016) Hydrothermal fluid venting in the offshore sector of Campi Flegrei caldera: a geochemical, geophysical, and volcanological study. Geochem Geophys Geosyst 17:4153–4178

Fischer TP, Hilton DR, Zimmer MM, Shaw AM, Sharp ZD, Walker JA (2002) Subduction and recycling of nitrogen along the Central American Margin. Science 297:1154–1157

Giggenbach WF (1987) Redox processes governing the chemistry of fumarolic gas discharges from White Island. New Zealand. Appl Geochem 2(2):143–161

Giggenbach WF (1992) Isotopic shifts in waters from geothermal and volcanic systems along convergent plate boundaries and their origin. Earth Planet Sci Lett 113(4):495–510

Giudicepietro F, Macedonio G, Martini M (2017) A physical model of sill expansion to explain the dynamics of unrest at calderas with application to Campi Flegrei. Front Earth Sci 5:54

Giudicepietro F, Chiodini G, Caliro S, De Cesare W, Esposito AM, Galluzzo D, Lo Bascio D, Macedonio G, Orazi M, Ricciolino P, Vandemeulebrouck J (2019) Insight into Campi Flegrei caldera unrest through seismic tremor measurements at Pisciarelli fumarolic field. Geochem Geophys Geosys 20:5544–5555

Guglielminetti M (1986) The Mofete geothermal field. Geothermics 15:781–790

Gresse M, Vandemeulebrouck J, Byrdina S, Chiodini G, Revil A, Johnson TC, Ricci T, Vilardo G, Mangiacapra A, Lebourg T, Grangeon J, Bascou P, Metralet L (2017) Three-dimensional electrical resistivity tomography of the Solfatara crater (Italy): Implication for the multiphase flow structure of the shallow hydrothermal system. J Geophys Res 122:8749–8768

Gresse M, Vandemeulebrouck J, Byrdina S, Chiodini G, Roux P, Rinaldi AP, Wathelet M, Ricci T, Letort J, Petrillo Z, Tuccimei P, Lucchetti C, Sciarra A (2018) Anatomy of a fumarolic system inferred from a multiphysics approach. Sci Rep 8(1):7580

Horita J (2001) Carbon isotope exchange in the system CO_2-CH_4 at elevated temperatures. Geochim Cosmochim Acta 65:1907–1919

Isaia R, Vitale S, Di Giuseppe MG, Iannuzzi E, D'Assisi Tramparulo F, Troiano A (2015) Stratigraphy, structure, and volcano-tectonic evolution of Solfatara maar-diatreme (Campi Flegrei, Italy). Geol Soc Am Bull 127:1485–1504

Martini M, Giannini L, Buccianti A, Prati F, Legittimo PC, Iozzelli P, Capaccioni B (1991) 1980–1990: ten years of geochemical investigation at Phlegrean Fields (Italy). J Volcanol Geotherm Res 48(1):161–171

Martelli M, Nuccio PM, Stuart FM, Burgess R, Ellam RM, Italiano F (2004) Helium strontium isotopic constrains on mantle evolution beneath the Roman Comagmatic Province, Italy. Earth Planet Sci Lett 224:295–308

Moretti R, De Natale G, Troise C (2017) A geochemical and geophysical reappraisal to the significance of the recent unrest at Campi Flegrei caldera (Southern Italy): Geochem Geophys Geosys 18:1244–1269

Panichi C, Volpi G (1999) Hydrogen, oxygen and carbon isotope ratios of Solfatara fumaroles (Phlegrean Fields, Italy): further insight into source processes. J Volcanol Geotherm Res 91(2–4):321–328

Peccerillo A, Frezzotti ML (2015) Magmatism, mantle evolution and geodynamics at the converging plate margins of Italy. J Geol Soc 172:407–427

Penta F (1954) Ricerche e studi sui fenomeni esalativo idrotermali ed il problema delle forze endogene. Ann Geophys 8(3):1–94

Petrillo Z, Chiodini G, Mangiacapra A, Caliro S, Capuano P, Russo G, Cardellini C, Avino R (2013) Defining a 3D physical model for the hydrothermal circulation at Campi Flegrei caldera (Italy). J Volcanol Geotherm Res 264:172–182

Rosi M, Sbrana A (1987) Phlegraean Fields. Quaderni de "La Ricerca Scientifica" 114, vol 9. CNR, Rome, Italy, pp 1–175

Siniscalchi A, Tripaldi S, Romano G, Chiodini G, Improta L, Petrillo Z, D'Auria L, Caliro S, Avino R (2019) Reservoir structure and hydraulic properties of the Campi Flegrei geothermal system inferred by audiomagnetotelluric, geochemical, and seismicity study. J Geophys Res 124:5336–5356

Smith VC, Isaia R, Pearce NJG (2011) Tephrostratigraphy and glass compositions of post-15 kyr Campi Flegrei eruptions: implications for eruption history and chronostratigraphic markers. Quat Sci Rev 30:3638–3660

Tamburello G, Caliro G, Chiodini G, DeMartino P, Avino R, Minopoli C, Carandente A, Rouwet D, Aiuppa A, Costa A, Bitetto A, Giudice G, Francofonte V, Ricci T, Sciarra A, Bagnato E, Capecchiacci F (2019) Escalating CO_2 degassing at the Pisciarelli fumarolic system, and implications for the ongoing Campi Flegrei unrest. J Volcanol Geotherm Res 384:151–157

Tedesco D (1997) Systematic variations in the $^3He/^4He$ ratio and carbon of fumarolic fluids from active volcanic areas in Italy: Evidence for radiogenic ^4He and crustal carbon addition by the subducting African Plate? Earth Planet Sci Lett 151(3–4):255–269

Tedesco D, Scarsi P (1999) Chemical (He, H_2, CH_4, Ne, Ar, Na) and isotopic (He, Ne, Ar, C) variations at the Solfatara crater (southern Italy): mixing of different sources in relation to seismic activity. Earth Planet Sci Lett 171(3):465–480

Tedesco D, Allard P, Sano Y, Wakita H, Pece R (1990) ^3He in Subaerial and Submarine Fumaroles of Campi-Flegrei Caldera. Italy. Geochim Cosmochim Acta 54 (4):1105–1116

Valentino GM, Stanzione D (2003) Source processes of the thermal waters from the Phlegrean Fields (Naples, Italy) by means of the study of selected minor and trace element distribution. Chem Geol 194:245–274

Valentino GM, Stanzione D (2004) Geochemical monitoring of the thermal waters of the Phlegrean Fields. J Volcanol Geotherm Res 133:261–289

Valentino GM, Cortecci G, Franco E, Stanzione D (1999) Chemical and isotopic compositions of minerals and waters from the Campi Flegrei volcanic system, Naples, Italy. J Volcanol Geotherm Res 91:329–344

Historic Unrest of the Campi Flegrei Caldera, Italy

Roberto Scarpa, Francesca Bianco, Paolo Capuano, Mario Castellano, Luca D'Auria, Bellina Di Lieto, and Pierdomenico Romano

Abstract

After some centuries of subsidence, following the AD 1538 Monte Nuovo last eruption, the Campi Flegrei caldera has shown unrest episodes since at least 1950. The first uplift episode dates back to 1950–1952 and amounted to 73 cm, without any report or record of seismic activity. Two strong inflation episodes occurred in 1970–1972 and 1982–1984. The first accompanied by moderate low seismicity, with only few events felt by the population, whereas the second was accompanied by relatively intense swarms of volcano-tectonic earthquakes, reaching up to magnitude 4. The seismic activity caused alarm in the population and a spontaneous nightly evacuation of part of the town of Pozzuoli (44,000 inhabitants). Since this last episode, subsidence has been recorded for several years, interrupted by some mini-uplift events, lasting several weeks and accompanied by seismic swarms of low-magnitude volcano-tectonic events. In recent years, high sensitivity instruments have been installed to detect slow earthquake transients and other mechanical/temperature low-intensity precursory signals. Since late 2004 another moderate uplift is occurring at very small rate, amounting to about 1–2 cm/year, accompanied by long-period events. This uplift is different from the past mini-uplift events due to its duration. This work summarises all seismic and ground deformation data as well as the models proposed to interpret these phenomena, suggesting possible methods for detecting precursors of future eruptive activity in the area.

1 Background

The Campi Flegrei caldera (CFc) is a nested and resurgent structure resulting from two major collapses related to the Campanian Ignimbrite (~39 ka) and the Neapolitan Yellow Tuff (~15 ka) catastrophic eruptions (Orsi et al. 1992, 1996; Chap. Volcanic and Deformation History of the Campi Flegrei Volcanic Field, Italy). The volcanism has been fed by a complex magmatic system. This system includes a deep (~8 km depth) and widespread reservoir (e.g., Zollo et al. 2008) that is fed by mantle derived magmas and, in turns, feeds shallower multiple magma pockets tapped by the eruptions of the past 15 kyrs

R. Scarpa (✉) · P. Capuano
Dipartimento di Fisica, Università degli Studi di Salerno, Salerno, Italy
e-mail: r.scarpa@unisa.it

F. Bianco · M. Castellano · B. Di Lieto · P. Romano
Istituto Nazionale di Geofisica e Vulcanologia, Sezione Osservatorio Vesuviano, Napoli, Italy

L. D'Auria
Instituto Volcanológico de Canarias (INVOLCAN), Tenerife, Spain

© Springer-Verlag GmbH Germany, part of Springer Nature 2022
G. Orsi et al. (eds.), *Campi Flegrei*, Active Volcanoes of the World,
https://doi.org/10.1007/978-3-642-37060-1_10

(Arienzo et al. 2010; Di Renzo et al. 2011; Moretti et al. 2013; Chaps. An Evolutionary Model for the Magmatic System of the Campi Flegrei Volcanic Field (Italy) Constrained by Petrochemical Data; Origin and Differentiation History of the Magmatic System Feeding the Campi Flegrei Volcanic Field (Italy) Constrained by Radiogenic and Stable Isotope Data). Above the magmatic system, a hydrothermal system has developed and is testified by diffuse soil degassing, and fumarole and hot spring activity (e.g., Chiodini et al. 2010; Chap. The Hydrothermal System of the Campi Flegrei Caldera, Italy).

The youngest caldera has been the site of intense volcanism and deformation that have generated about 70 eruptions and a resurgence with maximum net uplift of the La Starza marine terrace of about 90 m (Cinque et al. 1985; Di Vito et al. 1999; Orsi et al. 2004). The last eruption occurred in AD 1538 and generated the Monte Nuovo cone. Historical documents recently revised by Guidoboni and Ciuccarelli (2011), highlight the occurrence of felt earthquakes since about 70 years before the eruption. The caldera is the site of short-term deformation episodes, known also as bradyseismic events, which have been documented since the fifties of the last century (Del Gaudio et al. 2010) and have been interpreted as transient episodes within the long-term deformation related to the resurgence of the caldera (Orsi et al. 1999). However, no seismographs have been installed in the region before the uplift episode recorded in 1970. The renewal of ground uplift occurred in early 1970 caused a strong debate between Italian and French volcanologists, due to the forced evacuation of a part of Pozzuoli (Imbò 1977, 1980; Tazieff 1971, 1977). Parts of the documentation of the debate, which can be found in the Smithsonian Bulletin Reports (https://volcano.si.edu/volcano.cfm?vn=211010), are reported in the following.

On March 4, 1970, Imbò stated: "Ten new volcanic fissures have opened in the "Campi Flegrei" near the city of Pozzuoli (a suburb of Naples), which is being pushed upwards. The fissures showed a recrudescence of the volcanic activity which has always gone on in the area. The sudden raising of the Pozzuoli area has caused damage and fracturing of houses and walls. One church and two schools were recently closed. A levelling carried out on 6 February 1970 has shown that the floor of Serapaeum at Pozzuoli has risen about 80 cm with respect to the former levelling carried out in October 1968. The floor of Serapaeum had been sinking since the last eruption in the "Campi Flegrei" occurred in 1538, the average rate of sinking being 13 mm per year. Plans are being made to evacuate the inhabitants of the city if necessary." Statements concerning "new volcanic fissures", "recrudescence of volcanic activity", and "15 cm sinking of parts of the Isle of Ischia" were attributed to Giuseppe Imbò, director of the Osservatorio Vesuviano, by United Press International (UPI). Imbò claimed that these statements were "not entirely correct".

Haroun Tazieff (IPG, Paris) led a team which operated in the area between March 6 and 15, 1970 with the aim of investigating the Pozzuoli uplift. The results of the investigations of those researchers contradicted several important statements made by sources in the event area. The following was declared by Tazieff on April 21, 1970. "No volcanic fissures did open in Campi Flegrei. No recrudescence of the volcanic activity was actually detectable in this area". Tazieff's team could not obtain any valuable proof of the 20 cm uplift of Pozzuoli that allegedly occurred between March 4 and 5 and did not find any seismogram evidence of the ten shocks that allegedly occurred off the Pozzuoli coast. In fact, during their ten-day investigation not a single seismic event was recorded by their network of sensitive seismographs. Inquiry of local fishermen proved that no cooked fish or burnt nets were found in the area.

"Absolutely no other unusual phenomenon could be detected, neither by seismology, nor geochemistry, geomagnetism, heat, or temperature measurements. An important set of geophysical and geochemical investigations, both on land and at sea, are underway since late March and shall be carried on as long as necessary. A permanent station will be installed. Although an eruption can happen at any time and at any place both in Phlegrean Fields and Vesuvius area, absolutely no preliminary manifestations appeared to date in this area."

The reading of these documents well expresses now, after more than 50 years, the uncertainty in the interpretation of the phenomenon, which appeared quite contradictory and confuse, in spite of the relevant uplift. The absence of a monitoring network and a well-established group of volcanologists contributed to further increase of the uncertainty and led to the forced evacuation of a part of the town of Pozzuoli.

2 Seismicity at Campi Flegrei Caldera

2.1 The 1970 Seismic Activity

This activity has been documented by two small nets of seismographs installed in 1970 by the Osservatorio Vesuviano and the GEOLAB of the Italian Consiglio Nazionale delle Ricerche since the earliest evidence of uplift. The recorded activity showed swarms of volcano-tectonic (VT) earthquakes, accompanied by other events of artificial origin. In fact, in that period the seismic activity appeared to be dominated by events occurring in the Pozzuoli Bay, and certainly due to use of explosives for illegal fishing. This was indicated by: (i) the uneven hourly distribution of earthquakes, which was peaked during the night; (ii) the location of the earthquakes generally in the Pozzuoli Bay, close to the piers; (iii) the typical waveforms with compressional onsets containing surface waves trapped in the water (R. Scarpa, internal report of the Osservatorio Vesuviano). The total number of VT events and their energy was moderate, amounting to few hundred earthquakes, with maximum magnitudes smaller than 2. Reliable faulting mechanisms estimates are not available for these events. The activity reported after the uplift episode is dominated by artificial blasts until the onset of the next inflation event occurred at the end of 1982. The earthquakes reported as B-type, according to Minakami (1960), which correspond to those presently defined as long-period (LP) events, also seem to have an artificial origin. This activity was only recorded by the station in the area of the Pozzuoli harbour and was probably linked to oscillation of a nearby underwater pipeline forced by sea waves during storms. However, hypocentre locations reported by Corrado et al. (1977) show few VT earthquakes near the Solfatara crater. The mapped epicentres are quite few, with less than 50 located events (Orsi et al. 1999). Unfortunately, all the records, stored on magnetic tapes with the frequency-modulation system have been irremediably lost and no backup on other devices exist.

2.2 The 1982–1984 Seismic Activity

The 1982–1984 seismic sequence (Fig. 1) occurred during the so called "1982–1984 bradyseismic crisis" is the most important among those recorded in the last 70 years due to its energy that dominates the entire dataset. It was initially documented by the seismic network of the Osservatorio Vesuviano, consisting of up to 8 short-period seismometers and by a network of up to 14 seismometers, maintained by Aquater to monitor the geothermal field that was under geophysical exploitation with the drilling of some deep boreholes. Due to the increase of the seismicity in 1993, the seismic network operating in the Campi Flegrei area was improved up to 25 analog short-period seismic stations (11 operated by the Osservatorio Vesuviano and 14 by the Aquater). This analog seismic network was further improved with a temporary digital network operated by the University of Wisconsin in cooperation with the Osservatorio Vesuviano (Aster and Meyer 1988). This digital network was composed of 13 three-component seismic stations and operated from mid-January to April 1984 (Capuano et al. 2006). It produced high quality seismic data, which presently constitute the only dataset still available for analysis (Barberi et al. 1984; Aster et al. 1992). This most relevant 3D seismic waveforms database for the CFc was recently reconstructed (Capuano et al. 2006). It contains more than one thousand microearthquakes that occurred in 1984, and the whole waveform database has been re-picked. Most of the scientific literature on CFc still refers to these data for the analysis of source mechanisms of local earthquakes (De Natale et al. 1995), and for the generation of tomographic images of the velocity and attenuation structure (Vanorio et al. 2005; De Siena et al. 2010). Almost all the seismic activity consisted of VT events. A search for LP events is presently carried out but their number, if proven, will be negligible.

In 1982 this activity was characterised by few low-magnitude events until November 2 when a short swarm of 17 events with M_D max = 2.3, occurred. During 1983, the seismic activity accompanying the ground uplift became continuous and intense with several swarms of micro-earthquakes (Fig. 1b). The first "more energetic" earthquake occurred on May 15 (M_D = 3.4). Since then, the pattern of seismicity changed remarkably with a sequence of seismic swarms in September (more than 1,000 events with M_D > 0.2), October (more than 300 earthquakes with M_D between 0.2 and 3.0) and December (four events with M_D > 3.0 and M_D max = 3.8) of the same year. A shock of M_D = 4.0 was recorded on October 4, 1983 (Orsi et al. 1999). In 1984, the seismic activity increased. A swarm of more than 500 events (M_D max = 3.3) occurred on

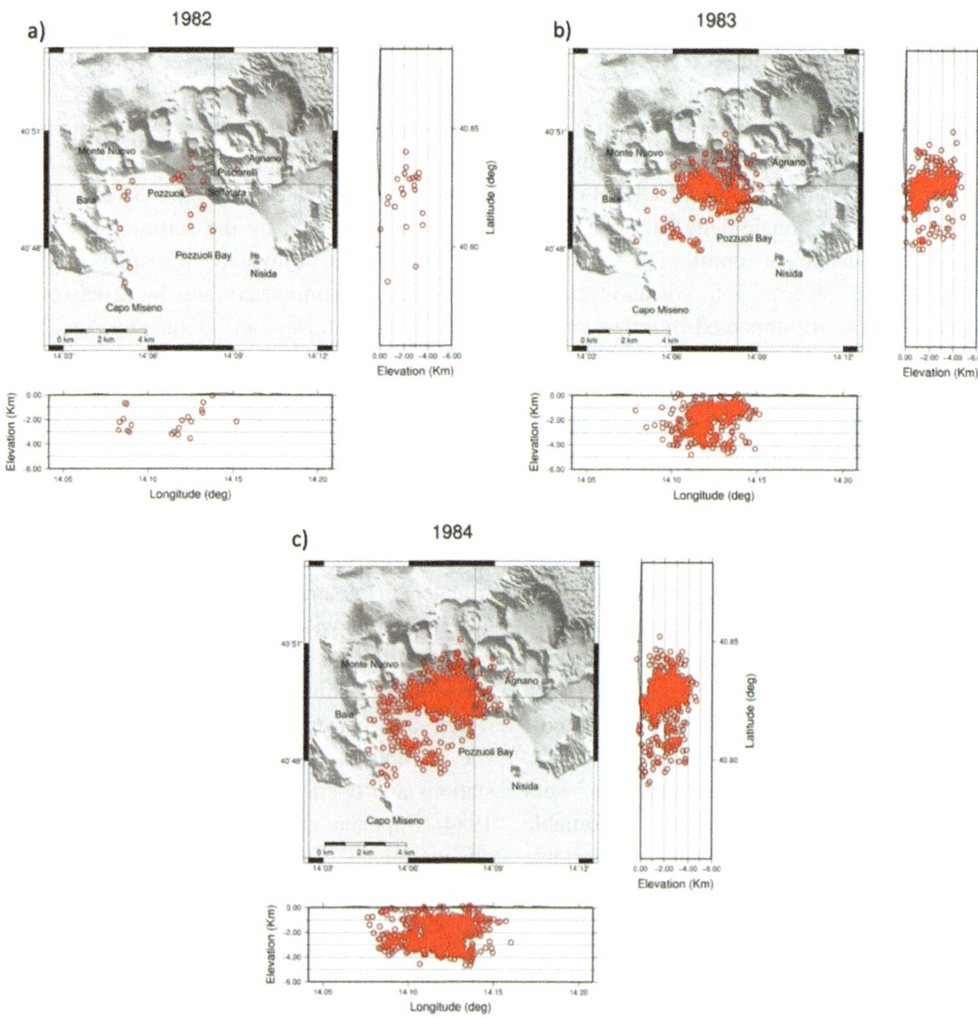

Fig. 1 Location of the seismic activity recorded during 1982, 1983 and 1984 at the Campi Flegrei caldera

April 1 in about 6 h. The last high energy event (M_D = 4.2) occurred on December 8, 1984 and was followed by a rapid decline of the seismic activity.

2.3 The Seismic Activity Accompanying the Most Recent Mini-Uplift Episodes

A new unrest episode at the CFc began on March 2000, after more than 15 years from the end of the 1982–1984 bradyseismic crisis characterised by weak local seismicity (Saccorotti et al. 2001; Chaps. The Permanent Monitoring System of the Campi Flegrei Caldera, Italy; The Hydrothermal System of the Campi Flegrei Caldera, Italy; Source Modelling from Ground Deformation and Gravity Changes at the Campi Flegrei Caldera, Italy). On July–August 2000, the ground uplift reached the maximum value of about 4 cm and was accompanied by two swarms of low-magnitude earthquakes (M_D < 2.4), one in July and one in August, that generated about sixty earthquakes.

Following the ground deformation renewal in November 2004 (Del Gaudio et al. 2010), additional swarms of microearthquakes occurred (Saccorotti et al. 2007). Accordingly, the seismic monitoring system was improved with the deployment of a temporary seismic network composed of 6 digital three-component seismic stations equipped with broadband seismometer, 2 digital six-channel seismic stations equipped with broadband seismometer and accelerometer, 1 seismic array based on five short-period three-component seismometer and one accelerometer, 1 hydrophone station (Castellano et al. 2012; Saccorotti et al. 2007). The events clustered in three separate episodes that occurred in October 2005, October 2006 and December 2006, respectively. The October 2005 swarm consisted of about 90 events (M_D max = 1.1), the October 2006 swarm included about 160 low-magnitude microearthquakes (M_D max = 0.8), whereas during the December 2006 swarm there were only 8 events, one of which had the highest magnitude (M_D max = 1.4; Saccorotti et al. 2007). After this last swarm, the seismic activity remained at low levels for both number of events and energy released until the swarm of March 30, 2010 composed of 147 events (M_D max = 1.2). Hereinafter, the seismicity was characterised by isolated low-energy events (M_D max < 1.0) culminating with a swarm occurred on September 7, 2012 and consisting of about 200 micro-earthquakes (M_D max = 1.7). At the end of 2012 the seismic monitoring system consisted of a Permanent and a Temporary Seismic Network. The Permanent Seismic Network was composed of 8 analog short-period seismic stations, 8 digital broadband stations (2 equipped also with an accelerometer and 3 with an infrasonic sensor). The Temporary Seismic Network included 9 digital three-component seismic stations equipped with a broadband seismometer, 3 six-channel seismic stations equipped with a broadband seismometer and an accelerometer, 1 new array based on 6 short-period three-component seismometers (La Rocca and Galluzzo 2012). The network has continuously monitored the seismic activity in the area allowing to record the earthquakes occurred so far (see Chap. The Permanent Monitoring System of the Campi Flegrei Caldera, Italy).

2.4 Source Parameters

2.4.1 Earthquake Locations

We have relocated the seismicity using an accurate 3D P-wave velocity model for the CFc and surroundings, centred on the Pozzuoli Bay. This model was developed by Battaglia et al. (2008) starting from previous 3D tomography images (Zollo et al. 2003; Judenherc and Zollo 2004). A probabilistic earthquake location based on a non-linear global search procedure (Lomax et al. 2000) has been performed by merging the data from the University of Wisconsin, the Osservatorio Vesuviano and the Aquater. The solution expressed as a probability density function, provides the uncertainty in the location due to the picking and travel-times calculation error as well as the network-event geometry. The maximum likelihood point of the complete non-linear probability density function is selected as an "optimal" hypocentre.

Figure 1 shows the best located events of the 1982–1984 bradyseismic crisis. The swarm of November 2, 1982 occurred in the Cimitero area, northwest of Pozzuoli, with hypocentres at maximum depth of 3.5 km b.s.l. (Orsi et al. 1999) (Fig. 1a). Other occasional events occurred in the Pozzuoli Bay. The epicentre of the earthquake occurred on May 15, 1983 was located near the eastern rim of the Solfatara crater (Fig. 1b). Those of the activity from September through December 1983 were located in the area between Pozzuoli and Solfatara, with hypocentral depth ranging from 0.5 to 3.0 km b.s.l. (Fig. 1b). In addition to these events, some earthquakes were in the Pozzuoli Bay, following a southeast-northwest alignment. The epicentral distribution of the earthquakes of the 1984 seismic activity delimited two seismogenetic areas (Fig. 1c). The highest concentration of events was in the Pozzuoli–Solfatara area. Moreover, as in 1983, an activity along a southeast-northwest alignment was detected and located in the western sector of the Pozzuoli Bay. The December 8, 1984 event was located in the Solfatara area at depth of about 3 km b.s.l.

The about sixty earthquakes of the two swarms occurred in July–August 2000 were located to the east of the Solfatara crater with hypocentral depth of about 2 km b.s.l. (Fig. 2a).

The earthquake hypocentres of two separate swarms occurred in October 2005 and October 2006 were mainly concentrated in the Solfatara area with depths between 1 and 3 km b.s.l. (Fig. 2b). The earthquakes of the December 2006 swarm were located in the Astroni area (about 2 km north of the Solfatara) with hypocentres up to 1.5 km b.s.l. They were the first events after the 1982–1984 crisis to occur outside the Pozzuoli-Solfatara area.

The March 30, 2010 swarm was located in the Pisciarelli area (between the Solfatara and Astroni volcanoes) with depths between 1 and 2 km b.s.l. (Fig. 2c). The microearthquakes of the seismic swarm occurred on September 7, 2012 were mainly located northwest of the town of Pozzuoli, with depth up to 2 km b.s.l. (Fig. 2c). In Fig. 2d is reported the activity detected between 2013 and 2019.

2.4.2 Focal Mechanisms

Fault-plane solutions have been analysed for many local earthquakes by using first P-wave onsets, polarisation analysis and P/S amplitude ratio (e.g., Gaudiosi and Iannaccone 1984; Aster et al. 1992; De Natale et al. 1995; Orsi et al. 1999). Gaudiosi and Iannaccone (1984), using a 1D velocity model to locate the earthquakes, reported 15 focal mechanisms with mainly normal faulting and various fault orientations. Aster et al. (1992) reported focal mechanisms, for data grouped into three different subzones of the caldera, computed using P wave polarities and illustrating the data scattering for one group (including the April 1, 1984 swarm) respect the other two. De Natale et al. (1995) reported 42 focal mechanisms for events occurred during the April 1, 1984 swarm, computed by a probability method using P-wave polarities and S-wave polarization data, whose main result is the evidence for a spatial change of the compressive stress axes inside the caldera. Orsi et al. (1999) reported 60 focal mechanisms computed using P-wave polarities, predominantly normal, normal/strike-slip.

The focal parameters for a selected subset (more than 200 earthquakes) of the 1982–1984 sequence estimated by using the grid search mechanism determination algorithm *FPFIT* (Reasenberg and Oppenheimer 1985) is presented here. The earthquake location parameters have been inferred from the maximum likelihood hypocentre determined by global search earthquake locations, and the corresponding ray take-off angles in the 3D velocity model have been used. The relevant feature for this set of earthquakes is the strong coherence of P-polarities at the same station for different events, thus a similar location and source mechanism is suggested for the different analysed earthquakes. Distribution of nodal planes, strike and rake angles (Fig. 3a) shows a preferred N150°E orientation with two secondary clusters striking N120°E and N30°E with all solutions showing a main normal component of the fault slip. The

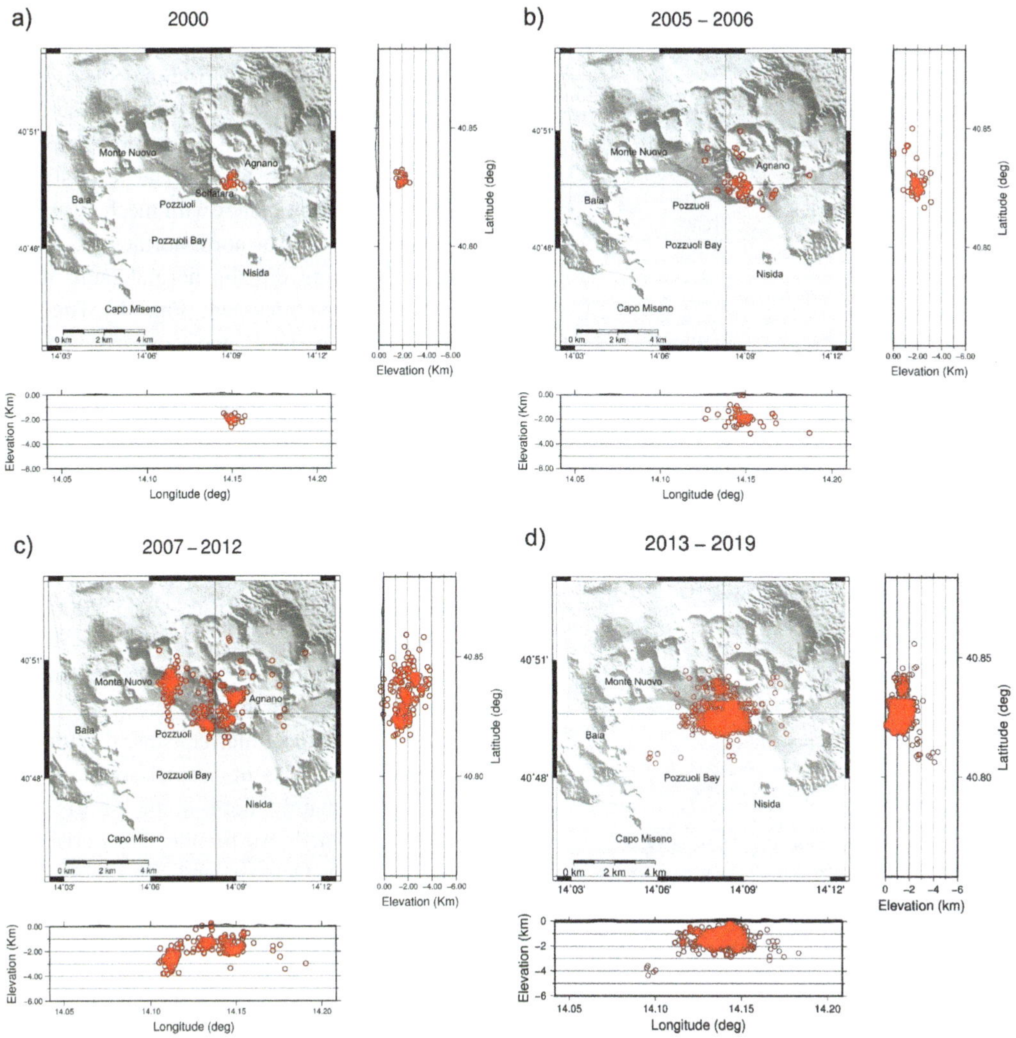

Fig. 2 Location of the seismic activity recorded during the recent mini-uplift episodes, and subsequently. **a** Seismic swarm of July–August 2000; **b** Seismic activity occurred in the 2005–2006 period; **c** Seismic activity occurred in the 2007–2012 period; **d** Seismic activity occurred in the 2013–2019 period

distribution of P-axis and T-axis azimuth (Fig. 3b) for the preferred fault plane solutions shows, in spite of a variability in the orientation of the axes, a cluster of roughly N20°E to near-vertical directions for compression axes, and roughly east–west to sub-horizontal directions for tension axes, confirming a dominant normal component of the fault movements. In fact, only few solutions show a reverse, reverse/strike-slip motion, the solution residuals likely reflect structural heterogeneities at a smaller scale within the volcanic area. The P-plunge vs. T-plunge diagram (Fig. 3c) clearly shows normal and normal/strike-slip main solutions, with a secondary cluster in the strike-slip range. Only few solutions show reverse, reverse/strike-slip displacements, with the presence of a number of odd-type solutions.

As regards the 2000 seismicity, the data quality of the July swarm was insufficient for any source mechanism estimate, due to the emergent first P-pulse of most of the recordings.

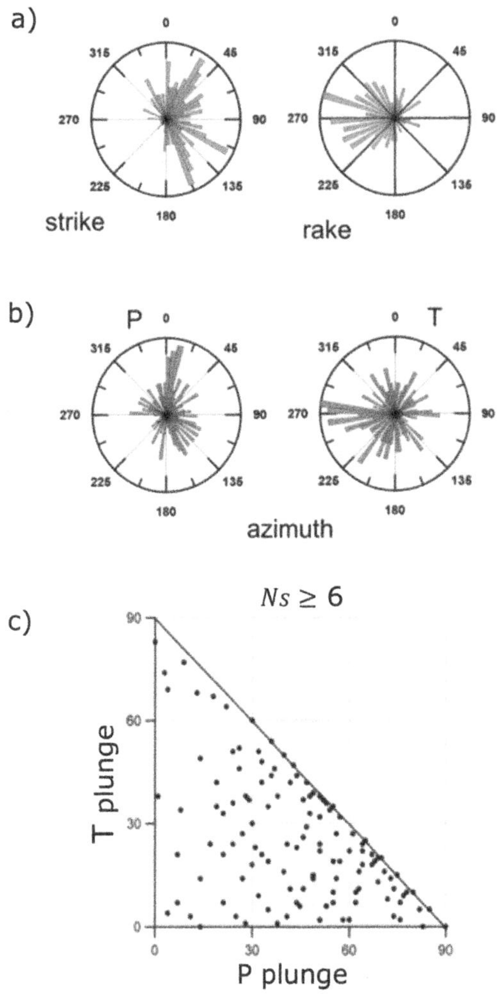

Fig. 3 Focal parameters evaluated for a selected dataset for earthquakes recorded at least by 6 stations. **a** Stereographic views of nodal planes (strike and rake); **b** Stereographic views of P-axis and T-axis azimuth; **c** P-plunge vs T-plunge diagram

Conversely, the better signal-to-noise ratio for the August events allowed us to obtain well-constrained focal mechanisms. Under the assumption of a double-couple mechanism, the fault plane solutions for this seismicity were obtained applying the grid-search algorithm *FPFIT* to the P-wave first motion polarities for a selected subset of the best-located events (Fig. 4a). Strike-slip focal mechanisms dipping in a narrow range are evident; the solutions are right-lateral, strike slip on vertical faults. Saccorotti et al. (2001) calculated fault plane solutions for the master events of the two clusters they found in the August 2000 swarm. They applied a grid search algorithm (Snoke et al. 1984), using as observables data both body-wave polarities (P, SV, SH) and body-wave amplitude ratios (SV/P, SV/SH, SH/P). The results depict the suite of solution planes with mechanisms that are compatible with a north-northeast to south-southwest striking dip-slip normal fault, with a minor strike-slip component (Fig. 4b). The strike of the retrieved nodal planes is in agreement with the alignment inferred by the relative locations of the epicentres found by the same authors. Saccorotti et al. (2001), based on the previous observation, suggested that the fault mechanism remained quite constant throughout each cluster and was representative of the rupture planes activated during the two earthquake clusters.

Fault plane solutions for the 2005–2006 crisis were obtained by Saccorotti et al. (2007) assuming a double couple mechanism and applying the grid-search.

The resulting focal mechanisms, illustrated in Fig. 4c, show a class of normal solutions with nodal planes mainly north-northeast to south-southwest oriented. As for the 2000 crisis, the events of both October and December 2006 swarms, with similar waveforms, have similar fault plane solutions.

2.4.3 Magnitude Statistics

The magnitude statistics of the CFc earthquakes were first studied by De Natale and Zollo (1986). They computed a b-value of 0.98 for the period May 1983–May 1984. D'Auria et al. (2011) computed for the same period a b-value of 0.72 ± 0.04 and for the period 1989–2010 an average b-value of 0.92 ± 0.25. The temporal variation of the duration magnitude (Md), the b-value and the magnitude completeness (Mc) for the 1982–1984 earthquakes are shown in Fig. 5. B-values and Mc have been computed using groups of 100 consecutive events without overlap. The b-value oscillated between 0.75 and 1.0 since April 1983. Then it showed three distinct peaks, the second of which, occurred at the end of June 1984, reached a maximum value of about 1.4. Mc, after a rapid decrease in the first few

Fig. 4 Fault plane solutions (lower hemisphere, equal area projection); open and solid circles denote the dilatational and compressional P-wave first motion, respectively. P and T axes are also shown. **a** The August 2000 events showing at least 8 reliable P-wave polarity readings; **b** Suite of solutions obtained using Focmec algorithm for the master event of the 2 multiplets recognised in the 2000 swarms (modified after Saccorotti et al. 2001); **c** 10 events of the 2006 swarm exhibiting the largest number of reliable P-wave polarity readings (modified after Saccorotti et al. 2007)

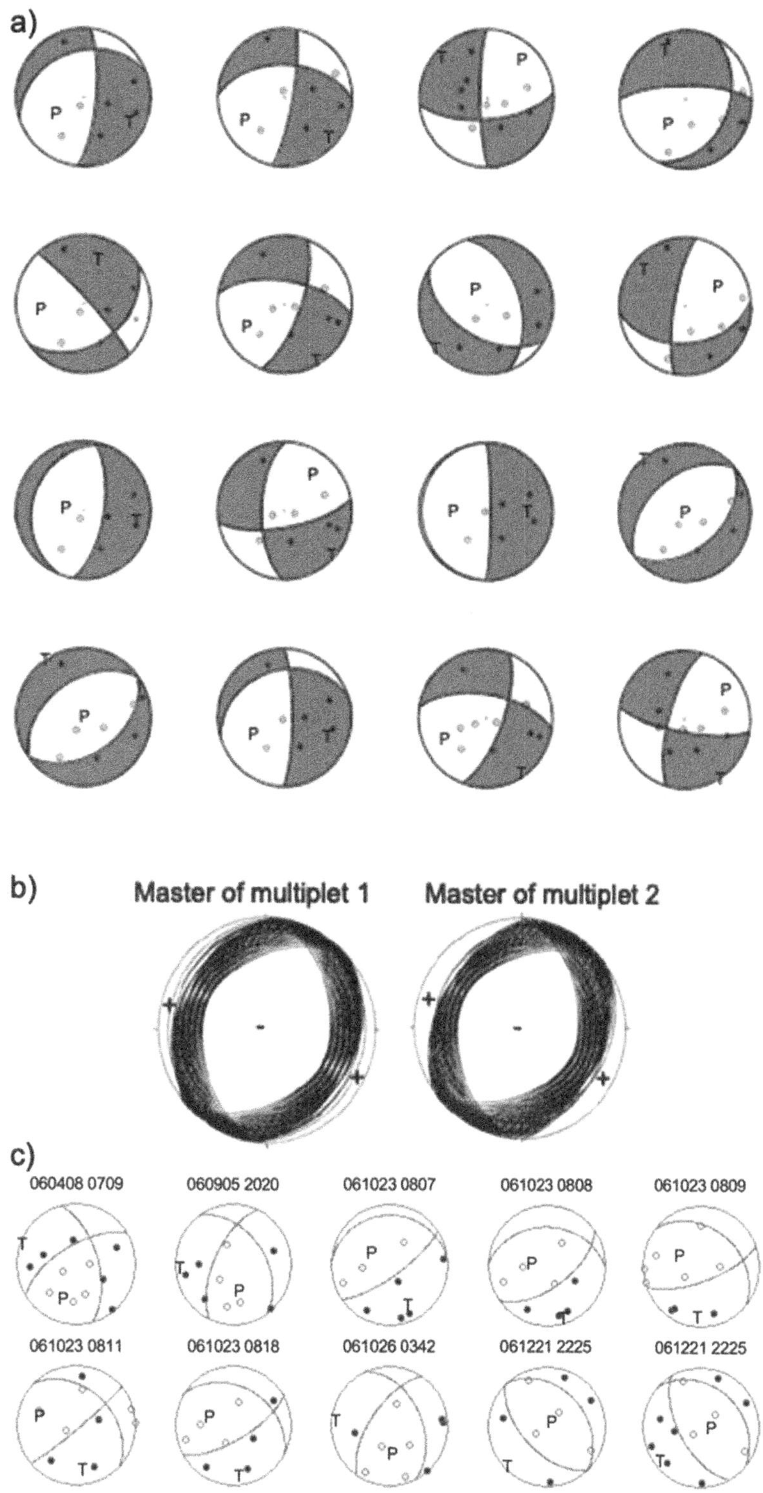

Fig. 5 Magnitude statistics for the 1982–1985 interval. The top panel shows duration magnitudes (Md) of the detected events. The intermediate and lower panels show the b-value and the magnitude completeness (Mc) with their root mean square, respectively. Each value is computed over sequences of 100 consecutive events

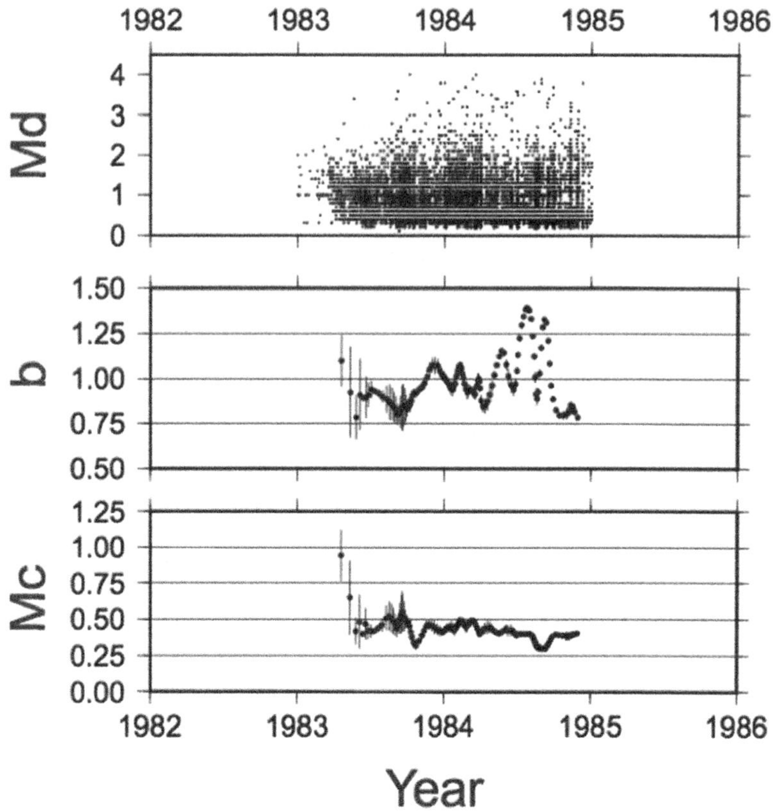

Fig. 6 Magnitude statistics for the 1989–2020 interval. The top panel shows duration magnitudes (Md) of the detected event. The intermediate and lower panels show the b-value and the magnitude completeness (Mc) with their root mean square, respectively. Each value is computed over sequences of 100 consecutive events

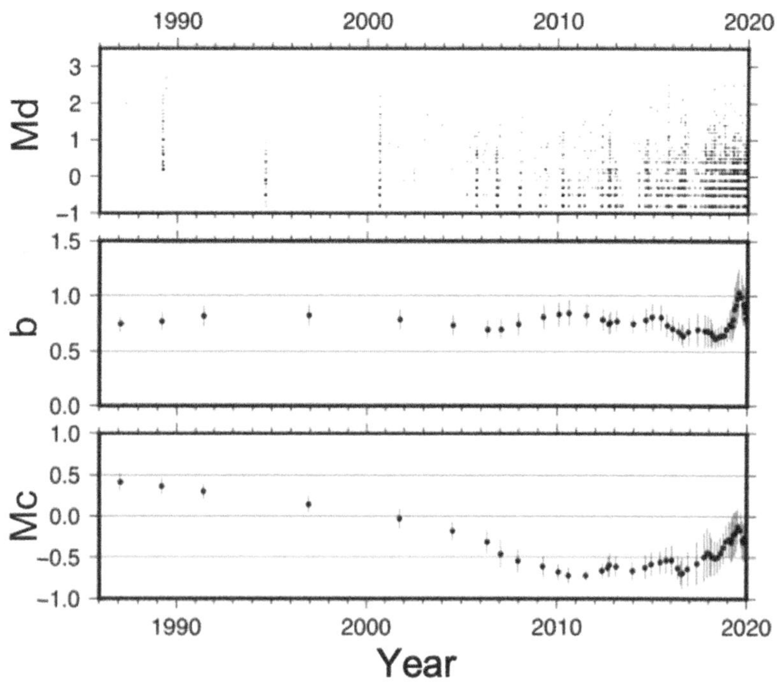

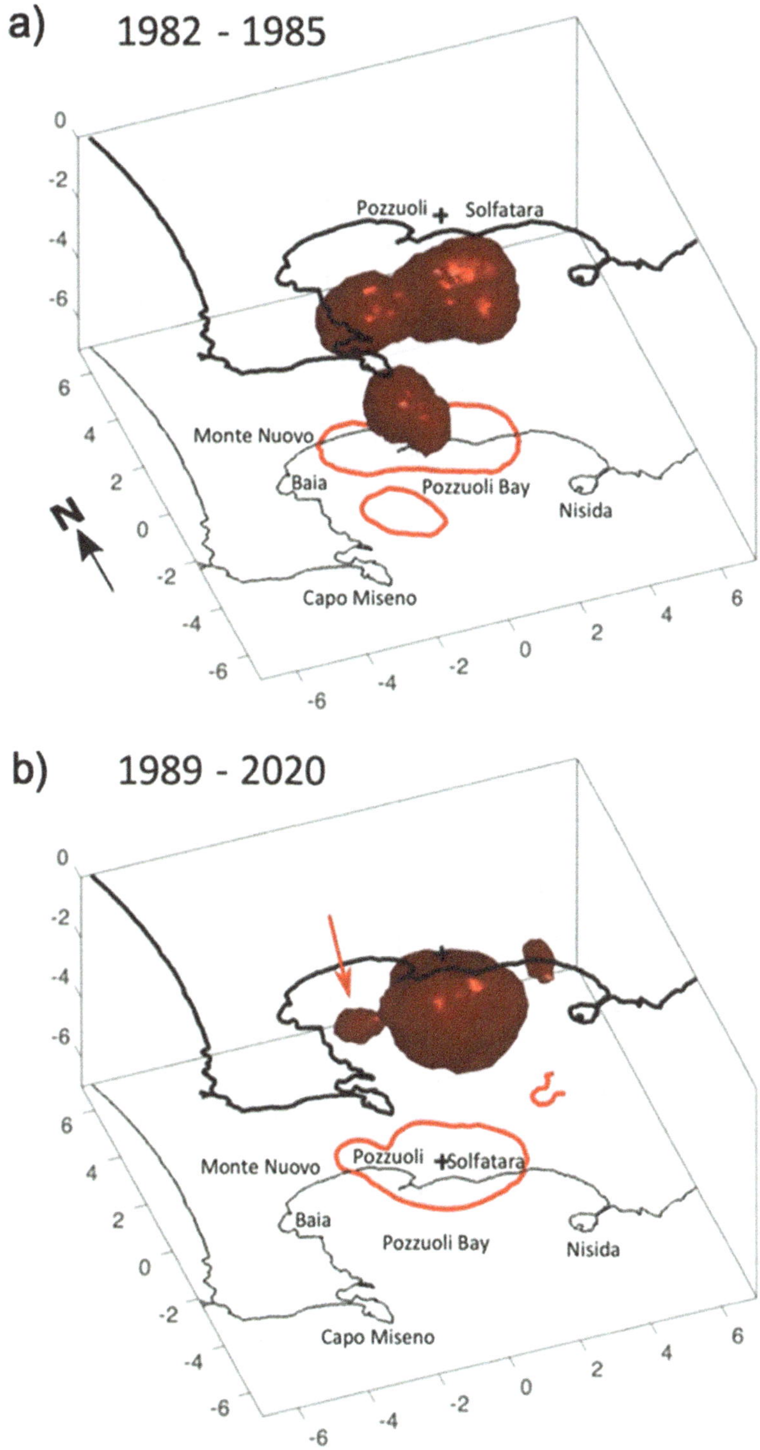

Fig. 7 Spatial pattern of the moment release. Red isosurfaces contain 90% of the total moment release for each of the two intervals. The horizontal projection is depicted by the red contour on the bottom of each panel. Black crosses indicate the position of the Solfatara crater. The arrow on the lower panel indicates the volume which is related to the September 7, 2012 seismic swarm

months of 1983, remained fairly constant around an average value of 0.75, for the rest of the period. Figure 6 shows the same parameters as Fig. 5 for the events occurred in the 1986–2020 time interval. The b-value was computed on groups of 50 events with an overlap of 10 events between adjacent groups. Because of the limited amount of data, a single value of Mc = 0.12 ± 0.43 has been determined. The b-values ranged mostly between 0.5 and 0.8, with the exception of a peak value of 1.1 reached in 2019.

2.4.4 Moment Release Patterns

The spatial and temporal distribution of the seismic moment release is characterised by using the relationship log $M_0 = 1.2\ M + 10$ (Bakun 1984) valid for small magnitude ($M < 3.5$) earthquakes. The moment release pattern shows a significant difference in the temporal distribution between the 1982–1985 time interval and the following years. Figure 7 represents the tridimensional distribution of the seismic moment release in these two intervals. The red isosurfaces contain 70% of the total moment released in each interval. The moment release values have been smoothed using a filter with a radius of 1 km. For the 1982–1985 period the moment was released mostly within two separate volumes (Fig. 7a). One extends from the Solfatara crater (indicated by a small cross) to the town of Pozzuoli toward the west. Its depth range is between 0 and 3 km. The other volume is located beneath the Pozzuoli Bay within thc depth range 3–4 km. For the 1989–2020 interval the moment release was mostly concentrated within two small volumes (Fig. 7b). The first one was located east of the Solfatara crater (Agnano-Pisciarelli area) in the depth range 0–3 km, while the other was beneath the town of Pozzuoli. The moment within this second volume was mostly released during the September 7, 2012 seismic swarm.

The moment release rate shows a peculiar temporal pattern for the interval 1982–1984 (Fig. 8a): at least four major peaks (September–October 1983, March–April 1984, August 1984 and November 1984) can be recognised. The

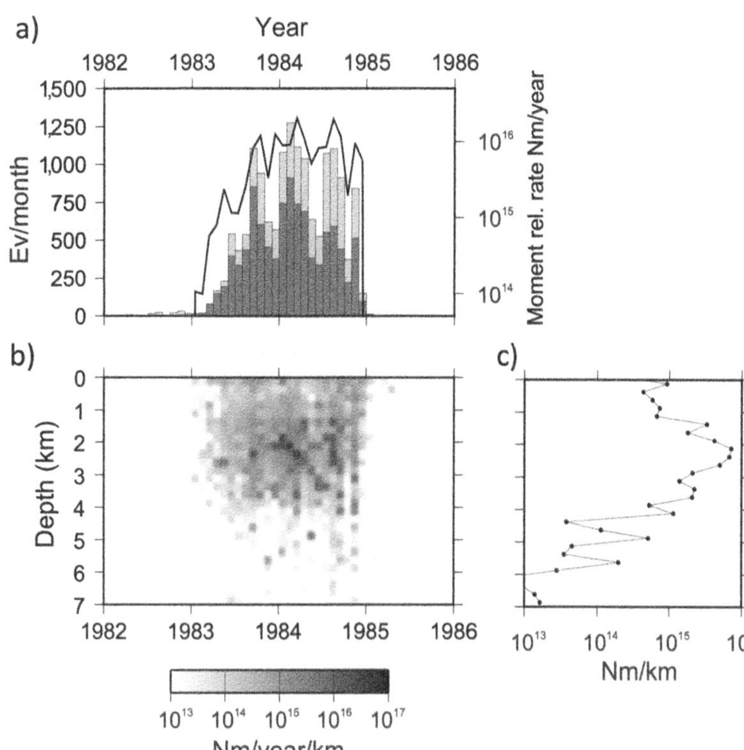

Fig. 8 Time-depth pattern of the moment release for the 1982–1985 interval. **a** The light grey bars indicate the number of detected events for each month, the dark grey bars indicate the number of events having a Md greater or equal to 0.5, the solid black line represents the moment release rate for the same interval; **b** Distribution of the moment release as a function of both time and depth: the size of the pixels is one month for the time and 0.25 km for the depth; **c** Distribution of the moment release as a function of depth

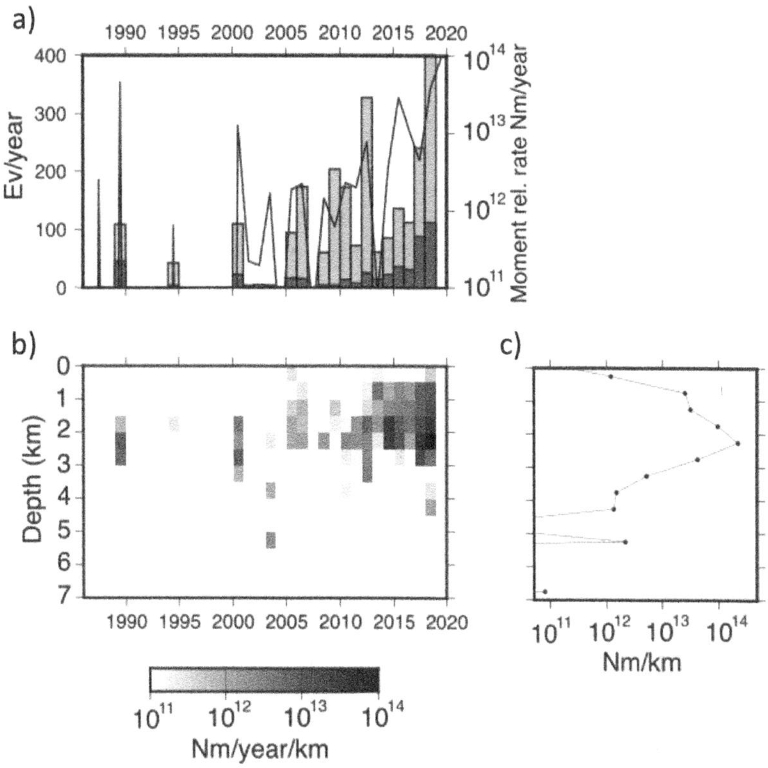

Fig. 9 Time-depth pattern of the moment release for the 1986–2020 interval. The meaning of the plots is the same as for Fig. 8. The size of the pixels in panel b is one year for the time and 0.5 km for the depth

second peak was the strongest and reached a value of 2.05×10^{16} Nm/year. A similar pattern was followed by the event occurrence rate (Fig. 8a), which reached a peak of about 1,300 events/month in February 1984. Furthermore, the whole temporal pattern is markedly asymmetric, with a smooth onset in 1982–1983 and an abrupt end at the end of December 1984. The total cumulative moment released during this interval was about 1.37×10^{16} Nm. Most of the moment was released at depths shallower than 3 km b.s.l., with a peak rate at 2 km b.s.l. without any time-depth distribution systematic trends (Fig. 8b, c) in the depth distribution with time. The time distribution of the moment release pattern for the time interval 1986–2020 has a highly non-stationary behaviour (Fig. 9a). Until 2005, the seismicity was dominated by isolated short duration seismic swarms (D'Auria et al. 2011), while since 2007 (and more markedly since 2010) a persistent background microseismicity appeared (see also Fig. 6). The moment shows a positive trend. The total cumulative moment released within this interval amounts to 8.27×10^{13} Nm. The depth distribution of the moment (Fig. 9c) within this interval is similar to the pattern found in the 1982–1985 interval (Fig. 8c)

Del Pezzo et al. (1987) calculated seismic source parameters (seismic moments, source radii and static stress drop) for a set of 181 seismograms recorded by the three-component digital stations network of the University of Wisconsin. The local magnitude of the analysed seismicity was in the range 0.7–3.2. The authors developed an inverse method to fit displacement spectra to the theoretical Brune's source model that permitted also to obtain error estimates for the computed spectral parameters. They obtained displacement S-wave spectra by log-averaging the smoothed spectra of the east–west and north–south components calculated over a 2.5 s time window starting at the S-wave arrival, thus fitting the spectra to the theoretical Brune's source model using S-wave velocity of 1.5 km/s, Qs = 110 and a density of 2,200 kg/m^3. The retrieved focal dimensions were in the range 43–131 m; the stress drops were mainly in the range 1–10 bar and seismic moments ranged between 10^{17} and 10^{20} dyne-cm. Saccorotti et al. (2001, 2007) adopted the method developed by Del Pezzo et al. (1987) to evaluate the source parameters for the 2000 and 2005–2006 crises. For the events of the 2000 crisis with magnitude between 1.6 and 2.2, the authors obtained focal dimensions in the order of 100 m, related stress drops in the range 1–10 bars and seismic moments in the range 10^{17}–10^{19} dyne-cm. Both the waveforms and the spectral features of these events are very similar to those of the events occurred during the 1982–1984 crisis and indicative of brittle shear failure. For the events of the 2005–2006 crisis with magnitude in the -0.5–1.4 range, Saccorotti et al. (2007) retrieved source radii in the range 30–60 m, stress drop between 1 and 2 bar, and seismic moments in the range 10^{16}–10^{18} dyne-cm.

In Fig. 10 seismic moment vs. Brune's source radius is shown for 1982–1984, 2000 and 2005–2006 episodes. On average, the 2005–2006 moments appear slightly lower than those of the other crises, whilst the range of seismic radii as well as that of the stress drops are similar. However, source dimensions, stress-drop and moments are of the same order for the 1982–1984 seismic swarm and the 2000 and 2005–2006 crises, suggesting that all the recent major seismic sequences at the CFc seem to share the same mechanism.

3 Ground Deformation

The earliest measurements of ground displacement at the Campi Flegrei were performed by the Istituto Geografico Militare, and date back to the beginning of the twentieth century (Del Gaudio et al. 2010; Chaps. The Permanent Monitoring System of the Campi Flegrei Caldera, Italy; The Hydrothermal System of the Campi Flegrei Caldera, Italy; Source Modelling from Ground Deformation and Gravity Changes at the Campi Flegrei Caldera, Italy). The renewal of uplift in early'70 promoted the intensification of levelling line measurements, some Electronic Distance Measurement analyses and the setting up of a tide gauge network. Since 1991, Interferometric synthetic aperture radar (InSar) images are available and since 2001 a network of continuous Global Positioning Survey (GPS) measurements at 13 benchmarks is available. Some general papers have been published in recent years on these data (See references in Chaps. The Permanent Monitoring System of the Campi Flegrei Caldera, Italy; The Hydrothermal System of the Campi Flegrei Caldera, Italy; Source

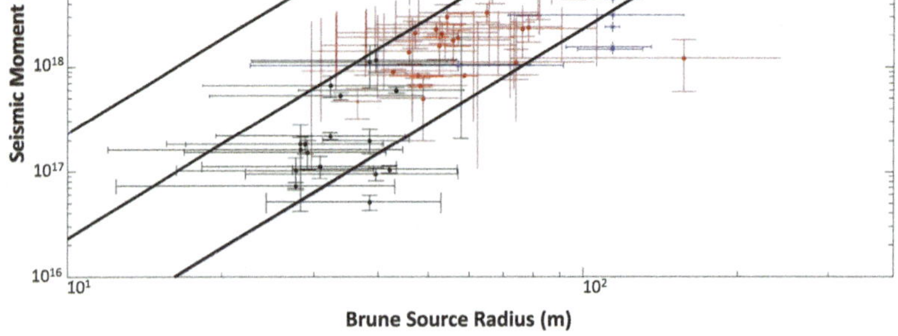

Fig. 10 Seismic moment vs. Brune's source radius: red colour for the 1982–1984 crisis, blue colour for the 2000 crisis and black colour for the 2005–2006 crisis

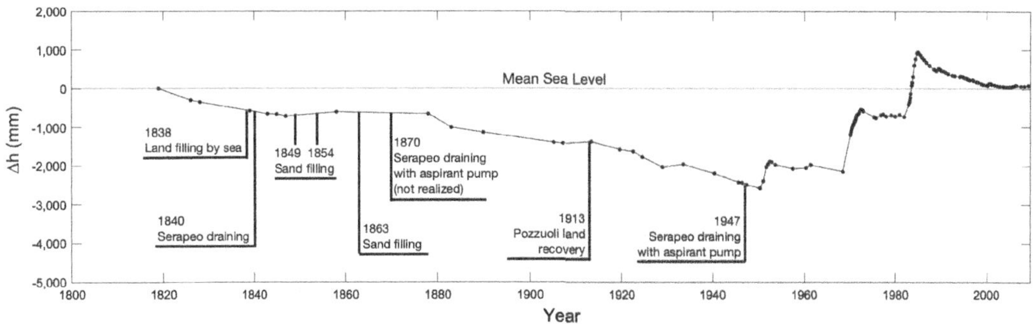

Fig. 11 Height variation of the Serapeo marble floor in the period 1800–2013. Figure modified after Del Gaudio et al. (2010)

Modelling from Ground Deformation and Gravity Changes at the Campi Flegrei Caldera, Italy), showing the very long-time series of vertical displacement within the CFc as well as the puzzling features arising from the time and space scale of this unrest. Recently another uplift episode during 1950–1952 was recognised from the analysis of documents reporting the damage of some pipelines (Del Gaudio et al. 2010). A summary of these vertical displacements in the last two centuries is reported in Fig. 11. The inflation rate in the 1982–1984 period is particularly high, and the cumulative uplift, including the 1950 and 1970 events, is in the same order of magnitude of the AD 1538 pre-eruptive stage (~10^2 cm/year) (Guidoboni and Ciuccarelli 2011; Di Vito et al. 2016).

3.1 Borehole Strainmeter Observations

Since spring 2004 an Italian research project has been developed to install borehole Sacks-Evertson strainmeters aimed at enhancing the monitoring systems of the Italian volcanoes. The Sacks-Evertson strainmeters (cylinders filled with degassed silicone oil, about 7 cm in diameter and 4 m in length) provide two signal outputs, obtained by two different hydro-mechanical amplification systems. The high-sensitivity output integrates the volumetric change in the strained reservoir. The low-sensitivity one is connected to the strained reservoir only when the instrument is sensing a rapid and strong strain change, thus measuring strain. Usually the low-sensitivity channel measures the pressure in a closed cell, that is proportional to local temperature. The temperature resolution is a few microdegrees. Air pressure is also measured at the surface, in order to correct for dilational strain resulting from atmospheric pressure loading. Nominal resolution of the Sacks-Evertson strainmeter is about 10^{-12} strain units, with a nominal dynamic range of 10^{-11}–10^{-3}. Low-frequency calibration of installed strainmeters is obtained by comparison with Earth tides (Hart et al. 1996; Amoruso et al. 2000), while surface waves caused by earthquakes can be used to calibrate the sensors in higher frequency bands (Currenti et al. 2017). A prototypal strainmeter was installed in October 1998 in the Vesuvio region (Scarpa et al. 2000). Six borehole dilatometers were installed at Campi Flegrei and Vesuvio in 2004–2005 (Scarpa et al. 2007; see Fig. 12), and two at the Stromboli volcano in 2006.

For the three instruments installed at the CFc the comparison with Earth tides and oceanic loading provided a sensitivity of 5×10^{-11} strain. High- and low-sensitivity strain signals, continuously recorded and sampled at 50 Hz, were sent to the Istituto Nazionale di Geofisica e Vulcanologia–Osservatorio Vesuviano and retrieved and processed at the University of Salerno. Presently this network is under revision and not operating. In November 2004, after the installation of the network, the last uplift at the

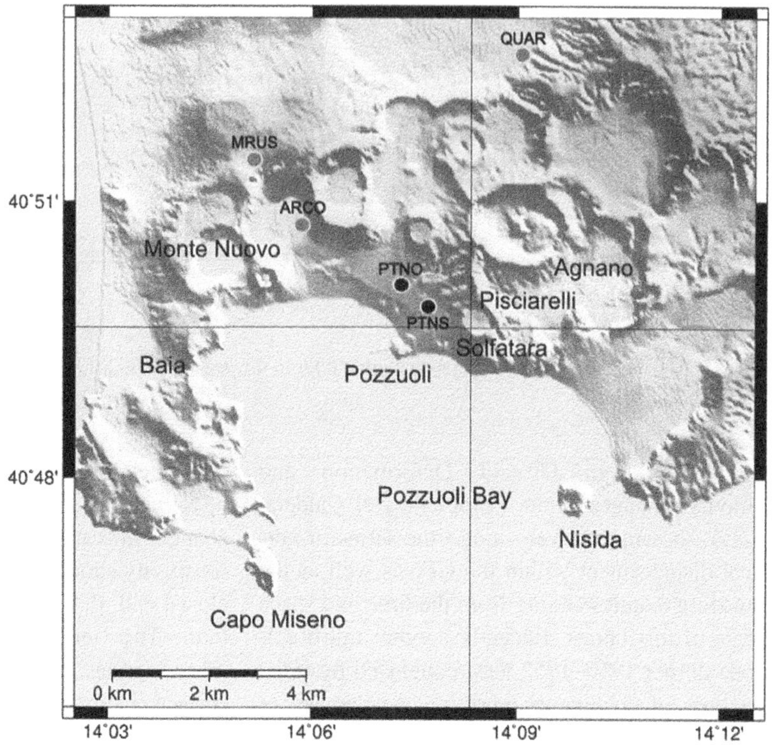

Fig. 12 Location of the borehole dilatometers (grey dots) and underground tiltmeters (black dots) installed in the Campi Flegrei caldera during 2004–2005 and 2008–2009, respectively

CFc started at various low rates, and, after a lower rate begun in December 2006, and it accelerated in 2011. The uplift currently continues at a rate of 5–7 cm/year, as documented by GPS data (Bevilacqua et al. 2020; De Martino et al. 2020). During the 2004–2006 mini-uplift episode, characterised by maximum vertical displacement of 4 cm, an anomalous strain was released in summer 2006, in correspondence of an anomalous CO_2 release and an increase in ground displacement measured by tiltmeters and GPS transducers. The strain episode preceded the CFc microseismic activity by a few months, similarly to what occurred during the 1982 last large uplift event. Other transient strain episodes occurred in October 2006, before the seismic swarm of LP events ($M_{max} = 0.8$; Saccorotti et al. 2007). In March 2010 other transient strain episodes were recorded, before the swarms of VT and LP events (D'Auria et al. 2011). The time scale of the transient strain episodes was less than one hour, placing a tight constraint on the origin of the ground uplift at the CFc. The locus of this deformation is close to a source inferred from long-term deformation signals, at about 4 km depth beneath Pozzuoli (Amoruso et al. 2007, 2014a, b).

The time series of the strainmeter signals at the CFc, starting from their installation, are shown in Fig. 13a and highlight overall stable trends. Data recorded during the first few months after installation are heavily affected by curing cement, and some interruptions in data acquisition are attributable to electronics issues. The ARCO borehole dilatometer station (Fig. 12) has the strongest signals and is the nearest (about 3 km) to the area of maximum uplift. Strain records at this station (Fig. 13a) show some slope change in temporal agreement with mini-uplift episodes recorded in the same period by GPS and levelling, and may possibly be related to their source. Changes in strain rates occurred in October 2004, May 2005 and July 2006, and preceded the three seismic swarms occurred in March and October 2005, and in October 2006 by about 4 months, consistently with the observations of the 1982–1984 crisis and the latest mini-uplift episodes. The changes in temperature rate are quite intriguing (Fig. 13 b). At the ARCO station increase rate was about 15 millidegrees/month in the first months of 2006. Thermal stability at the QUAR station is in the order of millidegrees, while temperature rate at MRUS station was about 10 millidegrees/month starting from June 2005 until December 2006. Variation in corrected strain and temperature time series at the ARCO station are similar but not identical: the differences may provide insights into the source processes. Corrected strain changes should give a "pure" tectonic signal. After May 2005 there was a strain change, consistent with both levelling and GPS data. However, the most interesting signals were recorded during summer 2006 (Fig. 13a). In this period a sudden change of strain occurred at the ARCO and MRUS stations, whereas the more distant QUAR station showed minimal changes involving only slight subsidence. The amount of the strain release was consistent with the tiltmeter data (Ricco et al. 2007). The strain changes preceded by a few months the October 2006 seismic swarm. Saccorotti et al. (2007), based also on an increase in CO_2 emission, interpreted this seismicity as due to an increase in the hydrothermal activity. Anyhow, the underground temperature variation in the geothermal reservoir seems to be the most important factor influencing the detected strain variations (Scarpa et al. 2007; Di Lieto et al. 2021).

In order to reveal anomalous but somehow smaller amplitude deformation events, the effects of Earth tides and atmospheric pressure variations were removed by using a Bayesian approach, as proposed by Tamura et al. (1991).

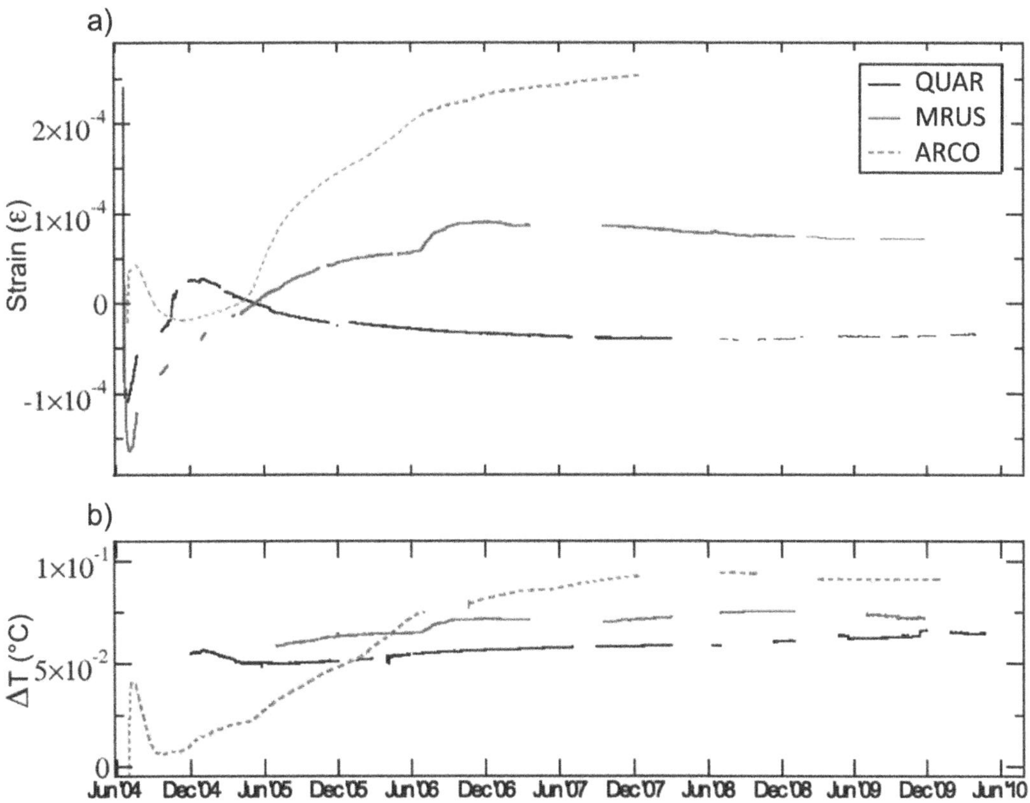

Fig. 13 Time series of strain and temperature signals recorded until 2010, at the QUAR, MRUS and ARCO stations. The station locations are shown in Fig. 12. Figure modified after Di Lieto et al. (2021)

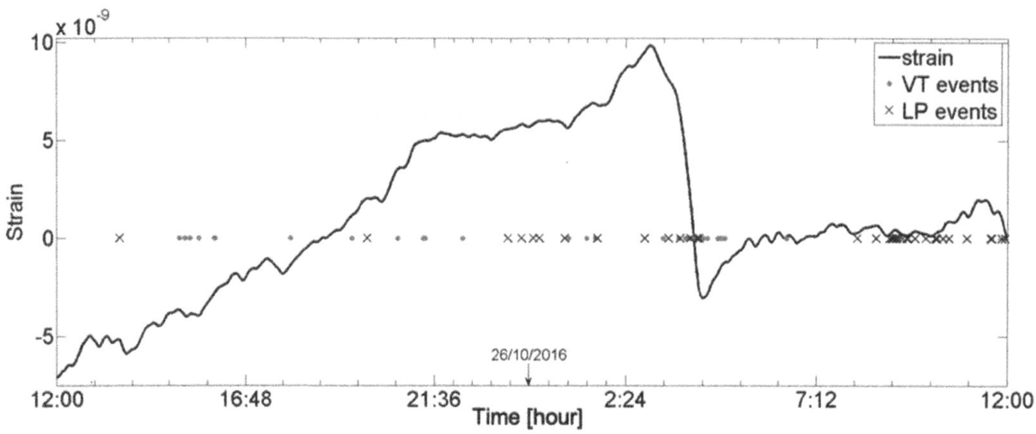

Fig. 14 Strainmeter signal detected at the ARCO station (black line). Crosses and dots on the zero axis show the time of occurrence of VT and LP events recorded near Solfatara. Figure modified after Di Lieto et al. (2021)

The obtained results have shown stronger, high-frequency content of strain signal, not correlated with rain or air pressure changes during the October 2006 seismic swarm. Two days of corrected data from the ARCO station (Fig. 14) include a dilatational strain contraction transient with a duration of approximately 1 h associated with increased volcanic tremor activity.

One of the largest transient episodes was recorded by the strainmeters on March 30, 2010. This episode started the day preceding a seismic swarm of about 100 VT earthquakes with small magnitudes (M $\leq$ 1.2), located in the Solfatara-Astroni area. It showed a small increase of the strain the day before the swarm, followed by a reversal and accelerated dilatation, preceding the

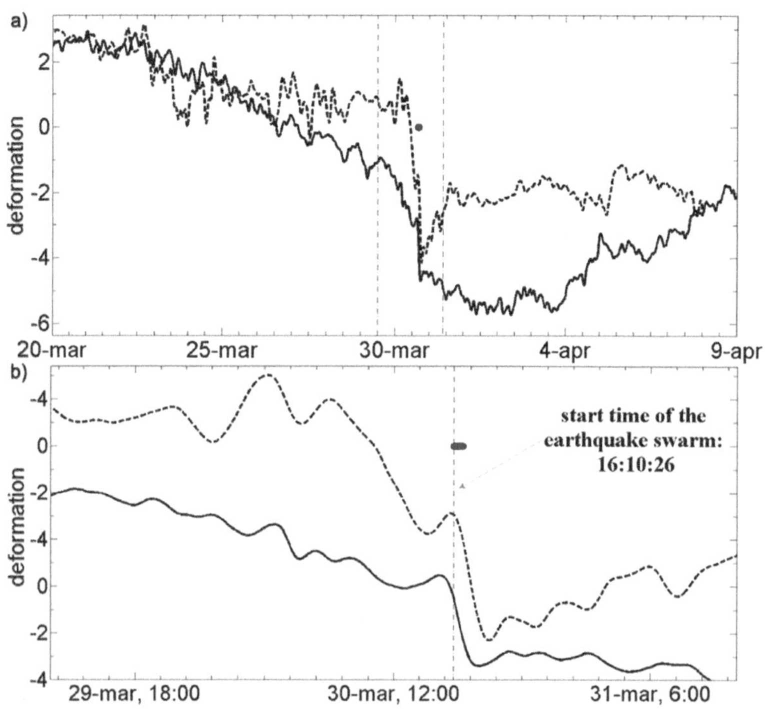

Fig. 15 **a** Strain (ARCO station, continuous line) and tilt (SOUTH station, dashed line) preceding and accompanying the March 2010 Astroni-Pisciarelli seismic swarm. The tilt amplitude unit is μrad and the strain unit multiplier is 10^{-8}. Blue dots represent the seismic swarm occurred on March 30. **b** Expanded view of the strain. Figure modified after Di Lieto et al. (2021)

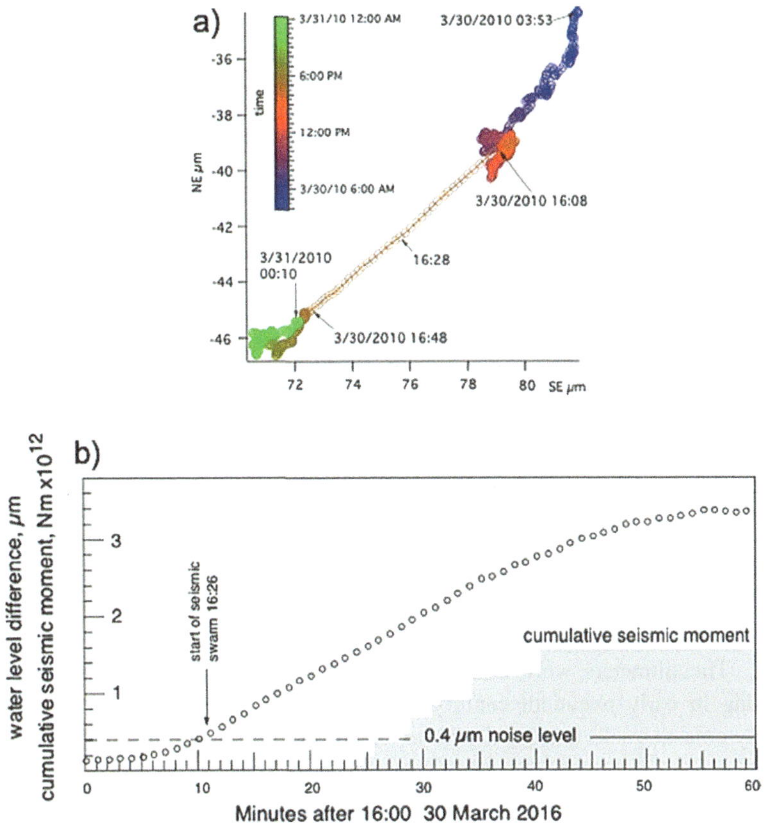

Fig. 16 **a** and **b** The tilt vector at minute intervals, compared to the seismic moment release, the bulk of which occurs 15–30 min after the first microearthquake in the swarm. The dashed line in **b** shows the 10–30 min RMS noise-level threshold in the raw data. The first microearthquake occurs approximately when the tilt signal exceeds the short-term noise level. However, the microseismic energy release starts 15 min after this threshold is exceeded, suggesting its utility in early warning methods. Figure modified after Di Lieto et al. (2021)

swarm by about 20 min (Figs. 15 and 16). The source of the process is compatible with the location of the recent inflation and deflation episodes analysed by Amoruso et al. (2014a, b, 2015).

3.2 Tiltmeter Observations

During the project UNREST, supported by the Italian Istituto Nazionale di Geofisica e Vulcanologia and the Dipartimento della Protezione Civile, two multiple-sensor, high-sensitivity long-baseline Michelson water pipe tiltmeters were installed at the CFc, in collaboration with Roger Bilham of the University of Colorado. Tiltmeters monitor ground deformation relative to a gravitational equipotential surface defined by gravity. A summary of the several tiltmeter programs focussed on the CFc in the past few decades can be found in Ricco et al. (2003). Prior to the installations described in this Chapter, the tilt instruments were all compact sensors using horizontal pendulums and more recently, biaxial electrolytic bubble sensors. These sensors, forming an array around the Pozzuoli Bay, are in tunnels, or castles, or in other underground sites far from large daily thermal signals. The tiltmeters are packaged in a 10–20 cm housing, fastened to concrete floor by bolts. The achieved data are recorded to 12-bit precision at rates of several samples per hour, permitting a resolution of approximately 0.1 μrad over a range of ±400 μrad with a linearity of 1%. Ricco et al. (2003) discuss the methods to optimise the measured signal-to-noise ratio of these sensors, especially the suppression of thermoelastic signals at daily and yearly intervals, and the effects of surface and subsurface hydrology.

The 12-cm-diameter horizontal pipes of the Michelson tiltmeters (Fig. 17) are half-filled with water and terminated by 20-cm-diameter reservoirs, fastened to concrete floor by steel pillars. Heated floating sensors support the core of a linear variable displacement transducer whose output is recorded to a 16-bit precision once per minute with a resolution of approximately 0.02 μm, and tilt is given by the ratio of the difference in height divided by the total length of the pipes. The tilt resolution in the longest water-pipes reaches a value of 0.08 nrad, that is approximately three orders of magnitude more sensitive than the electrolytic bubble sensors hitherto operating near Pozzuoli. Depending on the length of each pipe, the range obtainable without mechanical adjustment is ±9 to ±100 μrad, and the dynamic range before requiring pipe re-alignment is ±70 to ±700 μrad. Although water levels can be monitored at a higher order of magnitude, thermal noise in the tunnel environment increasingly dominates the signal at sensitivities of 1 nm.

The tiltmeters were installed in two tunnels dug in early twentieth century through mostly pyroclastic rocks and lined with tephra blocks or concrete (Fig. 17). The tunnels, named the North and the South Pozzuoli tunnel, are inclined at low gradients to permit drainage, but because the water-pipe must be installed within a few millimetres from the horizontal, the lengths of the tiltmeters in certain azimuths are restricted by these gradients and by the floor-to-ceiling height of the tunnels. Pipes length ranges from 28 m up to 285 m, and were installed at various azimuths in the two tunnels. The tunnels are linked to the atmosphere by openings and, although they afford good stability and immunity from surface-induced thermoelastic noise and precipitation, they are environmentally less stable than a sealed tunnel. Air currents passing in the tunnels result in thermal fluctuations of up to 3 °C in periods of minutes, especially at the passage of atmospheric pressure fronts.

Although the data could have been uploaded every minute, the time constants of the tiltmeters are sufficiently long period that a shorter latency is unjustified. In the 285-m-long water pipe the fundamental mode is approximately 8 min and

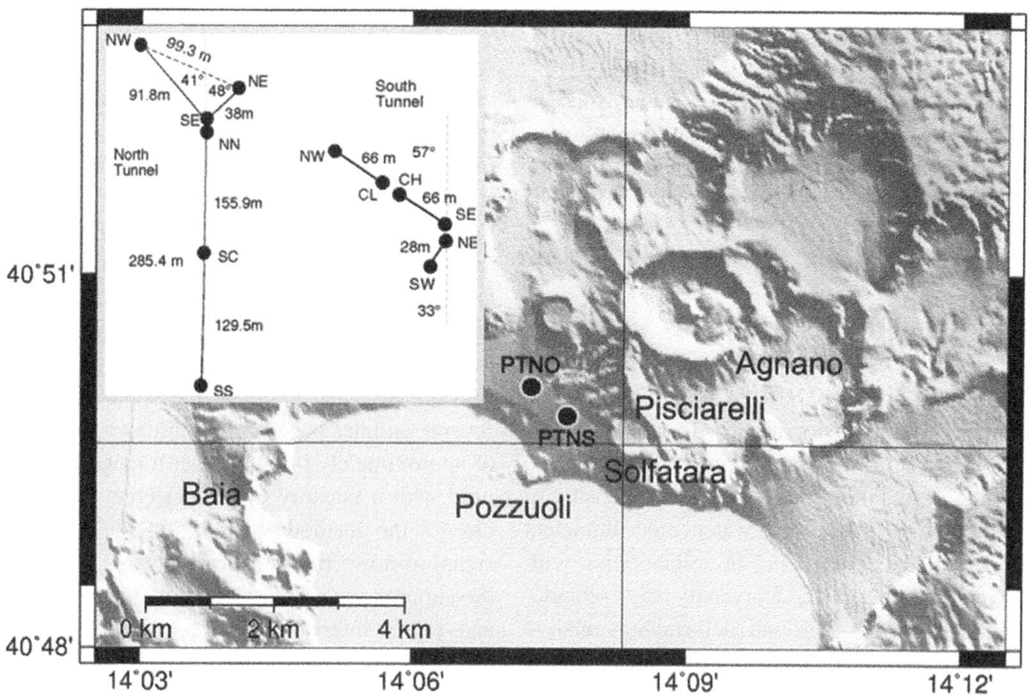

Fig. 17 Long baseline tiltmeters network: filled circles indicate the location of the sensors. The inset shows the geometric characteristics of both tiltmeters. Figure modified after Di Lieto et al. (2021)

we noted that resonance was rarely stimulated, damping being close to critical. Initially, also a 12-bit recording system in which the data were relayed as analog voltages to a central 8-channel data logger using isolating line drivers, was operated. A digital average on 30 samples was performed before uploading the data on a publicly accessible web-page. Despite the circuit breakers, the long data cables in the system (up to 300 m), proved to be vulnerable to lightning damage and this second form of transmission was discontinued.

A large inflation event, such as the 1982–1984 bradyseismic crisis, could produce an elevation change of 12 mm between the ends of the north–south pipe. Each transducer has a range of approximately 3 mm before mechanical adjustment is needed, although the linear variable displacement transducer sensor could be replaced with one of lower sensitivity if significant inflation is anticipated (6 mm, 12 mm or 25 mm). The 3 mm sensor results in a least-count resolution to water level changes of ~ 20 nm corresponding to ~ 0.1 nrad in the 285 m water pipe and ~ 1 nrad in the 28 m water pipe. Multiple sensors could also be arranged to monitor water level changes at each end of the water pipe although in the present installations we have not undertaken this. Should inflation exceed the double of that occurred in 1980 (maximum uplift of 1.7 m), the position of the longest pipe would need to be readjusted.

The North Pozzuoli tunnel tiltmeters, the longest installed in CFc, was dismissed in 2013. The array of tiltmeter sensors in the South Pozzuoli tunnel provided continuous time series from its installation in 2008 until 2015. Since September 2009 an inflation episode was recorded, interrupted by a small deflation, amounting to 341 nrad in the north–south components, which has been recorded coincident with a seismic swarm, lasting 40 min (Fig. 18). This observation provides a strong evidence that VT activity, which is related to normal faulting, occurs during small deflations in the background of the inflation. It may be related to the geometrical complex architecture of the magmatic feeding system and to the role of the hydrothermal fluids.

4 Conclusive Remarks

An ongoing debate on the origin of the unrest at the CFc has developed in the scientific literature since the 1970 event (Yokoyama 1971; Casertano et al. 1976; for a synthesis see Chaps. The Hydrothermal System of the Campi Flegrei Caldera, Italy; Source Modelling from Ground Deformation and Gravity Changes at the Campi Flegrei Caldera, Italy). The interpretation of the uplift has been attributed to interconnected magmatic and hydrothermal mechanisms. The modelling of recent unrest episodes indicates a variety of scenarios for the reappraisal of future volcanic activity. The extreme models, such as that proposed by Lima et al. (2009), suggest that the inflation episodes are related to the activity of the hydrothermal system due to the connection of reservoirs located at different depth. Volcanic activity would occur only if a new batch of magma would rise from depth. In contrast, many other models, such as that proposed by Gottsmann et al. (2006), explain the recent unrest episodes as due to magmatic intrusion with no significant role for the hydrothermal system.

Comparison of the historical activity of the CFc with that of other unrest calderas highlights a variety of scenarios. Large calderas, such as Yellowstone and Long Valley, recently have shown many unrest episodes not followed by an eruption (e.g., Acocella et al. 2015). The magmatic reservoirs of these calderas, however, are deeper and larger than that of the CFc (Acocella et al. 2015). Rabaul and Iwo Jima calderas, instead have generated seismic activity and unrest episodes culminated in volcanic eruptions (Acocella et al. 2015). CFc is in an intermediate state, and, while the probability of an eruption is small, the dense population makes this caldera one of the most risky in the world, even for a small eruption, which represents the most likely event, based on the volcanic activity of the last 5

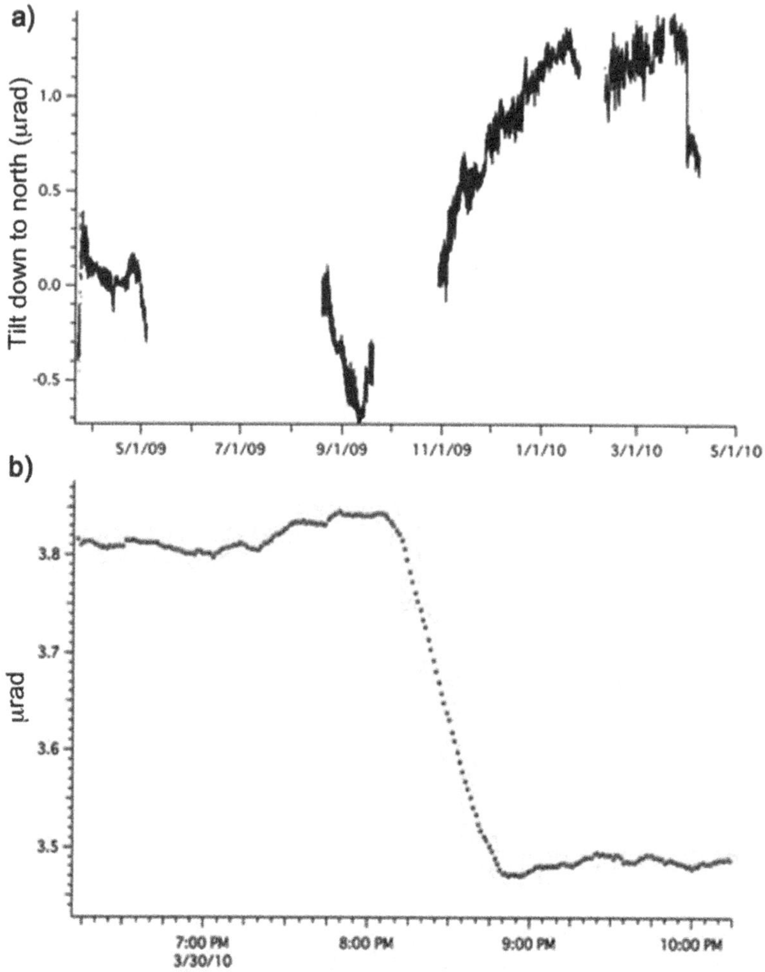

Fig. 18 a North–South tilt recorded in the South tunnel during the summer 2009 until May 2010. **b** Enlarged view of the deflation recorded at the South tunnel array during March 30, 2010

kyrs (Orsi et al. 2009; Chap. Volcanic Hazard Assessment at the Campi Flegrei Caldera, Italy).

In general, the inversion of ground deformation data can permit to infer some geometrical parameters (i.e., depth of the source and its extent) but not the dynamical ones (i.e., physical factors) controlling the process. From the published results it is evident that the inflation/deflation episodes share the same source, although the shape of the deformation fields are different. In particular, the 1969–1972 episode is different from the 1982–1984 for the absence of significant deflation (e.g., Troiano et al. 2011 who favour the hydrothermal explanation). In contrast, the petrologic models of the CFc (e.g., Arienzo et al. 2010) and the presence of shallow magmatic bodies as inferred by attenuation tomography (De Siena et al. 2010) favour a diffused field of small magma chambers located above a larger reservoir located at a 7–8 km depth. This dualism of hypotheses may be only an academic question even though cannot be neglected the possibility that a hydrothermal event may trigger a magmatic episode in an unstable system. However, at least since 1950, the CFc is in an unrest phase, with most of the energy released essentially aseismically. The low seismic efficiency has been already shown by Yokoyama (1988) who explained it as due to the relatively compliant rocks filling the central part of the caldera. The recent ground deformation data clearly indicate that the seismic swarms occur during inflation/deflation episodes. The subsidence is accompanied by seismic activity associated with normal faulting. The hypothesis that these faults are not free to slip (Troise et al. 2003) explains why the ground deformation episodes seem to anticipate the seismic activity by some months. However detailed observations performed during the more recent mini-uplift episodes (e.g., the 2005–2006 event) showed also that seismicity occurs with each variation in the rate of the ground uplift (e.g., Saccorotti et al. 2007).

Concerning the source process, the most important evidence of a magmatic origin for the actual unrest phase is the composition of the emitted gas species (Chiodini et al. 2012). The seismological data suggest that the October 2006 swarm was dominated by very shallow LP events, having hydrothermal origin (Saccorotti et al. 2007). These events, however, are quite rare. The seismic activity mostly occurs as VT events, consistent with double-couple focal mechanisms with low stress drops, ranging from 1 to 10 bars. There is no need to introduce additional and more complex seismic source models, and other proposed interpretations can be justified by the lateral inhomogeneities present within the central part of the caldera.

Consequently, the most important objective is to have a realistic picture of the possible precursory events. The recent unrest episodes have shown that the deformation related to the ground uplift is more energetic than the seismicity, which is due mostly to local VT events. But the most important observation is that the uplift, as demonstrated for the 1982–1984 sequence, preceded the seismicity by a time interval of some months. Seismic activity seems to be explained mostly by deflation, as indicated by the normal faulting and tiltmeters. LP events were recorded only during 2006 and have a quite shallow origin. Modelling of the convection in a magma chamber, as performed by Longo et al. (2011), indicate that this system should generate Ultra Long Period events, reflecting pressure transients in the frequency band of minutes with 10^{-4}–10^{-2} m displacement amplitude levels. If this should be the case particular relevance should be dedicated to very long period and high-sensitive instruments such as borehole strainmeters, long baseline tiltmeters and continuous microgravimetric measurements. This research has been recently initiated with few instruments and should be further potentiated.

Two workshops have been organised in 2006 and 2007 for checking the possibility to include the CFc in the Integrated Ocean Drilling Program and International Continental Scientific Drilling Program projects. The results of discussions stimulated an international debate on the need to continue the deep investigation at the CFc, not only for geothermal research but also for improving basic researches and monitoring. This debate has been intense, but the delay on this topic is mainly due to the lack of consensus, the small group of researchers still working on this issue, and the related organisational problems. In addition, the theme has been complicated for the urban environment and the risk of producing unexpected events such as those occurred at Lusi volcano (Indonesia) in 2006 (Mazzini et al. 2007). The relevance and costs of such a project requires the interest of a very large scientific community. Deep monitoring of strain, temperature and geochemical parameters are probably the best solution for providing a reliable early warning system for predicting future activity.

The presence of an active caldera in a densely populated area, may stimulate the need of creating an efficient early warning system in the region, merging all the ground and satellite data in a real-time system, linked to an international advisory panel, able to use statistical tools and procedures for prompt decision making. However, this is a very difficult task to realise since the area has experienced a growth in population and a significant urban expansion in the last 25 years, in spite of the recent unrest episodes.

References

Acocella V, Di Lorenzo R, Newhall C, Scandone R (2015) An overview of recent (1988 to 2014) caldera unrest: knowledge and perspectives. Rev Geophys 53 (3):1–60. https://doi.org/10.1002/2015RG000492

Amoruso A, Crescentini L, Scarpa R (2000) Removing tidal and atmospheric effects from Earth deformation measurements. Geophys J Int 140:493–505. https://doi.org/10.1046/j.1365-246X.2000.00052.x

Amoruso A, Crescentini L, Linde AT, Sacks IS, Scarpa R, Romano P (2007) A horizontal crack in a layered structure satisfies deformation for the 2004–2006 uplift of Campi Flegrei. Geophys Res Lett 34: L22313. https://doi.org/10.1029/2007GL031644

Amoruso A, Crescentini L, Sabbetta I (2014a) Paired deformation sources of the Campi Flegrei caldera (Italy) required by recent (1980–2010) deformation history. J Geophys Res Solid Earth 119:858–879. https://doi.org/10.1002/2013JB010392

Amoruso A, Crescentini L, Sabbetta I, De Martino P, Obrizzo F, Tammaro U (2014b) Clues to the cause of the 2011–2013 Campi Flegrei caldera unrest, Italy, from continuous GPS data. Geophys Res Lett 41:3081–3088. https://doi.org/10.1002/2014GL059539

Amoruso A, Crescentini L, Scarpa R, Bilham R, Linde AT, Sacks IS (2015) Abrupt magma chamber contraction and microseismicity at Campi Flegrei, Italy: Cause and effect determined from strainmeters and tiltmeters. J Geophys Res Solid Earth 120:1–12. https://doi.org/10.1002/2015JB012085

Arienzo I, Moretti R, Civetta L, Orsi G, Papale P (2010) The feeding system of Agnano-Monte Spina eruption (Campi Flegrei, Italy): Dragging the past into present activity and future scenarios. Chem Geol 270:135–147. https://doi.org/10.1016/j.chemgeo.2009.11.012

Aster R, Meyer R (1988) Three-dimensional velocity structure and hypocenter distribution in the Campi Flegrei caldera, Italy. Tectonophysics 149:195–218. https://doi.org/10.1016/0040-1951(88)90173-4

Aster RC, Meyer RP, De Natale G, Zollo A, Martini M, Del Pezzo E, Scarpa R, Iannaccone G (1992) Seismic investigation of the Campi Flegrei: a summary and synthesis of results. In: Gasparini P, Scarpa R, Aki K (eds) Volcanic seismology. IAVCEI Proceedings in Volcanology, vol 3, Springer, Berlin, Heidelberg, Germany. https://doi.org/10.1007/978-3-642-77008-1_28

Bakun WH (1984) Seismic moments, local magnitudes, and coda-duration magnitudes for earthquakes in central California. B Seismol Soc Am 74(2):439–458

Barberi F, Corrado G, Innocenti F, Luongo G (1984) Phlegraean fields 1982–1984: Brief chronicle of a volcano emergency in a densely populated area. Bull Volcanol 47:175–185. https://doi.org/10.1007/BF01961547

Battaglia J, Zollo A, Virieux J, Dello Iacono D (2008) Merging active and passive data sets in travel-time tomography: the case study of Campi Flegrei caldera (Italy). Geophys Prospect 56:555–573. https://doi.org/10.1111/j.1365-2478.2007.00687.x

Bevilacqua A, Neri A, De Martino P, Isaia R, Novellino A, D'Assisi Tramparulo F, Vitale S (2020) Radial interpolation of GPS and leveling data of ground deformation in a resurgent caldera: application to Campi Flegrei (Italy). J Geodyn 94(24):1–27. https://doi.org/10.10007/s00190-020-01355-x

Capuano P, Russo G, Vanorio T, Prevete R, Auger E, Bonagura M, Caielli G, Convertito V, Damiano N, D'Auria L, Emolo A, Lovisa L, Moretti M (2006) 1984 Campi Flegrei seismic waveforms compilation. In: Zollo A et al (eds) Geophysical exploration of the Campi Flegrei (Southern Italy) caldera' interiors: data, methods and results. Doppia Voce editore, Napoli, Italy, pp 15–24. ISBN: 88–89972–04–1

Casertano L, Oliveri Del Castillo A, Quagliariello MT (1976) Hydrodynamics and geodynamics in the Phlegrean Fields area of Italy. Nature 264:161–164. https://doi.org/10.1038/264161a0

Castellano M, Galluzzo D, La Rocca M, Capello M (2012) Lo studio dei vulcani attivi e delle strutture crostali con reti sismiche temporanee: storia, evoluzione e prospettive della Rete Sismica Mobile dell'Osservatorio Vesuviano (INGV). In: INGV (ed) Quaderni di Geofisica, vol 97. Rome, Italy, 51 pp. ISSN:1590–2595

Chiodini G, Caliro S, Cardellini C, Granieri D, Avino R, Baldini A, Donnini M, Minopoli C (2010) Long-term variations of the Campi Flegrei, Italy, volcanic system as revealed by the monitoring of hydrothermal activity. J Geophys Res 115:B03205

Chiodini G, Caliro C, De Martino P, Avino R, Gherardi F (2012) Early signals of new volcanic unrest at Campi Flegrei caldera? Insights from geochemical data and physical simulations. Geology 40:943–946. https://doi.org/10.1130/G33251.1

Cinque A, Rolandi G, Zamparelli V (1985) L'estensione dei depositi marini olocenici nei Campi Flegrei in relazione alla vulcano-tettonica. Boll Soc Geol Ital 104(2):327–348, ISSN 0037–8763

Corrado G, Guerra I, Lo Bascio A, Luongo G, Rampoldi F (1977) Inflation and microearthquake activity of Phlegraean Fields. Italy. Bull Volcanol 40(3):169–188. https://doi.org/10.1007/BF02596998

Currenti G, Zuccarello L, Bonaccorso A, Sicali A (2017) Borehole volumetric strainmeter calibration from a nearby seismic broadband array at Etna volcano. J Geophys Res 122:7229–7738. https://doi.org/10.1002/2017JB014663

D'Auria L, Giudicepietro F, Aquino I, Borriello G, Del Gaudio C, Lo Bascio D, Martini M, Ricciardi GP, Ricciolino P, Ricco C (2011) Repeated fluid transfer episodes as a mechanism for the recent dynamics of Campi Flegrei caldera (1989–2010). J Geophys Res 116:B04313. https://doi.org/10.1029/2010JB007837

De Martino P, Guardato S, Donnarumma GP, Dolce M, Trombetti T, Chierici F, Macedonio G, Beranzoli L, Iannaccone G (2020) Four years of continuous seafloor displacement measurements in the Campi Flegrei caldera. Front Earth Sci-Switz 8:615168. https://doi.org/10.3389/feart.2020.615178

De Natale G, Zollo A (1986) Statistical analysis and clustering features of the Phlegrean Fields earthquake sequence (May 1983–May 1984). B Seismol Soc Am 76(3):801–814

De Natale G, Zollo A, Ferraro A, Virieux J (1995) Accurate fault mechanism determinations for a 1984 earthquake swarm at Campi Flegrei caldera (Italy) during an unrest episode: implications for volcanological research. J Geophys Res 100(B12):24167–24185. https://doi.org/10.1029/95JB00749

De Siena L, Del Pezzo E, Bianco F (2010) Seismic attenuation imaging of Campi Flegrei: evidence of gas reservoirs, hydrothermal basins, and feeding systems. J Geophys Res 115:B09312. https://doi.org/10.1029/2009JB006938

Del Gaudio C, Aquino I, Ricciardi GP, Ricco C, Scandone R (2010) Unrest episodes at Campi Flegrei: A reconstruction of vertical ground movements during 1905–2009. J Volcanol Geotherm Res 195:48–56. https://doi.org/10.1016/j.jvolgeores.2010.05.014

Del Pezzo E, De Natale G, Martini M, Zollo A (1987) Source parameters of microearthquakes at Phlegrean Fields (Southern Italy) volcanic area. Phys Earth Planet in 47:25–42. https://doi.org/10.1016/0031-9201(87)90064-1

Di Lieto B, Romano P, Bilham R, Scarpa R (2021) Aseismic strain episodes at Campi Flegrei caldera, Italy. Adv Geosci 52:119–129. https://doi.org/10.5194/adgeo-52-119-2021

Di Renzo V, Civetta L, D'Antonio M, Tonarini S, Di Vito MA, Orsi G (2011) The magmatic feeding system of the Campi Flegrei caldera: architecture and temporal evolution. Chem Geol 281:227–241. https://doi.org/10.1016/j.chemgeo.2010.12.010

Di Vito MA, Isaia R, Orsi G, Southon J, de Vita S, D'Antonio M, Pappalardo L, Piochi M (1999) Volcanism and deformation since 12,000 years at the Campi Flegrei caldera. J Volcanol Geotherm Res 91:221–246. https://doi.org/10.1016/S0377-0273(99)00037-2

Di Vito M, Acocella V, Aiello G, Barra D, Battaglia M, Carandente A, Del Gaudio C, de Vita S, Ricciardi GP, Ricco C, Scandone R, Terrasi F (2016) Magma transfer at Campi Flegrei caldera (Italy) before the 1538 AD eruption. Sci Rep 6:32245. https://doi.org/10.1038/srep32245

Gaudiosi G, Iannaccone G (1984) A preliminary study of stress pattern at Phlegrean Fields as inferred from focal mechanism. Bull Volcanol 47:225–232. https://doi.org/10.1007/BF01961552

Gottsmann J, Folch A, Rymer H (2006) Unrest at Campi Flegrei: A contribution to the magmatic versus hydrothermal debatefrom inverse and finite element modeling. J Geophys Res 111:B07203. https://doi.org/10.1029/2005JB003745

Guidoboni E, Ciuccarelli C (2011) The Campi Flegrei caldera: historical revision and new data on seismic crises, bradyseisms, the Monte Nuovo eruption and ensuing earthquakes (twelfth century 1582 AD). Bull Volcanol 73:655–677. https://doi.org/10.1007/s00445-010-0430-3

Hart RHG, Gladwin MT, Gwyther RL, Agnew DC, Wyatt FK (1996) Tidal calibration of borehole strain meters: removing the effect of small-scale inhomogeneity. J Geophys Res 101(B11):25,553–25,571. doi.org/https://doi.org/10.1029/96JB02273

Imbò G (1977) Seismic prediction at Pozzuoli. Nature 227:211. https://doi.org/10.1038/277511b0

Imbò G (1980) Ancora sulla crisi bradisismica puteolana del 1970. Acc Pontaniana 29:53–83

Judenherc S, Zollo A (2004) The Bay of Naples (Southern Italy): Constraints on the volcanic structures inferred from a dense seismic survey. J Geophys Res 109: B10312. https://doi.org/10.1029/2003JB002876

La Rocca M, Galluzzo D (2012) A seismic array in the Town of Pozzuoli in Campi Flegrei (Italy). Seism Res Lett 83(1):86–96. https://doi.org/10.1785/gssrl.83.1.86

Lima A, Vivo BD, Spera FJ, Bodnar RJ, Milia A, Nunziata C, Belkin HE, Cannatelli C (2009) Thermodynamic model for uplift and deflation episodes (bradyseism) associated with magmatic–hydrothermal activity at the Campi Flegrei (Italy). Earth-Sci Rev 97:44–58. https://doi.org/10.1016/j.earscirev.2009.10.001

Lomax A, Virieux J, Volant P, Berge C (2000) Probabilistic earthquake location in 3D and layered models: introduction of a Metropolis-Gibbs method and comparison with linear locations. In: Thurber CH, Rabinowitz N (eds) Advances in seismic event location. Kluwer, Amsterdam, The Netherlands, pp 101–134

Longo A, Papale P, Vassalli M, Saccorotti G, Montagna CP, Cassioli A, Giudice S, Boschi E (2011) Magma convection and mixing dynamics as a source of Ultra-Long-Period oscillations. Bull Volcanol. https://doi.org/10.1007/s00445-011-0570-0

Mazzini A, Svensen H, Akhmanov GG, Aloisi G, Planke S, Melthe-Sorenssen A, Istadi B (2007) Triggering and dynamic evolution of the LUSI mud volcano, Indonesia. Earth Planet Sci Lett 261:375–388. https://doi.org/10.1016/j.epsl.2007.07.001

Minakami T (1960) Fundamental research for predicting volcanic eruptions. Part I. Bull Earthq Res Inst, Tokyo University 38:497–544

Moretti R, Arienzo I, Orsi G, Civetta L, D'Antonio M (2013) The deep plumbing system of the Ischia island (southern Italy): a physico-chemical and geodynamic window on the fluid-sustained and CO_2-dominated magmatic source of Campanian volcanoes. J Petrol 54(5):951–984. https://doi.org/10.1093/petrology/egt002

Orsi G, D'Antonio M, de Vita S, Gallo G (1992) The Neapolitan Yellow Tuff, a large-magnitude trachytic phreatoplinian eruption:eruptive dynamics, magma withdrawal and caldera collapse. J Volcanol Geotherm Res 53:275–287.https://doi.org/10.1016/0377-0273(92)90086-S

Orsi G, de Vita S, Di Vito M (1996) The restless, resurgent Campi Flegrei nested caldera (Italy): constraints on its evolution andconfiguration. J Volcanol Geotherm Res 74(3–4):179–214. https://doi.org/10.1016/S0377-0273(96)00063-7

Orsi G, Civetta L, Del Gaudio C, de Vita S, Di Vito MA, Isaia R, Petrazzuoli SM, Ricciardi GP, Ricco C (1999) Short-term deformations and seismicity in the resurgent Campi Flegrei caldera (Italy): an example of active block-resurgence in a densely populated area. J Volcanol Geotherm Res 91:415–451. https://doi.org/10.1016/S0377-0273(99)00050-5

Orsi G, Di Vito MA, Isaia R (2004) Volcanic hazard assessment at the restless Campi Flegrei caldera. Bull Volcanol 66:514–530. https://doi.org/10.1007/s00445-003-0336-4

Orsi G, Di Vito M, Selva J, Marzocchi W (2009) Long-term forecast of eruption style and size at Campi Flegrei caldera (Italy). Earth Planet Sci Lett 287:265–276. https://doi.org/10.1016/j.epsl.2009.08.013

Reasenberg P, Oppenheimer D (1985) FPFIT, FPPLOT and FPPAGE: Fortran computer programs for calculating and displaying earthquake fault-plane solutions. US Geol Surv Open File Rep 739:109

Ricco C, Aquino I, Del Gaudio C (2003): Ground tilt monitoring at Phlegraean Fields (Italy): a methodological approach. Ann Geophys 46(6):1297–1314

Ricco C, Aquino I, Borgstrom S, Del Gaudio C (2007) A study of tilt change recorded from July to October 2006 at the Phlegrean Fields (Naples, Italy). Ann Geophys 50(5):661–674. https://doi.org/10.4401/ag-3059

Saccorotti G, Bianco F, Castellano M, Del Pezzo E (2001) The July-August 2000 seismic swarms at Campi Flegrei volcanic complex. Italy. Geophys Res Lett 28(13): 2525–2528. https://doi.org/10.1029/2001GL013053

Saccorotti G, Petrosino S, Bianco F, Castellano M, Galluzzo D, La Rocca M, Del Pezzo E, Zaccarelli L, Cusano P (2007) Seismicity associated with the 2004–2006 renewed round uplift at Campi Flegrei caldera, Italy. Phys Earth Planet in 165:14–24. https://doi.org/10.1016/j.pepi.2007.07.006

Scarpa R, Amoruso A, Crescentini L, Linde AT, Sacks IS, Del Pezzo E, Martini M (2000) Forecasting volcanic eruptions: the case of Vesuvius and Campi Flegrei. In: Boschi E, Ekstrom G, Morelli A (eds) Problems in geophysics for the new millennium. Compositori, Bologna, Italy, pp 13–26

Scarpa R, Amoruso A, Crescentini L, Romano P, De Cesare W, Martini M, Scarpato G, Linde AT, Sacks SI (2007) New borehole strain system detects uplift at Campi Flegrei. EOS Trans AGU 88(18):197–203. https://doi.org/10.1016/j.pepi.2007.07.006

Snoke JA, Munsey JW, Teague AG, Bollinger GA (1984) A program for focal mechanism determination by combined use of polarity and Sv-P amplitude ratio data. Earthquake Notes 55(3):15–20

Tamura Y, Sato T, Ooe M, Ishiguro M (1991) A procedure for tidal analysis with a Bayesian information criterion. Geophys J Int 104(3):507–516. https://doi.org/10.1111/j.1365-246X.1991.tb05697.x

Tazieff H (1971) Petit discourse à propos de la panique qui eut lieu a Pozzuoli a l'occasion du Bradisismo Flegreo. Stromboli, Italy

Tazieff H (1977) La Soufrière, volcanology and forecasting. Nature 269:9. https://doi.org/10.1038/269096a0

Troiano A, Di Giuseppe MG, Petrillo Z, Troise C, De Natale G (2011) Ground deformation at calderas driven by fluid injection: modelling unrest episodes at Campi Flegrei (Italy). Geophys J Int 187:833–847. https://doi.org/10.1111/j.1365-246X.2011.05149.x

Troise C, Pingue F, De Natale G (2003) Coulomb stress change at calderas: modelling the seismicity of Campi Flegrei (Southern Italy). J Geophys Res 108:2292. https://doi.org/10.10291/2002J13002006

Vanorio T, Virieux J, Capuano P, Russo G (2005) Three-dimensional seismic tomography from P wave and S wave microearthquake travel times and rock physics characterization of the Campi Flegrei caldera. J Geophys Res 110:B03201. https://doi.org/10.1029/2004JB003102

Yokoyama I (1971) Pozzuoli event in 1970. Nature 229:532–533. https://doi.org/10.1038/229532a0

Yokoyama I (1988) Seismic energy release from volcanoes. Bull Volcanol 50:1–13. https://doi.org/10.1007/BF01047504

Zollo A, Judenherc S, Virieux J, Makris J, Auger E, Capuano P, Chiarabba C, De Franco R, Michelini A, Musacchio G (2003) Evidence for the buried rim of Campi Flegrei caldera from 3-d active seismic imaging. Geophys Res Lett 30(19), doi:https://doi.org/10.1029/2003GL018173

Zollo A, Maercklin N, Vassallo M, Dello Iacono D, Virieux J, Gasparini P (2008) Seismic reflections reveal a massive melt layer feeding Campi Flegrei caldera. Geophys Res Lett 35:L12306. https://doi.org/10.1029/2008GL034242

Source Modelling from Ground Deformation and Gravity Changes at the Campi Flegrei Caldera, Italy

Maurizio Bonafede, Antonella Amoruso, Luca Crescentini, Joachim H. Gottsmann, Micol Todesco, and Elisa Trasatti

Abstract

The deformation history of the Campi Flegrei caldera during the last decades consists of two large uplift events in 1970–1972 and 1982–1984, with ~3.5 m cumulative uplift, occurring at a rate of ~1 m/yr. Both events were accompanied by seismic activity, gravity changes and compositional variations of volcanic gases but no eruption took place. During the following decades, the area has been slowly subsiding but minor uplift episodes (~a few cm), seismic swarms and changes in degassing activity took place, showing that the area persisted in a near-critical state. Since November 2005, ground deformation resumed, although at a slower rate, totalling a displacement of ~0.77 m (February 2021). In this chapter, we present a retrospective analysis of ground deformation leading to a critical re-evaluation of the 1982–1984 uplift and of the following deflation, employing the most updated modelling techniques. Deformation and gravity data provide important constraints on depth, volume, mass density and dislocation mechanisms accompanying magma emplacement processes. The correct evaluation of these parameters is strongly conditioned by simplifying assumptions built in the different inversion procedures: in particular, the inferred source depth ranges from 5.5 km to less than 3 km and the intrusion density ranges from values pertinent to aqueous fluids to typical magmatic values. This review depicts the salient phases of the deformation history of this densely populated and high-risk volcanic area, helping to address debated issues, such as the role of the magmatic system, and their interaction with the shallower hydrothermal system. In spite of the mentioned difficulties, the following conclusion may be considered as firmly established: during 1982–1984 a magmatic intrusion took place, and the subsequent complex deformation history (1985–2010) was mainly controlled by the exsolution of volatiles with magmatic origin and their interaction with a shallow hydrothermal system. However, data collected after 2011 were modelled in terms of

M. Bonafede (✉)
Dipartimento di Fisica e Astronomia, Università degli Studi di Bologna, Bologna, Italy
e-mail: maurizio.bonafede@unibo.it

A. Amoruso · L. Crescentini
Dipartimento di Fisica, Università degli Studi di Salerno, Salerno, Italy

J. H. Gottsmann
Department of Earth Sciences, University of Bristol, Bristol, UK

M. Todesco
Istituto Nazionale di Geofisica e Vulcanologia, Sezione di Bologna, Bologna, Italy

E. Trasatti
Istituto Nazionale di Geofisica e Vulcanologia, Sezione Osservatorio Nazionale Terremoti, Roma, Italy

© Springer-Verlag GmbH Germany, part of Springer Nature 2022
G. Orsi et al. (eds.), *Campi Flegrei*, Active Volcanoes of the World,
https://doi.org/10.1007/978-3-642-37060-1_11

a deep inflating source of deformation, possibly resulting from a resumed magmatic recharge at depth.

1 Introduction

Campi Flegrei caldera (CFc) is a complex, nested and resurgent structure including a submerged and a subaerial part, at the western edge of the Naples Bay (Orsi et al. 1992, 1996). The last eruption took place in AD 1538 and since then intense degassing, seismic swarms and several episodes of ground uplift have been observed. The eruptive history and the structural setting of the area are reconstructed, among others, by Rosi et al. (1983), Orsi et al. (1996), De Vivo et al. (2001) and synthetised in Chap. Volcanic and Deformation History of the Campi Flegrei Volcanic Field, Italy. The caldera floor has been generally sinking in the last 20 centuries, at an average rate of 1.3 cm/yr at Serapeo, an ancient Roman market in the caldera centre, near the harbour of Pozzuoli (Parascandola 1947; Berrino et al. 1984), but large uplift events took place episodically (e.g., Morhange et al. 2006; Del Gaudio et al. 2010; Todesco et al. 2014; see also Chaps Volcanic and Deformation History of the Campi Flegrei Volcanic Field, Italy; The Permanent Monitoring System of the Campi Flegrei Caldera, Italy; The Hydrothermal System of the Campi Flegrei Caldera, Italy; Historic Unrest of the Campi Flegrei Caldera, Italy). Recently, two large uplift episodes occurred in 1970–1972 and 1982–1984, with about 3.5 m cumulative uplift in the city of Pozzuoli (Fig. 1). In both cases, the deformation field gradually vanished 6–7 km from the town, just beyond the caldera rim. These uplift episodes were very rapid, occurring at a rate of $\sim$1 m/yr and were accompanied by diffuse seismic activity, microgravity changes and compositional variations of geochemical emissions but no eruption took place. During the following decades, the area has been generally subsiding but minor uplift episodes, seismic swarms and changes in degassing activity took place showing that the area is in a critical state on the verge of instability. In 2004, a new phase of ground uplift began (Fig. 1b), characterised by a slower rate and still ongoing (Iannaccone et al. 2018). A retrospective analysis leading to a critical re-evaluation of the previous events, employing the most updated modelling techniques, may help solving unresolved problems and to depict more precisely the recent history of this densely populated and thus high risk active volcanic area (for a review see Chap. Volcanic Hazard Assessment at the Campi Flegrei Caldera, Italy).

2 Summary of Geodetic and Gravimetric Data

Since March 1970 levelling surveys were regularly carried out to monitor the elevation changes at the CFc (see Del Gaudio et al. 2010). The establishment of additional levelling benchmarks during the 1982–1984 episode improved the spatial definition of the uplift pattern. Furthermore, the vertical deformation was periodically monitored through 5 tide gauges placed in the harbour of Pozzuoli (close to the site of the detected maximum uplift), along the coastline of the Pozzuoli Bay and one in Naples (Fig. 2a). Levelling data for the 1982–1984 episode evidenced a maximum uplift of 1.8 m in the city of Pozzuoli (Fig. 2b).

The relative pattern of uplift (normalised to the maximum value) remained practically unchanged along the 2 years of unrest. Horizontal deformation was monitored through Electro-optical Distance Measuring (EDM) surveys during and after the main phase of unrest. However, only in June 1980 and in June 1983 measurements were computed in a large number of benchmarks, allowing to map changes of horizontal distances during the unrest. The direction of horizontal displacement is nearly radial if the reference point is suitably chosen (Fig. 2c) but EDM data show some asymmetry, the eastern sector of the caldera being characterised by larger displacements with respect to the western and northern sectors. Gravity data were also recorded regularly at a few benchmarks (Berrino 1994). During the uplift phase, the ratio

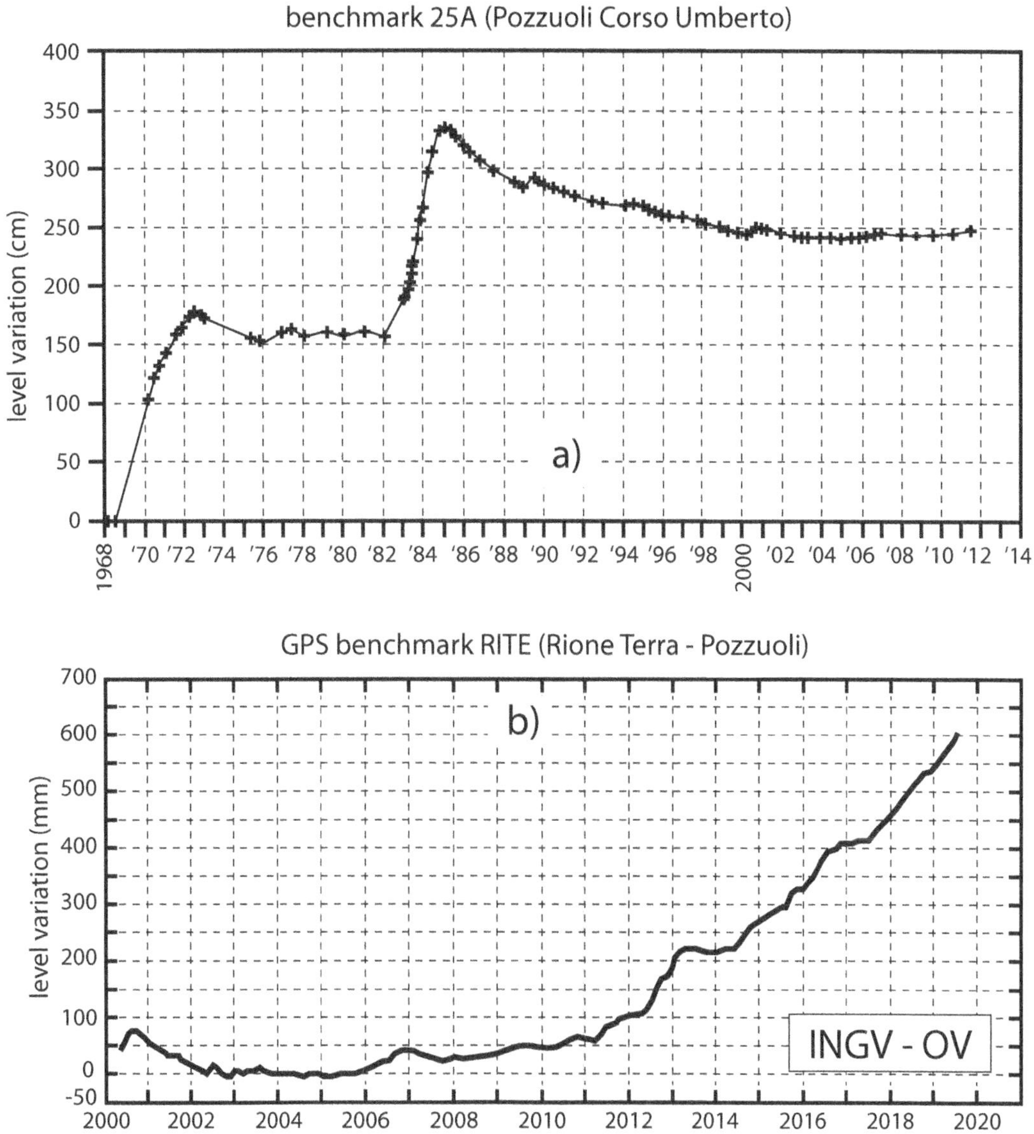

Fig. 1 **a** Height variation since 1968 at the Campi Flegrei caldera centre, Corso Umberto, Pozzuoli (modified after http://www.ov.ingv.it/). **b** Time series of weekly elevation changes at the RITE GPS benchmark, at Rione Terra in Pozzuoli, July 2000 to July 2019 (modified after http://www.ov.ingv.it/ov/bollettini-mensili-campania/Bollettino_Mensile_Campi_Flegrei_2019_07.pdf)

of gravity change to uplift, was -216 ± 6 μGal/m at Serapeo, in good agreement with the average of all the stations -213 ± 6 μGal/m (Fig. 2d).

Unfortunately, the CFc is partially submerged and underwater gravity changes are not measured, while offshore geodetic data only became available since 2011 (De Martino et al. 2014; Iannaccone et al. 2018). Therefore, in the preceding period only about 2/3 of the caldera deformation domain was imaged.

Since 1985 a deflation phase began and about 30–40% of the uplift was recovered in about 10 years (Fig. 1). Superimposed to the deflation trend since 1985, minor uplift episodes (termed "miniuplifts") of the order of few cm each, took

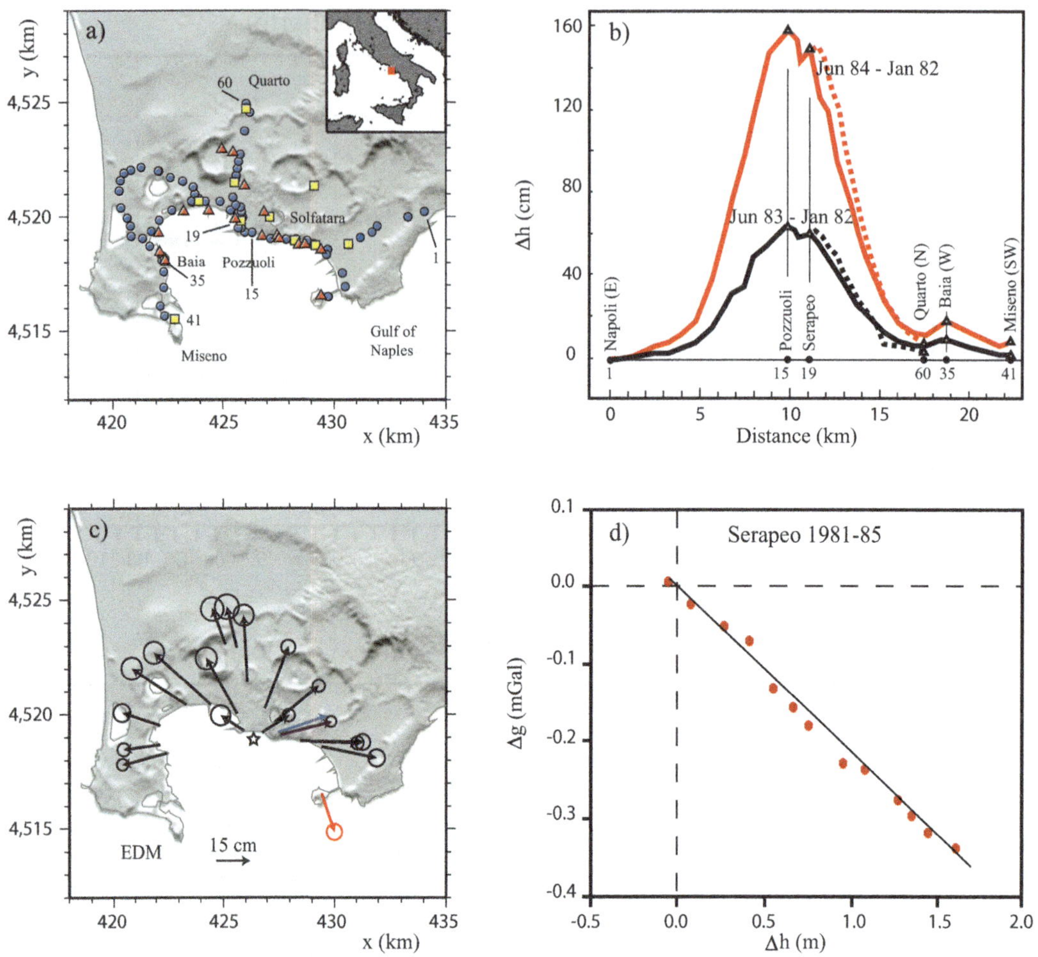

Fig. 2 Sketch of the Campi Flegrei caldera geodetic dataset for the 1982–1984 unrest. **a** Geodetic and gravity benchmarks surveyed during the 1982–1984 crisis: levelling (blue circles), EDM (red triangles) and gravity (yellow squares) stations. **b** Spatial pattern of uplift measured in June 1983 (black) and in June 1984 (red) w.r.t. to January 1982; the dotted lines are relative to the Pozzuoli-Quarto baseline: the maximum uplift was always found at benchmark no. 15 close to the centre of Pozzuoli. **c** EDM displacements between June 1980 and June 1983: all displacements but one (red arrow) are approximately radial when assuming the black star as the reference point. No error is given to the Accademia displacement (blue arrow) as it was originally used as reference for the EDM data set. **d** Gravity change Δg vs. uplift Δh at Serapeo of Pozzuoli benchmark. Figure modified after Trasatti et al. (2011), Amoruso et al. (2014a) and Berrino et al. (1994)

place in 1989, 2000–2001 and 2004–2006 (Chiodini 2009) often accompanied by peculiar seismic activity suggesting the involvement of hydrothermal fluids; the major phase of deflation apparently stopped after the year 2000 (Fig. 1).

A new phase of unrest started in November 2005, with the onset of a slower and longer-lasting uplift (De Martino et al. 2014). Since then, the uplift recorded at the Rione Terra GPS station, in Pozzuoli, has reached 60 cm (Fig. 1b). Global Positioning System (GPS) monitoring of ground deformation at the CFc began in year 2000.

Since 1992, the ground deformation at CFc has been monitored also by Synthetic Aperture Radar (SAR). Velocity time series from SAR analysis show that the deformation also extends outside the volcanological limits of the

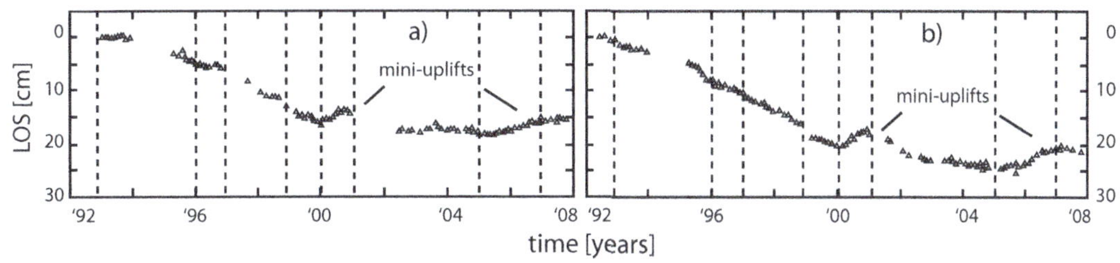

Fig. 3 Time series for the displacement along the line of sight in Pozzuoli for the **a** ascending and **b** descending orbits. Figure modified after Manconi et al. (2010)

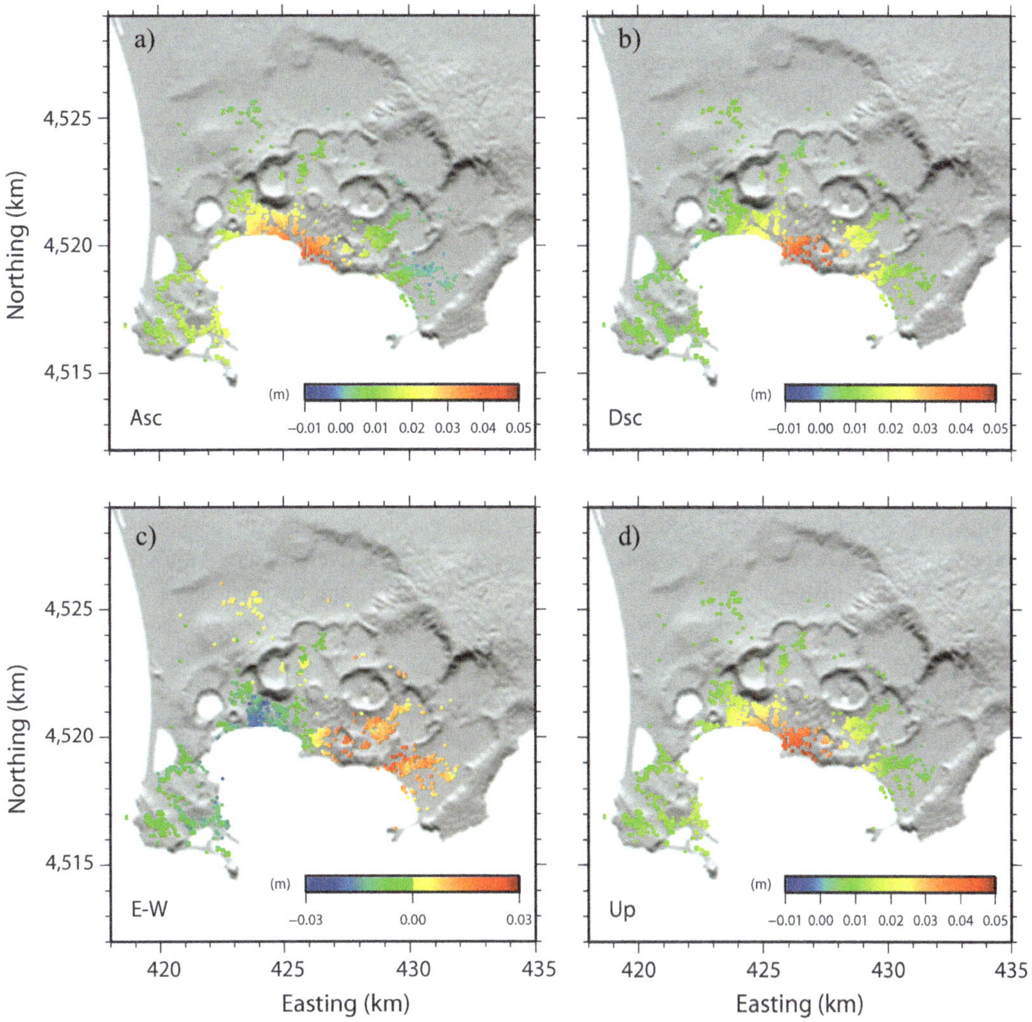

Fig. 4 SAR results for the 2004–2006 unrest. Total displacements from mean deformation velocity maps computed from the **a** ascending and **b** descending data acquired between summer 2004 and November 2006. Mean **c** east–west and **d** vertical deformation maps computed from both SAR geometries. Figure modified after Trasatti et al. (2008b)

Neapolitan Yellow Tuff caldera (see Chap. Volcanic and Deformation History of the Campi Flegrei Volcanic Field, Italy), without significant discontinuities (Manconi et al. 2010). Time series of the displacement at Pozzuoli between 1992 and 2008 clearly evidence a sub-linear subsidence trend of 2–3 cm/yr during 1992–2000, interrupted by two uplift phases in 2000–2001 and 2004–2006 (Fig. 3). The mini-uplift episodes are also monitored in detail by SAR interferometry. In particular, during 2004–2006, the total uplift amounted to 4 cm, as documented by SAR time series (Fig. 4; Trasatti et al. 2008b). The maximum deformation rate (up to 2 cm/yr) was in the area of the Pozzuoli harbour. After a deflation phase in 2007, following the 2004–2006 crisis, the ground uplift at the caldera centre evidenced a progressive increase until today, with a rate of a few cm/yr. Occasionally, the displacement rate increased up to 1 cm/month (e.g., between April and June 2011), going back to the background rate (from http://www.ov.ingv.it/). This irregular uplift rate reached a peak value of 3 cm/month during December 2012, when the volcano alert was shifted to the attention level (D'Auria et al. 2015a; www.protezionecivile.it).

Thus, the developments of modern volcano geodesy have clearly revealed a complex short-term deformation history at the CFc, renewing the concern following the 1982–1984 unrest episode. However, the interpretative models suggested so far lead to explanations not always in agreement with each other.

As discussed in the following, deformation and gravity changes may be interpreted as the result of instabilities of the hydrothermal system as well as due to magmatic intrusions. Since the controversy between magmatic and hydrothermal origin of deformation is of major importance for volcanic hazard assessment, the more realistic is the model (with parameters constrained by multidisciplinary data), the more reliable are the results and the ensuing interpretations. The first papers on the CFc unrest ascribed the ground deformation to magmatic intrusion, and the initial interpretation of measured gravity/elevation ratios ($\Delta g/\Delta h$) confirmed this hypothesis (e.g., Berrino et al. 1984; Bianchi et al. 1987; Berrino 1994). However, the same pattern of deformation may be produced by a pressurised hydrothermal source (Bonafede 1991) so that the answer to the main question is not univocal. In the following sections of the present chapter, we illustrate the advances in source modelling at the CFc, showing the progress in the definition of the physical model of its volcanic source. The deformation history of the last decades and the main modelling issues are discussed, focussing on the 1982–1984 and on the subsequent minor unrest episodes.

3 Inverse Problem for a Single Source: The 1982–1984 Uplift

The deformation source in a volcanic area is generally modelled as a cavity with pre-assigned shape, inflating under the increasing pressure due to the intrusion of fluid magma. Ground deformation is considered a reliable indicator of the location, depth, shape and incremental volume of the magmatic intrusion. From the knowledge of these parameters, the density of the emplaced material can be estimated, in principle, if gravity changes are available to constrain mass variations, and the overpressure within the source can be inferred if the dimensions of the source are known and elastic rheology is assumed.

3.1 Isotropic Point-Source in a Homogeneous Medium

Source parameters are typically inferred employing analytical solutions in elastic, homogeneous half-space models, under the point-source approximation (source dimensions much less than its depth). The most popular analytical model is the Mogi source (Mogi 1958), which describes the deformation due to a small spherical cavity, expanding isotropically in a homogeneous and elastic half-space. The bell-shaped vertical pattern of levelling measurements at the CFc during the 1982–1984 unrest (Fig. 2) is nicely fitted by a Mogi source located at ~ 3 km depth beneath Pozzuoli, at the caldera centre (Berrino et al. 1984; Berrino 1994; Trasatti et al. 2005), with an inferred source density of $\sim 2{,}500$ kg/m^3 (Table 1) which is compatible with a magmatic intrusion (Lesher and Spera 2015). Gottsmann et al. (2006a) proposed a prolate spheroid source undergoing a volume change of 71×10^6 m^3 as the best-fitting solution for a homogeneous, isotropic, elastic half-space model. Density values from gravimetric data inversion ($\sim 1{,}600$ kg/m^3) allowed these authors to suggest a hybrid nature for the causative source, with both magmatic and hydrothermal components.

It is important to point out that an isotropic (Mogi) source has some important peculiarities when the incremental volume and the density of the intrusion are considered. As discussed in Bonafede and Ferrari (2009), the volume ΔV_m of magma emplaced within the source boundary is typically larger than the outward volume expansion ΔV_{out} of the source, since the overpressure also contracts the material already resident within the source by ΔV_{in} so that $\Delta V_m = \Delta V_{out} + \Delta V_{in}$ and $\Delta V_m = 1.8 \Delta V_{out}$ if the material within the source has the same bulk modulus of the surrounding rock (assuming a Poisson ratio $\nu = \frac{1}{4}$). Furthermore, the gravity change Δg measured at the ground surface, due to the inflation of a buried pressurised cavity, can be generally split into 4 different contributions. One contribution, Δg_L, is due to the displacement of density boundaries (in a homogeneous medium it is simply due to the ground uplift Δh, it is positive if $\Delta h > 0$ and can be computed from the Bouguer formula). Another contribution, Δg_V, results from the density changes of the rock volume surrounding the deformation source (positive if the volume is compressed). A third contribution, Δg_S, is due to the normal displacement of the cavity boundary (negative for the expansion of a massless cavity). A final forth contribution, Δg_m, arises from the input of new mass within the source. The Mogi source has the peculiar property that $\Delta g_L + \Delta g_V + \Delta g_S = 0$ so that the residual gravity change, after free air correction, can be directly interpreted in terms of the mass added to the source interior: $\Delta g = \Delta g_m$. This simple result is not valid for other pressurised cavities, so that changing the source shape entails drastically different estimates of the density, as will be shown in the next section.

One of the problems regarding the employment of Mogi models in a homogeneous elastic medium at the CFc is that these models require unrealistic overpressures as high as 500 MPa in order to reproduce the detected uplift magnitude (Trasatti et al. 2005; Gottsmann et al. 2006a). Similar overpressure values exceed by 1 or 2 orders of magnitude the maximum shear stress sustainable by a typical basaltic rock. This problem can be overcome if a plastic (Trasatti et al. 2005) or a visco-elastic (Bonafede and Ferrari 2009) rheology is assumed for the embedding medium. However, the inferred source depth at ~ 3 km seems too shallow to be reconciled with the brittle behaviour of rocks at 4.5 km depth (as inferred from seismic hypocentres) and with deep geothermal drillings showing temperatures of ~ 420 °C at 2.7 km depth. Finally, since the gravity changes measured at the Serapeo benchmark during the deflation phase are very similar (224 ± 24 μGal/m) to those measured during the uplift phase (216 ± 7 μGal/m), a large fraction of magma ought to be removed from a Mogi source during the deflation phase after 1984, what seems implausible in the absence of an eruption. Therefore, more sophisticated models are required to interpret correctly experimental data and to understand the magmatic dynamics at CFc.

3.2 Non Isotropic Sources in a Homogeneous Half-Space

The a priori assumption of the shape of the source may influence not only the incremental volume expected to be involved in the unrest but, more important, leads to very different estimates of its density, affecting the interpretation of the physical origin of the unrest. By inverting

Table 1 Summary of the main results related to the source modelling of uplift and deflation episodes at Campi Flegrei since 1982

Inflation 1982–1984

Single source models

Source model	Depth (km)	Semi-major axis (km)	Semi-minor axis (km)	$\Delta V \times 10^6$ (m^3)	$\rho \times 10^3$ (kg/m^3)	Notes
Mogi elastic[a, b]	3			58–80	2.5	Homogeneous halfspace
Mogi viscoelastic[c]	3			$\Delta V_0 = 68$ $\Delta V_m = 120$	1.47–1.76	""
Sphere elastic[d]	2.7	0.7		64		""
Penny-shaped crack[d, e]	3.0	1	0	87	0.6	""
Prolate spheroid[d]	2.9	2.2	1.4	71	1.63	""
Tri-axial moment HOM1[f]	3.9	$M_2/M_1 \approx 0.3$	$M_3/M_1 \approx 0$		NA	""
Sphere plastic[g]	4–5	1			NA	Axially homogeneous
Triaxial moment HET1[f]	5.2	$M_2/M_1 \approx -0.35$	$M_3/M_1 \approx -0.25$		2–3	Laterally heterogeneous elastic
Horizontal circular crack[h]	3	2.75	2.75	22	2–3	Horizontally layered
Finite sill-like triaxial ellipsoid[m]	3.6	2.3	2.0	17	NA	Horizontally layered

Multiple source models

Source model	z_1 (km)	R_1 (m)	$\rho_1 \times 10^3$ (kg/m^3)	z_2 (km)	R_2 (m)	$\rho_2 \times 10^3$ (kg/m^3)	Horizontal distance (km)	Notes
Deep sill + triaxial moment HET2 [f]	7.5	horizontal crack (Point-source)	$\rho_1 V_1 = \rho_2 V_2$	5.5	$M_2/M_1 \approx 3.2$ $M_3/M_1 \approx 0.14$	2.4	0 (imposed)	Mass conservation—Laterally heterogeneous elastic halfspace
2 pressurised spheres [i]	5.3	1,286	3.6	2.0	1,374	1.4	2.1	Interacting spheres

(continued)

Table 1 (continued)

Deflation						
Deflation						
Single source models						
Source model	Depth z (km)	Semi-major axis (km)	Semi-minor axis (km)	$\Delta V \times 10^6$ (m^3)	$\rho \times 10^3$ (kg/m^3)	Notes
Spheroid [d]	2.9	2.2	1.4	-19	2.95 (h)	Rapid deflation 1985–87
Spheroid [d]	2.9	2.2	1.4	-19.8	1.87 (h)	Slow deflation 1988–2001
Finite sill-like triaxial ellipsoid [m]	3.6	2.3	2.0	-13.2	NA	Horizontally layered 1985–1988
Finite sill-like triaxial ellipsoid [m]	3.6	2.3	2.0	-5.9	NA	Horizontally layered 1989–1992
Finite sill-like triaxial ellipsoid [m]	3.6	2.3	2.0	-4.9	NA	Horizontally layered 1995–2000

Multiple source models								
Source model	z_1	$\Delta V_1 \times 10^6$ (m^3)	ρ_1	z_2	$\Delta V_2 \times 10^6$ (m^3)	ρ_2	Δr(km) (horizontal distance)	Notes
2 spherical [i]	5.0	-9.6	3.8	2.0	-7.3	1.60	2.2	Rapid deflation 1985–87
2 spherical [i]	1.1	-1.7	3.8	1.8	-7.3	1.60	2.9	Slow deflation 1988–2001

Mini-uplift 2004–2006				
Single source model				
Source model	z (km)	Aspect ratio	$\Delta V \times 10^6$ m^3	Notes
Spherical and spheroidal [j]	3	1 1.5	1.1	Laterally heterogeneous elastic
Penny-shaped crack [k]	3	>>1 (∞)	1.4	Horizontally layered elastic

NA = not available. [a]Berrino et al. (1984), [b]Berrino (1994), [c]Bonafede and Ferrari (2009), [d]Gottsmann et al. (2006a), [e]Battaglia et al. (2006), [f]Trasatti et al. (2011), [g]Trasatti et al. (2005), [h]Amoruso et al. (2008), [i]Gottsmann et al. (2006b), [j]Trasatti et al. (2008a), [k]Amoruso et al. (2007), [m]Amoruso et al. (2014a) geodetic and gravity data, Battaglia et al. (2006) found that levelling and gravimetric data are better fitted by a horizontal penny-shaped crack, as formerly considered by Dvorak and Berrino (1991). In this case, the source is placed at a similar depth (2.5–3.0 km), but the inferred volume variation is lower, being 20×10^6 m^3 instead of 60–70 $\times 10^6$ m^3, and the resulting density is as low as 600 ± 500 kg/m^3. This result must be ascribed to the rock on both sides of the crack being strongly compressed so that Δg_V accounts for a larger fraction of Δg. Since near-critical water densities are in the order of a few hundred kg/m^3, this value would indicate a source driven by hydrothermal instability rather than a magmatic intrusion. However, if the source shape is assumed as a prolate spheroid, the density inferred is quite high, being about 3,300 kg/m^3 a value even higher than that pertinent to trachybasalts at the CFc (2,300–2,500 kg/m^3; Battaglia et al. 2006). Similar inferences were confirmed by Trasatti and Bonafede (2008) employing a numerical procedure based on the Finite Element method (deployed to consider heterogeneous media and applied to a homogeneous half-space for comparison tests). An alternative finding for a finite spheroid centred at 2.9 km depth (vertical semi-major axis 2.2 km, semi-minor 1.4 km; Table 1) was found by Gottsmann et al. (2006b), estimating a volume variation of 70×10^6 m^3 (similar to Mogi model), but the density, obtained inverting gravity data separately, amounted to 1,600 kg/m^3, too low to be magma, but too high to be aqueous fluids (hence they suggested a hybrid system of magmatic/hydrothermal origin).

The contrasting results obtained employing different pressurised cavities (Trasatti and Bonafede 2008) show how sensitive density estimates are to the source shape, suggesting that the shape should not be assigned a priori but should be parameterised in such a way as to be recoverable from data inversion. Unfortunately, no convenient scheme exists to date to infer the shape of the surface over which overpressure boundary conditions are imposed: the most general geometric source is a tri-axial point-like pressurised ellipsoids (Davis 1986). However, a very general dislocation source model, accounting for tensile and shear contributions to deformation, is a moment tensor source. The moment tensor is a symmetric tensor describing the system of equivalent body forces producing a given deformation field. Its components may be retrieved by data optimisation. An interpretation of the moment tensor in terms of physically plausible source models (e.g., a pressurised cavity or a shear fault) must be performed "a posteriori". Trasatti et al. (2011) showed that inverting the vertical and horizontal deformation data (1982–1984) in terms of a general moment tensor, the source mechanism (HOM1) is located at 3.9 km depth but cannot be simply interpreted in terms of a pressurised ellipsoid since a significant shear component is needed (Fig. 5).

Finally, most of the source models mentioned so far face the same difficulties (the relatively low temperatures and the brittle behaviour at the inferred shallow depths) already mentioned for the isotropic source. This suggests the need to reconsider the commonly made assumption of a homogeneous elastic medium.

3.3 Finite Dimension Sources

Whatever picture of an unrest source is considered (e.g., a pressurised cavity, like a fluid-filled magma chamber, or a distribution of small expanding sources, like fluid-filled cracks) any source is finite. A finite source can be considered as a superposition of many (theoretically a continuum of) interacting point sources. Surface deformation data can provide clues on source morphology and dynamics, but data modelling is constrained by problems related to model complexity and computational cost; furthermore, any set of data can be almost perfectly fit by using a sufficiently complicated model, but it might be unrealistic and/or originate over-modelling. Simple analytical or semi-analytical models characterised by few parameters are thus convenient. The few available analytical (approximate or exact) solutions for superficial displacements

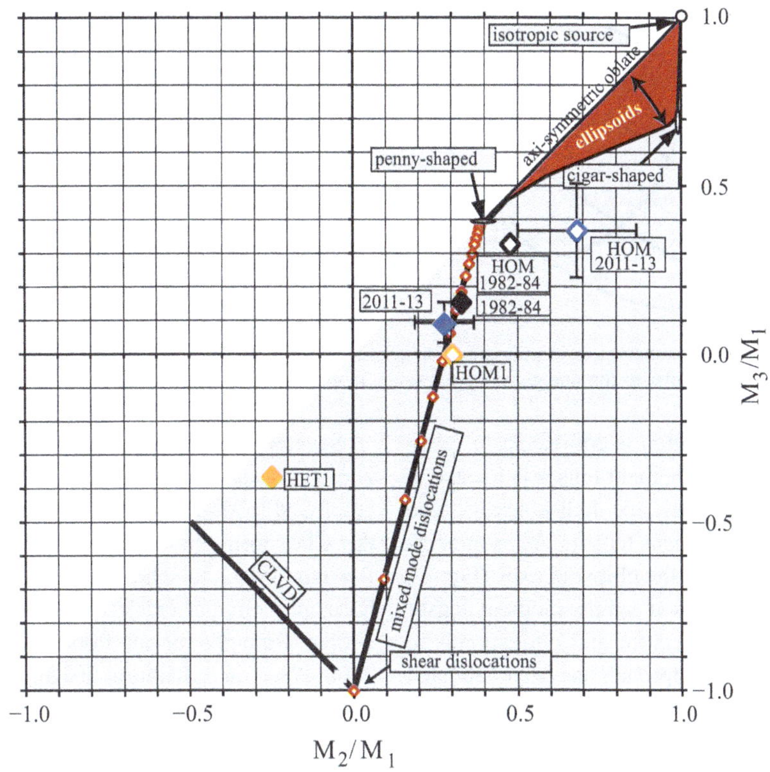

Fig. 5 Moment tensor eigenvalue ratios (assuming $M_1 \geq M_2 \geq M_3$, light grey area) and their relation with pressurised point-like ellipsoidal cavities (Davis 1986, red area). Mixed mode (tensile & shear) cracks are found along the oblique straight line on the right: the angle between the Burger's vector (opening direction) and the normal to the crack plane is noted (small red diamonds) in 5° steps from 0° (penny-shaped cracks) to 90° (pure shear dislocations). CLVD sources (sources with vanishing incremental volume) are found along the left line. Results for the moment tensor source of the 1982–1984 unrest are indicated in orange (full diamond—HET1 is in the heterogeneous medium, while the open diamond—HOM1 is for the homogeneous medium). The solutions are improved using a deeper feeding source, resulting as the final solutions, here reported in black (full diamond, heterogeneous medium, open diamond, HOM—homogeneous medium). Results for the 2011–2013 unrest are reported in blue (full diamond, heterogeneous medium, open diamond, HOM—homogeneous medium). A Poisson ratio $\nu = 0.28$ is assumed, as inferred from seismic tomography (the overall picture would change slightly if the standard value 0.25 were employed). Figure modified after Trasatti et al. (2011, 2015)

from finite deformation sources are generally restricted to simple geometries in a homogeneous elastic half-space.

Segall (2010) gives a comprehensive review of available models, including those for pressurised sources, like a point (Mogi 1958) or extended (McTigue 1987) sphere, a very small (with comparison to depth) triaxial ellipsoid (Davis 1986), an extended prolate spheroid (Yang et al. 1988), and a horizontal pressurised circular crack (Fialko et al. 2001). More recently, Cervelli (2013) gave analytical expressions for an extended oblate spheroid.

Amoruso and Crescentini (2011) used the multipole expansion method up to the quadrupole term to compute surficial deformations from an extended arbitrarily oriented triaxial ellipsoid; the same approach can be followed in the case of an arbitrary distribution of isotropic sources, like those due to thermal expansion, pore pressure changes, pressurisation of randomly oriented cracks. The quadrupole approximation includes only the monopole and quadrupole terms, as the dipole term is null because of symmetry. In the case of a pressurised ellipsoidal cavity or an ellipsoidal distribution of isotropic sources, the

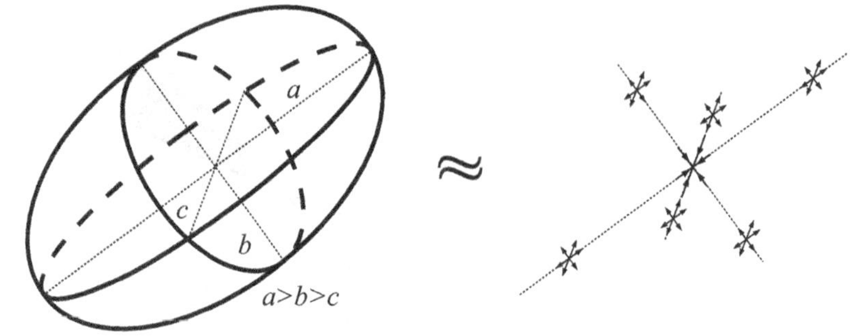

Fig. 6 Quadrupole approximation of a pressurised ellipsoidal cavity: the triaxial ellipsoid at left gives a ground deformation field very similar to the one given by the seven moment tensors at right

quadrupole approximation is equivalent to seven moment tensors: one moment tensor is located at the source centre and six half-potency (and opposite in sign) moment tensors are symmetrically distributed along the ellipsoid axes (Fig. 6). The monopole term is a single moment tensor located at the source centre and corresponds to Davis' solution. The approach has been validated against Finite Element Modelling for a layered half-space (Luongo et al. 2015).

Commonly used finite expansion sources include also tensile faults or, more generally, mixed-mode faults (Okada 1985), that do not fulfil the uniform-pressure boundary condition required in case of pressurised cavities. In this framework a mention is due to D'Auria et al. (2015b) who perform a joint inversion of geodetic data and seismic mechanisms, assuming a homogeneous elastic half space and inferring a nearly horizontal tensile dislocation 4 × 4 km wide, at 2,600 m depth for the 1982–1984 unrest.

Hydrothermal systems and aggregates of fluid-filled cracks and pores of general geometric configuration can be modelled as distributions of isotropic elementary sources. The ground displacement field is given by the convolution of the potency distribution of the elementary sources and the appropriate Green's functions. The potency distribution is made "smooth" (i.e., not too complex) by means of regularisation (Vasco et al. 2002; Camacho et al. 2011), cylindrical harmonics expansion (D'Auria et al. 2012) or, as shown above, multipole expansion.

Several finite-source models, embedded in homogeneous or layered media, have been employed to study the 1982–1984 bradyseism at the CFc, from the tensile fault in Dvorak and Berrino (1991), to spheres, prolate spheroids and horizontal circular cracks (e.g., Gottsmann et al. 2006a; Battaglia et al. 2006; Amoruso et al. 2008; Woo and Kilburn 2010), arbitrarily oriented triaxial ellipsoids (e.g., Amoruso et al. 2014a), injections of fluid batches (e.g., D'Auria et al. 2012).

4 Role of Medium Heterogeneities

Apart from avoiding unwarranted a priori assumptions regarding the source shape or the deformation mechanism, a reliable inference of source parameters in volcanically active areas should take into account a realistic description of the medium embedding the source. As shown numerically by Dieterich and Decker (1975), the vertical deformation field has little resolving power among different source geometries, is fairly symmetric and vanishes faster with the distance from the source, while the horizontal deformation is more discriminating and vanishes further away from the source. In order to exploit fully this property to infer source properties, anomalies in the deformation field arising from medium heterogeneities must be computed in order to extract signatures of the source mechanism itself. In a laterally heterogeneous medium, numerical models (such as finite element models) may account for complex medium structure (Beauducel et al. 2004; Folch and Gottsmann 2006), with the drawback that they are not easily implemented in inversion strategies. An active seismic survey (SERAPIS) was carried out in the submerged part of the caldera (the bays of Naples and Pozzuoli), with the aim of investigating its shallow structure (Chiarabba and Moretti 2006; Zollo et al. 2008). Seismic velocities increase mostly along the vertical, although significant lateral heterogeneities, related to the buried caldera rim, are also detected. The modelled deformation pattern depends strongly on the elastic structure assumed for the medium. Therefore, computing elastic solutions in a homogeneous half-space inevitably introduces a systematic bias in the interpretation of data collected in strongly heterogeneous regions. In particular, neglecting the elastic heterogeneities while inverting deformation data results in considerably inaccurate estimates of source depth because of the low flexural rigidity of the soft shallow layers (Trasatti et al. 2005; Crescentini and Amoruso 2007). Major elastic heterogeneities in the CFc are horizontally layered; extensions of few source models to layered elastic half-spaces are, e.g., in Wang et al. (2006), Crescentini and Amoruso (2007), and Amoruso and Crescentini (2011).

Trasatti et al. (2005) investigated the role of a plastic rheology for the Neapolitan Yellow Tuff caldera ($\sim$15 ka; Chap. Volcanic and Deformation History of the Campi Flegrei Volcanic Field, Italy and references therein) domain, within a layered structure of the surrounding medium. The authors found that the levelling data are reproduced by a deeper source (4–5 km) than that estimated from homogeneous elastic models employing a spherical source. Semi-analytic layered elastic and visco-elastic models, that typically provide deeper source locations with respect to homogeneous half-spaces, have also been used. Crescentini and Amoruso (2007) and Amoruso et al. (2008) employing a semi-analytic deformation model in a layered elastic medium (Wang et al. 2006) showed that the low density ($\sim$600 kg/m^3) inferred by Battaglia et al. (2006) assuming a horizontal penny-shaped crack, is mostly due to the homogeneous half-space assumption. Imposing an axisymmetric point-source, they inferred a depth of 4.7–5.0 km for the 1982–1984 uplift source, with an ill-constrained density of 2,000–3,000 kg/m^3, while a shallower depth (3.0 $\pm$ 0.5 km) with similar density was obtained assuming a finite horizontal circular crack with 2,750 m radius (Table 1).

Trasatti et al. (2011) employed a general moment tensor source within a Finite Element inversion scheme that can account for the vertical and lateral heterogeneities inferred from seismic tomography. In this way, they obtained a best fit moment tensor estimate that improves considerably data fitting (40% variance reduction with respect to the best moment source HOM1 in a homogeneous medium). The source depth was inferred at 5 km and the resulting mechanism HET1 contained a large deviatoric component that cannot be interpreted in terms of any pressurised ellipsoid (Fig. 5), nor it may be explained in terms of a mixed mode (tensile-and-shear) crack. The overall deformation mechanism may be interpreted splitting it in an expanding source and a purely deviatoric source, namely a pressurised cavity where magma intrusion takes place, surrounded by ring faults accommodating the intrusion volume through aseismic slippage with reverse dip-slip mechanism, as originally suggested by De Natale et al. (1997). The problem with such an interpretation of the moment tensor is that its decomposition is not unique, it cannot be reconciled with gravity data (the intrusion density should be unrealistically high) and the deviatoric moment driving the deformation would be hundreds of times greater than the observed seismicity.

5 Multiple Magmatic Source Models

A basic problem with all the above illustrated models is that they do not account for mass conservation: the mass emplaced within the source must come from somewhere else. Neglecting the effects of the disappearing mass is acceptable only if it comes from infinite distance.

A laterally extended magma reservoir has been detected at 7–8 km depth below the CFc from seismic studies (Zollo et al. 2008): this depth is too shallow to allow neglecting its effect on surface deformation and gravity changes. Hence, at the CFc we must account for both a shallow inflating source and a deep deflating origin source while inverting deformation and gravity data, as seldom acknowledged (e.g., Okubo and Watanabe 1989). Trasatti et al. (2011) have assumed that a deep deflating sill-like source is present at 7.5 km depth, as suggested from seismic inversions, vertically below a shallower moment source embedded in a laterally heterogeneous medium and inflating by the same mass, whose parameters are left to be inferred from a Bayesian inversion procedure. In this way, a variance reduction of 50% is obtained with respect to the single source model HET1, the shallow source depth is inferred at 5.5 km and its mechanism (indicated with a full black diamond in Fig. 5) fits extremely well an obliquely dipping dike (a pre-existing fault?) with mixed opening and reverse dip-slip dislocation mechanism. The inferred intrusion volume is $60–70 \times 10^6$ m^3 and its density ($\sim$2,400 kg/m^3) is compatible with trachytic magma (Lesher and Spera 2015), even if its value is ill constrained. The presence of a significant shear component implies that the intrusion took place across a prestressed solid medium, suggesting that the new magma emplaced at 5.5 km depth was responsible for fracture propagation beyond the previously emplaced fluid magma reservoir, with important implication for volcanic hazard assessment. An interesting point to be noted is that even homogeneous half-space models accounting for mass conservation provide better fit than the single source model HET1, if data are inverted by taking into account a deep deflating and a shallow inflating source (open black diamond in Fig. 5).

It is interesting to note that the deformation time series of the recent 2011–2013 uplift episode are best-fitted employing a moment source nearly identical to that of 1982–1984 (Trasatti et al. 2015 and blue solutions in Fig. 5). The apparent paradox of a reverse shear dislocation within a dilatational strain environment may be simply resolved in terms of the thermo-poro-elastic constitutive relation governing the stress and strain field inside a region undergoing large pore pressure and temperature increases as due to the exsolution of volatiles from a magmatic source (Belardinelli et al. 2019). Mantiloni et al. (2020) further developed this model to show that the 1982–1984 ground displacement and the extreme heterogeneity of fault mechanisms may be well explained by a thin disk-shaped thermo-poro-elastic inclusion at shallow (1.9 km) depth.

6 Deflation After 1984 and Mini Uplift Episodes

Since the beginning of 1985 the ground of the CFc began to subside (Fig. 1). In particular, during 1985–1987 a “fast” deflation took place, accompanied by significant gravity changes (Berrino 1994). In this time lapse, about 20% of the total amount of uplift was recovered. The post-1987 deflation is often referred to as “slow”, since the ground subsided at a slower rate. Indeed, the amount of deflation occurred between 1985 and 1987 is comparable to that registered during the period 1988–2000. The controversy among the nature of the volcanic source at the CFc accompanies also the interpretation of the fast/slow deflation and the associated gravity variations. These changes were initially interpreted as the result of magmatic changes, such as mass withdrawal or overpressure exhaustion (e.g., Berrino et al. 1984), and more recently ascribed to subsurface hydrothermal processes (e.g., Gottsmann et al. 2006b).

Since 1992, the deflation has been monitored by satellite techniques. A SAR interferogram covering the time interval 1993–1996 was modelled by Avallone et al. (1999) using a point source deflating by 2.5×10^6 m^3, placed at 2.7 km depth at 800 m southeast of Pozzuoli, in the seaside. Lundgren et al. (2001) inverted pairs of ascending and descending SAR images spanning 1993–1998. The joint inversion fits data with a sub-horizontal rectangular contracting tensile dislocation with dimensions $\sim 4 \times 2$ km and located beneath the town of Pozzuoli at 2.5–3 km depth. They pointed out that inversion for a spheroidal or Mogi point source also produced reasonable fits but with progressively poorer overall fits to the data. They suggested that the subsidence is due to hydrothermal diffusion as the primary deformation mechanism during this phase of caldera deflation. In a more complete data analysis, Manconi et al. (2010) computed the 1992–2008 mean velocities from SAR images. The contracting Mogi source used to fit the data was located in the centre of the caldera at about 3 km depth. The estimated variable overpressure of the small isotropic source is likely related to hydrostatic pressure variations occurring within the hydrothermal reservoir of the CFc system.

The long subsidence trend since 1985 was interrupted by several inflation episodes (see Chaps. Volcanic and Deformation History of the Campi Flegrei Volcanic Field, Italy; The Hydrothermal System of the Campi Flegrei Caldera, Italy; Historic Unrest of the Campi Flegrei Caldera, Italy), in particular, early in 1989 and during the summer 2000, when a maximum uplift of about 5 cm, associated with a deformation rate of about 10 cm/yr, was detected (Lanari et al. 2004). A new uplift episode started in the second half of 2004 and was analysed by using data from tiltmeters, precision levelling and the 5 continuous GPS stations located in the area (Troise et al. 2007). The uplift pattern between the summer 2004 and November 2006 was also clearly detected by SAR deformation time series (Fig. 4). The inversion of the SAR measurements was performed by attempting a spherical and a spheroidal source (Trasatti et al. 2008b). Both sources were located at 3 km depth, had a volume variation of 1.1×10^6 m^3, and showed comparable performances since the spheroid was quasi-spherical (mean axes ratio 1.5).

Previously, Amoruso et al. (2007) modelled the 2004–2006 episode using GPS and levelling data by means of a penny-shaped crack embedded in a stratified medium, retrieving a source depth of 3 km and volume variation of 1.4×10^6 m^3. It may be reminded that the inferred source location (3 km depth) is similar to that estimated by some authors for the 1982–1984 large unrest (e.g., Berrino et al. 1984; Amoruso et al. 2008) but much shallower than those inferred by others (Gottsmann et al. 2006b; Trasatti et al. 2011). However, if the hydrothermal hypothesis is nowadays ruled out as the primary cause of the 1982–1984 unrest from joint inversion of deformation and gravity, this is not the case for the mini uplifts. Indeed, the small reservoir responsible for the recent miniuplifts (Sect. 2) has been interpreted by Manconi et al. (2010) as a rock volume filled by pressurised gas and fluids, probably related to the hydrothermal reservoir at the CFc, hypothesised to react to the activity of a deeper magmatic body, in accordance with geochemical evidences (e.g., De Natale et al. 2006; Zollo et al. 2008; Chiodini 2009). A similar interpretation is given by D'Auria et al. (2012) since their sources are considered as hot fluid batches injected at the bottom of the geothermal reservoir. They argue that in 2000 and 2004–2006, the injected fluids migrated towards the centre of the caldera.

However, if we accept that more than one deformation mechanism is required to explain the unrest, other mechanisms might be taken into account, based on physical considerations. Thus, if the post-1984 deflation is interpreted in terms of depressurisation of a magma reservoir releasing hot and pressurised volatiles toward the surface, the interaction between the magma reservoir and the shallow hydrothermal system should be considered.

7 Physical Modelling of the Hydrothermal Instability

CFc hosts a large hydrothermal system that daily discharges thousands of tons of volcanic gases through diffuse degassing and fumarolic activity (e.g., Chiodini et al. 2001; Aiuppa et al. 2013; Pedone et al. 2014; Cardellini et al. 2017; Chap. The Hydrothermal System of the Campi Flegrei Caldera, Italy). Un-quantified gas emissions also take place offshore, venting from fractures on the seabed. Given the magnitude of these discharges, it is not surprising that aqueous fluids play an important role in the definition of the present state of the caldera and its evolution. As mentioned above and discussed below, several authors suggested the possibility that hydrothermal instabilities may be responsible for a significant fraction of caldera deformation and gas emissions (Bonafede 1991; De Natale et al. 1991; Orsi et al. 1996; Chiodini et al. 2003). Clearly, assessing the density of the fluids (whether aqueous or magmatic) responsible for the deformation episodes is a crucial issue to be addressed to perform a sound volcanic hazard assessment at the CFc.

Thermal and pore pressure changes in poroelastic media are important mechanisms of deformation (e.g., Todesco et al. 2010) and several authors suggested that hydrothermal instability might even be the main driving mechanisms of ground deformation (Casertano et al. 1976; Bonafede 1991; Gaeta et al. 1998; Orsi et al. 1999; Chiodini et al. 2003). The presence of gases and aqueous fluids alter the mechanical properties of rocks (Heap et al. 2014) and may enhance the deformation driven by other phenomena, such as magmatic or tectonic processes. The presence of fluids can be directly responsible for rock deformation: circulating fluids may cause pressure and temperature perturbations that act on the solid rock grains by releasing the effective intergranular stress, or by causing dilatancy and permanent alteration of the rock fabric (e.g., Todesco et al. 2010). The linear theory of poro-elasticity describes the mechanical interaction between interstitial fluid and the porous rock: an increase in pore pressure causes the dilation of the rock, while a compression of the rock affects the fluid, either causing its flow (drained conditions) or increasing its pore pressure (undrained conditions) (Biot 1956; Rice and Cleary 1976). When the circulating fluids are effective heat carriers, as in geothermal systems, the thermal perturbation associated with fluid flow may cause an additional thermo-elastic response of the porous medium (McTigue 1986). Coupling of thermal gradients, hydrologic flow, and mechanical deformation of rocks has been recognised in a variety of geological contexts (Tsang 1999). Hydromechanical effects were considered a possible trigger for the crustal deformation and the seismicity during the 1965–1967 Matsushiro earthquake swarm. During the two years of the strike-slip sequence, a large crustal uplift (up to 75 cm) was accompanied by enhanced spring discharge and CO_2 diffuse degassing. Hydro-mechanical modelling confirmed the link between fluid flow, seismicity and deformation (Cappa et al. 2009). Fluid pressure can also be focussed by the presence of impermeable fault planes and can act to weaken the fault (Fulton and Saffer 2009).

All these considerations led many scientists to investigate the role of fluids in the evolution of the bradyseismic activity at the CFc. Casertano et al. (1976) were among the first ones to suggest that uplift could be associated with the response of the fluid-saturated porous rocks to changes in the basal heat flux. They noted that the 1969–1972 uplift phase was accompanied by a remarkable increment of the fumarolic activity and by widening of the Fangaia mud pool within the Solfatara crater. They also highlighted the correlation between the amount of ground uplift and the thickness of the caldera fill deposits, through which volcanic fluids are expected to circulate. These authors also suggested a mechanism of tidal pumping by which the sea motion would affect the hydrodynamics within the permeable pyroclastic layers of the caldera fill. An active role of the hydrothermal fluids in the unrest episodes was suggested also by De Natale et al. (1991) based on geochemical and geophysical evidences. Bonafede (1991) calculated the effects of thermo-elastic expansion considering that heat is effectively transported by the circulation of hot hydrothermal fluids. The arrival of hot, pressure-driven fluids into a shallow and colder aquifer was shown to be an effective ground deformation mechanism providing the same deformation pattern of a Mogi source. The link between ground level and the conditions of the fluid-saturated porous rock is highlighted also by Gaeta et al. (1998), who suggested that subsidence is governed by a slower horizontal propagation of fluids, away from the caldera centre. A similar approach is followed in later works (Castagnolo et al. 2001; De Natale et al. 2001). An analysis of the energy required to uplift the ground showed that the pressure perturbation of the geothermal system is consistent with a few centimetres of ground deformation (Gaeta et al. 2003). This deformation is consistent with the ground displacement measured during the mini-uplift episodes. The involvement of hydrothermal fluids in the episodes of unrest provides an opportunely shallow and low-density source of ground deformation, required by the coupled inversion of gravity and ground deformation data (Battaglia et al. 2006; Bonafede and Mazzanti 1998; Gottsmann et al. 2006a).

Further insights on the fluid-related ground displacement derive from the application of coupled thermo-hydro-mechanical models (Rutqvist et al. 2002). Unrest episodes at the CFc have always been accompanied by remarkable changes in the composition of fumarolic gases. These changes have been interpreted as due to periods of increased magmatic degassing, when large amounts of CO_2-enriched volcanic gases are injected into the geothermal system (Chiodini et al. 2003; Todesco et al. 2003). The coupled thermo-hydro-mechanical model was applied here to compute the deformation arising from the changes in pore pressure and temperature associated with fluid circulation, neglecting the changes in porosity and permeability caused by rock deformation (a simplification often referred to as one-way coupling). Results showed that an increased injection of gas-rich, magmatic fluids may cause a quick uplift, of the order of tens of centimetres that is followed by a slower subsidence, when the injection rate is reduced and the hot fluids are released at the surface (Todesco et al. 2004). The amount of deformation depends on the system geometry, the assigned rock properties, and the duration of the simulated unrest. A similar approach was followed by Rinaldi et al. (2010) who first simulated the hydrothermal circulation undergoing a period of unrest, and then computed the displacement of the free surface, arising from the thermal and pressure perturbation of an elastic half-space, according to the theory of thermo-poro-elasticity. Although heterogeneities in the hydraulic properties of the rock may focus the fluid flow and enhance the resulting deformation (Todesco et al. 2010), the vertical displacement arising from the simulated unrest periods never exceeds a few tens of centimetres. These values are comparable with those of the mini-uplift episodes at the CFc, but they are far from the largest displacements detected in the caldera (e.g., 1.8 m over a similar time interval). Similar results were obtained by other authors, who achieved larger displacement but only considering a perturbation acting over a time scale of thousands of years (Hurwitz et al. 2007; Hutnak et al. 2009). Further developments allowed to introduce heterogeneities in the properties of the elastic medium and highlighted the effects of fluid flow rate and structural discontinuities on the maximum value and spatial distribution of ground displacement (Coco et al. 2016a, b).

The analysis was also extended to include both the magmatic system and the hydrothermal fluids. The emplacement of magmatic bodies at shallow crustal levels is accompanied by cooling and fractional crystallisation. This causes the exsolution of volcanic gases, which tend to rise toward the surface by buoyancy. Their propagation, however, is hindered by the extreme conditions that promote the closure of pore space and fractures and favour a plastic rock behaviour in the region surrounding the magma chamber. Therefore, the large volume increment associated with the crystallisation process results in a strong pressurisation of the fluid phase that accumulates at the top of the magma body, at near-lithostatic pressure (Bodnar et al. 2007; Lima et al. 2009). Ground deformation then results from the combined effect of the shallow, cooling intrusion and the volatile pressurisation. Subsidence may be due to cooling, that causes both the contraction of the intrusive body and the condensation of steam. Alternatively, fluid pressure may become high enough to open short-lived fractures, through which volatiles can escape, dissipating the overpressure (Dutrow and Norton 1995; Fournier 2006). If volatile exsolution continues, pore pressure may build up again after some time, leading to alternating phases of uplift and subsidence commonly observed in volcanic systems (Chiodini et al. 2003; Hurwitz et al. 2007).

The available literature confirms that the circulation of hydrothermal fluids is responsible for a fraction of the detected deformation (Todesco et al. 2004; Rinaldi et al. 2010). The available estimates of fluid-driven ground deformation are biased by the limits of the physical models that are used to represent the process. These models usually describe a shallow portion of the volcanic system, where temperatures are expected to be lower than the critical temperature of water (374 °C). By neglecting the deeper roots of the hydrothermal circulation, these models do not account for the entire range of pressure and temperature, which span from near-magmatic to atmospheric values. As the aqueous fluids circulate in the natural system, they are expected to approach near-critical conditions, where their thermodynamic and transport properties are known to be singular. Under such conditions, the feedbacks that characterise coupled processes are magnified (Norton and Dutrow 2001). The intrinsic physical instability of fluids near the critical point is well known and, although its effects in porous media are not completely assessed (Gaeta et al. 2003), the propagation of thermal fronts is enhanced in near-critical fluids. The transition between supercritical and subcritical conditions may occur in natural systems at different points in time and space, as the fluids move from one region to another, or as they are cooled or heated by the contact with surrounding rocks or other fluids. This certainly affects the complex pattern of ground motion detected at the surface, to an extent that has not been assessed yet.

The full coupling between thermo-hydro-mechanical processes, in which changes of hydraulic properties associated with rock deformation are accounted for, has not been implemented for volcanic systems. As a rock deforms, its porosity and permeability change affecting the fluid circulation. Where this leads to higher permeability, fluid pressure can be released, but the hot fluids may reach previously undisturbed regions. A reduction of permeability, on the other hand, may hinder fluid flow, favouring pressure build-up. How these different effects may add to the overall ground displacement is difficult to predict without further studies.

Additional complications may arise from the chemical evolution of the multi-component fluids: the deposition and dissolution of mineral phases modifies porosity and permeability of the rock and may also alter their mechanical and thermal properties. The effects of these changes in terms of ground deformation have not been quantified, although they probably act on long time scales.

8 Multiple Magmatic and Hydrothermal Source

The controversy between magmatic and hydrothermal origin of deformation is of major importance for assessing the CFc volcanic hazard. However, the two mechanisms are not mutually exclusive: the emplacement of a shallow magma body is necessarily accompanied by the release of hot magmatic gases, which in turn perturb the hydrothermal system. Some of the most recent works addressed the relation between these different deformation sources.

8.1 Interactions Between Magmatic and Hydrothermal Sources

Instead of discerning the origin of the source, mechanical coupling among different sub-surface pressure sources may be a plausible explanation for the observed rapid ground uplift of the 1970s and the 1980s (Fig. 1a). A viable scenario involves interactions between a deeper-seated magmatic source and a shallower hydrothermal source.

The conceptual model behind this is the arrival of a fresh batch of magma in the deeper reservoir, which triggers the release of fluids into an overlying hydro-thermal reservoir. Geodetic surface observables (e.g., ground deformation and gravity changes) must then be explained to result from a combination of mass and density changes in either sources with the possible contribution of medium heterogeneities such as the presence of shallow-seated near vertical faults. These scenarios were explored in detail by Folch and Gottsmann (2006) and Gottsmann et al. (2006a, b, c).

The potential role of multiple sources at the CFc has been explored using both analytical and numerical modelling approaches. The major limitation of the use of analytical models for this problem is their limited applicability for stacking of sources in the framework of an isotropic and homogeneous half-space due to the mechanical interaction of the stress fields induced by pressure variations in either. Furthermore, the role of faults or other mechanical heterogeneities in crustal rocks on stress evolution around multiple pressurised sources cannot be explored analytically. The key findings from the different approaches are summarised in the following.

Following the indication of a hybrid source, with mixed magmatic and hydro-thermal properties (Gottsmann et al. 2006a), Gottsmann et al. (2006b) employed two vertically stacked spherical sources representing a deep magmatic and a shallow hydrothermal reservoir using a 3D numerical solver within a finite element approach. These solutions accounted for both presence of the surface topography at the CFc and elastic rheology of sub-surface rocks. Deduced source parameters for this model are given within Table 1 in the Conclusive Remarks section. Assuming a rock rigidity of 10 GPa and a Poisson ratio of 0.25, the authors deduced pressure increases of 33 and 42 MPa for the deep and the shallow source, respectively. Permissible excess pressures were scaled with the tensile strength of the encasing rocks that was assumed at a maximum of 40 MPa. The results showed that ground inflation of the 1982–1984 unrest was predominantly caused by pressure increases in the shallow hydrothermal source, whereas the detected gravity changes were caused by practically equal contributions from both sources. The subsequent rapid deflation between 1985 and 1987 was again predominantly controlled by the shallow source, whereas changes in the deeper source produced the gravity variations. The shallow source then dominated both ground deformation and gravity variations during the 1988 to 2001 period of slow deflation.

When considering the presence of faults (a caldera-boundary fault or an inner caldera fault similar to the fault presented by Beauduce et al. 2004) two implications are important (Folch and Gottsmann 2006): the required pressure changes to fit the observed data are reduced by about 25% in the shallow source and 10% in the deeper source and gravity measurements need to be corrected for Bouguer effects and to account for the displacement of density layers (Bonafede and Mazzanti 1998). Assuming a Bouguer density of 2,000 kg/m^3, the model requires an input of material with a density ρ = 2,400 kg/m^3 into the deep source while the shallow pressure-dominated source undergoes a negligible mass input of material with density of 400 kg/m^3). A geologically plausible explanation for this modelling scenario of the 1982–1984 uplift is a recharge of magma at depth, perturbing the shallower hydrothermal system by inducing volume expansion of magmatic fluids. As such, the observed inflation may be interpreted to result from a combination of deep-seated magmatic and shallow hydrothermal processes.

The interaction between different sub-surface reservoirs as a source for observed inflation and deflation unrest at the CFc was also investigated using vertical residual gravity change over height change gradients $\Delta_r = \Delta_S + \Delta_{FA}$ where Δ_{FA} is the local free-air gravity gradient (−290 ± 5 μGal/m; Berrino et al. 1984) and Δ_S is the vertical surface gravity gradient expressed as the measured gravity change over the vertical deformation ($\Delta g_S/\Delta h$) at each observation point. The residual gravity gradient bears information on the dominant causative subsurface process. A pressure change caused by a density change (e.g., phase change) in a centre of expansion results in $\Delta_S \simeq \Delta_{FA}$. As a consequence, Δ_r values are around 0 (Rundle 1978; Walsh and Rice 1979). Evaluating data comprising the rapid uplift, rapid deflation and slow deflation between 1982 and 2001, Gottsmann et al. (2006c) showed that spatio-temporal variations in the gradiometric data is inconsistent with the idea of a single source scenario in particular for the period of deflation after 1988, requiring the adoption of multiple sources. These authors also inverted the gravity and deformation data between 1988 and 2001 using a random search approach for an elastic half-space model and a genetic algorithm invoking elasto-gravitational coupling (Rundle 1980, 1982); i.e., the mutual interaction between a subsurface mass change, the gravity field, and the effect of pressure changes within a spherical reservoir, which is mathematically approximated by a point source.

The use of the genetic algorithm was found to be superior over the random search approach in delineating source parameters for multiple sources by drastically reducing the uncertainty in the location of secondary sources. More confidence can be attributed to the period 1988 onwards due to a substantially spatially denser coverage of elevation and gravity changes in the CFc.

As a general outcome, Gottsmann et al. (2006a, b) identified two stable deformation sources beneath the Pozzuoli area: a deep-seated source at $\sim$2,700 m and a shallower one at $\sim$700 m depth. The deeper source undergoes pressure changes with negligible mass changes, whereas the shallower source is dominated by mass changes. Ephemeral tertiary sources are also identified at the periphery of the CFc and appear to vary both spatially and temporally during the survey periods. A key characteristic of these sources is their dominant mass change character rather than a density change. These sources are not associated with significant ground deformation, but their gravimetric signature seems statistically significant.

These models also highlight the contribution of both magmatic and hydrothermal sources to rapid inflation and deflation between 1982 and 1987 and a dominance of shallow-seated hydrothermal processes from 1988 onwards (Folch and Gottsmann 2006; Gottsmann et al. 2006a). The observed spatio-temporal variations in both gravity and deformation data between 1988 and 2001 can be best explained by fluid migration through a central hydrothermal system as well as along the caldera boundary faults.

Interactions between multiple sources has also been invoked. Using multi-parametric data, D'Auria et al. (2010) proposed the interplay between a deep and a shallow source as the dominant cause for miniuplifts during background surface deflation from 1988 onwards. Their model involves the release of fluids from a deep reservoir upon exceeding a pressure threshold to feed a shallow hydrothermal system. In another paper, D'Auria et al. (2012) highlighted a peripheral injection of fluids potentially along the western marginal faults of the Neapolitan Yellow Tuff caldera collapse (see Chap. Volcanic and Deformation History of the Campi Flegrei Volcanic Field, Italy) from Differential Interferometric Synthetic Aperture Radar data. Their source location matches the position of a pronounced gradiometric anomaly in the area of Capo Miseno and Arco Felice and a source derived by the integrated geodetic data inversion presented in Gottsmann et al. (2006b, c). Overall, there appears to be good evidence for the interaction of multiple subsurface sources beneath the CFc to generate the spatially and temporally complex deformation and mass/density change histories observed over the past few decades.

8.2 The Role of a Stable (~4 km Deep) Sill-Like Source of Strain

Amoruso et al. (2014a, b) have shown that the CFc ground displacement field from 1980 to 2013 can be decomposed into two components, related to large-scale and local deformation, respectively; local deformation is confined to the Solfatara fumarole field. Each displacement field component can be obtained by multiplying a function of the sole spatial coordinates (i.e., the component pattern) by a function of time (i.e., the component time evolution). By using the quadrupole approximation of Amoruso and Crescentini (2011) to compute surface deformation from a triaxial ellipsoid embedded in a layered half-space, large-scale deformation can be explained by a quasi-horizontal source northwest

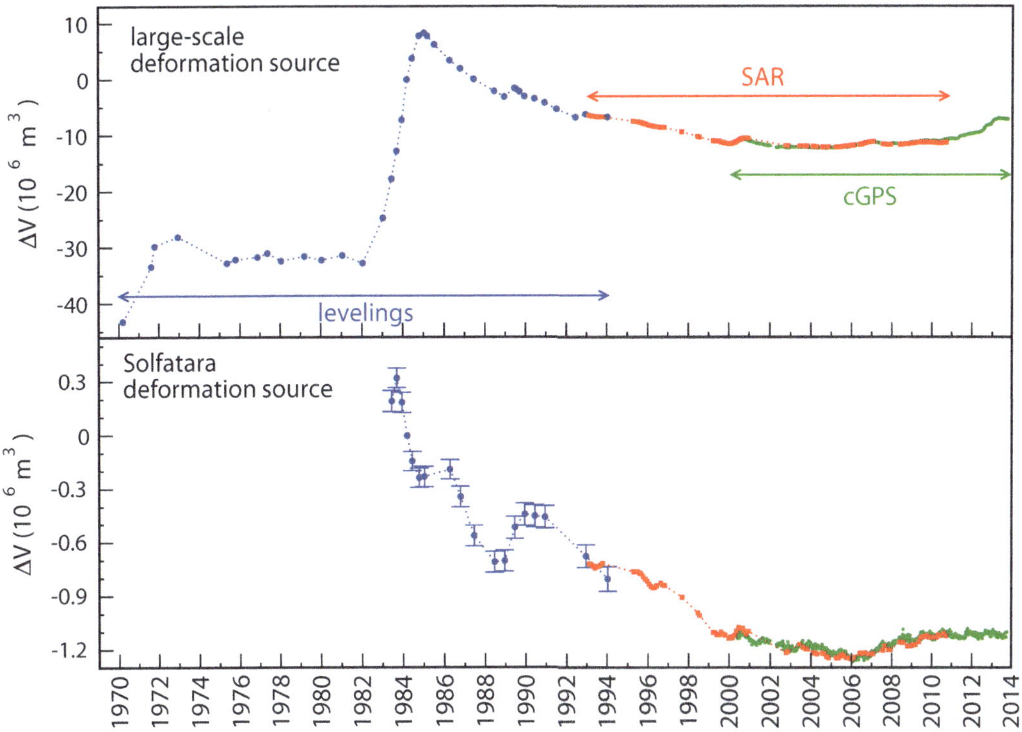

Fig. 7 Time histories of the pressurised triaxial ellipsoid (upper plot) and pressurised spheroid (lower plot) potencies from 1970 to end 2013, given with respect to March 1984 (data from Amoruso et al. 2014a, b). The pressurised spheroid potency is not given before June 1983 and for some other dates (e.g., July 1991 and June 1992) because of lack of levelling data in the Solfatara area

to southeast oriented and represented by a pressurised triaxial ellipsoid about 3,500 m deep, maybe related to the injection of magma and/or magmatic fluids from a deeper magma chamber into a sill. Residual (with respect to pressurised triaxial ellipsoid) deformation is confined to the Solfatara fumarole field and can be explained by a small pressurised spheroid at about 2,000 m depth beneath Solfatara. It may be related to the poro-elastic response of the substratum to pore pressure increases near the injection point of hot magmatic fluids into the hydrothermal system. All source parameters but potency (volume change) are constant over the whole investigated period (1980–2013) for both sources. Potency time histories of pressurised triaxial ellipsoid and pressurised spheroid (which the time evolutions of the two components of the ground displacement field are related to) are somewhat similar but not identical (Fig. 7), supporting the existence of a genuine local deformation source at the Solfatara against the emergence of a mere distortion of large-scale deformation. In particular, the Solfatara deformation source appears inactive during the whole 2011–2013 uplift.

This two-source model of the recent CFc ground deformation proved consistent with geochemical models and seismic tomography (e.g., Chiodini et al. 2015; De Siena et al. 2017a, b; Calò and Tramelli 2018; Akande et al. 2019). The pressurised triaxial ellipsoid has also been identified as the source for the ground deformation that preceded the AD 1538 Monte Nuovo eruption (Guidoboni and Ciuccarelli 2011; Di Vito et al. 2016; Amoruso et al. 2017) and the ground deformation which occurred during the last epoch of the post-NYT activity (~5.5 to ~3.5 ka; Di Vito et al. 1999; Smith et al. 2011; Marturano et al. 2018; Chap. Volcanic and Deformation History of the Campi Flegrei

Volcanic Field, Italy), suggesting the presence of a stable and relatively shallow (~4 km deep) source of strain like a sill.

A first model of isothermal sill propagation at Campi Flegrei was given by Macedonio et al. (2014) in the case depth-to-radius ratio is much smaller than 1, while non-isothermal sill propagation has been modelled by Amoruso and Crescentini (2019) in the case depth-to-radius ratio is larger than or around 1. Multiple about-equally-large sill emplacements at low injection rates, like those possibly occurred in the last centuries, might have been permitted by a favourable thermal structure related to a flat kilometre-sized relict thermal anomaly (Amoruso et al. 2017) dating from the last epoch of the post-NYT activity.

9 Conclusive Remarks

Ground deformation and gravity data provide important inferences on depth, volume, mass density and source mechanisms of magma emplacement processes. The correct evaluation of these parameters is extremely important in assessing volcanic hazard in a densely populated volcanic area, such as the CFc, but it is strongly conditioned by assumptions built in the inversion procedures. For instance, the depth of the inflating source of the CFc is inferred at 3 km if an isotropic point source in a homogeneous elastic half-space (Mogi model) is assumed, at 4 km if the layered elastic structure of the caldera is taken into account, and at ~5.5 km assuming a completely general moment tensor source in the point source approximation. The depth goes up again to ~3 km if the source is assumed as a circular horizontal crack with finite radius. The inferred density of the intrusion also varies from the values typical of basaltic magma to those of supercritical aqueous fluids, depending on the assumptions of the different models. Furthermore, the interaction of the volcanic source with the hydrothermal system provides additional changes which overlap onto the truly magmatic signals.

The non-uniqueness of inversions based only on geodetic and gravimetric data may be constrained by data of different origin, among which an important role is played by temperatures measured along deep drillings, the distribution of seismic hypocentres and their focal mechanisms (e.g., D'Auria et al. 2015b). Finally, the constitutive relation employed to describe the dynamical response of the medium should take into account the effect of temperature and pore pressure migration (e.g., Belardinelli et al. 2019; Mantiloni et al. 2020).

The development of new modelling methods and inversion strategies, as well as the advent of new data on the dynamics and structure of the caldera have given rise to wide range of source models over the years (including those proposed by the authors of this chapter) to explain the unrest dynamics of the CFc. In spite of the discrepancies in the modelling approach and the interpretation of the findings, a number of firm conclusions can be drawn:

1. The 1982–1984 inflation cannot be explained in terms of hydrothermal instability only: once the layered structure of the CFc is accounted for, the inferred intrusion densities are definitely greater than aqueous fluids.
2. If mass conservation is taken into account, with magma being transferred from a deep reservoir (detected from seismic data at 7.5 km) into a shallower general point source, the depth of magma emplacement is inferred as 5.5 km. Furthermore, its source mechanism is found consistent with an inclined tensile/shear crack inflating under input of material with typical magmatic density.
3. It is extremely important to account for medium heterogeneity: typically, the inferred source depth is much shallower if a homogeneous medium is assumed, but also the source mechanism, geometry and volume variation may be biased.
4. Inversion procedures are being devised to relax the point-source approximation, employing source mechanisms such as multipole expansions. This is an important step since point-sources such as moment tensors typically tend to overestimate depth.
5. A stable sill-like expansion source (at ~4 km depth) may be the main cause of recent

(1970–2013) ground deformation episodes, as well as the ground deformation which preceded the AD 1538 Monte Nuovo eruption and the ground deformation occurred during the last epoch of the post-NYT volcanism ($\sim$5.5–3.5 ka; see Chap. Volcanic and Deformation History of the Campi Flegrei Volcanic Field, Italy).

6. The role of instabilities of the shallow hydrothermal system is dominant in the post-1984 deflation and in the intervening mini-uplift episodes. If the deformation source is assumed as a prolate spheroid, the density of the emplaced material is $\sim$1,600 kg/m^3 and a mixed role between the shallow hydrothermal system and a deeper magmatic reservoir is suggested.

The main results obtained by the different authors are summarised in Table 1.

References

Aiuppa A, Tamburello G, Di Napoli R, Cardellini C, Chiodini G, Giudice G, Grassa F, Pedone M (2013) First observations of the fumarolic gas output from a restless caldera: Implications for the current period of unrest (2005–2013) at Campi Flegrei. Geochem Geophys Geosys 14:4153–4169. https://doi.org/10.1002/ggge.20261

Akande WG, De Siena L, Gan Q (2019) Three-dimensional kernel-based coda attenuation imaging of caldera structures controlling the 1982–84 Campi Flegrei unrest. J Volcanol Geotherm Res 381:273–283. https://doi.org/10.1016/j.jvolgeores.2019.06.007

Amoruso A, Crescentini L (2011) Modelling deformation due to a pressurized ellipsoidal cavity, with reference to the Campi Flegrei caldera, Italy. Geophys Res Lett 38:L01303. https://doi.org/10.1029/2010GL046030

Amoruso A, Crescentini L (2019) An approximate approach to nonisothermal emplacement of kilometer-sized kilometer-deep sills at calderas. J Geophys Res: Solid Earth 124:1236–1253. https://doi.org/10.1029/2018JB016254

Amoruso A, Crescentini L, Linde AT, Sacks IS, Scarpa R, Romano P (2007) A horizontal crack in a layered structure satisfies deformation for the 2004–2006 uplift of Campi Flegrei. Geophys Res Lett 34: L22313. https://doi.org/10.1029/2007GL031644

Amoruso A, Crescentini L, Berrino G (2008) Simultaneous inversion of deformation and gravity changes in a horizontally layered half-space: evidences for magma intrusion during the 1982–1984 unrest at Campi Flegrei caldera (Italy). Earth Planet Sci Lett 272:181–188. https://doi.org/10.1016/j.epsl.2008.04.040

Amoruso A, Crescentini L, Sabbetta I (2014a) Paired deformation sources of the Campi Flegrei caldera (Italy) required by recent (1980–2010) deformation history. J Geophys Res: Solid Earth 119:858–879. https://doi.org/10.1002/2013JB010392

Amoruso A, Crescentini L, Sabbetta I, De Martino P, Obrizzo F, Tammaro U (2014b) Clues to the cause of the 2011–2013 Campi Flegrei caldera unrest, Italy, from continuous GPS data. Geophys Res Lett 41:3081–3088. https://doi.org/10.1002/2014GL059539

Amoruso A, Crescentini L, D'Antonio M, Acocella V (2017) Thermally-assisted magma emplacement explains restless calderas. Sci Rep 7(1):7948. https://doi.org/10.1038/s41598-017-08638-y

Avallone A, Briole P, Delacourt C, Zollo A, Beauducel F (1999) Subsidence at Campi Flegrei (Italy) detected by SAR interferometry. Geophys Res Lett 26(15):2303–2306

Battaglia M, Troise C, Obrizzo F, Pingue F, De Natale G (2006) Evidence for fluid migration as the source of deformation at Campi Flegrei Caldera (Italy). Geophys Res Lett 33:L01307. https://doi.org/10.1029/2005GL024904

Beauducel F, De Natale G, Obrizzo F, Pingue F (2004) 3-D modelling of Campi Flegrei ground deformations: Role of caldera boundary discontinuities. Pure Appl Geophys 161(7):1329–1344

Belardinelli ME, Bonafede M, Nespoli M (2019) Stress heterogeneities and failure mechanisms induced by temperature and pore-pressure increase in volcanic regions, Earth Planet Sci Lett 525. https://doi.org/10.1016/j.epsl.2019.115765

Berrino G (1994) Gravity changes induced by height–mass variations at the Campi Flegrei caldera. J Volcanol Geotherm Res 61:293–309

Berrino G, Corrado G, Luongo G, Toro B (1984) Ground deformation and gravity change accompanying the 1982 Pozzuoli uplift. Bull Volcanol 47:187–200

Bianchi R, Coradini A, Federico C, Giberti G, Lanciano P, Pozzi JP, Sartoris G, Scandone R (1987) Modeling of surface deformation in volcanic areas: the 1970–72 and 1982–84 crises at Campi Flegrei, Italy. J Geophys Res 92(14):139–150

Biot MA (1956) General solution of the equations of elasticity and consolidation for a porous material. J Appl Mech 23:91–96

Bodnar RJ, Cannatelli C, De Vivo B, Lima A, Belkin HE, Milia A (2007) Quantitative model for magma degassing and ground deformation (bradyseism) at Campi Flegrei, Italy: Implications for future eruptions. Geology 35:791–794. https://doi.org/10.1130/G23653A.1

Bonafede M (1991) Hot fluid migration: an efficient source of ground deformation: application to the 1982–1985 crisis at Campi Flegrei-Italy. J Volcanol Geotherm Res 48:187–198

Bonafede M, Ferrari C (2009) Analytical models of deformation and residual gravity changes due to a

Mogi source in a viscoelastic medium. Tectonophys 471:4–13. https://doi.org/10.1016/j.tecto.2008.10.006
Bonafede M, Mazzanti M (1998) Modelling gravity variations consistent with ground deformation in the Campi Flegrei caldera (Italy). J Volcanol Geotherm Res 81:137–157
Calò M, Tramelli A (2018) Anatomy of the Campi Flegrei caldera using Enhanced Seismic Tomography Models. Sci Rep 8(1):16254. https://doi.org/10.1038/s41598-018-34456-x
Camacho AG, Gonzalez PJ, Fernandez J, Berrino G (2011) Simultaneous inversion of surface deformation and gravity changes by means of extended bodies with a free geometry: Application to deforming calderas. J Geophys Res 116:B10401. https://doi.org/10.1029/2010JB008165
Cappa F, Rutqvist J, Yamamoto K (2009) Modeling crustal deformation and rupture processes related to upwelling of deep CO_2-rich fluids during the 1965–1967 Matsushiro earthquake swarm in Japan. J Geophys Res 114:10304. https://doi.org/10.1029/2009JB006398
Cardellini C, Chiodini G, Frondini F, Avino R, Bagnato E, Caliro S, Lelli M, Rosiello A (2017) Monitoring diffuse volcanic degassing during volcanic unrests: the case of Campi Flegrei (Italy). Sci Rep 7(1):6757. https://doi.org/10.1038/s41598-017-06941-2
Casertano L, Oliveri del Castillo A, Quagliariello MT (1976) Hydrodynamics and geodynamics in the Phlegraean fields area of Italy. Nature 264:161–164. https://doi.org/10.1038/264161a0
Castagnolo D, Gaeta FS, De Natale G, Peluso F, Mastrolorenzo G, Troise C, Pingue F, Mita DG (2001) Campi Flegrei unrest episodes and possible evolution towards critical phenomena. J Volcanol Geotherm Res 109:13–40. https://doi.org/10.1016/S0377-0273(00)00302-4
Cervelli PF (2013) Analytical Expressions for Deformation from an Arbitrarily Oriented Sphe-roid in a Half-Space. American Geophysical Union, Fall Meeting 2013, abstract id. V44C-06
Chiarabba C, Moretti M (2006) An insight into the unrest phenomena at the Campi Flegrei caldera from Vp and Vp/Vs tomography. Terra Nova 18:373–379. https://doi.org/10.1111/j.1365-3121.2006.00701.x
Chiodini G (2009) CO_2/CH_4 ratio in fumaroles a powerful tool to detect magma degassing episodes at quiescent volcanoes. Geophys Res Lett 36:L02302. https://doi.org/10.1029/2008GL036347
Chiodini G, Frondini F, Cardellini C, Granieri D, Marini L, Ventura G (2001) CO_2 degassing and energy release at Solfatara volcano, Campi Flegrei, Italy. J Geophys Res 106(B8):16,213–16,221. https://doi.org/10.1029/2001JB000246
Chiodini G, Todesco M, Caliro S, Del Gaudio C, Macedonio G, Russo M (2003) Magma de-gassing as a trigger of bradyseismic events: The case of Phlegrean Fields (Italy). Geophys Res Lett 30:1434. https://doi.org/10.1029/2002GL016790
Chiodini G, Vandemeulebrouck J, Caliro S, D'Auria L, De Martino P, Mangiacapra A, Petrillo Z (2015) Evidence of thermal-driven processes triggering the 2005–2014 unrest at Campi Flegrei caldera. Earth Planet Sci Lett 414:58–67. https://doi.org/10.1016/j.epsl.2015.01.012
Coco A, Currenti G, Gottsmann J, Russo G, Del Negro C (2016a) A hydro-geophysical simulator for fluid and mechanical processes in volcanic areas. J Math Ind 6 (1):6. https://doi.org/10.1186/s13362-016-0020-x
Coco A, Gottsmann J, Whitaker F, Rust A, Currenti G, Jasim A, Bunney S (2016b) Numerical models for ground deformation and gravity changes during volcanic unrest: simulating the hydrothermal system dynamics of a restless caldera. Solid Earth 7(2):557–577. https://doi.org/10.5194/se-7-557-2016
Crescentini L, Amoruso A (2007) Effects of crustal layering on the inversion of deformation and gravity data in volcanic areas: an application to the Campi Flegrei caldera, Italy. Geophys Res Lett 34:L09303. https://doi.org/10.1029/2007GL029919
D'Auria L, Giudicepietro F, Martini M, Orazi M, Peluso R, Scarpato G (2010) Polarization analysis in the discrete wavelet domain: An application to volcano seismology. Bull Seismol Soc Am 100 (2):670–683. https://doi.org/10.1785/0120090166
D'Auria L, Giudicepietro F, Martini M, Lanari R (2012) The 4D imaging of the source of ground deformation at Campi Flegrei caldera (southern Italy). J Geophys Res 117:B08209. https://doi.org/10.1029/2012JB009181
D'Auria L, Massa B, Cristiano E, Del Gaudio C, Giudicepietro F, Ricciardi G, Ricco C (2015a) Retrieving the stress field within the Campi Flegrei caldera (southern Italy) through an integrated geodetical and seismological approach. Pure Appl Geophys 172(11):3247–3263
D'Auria L, Pepe S, Castaldo R, Giudicepietro F, Macedonio G, Ricciolino P, Tizzani P, Casu F, Lanari R, Manzo M, Martini M, Sansosti E, Zinno I (2015b) Magma injection beneath the urban area of Naples: a new mechanism for the 2012–2013 volcanic unrest at Campi Flegrei caldera. Sci Rep 5(1):343. https://doi.org/10.1038/srep13100
Davis PM (1986) Surface deformation due to inflation of an arbitrarily oriented triaxial ellip-soidal cavity in an elastic half-space, with reference to Kilauea Volcano. Hawaii. J Geophys Res 91:7429–7438
De Martino P, Guardato S, Tammaro U, Vassallo M, Iannaccone G (2014) A first GPS measurement of vertical seafloor displacement in the Campi Flegrei caldera (Italy). J Volcanol Geotherm Res 276:145–151. https://doi.org/10.1016/j.jvolgeores.2014.03.003
De Natale G, Pingue F, Allard P, Zollo A (1991) Geophysical and geochemical modelling of the 1982 1984 unrest phenomena at Campi Flegrei caldera (southern Italy). J Volcanol Geotherm Res 48:199–222. https://doi.org/10.1016/0377-0273(91)90043-Y
De Natale G, Petrazzuoli S, Pingue F (1997) The effect of collapse structures on ground deformation in calderas. Geophys Res Lett 24:1555–1558
De Natale G, Troise C, Pingue F (2001) A mechanical fluid-dynamical model for ground movements at Campi Flegrei caldera. J Geodyn 32:487–517

De Natale G, Troise C, Pingue F, Mastrolorenzo G, Pappalardo L, Battaglia M, Boschi E (2006) The Campi Flegrei caldera: unrest mechanisms and hazards. Geol Soc London Sp Publ 269:25–45. https://doi.org/10.1144/GSL.SP.2006.269.01.03

De Siena L, Amoruso A, Del Pezzo E, Wakeford Z, Castellano M, Crescentini L (2017a) Space-weighted seismic attenuation mapping of the aseismic source of Campi Flegrei 1983–1984 unrest. Geophys Res Lett 44:1740–1748. https://doi.org/10.1002/2017GL072507

De Siena L, Chiodini G, Vilardo G, Del Pezzo E, Castellano M, Colombelli S, Tisato N, Ventura G (2017b) Source and dynamics of a volcanic caldera unrest: Campi Flegrei, 1983–84. Sci Rep 7(1):8099. https://doi.org/10.1038/s41598-017-08192-7

De Vivo B, Rolandi G, Gans PB, Calvert A, Bohrson WA, Spera FJ, Belkin HE (2001) New constraints on the pyroclastic eruptive history of the Campanian Volcanic Plain (Italy). Miner Petrol 73:47–65

Del Gaudio C, Aquino I, Ricciardi GP, Ricco C, Scandone R (2010) Unrest episodes at Campi Flegrei: a reconstruction of vertical ground movements during 1905–2009. J Volcanol Geotherm Res 195:48–56

Dieterich J, Decker R (1975) Finite element modeling of surface deformation associated with volcanism. J Geophys Res 80(29):4094–4101

Di Vito MA, Isaia R, Orsi G, Southon J, de Vita S, D'Antonio M, Pappalardo L, Piochi M (1999) Volcanic and deformation history of the Campi Flegrei caldera in the past 12 ka. J Volcanol Geotherm Res 91:221–246

Di Vito A, Acocella V, Aiello G, Barra D, Battaglia M, Carandente A, Del Gaudio C, de Vita S, Ricciardi GP, Ricco C, Scandone R, Terrasi F (2016) Magma transfer at Campi Flegrei caldera (Italy) before the 1538 AD eruption. Sci Rep 6:32245. https://doi.org/10.1038/srep32245

Dutrow B, Norton D (1995) Evolution of fluid pressure and fracture propagation during contact metamorphism. J Metam Geol 13:677–686

Dvorak J, Berrino G (1991) Recent ground movement and seismic activity in Campi Flegrei, southern Italy, episodic growth of a resurgent dome. J Geophys Res 96:2309–2323

Fialko Y, Khazan Y, Simons M (2001) Deformation due to a pressurized horizontal circular crack in an elastic half-space, with applications to volcano geodesy. Geophys J Int 146:181–190

Folch A, Gottsmann, JH (2006) Volcanic unrest: Faults and ground uplift at active calderas. In: Troise C, De Natale G, Kilburn CRJ (eds) Mechanisms of activity and unrest at large calderas. Geol Soc London, pp 109–120. ISBN:1862392110

Fournier RO (2006) Hydrothermal systems and volcano geochemistry. In: Dzurisin D (ed) Volcano deformation. Springer, pp 323–341

Fulton PM, Saffer DM (2009) Potential role of mantle-derived fluids in weakening the San Andreas Fault. J Geophys Res 114:B07408. https://doi.org/10.1029/2008JB006087

Gaeta FS, De Natale G, Peluso F, Mastrolorenzo G, Castagnolo D, Troise C, Pingue F, Mita DG, Rossano S (1998) Genesis and evolution of unrest episodes at Campi Flegrei caldera: The role of thermal fluid-dynamical processes in the geothermal system. J Geophys Res 103:20921–20933. https://doi.org/10.1029/97JB03294

Gaeta F, Peluso F, Arienzo I, Castagnolo D, De Natale G, Milano G, Albanese C, Mita DG (2003) A physical appraisal of a new aspect of bradyseism: The mini-uplifts. J Geophys Res 108:2363

Gottsmann JH, Camacho AG, Tiampo KF, Fernández J (2006a) Spatiotemporal variations in vertical gravity gradients at the Campi Flegrei caldera (Italy): a case for source multiplicity during unrest? Geophys J In 167(3):1089–1096. https://doi.org/10.1111/j.1365-246X.2006.03157.x

Gottsmann J, Folch A, Rymer H (2006b) Unrest at Campi Flegrei: A contribution to the mag-matic versus hydrothermal debate from inverse and finite element modeling. J Geophys Res 111:B07203. https://doi.org/10.1029/2005JB003745

Gottsmann J, Rymer H, Berrino G (2006c) Caldera unrest at the Campi Flegrei: A critical evaluation of source parameters from geodetic data inversion. J Volcanol Geotherm Res 150:132–145. https://doi.org/10.1016/j.jvolgeores.2005.07.002

Guidoboni E, Ciuccarelli C (2011) The Campi Flegrei caldera: historical revision and new data on seismic crises, bradyseisms, the Monte Nuovo eruption and ensuing earthquakes (twelfth century 1582 AD). Bull Volcanol 73:655–677

Heap MJ, Baud P, Meredith PG, Vinciguerra S, Reuschle T (2014) The permeability and elastic moduli of tuff from Campi Flegrei, Italy: implications for ground deformation modelling. Solid Earth 5 (1):25–44. https://doi.org/10.5194/se-5-25-2014

Hurwitz S, Christiansen LB, Hsieh PA (2007) Hydrothermal fluid flow and deformation in large calderas: Inferences from numerical simulations. J Geophys Res 112:1–16. https://doi.org/10.1029/2006JB004689

Hutnak M, Hurwitz S, Ingebritsen SE, Hsieh PA (2009) Numerical models of caldera defor-mation: Effects of multiphase and multicomponent hydrothermal fluid flow. J Geophys Res 114:B04411

Iannaccone G, Guardato S, Donnarumma GP, De Martino P, Dolce M, Macedonio G, Chierici F, Beranzoli L (2018) Measurement of seafloor deformation in the marine sector of the Campi Flegrei caldera (Italy). J Geophys Res 123:66–83. https://doi.org/10.1002/2017JB014852

Lanari R, Berardino P, Borgstrom S, Del Gaudio C, De Martino P, Fornaro G, Guarino S, Ric-ciardi GP, Sansosti E, Lundgren P (2004) The use of IFSAR and classical geodetic tech-niques for caldera unrest episodes: application to the Campi Flegrei uplift event of 2000. J Volcanol Geotherm Res 133:247–260

Lesher CE, Spera FJ (2015) Chapter 5—Thermodynamic and transport properties of silicate melts and magma. In: Sigurdsson H. (ed) The Encyclopedia of Volcanoes (2nd ed), pp 113–141. https://doi.org/10.1016/B978-0-12-385938-9.00005-5

Lima A, Vivo BD, Spera FJ, Bodnar RJ, Milia A, Nunziata C, Belkin HE, Cannatelli C (2009) Thermodynamic model for uplift and deflation episodes (bradyseism) associated with magmatic–hydrothermal activity at the Campi Flegrei (Italy). Earth-Sci Rev 97: 44–58. https://doi.org/10.1016/j.earscirev.2009.10.001

Lundgren P, Usai S, Sansosti E, Lanari R, Tesauro M, Fornaro G, Berardino P (2001) Modeling surface deformation observed with synthetic aperture radar interferometry at Campi Flegrei caldera. J Geophys Res 106(B9):19355–19366. https://doi.org/10.1029/2001JB000194

Luongo A, Amoruso A, Crescentini L (2015) Finite element modeling tests of the seven moment tensor approximation of ground displacement from tri-axial pressurized ellipsoids. Geophys Res Abstr 17, EGU2015-10574

Macedonio G, Giudicepietro F, D'Auria L, Martini M (2014) Sill intrusion as a source mecha-nism of unrest at volcanic calderas. J Geophys Res: Solid Earth 119:3986–4000. https://doi.org/10.1002/2013JB010868

Manconi A, Walter TR, Manzo M, Zeni G, Tizzani P, Sansosti E, Lanari R (2010) On the effects of 3-D mechanical heterogeneities at Campi Flegrei caldera, southern Italy. J Geophys Res 115:B08405. https://doi.org/10.1029/2009JB007099

Mantiloni L, Nespoli M, Belardinelli ME, Bonafede M (2020) Deformation and stress in hydrothermal regions: The case of a disk-shaped inclusion in a half-space. J Volcanol Geotherm Res 403:1–12. https://doi.org/10.1016/j.jvolgeores.2020.107011

Marturano A, Isaia R, Aiello G, Barra D (2018) Complex dome growth at Campi Flegrei caldera (Italy) in the last 15 ka. J Geophys Res: Solid Earth 123. https://doi.org/10.1029/2018JB015672

McTigue D (1986) Thermoelastic response of fluid-saturated porous rock. J Geophys Res 91:9533–9542

McTigue DF (1987) Elastic stress and deformation near a finite spherical magma body: Resolution of the point source paradox. J Geophys Res 92:12931–12940

Mogi K (1958) Relations of the eruptions of various volcanoes and the deformation of ground surfaces around them. Bull Earthq Res Inst Univ Tokyo 36:94–134

Morhange C, Marriner N, Laborel J, Todesco M, Oberlin C (2006) Rapid sea-level movements and noneruptive crustal deformations in the Phlegrean Fields caldera. Italy. Geology 34(2):93–96. https://doi.org/10.1130/G21894.1

Norton DL, Dutrow BL (2001) Complex behavior of magma-hydrothermal processes: role of supercritical fluid. Geochim Cosmochim Acta 65:4009–4017

Okada Y (1985) Surface deformation due to shear and tensile faults in a half-space. Bull Seismol Soc Am 75:1135–1154

Okubo S, Watanabe H (1989) Gravity change caused by a fissure eruption. Geophys Res Lett 16:445–448

Orsi G, D'Antonio M, de Vita S, Gallo G (1992) The Neapolitan Yellow Tuff, a large-magnitude trachytic phreatoplinian eruption: eruptive dynamics, magma withdrawal and caldera collapse. J Volcanol Geotherm Res 53:275–287

Orsi G, de Vita S, Di Vito M (1996) The restless, resurgent Campi Flegrei nested caldera (Italy): constraints on its evolution and configuration. J Volcanol Geotherm Res 74(3–4):179–214

Orsi G, Petrazzuoli SM, Wohletz K (1999) Mechanical and thermo-fluid behaviour during unrest at the Campi Flegrei caldera (Italy). J Volcanol Geotherm Res 91 (2–4):453–470

Parascandola A (1947) I fenomeni bradisismici del Serapeo di Pozzuoli. Boll Soc Natur Napoli, Italy 56:115

Pedone M, Aiuppa A, Giudice G, Grassa F, Cardellini C, Chiodini G, Valenza M (2014) Volcanic CO_2 flux measurement at Campi Flegrei by tunable diode laser absorption spectros-copy. Bull Volcanol 76:812. https://doi.org/10.1007/s00445-014-0812-z

Rice JR, Cleary MP (1976) Some basic stress diffusion solutions for fluid-saturated elastic porous media with compressible constituents. Rev Geophys Space Phys 14:227–241

Rinaldi A, Todesco M, Bonafede M (2010) Hydrothermal instability and ground displacement at the Campi Flegrei caldera. Phys Earth Plan Inter 178:155–161

Rosi M, Sbrana A, Principe C (1983) The Phlegraean Fields; structural evolution, volcanic history and eruptive mechanisms. J Volcanol Geotherm Res 17:273–288

Rundle JB (1978) Gravity changes and the Palmdale uplift. Geophys Res Lett 5:41–44

Rundle JB (1980) Static elastic—gravitational deformation of a layered half—space by point couple stresses. J Geophys Res 85:5355–5363

Rundle JB (1982) Deformation, gravity and potential change due to volcanic loading of the crust. J Geophys Res 87:10724–10744

Rutqvist J, Wu YS, Tsang CF, Bodvarsson G (2002) A modeling approach for analysis of coupled multiphase fluid flow, heat transfer, and deformation in fractured porous rock. Intern J Rock Mech Mining Sci 39:429–442

Segall P (2010) Earthquake and volcano deformation. Princeton University Press, Princeton, 456 pp, ISBN 9780691133027

Smith VC, Isaia R, Pearce NJG (2011) Tephrostratigraphy and glass compositions of post-15 kyr Campi Flegrei eruptions: implications for eruption history and chronostratigraphic markers. Quaternary Sci Rev 30:3638–3660

Todesco M, Chiodini G, Macedonio G (2003) Monitoring and modelling hydrothermal fluid emission at La Solfatara (Phlegrean Fields, Italy). An interdisciplinary approach to the study of diffuse degassing. J Volcanol Geotherm Res 125:57–79

Todesco M, Rutqvist J, Chiodini G, Pruess K, Oldenburg CM (2004) Modeling of recent volcanic episodes at Phlegrean Fields (Italy): geochemical variations and ground deformation. Geothermics 33:531–547. https://doi.org/10.1016/j.geothermics.2003.08.014

Todesco M, Rinaldi AP, Bonafede M (2010) Modeling of unrest signals in heterogeneous hydrothermal systems. J Geophys Res 115:B09213. https://doi.org/10.1029/2010JB007474

Todesco M, Costa A, Comastri A, Colleoni F, Spada G, Quareni F (2014) Vertical ground displacement at Campi Flegrei (Italy) in the fifth century: rapid subsidence driven by pore pressure drop. Geophys Res Lett 41:1471–1478. https://doi.org/10.1002/2013GL059083

Trasatti E, Bonafede M (2008a) Gravity changes due to overpressure sources in 3D heterogeneous media: application to Campi Flegrei caldera, Italy. Ann Geophys 51:121–135

Trasatti E, Giunchi C, Bonafede M (2005) Structural and rheological constraints on source depth and overpressure estimates at the Campi Flegrei caldera, Italy. J Volcanol Geotherm Res 144:105118

Trasatti E, Casu F, Giunchi C, Pepe S, Solaro G, Tagliaventi S, Berardino P, Manzo M, Pepe A, Ricciardi GP, Sansosti E, Tizzani P, Zeni G, Lanari R (2008b) The 2004-2006 uplift episode at Campi Flegrei caldera (Italy): constraints from SBAS-DInSAR ENVISAT data and Bayesian source inference. Geophys Res Lett 35. https://doi.org/10.1029/2007GL033091

Trasatti E, Bonafede M, Ferrari C, Giunchi C, Berrino G (2011) On deformation sources in volcanic areas: modeling the Campi Flegrei (Italy) 1982–84 unrest. Earth Planet Sci Lett 306:175–185. https://doi.org/10.1016/j.epsl.2011.03.033

Trasatti E, Polcari M, Bonafede M, Stramondo S (2015) Geodetic constraints to the source mechanism of the 2011–2013 unrest at Campi Flegrei (Italy) caldera. Geophys Res Lett 42(10):3847–3854

Troise C, De Natale G, Pingue F, Obrizzo F, De Martino P, Tammaro U, Boschi E (2007) Renewed ground uplift at Campi Flegrei caldera (Italy): New insight on magmatic processes and forecast. Geophys Res Lett 34:L03301. https://doi.org/10.1029/2006GL028545

Tsang CF (1999) Linking thermal, hydrological, and mechanical processes in fractured rocks 1. Annu Rev Earth Plan Sci 27:359–384

Vasco DW, Wicks C, Karasaki K, Marques O (2002) Geodetic imaging: Reservoir monitoring using satellite interferometry. Geophys J Int 149:555–571

Walsh J, Rice J (1979) Local changes in gravity resulting from deformation. J Geophys Res 84(B1). https://doi.org/10.1029/JB084iB01p00165

Wang R, Lorenzo Martin F, Roth F (2006) PSGRN/PSCMP a new code for calculating co- and post-seismic deformation, geoid and gravity changes based on the viscoelastic-gravitational dislocation theory. Comp Geosci 32:527–541

Woo JYL, Kilburn CRJ (2010) Intrusion and deformation at Campi Flegrei, southern Italy: Sills, dikes, and regional extension. J Geophys Res 115:B12210. https://doi.org/10.1029/2009JB006913

Yang X, Davis PM, Dieterich JH (1988) Deformation from inflation of a dipping finite prolate spheroid in an elastic half-space as a model for volcanic stressing. J Geophys Res 93:4249–4257

Zollo A, Maercklin N, Vassallo M, Iacono DD, Virieux J, Gasparini P (2008) Seismic reflections reveal a massive melt layer feeding Campi Flegrei caldera. Geophys Res Lett 35:L12306. https://doi.org/10.1029/2008GL034242

Volcanic Hazard Assessment at the Campi Flegrei Caldera, Italy

Andrea Bevilacqua, Giovanni Macedonio, Augusto Neri, Giovanni Orsi, and Paola Petrosino

Abstract

The Campi Flegrei caldera is a complex and resurgent structure, and its active portion has been the site of an intense volcanism with eruptions concentrated in temporal clusters, called epochs. The caldera is home to about 350,000 people. In the past decades, many scientific studies were aimed at both defining fundamental parameters of a future eruption, and hazard zoning of the territory. The approach to the volcanic hazard assessment of the Campi Flegrei caldera has progressively changed through time from a "deterministic" interpretation of the available information to a quantitative probabilistic elaboration of the main sources of uncertainty. In particular, on the long-term, Astroni-Agnano-Solfatara is recognised to be the most probable area in which a vent will likely open, while the Averno-Monte Nuovo is the second most probable one. Significant uncertainty affects these results, and a non-negligible vent opening probability spreads over the whole caldera. The inter-event times of volcanic eruptions span from a few years to thousands of years. Within the epochs of activity, the mean recurrence time is tens of years, and intra-epoch temporal groups are evident. The probability that the next eruption will be of < 0.1 km^3 volume of magma DRE is $\sim$95%. Larger size eruptions tend to be localised in the central-eastern sector of the caldera. Expected hazards from renewal of explosive volcanism result from pyroclastic fallout and pyroclastic density currents. Pyroclastic fallout deposits in excess of 300 kg/m^2 can affect most of the caldera and the city of Naples with more than 10% probability. The entire caldera has the potential to be affected by pyroclastic density currents invasion with mean probability above 30% in its central-eastern portion, and above 50% in the Agnano area. Invasion probabilities of $\sim$5–10% have been estimated for the urbanised areas along the eastern slopes of the Posillipo Hill. In summary, probabilistic hazard assessment is particularly important in Campi Flegrei caldera, due to the uncertain location for potential vents and the large variability of eruption styles and sizes that cannot be, at the present time, effectively constrained by monitoring signals.

A. Bevilacqua (✉) · A. Neri
Istituto Nazionale di Geofisica e Vulcanologia, Sezione di Pisa, Pisa, Italy
e-mail: andrea.bevilacqua@ingv.it

G. Macedonio
Istituto Nazionale di Geofisica e Vulcanologia, Sezione Osservatorio Vesuviano, Napoli, Italy

G. Orsi · P. Petrosino
Dipartimento di Scienze della Terra, dell'Ambiente e delle Risorse, Università degli Studi di Napoli Federico II, Napoli, Italy

© Springer-Verlag GmbH Germany, part of Springer Nature 2022

G. Orsi et al. (eds.), *Campi Flegrei*, Active Volcanoes of the World,
https://doi.org/10.1007/978-3-642-37060-1_12

1 Introduction

At least 575 volcanoes, out of the over 1,250 that have been active in the Holocene, have erupted in historical times (Simkin and Siebert 1994; Brown et al. 2014; Siebert et al. 2015), and 97 out of 446 calderas are in unrest as of December 2020 (Newhall and Dzurisin 1988a, b; Geyer and Marti 2008; Siebert et al. 2010; Acocella et al. 2015; https://volcano.si.edu/). About 11% of the Earth population lives in a radius of 100 km around volcanoes that in the last 10,000 years have generated hazardous phenomena, able to adversely affect people, structures and infrastructures (Blong 1984; Small and Naumann 2001; Heiken et al. 2003; Marti and Ernst 2005; Aspinall and Blong 2015).

Modern volcanology tackled the challenge of eruption forecasting, assessing volcanic hazard, and zoning the exposed territory in relation to the likely effects produced by the expected volcanic phenomena (Sparks 2003; Aspinall et al. 2003; Connor et al. 2015; National Academy of Science 2017). A thorough volcanic hazard assessment has to answer three basic questions: when, where and how will next eruption occur. In a caldera or a volcanic field, the spatial assessment of future vent opening typically comes first. Therefore, three key forecasting issues naturally arise: (i) the spatial location of a new vent opening; (ii) the onset time, that is, the expected time before the beginning of the next eruption; (iii) the style and size of the expected hazardous phenomena, and the affected areas.

Volcanic hazard assessments are typically classified in two groups: long-term assessments and short-term assessments (Banks et al. 1989; Marzocchi and Bebbington 2012; Poland and Anderson 2020). Long-term hazard assessments are based on the assumption that the past is a key to understanding the future, implying that the expected phenomena will be similar to those generated by that specific volcano in the past and affect similar areas. To do so, it is fundamental to reconstruct the eruptive history of that volcanic system. Another assumption is the portion of the entire history of the volcano to be taken into account. In a caldera, this portion generally post-dates the last event that has generated a dramatic and significant change in the structural setting of the entire volcanic system, including its magmatic feeding system, so as to determine a change in its behaviour. From a practical point of view, a long-term hazard assessment is a preparatory approach valid on a long-term perspective, which allows being ready in case of an impending eruption.

Instead, short-term hazard assessments are based on the investigation of phenomena acknowledged as precursors of past eruptions and on the definition of the current state of the volcano through monitoring data. They will complement existing hazard models during an eruptive crisis, given enough time to react. In closed conduit volcanoes, type, size and time-evolution of precursors of past eruptions are defined through geological and historical-documentary studies. The current state of a volcano is assessed by analysing the data collected by geophysical and geochemical monitoring networks and geological-structural surveys (McGuire et al. 1995; Scarpa and Tilling 1996; Sparks et al. 2012; Dempsey et al. 2020).

The knowledge of when, where, and how a future eruption will occur, along with socio-economic evaluations, is propaedeutic to the assessment of the risk related to the activity of a volcanic system (Fournier d'Albe 1979; Tilling 2001; ISDR 2005; Aspinall and Blong 2015). Volcanic hazard and risk assessment, as well as the zoning of the territory in relation to the expected hazards and related risks, are the essential requirements for a sound territorial planning and for actions aimed at volcanic risk mitigation by the civil authorities.

The Campi Flegrei caldera (CFc) is an active, complex and resurgent structure whose volcanic and deformation history is outlined in Chap. Volcanic and Deformation History of the Campi Flegrei Volcanic Field, Italy. It is the site of widespread fumarole and thermal spring activity (see Chap. The Hydrothermal System of the Campi Flegrei Caldera, Italy) that has persisted for millennia, as testified by the name Campi Flegrei (Burning Fields in English), that was given to the area by the earliest Hellenic colonists in the eighth century BC (see Chap. The Urban Development of Campi Flegrei, Italy). The caldera floor is also affected by an ongoing resurgence through a series of short-term events, also known as bradyseism, that have occurred at least since the 1950s (Del Gaudio et al. 2010; Ricco et al. 2019; Chaps. The Permanent Monitoring System of the Campi Flegrei Caldera, Italy; The Hydrothermal System of the Campi Flegrei Caldera, Italy; Historic Unrest of the Campi Flegrei Caldera, Italy; Source Modelling from Ground Deformation and Gravity Changes at the Campi Flegrei Caldera, Italy), and that determine the current unrest state of the caldera. All these characteristics highlight that the magmatic system is active.

CFc is in unrest since decades, and its volcanism had a prevailing explosive character, therefore volcanic hazard potential is high. The active portion of the caldera, hosting a densely inhabited conurbation and the western neighbourhoods of the city of Naples, is home to about 350,000 people, meaning that volcanic risk is also very high. In such a situation, a volcanic hazard assessment and a zoning of the territory in relation to the expected hazardous phenomena in case of renewal of the activity in short- or mid-terms, is a must for the scientific community. In fact, in the past decades several scientific studies focussed on these topics have been published. The approach to the volcanic hazard assessment of the CFc has progressively changed through time from a "deterministic" interpretation of geological, volcanological, geochemical and geophysical data to a quantitative probabilistic elaboration of these data.

This Chapter presents the results achieved by the scientific community over the years and synthesises the current state of knowledge on the main topics needed for a comprehensive volcanic hazard assessment.

2 Volcanological Outlines

The CFc, the dominant feature of the CF volcanic field, was first interpreted by Orsi et al. (1992, 1996) as a nested, resurgent and restless structure resulting from two cataclysmic eruptions (see also Chap. Volcanic and Deformation History of the Campi Flegrei Volcanic Field, Italy and references therein; Fig. 1). The youngest caldera is believed to be the only portion still active of the entire structure (Capuano et al. 2013).

The earliest caldera-forming event was the Campanian Ignimbrite (CI) eruption occurred around 40 ka (Barberi et al. 1978; Fisher et al. 1993; Orsi et al. 1996; Rosi et al. 1996; Civetta et al. 1997; Fedele et al. 2008; Perrotta et al. 2010; Giaccio et al. 2017; Silleni et al. 2020). The second event, with a collapse nested within the earlier one, was the Neapolitan Yellow Tuff (NYT) eruption that took place at about 15 ka (Orsi et al. 1992, 1995; Scarpati et al. 1993; Wohletz et al. 1995; Deino et al. 2004). A caldera collapse between these two has been recently hypothesised (Albert et al. 2019). Prior to the CI caldera collapse, the volcanism extended far outside the margins of the CFc, while after each collapse, it has been concentrated within the collapsed area. The stratigraphic sequence of the deposits of the pre-NYT activity and, thus, the volcanic history until the collapse of the NYT caldera is difficult to reconstruct in detail because of the large area affected by the CI caldera collapse, the huge amount of material deposited by the CI eruption, and the occurrence of the NYT caldera collapse over a large portion of the western sector of the CI caldera floor.

The NYT caldera collapse occurred through a piecemeal mechanism mostly along northeast-southwest and northwest-southeast, and subordinately north–south trending faults, likely reactivating portions of pre-existing regional structures (Capuano et al. 2013). The NYT eruption and related caldera collapse have produced such a remarkable change in the structural setting of the volcanic system to significantly modify its behaviour. The floor of the NYT caldera has been the site of an intense and interconnected deformation and volcanic activity. The main display of the deformation is an ongoing resurgence of the central portion of the caldera floor, while the volcanism has generated at least 70 eruptions in the last 15 kyrs.

Within the NYT caldera resurgence, a long-term and a short-term deformation have been distinguished; the long-term deformation sums up the effects of the short-term deformation, which occurs over years to tens of years, and accounts for

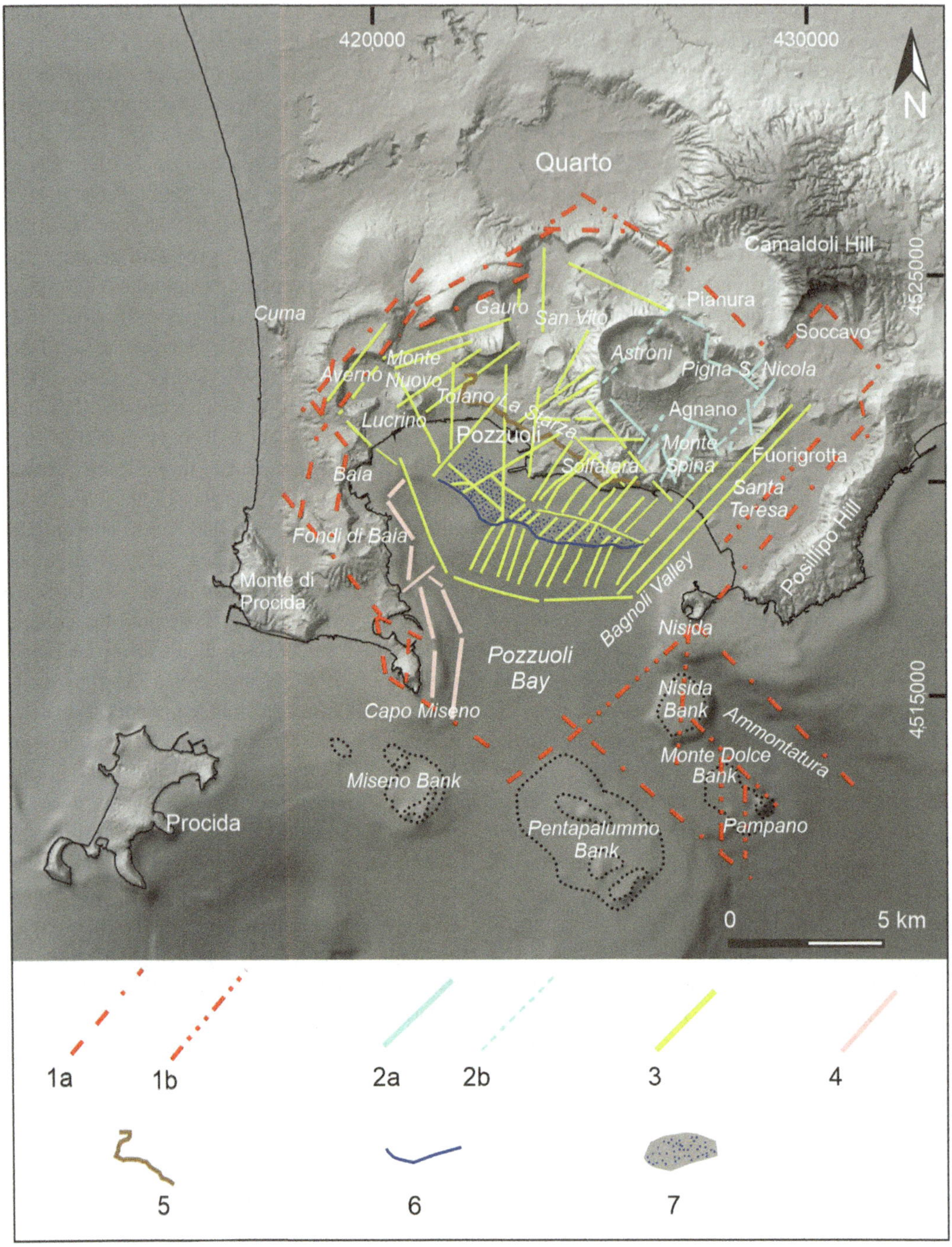

Fig. 1 Structural sketch map of the Neapolitan Yellow Tuff caldera. **1a** Faults active during the Neapolitan Yellow Tuff eruption. **1b** Faults likely active during the Neapolitan Yellow Tuff eruption. **2a** Faults active during the Agnano-Monte Spina eruption. **2b** Faults likely active during the Agnano-Monte Spina eruption. **3** Faults active during resurgence of La Starza block. **4** Faults active during subsidence of the Neapolitan Yellow Tuff caldera floor. **5** Rim of La Starza marine terrace. **6** Rim of the post-Würmian level surface. **7** Post-Würmian level surface. Figure modified after Chap. Volcanic and Deformation History of the Campi Flegrei Volcanic Field, Italy

the general process of resurgence operating since the NYT caldera collapse (Orsi et al. 1996, 1999). The long-term deformation has disjointed the caldera floor generating the uplifting of blocks in its central portion (Acocella 2010; Natale et al. 2020). The main expressions of this uplift are two parallel morphostructural features (Chap. Volcanic and Deformation History of the Campi Flegrei Volcanic Field, Italy; Fig. 1). One is a northwest-southeast oriented on-land feature topped by the La Starza marine terrace for which a maximum net uplift of 85–100 m has been evaluated (Cinque et al. 1985; Sacchi et al. 2014; Isaia et al. 2019). A hypothetical maximum cumulative uplift of ~ 180 m has been estimated for the entire structure after the NYT caldera collapse (Sacchi et al. 2014). The second is the Lucrino-Bagnoli Valley morphostructural high, delimited towards the southwest by a roughly northwest-southeast trending alignment of high-angle morphostructural slopes and truncated by a post-Würmian level surface in the Pozzuoli Bay (Chap. Volcanic and Deformation History of the Campi Flegrei Volcanic Field, Italy; Fig. 1).

The short-term deformation (i.e., the bradyseism) is interpreted as transient episodes during the general resurgence process (Orsi et al. 1999; Del Gaudio et al. 2010). These events generate ground deformation, seismicity, gravity changes and compositional variations of fluid effluents (for reviews see Chaps. The Permanent Monitoring System of the Campi Flegrei Caldera, Italy; The Hydrothermal System of the Campi Flegrei Caldera, Italy; Historic Unrest of the Campi Flegrei Caldera, Italy; Source Modelling from Ground Deformation and Gravity Changes at the Campi Flegrei Caldera, Italy, and references therein). In 2012 values and trends of these phenomena induced the Italian Dipartimento della Protezione Civile to raise the alert level from "base" to "attention" (www.protezionecivile.it).

2.1 Post-Neapolitan Yellow Tuff Volcanism

The post-NYT volcanism was fed by a complex feeding system, and extruded magmas varying in composition from trachybasalt to phonolite (Chaps. An Evolutionary Model for the Magmatic System of the Campi Flegrei Volcanic Field (Italy) Constrained by Petrochemical Data; Origin and Differentiation History of the Magmatic System Feeding the Campi Flegrei Volcanic Field (Italy) Constrained by Radiogenic and Stable Isotope Data; Tephrochronology and Geochemistry of Tephra from the Campi Flegrei Volcanic Field, Italy). Their time sequence, reported in Table 1, is derived from the current literature (e.g., Orsi et al. 2004; Smith et al. 2011; Fedele et al. 2011, 2012; Isaia et al. 2012; Bevilacqua et al. 2015, 2016). The eruptions over the past 15 kyrs occurred in three clusters of activity, called epochs, followed by periods of quiescence (Di Vito et al. 1999). After more than 3 kyrs of quiescence following the end of the last epoch, the Monte Nuovo last eruption occurred as a solitary event in AD 1538 (Guidoboni and Ciuccarelli 2011; Di Vito et al. 2016 and references therein).

During each epoch of activity, the eruptions followed each other with time intervals of tens to hundreds of years. The distribution of the active vents through time has followed a specific pattern in each epoch of activity (Figs. 2 and 3) and has been considered as a tracer of the structures acting as feeding features for volcanism (Orsi et al. 1996, 1999, 2004).

The eruptions of the past 15 kyrs of the CFc were predominantly explosive, with only four effusive events (Table 1). The effusive eruptions mostly extruded dense trachytic magmas that formed lava domes and subordinately short and stubby lava flows. The explosive eruptions were characterised by phreatomagmatic phases prevailing over purely magmatic. Contemporaneous magmatic and phreatomagmatic explosions have also been documented (Dellino et al. 2004; Romano et al. 2020). The phreatomagmatic explosions generated mostly dilute pyroclastic density currents (PDCs) that flowed within the floor of the NYT caldera and in some cases overcame its high-angle morphological boundaries to spread over the surrounding plain (Fig. 4). The magmatic explosions gave rise to

Table 1 Post-NYT eruptions

ID	Eruption	Epoch	Age (cal. years BP)		Magma Volume (DRE) (km^3)			Area PDC (km2)
			Percentile		Percentile			
			2.5th	97.7th	5th	50th	95th	
70	Monte Nuovo	AD 1538			0.03	0.03	0.03	5.7
69	Nisida	3b	3,213	4,188	0.01	0.02	0.03	4.7
68	Fossa Lupara	3b	3,978	4,192	0.01	0.02	0.03	8.9
67	Astroni 7	3b	4,098	4,297	0.04	0.07	0.11	10.2
66	Astroni 6	3b			0.06	0.12	0.18	26.9
65	Astroni 5	3b			0.05	0.10	0.15	29.1
64	Astroni 4	3b			0.07	0.14	0.21	60.4
63	Astroni 3	3b			0.08	0.16	0.24	41.1
62	Astroni 2	3b			0.01	0.02	0.03	19.1
61	Astroni 1	3b	4,153	4,345	0.03	0.06	0.09	39.7
60[?]	Capo Miseno	3b	3,200	4,200	0.01	0.02	0.03	1.1
59[a]	Averno 2	3b			0.04	0.07	0.11	24.8
58	Solfatara	3b	4,181	4,386	0.02	0.03	0.05	8.7
57[b]	Accademia Lava Dome	3b			0.00	-	0.01	
56	Monte Olibano Tephra	3b			0.01	-	0.10	
55	Solfatara lava dome	3b			0.00	-	0.01	
54	Paleo-Astroni 3	3b			0.01	0.02	0.03	
53[b]	M.te Olibano Lava Dome	3b			0.00	-	0.01	
52	S.ta Maria delle Grazie	3b	4,382	4,509	0.01	-	0.10	
51	Agnano-Monte Spina	3a	4,482	4,625	0.43	0.85	1.28	312.5
50	Paleo-Astroni 2	3a	4,712	4,757	0.10	-	0.30	5.4
49	Paleo-Astroni 1	3a	4,745	4,834	0.03	0.05	0.08	18.1
48[b]	Monte Sant'Angelo	3a	4,832	5,010	0.10	-	0.30	43.8
47	Pignatiello 2	3a			0.01	0.02	0.03	
46	Cigliano	3a			0.03	0.05	0.08	28.3
45	Agnano 3	3a			0.10	0.19	0.29	68.0
44	Averno 1	3a	5,064	5,431	0.01	-	0.10	27.0
43	Agnano 2	3a			0.01	0.01	0.02	17.1
42	Agnano 1	3a	5,266	5,628	0.01	0.02	0.03	
41	San Martino	2	9,026	9,370	0.03	0.05	0.08	19.7
40	Sartania 2	2			0.01	-	0.10	27.0
39	Pigna San Nicola	2	9,201	9,533	0.10	-	0.30	8.0
38	Costa San Domenico	2			0.01	-	0.10	16.9
37	Monte Spina Lava Dome	2			0.00	-	0.01	
36	Sartania 1	2	9,500	9,654	0.01	-	0.10	40.7
35	Fondi di Baia	2	9,525	9,695	0.02	0.04	0.06	15.7
34	Baia	2			0.00	-	0.01	
33[?]	Porto Miseno	1	10,347	12,860	0.01	-	0.10	0.7
32[?]	Bacoli	1	11,511	14,154	0.10	0.20	0.30	1.1

(continued)

Table 1 (continued)

ID	Eruption	Epoch	Age (cal. years BP)		Magma Volume (DRE) (km^3)			Area PDC (km2)
			Percentile		Percentile			
			2.5th	97.7th	5th	50th	95th	
31	Casale	1			0.01	-	0.10	
30	Pisani 3	1	10,516	10,755	0.01	-	0.10	3.0
29	Pignatiello 1	1			0.01	-	0.10	
28	Montagna Spaccata	1			0.01	0.02	0.03	3.0
27	Concola	1			0.00	-	0.01	
26	Fondo Riccio	1			0.00	-	0.01	
25	Pisani 2	1			0.10	-	0.30	21.1
24	Pisani 1	1			0.10	-	0.30	
23	Soccavo 5	1			0.01	-	0.10	66.2
22	Minopoli 2	1			0.01	-	0.10	113.6
21	Paleo-San Martino	1			0.01	-	0.10	37.3
20	Soccavo 4	1			0.10	-	0.30	180.2
19	S4s3_2	1			0.01	-	0.10	
18	S4s3_1	1			0.10	-	0.30	
17	Soccavo 3	1			0.01	-	0.10	147.5
16	Soccavo 2	1			0.01	-	0.10	75.8
15	Paleo-Pisani 2	1			0.10	-	0.30	37.7
14	Paleo-Pisani 1	1			0.01	-	0.10	190.5
13	Pomici Principali	1	11,915	12,158	0.43	0.85	1.28	129.2
12	Gaiola	1			0.01	-	0.10	
11	Soccavo 1	1			0.25	0.50	0.75	190.5
10	Paradiso	1			0.01	-	0.10	
9	Minopoli 1	1			0.01	-	0.10	
8	Torre Cappella	1			0.01	-	0.10	1.0
7	La Pigna 2	1			0.01	-	0.10	
6	La Pigna 1	1	12,749	13,110	0.01	-	0.10	
5	La Pietra	1			0.01	-	0.10	2.6
4	Santa Teresa	1			0.01	-	0.10	0.9
3	Gauro	1	12,721	15,511	0.25	0.50	0.75	16.1
2	Mofete	1			0.01	-	0.10	2.1
1	Bellavista	1			0.01	-	0.10	3.9

Legend: [a]coeval with the previous eruption; [b]uncertain stratigraphic order with the previous eruption; [?]unconstrained stratigraphic order. The eruptions record is based on Di Vito et al. (1999), de Vita et al. (1999), Isaia et al. (2004, 2009), Orsi et al. (2004, 2009), Di Renzo et al. (2011), Fedele et al (2011), Smith et al. (2011), Neri et al. (2015a, b) and Bevilacqua et al. (2016). The Age and Magma Volume estimates are from Smith et al. (2011) and references therein. The Age of Capo Miseno is from Di Renzo et al. (2011). The Magma Volume uncertainty ranges are from Bevilacqua et al. (2016). The Area PDC estimates are from Orsi et al. (2004) but those of events 33, 15, 8, 4 that are from Neri et al. (2015a, b), and the one of event 51 that is from de Vita et al. (1999)

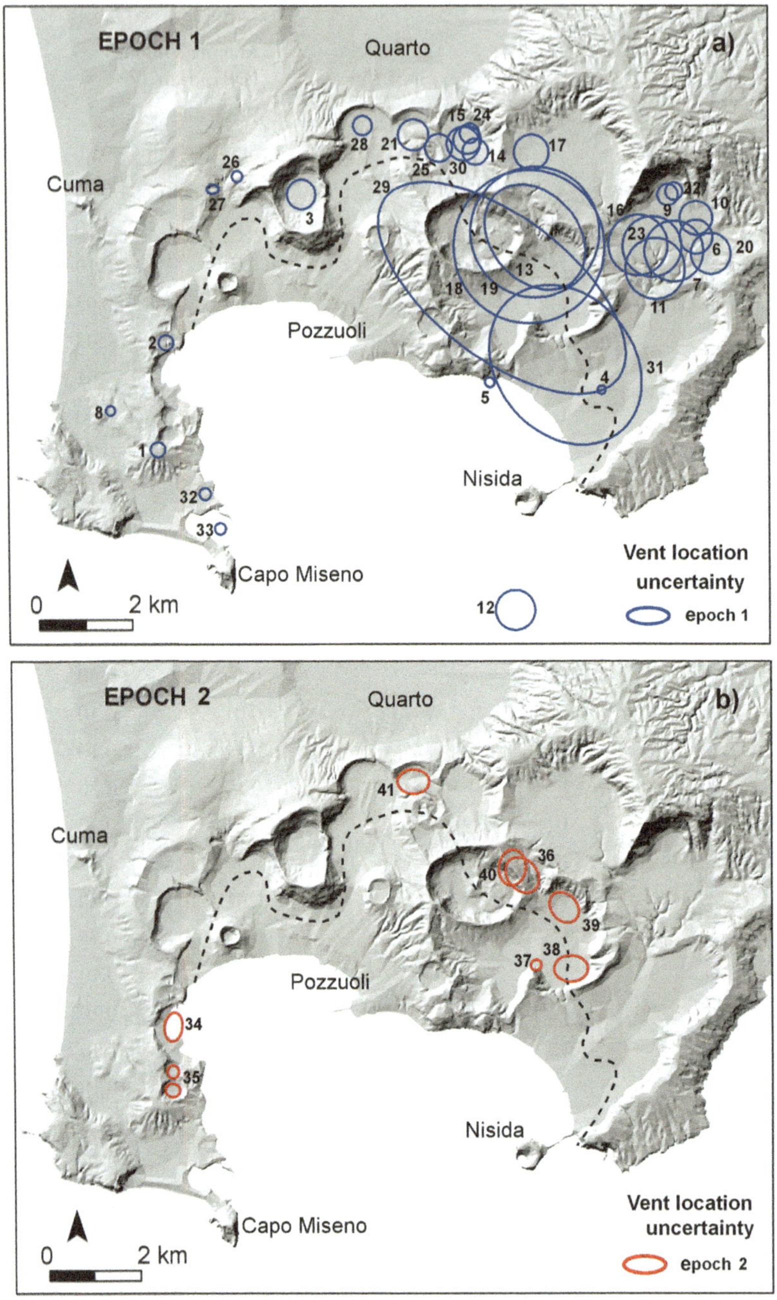

Fig. 2 Location of the vents of the eruptions of **a** epoch 1 and **b** epoch 2. Numbering of the events follows Table 1. The dashed line indicates the likely location of the coast line between epochs 2 and 3 (modified after Orsi et al. 2004). Figure modified after Bevilacqua et al. (2015)

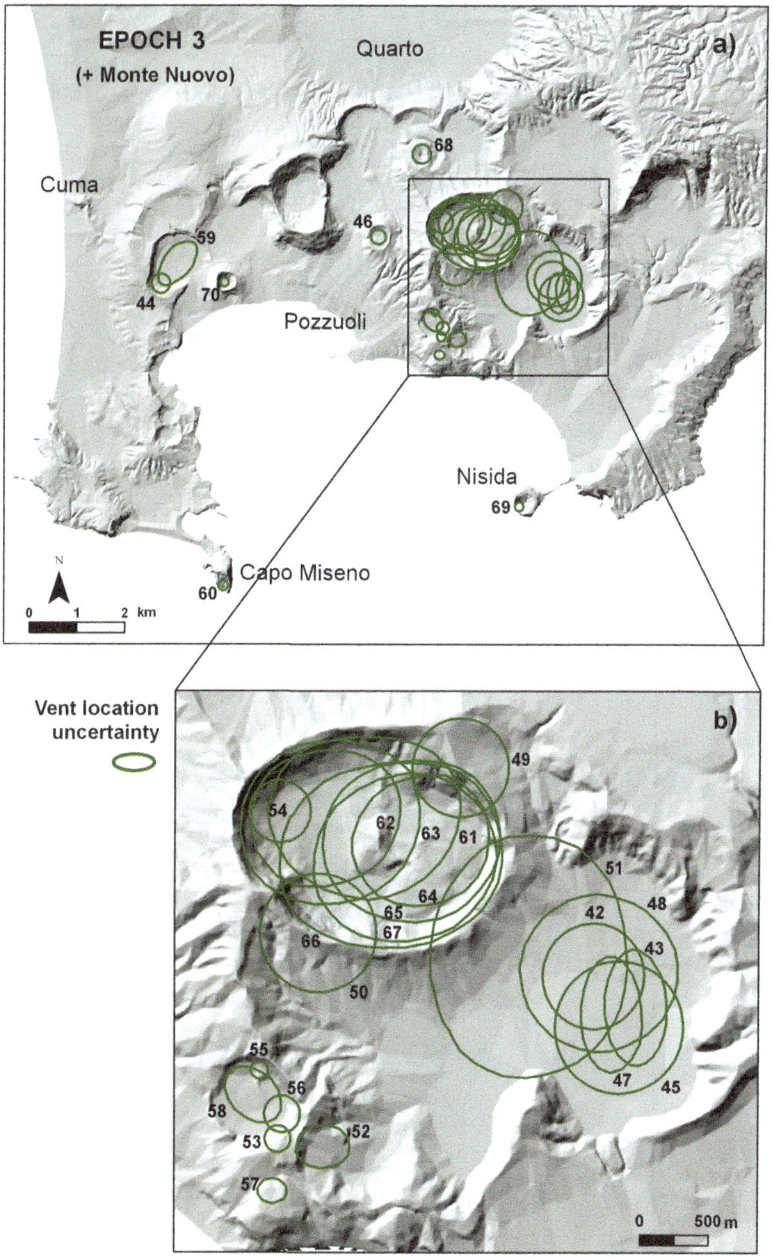

Fig. 3 **a** Location of the vents of the eruptions of epoch 3. **b** zoom of **a**. Numbering of the events follows Table 1. Figure modified after Bevilacqua et al. (2015)

sustained columns that laid down pyroclastic fallout (PF) beds mostly easterly dispersed (Fig. 5). In a few cases, they also generated dense PDCs.

Most of the explosive eruptions varied from low- to medium-magnitude events (Table 1); only three in epoch 1 and one in epoch 3 had a magnitude significantly higher. There is no clear relationship between the length of a period of quiescence and the size of the first eruption of the following epoch. Furthermore, the first eruption has never been the highest magnitude event of any of the three epochs (Orsi et al. 2003, 2004). An estimate of the temporal rate of the erupted volume in the epochs of activity was obtained, with uncertainty quantification (Fig. 6; Bevilacqua et al. 2016). An intensification of the activity rate with time was remarked during epochs 1 and 3, an

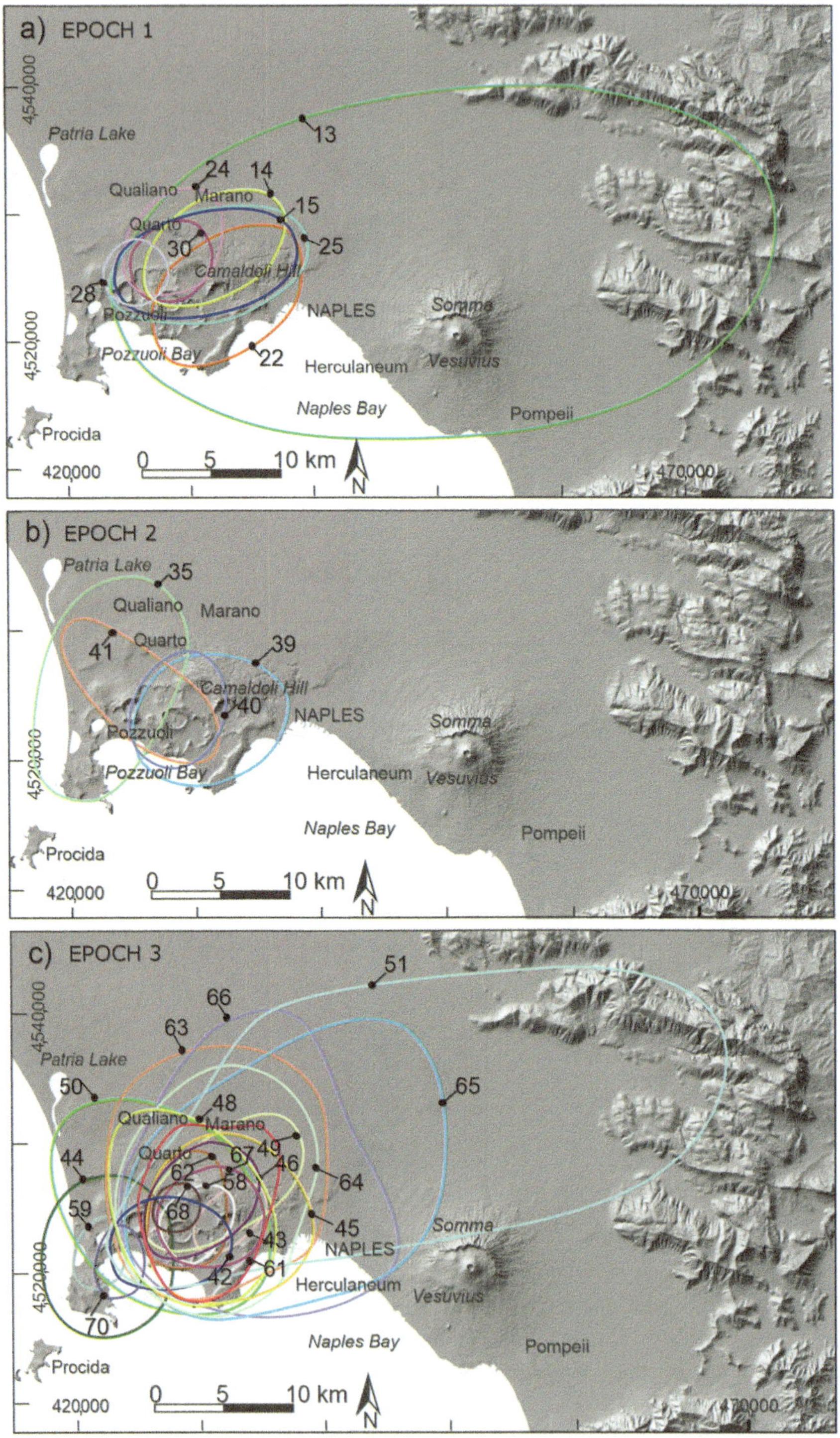

Fig. 4 10-cm isopachs of the pyroclastic fallout deposits of **a** epoch 1, **b** epoch 2, and **c** epoch 3 and Monte Nuovo. **d** zoom of the Campi Flegrei area in **c**. Numbering of the events follows Table 1. Figure modified after Orsi et al. (2004)

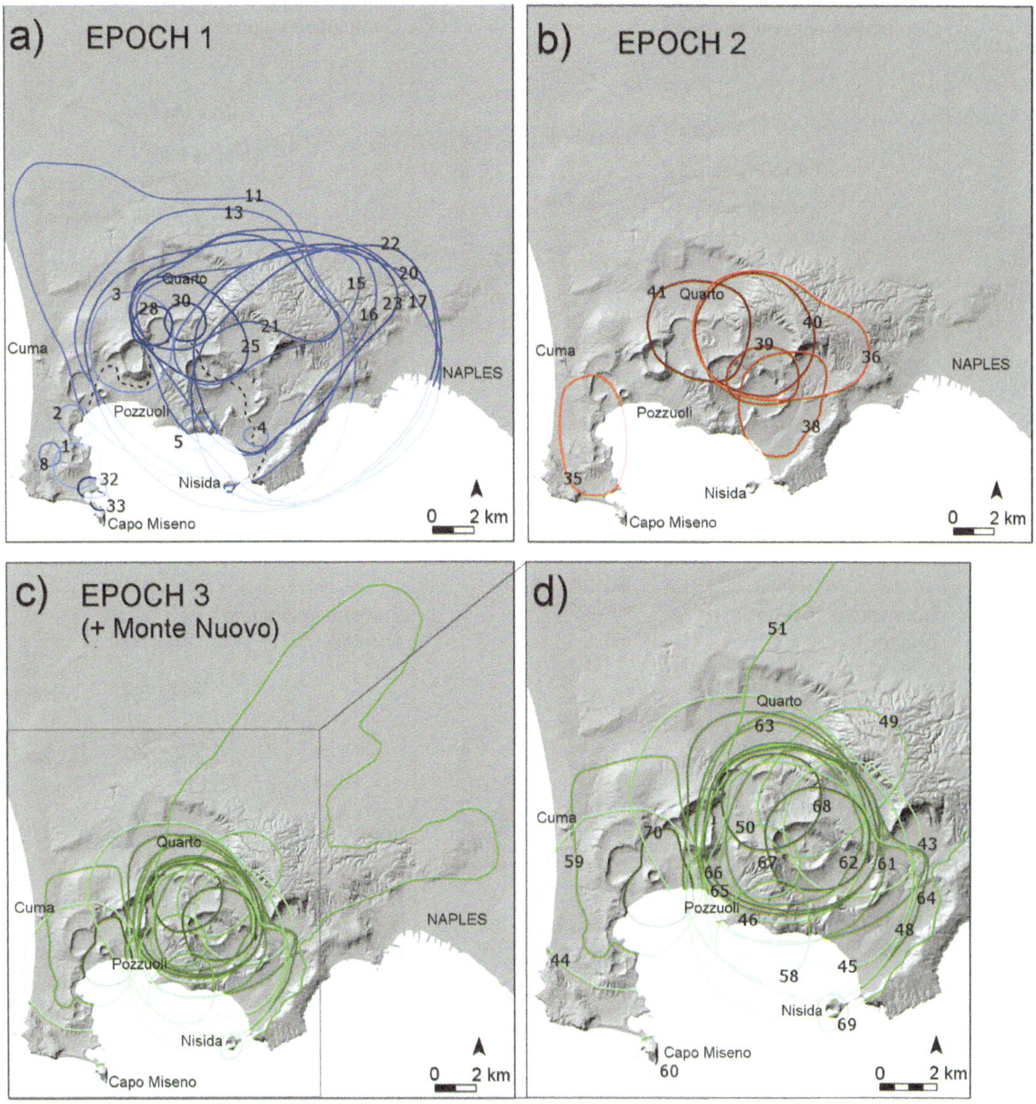

Fig. 5 Distribution of pyroclastic density current deposits of the eruptions of **a** epoch 1, **b** epoch 2, and **c** epoch 3 plus the Monte Nuovo event. Numbering of the events follows Table 1. Different colour tones indicate different events. Reported deposit boundaries were extended overseas to allow estimation of reasonable values for pyroclastic density current invasion area. Figure modified after Neri et al. (2015a, b)

apparent feature despite the large uncertainties affecting the age of the oldest events of epoch 1. It was not possible to clearly identify similar features in the much shorter epoch 2, due to the uncertainty affecting time and size of the eruptions.

Some eruptions have shown peculiar characteristics, such as contemporaneous events in distinct portions of the caldera floor, migration of the vent during a single event, building of a volcano through several eruptions close in time and space, and occurrence of a structural collapse during the event.

2.1.1 Epoch 1

Epoch 1, lasted from the NYT eruption to ~10.6 ka, generated at least 33 eruptions largely varied in size that followed each other at mean time intervals of about 80–140 years, and erupted a volume of magma of 4.2 ± 0.7 km^3 DRE (Table 1; Fig. 2c; Bevilacqua et al. 2016; see

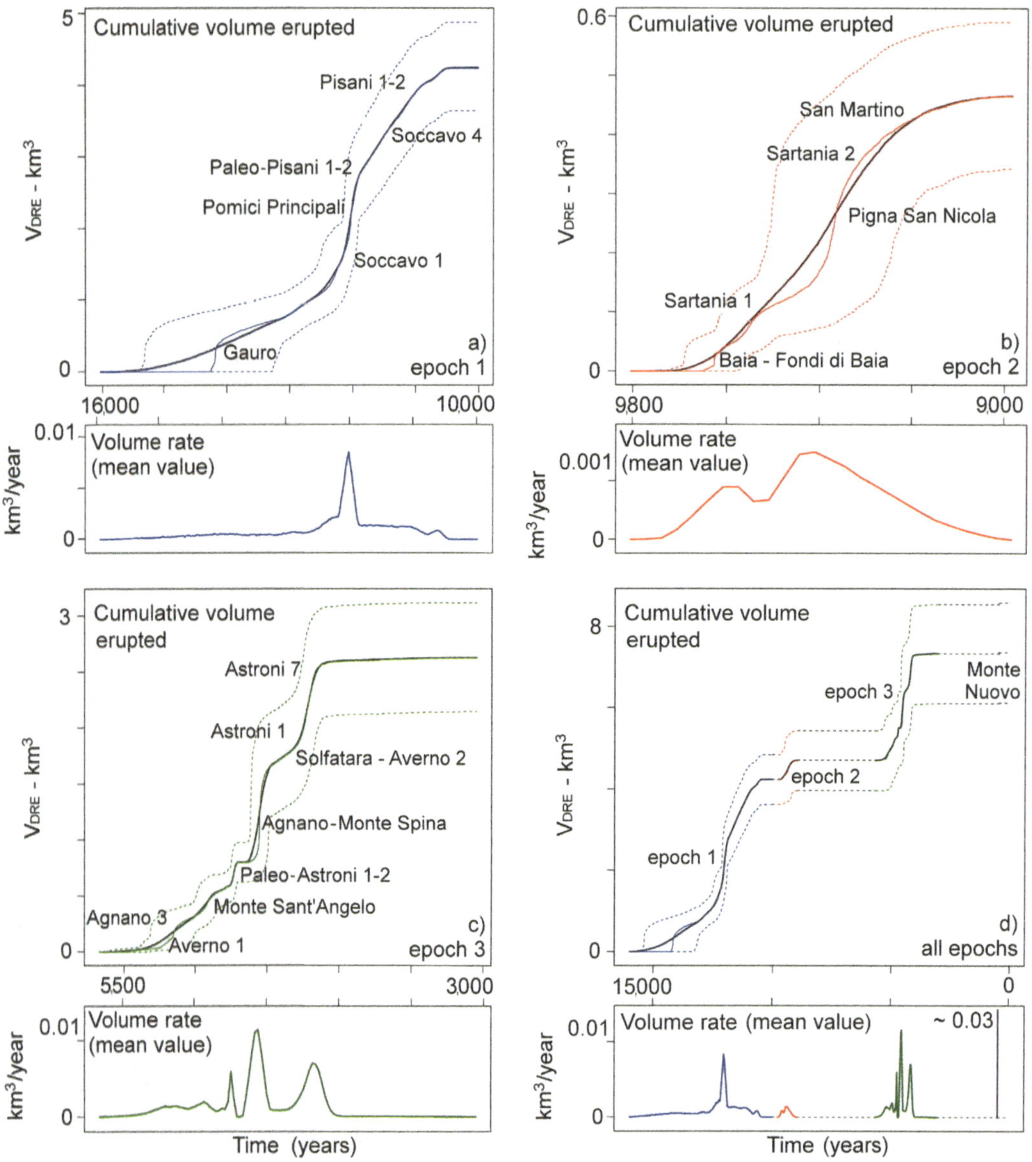

Fig. 6 Cumulative volume erupted as a function of time during **a** epoch 1 in blue, **b** epoch 2 in red, and **c** epoch 3 in green, and **d** the entire post-NYT activity (including Monte Nuovo in dark violet), assuming the described probability model. The bold line is the mean value, the narrow line is the 50th percentile and the dashed lines are the 5th and 95th percentiles of the epistemic uncertainty. The event labels correspond to the largest eruptions of each epoch. Under the main plots are reported the time derivatives of the mean graphs, which represent the mean erupted volumes rate per year. It should be noted that such mean eruption volume rates are directly affected by the uncertainty associated to each event age. Figure modified after Bevilacqua et al. (2016)

Chap. Volcanic and Deformation History of the Campi Flegrei Volcanic Field, Italy and references therein). During this epoch, the vents were located along the structural boundary of the NYT caldera, with an apparent concentration in the northern and north-eastern sectors. The earliest known activity of this epoch built up the Bellavista tuff cone through phreatomagmatic explosions that generated a zeolitised sequence of fine- to coarse-ash, mostly plain-parallel to cross-laminated beds (Di Vito et al. 1999). The largest eruptions were Pomici Principali, Gauro, and Soccavo 1. The maximum rate of activity appeared to be reached with the Pomici Principali event and maintained for most of the second part of this epoch. The Pomici Principali event laid down a sequence, from base upward, of fine-ash fallout beds, pumice lapilli layers with subordinate ash beds that in proximal area show sedimentological characteristics of surge deposits, and fine-ash surge beds rich in accretionary lapilli. The three eruptions of Montagna Spaccata, Fondo Riccio, Concola in epoch 1 are suspected to have been contemporaneous (Rosi and Sbrana 1987). A few post-NYT events occurred in the submerged portion of the caldera (e.g., Nisida Bank) and their stratigraphic correlations are the subject of recent research (e.g., Steinmann et al. 2018; Chap. Volcanic and Deformation History of the Campi Flegrei Volcanic Field, Italy). This epoch likely ended with the growth of the Porto Miseno tuff ring, composed of a sequence of zeolitised plain parallel to cross-laminated, fine- to coarse-ash beds (Fedele et al. 2011). During the quiescence between epochs 1 and 2, which lasted between 0.8 and 1.1 ka, Paleosol A formed. It is largely exposed in the Neapolitan-Phlegraean area with variable thickness, and ranges in colour from light- to dark-brown.

2.1.2 Epoch 2

Epoch 2, lasted from $\sim$9.6 to $\sim$9.2 ka, was characterised by a minimum of 8 eruptions which occurred at mean time intervals of 35–75 years (Table 1; Fig. 2b; see Chap. Volcanic and Deformation History of the Campi Flegrei Volcanic Field, Italy and references therein). All eruptions but one were explosive, low magnitude events characterised by phreatomagmatic dominating over magmatic phases. They extruded a volume of magma of 0.5 ± 0.1 km^3 DRE (Bevilacqua et al. 2016). The only effusive eruption generated the Monte Spina lava dome. The few vents active during epoch 2 were mostly located along the north-eastern and the western portions of the structural boundary of the NYT caldera. The epoch began with the Fondi di Baia eruption that formed an eight-shaped tuff cone (Di Vito et al. 1999; Pistolesi et al. 2017; Voloschina et al. 2018). This eruption produced mostly phreatomagmatic and subordinately magmatic explosions that generated a sequence of sandwave, ash surge beds with coarse fallout layers in its basal portion. The highest magnitude event of this epoch was the Pigna San Nicola eruption (Di Vito et al. 1999). It was characterised by alternating magmatic and phreatomagmatic explosions, and formed a tuff cone, only partly preserved today. The deposits are a sequence of pumice lapilli and subordinate ash fallout beds topped by sandwave, fine- to coarse-ash surge deposits. The last event was the San Martino eruption (Di Vito et al. 1999), which alternated magmatic and phreatomagmatic explosions. This activity produced a sequence of fallout layers alternating to thin, mostly ash surge beds, with the pumice fallout dominating over the PDC deposits within the mid-lower portion of the sequence. A quiescence lasting between 3.5 and 4.0 kyrs followed this epoch of activity during which the widely exposed yellowish-brown Paleosol B formed.

2.1.3 Epoch 3

Epoch 3, lasted from $\sim$ 5.5 to $\sim$ 3.5 ka, produced at least 28 eruptions at mean time intervals of 50–80 years (Table 1; Fig. 2a; see Chap. Volcanic and Deformation History of the Campi Flegrei Volcanic Field, Italy and references therein). These eruptions were fed by 2.6 ± 0.5 km^3 of magma DRE (Bevilacqua et al. 2016). A further twofold subdivision of this epoch in relation to a second-order pause in the activity, at $\sim$4.4 ka, has also been suggested (Isaia et al. 2009). Most of epoch 3 vents were concentrated in the central-eastern portion of the NYT caldera floor, but the vents of the two Averno and the Monte Nuovo eruptions were in the north-western sector of the caldera, while those of the Nisida and Capo Miseno eruptions were at the south-western and south-eastern edges of the emerged portion of the caldera. The large majority of the eruptions of this epoch were explosive, with only 3 effusive events: namely Monte Olibano, Solfatara, and Accademia lava domes, from the oldest to the youngest. The epoch began with the Agnano 1 eruption characterised by phreatomagmatic and magmatic explosions (Di Vito et al. 1999). These explosions produced a sequence of ash and pumice-lapilli fallout beds containing abundant ash particles, which becomes a thin ash layer in distal areas. The highest-magnitude event was the Agnano-Monte Spina (A-MS) eruption, accompanied by a volcano-tectonic collapse in the north-eastern portion of the NYT caldera (de Vita et al. 1999). The change in the eruption rate appears to coincide with this eruption, which somehow marks the subdivision of the epoch in two sub-epochs (Isaia et al. 2009). A-MS was characterised by phreatomagmatic and magmatic, sometimes contemporaneous, explosions that produced PDCs at variable particle concentrations and two sustained columns. The latter were generated during the eruption phases named B1 and D1, and reached height of 23 and 27 km, respectively (de Vita et al. 1999). The high particle concentration PDCs were confined within the caldera depression, while the more diluted ones overtopped the caldera boundaries and flowed for about 15 km within the surrounding plain. The sustained columns laid down easterly-to-northeasterly dispersed pumice fall deposits not less than 10 cm thick at about 42 km from the vent. The sequence is cut by an erosional unconformity, suggesting a time break in the eruption. A volcano-tectonic collapse likely occurred during this pause in the eruption. It affected an area of about 6 km^2, polygonal in shape, roughly northwest-southeast oriented, and located towards the north-eastern border of the NYT caldera floor.

The eruptions of Averno 2 and Solfatara, postdating the A-MS eruption and volcano-tectonic collapse, likely occurred contemporaneously (Isaia et al. 2009; Pistolesi et al. 2016). Their vents were located about 5 km apart, at the opposite sides of the north–south Pozzuoli structure that affects the NYT caldera floor (Capuano et al. 2013). The Averno 2 eruption (Di Vito et al. 2011) was characterised by a sequence of magmatic and phreatomagmatic explosions generating sustained and collapsing columns that laid down PF and PDC deposits, prevailing during the first and the second part of the eruption, respectively. The maximum height reached by the sustained columns was about 10 km, and the vent migrated over a limited area during the eruption. The Solfatara eruption was characterised by phreatomagmatic with subordinate magmatic explosions that generated a sequence of surge beds with minor PF deposits (Rosi and Sbrana 1987; Isaia et al. 2015; Pistolesi et al. 2016). The Astroni volcano (Di Vito et al. 1999; Isaia et al. 2004; Tonarini et al. 2009) grew through 7 distinct eruptions occurred in a time span not longer than 300 years. The eruptions were dominantly phreatomagmatic, with generation of ashy PDCs; a sustained column formed only during eruption 6 and laid down a sub-Plinian pumice fall deposit. The eruption vent of each of the seven events, although confined within the 3 km^2 area of the present crater, migrated during the activity roughly from southeast to northwest. The vent of the Averno 2 eruption also migrated along a 2-km-long northeast-southwest oriented system of fractures.

3 Long-Term Assessment of Volcanic Hazard

Pioneering hazard assessment studies at the CFc were based on the "deterministic" reconstruction of reference events from the past. In particular, during the 1982–1984 bradyseismic crisis, Rosi and Santacroce (1984) considered three reference scenarios: a phreatic explosion like the AD 1198 event at the Solfatara volcano (Scandone et al. 2010), a low-magnitude explosive eruption like the AD 1538 Monte Nuovo event (Guidoboni and Ciuccarelli 2011; Di Vito et al. 2016 and references therein), and a maximum expected eruption like the A-MS event (de Vita et al. 1999). The spatial definitions of the hazardous phenomena were based on the areal distribution of the deposits of the reference events. Orsi et al. (2004) performed the first comprehensive eruption forecasting and hazard assessment at the restless CFc using stratigraphic, volcanological, structural and petrological data. Due to the remarkable change in the structural setting of the CFc system produced by the NYT eruption and related caldera collapse, the authors assumed the entire history of the CFc system subsequent to these events as reference for volcanic hazard assessment. More recently, the approach to the volcanic hazard assessment has changed to include probabilistic forecasts that incorporate a variety of numerical modelling of past hazardous phenomena, along with the uncertainty quantification.

The expected phenomena in case of renewal of the volcanic activity of the CFc have various intensities and impacts depending on the style and size of the next eruption. This section describes the researches developed in the last two decades aimed at both defining fundamental parameters of a future eruption of the CFc and zoning of the territory in relation to the expected volcanic hazards. These results include vent location, time, size, and style of a future eruption.

3.1 Site of a Future Eruption and Vent Opening Probability Maps

Alberico et al. (2002) were the first to quantify potential new vent openings at CFc. They subdivided the subaerial CF area into a regular grid of 1 km cells and evaluated the relative spatial probability of vent opening based on geological, geophysical and geochemical data. In particular, they included earthquake epicentres, gravity anomalies, along with location of past vents, recent faults, helium concentration anomalies, and fumaroles. The number of these elements present in a cell empirically defined a score for each cell of the grid. After a normalisation step, this score provided the probability of vent opening in each cell. The probability distribution spread over the whole NYT caldera, and had maximum values (above 1.6% over km^2) in the central part, near Solfatara and Monte Nuovo.

Orsi et al. (2004) applied a different approach taking into consideration both history and current dynamics of the system. They inferred that if the next eruption is triggered by the ongoing dynamics of the NYT caldera, the most likely new vent location would be in the north-eastern sector of the caldera. They correlated this area to the portion of the resurgent block under extension that has been the site of the majority of the eruptions of the past 5 kyrs (Fig. 2a). However, they did not exclude that a future vent could open in the Monte Nuovo-Averno Lake area, at the intersection of faults within the resurgent block with those bordering the block towards the northwest. They finally delimited one area of "higher" and one of "lower" probability of vent opening (Fig. 7a).

A new spatial probability map of vent opening was developed by Selva et al. (2012a) following a Bayesian approach based on Dirichlet random variables corresponding to 700 cells with 500-m sides in a rectangular grid covering the NYT caldera. A prior distribution was defined using an empirical procedure based on the association of unequal scores to the geological features present in a given spatial cell. These features included, with score points in parentheses: past vents (3 points), faults (2 points), and location within the NYT caldera (1 point). This procedure was simplified with respect to that of Alberico et al. (2002), in order to avoid the influence of modern unrest signals on a long-term, vent-opening probability map. The prior model was then combined with the likelihood of the number of epoch 3 vents in each cell. Vent migration during an eruption was not considered as indicative of multiple events, but the seven eruptions building up the Astroni volcano were considered as distinct events. This likelihood was defined according to a multinomial distribution, and the resulting probability was smoothed by using a Gaussian filter with a 2σ radius of 1 km, in order to reduce scatter and to cope with the spatial uncertainty affecting the data and the prior distribution. These maps (Fig. 7b) showed a sickle-shaped region of increased probability of vent opening, following the inland portion of the caldera, from Agnano to Baia, and including the Solfatara crater. Maximum values of $\sim$2% probability over km^2 were calculated in the area of the Astroni volcano and the Agnano volcano-

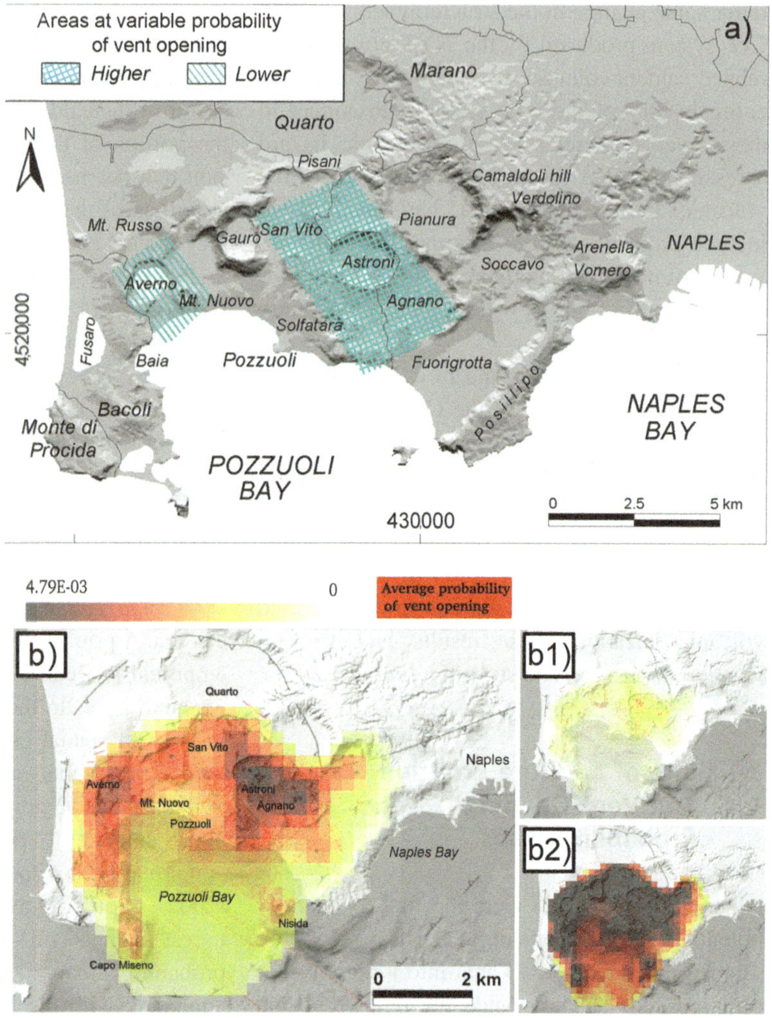

Fig. 7 **a** Areas at variable probability of vent opening in case of renewal of volcanism in short-mid-terms at the Campi Flegrei caldera (modified after Orsi et al. 2004). **b**, **b1**, **b2** Conditional probability of vent opening at the Campi Flegrei caldera of the posterior filtered model. The average (best guess) value is shown in **b**, while **b1** and **b2** show the 10th and 90th percentiles confidence interval, respectively (modified after Selva et al. 2012a)

tectonic depression. The authors provided additional maps of the 10th and 90th percentiles of the uncertain probability values over the cells (Fig. 7b1, b2). Uncertainty ranged from +180% to −90% of the local mean values.

Bevilacqua et al. (2015) developed a more comprehensive vent opening probability map for a future eruption within the CFc, through a weighted linear combination of five maps describing the spatial distribution of relevant volcanic features. Such features included the eruption vent locations of the past 15 kyrs (each epoch was separately considered; Fig. 2), and the maximum fault displacement and the surface fractures density from Vitale and Isaia (2014). A uniform probability map within the caldera was included to account for unconsidered or unidentified processes and variables. The past vent locations were represented by ellipses defining the associated uncertainty area, rather than by deterministic points. Their density distribution was processed using either Gaussian kernel functions or a partitioning of the caldera into sixteen homogeneous zones, with quite consistent results. An uncertain number of "lost vents" with unconstrained location in the subaerial part of the caldera was also included in the analysis. The uncertain weights to be associated with the features that contributed to the definition of the vent opening map relied on expert judgment techniques (see also Aspinall and Cooke 2013; Bevilacqua 2016). Based on the expert

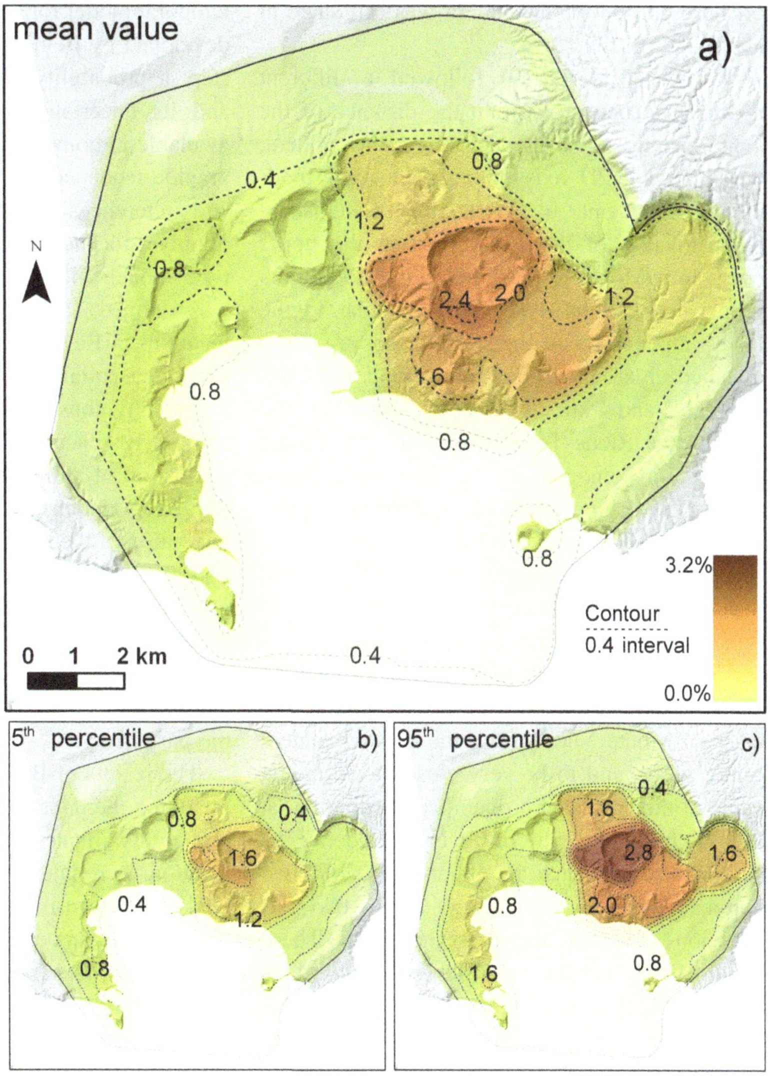

Fig. 8 Probability maps of new vent opening as obtained weighting the six variable distributions considered in Bevilacqua et al. (2015). Contours and colours indicate the percentage probability of vent opening per km^2 (conditional on the occurrence of an eruption). **a** Mean values, **b** and **c** 5th and 95th percentile values, respectively. Figure modified after Bevilacqua et al. (2015)

elicitation outcomes, areal distribution of previous vents was judged the most important variable for quantifying the vent opening probability map, with a total contribution weight of about 47% (mean value). The distributions of the maximum fault displacement and surface fracture density were weighted about 16% and 12%, respectively. About 25% weight was assigned to the uniformly homogeneous spatial vent opening map (i.e., to the possibility that the next vent could open anywhere inside the NYT caldera). The results, expressed as a mean probability map and 5th and 95th uncertainty percentiles of the probability values, outlined a main, quite wide, region in the central-eastern part of the caldera characterised by the highest probabilities of vent opening (Fig. 8). The maximum probabilities were between 1.6%/km^2 and 3.2%/km^2 with a mean value of about 2.4%/km^2. Secondary maxima were obtained in the western part of the caldera, with mean values of about 1–1.2%/km^2. However, the probability of vent opening was not confined to these areas and, with mean values everywhere above 0.4%/km^2, the possibility is widespread over the caldera. Uncertainty ranged about ±30% of the mean value, with variations from ±10% to ±50% in different areas of the caldera. The probability of vent opening in the offshore portion of the caldera was estimated at about 25% ± 5%.

Rivalta et al. (2019) followed a different approach to better constrain the distance of the vent opening from the centre of the caldera, based on the 2D axisymmetric physical modelling of magma transport after a statistical reconstruction of the stress field in the upper crust. In a Monte Carlo simulation, the authors propagated magma dykes perpendicular to the lowest eigenvector of the stress tensor σ_3. They assumed dyke nucleation at a random depth between 3 and 4 km, placed at the caldera centre, beneath the Rione Terra (Pozzuoli). The model considers a homogeneous tectonic stress and a uniform unloading stress applied to the caldera floor, the latter affected by an axisymmetric topography. The stress coefficients were estimated through a Bayesian technique that considered the past vent locations. The model performance was tested on the events of epoch 3 and the Monte Nuovo eruption. The radial distance of Monte Nuovo from the defined caldera centre was found to be very close to the modal value of the estimated probability distribution, at 2–3 km from the centre of the NYT caldera. The authors also evaluated the effects of a simplified non-axisymmetric topography along two 1D directions, towards northeast and south. These results were broadly consistent with previous maps and indicated a higher probability of vent opening in the north-eastern portion of the caldera.

3.2 Timing of a Future Eruption

The long-term temporal forecasting of volcanic eruptions in the CFc is particularly complex and very few studies have been devoted to its solution. Indeed, constant rates based on the ratio of number of eruptions over time turn out to be inadequate due to the strong temporal clustering occurring at various time scales. Spatial locations and temporal rates also appear to be related. Uncertainty affecting the past record cannot be neglected when the spatial and temporal clustering properties are evaluated. A spatial–temporal model for eruption rates at the CFc was developed by Bevilacqua et al. (2016). As a first step, a probability model of the past eruptions and its uncertainty was developed, using the available radiometric ages and the known stratigraphic sequence of the events (Smith et al. 2011 and references therein). Then, a space–time doubly stochastic non-homogeneous Poisson-type model with a local self-excitation feature was defined. This model, based on the vent locations in Bevilacqua et al. (2015), was able to generate spatial–temporal clusters of eruptions statistically consistent with the CF reconstructed record (Bevilacqua 2016). Results allowed the evaluation of similarities and differences among the three epochs of activity as well as the derivation of the eruptive base rate of the caldera (i.e., excluding the clusters) and its capacity to generate clusters. Thanks to this uncertainty assessment, it was possible to obtain a temporal probability model able to describe the temporal and spatial eruptive behaviour of the caldera over the past 15 kyrs.

The results of Bevilacqua et al. (2015, 2016), including the dependence of both scale and time probability of an eruption on the vent opening location, were later revisited by Bevilacqua et al. (2017). This resulted in separate vent opening probability maps conditional on the western and eastern sectors of the subaerial portion of the caldera, and a conditional version of the vent opening map was developed. This version enabled the authors to restrict the eruption vent sampling inside various subsets of the caldera (typically of 2–4 km in diameter) and to produce PDC invasion hazard maps under such conditions (see Sect. 3.4.2 and Fig. 14a, b).

3.3 Size of a Future Eruption

A first attempt to constrain the size of a future eruption of the CFc was made by Orsi et al. (2004), who developed a dataset of the areas covered by PF and PDC deposits of the eruptions of the past 15 kyrs, as proxies of the magnitudes of the eruptions. These areas were constrained by the 10 cm isopachs for fallout beds, and the presence of sedimentological features indicating a horizontal component in the transport of the particles for PDCs deposits. An analysis of the magnitude–frequency ratio of the PF deposits led Orsi et al. (2004) to infer that in epoch 3 the probability distribution had a polymodal profile, with one mode between 0 and 100 km^2, another between 100 and 500 km^2, and a third, represented by the A-MS outlier, at 800 km^2. Finally, the authors suggested that although A-MS could not be excluded as the maximum expected event, the most probable maximum event in the case of renewal of volcanism in the short- or mid-term was an eruption falling between 100 and 500 km^2 area for PF deposits. Very large-scale eruptions, such as the CI and NYT caldera forming events, were not included in the dataset due to their very low probability of occurrence, likely below 1% based on frequency of occurrence.

Orsi et al. (2009) built upon the above analysis by developing a quantitative long-term forecast of the size of the next eruption. Some physical parameters of 22 explosive eruptions of the CFc occurred during the past 5 kyrs were analysed. They included dispersal area, volume and density of the pyroclastic deposits, volume of erupted magma, total erupted mass, and eruption magnitude. This analysis resulted in a size classification of the explosive eruptions, which were grouped into three classes: small, medium, and large, each with a conditional probability of occurrence. They selected some reconstructed events as representative of each size class. A small-size explosive event (e.g., Monte Nuovo, Averno 2) is the most likely with a probability of about 60%; a medium-size (e.g., Astroni 6) has a probability of occurrence of about 25%; a large-size (A-MS) is the least likely with a probability of about 4%. An effusive eruption has a chance of occurrence of about 11%.

Neri et al. (2015b) designated the areas invaded by PDCs as a random variable representative of the uncertainty affecting the scale of the next eruption with the aim to quantifying the PDC hazard. This approach was based on a continuous statistics of the areas expected to be inundated by a PDC, rather than on discrete classes of the total erupted volume. Based on the available field datasets and estimates for poorly exposed old deposits, and using alternative expert judgment procedures (Bevilacqua et al. 2015), the authors considered that a radial underestimation error of deposit boundaries (treated as a source of uncertainty) varied between about 150 and 1,000 m, with a mean value of about 500 m. Thus, they generated log-normal probability density functions of spatial extent distribution considering either the past 5 kyrs or the past 15 kyrs datasets. The presence of several intermediate data points between the main empirical distribution for the 15 kyrs dataset and the A-MS event allowed a quasi-continuous distribution of the PDC inundation areas to be hypothesised. The curves were very similar, although the one related to the longer time interval had a slightly higher number of smaller and larger events, and less medium scale events.

Bevilacqua et al. (2016) developed a probability model of the post-NYT caldera eruptions record, which included some time-volume statistics of the sum of PF and PDC deposits (see also Sect. 2.1). Uncertain volumes were uniformly sampled inside three separate intervals associated with different volumes of erupted magma (DRE): 0–0.01 km^3 for very small explosive/effusive eruptions, 0.01–0.1 km^3 for small explosive eruptions, and 0.1–0.3 km^3 for medium explosive eruptions (similarly to Orsi et al. 2009). If volume estimates were available 5th and 95th percentiles were adopted, corresponding to ±50% relative errors (Fig. 6). The analysis included a comparison of the volumes of magma erupted in the eastern and the western sectors, showing that the latter were significantly lower. A first attempt to find a correlation between eruption size and vent location was made by Bevilacqua et al. (2017) in the framework of a more general study mainly aimed at producing PDC hazard maps. They showed that the western sector of the caldera was characterised by smaller size events, as well as by a less frequent and less clustered activity (Fig. 9). This resulted in significantly different probability distributions for the PDC sizes according to the specific sector of vent opening considered.

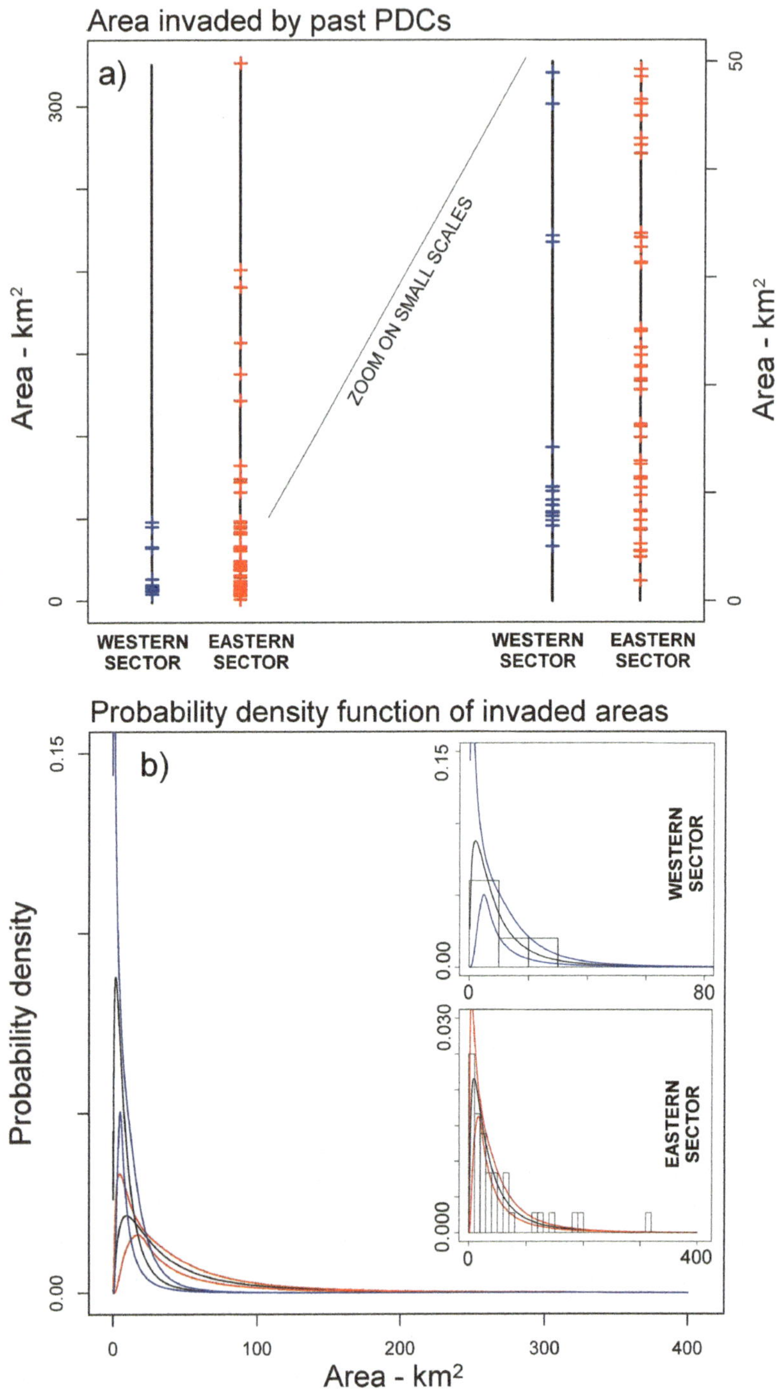

Fig. 9 **a** Representation of the areas invaded by pyroclastic density current originating in the western (blue) or eastern (red) sectors of the caldera. Each coloured dash represents an event. The sequence is ordered increasingly as a function of the areal extents. On the right of the plot is displayed a zoom of the small scales data. **b** Probability density functions (log-normal class) for the invasion areas originating in the western (blue) or eastern (red) sectors of the caldera. The black curve is the mean value and the coloured curves are the 5th and 95th uncertainty bounds. In the small boxes the two estimates are displayed at different scales and the histogram data are included. Figure modified after Bevilacqua et al. (2017)

3.4 Style of a Future Eruption

In addition to the previously presented definition of the size classes of the past eruptions of the CFc and the possibility for each of them to occur in case of renewal of volcanism in short- to mid-terms, Orsi et al. (2009) also discussed the likely style of a future eruption that will be either effusive or explosive of three different sizes. The possibility that the next eruption will be the first of a series of events following each other at short time intervals (years or tens of years), that more than one eruption will occur in the same vent area, and that two contemporaneous eruptions will occur at vents located in different portions of the NYT caldera cannot be ruled out. An effusive eruption will very likely extrude viscous magma that will generate either lava flows travelling short distances or a lava dome. An explosive eruption, regardless of its size, will likely alternate magmatic and phreatomagmatic phases that will generate particle fallout and PDCs. The detailed sequential development of the eruption dynamics and the related phenomena expected during the course of the eruption is rather difficult to predict. The areas affected by accumulation of falling pyroclastic fragments will be exposed to different effects (e.g., Wilson et al. 2014, for a review) including roofs collapse (Blong 1981, 2003; Macedonio and Costa 2012), disruption to aviation (Casadevall 1994; Folch and Sulpizio 2010), water pollution (Blong 1984; Wilson et al. 2010), damage to electric power supply and radio communication (Wilson et al. 2014), health problems to humans and animals (Baxter and Horwell 2015; Armienta et al. 2011; Flueck 2016), lahars generation and damages to agriculture (Cronin et al. 2014). Fatalities will occur in the proximal areas, through roof collapse, asphyxiation and burial (Spence et al. 2005; Brown et al. 2017). PDCs will spread either 360° from the vent or over preferential sectors. Their runout will be significantly affected by the topography surrounding the vent; higher elevations can divert the currents, valleys can channelise them. PDCs entering the sea will have the potential to produce vast clouds of water vapour bearing accretionary lapilli, and to generate tsunami waves. Due to their velocity, temperature, and particle concentrations, PDCs can produce heavy damage to urban structures and lethal conditions for human beings (Baxter et al. 2005; Neri et al. 2015a). Mele et al. (2015) reconstructed the dynamics of the PDCs generated by some phases of the Averno 2 (small scale), Astroni (intermediate scale) and A-MS (large scale) eruptions. They combined facies architecture with laboratory analysis and physical modelling to evaluate the dynamic characteristics of the original currents. Facies analysis indicated that the deposits of the small and intermediate events resulted from pyroclastic surges fed by multiple explosions of phreatomagmatic origin. Instead, in the large-scale event some of the currents started as a concentrated flow near the vent and then evolved laterally into expanded flows by the propagation of the basal shear current. Finally, the authors provided probability density functions for the dynamic pressure and the particle volumetric concentration.

Esposti Ongaro et al. (2006) explored the effects of the presence of different amounts of CO_2 in the 4 wt% H_2O-bearing A-MS magma, by simulating the eruption dynamics from the base of the volcanic conduit up into the atmosphere. Within the range of conditions explored, the eruption style and dynamics in the atmosphere were mainly controlled by the H_2O content. Nevertheless, the partial substitution of H_2O with CO_2 in the erupted magma produced a substantially more collapsing behaviour of the resulting column.

At the lower range of the size spectrum of explosive eruptions in the CFc there are phreatic explosions, i.e., driven by over pressurised geothermal fluids, such as H_2O, CO_2 (Barberi et al. 1984; Macedonio et al. 2013). Phreatic explosions can have various scale. The most energetic can eject ballistics at kilometres distance (Neri et al. 1999; Houghton et al. 2015) and produce great amounts of fine dust of clay and other hydrothermally altered minerals, generating 'soap-like' wet deposits (Montanaro et al. 2016). Small convective plumes and pyroclastic surges can also be generated. They can precede a magmatic or phreatomagmatic eruption, although often they occur as isolated events (Barberi et al. 1993). Phreatic explosions produce less evident precursor signals than magmatic explosions and can unexpectedly occur in hydrothermal areas, like Solfatara-Pisciarelli in the CFc (Mayer et al. 2016).

In the following, we summarise both the evolution and the state of the art in development of PF and PDC hazards maps at the CFc.

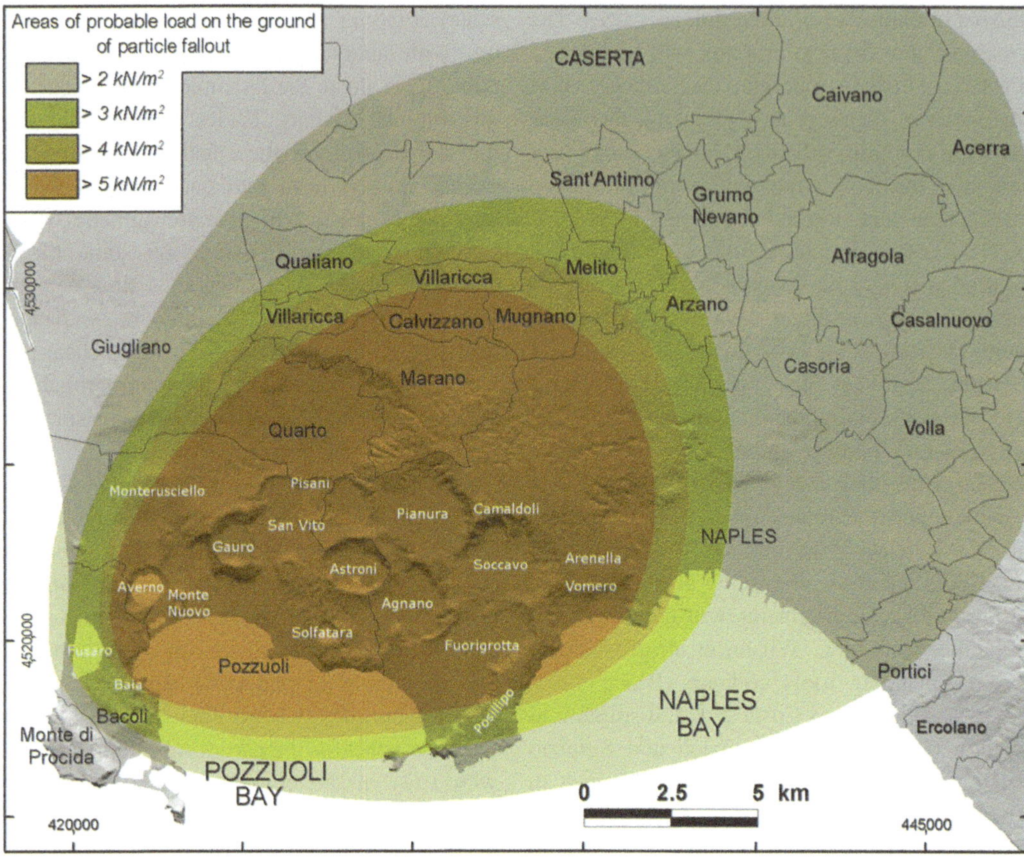

Fig. 10 Tephra fallout hazard map of the Campi Flegrei caldera. Different colours mean different tephra loads. Figure modified after Orsi et al. (2004)

3.4.1 Pyroclastic Fallout Hazard Maps

Orsi et al. (2004) constructed the first PF hazard map for the CFc (Fig. 10). The map was based on the frequency of deposition of fallout beds thicker than 10 cm (Fig. 5), their load on horizontal surfaces, and the direction of their dispersal axes. It delimits the areas that could be affected by load on the ground in excess of 2.0, 3.0, 4.0 and 5.0 kN/m^2.

Mastrolorenzo et al. (2006, 2008) first produced PF hazard maps by using numerical modelling, for three different scenarios, and sampling empirical probability distributions of the input parameters such as column height, erupted mass, initial velocity and grain size. Average wind direction was obtained from historical data. The PF hazard level was found to depend strictly on the scenario being considered.

Orsi et al. (2009), as already reported in Sect. 3.3, evaluated the most relevant physical parameters of the 22 explosive eruptions recognised at the CFc over the past 5 kyrs. These parameters include dispersal, volume and density of the pyroclastic deposits, volume of erupted magma, total erupted mass, and eruption magnitude, and permitted to group the fallout deposits in three classes. Costa et al. (2009) simulated and analysed the PF of the three eruption scenarios of Orsi et al. (2009) using the HAZMAP computational model (Macedonio et al. 2005). This model is based on a semi-analytical solution of the two-dimensional advection–diffusion-sedimentation equation for volcanic tephra and is able to perform many simulations in a short time. The volcanological model parameters were estimated from best fit of the model output with the field data of the eruption representative of each size class.

These parameters included total mass, eruption column height, total grain size distribution or settling velocity distribution of the particles, column shape and meteorological parameters, such as wind velocity and atmospheric diffusion coefficient. After the best-fit phase, a statistical analysis was performed by varying the meteorological conditions, namely the wind velocity profile, using a statistical set of wind profiles obtained from National Oceanic and Atmospheric Administration (NOAA) reanalysis. Probability maps, relative to the considered scenarios, were also constructed for tephra loads of 200, 300 and 400 kg/m^2.

Selva et al. (2010) explored the joint effect of two great uncertainties (vent location, and eruption size) affecting PF hazards at CFc by using the tool BET_VH of Marzocchi et al. (2010). The study did not produce new simulations but relied on the data in Orsi et al. (2004) and Costa et al. (2009). The results showed that volcanic hazard based on the weighted average of all possible eruptive scenarios is very different from the analysis of single reference scenarios.

Lirer et al. (2010) and Alberico et al. (2011) produced multi-source (CFc, Somma-Vesuvio, Ischia) PF hazard maps by combining the dispersal of past eruption products as reconstructed by field surveys and their recurrence over the whole area. They included a semi-empirical processing of the results of numerical simulation in Costa et al. (2009). Later, through numerical modelling within the BET_VH tool, a multi-source PF hazard map was obtained by Marzocchi et al. (2015b), focussing on the city of Naples. Although an underwater eruption may not generate any eruptive column, Tonini et al. (2015) focussed on the chance of PF source opening in the Pozzuoli Bay, and showed that this assumption enhances the hazard in the city of Naples.

Sandri et al. (2016) performed further uncertainty quantification on the PF hazard maps, both

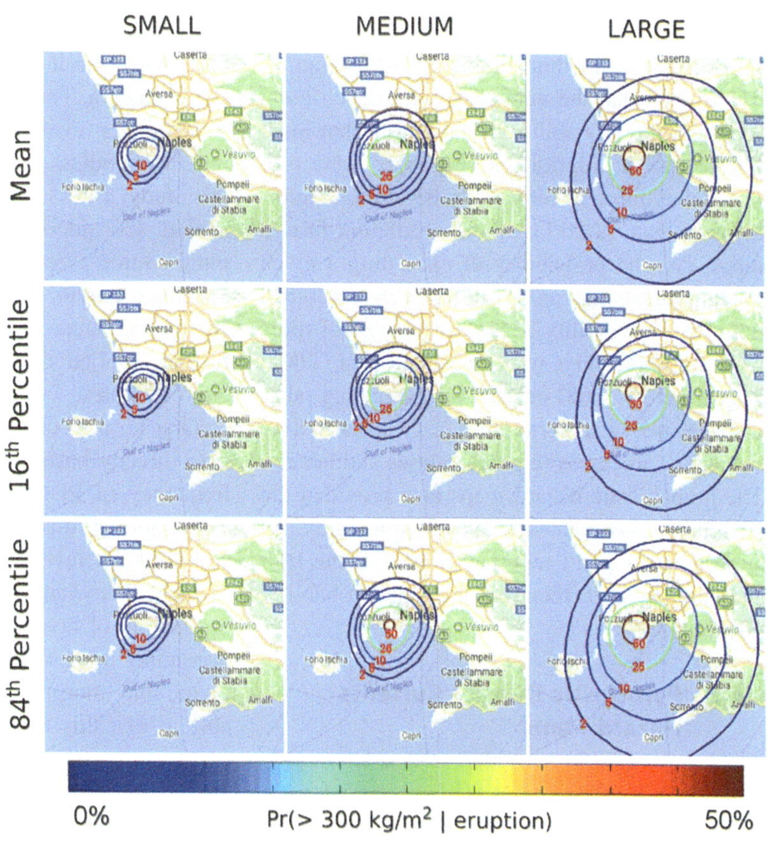

Fig. 11 Conditional probability map for volcanic ash loading greater than 300 kg/m^2 in response to an eruption of the Campi Flegrei caldera. The contour lines, from the outer to the inner, refer to 2, 5, 10, 25 and 50%. In some panels the highest values are not reached and the corresponding curves are missing. From left to right, the hazard maps corresponding to the three different eruption scenarios (small, medium and large) are shown. The top panels report the mean values, whereas the other panels report the 16th and the 84th percentiles. Figure modified after Selva et al. (2018)

for CFc and Somma-Vesuvio volcano. The possible wind directions and speed were explored by using real winds based on model reanalysis provided by the European Centre for Medium-Range Weather Forecasts, freely available in the internet. A period of about ten years (2001–2010) was considered sufficient to obtain a robust statistical analysis (Macedonio et al. 2016). In this study, the authors first explored extra and intra-class size variability (i.e., size difference within an eruptive size class) by using a power law for the magnitude-frequency modelling. They showed that also the effects of intra-class variability within a single scenario can be relevant. Secondly, they tested both the semi-analytic 2D model HAZMAP, and the fully 3D model FALL3D (Folch et al. 2009). They found that the influence of the simulator on the statistical results is relatively small.

Selva et al. (2018) presented a state-of-the-art analysis of the hazard maps for PF at the CFc by exploring the possible eruption size, total grain size distribution of products, wind field, vent location, also accounting for their respective uncertainties. The hazard map reported in Fig. 11 is relative to a loading threshold of 300 kg/m^2. This means that the values shown in each point represent the probability of exceeding that loading in case of eruption. Similar maps were produced for different thresholds between 1 kg/m^2 to more than 1,000 kg/m^2. The different hazard maps, i.e., the probability of exceeding a given loading threshold, can be successively combined with the vulnerability of the given types of roofs present in the area (e.g., Spence et al. 2005; Zuccaro et al. 2008) or used for estimating other damage such as to agriculture or those caused by secondary processes (e.g., landslides and lahars). The probabilistic hazard maps also provided the uncertainties associated with each probability. As an example, Fig. 11 reports the median, the 16th and the 84th percentiles of the probability distribution.

3.4.2 Pyroclastic Density Currents Hazard Maps

Lirer et al. (2001) provided a detailed reconstruction of the PDC deposits of six characteristic eruptions, ranging from the Monte Nuovo small event to the Pomici Principali large event. In particular, the authors widely discussed the interaction between the flow capacity and the topography. They concluded that both Posillipo and Camaldoli hills act as effective obstacles for PDCs flowing towards the centre of the city of Naples, while Bagnoli and Fuorigrotta plains are significantly exposed to the PDC hazard.

A first qualitative PDC hazard map was developed by Alberico et al. (2002) on the basis of their vent opening map (see Sect. 3.1). They used a Monte Carlo application of the energy line model (Sheridan and Malin 1983), assuming a 6° friction angle φ, for a VEI 3 and a VEI 4 eruption with column collapse height of 100 and 300 m, respectively (Fig. 12). The resulting map for a VEI 3 eruption indicates that only limited areas have a cumulative 3% chance of being hit by PDCs erupted from a distal vent, as their propagation does not exceed a distance of 1 km. Thus, the inundation area is strongly dependent on the vent location. Instead, in case of a VEI 4 eruption, the central-eastern portion of the caldera has a 25–50% probability of being invaded by PDCs, even if they originate in other portions of the caldera. The authors also stated that a VEI 5 or higher event would inundate the whole region regardless of the vent location. Based on this study, Lirer et al. (2010) constructed multi-hazard (considering both PF and PDC) and multi-source (considering Campi Flegrei, Somma-Vesuvio, and Ischia) maps for the Neapolitan area. Later, Alberico et al. (2011) further detailed these results, focussing on the city of Naples.

Orsi et al. (2004) constructed a PDC hazard map directly based on the areal distribution and frequency of PDC deposits of the past 5 kyrs. They argued that PDCs of a future eruption would likely travel at high speed within portions of the CF lowland, defined as the area at high probability of invasion by PDCs. Some of them, depending upon vent location and energy, and in favourable morphological conditions could be able to exit this lowland towards the north and northeast, invading an area defined as the area at lower probability of invasion by PDCs.

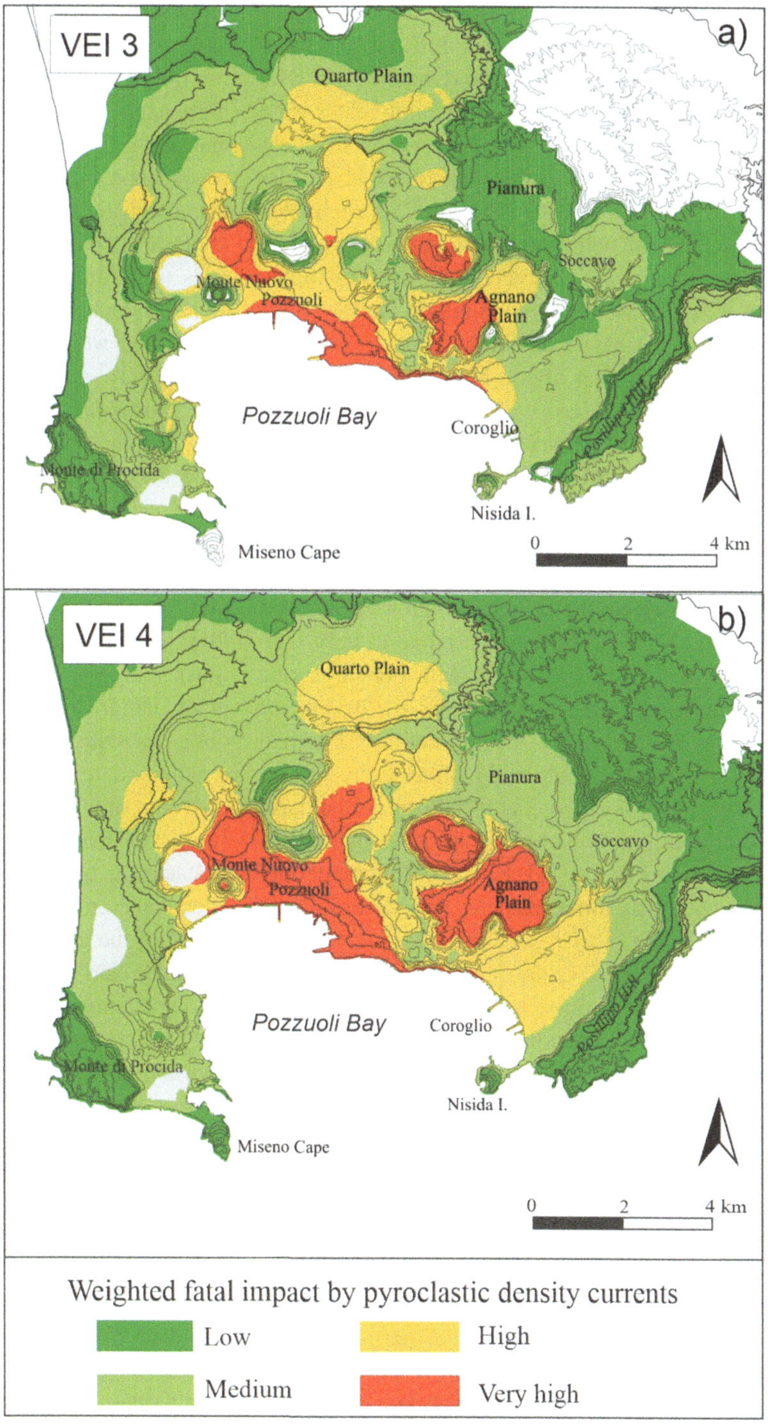

Fig. 12 Composite probability maps of weighted fatal impact from pyroclastic flows by any eruption in any of the cells for **a** VEI 3 and **b** VEI 4 eruptions. Figure modified after Alberico et al. (2002)

Other PDC hazard maps were developed by Rossano et al. (2004) using the Monte Carlo application of a 1D physical model, while varying 14 vent locations and 1,200 combinations of input parameters (e.g., initial velocity, density, viscosity, thickness, yield strength) of the PDCs. The flow was assumed to be incompressible with a Newtonian or Bingham rheology, in contrast with the real dynamics of PDCs (e.g., Sulpizio et al. 2014). The input values were empirically weighted and assumed as independent. They computed the trajectories of an idealised, 1D, mass-independent flow, approximated as a material point, representing the flow plug, with an initial radial direction uniformly sampled over 360°. The results highlighted that the most mobile flows have a maximum range of ~20 km and can pass over 400-m-high topographic barriers. Otherwise, most of the flows with intermediate mobility were affected by the main topographic barriers, in particular the western slopes of the Posillipo Hill. Instead, slow-moving flows were strongly controlled by the rugged topography of the area, and stopped within few kilometres from the vent. The authors concluded that the NYT caldera floor is prone to very high inundation hazard by flows with high dynamic overpressure, whereas the centre of the city of Naples and the plain to the north of the CF have a lower probability of being affected by PDCs. Later, Mastrolorenzo et al. (2006) included these results in a multi-hazard perspective, by considering both PDC and PF hazards. This modelling approach was further detailed by Mastrolorenzo et al. (2017).

Todesco et al. (2006) performed the first 2D axisymmetric numerical simulations of PDCs generated by column collapse at the CFc, describing the transient, multiphase flow dynamics of a mixture of three solid particulate phases with different densities and dimensions in a continuous gas phase. The particulate phases were used to represent ash, pumice and lithic solid components, whereas water vapour was the only volatile component assumed in the eruptive mixture. Grain-size, density, and mass fraction of the different classes of solid particles were obtained from the analyses of the eruption deposits. The simulations provided the spatial and temporal distributions of some variables essential for PDC hazard assessment, such as flow temperature and concentration of ash in air. The simulations considered two different radial topographic profiles representative of the south-eastern sector of the NYT caldera and accounting for different vent positions with respect to the high angle slopes bordering the Agnano Plain and the Posillipo Hill. The simulations were based upon the A-MS eruption, in particular to the magmatic phases of the eruption producing the Plinian columns that generated the B1 and D1 beds (de Vita et al. 1999). Largest events generated pyroclastic flows which were thicker and faster, and which overran topographic obstacles more easily, covering longer distances. In contrast, flows generated by small-scale events were fully confined by the distal ridge. Every time the flow decelerated, hot gases and light particles decoupled from the flow and rose to form phoenix clouds, thereby contributing to flow deceleration. As the flow propagated uphill, a portion of material segregated at the base of the flow and eventually slid back, generating back-flows that propagated toward the vent. On the other hand, the presence of two concentric obstacles reduced dynamic pressures in the distal region and prevented flow propagation in the case of small eruptions. The authors finally suggested that in the case of large events, distal obstacles as high as the Posillipo Hill could not effectively protect the area beyond, that is the centre of the city of Naples.

Then, Neri et al. (2015b) produced, through the application of a doubly stochastic approach, fully developed probabilistic maps of PDC invasion able to incorporate some of the main sources of epistemic uncertainty that influence the models. In particular, the new method combined the vent opening map of Bevilacqua et al. (2015) (Fig. 8), the density distribution of PDC invasion areas described in the previous section (Fig. 4), and a PDC integral model able to describe the dynamics of inertial dilute fully-turbulent density currents (the so-called "box-model"; Huppert and Simpson 1980). The integral model had been validated and calibrated through extensive comparison with 2D numerical simulations also on the CF topography, as detailed in Esposti Ongaro et al. (2016). It assumed that the current was vertically homogeneous and deposited particles during propagation as a function of their (constant) sedimentation velocity. A single particle size, representative of the mean Sauter diameter of the grain-size distribution of the mixture, was adopted. Bevilacqua (2016) provided further details on the propagation model and the Monte Carlo algorithm adopted to produce the PDC hazard maps. In order to quantify the main effects of topography on the propagation of a PDC, the flow kinetic energy was compared to the potential energy associated with any topographical relief that the flow encountered (Biagioli et al. 2019). The study suggested that the entire caldera has the potential to be affected, with a mean probability of flow invasion higher than about 5% and the central-eastern area of the caldera (i.e., Agnano-Astroni-Solfatara) having invasion probabilities above about 30%, with local peaks at or above 50% in Agnano (Fig. 13). Significant mean probabilities, up to values of ~10%, were also computed in some areas outside the caldera (i.e., over the Posillipo Hill and in some neighbourhoods of Naples). Considering the density distribution of the PDC invasion areas over the past 15 kyrs, instead of the past 5 kyrs, slightly extended the area enclosed by low-probability isolines. If the probability distribution of the PDC invasion areas excluded the occurrence of large-scale events (top 5%), the computed distribution of probability showed a general decrease in mean values of about 2%. Finally, the chance of a simultaneous activation of two separate vents during the same eruptive event was explored. Assuming that this scenario could occur in 10% of all eruption episodes, with a credible range between about 5% and 25%, the resulting mean invasion map produced slightly

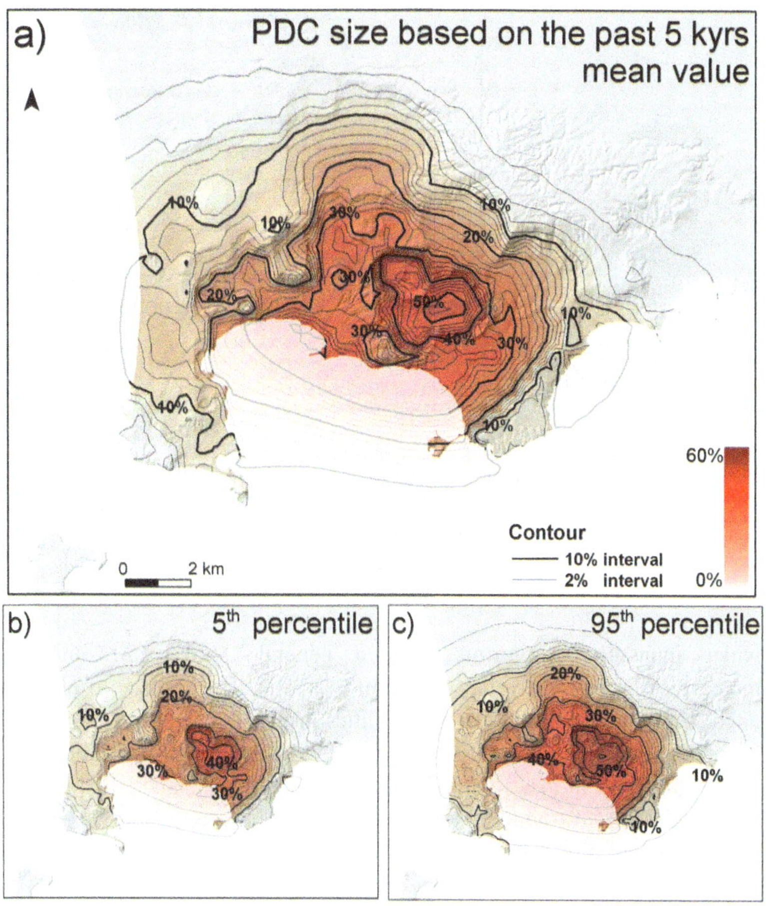

Fig. 13 PDC invasion probability maps computed by assuming the vent opening distribution of Bevilacqua et al. (2015), and the spatial density distribution of invasion areas of the last 5 kyrs. The maps assume that PDCs originate from a single vent per eruption and that the vent is located in the inland part of the caldera. Contours and colours indicate the percentage probability of PDC invasion conditional on the occurrence of an explosive eruption. The maps relate to **a** the mean spatial probability, **b** the 5th percentile and **c** the 95th percentile, respectively. Figure modified after Neri et al. (2015a, b)

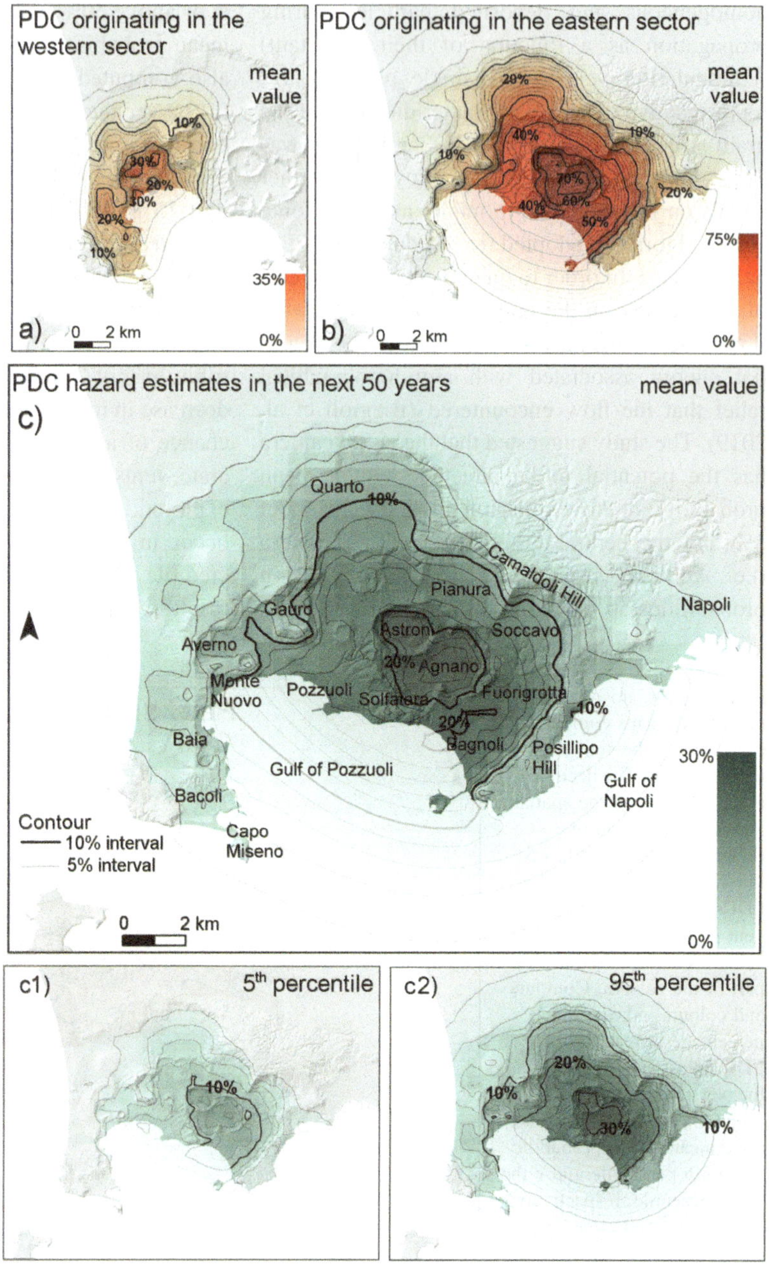

Fig. 14 PDC invasion hazard maps based on the areal size distributions and vent opening probability maps of Bevilacqua et al. (2017). Contours and colours indicate the mean percentage probability of PDC invasion conditional on the occurrence of an explosive event **a** originating inland in the western sector, **b** originating inland in the eastern sector. **c**, **c1**, **c2** show the temporal PDC invasion hazard map assuming that in AD 1538 the volcano entered a new eruptive epoch and including the possibility of a sequence of multiple events in this time period. Maps in **c1** and **c2** show the 5th and 95th uncertainty percentiles of the distribution by using the same colour scales. Figure modified after Bevilacqua et al. (2017)

wider inundation footprints with a general increase of probability values of about 2%. The uncertainty range on invasion probabilities inside the caldera typically was between ±15 and ±35% of the local mean value, with an average of about ±25%; wider uncertainties were found outside the caldera, with an average above ±50% and a significantly larger variability from place to place (Fig. 13).

Tierz et al. (2016) compared eighteen PDC deposits of CFc with a statistical sample of currents modelled testing again with the energy cone model, following a similar approach to Lirer et al. (2001). In particular, the authors compared invaded area, maximum runout, and frequency of PDC arrival. The initial height and the energy cone angle were sampled according to a truncated exponential and a uniform probability distribution, respectively. These distributions were assumed to be independent. The average initial height was obtained from models of eruptive column height, assuming that the 95th percentile of the exponential was equal to the gas thrust height. The energy cone angle range was collected from values found in literature. The authors concluded that, although the energy cone model seemed able to capture the maximum runout for several events, there were circumstances in which it struggled to accurately reproduce past PDC deposits. This was usually related to PDC that became strongly channelled despite modest topography (Aravena et al. 2020). It is worth noting that fully 3D simulations of the multiphase flow model clearly showed that column collapse height is almost irrelevant in the determination of PDCs runout and intensity (Esposti Ongaro et al. 2007, 2008, 2020; Neri et al. 2007, 2021). This observation is consistent with results of integral inertial models and fieldwork evidence (Tadini et al. 2021), but it contrasts the results of the models based on the energy-line approach.

New long-term PDC invasion hazard maps for the CFc (Fig. 14) were produced by Bevilacqua et al. (2017), based on the vent opening probability of Bevilacqua et al. (2015) (Fig. 8), the density distribution of PDC invasion areas (Fig. 9), the integral inertial PDC propagation model of Neri et al. (2015a, b), and the temporal model of Bevilacqua et al. (2016). The authors further evaluated the effects of key epistemic uncertainty sources affecting the hazard assessments. This study had several results. (i) The separate analysis of the eruptive record in the western and eastern sectors of the caldera produced a significant shift of the higher probability values toward the east, with an increase of $\sim 5\%$ of the peak probability and of $\sim 3\%$ of the chance for a PDC to overcome the NYT caldera boundary toward the city centre of Naples. Vice versa, the hazard levels were significantly reduced on the western sector of the caldera (Fig. 14a, b). (ii) Major differences were highlighted between a map generated by considering all the scales up to a given PDC invasion area, and one generated assuming a specific PDC invasion area (i.e., conditional to the occurrence of a PDC of a given size). (iii) Given the assumption that each vent was inside one of the zones in which the CFc was subdivided, a PDC originated in the Agnano zone would have about three times more chance to overcome the Posillipo Hill than one starting in the Astroni zone. In contrast, the area subject to significant PDC invasion probability was strongly reduced when assuming a PDC originating from the Averno zone. (iv) The temporal model enabled the production of hazard maps under the assumption that in AD 1538 the volcano entered a new epoch of activity (Fig. 14c). The effect of a sequence of multiple events in the given time frame was explored. The outcomes showed that the entire caldera had a mean hazard of flow invasion above 1% in the next 10 years and above 5% in the next 50 years (although significantly larger values were estimated in the central-eastern part of the caldera, i.e., 5% and 25%, respectively). Without the assumption that the CFc in AD 1538 entered a new eruptive epoch, the hazard estimates were about three to four times lower. These results are being currently improved by using more advanced numerical models and Monte Carlo sampling techniques (Bevilacqua et al. 2019a).

The most recent attempt to construct a PDC hazard map of the CFc has been performed by Sandri et al. (2018) who developed a multi-source hazard mapping for PDC invasion over the metropolitan area of Naples. The authors considered both Campi Flegrei and Somma-Vesuvio, and accounted for both aleatory variability and epistemic uncertainty, similarly to Neri et al. (2015b) and Bevilacqua et al. (2017), but using the energy cone approach. In detail, they implemented a complete probability assessment, with a Bayesian Event Tree structure. They based the temporal rate on the model of Bevilacqua et al. (2016), but re-ran it on a reconstruction of the last 40 kyrs record. Concerning the eruption scale, the authors adopted, like Sandri et al. (2016), a power law for the total mass erupted by explosive events, constrained with the size classes of Orsi et al. (2009). The uncertainties on column collapse height and energy cone angle were consistent with the evaluations of Tierz et al. (2016), but the authors explored various correlation structures linking the two parameters, instead of assuming them as independent variables. The authors found that the uncertainty associated with this correlation (i.e., direct or inverse) was the major source of uncertainty in their approach to produce hazard maps. According to this study, the maximum probability of PDC invasion in the next 50 years was about 7% (mean value) and ranged between 4 and 9% (assumed as 80% confidence interval), in the central-eastern part of the caldera. The mean probabilities of PDC arrival in the centre of the city of Naples were around 1% and 0.01–2%. The authors also tested the chance of arrival of a PDC originating under the sea, following a linearly decreasing probability of occurrence given an eruption, similar to Tonini et al. (2015). The obtained results showed that the changes in the probability maps were negligible.

4 Hazards Indirectly Related to Magmatism and Volcanism

In addition to the hazards directly related to volcanism, summarised in Sect. 3.4, geological and geomorphologic setting, along with the consequences of millenary human settlement (Orsi et al. 2003; Chap. The Urban Development of Campi Flegrei, Italy) expose the CFc to several hazards indirectly induced by magmatism and volcanism. The recent and ongoing short-term unrest episodes are also accompanied by seismicity as synthesised in Chaps. The Permanent Monitoring System of the Campi Flegrei Caldera, Italy; Historic Unrest of the Campi Flegrei Caldera, Italy; Source Modelling from Ground Deformation and Gravity Changes at the Campi Flegrei Caldera, Italy. Seismicity occurs during the uplifting phases of these episodes, while is almost absent during subsidence. The higher the uplift rate, the stronger and more frequent the seismicity has been in the last decades. The most intense detected shocks since the 1960s occurred in 1983–1984 and had maximum magnitude of 4.0–4.2 according to different estimates (Branno et al. 1984; Barberi et al. 1984; De Natale and Zollo 1986; Orsi et al. 1999; Chap. Historic Unrest of the Campi Flegrei Caldera, Italy). Their intensity reached maximum values in the Pozzuoli area, and decreased within the caldera lowland that includes the western portion of Naples, and the towns of Pozzuoli, Quarto and Bacoli.

Soft-sediment deformation is reported at various times in the history of the caldera. Orsi et al. (1992) report diapir-like deformation with intrusion of several metres of ash beds within the NYT sequence up to 15 km from Campi Flegrei. They interpreted these features as due to liquefaction of ash layers into fluid masses generated by syn-eruptive earthquakes likely associated to activation of the caldera faults. Vitale et al. (2019) reported evidence of seismically induced soft-sediment deformation in the central area of the CFc. They surveyed several soft-sediment structures including sand dikes and sand volcanoes, largely dated between 4.6 and 4.3 ka. They envisaged that the marine-transitional sands of the La Starza unit, when subject to seismic shaking, could go through liquefaction processes. The evidence of liquefaction processes increases the seismic hazard during pre-eruptive, syn-eruptive and, if any, post-eruptive phases even at a long distance from CF.

Episodes of increased unrest at calderas could be accompanied by release of great quantities of CO_2 and other toxic gases, either impulsively or for a prolonged time (Le Guern et al. 1982; Williams-Jones and Rymer 2015). Cold CO_2 is heavier than atmospheric air, and its dispersion is related to both relative temperature and weather conditions (Costa et al. 2005). In CFc, the Solfatara crater is a main source of diffuse CO_2 degassing (Chiodini et al. 2001, 2003, 2010; Caliro et al. 2007; Cardellini et al. 2017). Granieri et al. (2013) estimated that the annual emission of natural CO_2 from the Solfatara crater led to a significant air CO_2 concentration in part of the urban area of Naples, although lower than recommended health protection thresholds. The current unrest was accompanied by a significant increase in CO_2 flux localised in Pisciarelli area (Queißer et al. 2017; Tamburello et al. 2019).

Although the case of a strong explosive scenario in deep water conditions is unlikely, the opening of an eruption vent underwater has been considered a possible event at CFc by Selva et al. (2012a) and Bevilacqua et al. (2015). Paris et al. (2019) produced a probabilistic hazard analysis of tsunami generated by subaqueous volcanic explosions related to the opening of a new eruption vent offshore. They considered a 1.5 km spaced grid of 17 different sites in the underwater part of the caldera, weighted according to the map of Selva et al. (2012a). Thus, they simulated (with a fully nonlinear Boussinesq model) the generation and propagation of the consequent tsunami waves able to reach the coasts of the Campania region for all the combinations of tsunami-generating vent sites and sizes. Results of the study showed highest hazard along the coasts of the Pozzuoli Bay that are in the imminent vicinity of the explosion centres. In a scenario with 650 m vent radius, maximum wave heights at the coast exceeded 3 m only in the Pozzuoli Bay and the eastern coast of the Procida Island. Secondary peaks of wave heights are obtained on the western coast of the Sorrento Peninsula and eastern coast of Capri Island. Instead, with a vent radius of 900 m, wave heights were in excess of 10 m, and with a vent radius of 200 m, they were 1–2 m high at most.

The excavation of many quarries adds further complexity to the geomorphic setting of the Neapolitan-Phlegraean area, with its steep caldera walls and variably preserved monogenetic volcano morphologies within a flat area inside the NYT caldera. These facilitate mass movements along slopes. A synthesis of these mass movements and relative hazard and risk is presented in Chap. Landslide Hazard and Risk in the Campi Flegrei Caldera, Italy. The various types of movements depend upon the lithological characteristics of the slopes along which they occur. Falls and topples occur along slopes made up of tuffs or lavas, while slides and flows develop along slopes composed of loose pyroclastic deposits. The former two phenomena mostly occur along the high-angle scarps of the CFc, the wave-cut cliffs of the Posillipo Hill, and the vertical walls of quarries. These gravitational movements are favoured by wind and sea erosion on highly fractured tuffs. Slides and flows mobilise the loose pyroclastic deposits mantling the high-angle slopes of the caldera walls and the flanks of both volcanic edifices and valleys. They occur during heavy rain periods and mobilise loose material blankets generally thinner than 1 m. The volume of slide material generally does not exceed few hundreds cubic metres.

5 Short-Term Assessment of Volcanic Hazard

CFc is one of the most monitored volcanoes in the world (see Chap. The Permanent Monitoring System of the Campi Flegrei Caldera, Italy), and the main monitored parameters can be divided in three groups: seismic, geodetic, geochemical. A significant uncertainty affects any forecasting effort based on the registered information, because previous pre-eruptive data for CFc to train with are very scarce. Few attempts have been made for the only historical eruption of the caldera: Guidoboni and Ciuccarelli (2011) systematically analysed historical documents; Di Vito et al. (2016) exploited the historical, archaeological and geological record to estimate the ground deformation before the eruption; Di Napoli et al. (2016) investigated the related offshore degassing. Interaction of the magmatic and hydrothermal systems complicates the interpretation of detected signals. Finally, caldera unrest may produce intense signals not followed by an eruption, and, at the same time, the final precursory signals before the eruption may be significantly weaker than those registered during the unrest phase (e.g., Newhall and Dzurisin 1988a, b; Acocella et al. 2015).

Crucial observations in the final pre-eruptive phase include the acceleration in the seismic count, and in the rate and pattern of ground deformation. The Real-time Seismic Amplitude Measurement may become particularly relevant when intense seismic activity hinders event counting.

Moreover, the detection of acid gasses like HF, HCl, SO_2 in the fumaroles, and the opening of new hydrothermal vents is an additional critical element (Caliro et al. 2014; Chap. The Hydrothermal System of the Campi Flegrei Caldera, Italy). Variation in the geometry of the deformation, such as migration of the zone of maximum uplift leading to a bimodal profile, would represent a fundamental clue of the magmatic intrusion at shallow depth, but it was never detected in the CFc (Corrado et al. 1977; Berrino et al. 1984; Orsi et al. 1999; Del Gaudio et al. 2010; De Martino et al. 2014, 2020; Bevilacqua et al. 2020a).

The temporal evolution of the system is characterised by the greatest uncertainty. Unrest signals may persist for months/years before an eruption, but clear evidence of an incoming eruption may become detectable only few days/hours before the actual event. Large-scale variations historically recorded before the AD 1538 eruption suggest decade-long unrest, with a final phase of a couple of months of strong seismicity and a rapid uplifting of the vent opening area only a few hours before the eruption onset. This extremely long and intense unrest resulted in a relatively small eruption (Guidoboni and Ciuccarelli 2011; Di Vito et al. 2016 and references therein). Evidence of fast magma mixing and ascent has been found through geochemical and isotope-geochemical investigations of the products of recent eruptions (i.e., Monte Nuovo, Astroni 6, Averno 2). The results of these studies also suggest time intervals of tens of hours to few days from injection of a new magma batch in the shallow reservoir, with related mixing, to the eruption (e.g., Rutherford 2004; Perugini et al. 2010; Chaps. An Evolutionary Model for the Magmatic System of the Campi Flegrei Volcanic Field (Italy) Constrained by Petrochemical Data; Origin and Differentiation History of the Magmatic System Feeding the Campi Flegrei Volcanic Field (Italy) Constrained by Radiogenic and Stable Isotope Data; Rheological Properties of the Magmas Feeding the Campi Flegrei Caldera (Italy) and Their Influence on Mixing Processes; Magma Chamber Dynamics at the Campi Flegrei Caldera, Italy).

Selva et al. (2012b) developed an event tree for the volcanic unrest of the CFc through a five-years long investigation that involved over 30 researchers, including experts in geophysical monitoring, fluid geochemistry, and in the volcanic history of the CFc. Node 1 of the tree evaluated if the caldera is in state of unrest or quiescence. Node 2 evaluated if the unrest is driven by a magmatic process. Node 3 evaluated the occurrence of a volcanic eruption in the following month. Different signals and Boolean observations were recognised to be crucial in the estimation of the probabilities at each node. Seismic signals were present in every node, the involvement of magma was recognised from the depth and waveforms of earthquakes, and the approaching of an eruption was associated with a significant acceleration in the seismic activity. Application of this event tree analysis resulted in $\sim 40\%$ probability of eruption in August 1983, during the major 1982–1984 unrest episode, and up to $\sim 10\%$ probability of eruption during the later minor uplift episodes. Currently efforts are being made for a real-time updating of vent opening maps through merging the long-term information with the short-term localisation and careful spatial-interpolation of the signals (e.g., Selva et al. 2015; Patra et al. 2019; Sandri et al. 2020).

Macedonio et al. (2014) hypothesised that the intrusion of a sill can be responsible for the dynamics observed during unrest at calderas, and developed a dynamic model of sill intrusion in a shallow volcanic environment. The model is based on the numerical solution of the equation for the elastic plate, coupled with a Navier–Stokes equation for simulating the dynamics. The stress field produced by the intrusion is mainly concentrated in a circular zone that follows the sill intrusion front (D'Auria et al. 2015; Giudicepietro et al. 2016, 2017; Chap. Source Modelling from Ground Deformation and Gravity Changes at the Campi Flegrei Caldera, Italy).

Amoruso et al. (2014a, b) interpreted the deformation history of the CFc in the periods 1980–2010 and 2011–2013 as a consequence of paired deformation sources. In particular, a

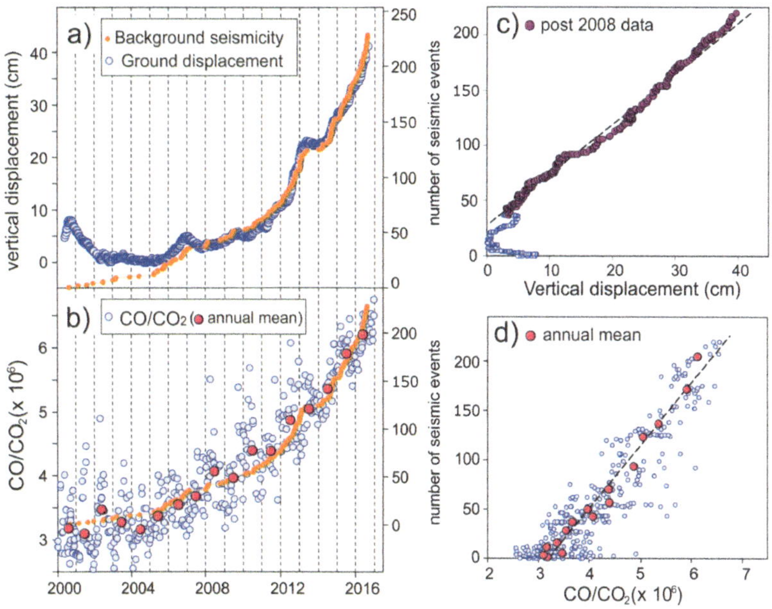

Fig. 15 Background seismicity compared with other observations. **a** Chronogram of the cumulative background seismicity (orange dots) and vertical ground displacement at the RITE GPS station at Rione Terra, Pozzuoli. **b** Chronogram of the cumulative background seismicity (orange dots) and fumarolic CO/CO_2 ratios. **c** Binary plot of the cumulative background seismicity vs. the vertical ground displacement at the RITE GPS station. **d** Binary plot of the cumulative background seismicity vs. the fumarolic CO/CO_2 ratio (the magenta dots refer to annual mean values of both CO/CO_2 ratio and seismicity). Figure modified after Chiodini et al. (2017a, b)

quasi-horizontal, northwest-southeast elongated crack, embedded in an elastic layered half-space at a depth of about 3,600 m, satisfies large-scale deformation. Residual deformation is confined to the area of the Solfatara fumarolic field and satisfied by a small spheroid located at about 1,900 m depth. All source parameters but volume change are constant over time. A synthesis of the hypotheses proposed on the source of the recent short-term deformation events is presented in Chap. Source Modelling from Ground Deformation and Gravity Changes at the Campi Flegrei Caldera, Italy.

In general, although uncertain until the last hours, the approach of an eruption is believed to be associated with a substantial acceleration in the unrest signals (Chiodini et al. 2016; Kilburn et al. 2017). Chiodini et al. (2017a) studied the inter-event times of the post-2000 seismicity at the CFc and distributed it into different populations. The low inter-event times population represented swarm events, while the high inter-arrival times population marked background seismicity. The background seismicity was found to increase at the same rate of the ground uplift and of the concentration of the fumarolic gas species more sensitive to temperature (Fig. 15). The authors proposed that the whole sequence of the CFc unrest episodes since 1950 belongs to a single evolutionary trend of accumulating stress and crustal damage, and that the continuation of the trend will favour the progressive approach to eruptive conditions. After 2016 the rate of seismicity, both in terms of event count and strain release, sped up faster than the rate of ground displacement (Bevilacqua et al. 2020c). De Siena et al. (2017) studied the seismic source locations and waveform attenuation of earthquakes in the CFc during the 1983–1984 unrest episode, constraining a 4–4.5 km deep northwest-southeast striking aseismic zone of high attenuation offshore Pozzuoli, a 3–4 km deep reservoir of supercritical fluids under Pozzuoli, and a shallower aseismic deformation source under Solfatara. They showed that the high attenuation domain controlled the largest monitored seismic, deformation, and geochemical unrest at the caldera (Chiodini et al. 2012, 2015).

Giudicepietro et al. (2021) analysed the two episodes of seismicity and gas emission that occurred on October 7, 2015 (M_d max = 2.5) and December 6, 2019 (M_d max = 3.1) also in terms of ground displacement rates before and after the episodes. They interpreted these as examples of repeated volcanic and (or) hydrothermal system pressurisation that culminate in injection of fluids along a conduit-like path (Chiodini et al. 2017b; Giudicepietro et al. 2019) which behaves as a valve that allows fluids discharge and temporary depressurisation of the source region. The CO_2 flux from the main Pisciarelli fumarolic field has increased by a factor > 3 since 2012, reaching in 2018–2019 levels that are comparable to those typical of a medium-sized erupting arc volcano (Queißer et al. 2017; Tamburello et al. 2019).

Bevilacqua et al. (2020b) described a first assessment of the "failure time" on present-day unrest signals at the CFc based on the horizontal deformation data collected between 2011 and April 2020 at eleven GPS stations. In particular, they applied a probabilistic approach that enhanced the well-established failure forecast method of Voight (1988) by incorporating a stochastic noise in the linearised equations and a mean-reversion property to constrain it (Bevilacqua et al. 2019b). The new formulation enabled the processing of decade-long time windows of data, including the effects of variable dynamics. The "failure time" is realised when the system will reach a critical state if the accelerating trend observed in the monitoring signals will not change in the future. The possibility for false alarms is not eliminated by this method, and included in this category is the "arrested" (or failed) eruption, in which the volcano displays the precursory symptoms typical of an eruption, but does not culminate with magma reaching the surface (Cornelius and Voight 1995). This is a phenomenon typical of restless calderas (Acocella et al. 2015). Based on the assumption that the trends observed in the last several years will continue in the future, Bevilacqua et al. (2020b) provided temporal forecasts, with uncertainty quantification, of a range of failure times (possibly indicative of eruption times). The probability density function of a failure time had peaks of about 12% mean probability per year, and 95th percentile values that can reach 25–30% probability per year (Fig. 16). The analysis of the data of the four GPS stations showing greatest displacement values produced failure time probabilities of 31–36% in 5 years, 60–64% in 10 years, 92–94% in 25 years. However, the interpretation of the failure time as the onset of a volcanic eruption is speculative in caldera systems (Kilburn 2018). Different types of signals can produce different forecasts, and the same type of signals recorded in different locations can produce different results (Bevilacqua et al. 2020c). Moreover, these results relied on long-term trends registered across multiple years. It is worth mentioning that future variations in monitoring data could either slow down the increasing trends so far observed, or speed them up. Moreover, to model any short-term trend, an appropriate time window should be selected.

6 Main Outcomes and Future Implications

The Campi Flegrei caldera (CFc) hazard assessment is complex, although the many studies carried out through time have produced some likely scenarios and probabilistic estimates in case of renewal of volcanism in short-mid-terms largely shared by the scientific community, and summarised in the following.

Astroni-Agnano-Solfatara is the most probable area in which a vent will open, while the Averno-Monte Nuovo is the second most probable one. In general, the vent could open over a region at 2–3 km from the centre of the NYT caldera, although not symmetrically distributed, but most likely occurring in its central-eastern portion. However, a significant uncertainty affects these results, and a non-negligible vent opening probability spreads over the whole caldera. Moreover, multiple vents opening at the same time cannot be ruled out. The information arising from the monitoring network will likely

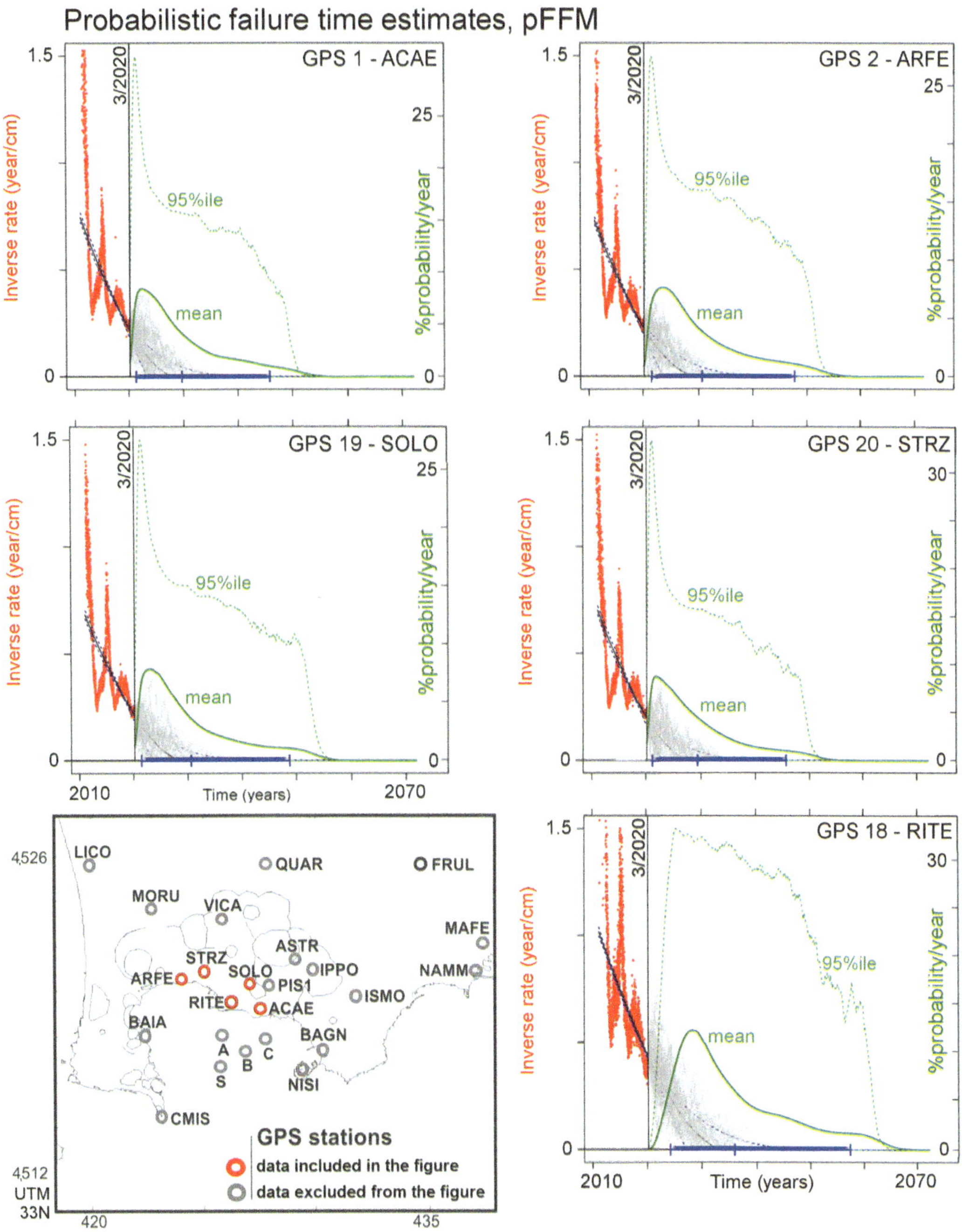

Fig. 16 Probability forecasts of the failure time of CFc using the 1/2011–3/2020 GPS data. Red points on the left are inverse-rate data. The green continuous line on the right is mean values of the annual probability of failure, dashed lines mark its 95th percentile values. A blue line bounds the 90% confidence interval of the forecast. Grey dotted lines display 50 stochastic solution paths. The GPS stations are mapped in the lower left corner (UTM 33 T coordinates). Figure modified after Bevilacqua et al. (2020b)

constrain the vent opening location to a specific area, although it is not possible to foresee how long before the eruption onset.

The eruptions of the CFc, excluding the caldera-forming events, have ejected volumes of magma up to 1 km^3 DRE. However, only four over a total of seventy events have been large eruptions, and the probability that the next eruption will be of medium or smaller size (i.e., < 0.1 km^3 volume of magma DRE) is ~95%. Larger size and more frequent eruptions tend to be localised in the central-eastern portion of the NYT caldera. The inter-event times of volcanic eruptions span from a few years to thousands of years between two subsequent epochs of activity. Inside the epochs, the mean recurrence time is tens of years, but their time sequence is not Poissonian and temporal clusters are evident. However, once these clusters are considered and modelled, the system appears to have followed a self-consistent behaviour in the last 15 kyrs with the eruptions of the past 5 kyrs as a representative sample of all those occurred after 15 ka.

The next eruption can be either effusive or, more likely, explosive, and the hazards that it will generate depend upon its style. An effusive eruption could produce extrusion of a lava dome and/or lava flows. An explosive eruption could be in the range from a phreatic to a Plinian event. The most expected hazards from renewal of explosive volcanism at CFc, corresponding to those with the highest impact on environment and humans, are therefore PF and PDCs. It is worth noting that the winds blow more frequently from west to east, with a more pronounced effect shown by the stratospheric winds above 10–11 km. This means that a volcanic eruption will have higher probability to produce a deposit towards east and this probability depends on the column height. PF deposits in excess of 300 kg/m^2 can affect most of the caldera and the city of Naples with more than 10% probability, as well as the territory that extends for tens of kilometres to the northeast of the CFc, with more than 1% probability, given an eruption.

The entire caldera has the potential to be affected by PDC invasion with mean probability above ~30% in the central-eastern, and above ~50% in Agnano area. However, the high-angle slopes of the caldera morphological boundary tend to act as barriers for small PDC. Nevertheless, these barriers can be overcome in case of large PDC and/or if the eruptive vent lies close to the caldera boundary. Thus, invasion probabilities of ~5–10% have been estimated for the urbanised areas along the eastern slopes of the Posillipo Hill.

In conclusion, probability hazard assessment is particularly complex for CFc due to the sparse location of potential vents and the large variability of eruption styles and sizes that cannot be, at the present time, effectively constrained by monitoring signals. CFc may persist in state of unrest for a long time (decades?), periodically displaying precursor signals that would likely lead a central volcano to eruption. At the same time the possibility of an eruption, with clear pre-eruptive signals only occurring short before the event cannot be ruled out. Nevertheless, volcano monitoring may likely improve the estimates of the probability of vent opening and weather forecasts can constrain the wind field in the near future. From this point of view, the PF and the PDC hazard maps should be regarded as a dynamical tool to be used by decision makers for both long-term land-use zoning of volcanic areas and short-term risk analysis, such as managing volcano crises. These reasons increase the importance of having a robust and flexible hazard model based on the past behaviour of the volcano as well as on all other monitoring and modelling information.

Large part of the analyses and thematic maps presented in this chapter have been used, together with vulnerability and socio-economical information, by the Italian Civil Protection Department of the Presidency of the Council of Ministers for the evaluation of risk scenarios and the elaboration of the National Emergency Plan of the CFc (www.protezionecivile.it).

Acknowledgements Valerio Acocella and Greg Valentine are gratefully thanked for reviewing a first version of the manuscript.

References

Acocella V (2010) Evaluating fracture patterns within a resurgent caldera: Campi Flegrei, Italy. Bull Volcanol 72:623–638. https://doi.org/10.1007/s00445-010-0347-x

Acocella V, Di Lorenzo R, Newhall C, Scandone R (2015) An overview of recent (1988 to 2014) caldera unrest: Knowledge and perspectives. Rev Geophys 53:896–955. https://doi.org/10.1002/2015RG000492

Alberico I, Lirer L, Petrosino P, Scandone R (2002) A methodology for the evaluation of long-term volcanic risk from pyroclastic flows in Campi Flegrei (Italy). J Volcanol Geotherm Res 116(1–2):63–78. https://doi.org/10.1016/S0377-0273(02)00211-1

Alberico I, Petrosino P, Lirer L (2011) Volcanic hazard and risk assessment in a multi-source volcanic area: the example of Napoli city (Southern Italy). Nat Hazard Earth Syst 11:1057–1070. https://doi.org/10.5194/nhess-11-1057-2011

Albert PG, Giaccio B, Isaia R, Costa A, Niespolo EM, Nomade S, Pereira A, Renne PR, Hinchliffe A, Mark DF, Brown RJ, Smith VC (2019) Evidence for a large-magnitude eruption from Campi Flegrei caldera (Italy) at 29 ka. Geology 47:1–5. https://doi.org/10.1130/G45805.1

Amoruso A, Crescentini L, Sabbetta I (2014a) Paired deformation sources of the Campi Flegrei caldera (Italy) required by recent (1980–2010) deformation history. J Geophys Res 119:858–879. https://doi.org/10.1002/2013JB010392

Amoruso A, Crescentini L, Sabbetta I, De Martino P, Obrizzo F, Tammaro U (2014b) Clues to the cause of the 2011–2013 Campi Flegrei caldera unrest, Italy, from continuous GPS data. Geophys Res Lett 41:3081–3088. https://doi.org/10.1002/2014GL059539

Aravena A, Cioni R, Bevilacqua A, de' Michieli Vitturi M, Esposti Ongaro T, Neri A (2020) Tree-branching based enhancement of kinetic energy models for reproducing channelization processes of pyroclastic density currents. J Geophys Res Solid Earth 125:e2019JB019271. https://doi.org/10.1029/2019JB019271

Armienta MA, De la Cruz-Reyna S, Cruz O, Ceniceros N, Aguayo A, Marin M (2011) Fluoride in ash leachates: environmental implications at Popocatépetl volcano, central Mexico. Nat Haz Earth Syst Sci 11:1949–1956. https://doi.org/10.5194/nhess-11-1949-2011

Aspinall W, Blong R (2015) Volcanic risk assessment. In: Sigurdsson H, Houghton B, McNutt S, Rymer H, Stix J (eds) Encyclopaedia of Volcanoes, 2nd ed. Academic Press, San Diego (CA), pp 1215–1230. https://doi.org/10.1016/B978-0-12-385938-9.00070-5

Aspinall WP, Cooke RM (2013) Quantifying scientific uncertainty from expert judgement elicitation. In: Rougier J, Sparks RSJ, Hill LJ (eds) Risk and uncertainty assessment for natural hazards. Cambridge University Press, New York, pp 64–99. www.cambridge.org/9781107006195

Aspinall WP, Woo G, Voight B, Baxter PJ (2003) Evidence-based volcanology: Application to eruption crises. J Volcanol Geotherm Res 128:273–285. https://doi.org/10.1016/S0377-0273(03)00260-9

Banks GN, Tilling RI, Harlow DH, Ewert JW (1989) Volcano monitoring and short-term forecasts. In: Tilling RI (ed) Volcanic hazards: short course in geology, vol. 1. Am Geophys Un, Washington (DC), pp 51–80. https://doi.org/10.1029/SC001p0051

Barberi F, Innocenti F, Lirer L, Munno R, Pescatore T, Santacroce R (1978) The Campanian Ignimbrite: a major prehistoric eruption in the Neapolitan area (Italy). Bull Volcanol 41:10–31. https://doi.org/10.1007/BF02597680

Barberi F, Corrado G, Innocenti F, Luongo G (1984) Phlegraean Fields 1982–1984: Brief chronicle of a volcano emergency in a densely populated area. Bull Volcanol 47:175–185. https://doi.org/10.1007/BF01961547

Barberi F, Bertagnini A, Landi P, Principe C (1993) A review on phreatic eruptions and their precursors. J Volcanol Geotherm Res 52:231–246

Baxter PJ, Horwell CJ (2015) Impacts of eruptions on human health. In: Sigurdsson H, Houghton B, McNutt S, Rymer H, Stix J (eds) Encyclopedia of volcanoes, 2nd ed. Academic Press, San Diego (CA), pp 1035–1047. https://doi.org/10.1016/B978-0-12-385938-9.00060-2

Baxter PJ, Boyd R, Cole P, Neri A, Spence R, Zuccaro G (2005) The impacts of pyroclastic surges on buildings at the eruption of the Soufriere Hills Volcano, Montserrat. Bull Volcanol 67:292–313. https://doi.org/10.1007/s00445-004-0365-7

Berrino G, Corrado G, Luongo G, Toro B (1984) Ground deformation and gravity changes accompanying the 1982 Pozzuoli uplift. Bull Volcanol 47:187–200. https://doi.org/10.1007/BF01961548

Bevilacqua A (2016) Doubly stochastic models for volcanic hazard assessment at Campi Flegrei caldera. Theses, 21, Edizioni della Normale, Birkhäuser/Springer, Pisa, Italy, 227 pp. https://doi.org/10.1007/978-88-7642-577-6. ISBN 978-88-7642-577-6

Bevilacqua A, Isaia R, Neri A, Vitale S, Aspinall WP, Bisson M, Flandoli F, Baxter PJ, Bertagnini A, Esposti Ongaro T, Iannuzzi E, Pistolesi M, Rosi M (2015) Quantifying volcanic hazard at Campi Flegrei caldera (Italy) with uncertainty assessment: I. Vent opening maps. J Geophys Res 120:2309–2329. https://doi.org/10.1002/2014JB011775

Bevilacqua A, Flandoli F, Neri A, Isaia R, Vitale S (2016) Temporal models for the episodic volcanism of Campi Flegrei caldera (Italy) with uncertainty quantification. J Geophys Res 121:7821–7845. https://doi.org/10.1002/2016JB013171

Bevilacqua A, Neri A, Bisson M, Esposti Ongaro T, Flandoli F, Isaia R, Rosi M, Vitale S (2017) The effects of vent location, event scale, and time forecasts on pyroclastic density current hazard maps at Campi

Flegrei caldera (Italy). Front Earth Sc-Switz 5:72. https://doi.org/10.3389/feart.2017.00072

Bevilacqua A, de' Michieli Vitturi M, Esposti Ongaro T, Neri A (2019a) Enhancing the uncertainty quantification of pyroclastic density current dynamics in the Campi Flegrei caldera. Workshop on Frontiers of Uncertainty Quantification in Fluid Dynamic FrontUQ19, 11–13 September 2019, Pisa, Italy. http://hdl.handle.net/2122/13832

Bevilacqua A, Pitman EB, Patra A, Neri A, Bursik M, Voight B (2019b) Probabilistic enhancement of the Failure Forecast Method using a stochastic differential equation and application to volcanic eruption forecasts. Front Earth Sc-Switz 7:135. https://doi.org/10.3389/feart.2019.00135

Bevilacqua A, Neri A, De Martino P, Isaia R, Novellino A, Tramparulo F, Vitale S (2020a) Radial interpolation of GPS and leveling data of ground deformation in a resurgent caldera: application to Campi Flegrei (Italy). J Geodyn 94(2):24. https://doi.org/10.1007/s00190-020-01355-x

Bevilacqua A, Patra A, Pitman EB, Bursik M, De Martino P, Giudicepietro F, Macedonio G, Vitale S, Flandoli F, Voight B, Neri A (2020b) Utilizzo preliminare del failure forecast method sui dati GPS di spostamento orizzontale registrati nella caldera dei Campi Flegrei dal 2011 al 2020. Misc INGV 57:135–139. doi:https://doi.org/10.13127/misc/57/25. ArXiv 2007.02756 (English version)

Bevilacqua A, Patra A, Pitman EB, Bursik M, Neri A, Voight B, Flandoli F, De Martino P, Giudicepietro F, Ricciolino P, Macedonio G, Vitale S (2020c) The Failure Forecast Method applied to the GPS and seismic data collected in the Campi Flegrei caldera (Italy) in 2011–2020, AGU Fall Meeting 2020, Abstract V029–07, San Francisco (CA). USA. https://doi.org/10.1002/essoar.10505832.1

Biagioli G, Bevilacqua A, Esposti Ongaro T, de' Michieli Vitturi M (2019) PyBox: a Python tool for simulating the kinematics of Pyroclastic density currents with the box-model approach reference and user's guide. Zenodo web platform, https://zenodo.org/record/2616551#.YGbw7T_OOUk. https://doi.org/10.5281/zenodo.2616551

Blong RJ (1981) Some effects of tephra falls on buildings. In: Self S, Sparks R (eds) Tephra studies. D. Reidel Publ Co, pp 405–420. https://doi.org/10.1007/978-94-009-8537-727

Blong RJ (1984) Volcanic Hazards: a Sourcebook of the Effects of Eruptions. Academic Press, Sydney, Australia, 424 pp, ISBN:9781483288208

Blong RJ (2003) Building damage in Rabaul, Papua New Guinea, 1994. Bull Volcanol 65:43–54. https://doi.org/10.1007/s00445-002-0238-x

Branno A, Esposito EGI, Luongo G, Marturano A, Porfido S, Rinaldis V (1984) The October 4th, 1983 —Magnitude 4 earthquake in Phlegraean Fields: Macroseismic survey. Bull Volcanol 47:233–238. https://doi.org/10.1007/BF01961553

Brown SK, Crosweller HS, Sparks RSJ, Cottrell E, Deligne NI, Guerrero NO, Hobbs L, Kiyosugi K, Loughlin SC, Siebert L, Takarada S (2014) Characterisation of the Quaternary eruption record: analysis of the Large Magnitude Explosive Volcanic Eruptions (LaMEVE) database. J Appl Volcanol 3:5. https://doi.org/10.1186/2191-5040-3-5

Brown SK, Jenkins SF, Sparks RSJ, Odbert H, Auker MR (2017) Volcanic fatalities database: analysis of volcanic threat with distance and victim classification. J Appl Volcanol 6:15. https://doi.org/10.1186/s13617-017-0067-4

Caliro S, Chiodini G, Moretti R, Avino R, Granieri D, Russo M, Fiebig J (2007) The origin of the fumaroles of La Solfatara (Campi Flegrei, South Italy). Geochim Cosmochim Acta 71(12):3040–3055. https://doi.org/10.1016/j.gca.2007.04.007

Caliro S, Chiodini G, Paonita A (2014) Geochemical evidences of magma dynamics at Campi Flegrei (Italy). Geochim Cosmochim Acta 132:1–15. https://doi.org/10.1016/j.gca.2014.01.021

Capuano P, Russo G, Civetta L, Orsi G, D'Antonio M, Moretti R (2013) The Campi Flegrei caldera structure imaged by 3-D inversion of gravity data. Geochem Geophys Geosyst 14:4681–4697. https://doi.org/10.1002/ggge.20276

Cardellini C, Chiodini G, Frondini F, Avino R, Bagnato E, Caliro S, Lelli M, Rosiello A (2017) Monitoring diffuse volcanic degassing during volcanic unrests: the case of Campi Flegrei (Italy). Sci Rep 7:6757. https://doi.org/10.1038/s41598-017-06941-2

Casadevall TJ (1994) The 1989–1990 eruption of Redoubt Volcano, Alaska: impacts on aircraft operations. J Volcanol Geotherm Res 62:301–316. https://doi.org/10.1016/0377-0273(94)90038-8

Chiodini G, Frondini F, Cardellini C, Granieri D, Marini L, Ventura G (2001) CO_2 degassing and energy release at Solfatara volcano, Campi Flegrei. Italy. J Geophys Res 106(B8):16213–16221. https://doi.org/10.1029/2001JB000246

Chiodini G, Todesco M, Caliro S, Del Gaudio C, Macedonio G, Russo M (2003) Magma degassing as a trigger of bradyseismic events: The case of Phlegrean Fields (Italy). Geophys Res Lett 30:1434. https://doi.org/10.1029/2002GL016790,8

Chiodini G, Caliro S, Cardellini C, Granieri D, Avino R, Baldini A, Donnini M, Minopoli C (2010) Long-term variations of the Campi Flegrei, Italy, volcanic system as revealed by the monitoring of hydrothermal activity. J Geophys Res 115:B03205. https://doi.org/10.1029/2008JB006258

Chiodini G, Caliro S, De Martino P, Avino F, Gherardi F (2012) Early signals of new volcanic unrest at Campi Flegrei caldera? Insights from geochemical data and physical simulations. Geology 40(10):943–946. https://doi.org/10.1130/G33251.1

Chiodini G, Vandemeulebrouck J, Caliro S, D'Auria L, De Martino P, Mangiacapra A, Petrillo Z (2015) Evidence of thermal-driven processes triggering the

2005–2014 unrest at Campi Flegrei caldera. Earth Planet Sci Lett 414:58–67. https://doi.org/10.1016/j.epsl.2015.01.012

Chiodini G, Paonita A, Aiuppa A, Costa A, Caliro S, De Martino P, Acocella V, Vandemeulebrouck J (2016) Magmas near the critical degassing pressure drive volcanic unrest towards a critical state. Nat Commun 7:1–9. https://doi.org/10.1038/ncomms13712

Chiodini G, Selva J, Del Pezzo E, Marsan D, De Siena L, D'Auria L, Bianco F, Caliro S, De Martino P, Ricciolino P, Petrillo Z (2017a) Clues on the origin of post-2000 earthquakes at Campi Flegrei caldera (Italy). Sci Rep 4472:2045–2322. https://doi.org/10.1038/s41598-017-04845-9

Chiodini G, Giudicepietro F, Vandemeulebrouck J, Aiuppa A, Caliro S, De Cesare W, Tamburello G, Avino R, Orazi M, D'Auria L (2017b) Fumarolic tremor and geochemical signals during a volcanic unrest. Geology 45(12):1131–1134. https://doi.org/10.1130/G39447.1

Cinque A, Rolandi G, Zamparelli V (1985) L'estensione dei depositi marini olocenici nei Campi Flegrei in relazione alla vulcano-tettonica. Boll Soc Geol Ital 104(2):327–348. ISSN 0037-8763

Civetta L, Orsi G, Pappalardo L, Fisher RV, Heiken GH, Ort M (1997) Geochemical zoning, mixing, eruptive dynamics and depositional processes—the Campanian Ignimbrite, Campi Flegrei, Italy. J Volcanol Geotherm Res 75:183–219. https://doi.org/10.1016/S0377-0273(96)00027-3

Connor C, Bebbington M, Marzocchi W (2015) Probabilistic volcanic hazard assessment, In: Sigurdsson H, Houghton B, McNutt SR, Rymer H, Stix J (eds) Encyclopedia of Volcanoes, 2nd ed. Academic Press, San Diego (CA), USA, pp 897–910. https://doi.org/10.1016/B978-0-12-385938-9.00051-1

Cornelius R, Voight B (1995) Graphical and PC-software analysis of volcano eruption precursors according to the Material Failure Forecast Method (FFM). J Volcanol Geotherm Res 64:295–320. https://doi.org/10.1016/0377-0273(94)00078-U

Corrado G, Guerra I, Lo Bascio A, Luongo G, Rampoldi R (1977) Inflation and microearthquake activity of phlegraean fields, Italy. Bull Volcanol 40:169–188. https://doi.org/10.1007/BF02596998

Costa A, Macedonio G, Chiodini G (2005) Numerical model of gas dispersion emitted from volcanic sources. Ann Geophys 48(4–5):805–815. https://doi.org/10.4401/ag-3236

Costa A, Dell'Erba F, Di Vito MA, Isaia R, Macedonio G, Orsi G, Pfeiffer T (2009) Tephra fallout hazard assessment at the Campi Flegrei caldera (Italy). Bull Volcanol 71:259–273. https://doi.org/10.1007/s00445-008-0220-3

Cronin SJ, Stewart C, Zernack AV, Brenna M, Procter JN, Pardo N, Christenson B, Wilson T, Stewart RB, Irwin M (2014) Volcanic ash leachate compositions and assessment of health and agricultural hazards from 2012 hydrothermal eruptions, Tongariro, New Zealand. J Volcanol Geotherm Res 286:233–247. https://doi.org/10.1016/j.jvolgeores.2014.07.002

D'Auria L, Pepe S, Castaldo R, Giudicepietro F, Macedonio G, Ricciolino P, Tizzani P, Casu F, Lanari R, Manzo M, Martini M, Sansosti E, Zinno I (2015) Magma injection beneath the urban area of Naples: a new mechanism for the 2012–2013 volcanic unrest at Campi Flegrei caldera. Sci Rep 5:13100. https://doi.org/10.1038/srep13100,2015

de Vita S, Orsi G, Civetta L, Carandente A, D'Antonio M, Di Cesare T, Di Vito M, Fisher RV, Isaia R, Marotta E, Ort M, Pappalardo L, Piochi M, Southon J (1999) The Agnano-Monte Spina eruption (4.1 ka) in the resurgent, nested Campi Flegrei caldera (Italy). J Volcanol Geotherm Res 91:269–301. https://doi.org/10.1016/S0377-0273(99)00039-6

De Natale G, Zollo A (1986) Statistical analysis and clustering features of the Phlegraean Fields earthquake sequence (May 1983–May 1984). Bull Seism Soc Am 76(3):801–814

De Martino P, Tammaro U, Obrizzo F (2014) GPS time series at Campi Flegrei caldera (2000–2013). Ann Geophys 57(2):S0213. https://doi.org/10.4401/ag-6431

De Martino P, Guardato S, Donnarumma GP, Dolce M, Trombetti T, Chierici F, Macedonio G, Beranzoli L, Iannaccone G (2020) Four years of continuous seafloor displacement measurements in the Campi Flegrei caldera. Front Earth Sc-Switz 8:615178.https://doi.org/10.3389/feart.2020.615178

De Siena L, Chiodini G, Vilardo G, Del Pezzo E, Castellano M, Colombelli S, Tisato N, Ventura G (2017) Source and dynamics of a volcanic caldera unrest: Campi Flegrei, 1983–84. Sci Rep 7:8099. https://doi.org/10.1038/s41598-017-08192-7

Deino AL, Orsi G, Piochi M, de Vita S (2004) The age of the Neapolitan Yellow Tuff caldera-forming eruption (Campi Flegrei caldera—Italy) assessed by $^{40}Ar/^{39}Ar$ dating method. J Volcanol Geotherm Res 133:157–170. https://doi.org/10.1016/S0377-0273(03)00396-2

Del Gaudio C, Aquino I, Ricciardi GP, Ricco C, Scandone R (2010) Unrest episodes at Campi Flegrei: a reconstruction of vertical ground movements during 1905–2009. J Volcanol Geotherm Res 195:48–56. https://doi.org/10.1016/j.jvolgeores.2010.05.014

Dellino P, Isaia R, La Volpe L, Orsi G (2004) Interaction between particles transported by fallout and surge in the deposits of the Agnano-Monte Spina eruption (Campi Flegrei, Southern Italy). J Volcanol Geotherm Res 133(1–4):193–210. https://doi.org/10.1016/S0377-0273(03)00398-6

Dempsey DE, Cronin SJ, Mei S, Kempa-Liehr AW (2020) Automatic precursor recognition and real-time forecasting of sudden explosive volcanic eruptions at Whakaari. New Zealand. Nat Commun 11:3562. https://doi.org/10.1038/s41467-020-17375-2

Di Napoli R, Aiuppa A, Sulli A, Caliro S, Chiodini G, Acocella V, Ciraolo G, Di Vito MA, Interbartolo F, Nasello C, Valenza M (2016) Hydrothermal fluid venting in the offshore sector of Campi Flegrei caldera: A geochemical, geophysical and volcanological study. Geochem Geophys Geosyst 17:4153–4178. https://doi.org/10.1002/2016GC006494

Di Renzo V, Arienzo I, Civetta L, D'Antonio M,Tonarini S, Di Vito MA, Orsi G (2011) The magmatic feeding system of the Campi Flegrei caldera: architecture and temporal evolution. Chem Geol 281:227–241

Di Vito MA, Isaia R, Orsi G, Southon J, de Vita S, D'Antonio M, Pappalardo L, Piochi M (1999) Volcanic and deformation history of the Campi Flegrei caldera in the past 12 ka. J Volcanol Geotherm Res 91:221–246. https://doi.org/10.1016/S0377-0273(99)00037-2

Di Vito MA, Arienzo I, Braia G, Civetta L, D'Antonio M, Di Renzo V, Orsi G (2011) The Averno 2 fissure eruption: a recent small-size explosive event at the Campi Flegrei caldera (Italy). Bull Volcanol 73:295–320. https://doi.org/10.1007/s00445-010-0417-0

Di Vito M, Acocella V, Aiello G, Barra D, Battaglia M, Carandente A, Del Gaudio C, de Vita S, Ricciardi GP, Ricco C, Scandone R, Terrasi F (2016) Magma transfer at Campi Flegrei caldera (Italy) before the 1538 AD eruption. Sci Rep 6:32245. https://doi.org/10.1038/srep32245

Esposti Ongaro T, Papale P, Neri A, Del Seppia D (2006) Influence of carbon dioxide on the large-scale dynamics of magmatic eruptions at Phlegrean Fields (Italy). Geophys Res Lett 33:L06318. https://doi.org/10.1029/2005GL025528

Esposti Ongaro T, Cavazzoni C, Erbacci G, Neri A, Salvetti MV (2007) A parallel multiphase flow code for the 3D simulation of volcanic explosive eruptions. Parallel Comput 33(7–8):541–560. https://doi.org/10.1016/j.parco.2007.04.003

Esposti Ongaro T, Neri A, Menconi G, de' Michieli Vitturi M, Marianelli P, Cavazzoni C, Erbacci G, Baxter PJ (2008) Transient 3D numerical simulations of column collapse and pyroclastic density current scenarios at Vesuvius. J Volcanol Geotherm Res 178(3):378-396.https://doi.org/10.1016/j.jvolgeores.2008.06.036

Esposti Ongaro T, Orsucci S, Cornolti F (2016) A fast, calibrated model for pyroclastic density currents kinematics and hazard. J Volcanol Geotherm Res 327:257–272. https://doi.org/10.1016/j.jvolgeores.2016.08.002

Esposti Ongaro T, Komorowski JC, Legendre Y, Neri A (2020) Modelling pyroclastic density currents from a subplinian eruption at La Soufrière de Guadeloupe (West Indies, France). Bull Volcanol 82:76. https://doi.org/10.1007/s00445-020-01411-6

Fedele L, Scarpati C, Lanphere M, Melluso L, Morra V, Perrotta A, Ricci G (2008) The Breccia Museo formation, Campi Flegrei, Southern Italy: geochronology, chemostratigraphy and relationship with the Campanian Ignimbrite eruption. Bull Volcanol 70:1189–1219. https://doi.org/10.1007/s00445-008-0197-y

Fedele L, Insinga DD, Calvert AT, Morra V, Perrotta A, Scarpati C (2011) $^{40}Ar/^{39}Ar$ dating of tuff vents in the Campi Flegrei caldera (Southern Italy): toward a new chronostratigraphic reconstruction of the Holocene volcanic activity. Bull Volcanol 73:1323–1336. https://doi.org/10.1007/s00445-011-0478-8

Fedele L, Insinga DD, Calvert AT, Morra V, Perrotta A, Scarpati C (2012) Reply to the comment on the article "$^{40}Ar/^{39}Ar$ dating of tuff vents in the Campi Flegrei caldera (southern Italy): toward a new chronostratigraphic reconstruction of the Holocene volcanic activity" by Isaia et al. Bull Volcanol 74:297-299. https://doi.org/10.1007/s00445-011-0560-2

Fisher RV, Orsi G, Ort M, Heiken G (1993) Mobility of large-volume pyroclastic flow—emplacement of the Campanian Ignimbrite, Italy. J Volcanol Geotherm Res 56:205–220. https://doi.org/10.1016/0377-0273(93)90017-L

Flueck WY (2016) Brief communication: extended chronology of the Cordón Caulle volcanic eruption beyond 2011 reveals toxic impacts. Nat Haz Earth Sys Sci 16:2351–2355. https://doi.org/10.5194/nhess-16-2351-2016

Folch A, Costa A, Macedonio G (2009) FALL3D: A computational model for transport and deposition of volcanic ash. Comput Geosci 35:1334–1342. https://doi.org/10.1016/j.cageo.2008.08.008

Folch A, Sulpizio R (2010) Evaluating the long-range volcanic ash hazard using supercomputing facilities. Application to Somma-Vesuvius (Italy), and consequences on civil aviation over the Central Mediterranean Area. Bull Volcanol 72:1039–1059. https://doi.org/10.1007/s00445-010-0386-3

Fournier d'Albe EM (1979) Objectives of volcano monitoring and prediction. J Geol Soc London 136:321–326. https://doi.org/10.1144/gsjgs.136.3.0321

Geyer A, Martí J (2008) The new worldwide collapse caldera database (CCDB): A tool for studying and understanding caldera processes. J Volcanol Geotherm Res 175(3):334–354. https://doi.org/10.1016/j.jvolgeores.2008.03.017

Giaccio B, Hajdas I, Isaia R, Deino A, Nomade S (2017) High-precision ^{14}C dating and $^{40}Ar/^{39}Ar$ dating of the Campanian Ignimbrite (Y-5) reconciles the time-scales of climatic-cultural processes at 40 ka. Sci Rep 7:45940. https://doi.org/10.1038/srep45940

Giudicepietro F, Macedonio G, D'Auria L, Martini M (2016) Insight into vent opening probability in volcanic calderas in the light of a sill intrusion model. Pure Appl Geophys 173:1703–1720. https://doi.org/10.1007/s00024-015-1190-y

Giudicepietro F, Macedonio G, Martini M (2017) A physical model of sill expansion to explain the dynamics of unrest at calderas with application to Campi flegrei. Front Earth Sc-Switz 5:54–65. https://doi.org/10.3389/feart.2017.00054

Giudicepietro F, Chiodini G, Caliro S, De Cesare W, Esposito AM, Galluzzo D, Lo Bascio D, Macedonio G,

Orazi M, Ricciolino P, Vandemeulebrouck J (2019) Insight into Campi Flegrei Caldera Unrest through seismic tremor measurements at Pisciarelli Fumarolic Field. Geochem Geophys Geosys 20:5544–5555. https://doi.org/10.1029/2019GC008610

Giudicepietro F, Chiodini G, Avino R, Brandi G, Caliro S, De Cesare W, Galluzzo D, Esposito A, La Rocca A, Lo Bascio D, Obrizzo F, Pinto S, Ricci T, Ricciolino P, Siniscalchi A, Tramelli A, Vandemeulebrouck J, Macedonio G (2021) Tracking episodes of seismicity and gas transport in Campi Flegrei caldera through seismic, geophysical, and geochemical measurements. Seismol Res Lett 92(2A):965-975. https://doi.org/10.1785/0220200223

Guidoboni E, Ciuccarelli C (2011) The Campi Flegrei caldera: historical revision and new data on seismic crises, bradyseisms, the Monte Nuovo eruption and ensuing earthquakes (twelfth century 1582 AD). Bull Volcanol 73:655–677. https://doi.org/10.1007/s00445-010-0430-3

Heiken G, Fakundiny R, Sutter J (eds.) (2003) Earth Sciences in the Cities: A Reader. Am Geophys Un, Sp Publ Series 56, 444 pp. https://doi.org/10.1029/SP056

Houghton B, White JDL, Van Eaton AR (2015) Phreatomagmatic and related eruption styles. In: Sigurdsson H, Houghton B, McNutt S, Rymer H, Stix J (eds) Encyclopaedia of Volcanoes, 2nd ed. Academic Press, San Diego (CA), USA, pp 537–552. https://doi.org/10.1016/B978-0-12-385938-9.00030-4

Huppert HE, Simpson JE (1980) The slumping of gravity currents. J Fluid Mech 99(4):785–799

Kilburn CRJ (2018) Forecasting volcanic eruptions: beyond the failure forecast method. Front Earth Sc-Switz 6:133. https://doi.org/10.3389/feart.2018.00133

Kilburn C, De Natale G, Carlino S (2017) Progressive approach to eruption at Campi Flegrei caldera in southern Italy. Nat Commun 8:15312. https://doi.org/10.1038/ncomms15312

Granieri D, Costa A, Macedonio G, Bisson M, Chiodini G (2013) Carbon dioxide in the urban area of Naples: Contribution and effects of the volcanic source. J Volcanol Geotherm Res 260:52–61. https://doi.org/10.1016/j.jvolgeores.2013.05.003

Isaia R, D'Antonio M, Dell'Erba F, Di Vito MA, Orsi G (2004) The Astroni volcano: the only example of close eruptions within the same vent area in the recent history of the Campi Flegrei caldera (Italy). J Volcanol Geotherm Res 133:171–192. https://doi.org/10.1016/S0377-0273(03)00397-4

Isaia R, Marianelli P, Sbrana A (2009) Caldera unrest prior to intense volcanism in Campi Flegrei (Italy) at 4.0 ka B.P.: Implications for caldera dynamics and future eruptive scenarios. Geophys Res Lett 36: L21303. https://doi.org/10.1029/2009GL040513

Isaia R, Di Vito MA, de Vita S, Rosi M, Sbrana A (2012) Comment on "^{40}Ar/^{39}Ar dating of tuff vents in the Campi Flegrei caldera (southern Italy): Toward a new chronostratigraphic reconstruction of the Holocene volcanic activity" by Fedele et al. [Bull Volcanol 73:1323-1336]. Bull Volcanol 74:293-296.https://doi.org/10.1007/s00445-011-0561-1

Isaia R, Vitale S, Di Giuseppe MG, Iannuzzi E, Tramparulo F, Troiano A (2015) Stratigraphy, structure, and volcano-tectonic evolution of Solfatara maar-diatreme (Campi Flegrei, Italy). Geol Soc Am Bull 127:1485–1504. https://doi.org/10.1130/B31183.1

Isaia R, Vitale S, Marturano A, Aiello G, Barra D, Ciarcia S, Iannuzzi E, D'Assisi Tramparulo F (2019) High-resolution geological investigations to reconstruct the long-term ground movements in the last 15 kyr at Campi Flegrei caldera (southern Italy). J Volcanol Geotherm Res 385:143–158. https://doi.org/10.1016/j.jvolgeores.2019.07.012

ISDR 2005 Hyogo Framework for Action 2005–2015: ISDR international strategy for disaster reduction: building the resilience of nations and communities to disasters. https://www.unisdr.org/2005/wcdr/intergover/official-doc/L-docs/Hyogo-framework-for-action-english.pdf. Last accessed February 8th 2020

Jolly G, De La Cruz S (2015) Volcanic crisis management. In: Sigurdsson H, Houghton B, McNutt S, Rymer H, Stix J (eds) Encyclopaedia of Volcanoes, 2nd ed. Academic Press, San Diego (CA), USA, pp 1188–1202. https://doi.org/10.1016/B978-0-12-385938-9.00068-7

Le Guern F, Tazieff H, Fivre-Pierret R (1982) An example of health hazard: people killed by gas during a phreatic eruption: Dieng Plateau (Java, Indonesia), Feb. 20 1979. Bull Volcanol 45:153–156. https://doi.org/10.1007/BF02600430

Lirer L, Petrosino P, Alberico I (2001) Hazard assessment at volcanic fields: the Campi Flegrei case history. J Volcanol Geotherm Res 112:53–74. https://doi.org/10.1016/S0377-0273(01)00234-7

Lirer L, Petrosino P, Alberico I (2010) Hazard and risk assessment in a complex multi-source volcanic area: the example of the Campania Region, Italy. Bull Volcanol 72:411–429. https://doi.org/10.1007/s00445-009-0334-2

Macedonio G, Costa A (2012) Brief Communication: Rain effect on the load of tephra deposits. Nat Haz Earth Sys Sci 12:1229–1233. https://doi.org/10.5194/nhess-12-1229-2012

Macedonio G, Costa A, Longo A (2005) A computer model for volcanic ash fallout and assessment of subsequent hazard. Comput Geosci 31:837–845. https://doi.org/10.1016/j.cageo.2005.01.013

Macedonio G, Martini M, Neri A, Papale P, Rosi M, Zuccaro G (2013) Rapporto Finale. Gruppo di Lavoro incaricato della definizione dello scenario di riferimento per il piano di emergenza dei Campi Flegrei per il rischio vulcanico. 122 pp. http://www.protezionecivile.gov.it/resources/cms/documents/rapporto_GdL_CampiFlegrei_art_70_def.pdf

Macedonio G, Giudicepietro F, D'Auria L, Martini M (2014) Sill intrusion as a source mechanism of unrest at volcanic calderas. J Geophys Res Solid Earth 119:3986–4000. https://doi.org/10.1002/2013JB010868

Macedonio G, Costa A, Scollo S, Neri A (2016) Effects of eruption source parameter variation and meteorological dataset on tephra fallout hazard assessment: an example from Vesuvius, Italy. J Appl Volcanol 5:1–19. https://doi.org/10.1186/s13617-016-0045-2

Mayer K, Scheu B, Montanaro C, Yilmaz T, Isaia R, Aßbichler D, Dingwell DB (2016) Hydrothermal alteration of surficial rocks at Solfatara (Campi Flegrei): Petrophysical properties and implications for phreatic eruption processes. J Volcanol Geotherm Res 320:128–143. https://doi.org/10.1016/j.jvolgeores.2016.04.020

Mastrolorenzo G, Pappalardo L, Troise C, Rossano S, Panizza A, De Natale G (2006) Volcanic hazard assessment at the Campi Flegrei caldera. In Troise C, De Natale G, Kilburn CRJ (eds). Mechanims of Activity and Unrest at Large Calderas. Geol Soc Lond Spec Publ 269:159–171. https://doi.org/10.1144/GSL.SP.2006.269.01.10

Mastrolorenzo G, Pappalardo L, Troise C, Panizza A, De Natale G (2008) Probabilistic tephra hazard maps for the Neapolitan area: Quantitative volcanological study of Campi Flegrei eruptions. J Geophys Res 113: B07203. https://doi.org/10.1029/2007JB004954

Mastrolorenzo G, Palladino DM, Pappalardo L, Rossano S (2017) Probabilistic-numerical assessment of pyroclastic current hazard at Campi Flegrei and Naples city: Multi-VEI scenarios as a tool for "full-scale" risk management. PLoS ONE 12(10):e0185756. https://doi.org/10.1371/journal.pone.0185756

Marti J, Ernst GGJ (2005) Volcanoes and the environment. Cambridge University Press, Cambridge, UK, pp 1-471

Marzocchi W, Bebbington MS (2012) Probabilistic eruption forecasting at short and long time scales. Bull Volcanol 74:1777–1805. https://doi.org/10.1007/s00445-012-0633-x

Marzocchi W, Sandri L, Selva J (2010) BET_VH: a probabilistic tool for long-term volcanic hazard assessment. Bull Volcanol 72:705–716. https://doi.org/10.1007/s00445-010-0357-8

Marzocchi W, Taroni M, Selva J (2015a) Accounting for epistemic uncertainty in PSHA. Logic tree and ensemble modeling. B Seismol Soc Am 104:2151–2159. https://doi.org/10.1785/0120140131

Marzocchi W, Selva J, Costa A, Sandri L, Tonini R, Macedonio G (2015b) Tephra fall hazard for the Neapolitan area. In: Loughlin SC, Sparks RSJ, Brown SK, Jenkins SF, Vye-Brown C (eds) Global volcanic hazards and risk, Cambridge University Press, Cambridge, UK, pp 239–247. https://doi.org/10.1017/CBO9781316276273.008

McGuire WJ, Kilburn CRJ, Murray JB (1995) Monitoring active volcanoes: strategies, procedures and techniques. Univ College London Press, London, UK, 420 pp. https://doi.org/10.1002/(SICI)1096-9837(199609)21:9<874::AID-ESP583>3.0.CO;2-I

Mele D, Dioguardi F, Dellino P, Isaia R, Sulpizio R, Braia G (2015) Hazard of pyroclastic density currents at the Campi Flegrei caldera (southern Italy) as deduced from the combined use of face architecture, physical modeling and statistics of the impact parameters. J Volcanol Geotherm Res 299:35–53. https://doi.org/10.1016/j.jvolgeores.2015.04.002

Montanaro C, Scheu B, Mayer K, Orsi G, Moretti R, Isaia R, Dingwell DB (2016) Experimental investigations on the explosivity of steam-driven eruptions: A case study of Solfatara volcano (Campi Flegrei). J Geophys Res Solid Earth 121:7996–8014. https://doi.org/10.1002/2016JB013273

Nairn IA, McKee CO, Talai B, Wood CP (1995) Geology and eruptive history of the 550 Rabaul Caldera area, Papua New Guinea. J Volcanol Geotherm Res 69:255–284. https://doi.org/10.1016/0377-0273(95)00035-6

Natale J, Ferranti L, Marino C, Sacchi M (2020) Resurgent-dome faults in the offshore of the Campi Flegrei caldera (Pozzuoli Bay, Campania): preliminary results from high-resolution seismic reflection profiles. Boll Geof Teor Appl 61(3):333–342. https://doi.org/10.4430/bgta0315

National Academies of Sciences, Engineering, and Medicine (2017) Volcanic eruptions and their repose, unrest, precursors, and timing. The National Academies Press, Washington, DC, USA, pp 1–122. https://doi.org/10.17226/24650

Neri A, Macedonio G, Gidaspow D (1999) Phreatic explosions hazard assessment by numerical simulation. Phys Chem Earth 24(11–12):989–995. https://doi.org/10.1016/S1464-1895(99)00147-7

Neri A, Esposti Ongaro T, Menconi G, de' Michieli Vitturi M, Cavazzoni C, Erbacci G, Baxter PJ (2007) 4D Simulation of explosive eruption dynamics at Vesuvius. Geophys Res Lett 34. https://doi.org/10.1029/2006GL028597

Neri A, Esposti Ongaro T, Voight B, Widiwijayanti C (2015a) Pyroclastic density current hazards and risk. In: Papale P, Schroder J (eds) Volcanic hazards, risks and disasters. Elsevier, London, UK, 5:109–140. https://doi.org/10.1016/B978-0-12-396453-3.00005-8

Neri A, Bevilacqua A, Esposti Ongaro T, Isaia R, Aspinall WP, Bisson M, Flandoli F, Baxter PJ, Bertagnini A, Iannuzzi E, Orsucci S, Pistolesi M, Rosi M, Vitale S (2015b) Quantifying volcanic hazard at Campi Flegrei caldera (Italy) with uncertainty assessment: 2. Pyroclastic density current invasion maps. J Geophys Res 120:2330–2349. https://doi.org/10.1002/2014JB011776

Neri A, Esposti Ongaro T, de' Michieli Vitturi M, Cerminara M. (2021) Multiphase flow modeling of explosive volcanic eruptions, In: Arastoopour H, Gidaspow D, Lyczkowski (eds) Transport phenomena in multiphase systems. Mechanical Engineering Series, Springer. ISBN 978–3–030–68578–2

Newhall CG, Dzurisin D (1988a) Historical Unrest at Large Calderas of the World, volume 1. USGS Bulletin, Washington, (DC). USA 1855:1–610

Newhall CG, Dzurisin D (1988b) Historical unrest at large calderas of the world, volume 2. USGS Bulletin, Washington, (DC). USA 1855:1–520

Orsi G, D'Antonio M, de Vita S, Gallo G (1992) The Neapolitan Yellow Tuff, a large-magnitude trachytic phreatoplinian eruption: eruptive dynamics, magma withdrawal and caldera collapse. J Volcanol Geotherm Res 53:275–287. https://doi.org/10.1016/0377-0273(92)90086-S

Orsi G, Civetta L, D'Antonio M, Di Girolamo P, Piochi M (1995) Step-filling and development of a three-layers magma chamber: the Neapolitan Yellow Tuff case history. J Volcanol Geotherm Res 67:291–312. https://doi.org/10.1016/0377-0273(94)00119-2

Orsi G, de Vita S, Di Vito M (1996) The restless, resurgent Campi Flegrei nested caldera (Italy): constraints on its evolution and configuration. J Volcanol Geotherm Res 74:179–214. https://doi.org/10.1016/S0377-0273(96)00063-7

Orsi G, Civetta L, Del Gaudio C, de Vita S, Di Vito MA, Isaia R, Petrazzuoli SM, Ricciardi G, Ricco C (1999) Short-term ground deformations and seismicity in the nested Campi Flegrei caldera (Italy): an example of active block-resurgence in a densely populate area. J Volcanol Geotherm Res 91:415–451. https://doi.org/10.1016/S0377-0273(99)00050-5

Orsi G, de Vita S, Di Vito M, Isaia R, Nave R, Heiken G (2003) Facing volcanic and related hazards in the Neapolitan area. In: Heiken G, Fakundiny R, Sutter J (eds) Earth sciences in the cities: a reader. Am Geophys Un, Sp Publ Series, 56:121–170, ISBN 0–87590–299–5

Orsi G, Di Vito MA, Isaia R (2004) Volcanic hazard assessment at the restless Campi Flegrei caldera. Bull Volcanol 66:514–530. https://doi.org/10.1007/s00445-003-0336-4

Orsi G, Di Vito MA, Selva J, Marzocchi W (2009) Long-term forecast of eruption style and size at Campi Flegrei caldera (Italy). Earth Planet Sci Lett 287:265–276. https://doi.org/10.1016/j.epsl.2009.08.013

Pallister J, McNutt SR (2015) Synthesis of Volcano Monitoring. In: Sigurdsson H, Houghton B, McNutt SR, Rymer H, Stix J (eds) Encyclopedia of Volcanoes, 2nd ed. Academic Press, San Diego (CA), USA, pp 1151–1171. https://doi.org/10.1016/B978-0-12-385938-9.00066-3

Paris R, Ulvrova M, Selva J, Brizuela B, Costa A, Grezio A, Lorito S, Tonini R (2019) Probabilistic hazard analysis for tsunamis generated by subaqueous volcanic explosions in the Campi Flegrei caldera, Italy. J Volcanol Geotherm Res 379:106–116. https://doi.org/10.1016/j.jvolgeores.2019.05.010

Patra A, Bevilacqua A, Pitman EB, Bursik B, Voight B, Neri A, Macedonio G, Flandoli F, De Martino P, Giudicepietro F, Vitale S (2019) A statistical approach for spatial mapping and temporal forecasts of volcanic eruptions using monitoring data. AGU 2019 Fall Meeting, San Francisco, CA, USA, 9–13 December 2019. https://doi.org/10.1002/essoar.10502434.1

Perrotta A, Scarpati C, Luongo G, Morra V (2010) Stratigraphy and volcanological evolution of the southern sector of CampiFlegrei and Procida Island, Italy. In: Groppelli G, Viereck-Goette L (eds) Stratigraphy and geology of volcanic areas. Geol SocAm Sp Paper 464:171–191. https://doi.org/10.1130/2010.2464(09)

Perugini D, Petrelli M, Poli G, De Campos CP, Dingwell DB (2010) Time-scales of recent Phlegrean Fields eruptions inferred by the application of the 'diffusive fractionation' model of trace elements. Bull Volcanol 72:431–447. https://doi.org/10.1007/s00445-009-0329-z

Pistolesi M, Isaia R, Marianelli P, Bertagnini A, Fourmentraux C, Albert P, Tomlinson E, Menzies M, Rosi M, Sbrana A (2016) Simultaneous eruptions from multiple vents at Campi Flegrei (Italy) highlight new eruption processes at calderas. Geology 44(6):487–490. https://doi.org/10.1130/G37870.1

Pistolesi M, Bertagnini A, Di Roberto A, Vona A, Cioni R, Giordano G (2017) The Baia-Fondi di Baia eruption at Campi Flegrei: stratigraphy and dynamics of a multi-stage caldera reactivation event. Bull Volcanol 79:67. https://doi.org/10.1007/s00445-017-1149-1

Poland M, Anderson K (2020) Partly cloudy with a chance of lava flows: Forecasting volcanic eruptions in the twenty-first century. J Geophys Res Solid Earth 125(1):e2018JB016974. https://doi.org/10.1029/2018JB016974

Queißer M, Granieri D, Burton M, Arzilli F, Avino R, Carandente A (2017) Increasing CO_2 flux at Pisciarelli, Campi Flegrei, Italy. Solid Earth 8:1017–1024. https://doi.org/10.5194/se-8-1017-2017

Ricco C, Petrosino S, Aquino I, Del Gaudio C, Falanga F (2019) Some investigations on a possible relationship between ground deformation and seismic activity at Campi Flegrei and Ischia volcanic areas (Southern Italy). Geosciences 9(5):222. https://doi.org/10.3390/geosciences9050222

Rivalta E, Corbi F, Passarelli L, Acocella V, Davis T, Di Vito MA (2019) Stress inversions to forecast magma pathways and eruptive vent location. Sci Adv 5:7. https://doi.org/10.1126/sciadv.aau97

Robertson RM, Kilburn CR (2016) Deformation regime and long-term precursors to eruption at large calderas: Rabaul, Papua New Guinea. Earth Planet Sci Lett 438:86–94. https://doi.org/10.1016/j.epsl.2016.01.003

Romano C, Vona A, Campagnola S, Giordano G, Arienzo I, Isaia R (2020) Modelling and physico-chemical constraints to the 4.5 ka Agnano-Monte Spina Plinian eruption (Campi Flegrei, Italy). Chem Geol 532:119301. https://doi.org/10.1016/j.chemgeo.2019.119301

Rosi M, Santacroce R (1984) Volcanic hazard assessment in the Phlegraean Fields: a contribution based on stratigraphic and historical data. Bull Volcanol 47:359–370. https://doi.org/10.1007/BF01961567

Rosi M, Sbrana A (eds) (1987) Phlegrean Fields. Quaderni de "La Ricerca Scientifica" CNR, Rome, Italy, 114:1–175. ISSN 0556–9664

Rosi M, Vezzoli L, Aleotti P, De Censi M (1996) Interaction between caldera collapse and eruptive dynamics during the Campanian Ignimbrite eruption, Phlegraean Fields, Italy. Bull Volcanol 57:541–554. https://doi.org/10.1007/BF00304438

Rossano S, Mastrolorenzo G, De Natale G (2004) Numerical simulation of pyroclastic density currents on Campi Flegrei topography: A tool for statistical hazard estimation. J Volcanol Geotherm Res 132:1–14. https://doi.org/10.1016/S0377-0273(03)00384-6

Rutherford MJ (2004) Experimental petrology studies. In: Papale, P. Rapporto finale del progetto GNV 2001–03 n. 17 "Simulation of Eruptive Scenarios at Phlegrean Fields Based on Field, Laboratory, and Numerical Studies, and Implications for Volcanic Hazard"

Sacchi M, Pepe F, Corradino M, Insinga DD, Molisso F, Lubritto C (2014) The Neapolitan Yellow Tuff caldera offshore the Campi Flegrei: Stratal architecture and kinematic reconstruction during the last 15 ky. Mar Geol 354:15–33. https://doi.org/10.1016/j.margeo.2014.04.012

Sandri L, Costa A, Selva J, Tonini R, Macedonio G, Folch A, Sulpizio R (2016) Beyond eruptive scenarios: assessing tephra fallout hazard from Neapolitan Volcanoes. Sci Rep 6:24271. https://doi.org/10.1038/srep24271

Sandri L, Tierz P, Costa A, Marzocchi W (2018) Probabilistic hazard from pyroclastic density currents in the Neapolitan area (Southern Italy). J Geophys Res Solid Earth 123:3474–3500. https://doi.org/10.1002/2017JB014890

Sandri, L, Bevilacqua A, Selva J, Neri A, Costa A, Macedonio G (2020) Eruption forecasting and hazard assessment at INGV during the 2019 crisis exercise at Campi Flegrei. 4th Rittmann Conference, Miscellanea INGV 52, ISSN 2039–6651, Catania, Italy, 12 - 14 February 2020, http://hdl.handle.net/2122/13862

Scandone R, D'Amato J, Giacomelli L (2010) The relevance of the 1198 eruption of Solfatara in the Phlegraean Fields (Campi Flegrei) as revealed by medieval manuscripts and historical sources. J Volcanol Geotherm Res 189(1–2):202–206. https://doi.org/10.1016/j.jvolgeores.2009.09.012

Scarpa R, Tilling RI (eds.) (1996) Monitoring and mitigation of volcano hazards. Springer-Verlag, Heidelberg, Germany, 841 pp, ISBN 978–3–642–80087–0

Scarpati C, Cole P, Perrotta A (1993) The Neapolitan Yellow Tuff - A large volume multiphase eruption from Campi Flegrei, Southern Italy. Bull Volcanol 55:343–356. https://doi.org/10.1007/BF00301145

Selva J, Costa A, Marzocchi W, Sandri L (2010) BET_VH: Exploring the influence of natural hazard uncertainties on long-term hazard from tephra fallout at Campi Flegrei, (Italy). Bull Volcanol 72:717–733. https://doi.org/10.1007/s00445-010-0358-7

Selva J, Marzocchi W, Papale P, Sandri L (2012a) Operational eruption forecasting at high-risk volcanoes: the case of Campi Flegrei. Naples. J Appl Volcanol 1:5. https://doi.org/10.1186/2191-5040-1-5

Selva J, Orsi G, Di Vito M, Marzocchi W, Sandri L (2012b) Probability hazard map for future vent opening at the Campi Flegrei caldera. Italy. Bull Volcanol 74(2):497–510. https://doi.org/10.1007/s00445-011-0528-2

Selva J, Marzocchi W, Sandri L, Costa A (2015) Operational short-term volcanic hazard analysis: methods and perspectives. In: Shroder JF, Papale P (eds) Volcanic Hazards, Risks and Disasters, Ch. 9. Elsevier, New York, USA, pp 233–259. https://doi.org/10.1016/B978-0-12-396453-3.00009-5

Selva J, Costa A, De Natale G, Di Vito MA, Isaia R, Macedonio G (2018) Sensitivity test and ensemble hazard assessment for tephra fallout at Campi Flegrei, Italy. J Volcanol Geotherm Res 351:1–28. https://doi.org/10.1016/j.jvolgeores.2017.11.024

Sheridan MF, Malin MC (1983) Application of computer-assisted mapping to volcanic hazard evaluation of surge eruptions:Vulcano, Lipari, and Vesuvius. J Volcanol Geotherm Res 17(1):187–202. https://doi.org/10.1016/0377-0273(83)90067-7

Siebert L, Simkin T, Kimberly P (2010) Volcanoes of the World, 3rd edn. Smithsonian Institution, Berkeley, CA, USA

Siebert L, Cottrell E, Venzke E, Andrews B (2015) Earth's Volcanoes and Their Eruptions: An Overview. In: Sigurdsson H, Houghton B, McNutt S, Rymer H, Stix J (eds) Encyclopaedia of Volcanoes, 2nd ed. Academic Press, San Diego, CA, USA, pp. 239–255. https://doi.org/10.1016/B978-0-12-385938-9.00012-2

Silleni A, Giordano G, Isaia R, Ort MH (2020) The Magnitude of the 39.8 ka Campanian Ignimbrite Eruption, Italy: Method, Uncertainties and Errors. Front Earth Sc-Switz 8:543399. https://doi.org/10.3389/feart.2020.543399

Simkin T, Siebert L (1994) Volcanoes of the World: A Regional Directory, Gazetteer, and Chronology of Volcanism During the Last 10,000 Years (Second Edition). Smithsonian Inst Geosci Press Inc., Tucson, AZ, USA, p 349

Small C, Naumann T (2001) The global distribution of human population and recent volcanism. Global Environ Chang Part b, Environ Haz 3:93–109. https://doi.org/10.1016/S1464-2867(02)00002-5

Smith VC, Isaia R, Pearce NJG (2011) Tephrostratigraphy and glass compositions of post-15 kyr Campi Flegrei eruptions: implications for eruption history and chronostratigraphic markers. Quaternary Sci Rev 30:3638–3660. https://doi.org/10.1016/j.quascirev.2011.07.012

Sparks RSJ (2003) Forecasting volcanic eruptions. Earth Planet Sci Lett 210:1–15. https://doi.org/10.1016/S0012-821X(03)00124-9

Sparks RSJ, Biggs J, Neuberg JW (2012) Monitoring Volcanoes. Science 335(6074):1310–1311. https://doi.org/10.1126/science.1219485

Spence RJS, Kelman I, Baxter PJ, Zuccaro G, Petrazzuoli S (2005) Residential building and occupant vulnerability to tephra fall. Nat Haz Earth Sys Sc 5:477–494. https://doi.org/10.5194/nhess-5-477-2005

Steinmann L, Spiess V, Sacchi M (2018) Post-collapse evolution of a coastal caldera system: Insights from a 3D multichannel seismic survey from the Campi Flegrei caldera (Italy). J Volcanol Geotherm Res 349:83–98. https://doi.org/10.1016/j.jvolgeores.2017.09.023

Sulpizio R, Dellino P, Doronzo DM, Sarocchi D (2014) Pyroclastic density currents: state of the art and perspectives. J Volcanol Geotherm Res 283:36–65. https://doi.org/10.1016/j.jvolgeores.2014.06.014

Tadini A, Bevilacqua A, Neri A, Cioni R, Biagioli G, de' Michieli Vitturi M, Esposti Ongaro T (2021) Reproducing pyroclastic density currents deposits of the AD 79 eruption of the Somma-Vesuvius volcano using the box-model approach. Solid Earth 12:119–139. https://doi.org/10.5194/se-12-119-2021

Tamburello G, Caliro S, Chiodini G, De Martino P, Avino R, Minopoli C, Carandente A, Rouwet D, Aiuppa A, Costa A, Bitetto M, Giudice G, Francofonte V, Ricci T, Sciarra A, Bagnato E, Capecchiacci F (2019) Escalating CO_2 degassing at the Pisciarelli fumarolic system, and implications for the ongoing Campi Flegrei unrest. J Volcanol Geotherm Res 384:151–157. https://doi.org/10.1016/j.jvolgeores.2019.07.005

Tierz P, Sandri L, Costa A, Zaccarelli L, Di Vito MA, Sulpizio R, Marzocchi W (2016) Suitability of energy cone for probabilistic volcanic hazard assessment: validation tests at Somma-Vesuvius and Campi Flegrei (Italy). Bull Volcanol 78:79. https://doi.org/10.1007/s00445-016-1073-9

Tilling RI (2001) Volcano monitoring and eruption warnings. Springer-Verlag, Berlin Heidelberg, Germany, pp 505–510. https://doi.org/10.1007/978-3-642-55903-766

Todesco M, Neri A, Esposti Ongaro T, Papale P, Rosi M (2006) Pyroclastic flow dynamics and hazard in a caldera setting: application to Phlegrean Fields (Italy). Geochem Geophys Geosys 7:11. https://doi.org/10.1029/2006GC001314

Tonarini S, D'Antonio M, Di Vito MA, Orsi G, Carandente A (2009) Geochemical and isotopical (B, Sr, Nd) evidence for mixing and mingling processes in the magmatic system feeding the Astroni volcano (4.1-3.8 ka) within the Campi Flegrei caldera (South Italy). Lithos 107:135–151. https://doi.org/10.1016/j.lithos.2008.09.012

Tonini R, Sandri L, Costa A, Selva J (2015) Brief Communication: The effect of submerged vents on probabilistic hazard assessment for tephra fallout. Nat Hazards Earth Syst Sci 15:409–415. https://doi.org/10.5194/nhess-15-409-2015

Vitale S, Isaia R (2014) Fractures and faults in volcanic rocks (Campi Flegrei, Southern Italy): Insights into volcano-tectonic processes. Int J Earth Sci 103:801–819. https://doi.org/10.1007/s00531-013-0979-0

Vitale S, Isaia R, Ciarcia S, Di Giuseppe MG, Iannuzzi E, Prinzi EP, Tramparulo FDA, Troiano A (2019) Seismically induced soft-sediment deformation phenomena during the volcano-tectonic activity of Campi Flegrei caldera (southern Italy) in the last 15 kyr. Tectonics 38:1999–2018. https://doi.org/10.1029/2018TC005267

Voight B (1988) A method for prediction of volcanic eruptions. Nature 332:125–130. https://doi.org/10.1038/332125a0

Voloschina M, Pistolesi M, Bertagnini A, Métrich N, Pompilio M, Di Roberto A, Di Salvo S, Francalanci L, Isaia R, Cioni R, Romano C (2018) Magmatic reactivation of the Campi Flegrei volcanic system: insights from the Baia-Fondi di Baia eruption. Bull Volcanol 80:75. https://doi.org/10.1007/s00445-018-1247-8

Wohletz K, Orsi G, de Vita S (1995) Eruptive mechanisms of the Neapolitan Yellow Tuff interpreted from stratigraphic, chemicaland granulometric data. J Volcanol Geotherm Res 67:263–290. https://doi.org/10.1016/0377-0273(95)00002-C

Williams-Jones G, Rymer H (2015) Hazards of Volcanic Gases. In: Sigurdsson H, Houghton B, McNutt S, Rymer H, Stix J (eds) Encyclopaedia of Volcanoes, 2nd ed. Academic Press, San Diego, CA, USA, pp 985–992. https://doi.org/10.1016/B978-0-12-385938-9.00057-2

Wilson G, Wilson TM, Deligne NI, Cole JW (2014) Volcanic hazard impacts to critical infrastructure: A review. J Volcanol Geotherm Res 286:148–182. https://doi.org/10.1016/j.jvolgeores.2014.08.030

Wilson T, Stewart C, Cole J, Johnston D, Cronin S (2010) Vulnerability of farm water supply systems to volcanic ash fall. Environ Earth Sci 61:675–688. https://doi.org/10.1007/s12665-009-0380-2

Zuccaro G, Cacace F, Spence RJS, Baxter PJ (2008) Impact of explosive eruption scenarios at Vesuvius. J Volcanol Geotherm Res 178:416–453. https://doi.org/10.1016/j.jvolgeores.2008.01.005

Landslide Hazard and Risk in the Campi Flegrei Caldera, Italy

Domenico Calcaterra and Diego Di Martire

Abstract

This chapter deals with the main characteristics of the mass movements affecting the slopes of the Campi Flegrei Volcanic Field. To this aim, giving due credit to previous works, an overview of the Phlegraean landslides is presented, taking into account other ongoing researches. The prevailing landslide types depend upon the lithological characteristics of the rocks involved: slides and flows dominate in the loose pyroclastic deposits, while falls and topples characterise the slopes where tuffs and lavas crop out. All the Phlegraean landslides show a moderate mobility, with a maximum runout distance lower than 350 m, and a mean angle of reach equal to 39°. More than 15% of the whole territory of the six Phlegraean towns is prone to landslide hazard, though the distribution of at-risk areas among the six towns is uneven (e.g., Monte di Procida, 31.7%; Quarto, 5%). Due to the intense urbanisation of the area, such a hazard results in a quite high risk. In recent years, a number of structural countermeasures, including low-impact soil bioengineering techniques, have been introduced as to mitigate the landslide risk in the Neapolitan-Phlegraean area. However, a paradigm shift is strongly needed, due in part to the persistent economic crisis, which should induce to emphasise the importance of non-structural preventive solutions, among which the improvement of the urban communities' societal resilience is of paramount importance.

1 Introduction

The main concern in a densely populated active volcanic area, such as the Campi Flegrei volcanic field (CFvf), is obviously and correctly addressed to the societal impacts deriving from a future eruption or volcanic-related events. However, volcanic eruptions, earthquakes and bradyseismic activity have continuously combined to create a unique geomorphologic scenario, characterised by a volcanic-derived hilly landscape, mostly carved into loose or variably consolidated pyroclastic deposits. For a reconstruction of the geological and volcanological evolution of the CFvf, including the Campi Flegrei caldera (CFc), as well as of the characteristics of the outcropping rocks, see Chap. Volcanic and Deformation History of the Campi Flegrei Volcanic Field, Italy. The petrophysical characteristics of most widespread rock-types of this volcanic field and their use are presented in Chap. Building Stones and

D. Calcaterra (✉) · D. Di Martire
Dipartimento di Scienze della Terra, dell'Ambiente e delle Risorse, Università degli Studi di Napoli Federico II, Napoli, Italy
e-mail: domenico.calcaterra@unina.it

© Springer-Verlag GmbH Germany, part of Springer Nature 2022

G. Orsi et al. (eds.), *Campi Flegrei*, Active Volcanoes of the World,
https://doi.org/10.1007/978-3-642-37060-1_13

Technological Materials of the Campi Flegrei Caldera, Italy. Such terrains, generally displaying poor geomechanical properties, are exposed to both natural and anthropogenic factors that together predispose and sometimes trigger slope instabilities.

The occurrence of both endogenous and exogenous factors controlling the everyday morphodynamics of the Phlegraean hills, makes them particularly vulnerable to mass movements, which only in the last decades have received the proper attention they deserve from the scientific community and public administrators.

In the following, data from the literature and from ongoing studies on the landslides affecting the CFvf, hereinafter called "Phlegraean landslides", will be described. Then the overall hazard and risk conditions will be critically commented on, with reference to the official documents recently produced by the competent authorities. Finally, some considerations will be given to the possible scenarios related to the landslide risk mitigation and management in the Neapolitan-Phlegraean area.

2 Previous Studies

Although landslides have been a serious geohazard in the CFvf since distant historical times, they have been a focus of a systematic scientific research only in the last few decades. In fact, Naples and the other Phlegraean towns of Pozzuoli, Bacoli, Monte di Procida, Quarto and Marano di Napoli (Fig. 1) were not cited in the first nation-wide report on landslides compiled in Italy by the geographer Roberto Almagià at the beginning of the twentieth century (Almagià 1907, 1910), through a questionnaire sent to local governing bodies, eventually integrated with field-checks.

Dell'Erba (1923) was probably the first author to highlight the importance of landslides in the Neapolitan-Phlegraean area. He described some landslides which involved the Neapolitan Yellow Tuff (NYT; ~15 ka; see Chap. Volcanic and Deformation History of the Campi Flegrei Volcanic Field, Italy) between 1886 and 1911, pointing out the close relationship existing between mass movements and quarrying activity. The following decades were again characterised by a notable absence of the CFvf from both technical and scientific documents. Such an absence is well testified by the volume published by the Italian Ministry for the Public Works (Ministero per i Lavori Pubblici) (1965), where, in a synopsis of the inventories performed in 1957 and 1963 through the local ministerial divisions, no landslide is mentioned for the Phlegraean towns.

A paper by Beneduce et al. (1988) can be considered as the first modern contribution aimed at a deeper comprehension of causes and features of the Phlegraean landslides. It reports the main results of a research carried out after a rainfall-induced event occurred on February 1986. Catenacci (1992), in a catalogue of the main hydraulic and geological disasters occurred in Italy from 1945 to 1990, pointed out three landslides at Naples. Pellegrino (1994) reviewed the landslides studied by Beneduce et al. (1988), in the framework of a wider study dealing with the landslides in the Neapolitan area. Some rainfall-induced landslides occurred at Naples in 1910, are mentioned by Migale and Milone (1998) in a historical research on debris flows in the Campanian pyroclastic deposits.

A specific archival research on the historical mass movements at Naples and the CFvf has been conducted by Calcaterra and co-workers and is still underway; the results so far obtained have been published in a number of papers (Calcaterra et al. 2002, 2003b; Calcaterra and de Luca Tupputi Schinosa 2006; Morra et al. 2010; Di Martire et al. 2012). The same research group has analysed a cluster of shallow landslides triggered by a rainfall event that occurred in Naples on January 1997 (Calcaterra and Guarino 1999a, b). The relationship between mass movements and rainfall at Naples has also been dealt with by Esposito et al. (2002).

The Camaldoli Hill, representing the main peak within the entire CFvf (Fig. 1), was treated in detail by Parise et al. (2004) and by Calcaterra et al. (2007a). The latter paper was centred on the flow-like movements in the post-NYT loose

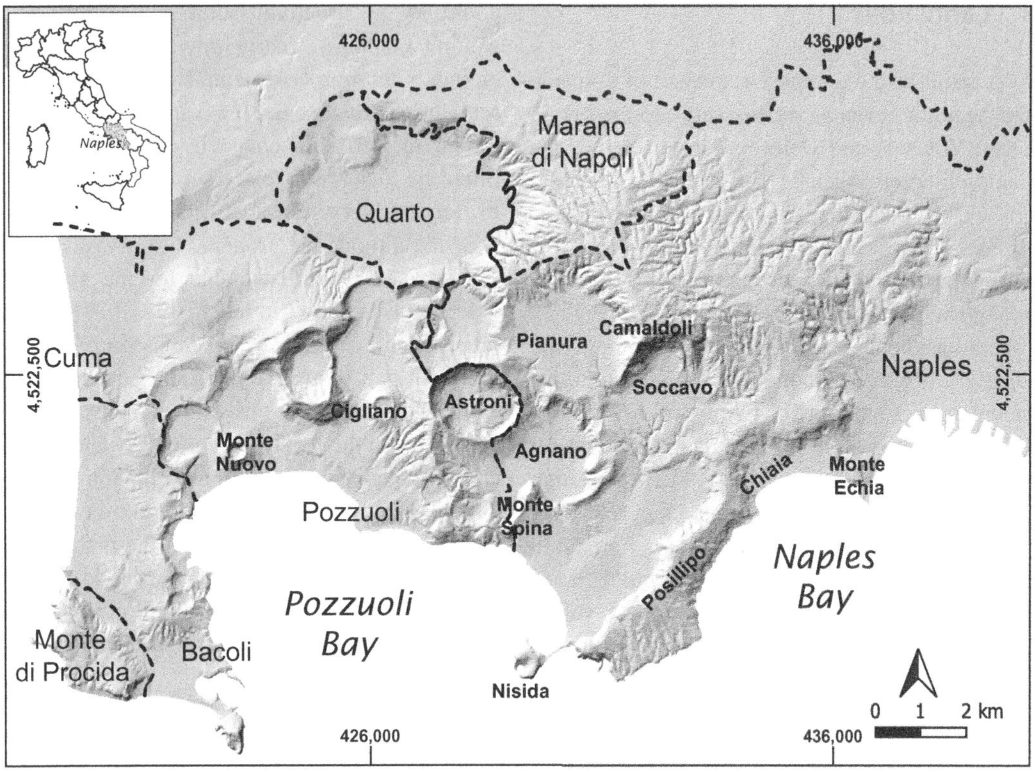

Fig. 1 Naples and the five Phlegraean towns within the Campania region (highlighted in the inset) that are the site of the landslides discussed in this chapter

pyroclastic terrains, while in the first contribution rockfall instabilities were analysed. The same hill has been the study area chosen by Calcaterra et al. (2007b) to highlight the role played by changes in land use and by wildfires as factors predisposing to landslides and erosion. The flow-like movements affecting the youngest pyroclastic deposits of the Phlegraean sequence have been further studied with the help of some physically-based models, aiming to assess the proneness of the Neapolitan slopes to the detachment events ranging from slides to flows (Calcaterra et al. 2004a, 2005, 2010; Calcaterra 2010). Taking advantage of the ongoing research on several areas in the Campania region, a number of papers have been published. In these papers the features of the Phlegraean landslides in pyroclastic loose terrains have been critically compared to similar phenomena that involve the pyroclastic cover, mainly of Vesuvian origin, mantling the carbonate ridges bordering the Campanian Plain (de Riso et al. 1999, 2004, 2007; Calcaterra et al. 2003b, 2004b; Palma et al. 2009; Calcaterra 2010).

The published data have been fruitfully used by the Basin Authority of the North-western Campania Region, a governmental agency in charge of land planning and management, to issue the Hydro-geomorphological Setting Plan in 2002 and to revise it in 2010 (Basin Authority of North-western Campania Region 2002, 2010). Such agency has been recently incorporated into a wider administrative body (*Autorità di Bacino Distrettuale dell'Appennino Meridionale*, www.distrettoappenninomeridionale.it), responsible for the whole southern Italy. The Plan is an act aimed at identifying landslide and flood risk, through a set of basic thematic maps, among which the Landslide-Inventory map plays a role of paramount importance as to define the areas prone to detachment, transit and runout of slope instabilities.

3 Landslides

As a result of the complex sequence of volcanic and tectonic events, extensively described in Chap. Volcanic and Deformation History of the Campi Flegrei Volcanic Field, Italy, the Phlegraean hillslopes are either remnants of volcanic edifices or derive from the inner walls of the CFc. Consequently, along these slopes a peculiar geological setting can be recognised (Fig. 2), with a backbone prevailingly made up of weak, jointed tuffs and subordinately of lavas, surmounted by some tens of metres of loose pyroclastic deposits. The main geomorphological feature of the Phlegraean hills is their quite high "relief energy". In fact, despite their relatively moderate altitude (max. elevation: Camaldoli Hill, 458 m a.s.l.), local hills show high slope angles (>30°), which reach the highest values where cliff-forming tuffs and lavas crop out. In addition, the drainage network is markedly controlled by structural factors, where low-order straight channels clearly prevail, usually coinciding with morphostructural lineaments. Consequently, small-scale fans are present at the mouth of the main channels only, while more often the material displaced from hillsides tends to remain very close to the foothill, hence showing a moderate proneness to invasion.

The data so far available for the six Phlegraean towns are reported in Fig. 3, where the inventoried events refer to a timespan of slightly less than two thousand years. In fact, the oldest evidence related to mass movements can be found within the Ancient Centre of Naples, in the archaeological site found under the San Lorenzo church. Here, 7 m below the present-day ground surface, the city market, built in the first century BC, is partly buried under a chaotic deposit ascribed to a flood which occurred at the end of the fifth century CE (Touring Club Italiano 2001; Morra et al. 2010).

Fig. 2 Typical Phlegraean slope (Posillipo Cape, Naples). **A** Loose pyroclastic terrains, younger than 15 ka; **B** Neapolitan Yellow Tuff; **C** Landslide debris. Photo by the authors

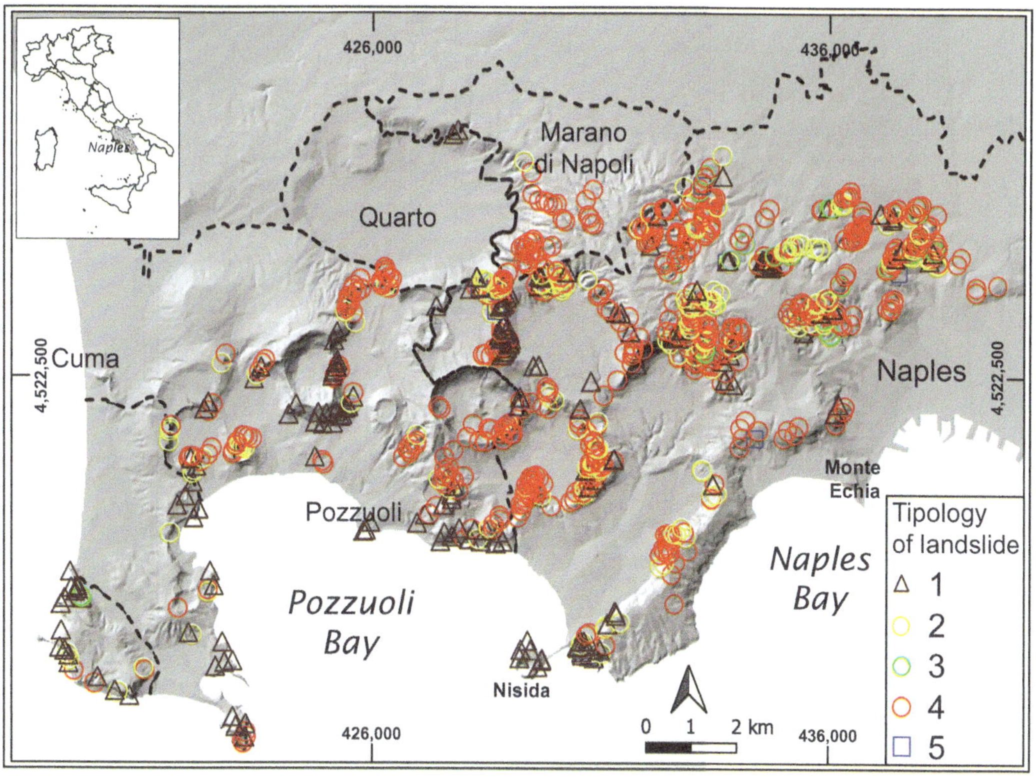

Fig. 3 Landslide-inventory map of the Campi Flegrei volcanic field. 1. Fall/topple; 2. Flow; 3. Slide; 4. Complex landslide; 5. Unclassified landslide. Author's unpublished data

The deadliest landslide among those known of the entire CFvf struck Naples on January 28, 1868 (Fig. 4). On that occasion 60 victims and tens of injuries were caused by a huge rockfall which detached from Monte Echia (Fig. 1), in the Chiaia borough, and reached the nearby Castel dell'Ovo (Calcaterra and de Luca Tupputi Schinosa 2006). After the event, a legal suit went on for about 40 years, and in 1908 the huge sum of 1,300,000 Italian lire (corresponding to a present-day value of about 5 million euro) was granted by the Italian government for the landslide repair project. Other relevant slope instabilities involved the urban area of Naples in the nineteenth century, such as a rockfall in NYT that took place on New Year's Eve of 1889 and disrupted an important road running along the coast of Posillipo (Gunther 1993) (Fig. 1). The occurrence of a reply in the same place on January 11, 1889, induced the authorities to remove some unstable blocks by bombarding them from the sea. The unusual intervention started on January 21 and ended on February 2, when 200 m^3 of NYT moved downslope, seriously damaging an edifice (Calcaterra et al. 2002).

The twentieth century CE started with a huge rainfall event, which, on October 24, 1910, affected vast portions of the Campania region, with the epicentral area located in the Amalfi coast, where 200 people lost their lives. However, Naples and the surrounding towns were also severely hit on that occasion, suffering serious material damage. Based on contemporary newspaper reports, about 150 mm of rainfall were discharged at Naples in about four hours.

Coming to more recent times, in the last decades of the past century some rainfall-induced episodes can be highlighted, among which those occurred in February 1986, January 1997, September 2001 and March 2005 are probably

Fig. 4 The January 28, 1868 Monte Echia landslide. Red dotted line refers to the crown of the mass movement. Figure modified after Emporio Pittoresco (1868)

the most important in terms of effects. The 1986 event produced the worst effects in the western part of Naples (Fig. 5), as well as in the remaining towns of the Phlegraean continental area (Beneduce et al. 1988). Naples was again the focus of the event that occurred on January 10–11, 1997, when 110 mm of rainfall fell in 40 h (maximum hourly intensity: 10 mm) and, preceded by a long rainy period, triggered a huge cluster of landslides. Calcaterra and Guarino (1999a, b) described the basic features of that event, when some hundred shallow landslides (Fig. 5), mostly of the slide-flow type, involved essentially the loose pyroclastic rocks younger than 15 ka, causing severe damage to man-made structures. A large majority of such phenomena took place along the slopes of the western sectors of the city (Camaldoli Hill, Posillipo Hill, Agnano Plain) (Fig. 1).

Two more instability episodes, characterised by highly variable values of rainfall duration and intensity, affected the Phlegraean area in September 2001 (183.4 mm in 4 h) and March 2005 (130.2 mm in 27 h). In the first case, water-driven debris travelled for quite long distances along the Camaldoli hillslopes (Fig. 1), sometimes reaching the foothill inhabited areas. The 2005 event mainly involved the territory of Pozzuoli (Figs. 1 and 5). Here the worst single movement was a small-scale translational slide which invaded the track of a usually heavily trafficked motorway (Fig. 5): luckily, only material damage was caused.

By integrating all the available data coming from different sources, slightly less than 1,500 landslides have been inventoried in the six towns of the study area, among which Naples is by far the leading one, with about 75% of the total (Fig. 6). Analysing the same data in terms of areal density, the highest value pertains to Naples (9.4 events/km^2), followed by the smallest town, Monte di Procida, where 7.4 landslides per square kilometre have occurred; the mean value is 7.1. All these values are notably higher than the one of the Campania region (1.71), which in turn is similar to that (1.6) established by the Italian Institute for Environmental Protection and Research (Istituto Superiore per la Protezione e la Ricerca Ambientale) (2008) for the whole Italian territory.

Complex landslides, following the Varnes (1978) classification scheme, are the prevailing type (Fig. 6) with single contributing movements depending upon the lithological characteristics of the involved rocks: slides and flows dominate in the loose pyroclastic deposits, while falls and topples characterise the slopes along which tuffs and lavas crop out.

As evident from Fig. 7, all the Phlegraean landslides show a moderate mobility, expressed in terms of H/L ratio or "angle of reach" (i.e., the angle of the line connecting the highest point of the landslide crown scarp—H—to the distal margin of the displaced mass; Heim 1932). In fact, the travel distance of all the main Phlegraean landslide types (slides, falls/topples and complex movements) show a runout distance lower than 350 m, with a mean value of the angle of reach equal to 39°. If the various types of movements are individually considered, no striking difference can be highlighted (Fig. 7).

The moderate invasion potential that characterises the Phlegraean landslides can be conveniently compared with the slope instabilities affecting the pyroclastic cover of the nearby carbonate ridges. In this case, the H/L ratio reaches values as low as 10–15°, corresponding to the channelled slide-flows occurred during the 1998 Sarno catastrophic event, with a typical higher bound at 25–30° (Calcaterra et al. 2004b).

4 Landslide Hazard and Risk

The subject of landslide/flood defence and soil conservation was treated in Italy for the first time in 1989, when a law was promulgated, at the end of a long process that started after the flooding of Florence in 1966. The main innovations introduced by this law were: the change in the frame of reference, from administrative (region, province or municipality) to geographical (river basin); the establishment of the River Basin Authorities, governmental agencies (either national, inter-regional or regional) in charge of land planning and management; the need for the River Basin Master Plan, a comprehensive planning act, dealing with different territorial and environmental issues, such as landslides, floods, groundwater, coastal erosion. However, it was only after nine years that the law was fully implemented, when, following the aftermath of the Sarno tragedy (160 people killed by flow-like movements in Sarno and four more Campanian towns—May 1998) a decree was issued (D.P.C. M. 29 September 1998). This decree, among other topics, defined the landslide and flood risk grades, from R4 (very high) to R1 (moderate), to be included in the Master Plans on behalf of the River Basin Authorities.

Fig. 5 Recent slope instabilities of the Neapolitan-Phlegraean area. **a** Post-fire shallow slides along the southern slope of the Camaldoli Hill (Naples, summer 1996); **b** Typical small-scale slide along the inner slope of the Agnano Plain (Naples, January 1997); **c** Cluster of slides at the border between Naples and Quarto (January 1997); **d** Translational slide that invaded the Naples bypass (Tangenziale) (Pozzuoli, March 4, 2005); **e** different generations of translational slides along the western slope of Monte Spina (Agnano Plain, Naples): 1. February 1986, 2. September 2002, 3. January 2003; **f** Slide at Cigliano (Pozzuoli, March 4, 2005); **g** Fall along the southern slope of the Camaldoli Hill (Naples, 2005—photo courtesy by A. Santo); **h** Column-like mass of Whitish Tuffs (see Chap. Volcanic and Deformation History of the Campi Flegrei Volcanic Field, Italy) prone to topple (Camaldoli Hill, Naples). For location of cited localities, see Fig. 1. Photos by the authors

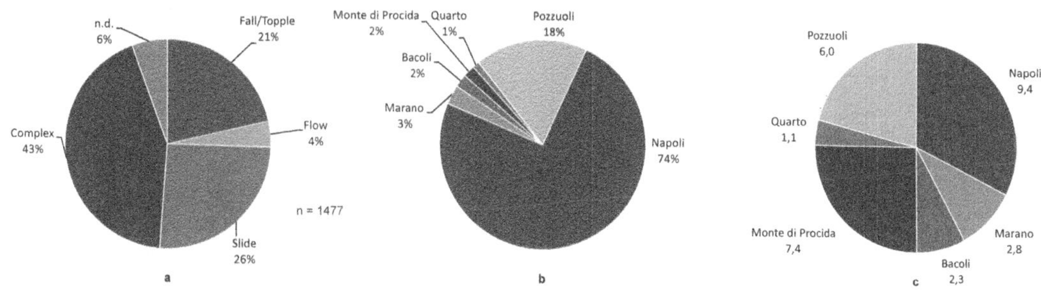

Fig. 6 Statistical features of the Phlegraean landslides. **a** Type of events; **b** Percentage of total events in each town; **c** Density of events (number/km^2) in each town. Author's unpublished data

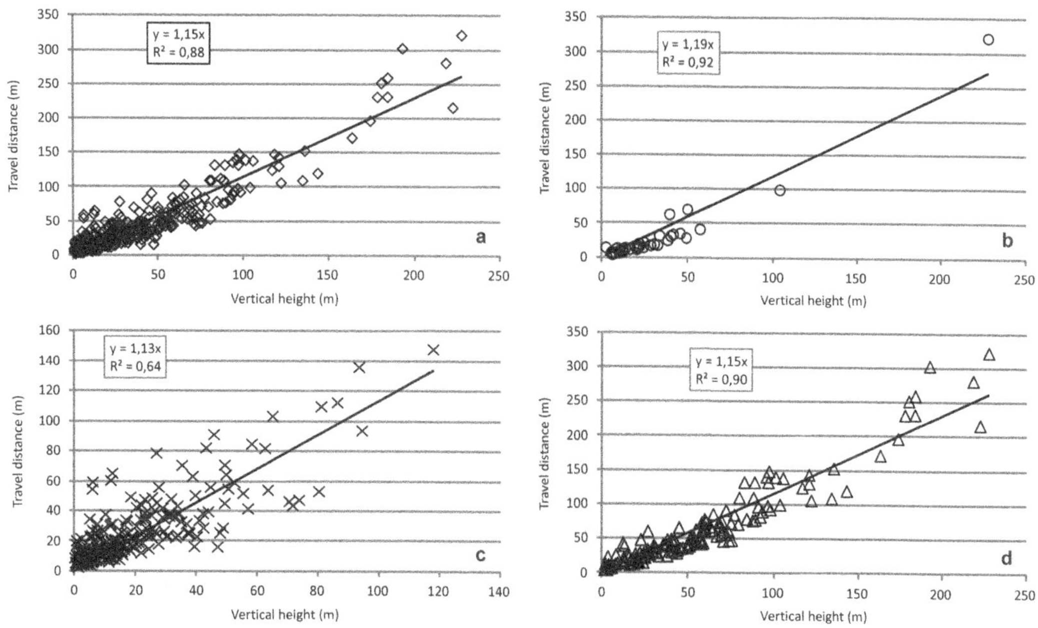

Fig. 7 Relationships between vertical height (H) and travel distance (L) for the Phlegraean landslides. **a** All types (n = 453); **b** Falls/topples (n = 45); **c** Slides (n = 262); **d** Complex (n = 146). Author's unpublished data

The Phlegraean towns fall under the jurisdiction of the Basin Authority of the North-western Campania Region that issued its Hydrogeomorphological Setting Plan in 2002 and revised it in 2010. Both versions of the Plan were released at a 1:5,000 scale. The procedure adopted to obtain the final landslide risk map has been extensively described by Calcaterra et al. (2003a) and Di Martire et al. (2012) and critically reviewed by Sorriso Valvo (2005).

Landslide risk, subdivided in the four official levels, affects more than 15% (31.9 km^2) of the whole territory of the six Phlegraean towns (Fig. 8). The figure is coherent with that referred to all the towns under the control of the Basin Authority of the North-western Campania Region (16%). However, the distribution of the at-risk areas among the six towns is uneven: Monte di Procida is, in fact, the town most exposed to landslides, with 31.7% of its territory, while at Quarto the percentage drops down to 5% (Figs. 1 and 8). A very high percentage (72%) of all the at-risk areas falls in the two highest risk grades, where loss of human lives (R4) or serious threat to public safety (R3) is expected.

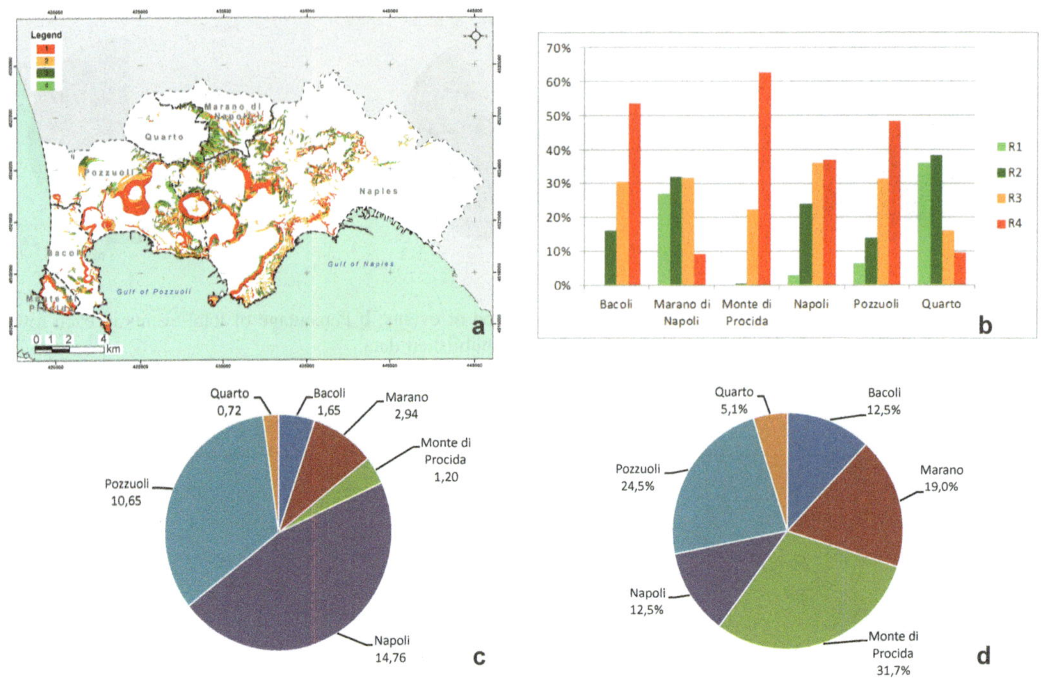

Fig. 8 Summary of the landslide risk conditions for the Phlegraean towns (Author's unpublished data). **a** Landslide risk map (modified after Basin Authority of the North-western Campania Region 2010): 1. Very high risk; 2. High risk; 3. Medium risk; 4. Moderate risk; **b** Distribution of landslide risk levels within each town; **c** Total area exposed to landslide risk within each town; **d** Percentage of areas exposed to landslide hazard within each town

5 Considerations on Landslide Hazard Mitigation and Risk Reduction

According to the results of the previously presented detailed studies performed over the last 20 years, landslides have to be considered one of the sources of natural hazards for the entire CFvf. Their typologies, geological and geomorphological features, triggering conditions and invasion potential are among the key-aspects, which have been so far treated in depth, thus allowing a reasonably good comprehension of the propensity to slope instabilities in the whole area.

With respect to the two main rock types erupted by the CFvf, post-NYT loose pyroclastic terrains and lithified tuffs, some basic similarities and differences in the Phlegraean landslides can be determined. Both rock types are in fact affected by small-scale phenomena, characterised by a relatively low mobility. On the contrary, the geological and geomechanical features of each of them favour different types of landslides. Loose pyroclastic rocks mostly generate open-slope slides that can evolve into debris flows; on the contrary, tuffs, due to their both typical columnar jointing and exposure in sub-vertical natural cliffs, cut slopes or quarry walls, induce falls and topples.

The Phlegraean landslides have been triggered by both natural physical processes and man-related activities. In fact, if rainfall represents the main triggering cause, anthropogenic actions significantly contributed to landslides generation (Calcaterra et al. 2002; Di Martire et al. 2012). In the past, especially at the turn of the twentieth century, quarrying activity and excavations played a significant role in causing landslides. More recently, the progressive abandonment of agricultural practices along with extensive urbanisation of the slopes (often illegal) and wildfires have assumed a greater relevance (Calcaterra et al. 2007b; Di Martire et al. 2012).

Despite their moderate mobility, landslides in the study area represent a serious threat for the local population due to the wild expansion of the urban areas (Chap. The Urban Development of Campi Flegrei, Italy). Especially after World War II, urbanisation has progressively expanded toward foothill zones, hence occupying relatively unsafe territories and determining a serious exposure of people and infrastructures to slope-related instabilities. Nowadays, some of the foothill boroughs show population densities among the highest in the western countries (e.g., Soccavo with 9,300 inhabitants/km^2) (Municipal Government of the City of Naples 2007).

As a result, landslide risk conditions of the entire CFvf are definitely troublesome, with a relevant percentage of territory exposed to a high to very high risk, which, following the Italian legislation, imply the possibility of loss of human life and irreparable damage to property. Therefore, the issue of landslide risk mitigation in the Phlegraean towns is of paramount importance, as a key-aspect in the attempt to improve life quality and territorial safety.

In the last 15 years, several remedial measures have been designed and partly implemented. To this respect, the turning point was represented, at least for the city of Naples, by the institution in 1997 of a special governmental agency (in Italian, *Commissariato Straordinario di Governo*) in charge of planning and managing structural and non-structural measures against landslides and floods. This agency benefited for over 10 years from extraordinary funds granted by the national government. In these years, some types of remedial measures have been more frequently adopted in the CFvf, inspired to both active and passive principles. In the first case, stabilisation works have been aimed at preventing future triggering of mass movements, usually leading to an increase of the safety factor in the slope sectors susceptible to detachment. Passive countermeasures have reduced the risk, protecting man-made works from impacts, for example controlling the distance and direction of landslide travel. Falls and topples affecting the weak tuffs, such as the NYT, have been usually stabilised through rock removal or scaling, bolts, wire mesh or woven wire-rope nets and, more rarely, passive solutions such as ditches, barriers or fences (Fig. 9). Fast-moving slide-flows involving the post-NYT loose pyroclastic deposits have been prevailingly mitigated by means of passive works such as check dams and retention basins, located in the lower reaches of the slope channels and at the mouths. However, the efficiency of the latter works is seriously hindered by a chronic lack of maintenance, basically represented by a periodic removal of any debris and silt build-up (Fig. 9). In the same years, a considerable effort has been put into an attempt to reduce the environmental impact of the remedial measures. To this aim, biotechnical slope stabilisation works have been adopted, such as vegetated timber walls, geogrids, contour wattling, brush layering, biotextile revetments, live wooden crib-walls (Fig. 9).

6 Conclusive Remarks

Naples and the Phlegraean towns are an iconic example of a densely populated urban territory highly vulnerable to recurrent landslides. Such recurrent threat has been, over the time, triggered by both natural and human-induced causal factors. In fact, if in recent decades all the main geomorphological crises (e.g., 1986, 1997, 2001) have been caused by intense and/or prolonged rainfall events, at the turn between nineteenth and twentieth century landslides were of a prevailing anthropogenic nature, due to a strong urban expansion which, at that time, involved some of the urban hillsides (e.g., Posillipo; Fig. 1).

However, even though 150 years of documented events clearly testify of a recurring hazard, it is not possible yet to assign a given "return period" to Phlegraean landslides, also considering the variety of instabilities (falls, topples, slides, complex) which affect both soils and rocks cropping out in the Phlegraean area. Therefore, taking into account the different predisposing factors recognised in the area, wide portions of its territory have been classified as

Fig. 9 Examples of structural remedial measures adopted to mitigate landslide risk in the Campi Flegrei volcanic field. **a** Camaldoli Hill (Naples): in the insets, some retention basins; **b** One of the Camaldoli retention basins completely filled with debris; **c** Posillipo Hill (Naples): integrated solution, through joint sealing and flexible fences; **d** Marano di Napoli: live double-layer wooden crib-wall. **e** Marano di Napoli: geogrid revetment and hexagonal double-torsion net coupled with hydroseeding; **f** Naples: rock nails and wire-mesh net. For location of cited localities, see Fig. 1. Photos by the authors

exposed to high or very high landslide risk (11%), corresponding to about 23 km^2.

From this perspective, the Phlegraean area is a sort of "litmus test", which well shows how critical is the Italian situation in terms of resources needed to mitigate landslide risk. In fact, a quite recent survey has shown that more than eleven thousand priority interventions are needed to significantly reduce the landslide and flooding risk over the Italian territory; the related costs have been evaluated at about forty billion Euros (Giannella and Guida 2010). On the other hand, from 1991 to 2008 the Italian government has invested about seven billion euros of ordinary funds for the mitigation of landslide and flooding risk. This implies that, maintaining the same average annual expenditure values, 100 years would be necessary to reach a reasonable level of structural protection against landslides and floods.

It is therefore evident, that, due to the chronic scarcity of resources allocated for the implementation of structural measures against geohazards, a fundamental commitment is to improve the Nations' understanding of where and how they are vulnerable to natural hazards and how to effectively respond so that recovery is rapid when catastrophes occur. To this aim, the World Conference on Disaster Reduction held in 2005 at Kobe, Japan, launched a worldwide campaign centred on societal resilience to natural hazards, a concept that implies exposure, sensitivity, and adaptive capacity of human-environmental systems. Since 2005, Italy participates to the Hyogo Framework for Action; some of the largest Italian cities, among which there is Naples, have joined the campaign of resilient cities, which aims to strengthen the capacity of disaster response at the local level. However, it must be stressed that no investments of specific resources to implement prevention campaigns or risk management have been made up to now.

Nowadays 50% of the world population is living in urban areas, a figure which is expected to double in the next 30 years. Recent catastrophic events in Italy and elsewhere have clearly demonstrated that the severity of damage largely depends on how much both high value assets and critical urban infrastructure are affected, either directly or indirectly. It is therefore hoped, especially in this persistent period of political and economic crisis, that governments could understand the importance of investing in non-structural preventive measures devoted to reducing natural hazards. Such an option, among other benefits, allows remarkable economic savings, if compared to the huge amounts of money usually needed in the post-emergency costs, and, above all, promotes a responsible behaviour in the public.

References

Almagià R (1907) Studi geografici sulle frane in Italia (Geographic studies on landslides in Italy). Mem Soc Geograf Ital 13

Almagià R (1910) Studi geografici sulle frane in Italia. L'Appennino centro meridionale (Geographic studies on landslides in Italy. The central and southern Apennines). Mem Soc Geograf Ital 14(2)

Basin Authority of North-western Campania Region (2002) Piano Stralcio per l'Assetto Idrogeologico (Hydro-geomorphological Setting Plan). 4 vol, TPS—SELCA, Naples, Italy (in Italian)

Basin Authority of North-western Campania Region (2010) Piano Stralcio per l'Assetto Idrogeologico (Hydro-geomorphological Setting Plan)—Upgrade year 2010. Basin Authority of North-western Campania Region, Naples, Italy, 2010 (in Italian)

Beneduce P, D'Elia G, Guida M (1988) Morfodinamica dei versanti dell'area flegrea (Campania): erosione in massa ed erosione lineare (Slope morphodynamics of the Phlaegraean area (Campania): mass movements and linear erosion). Mem Soc Geol Ital 41:949–961

Calcaterra D (2010) La pericolosità da frana nei distretti vulcanici e nelle aree peri-vulcaniche della Campania (Landslide hazard in the volcanic districts and in the peri-volcanic areas of Campania region). In Calcaterra D, Morra V (eds) Proc Intern Conf on "Montagne di Fuoco - Rischi e Risorse in aree vulcaniche - Vesuvio ed Etna", Naples - Nicolosi (CT), Italy, 27–28 April 2009, pp 85–108

Calcaterra D, Guarino PM (1999a) Fenomeni franosi recenti nell'area urbana napoletana: il settore centro-orientale (Recent landslides in the urban area of Naples: the middle-eastern sector). In: Proc Conf on "Geologia delle Grandi Aree Urbane", Bologna, Italy, 4–5 November 1997, pp 257–261

Calcaterra D, Guarino PM (1999b) Dinamica morfologica e fenomeni franosi recenti nell'area collinare napoletana (Settore occidentale) (Morphodynamics and recent landslides in the Neapolitan slopes - western sector). Geologia Tecnica Ed Ambientale 2(99):11–17

Calcaterra D, de Riso R, Nave A, Sgambati D (2002) The role of historical information in landslide hazard assessment of urban areas: the case of Naples (Italy). In: Stemberk J, Wagner P (eds) Rybar J. Proc 1st Eur Conf on Landslides, Prague, Czech Republic, 24–26 June 2002. Swets & Zeitlinger, Lisse, pp 129–135

Calcaterra D, de Riso R, Santo A (2003a) Landslide hazard and risk mapping: experiences from Campania, Italy. In: Picarelli L (ed) Proc Intern Conf "Fast Slope Movements: Prediction and Prevention for Risk Mitigation", Naples, Italy, 11–13(1), May 2003. Pàtron Editore, Bologna, pp 63–70

Calcaterra D, Del Prete S, Mele R (2003b) L'influenza dei fenomeni franosi sugli insediamenti costieri del distretto flegreo (Campania, Italia) (The influence of landslides on the coastal settlements of the Phlaegraean district - Campania region, Italy). Proceedings International Conference CITTAM 2003 on "The requalification of Mediterranean coasts among tradition, development and sustainability", Naples, Italy, 26–28 June 2003, 524–534

Calcaterra D, de Riso R, Di Martire D (2004a) Assessing shallow debris slide hazard in the Agnano Plain (Naples, Italy) using SINMAP, a physically based slope-stability model. In: Lacerda WA, Ehrlich M, Fontoura SAB, Sayão ASF (eds) Proceedings 9th International Symposium on Landslides, Rio de Janeiro, Brazil, 28 June–7 July 2004, 177–183, Taylor & Francis Group, London

Calcaterra D, de Riso R, Evangelista A, Nicotera MV, Santo A, Scotto Di Santolo A (2004b) Slope instabilities of the pyroclastic deposits in the Phlegraean district and in the carbonate Apennine (Campania, Italy). In: Picarelli L (eds) Proceedings International Workshop on "Occurrence and Mechanisms of Flows in Natural Slopes and Earthfills", Sorrento, Italy, 14–16 May 2003, 61–75, Pàtron Editore, Bologna

Calcaterra D, de Riso R, Di Martire D (2005) Valutazione della suscettibilità da frana nella Conca di Agnano (Napoli) mediante l'applicazione di un modello su base fisica (SHALSTAB) (Assessment of landslide susceptibility in the Agnano Plain (Naples) by means of a physically-based model (SHALSTAB)). In: Versace P (ed) Proceedings Workshop on "Modelli matematici per la simulazione di Catastrofi Idrogeologiche", Arcavacata di Rende, Italy, 30–31 March 2004, 355–368

Calcaterra D, de Luca Tupputi Schinosa F (2006) The February 28, 1868 Mt. Echia rockfall in the framework of the historical and present-day landslide activity at Naples, Italy. Geophys Res Abstr 8:04708

Calcaterra D, Coppin D, De Vita S, Di Vito MA, Orsi G, Palma B, Parise M (2007a) Slope processes in weathered volcaniclastic deposits within the city of Naples: the Camaldoli Hill case. Geomorphol 87:132–157

Calcaterra D, Parise M, Strumia A, Mazzella E (2007b) Relations between fire, vegetation and landslides in the heavily populated metropolitan area of Naples, Italy. In: Schaefer VR, Schuster RL, Turner KA (eds) Proc 1st North American on Landslides Conf., Vail, Colorado, 3–8 June 2007, A.E.G. Sp Publ 23:1448–1461

Calcaterra D, de Luca Tupputi Schinosa F, Evangelista A, Ruopolo S, Scotto di Santolo A (2010b) Modellazione di frane superficiali da scorrimento-colata ad innesco meteorico nei depositi piroclastici del cratere degli Astroni (Campi Flegrei) (Modeling of rainfall-induced surficial slide-flows in the pyroclastic deposits of the Astroni crater (Phlegrean Fields)). In: Picarelli L, Tommasi P, Urciuoli G, Versace P (eds) Proc 1st Italian Workshop on "Rainfall-induced landslides: mechanisms, monitoring techniques and nowcasting models for early warning systems", Naples, Italy, 8–10 June 2009, 2: 9–19

Catenacci V (1992) Il dissesto idrogeologico e geoambientale in Italia dal dopoguerra al 1990 (The geohydrologic and geo-environmental instability in Italy from the post-war period to 1990). Mem Descr Carta Geol d'Italia, 47, 301 pp

de Riso R, Budetta P, Calcaterra D, Santo A (1999) Le colate rapide in terreni piroclastici del territorio campano (Rapid flows in pyroclastic deposits of the Campania region). In: Peila D (ed) Proc Conf on "Previsione e prevenzione di movimenti franosi rapidi", Trento, Italy, 17–19 June 1999, 133–150, GEAM – Ass Georisorse e Ambiente, Torino

de Riso R, Budetta P, Calcaterra D, Santo A, Del Prete S, De Luca C, Di Crescenzo G, Guarino PM, Mele R, Palma B, Sgambati D (2004) Fenomeni di instabilità dei Monti Lattari e dell'area Flegrea (Campania): scenari di suscettibilità da frana in aree-campione (Landslides in the Lattari Mts. and in the Phlegraean area - Campania region): scenarios of landslide susceptibility in test-areas). Quad Geol Appl 11(1): 5–30

de Riso R, Budetta P, Calcaterra D, Santo A (2007) Riflessioni sul comportamento delle colate rapide non incanalate della Campania, alla luce delle conoscenze pregresse (Discussion on the behaviour of the unchannelled rapid flows of Campania region, on the basis of previous experiences). In: Versace P (ed) Proc Italian Conf on "La mitigazione del rischio da colate di fango a Sarno e negli altri Comuni colpiti dagli eventi del maggio 1998", Naples, Italy, 2–3 May 2005, 81–92

Dell'Erba L (1923) Il tufo giallo napoletano. Studio scientifico-tecnico esteso alle cave e alle frane (The Neapolitan Yellow Tuff. Scientific and technical study extended to quarries and landslides). R. Pironti Editore, Naples, Italy (in Italian)

Di Martire D, De Rosa M, Pesce V, Santangelo MA, Calcaterra D (2012) Landslide hazard and land management in high-density urban areas of Campania region, Italy. Nat Hazards Earth Syst Sci 12:905–926

Emporio Pittoresco (1868) La catastrofe di Pizzofalcone (The Pizzofalcone catastrophe). Vol 5(184):117, Sonzogno Editore, Milan, Italy (in Italian)

Esposito E, Porfido S, Tranfaglia G, Iaccarino G, Esposito G, Braca G (2002) Correlation between pluviometric data and sliding phenomena in Naples, Southern Italy. Proc Convegni Lincei 181:379–386

Giannella G, Guida T (2010) I costi delle frane e delle alluvioni in Italia (The costs of landslides and floods in Italy). In: Consiglio Nazionale dei Geologi (eds) Proc Forum on "Le frane in casa", Rome, Italy, 16 June 2010, available at http://www.geologilazio.it/ordine/forum-rischio-frana.aspx (in Italian)

Gunther RT (1993) Posillipo romana (Roman Posillipo). Electa, Naples, Italy (in Italian)

Hartlen J, Viberg L (1988) General report: evaluation of landslide hazard. In: Bonnard C (ed) Proc 5th Int Symp on Landslides, Lausanne, Switzerland, 10–15 July 1988, 2: 1037–1057

Heim A (1932) Landslides and human lives (Bergstürz and Menchen leben). Translated by N. Skermer, BiTech Publishers, Canada

Institute for Environmental Protection and Research (ISPRA) (2008) Frane in Italia (Landslides in Italy), Special report 2008. Report 83, 32 pp (in Italian)

Migale L, Milone A (1998) Ricerca storica sulle colate di fango in terreni piroclastici della Campania (Historical research on mud flows in Campania's pyroclastic terrains). Rassegna Storica Salernitana 30:235–271 (in Italian)

Ministry for the Public Works—Min LL PP (1965) Frane in Italia (Landslides in Italy). Italy (in Italian), Rome

Morra V, Calcaterra D, Cappelletti P, Colella A, Fedele L, de Gennaro R, Langella A, Mercurio M, de Gennaro M (2010) Urban geology: relationships between geological setting and architectural heritage of the Neapolitan area. In: Beltrando M, Peccerillo A, Mattei M, Conticelli S, Doglioni C (eds) The Geology of Italy. J Virtual Explor 36, paper 26. https://doi.org/10.3809/jvirtex.2010.00261

Municipal Government of the City of Naples (MGCN) (2007) La popolazione di Napoli dal 1951 al 2011 (The population of Naples from 1951 to 2001). SISTAN, Servizio Statistico Nazionale

Palma B, Calcaterra D, Parise M (2009) Modelli geologici e meccanismi di innesco di frane da scorrimento-colata rapida nei depositi vulcanoclastici della Campania (Geological models and triggering mechanisms of slide-flows in the volcaniclastic deposits of Campania region (Italy)). Geoing Ambient Min 46(1):21–48

Parise M, Calcaterra D, de Luca Tupputi Schinosa F, Palma B (2004) Rockfall stability assessment at the western slope of Camaldoli Hill (Naples, Italy). In: Lacerda WA, Ehrlich M, Fontoura SAB, Sayão ASF (eds) Proc 9th Intern Symp on Landslides, Rio de Janeiro, Brazil, 28 June – 7 July 2004, 257-263, Taylor & Francis Group, London

Pellegrino A (1994) I fenomeni franosi nell'area metropolitana napoletana (Landslides in the Province of Naples). In: CIRAM (ed) Proc Symp on "Rischi naturali ed impatto antropico nell'area metropolitana napoletana", Naples, Italy, 7–8 June 1991, Acta Neapolitana 18: 237–256, Guida ed, Naples

Sorriso Valvo M (2005) Landslide risk assessment in Italy. In Glade T, Anderson M, Crozier MJ (eds) Landslide Hazard and Risk. John Wiley and Sons, Chichester

Touring Club Italiano (2001) Napoli e dintorni (Naples and its surroundings). Touring Club Italiano, Milano

Varnes DJ (1978) Slope movement types and processes. In: Schuster RL, Krizek RS (eds), Landslides: analysis and control. US Nat Acad Sci, Transp Res Board Sp Rep 176:11–33

Building Stones and Technological Materials of the Campi Flegrei Caldera, Italy

Alessio Langella, Piergiulio Cappelletti, Mariano Mercurio, Domenico Calcaterra, and Maurizio de Gennaro

Abstract

This chapter highlights how, lacking high value rocks in the Campanian area (South Italy), those produced by the activity of the Campi Flegrei volcanic field, have represented a fundamental resource that has made possible to carry out important architectural works. These works can be admired in the numerous archaeological sites, hamlets and most notably throughout the city of Naples, whose Ancient Centre is a splendid example of a stratified city. Phlegraean Tuffs (i.e., Neapolitan Yellow Tuff, various facies of the Campanian Ignimbrite) and Lavas are the "stones" that mostly have marked the architectural history of the city. In some cases, thanks to the mastery of the *lapicidi* (Italian for stone craftsmen), these stones have been worked to obtain works of art of priceless value. The use of imported stones (marbles, granites, etc.) solely for particular architectural elements (i.e., church and building portals, columns) further corroborates the aforementioned. When the Phlegraean rocks were used for purely structural functions, they were covered with plaster made with Phlegraean sands and aggregates, or, less frequently, with stone slabs from other exploitation sites in the Campania region. However, these rocks have had an important role not only in the building industry. As demonstrated by numerous researches carried out in over sixty years, the Neapolitan Yellow Tuff and the yellow facies of the Campanian Ignimbrite with a high content of zeolites (phillipsite and chabazite), can be used, exploiting the peculiar properties of these minerals, in numerous technological sectors, in particular in those with environmental implications.

1 Introduction

A city and its environs, a people and its culture and art, are inextricably linked to the georesources present in their territory. Man has always turned to these resources to draw sustenance and uses them for his necessities and development. Great efforts are therefore required in the search for and the use of materials for the realisation of both everyday objects and engineering works, from the most modest constructions to architectural works of great value. This represents the close link between the geology of a territory and

A. Langella (✉) · P. Cappelletti · D. Calcaterra
Dipartimento di Scienze della Terra, dell'Ambiente e delle Risorse, Università degli Studi di Napoli Federico II, Napoli, Italy
e-mail: alessio.langella@unina.it

M. Mercurio
Dipartimento di Scienze e Tecnologie, Università degli Studi del Sannio, Benevento, Italy

M. de Gennaro
Olim Full Professor in Georesources, University of Naples Federico II, Naples, Italy

© Springer-Verlag GmbH Germany, part of Springer Nature 2022
G. Orsi et al. (eds.), *Campi Flegrei*, Active Volcanoes of the World,
https://doi.org/10.1007/978-3-642-37060-1_14

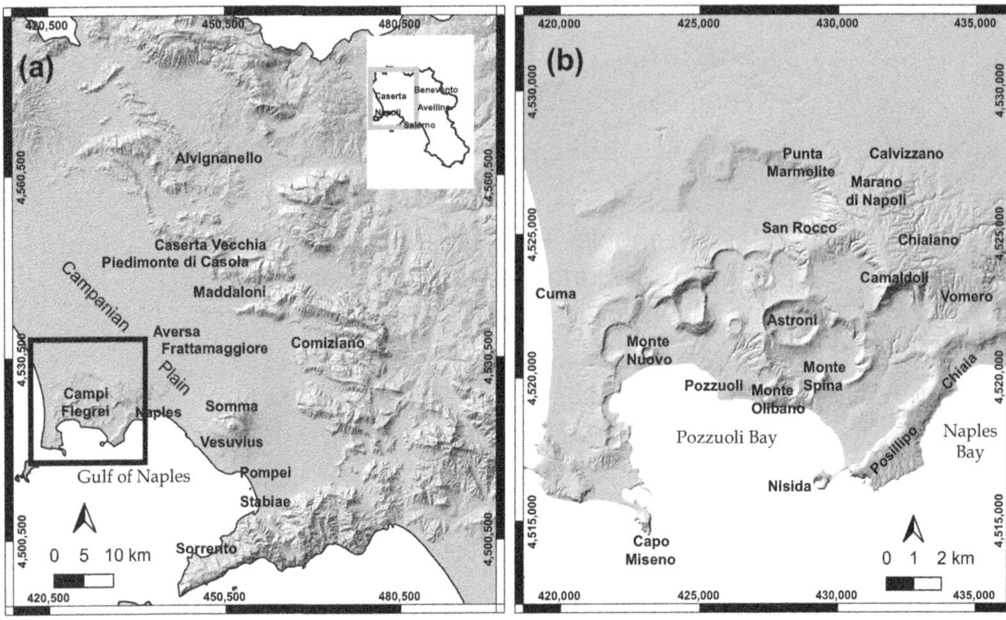

Fig. 1 **a** Schematic map of the investigated area. The white inset highlights this area within the Campania region. **b** Enlarged vision of Campi Flegrei (black frame in **a**). Toponyms of all the localities cited in this chapter are also reported

the constructions that over the centuries have been erected upon it. A good example of such a link is given by the Neapolitan area in South Italy (Fig. 1) and its volcanic rocks erupted by the Campi Flegrei caldera (CFc), the major structure of the Campi Flegrei volcanic field (CFvf). For a synthesis of the volcanic and deformation history of this field see Chap. Volcanic and Deformation History of the Campi Flegrei Volcanic Field, Italy. The urban centres, in particular the city of Naples, that over the course of history have risen up in the CFvf, have reached heights of great splendour as well as witnessed periods of decline and abandonment (Chap. The Urban Development of Campi Flegrei, Italy).

The CFvf has generated mostly explosive eruptions, including the two catastrophic caldera-forming events of the Campanian Ignimbrite (CI) and Neapolitan Yellow Tuff (NYT). Therefore, the rocks are predominantly volcanoclastites. The eruptions were fed by magmas ranging in composition from shoshonite to trachyte and evolving in a complex magmatic system (see Chaps. An Evolutionary Model for the Magmatic System of the Campi Flegrei Volcanic Field (Italy) Constrained by Petrochemical Data; Origin and Differentiation History of the Magmatic System Feeding the Campi Flegrei Volcanic Field (Italy) Constrained by Radiogenic and Stable Isotope Data; Tephrochronology and Geochemistry of Tephra from the Campi Flegrei Volcanic Field, Italy; Rheological Properties of the Magmas Feeding the Campi Flegrei Caldera (Italy) and Their Influence on Mixing Processes; Magma Chamber Dynamics at the Campi Flegrei Caldera, Italy). The CI is a complex sequence of variably consolidated pyroclastic rocks, including high dilute pyroclastic density current (PDC) deposits in distal areas, and a lithic breccia known as Museum Breccia (Johnston-Lavis 1888) and a succession in proximal areas of layers of eutaxitic tuff named Piperno (PI) (De Lorenzo 1904) (see Chap. Volcanic and Deformation History of the Campi Flegrei Volcanic Field, Italy). The deposits of the PDCs can occur in two variable facies according to the post-depositional lithification processes. One facies is characterised by a yellow colour while the other is grey. The yellow one is called Lithified Yellow Tuff (LYT) and the grey one

Welded Grey Ignimbrite (WGI) (Cappelletti et al. 2003; Langella et al. 2013b). The NYT includes a basal sequence of phreatoplinian and magmatic layers surmounted by deposits of variably diluted PDCs (see Chap. Volcanic and Deformation History of the Campi Flegrei Volcanic Field, Italy). Almost all the other explosive eruptions have generated sequences of loose pyroclastic deposits laid down by mostly dilute PDCs and particles fallout (see Chap. Volcanic and Deformation History of the Campi Flegrei Volcanic Field, Italy). All these deposits are known with the comprehensive Italian name of *pozzolana* (Scherillo 1955). These rocks have been important georesources over the course of history. Furthermore, due to their specific features, have been widely used not only in architecture but also in technological applications (Colella 1996; de Gennaro and Langella 1996; Pansini 1996; de Gennaro et al. 2000a, 2003, 2009; Colella et al. 2001; Coppola et al. 2003; Mercurio et al. 2010; Morra et al. 2010; Cappelletti et al. 2011). The volcanoclastites, such as PI, WGI and LYT facies of the CI, NYT as well as the *pozzolana*, for several reasons, such as their wide areal distribution, easy workability and good or discreet physico-mechanical characteristics, have been the subject of a very intense exploitation activity and utilisation, not necessarily local. This activity is one of the most important anthropogenic factors predisposing and even triggering slope instability (Chap. Landslide Hazard and Risk in the Campi Flegrei Caldera, Italy). The *pozzolana* is also ascribable to the category of technological materials due to its specific use in the building industry. The lavas (Langella et al. 2009), also on account of the limited occurrence, have had less fortune and have been mainly used whenever materials with better physico-mechanical properties were required. Lavas, PI and WGI are to be considered materials of a typically architectural interest. The NYT and the LYT, further than important building materials, have recently become of high technological interest. This is because the particular mineralogical composition permits their use in various technological sectors.

2 Ornamental Stones in Building Constructions

2.1 Lithified Volcanoclastites

The Phlegraean tuffs (NYT and CI, in both LYT and WGI facies) are macroporous and light rocks whose lithification process is the result of secondary minerogenetic processes. They have been widely used as building material in historic constructions due to their qualitative and quantitative characteristics attributable to their genesis and mineralogical composition.

de Gennaro et al. (2000b) suggested that the formation of the zeolites in the NYT occurred almost immediately after the eruption, in a sort of thermally insulated system. Physico-chemical conditions favourable to the zeolitisation process occurred immediately after the emplacement of the deposit and can be summarised as follows: (1) enough water of phreatomagmatic origin to enhance the hydration-dissolution processes of the volcanic glass shards; (2) emplacement temperatures at least close to that of water vapour condensation; (3) fine grained volcaniclastic components and highly reactive alkali-trachytic glass that quickly buffered the acid pH of the system; (4) considerable thickness of the deposit, favouring natural thermal insulation and lengthy persistence of high temperatures. These considerations demonstrate that the zeolitisation of the NYT required enough water to trigger secondary minerogenetic processes. The emplacement conditions and the consequent variable thermal dispersion patterns played a fundamental role, as testified by the vertical and lateral variations, in the lithification grade. Furthermore, the authors correlated the reduction in the lithification grade towards the base and top of the sequence to heat loss that could have affected the minerogenetic process, cooling off those parts of the deposit very quickly and thus preventing their zeolitisation. In fact, in distal areas, the NYT is less than 20–30 m thick, and never lithified. This evidence suggested the possible existence of a "thickness threshold value", below which thermal dispersion does not allow the full development of a zeolitisation process.

Unlike the NYT, the secondary minerogenetic processes affecting the CI are definitely very complex for a variety of reasons. With regards to the lithological facies, the huge PDCs deposits can be divided into: the welded and highly feldspathised (pyrogenic and authigenic feldspars) WGI facies and the highly zeolitised LYT facies, characterised by a large stratigraphic variability of the zeolite species (phillipsite, chabazite and analcime) (Cappelletti et al. 2003; Langella et al. 2013b). Detailed volcanological, chemical and mineralogical analyses of several stratigraphic sections allowed Langella et al. (2013b) to formulate a genetic model demonstrating that once again open and/or closed hydrologic systems (Sheppard and Hay 2001; Hay and Sheppard 2001) are not always strictly pertinent to the alteration of volcanic flow units. Temperature was the key factor in the entire process. High emplacement temperatures, mainly in the lower portions of the deposit ($\sim$600 °C; Incoronato and Di Girolamo, unpublished data), caused the sudden welding of the central part of the WGI and led to the formation (by devitrification of the glass shards) of authigenic feldspars. Physical features in the lower and upper portions of the WGI are quite different from those recorded in the welded portion, including the presence of uncollapsed scoriae. The WGI gently fades upwards into the LYT facies, defining a transition zone, where temperatures, although still high, were insufficient to allow feldspathisation processes. In addition, the amount of water of volcanic origin was inadequate to trigger new mineral-formation processes. The zeolitisation process leading to the lithification of the LYT only began after the infiltration of percolating meteoric waters. Subsequent glass hydrolysis caused smectite formation, followed by the crystallisation of phillipsite, chabazite, and analcime. The association of chabazite and analcime in the lower part of the deposit supports the hypothesis of a process governed by persistent high temperatures, but temperature cannot be considered the only element driving the process. In fact, some other factor(s) prevailing over temperature could explain the predominance of one of these two zeolites (phillipsite and chabazite), such as the composition of the interacting solutions. In particular, Na^+, K^+, and Ca^{2+} and their ratios are crucial in determining the growth of a specific zeolite (phillipsite, chabazite, and analcime). Waters interacting with fresh glass determine a subsequent enrichment in K^+ and Na^+ and an increase in the pH defining the condition suitable for the crystallisation of phillipsite (de Gennaro et al. 1988, 1992, 1993, 1999; de Gennaro and Colella 1991). The removal of K^+ and Na^+ via phillipsite precipitation indirectly caused a relative Ca^{2+} enrichment, thereby creating conditions suitable for chabazite crystallisation (Langella et al. 2013b).

The welding process of the PI was not different from that of the WGI. The high emplacement temperature, further than favouring the welding of the glassy component also determined an almost complete feldspathisation with sanidine content often reaching 99 wt% (Calcaterra et al. 2000).

2.2 Exploitation and Use

The areal diffusion of the NYT, that can be defined as the backbone of the city of Naples, has determined its use for more than two millennia. Mining was predominantly carried out underground (Fig. 2a), utilising very simple yet still functional tools. The consequence of this activity is the presence of a large and diffused network of quarries, cisterns and hypogea aqueducts that represent the so-called Underground Naples, a fascinating network of itineraries (Fig. 2b) (Calcaterra et al. 2018). The underground tuff quarries were developed following two mining methods: the gallery type and the bottle type.

Quarries dug by the gallery-type mining method had the so-called "flat sky" faces with almost parabolic transversal sections and downward concavity (Aveta 1987). In this case, the attack phase of the excavation began from the summit of a section previously traced on the front, proceeding downwards right down to the prefixed quota that generally coincided with the exploitation ground level. It was quite common

Fig. 2 **a** Trapezoidal section of a Neapolitan Yellow Tuff exploitation gallery intersected by a later open pit quarry (Savanelli quarry, Marano—Naples) (photo by P. Cappelletti); **b** central portion of the Bourbon Tunnel (Naples, from via Solaria to via Morelli) (photo by G. Minin) **c** Remains of the old Campania Ignimbrite exploitation gallery at Piedimonte di Casola (Caserta), intersected by the later open pit quarry, to extract the tuff used for the construction of the Medieval hamlet of Casertavecchia (photo by A. Langella)

to dig parallel galleries (principal galleries) connected to each other via transversal excavations thus creating a dense network of tunnels. This system provided two advantages: the possibility to excavate as much material as possible while still maintaining the stability of the excavation due to the presence of tuff portions that thus represented the supporting pillars of the gallery. A rare example of the gallery-type extraction is that of the Piedimonte di Casola (Caserta) quarry (Fig. 2c) where some of the stones used for the construction of the Medieval Hamlet of Casertavecchia originated.

The bottle-type mining method (Fig. 3a), widely used for a long period up until the late fifties of last century, envisaged the realisation of a pit, the so-called *occhio del monte* (Italian for "eye of the mountain"), necessary to overcome the incoherent deposits overlying the tuff bank. The latter was manually perforated for a few meters maintaining a reduced section of the hole in order to give greater stability to the roof of the excavation area. Then the tuff was excavated realising a cavity that, as it proceeded downwards, opened out into a bell shape. This technique permitted to get the tuff blocks directly where they had to be used for building construction. This activity generally continued as long as the need for material persisted or up until it was no longer possible to extract the blocks. At

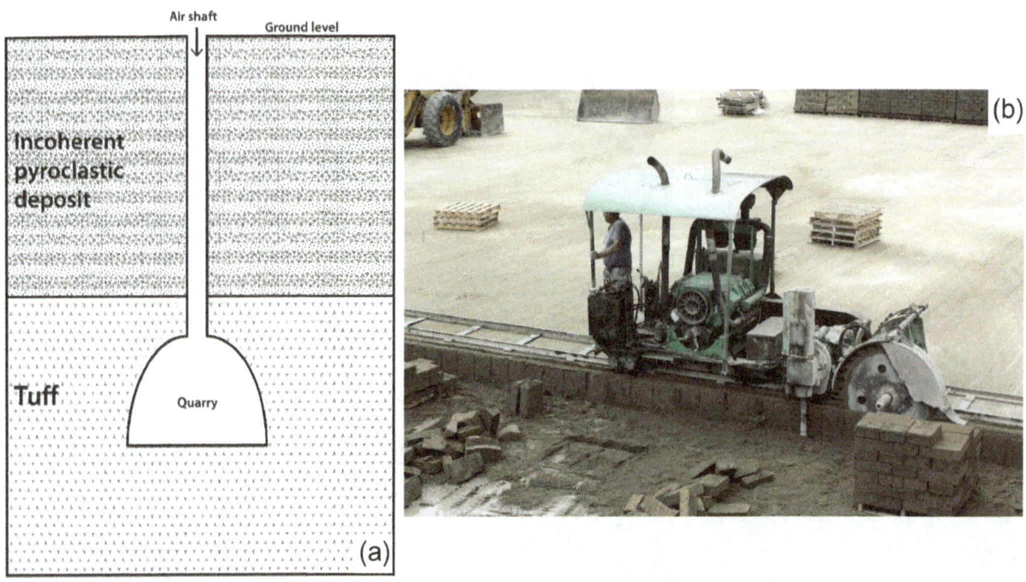

Fig. 3 **a** Sketch of an underground quarry with a vertical access well (modified after Galante 1971); **b** sewing machinery specialised for tuff dimension stones (Apostolico and Tanagro quarry, Comiziano, Naples) (photo by A. Langella)

the end of the mining work, either the hole was plugged and the cavity abandoned, or, more often, the cavity was used as a cistern for rainwater by connecting it to the roofs of the building through drainpipes. A particular significant case was brought to light during the restoration works on the eighteenth-century *Villa Floridiana* located on the Vomero Hill (Fig. 1b), undertaken in the early nineties of last century. On this occasion, workers revealed, in the subsoil of the Villa, the cavity originating as a consequence of tuff exploitation as well as important underlying service infrastructures. Actually, rainwater was channelled within the cavity that, passing via a series of decantation tanks, located at different depths, reached the final collection tank. This permitted having the necessary water available for the various needs of the household.

The improper and rash use of the tuff extracted from the subsoil of the city produced notable problems of stability. This induced Ferdinand IV King of Naples to promulgate in 1781 a building code and professional law that prohibited the excavation of tuff, *pozzolana* and lapilli in caverns within the urban perimeter and in the proximity of the access roads to Naples (Cardone and Papa 1993). Since then, and in particular from the end of the nineteenth and the beginning of the twentieth century, numerous open-pit quarries were dug at the foot of the tuffaceous hills adjoining the city, such as at Posillipo and Chiaia (Fig. 1b). The same type of quarries was excavated also northwest of Naples, such as those at San Rocco, Chiaiano, Marano and Calvizzano, and even in more distant areas (Fig. 1b). The high demand of construction materials along with the concomitant huge urban expansion of the city at the end of the nineteenth century, resulted in a flourishing mining activity in the Phlegraean area that in the early decades of the twentieth century numbered 80 quarries (Dell'Erba 1923). Starting from the sixties of the last century, the production of tuff blocks increased notably and assumed industrial proportions thanks to the introduction of mechanised mining. An open-pit quarry envisages as a first phase the removal of the loose pyroclastic cover that permits reaching the tuff bank. The cutting machine is equipped with a large vertically pronged wheel that cuts the quarry plane in parallel furrows ("lining" according to Colella et al. 2001) 40 cm deep at a constant distance of 25 cm. A second orthogonal furrow cuts the quarry plane at a standard distance of 20 cm along with a second circular horizontal cutting saw that isolates the block whose final dimension is standardised and measures 40 × 25 × 20 cm (Fig. 3b).

The exploitation activity for the CI slightly differs from the one previously described for the NYT. The great majority of the quarries of the last century within the Campanian Plain are pit quarries, while those on the surrounding carbonate ridges of the Apennines are slope quarries. The areal development of this material in flat zones resulted in a prevalent underground exploitation (bottle-type excavation method). In numerous town (e.g., Frattamaggiore, Comiziano, Alvignanello; Fig. 1), this exploitation technique has resulted in a dense network of underground tunnels, often forgotten. The presence of these tunnels has contributed to catastrophic events, especially in the cases in which, due to an intense urban expansion, the overlying territory has become the site of a high concentration of buildings. One of the most emblematic cases of this situation comes from the Municipality of Frattamaggiore documented in a monograph by the engineer in charge of the Municipality's Technical Office (Galante 1971).

The destination of the historic use of these tuffs is certainly the building construction activity where tuff was often used "nude" without any plaster covering, also in order to exploit its good qualitative characteristics, such as colour and design. The NYT is undoubtedly the most used building stone in all the Phlegraean area and mostly in the city of Naples, in both great architectural works and housing construction.

One needs only to remember the walls in *opus reticulatum* or *incertum* of the numerous archaeological sites from Cuma to Pompeii to Stabiae, the castles of Naples (Fig. 4a), the gothic Churches (Fig. 4b) and, more recently, the nineteenth century Academy of the Fine Arts (Fig. 4c). Despite being constituted in tuff, architectural works of lesser value in almost all cases were covered in plaster. A use, which most certainly is not suitable for a somewhat fragile stone, is that of the paving of some open areas of the city (the Public Garden, *Virgiliano* Park and *Tecchio* Square) (Fig. 4d).

The main use of the CI, mostly its WGI facies, is found in the areas of Caserta and Salerno where important works were realised in stone facing. One of the maximum expressions of this art is certainly the Medieval Hamlet of Casertavecchia (Fig. 1a) with its *San Michele Arcangelo* Cathedral (Fig. 5a), the bell tower and the Castle. By contrast, the "Mastio" (tower) of the Castle was realised with the LYT facies. Although hard to cite all the examples, it is sufficient to mention just a few, such as the impressive hydraulic work realised by the architect Luigi Vanvitelli (1700–1773) (*Ponti della Valle*, *Carolino* Aqueduct) by using the WGI facies and bricks (Fig. 5b), the *S. Maria delle Grazie* Church, Vaccheria, Caserta (early 1800) (Fig. 5c) and the Belltower of *S. Paolo* Cathedral (Aversa, tenth century) (Fig. 5d). In the historic buildings of Naples, the WGI has been used for architectural elements mainly to replace PI.

Differently from the Phlegraean tuffs, the exploitation of PI, occurring at the base of the Camaldoli Hill (Fig. 1b), presented severe difficulties such as a thick sequence of overlying volcanoclastites which required an underground mining activity. The exploitation presumably followed an "abandoned pillar" geometry, whose dimensions were probably defined based on the quarrymen's ability. The distribution of the pillars and their shape was irregular in each part of the hypogeum, perhaps due to the elevated resistance of the rock that obliged the quarrymen to follow the main discontinuities that crossed the deposit (Fig. 6). The total surface of these pillars did not reach 14% of the entire underground area, thus confirming the low safety margins that characterised the mining activity. Such a situation caused, and still today does, a continuous static stress not only on the pillars but also on some walls of the quarry (Calcaterra et al. 2007).

With regards to the presumed mining techniques, numerous marks left by the tools of the quarrymen of the time, seem to suggest that PI was mined using methods analogous to those used for the NYT. The blocks were shaped and moulded into the required dimensions directly in

Fig. 4 The Neapolitan Yellow Tuff in the historical buildings of Naples. **a** *Castel dell'Ovo* (photo by A. Langella); **b** *Santa Chiara* Cathedral: pronaos and main *façade* (photo by A. Langella); **c** *Accademia delle Belle Arti*, in English Academy of Fine Arts (photo by A. Langella); **d** flooring in tuff blocks of the Public Garden—Naples (photo by M. de Gennaro)

the quarry; the final shape and dimension were conferred to them by the *pipernieri* (Italian name for craftsmen working *Piperno* stone) according to the final use.

The mining difficulties and the scarcity of the produced material, along with the high demand resulted in not always sustainable costs. This led to the search for substitute materials such as the WGI and successively the *pietrarsa*, namely the lavas from Vesuvio.

The ability of the Neapolitan *lapicidi* along with the good properties of PI suitable to be cut and shaped, enhanced the use of this rock to decorate entire walls of notable architectural works. Four of the five towers as well as the southern defensive wall of the *Castel Nuovo* (New Castle) or, as it is more commonly known, the *Maschio Angioino* (twelfth–fifteenth century) (Fig. 7a) were constructed in squared blocks of PI, while the tower known as "San Giorgio" and the remaining defensive walls are in NYT. The *Filangieri* Museum (*Como* Palace, fifteenth–nineteenth century) (Fig. 7b), whose *façade* is distinguished by its smooth skirting, its first floor with rustication in the Florentine style and the second floor with a smooth rustication. Noteworthy for the simplicity of its style is the *Pontano* Chapel (Fig. 7c), built at the close of the fifteenth century and commissioned by the humanist Giovanni Pontano (1429–1503), one of the founders of the Neapolitan Academy; the diamond rustication of the *façade* of the *Gesù Nuovo* Church (sixteenth century; Fig. 7d), previously *Sanseverino* Palace; the moulded Portal of the *Tocco di Montemiletto* Palace (sixteenth century; Fig. 7e, f). Finally, worth of mention is the *Gravina* Palace (sixteenth–eighteenth century) where the PI constitutes the base, the cushion rustication up to the first floor and all the covering in smooth slabs of the superior sections.

Fig. 5 Campanian Ignimbrite in the architecture of the Campania Region. **a** Cathedral of *San Michele Arcangelo* and bell tower in Casertavecchia; **b** *Carolino* Aqueduct designed by Vanvitelli (Valle di Maddaloni, Caserta); **c** *Santa Maria delle Grazie* Church at Vaccheria - San Leucio (Caserta); **d** Bell tower of *San Paolo* Cathedral (Aversa, Caserta). Photos by A. Langella

2.3 Petrophysical Properties

The NYT is characterised by an elevated textural heterogeneity (presence of pumice of variable size, scoriae, crystal fragments, etc.) that influences its petrophysical characteristics. A large set of data permits defining the lithotype NYT as a macroporous light and soft rock (Zezza et al. 1985; Primavori 1999; Colella et al. 2017). Table 1 reports the main petrophysical features of the NYT measured on several rock samples collected in the most important and large outcrops of the Neapolitan area.

A similar situation is depicted for both WGI and LYT (Tables 2 and 3). It should be noted in particular that, notwithstanding the genetic and

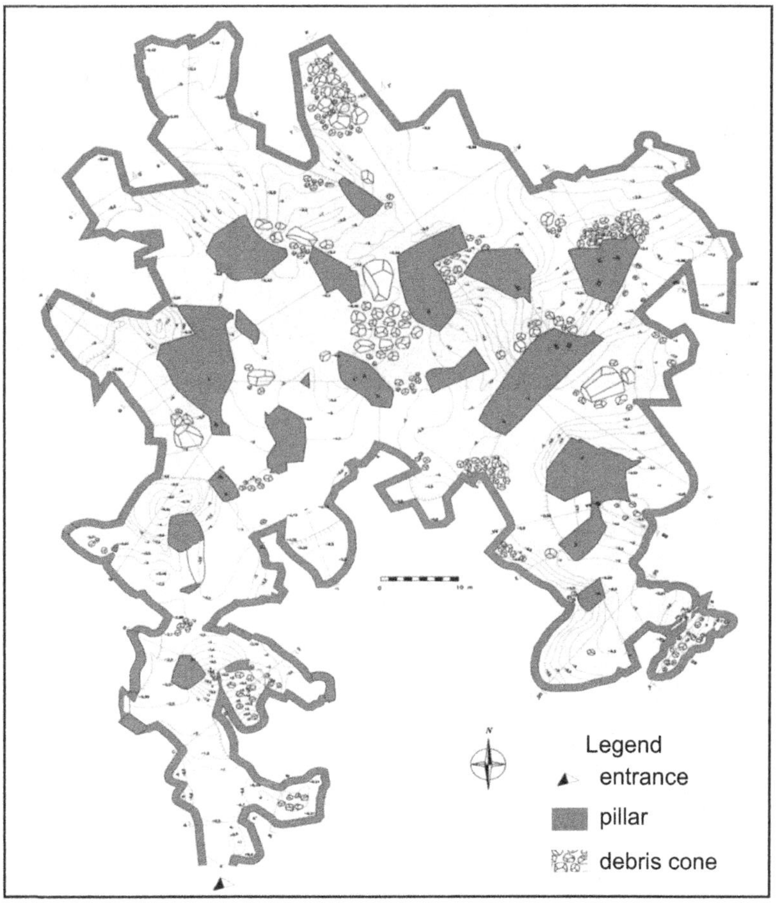

Fig. 6 Sketch map of the Piperno quarry in Masseria del Monte, Naples. Figure modified after Calcaterra et al. (2007)

compositional differences described in the previous paragraph, the two stones do not show significant variations, with the only exception of the thermal dilation coefficient that turns to be negative for the LYT and positive for the WGI. The negative anomaly of the LYT is probably related to the peculiar behaviour of the zeolites that dehydrate upon heating, thus defining the specimen contraction.

Notwithstanding NYT was the most utilised building stone in the architecture of the city of Naples and the majority of its Province, undoubtedly PI was the stone most used for the realisation of architectural works of notable value. This aspect is evidenced in the colonnades of the numerous cloisters of the Ancient Centre of Naples, where its aesthetic aspect and functional structure are emphasised, as well as the frequent decorations of basal sections or entire walls. In addition, the sculptured stone was also utilised for particular decorations (Fig. 7f). The use of PI as an ornamental stone is closely linked to its eutaxitic fabric characterised by black flattened scoriae (*fiammae*) set in a hard and light grey matrix that determines a particular and pleasant design when the stone is cut in the direction of the *contro* (that is to say parallel to the minor axis of the scoriae).

The size of the scoriae is variable, sometimes reaching 30–40 cm and an average flattening ratio of 1:10. The use of PI not only as an architectural material but also with a structural function arises from the fair or good petrophysical properties of the stone as highlighted by the data reported in Table 4.

Fig. 7 Piperno in the monumental architecture of Naples. **a** Southern *façade* of *Castel Nuovo* (known as *Maschio Angioino*); **b** *Como* Palace, today *Filangieri* Museum; **c** *Pontano* Chapel; **d** *Gesù Nuovo* Church: main *façade* characterised by the diamond ashlar; **e** Portal of *Tocco di Montemiletto* Palace; **f** Detail of the moulded portal. Photos by A. Langella

Table 1 Petrophysical parameters of the Neapolitan Yellow Tuff

	Unit	Samples number	Mean	Min	Max	Standard deviation
Dry density	(kN/m^3)	**181**	**10.80**	**8.14**	**14.65**	**1.27**
Specific gravity	(kN/m^3)	**181**	**22.74**	**21.59**	**24.60**	**0.79**
Compactness		**181**	**0.47**			
Capillary absorption	(g/cm^2 s$^{1/2}$)	**30**	**0.042**	**0.022**	**0.066**	**0.01**
Imbibition coefficient	(%)	**73**	**39.37**	**23.20**	**47.95**	**5.88**
Open porosity	(%)	102	51.81	39.50	63.28	6.62
Mean pore size	(μm)	13	5.27	0.03	16.08	6.20
Specific surface	(m^2/g)	13	16.93	13.02	21.54	2.39
Pore total volume	(cm^3/g)	13	0.35	0.27	0.45	0.06
Compressive strength	(MPa)	**93**	**5.99**	**1.20**	**38.83**	**6.44**
Thermal dilation coefficient	(°C^{-1})	**6**	**−17.65**	**−35.50**	**−4.67**	**8.02**
Volumetric deformation	(%)	26	0.54	0.23	1.15	0.20
Dry ultrasonic velocity	(m/s)	**265**	**1,760**	**1,373**	**2,617**	**220.29**
Saturation ultrasonic velocity	(m/s)	208	1,720	1,382	2,233	**148.02**

Values in bold from Colella et al. (2017); other data from Colella et al. (2013)

Table 2 Petrophysical parameters of the Welded Grey Ignimbrite facies of the Campanian Ignimbrite

		Samples number	Mean	Min	Max	Standard deviation
Dry density	(kN/m^3)	93	11.42	9.66	14.80	0.71
Specific gravity	(kN/m^3)	93	25.53	22.26	26.30	0.45
Compactness		93	0.45			
Capillary absorption	(g/cm^2 s$^{1/2}$)	4	0.165	0.160	0.169	0.003
Imbibition coefficient	(%)	16	41.61	38.17	52.28	3.38
Open porosity	(%)	93	55.24	42.06	62.79	0.03
Mean pore size	(μm)	3	7.62	6.84	8.26	0.59
Specific surface	(m^2/g)	3	1.57	0.64	2.95	1.00
Pore total volume	(cm^3/g)	3	0.48	0.47	0.48	0.00
Compressive strength	(MPa)	17	5.23	1.07	11.68	2.80
Thermal dilation coefficient	(°C^{-1})	3	0.78	0.67	1.00	
Dry ultrasonic velocity	(m/s)	89	1,691	1,221	2,379	297.92
Saturation ultrasonic velocity	(m/s)	49	1,491	1,004	2,153	380.25

Data from Langella et al. (2013a)

Table 3 Petrophysical parameters of the Lithified Yellow Tuff facies of the Campanian Ignimbrite

		Samples number	Mean	Min	Max	Standard deviation
Dry density	(kN/m^3)	40	10.97	9.85	12.10	0.52
Specific gravity	(kN/m^3)	40	22.68	21.51	23.13	0.29
Compactness		40	0.48			
Capillary absorption	(g/cm^2 s$^{1/2}$)	6	0.014	0.012	0.015	0.001
Imbibition coefficient	(%)	10	34.73	28.49	40.43	5.24
Open porosity	(%)	40	51.61	46.85	57.20	0.02
Mean pore size	(μm)	3	0.66	0.56	0.79	0.09
Specific surface	(m^2/g)	3	8.91	8.07	9.60	0.63
Pore total volume	(cm^3/g)	3	0.36	0.33	0.41	0.04
Compressive strength	(MPa)	13	6.45	4.11	8.02	1.01
Thermal dilation coefficient	(°C^{-1})	3	−25.4	−21.5	−28.50	
Volumetric deformation	(%)	9	0.49	0.3	1.01	0.21
Dry ultrasonic velocity	(m/s)	64	1,777	1,672	2,204	112.63
Saturation ultrasonic velocity	(m/s)	33	1,621	1,536	2,092	89.11

Data from Langella et al. (2013a)

Table 4 Petrophysical parameters of Piperno

		Samples number	Mean	Min	Max	Standard deviation
Dry density	(kN/m^3)	98	17.33	12.95	22.59	2.60
Specific gravity	(kN/m^3)	98	25.71	25.18	26.12	0.20
Capillary absorption	(gr/cm^2 s$^{1/2}$)	35	0.0246	0.0063	0.0434	0.0124
Imbibition coefficient	(%)	28	18.80	8.88	27.77	5.92
Water vapor permeability	(gr/m^2 24 h)	18	117.03	86.90	139.90	12.71
Open porosity	(%)	98	32.81	12.03	49.90	10.46
Mean pore size	(μm)	5	4.40	0.36	7.43	2.33
Specific surface	(m^2/g)	5	6.61	3.92	10.52	2.27
Pore total volume	(cm^3/g)	5	0.16	0.05	0.27	0.08
Compressive strength	(MPa)	42	20.95	4.75	67.47	13.84
Point load	(MPa)	46	1.31	0.28	7.43	1.23
Tangent elasticity module (50%)	(GPa)	24	2.54	0.86	6.49	1.47
Dry ultrasonic velocity	(m/s)	76	2,694	2,229	3,239	267.9
Saturation ultrasonic velocity	(m/s)	76	2,891	2,314	3,997	384.1

Data from Calcaterra et al. (2013)

2.4 Lava Stones

The Phlegraean lavas, despite being rocks with good technical qualities, were not widely used in historical architecture due to the exiguity of the deposits and the consequent scarcity of the resource. Indeed, as is well-known, the activity of the CFc (Orsi et al. 1996, 1999; Di Vito et al. 1999) has been characterised by rare effusive events that produced small volume lava flows and domes, presently exposed within the Astroni Crater, and at Monte Spina, Cuma, Punta Marmolite and Monte Olibano (Fig. 1). This latter (Fig. 8) is undoubtedly the most important deposit and was the site of an important mining activity, favoured by the ease of transport by sea (Cardone and Papa 1993). The mining activity of Phlegraean lavas has always been carried out in open pits.

All the lava samples show a porphyritic texture and a fluidal texture groundmass. The main phenocrysts are alkaline feldspar (sanidine) and subordinate clinopyroxene. Microphenocrysts are mainly constituted by magnetite and highly-zoned sodic plagioclase; rare olivine, apatite and biotite crystals also occur. The groundmass is prevailingly composed of alkaline feldspar, diopsidic pyroxene, brown biotite, magnetite and very rare plagioclase (Langella et al. 2009). Minerals present in decreasing order of abundance are: sanidine (more than 80 vol%), sodic plagioclase (6.9–4.5 vol%) and clinopyroxene (5.7–5.4 vol%); biotite, magnetite and amphibole are also present in subordinate amount. From a chemical point of view, no substantial differences were recorded between lavas from different deposits, all classified as trachytes.

Table 5 reports the main petrophysical parameters of the lavas of Monte Olibano Dome that are characterised by an open porosity close to 10% and maximum values of imbibition coefficient and capillary absorption perfectly concordant with those of the porosity. An evident variability can be noted in the compressive strength values, ranging between 138 and 208 MPa, that, according to the classification of Deere and Miller (1966), permit defining the rock as strong to very strong and stiff.

In old times, the lavas were mainly used in road paving for which purpose were cut in polygonal blocks with flat surfaces. However, lavas were also used in architecture and civil constructions as demonstrated by some structures in the Roman *Flavio* amphitheatre, in Pozzuoli.

Undoubtedly, the use of the Phlegraean lavas in the Neapolitan architecture saw its highest

Fig. 8 View of the Phlegraean lava quarry (Cava Regia) at Monte Olibano, Pozzuoli. Photo by M. d'Albora

Table 5 Petrophysical parameters of Phlegraean lavas

		Samples number	Mean	Min	Max	Standard deviation
Dry density	(kN/m^3)	11	23.83	23.05	24.52	2.60
Specific gravity	(kN/m^3)	11	26.39	26.06	26.58	0.20
Compactness		12	0.90	0.88	0.92	0.02
Capillary absorption	(g/cm^2 s$^{1/2}$)	9	1.6E^{-04}	1.1E^{-04}	2.3E^{-04}	0.5
Imbibition coefficient	(%)	9	1.84	1.43	2.51	0.83
Open porosity	(%)	11	9.82	7.67	11.39	1.57
Mean pore size	(μm)	9	0.25	0.005	0.69	0.31
Specific surface	(m^2/g)	9	6.61	3.92	10.52	2.27
Pore total volume	(cm^3/g)	9	0.16	0.05	0.27	0.08
Compressive strength	(MPa)	5	161	138	208	33
Tangent elasticity module (50%)	(GPa)	3	53.85	51.54	56.80	2.20
Dry ultrasonic velocity	(m/s)	8	4,398	4,255	4,494	103
Saturation ultrasonic velocity	(m/s)	8	4,735	4,580	4,927	143

Modified after Langella et al. (2009)

Fig. 9 The Phlegraean trachytic lava in the monumental architecture of Naples. **a** Colonnade of the *Santa Chiara* Monastery cloister (photo by M. d'Albora); **b** colonnade of *San Francesco di Paola* Church. (photo by A. Langella)

peak between the thirteenth and fourteenth century. Worth to be mentioned are the churches of *San Lorenzo Maggiore*, *San Domenico Maggiore*, *Santa Chiara* (Fig. 9a), *Santa Maria Donnaregina*, *San Giovanni a Carbonara*, and the central colonnade of *San Francesco di Paola* (Fig. 9b), the bases of the triumphant arch and the windows of the Barons' hall in the *Maschio Angioino* and, more recently, the *Palazzo del Banco di Napoli* at Spirito Santo (Penta 1935; Cardone and Papa 1993). The use of these lavas continued in the successive centuries. However, it witnessed a progressive decline, at least as a cutting stone, between the seventeenth century and the end of the nineteenth century, as PI became the most favoured choice (Aveta 1987; Cardone and Papa 1993). With the appearance of this last material, the Phlegraean trachytic lavas were marginalised to some specific uses such as road paving port and maritime works, dams and breakwater reefs. At the end of the seventeenth century, Phlegraean lavas were progressively replaced by the Vesuvian *pietrarsa* because of its higher availability and, in some cases, better performances (Cardone and Papa 1993).

3 Materials of Technological Relevance

3.1 Zeolite-Rich Tuffs

As previously stated, the NYT and the LYT facies of the CI are characterised by the presence of zeolites, such as chabazite and phillipsite. Zeolites are a group of alkaline and earth-alkaline hydrated aluminosilicates that differ from the other tectosilicates. In fact, they are characterised by some specific properties of relevant technological interest, that make them useful in processes such as cationic exchange, polar molecule adsorption, molecular sieving and catalysis. Numerous studies have been undertaken to individuate and characterise natural Phlegraean zeolite deposits, and above all to identify their possible uses in relation to their potential technological characteristics, and hence chemico-physical specifications. These include an excellent cation exchange capacity ($\sim$3.5 meq/g) and a marked selectivity towards polluting ions, present in urban and industrial wastewater (e.g., NH^{4+}, Pb^{2+}, Cr^{3+}) (Pansini et al. 1991, 1996, 2010; Colella et al. 1998; Caputo et al. 1999; Colella and de Gennaro 2001; Cappelletti et al. 2011). A recent research field investigates the surface modification of natural zeolites for a potential application as nanocomposite versatile materials for health and environment (Izzo et al. 2019; Mercurio et al. 2018; Cappelletti et al. 2017; de Gennaro et al. 2015, 2016, 2017; Markovic et al. 2016; Serri et al. 2016).

The interest for the Phlegraean zeolite-rich tuffs has traditionally been related to their use in the building construction activity. This activity has induced a continuous and growing exploitation of these tuffs starting from the end of World War II until the nineties of the last century. During this period, a huge number of tuff quarries were opened with a consequent strong environmental impact which determined the almost complete block of the exploitation. As far as the tuff exploitation is concerned, the Regional Plan for Mining Activities of the Campania Region identified a total of 9 exploitation areas with an overall potential of about 6 million tons of material to be extracted in ten years. No reliable data on the current production are available so far since only two areas have been authorised. Most of the production (likely few hundred thousand tons per year) is used by the building industry and only 10–15% in technological applications. A consolidated use of this raw material is as binding and anti-agglomerant agent in the preparation of animal food. However, the potential occurrence of undesirable substances in supplements for the animal diet imposes continuous quality controls on commercial lots, particularly on those addressed to the biomedical and veterinary sectors (Mercurio et al. 2012, 2016). It should be remarked that, since several years, on the European market a registered trademark product compliant with European legislation 2002/32/CE and subsequent modifications, is available.

In the last years, the potential use of these raw materials has been introduced, and largely consolidated, in agriculture. Natural zeolites are used as soil amendants or added to fertilisers. Ultimately, climate changes are stimulating their use as tropicalisation along with lasting drought periods are hard testing the agriculture. Zeolites provide a marked increase of biofertiliser efficiency as they control the release process of macroelements (nitrogen, phosphorous, potassium) thus limiting the loss by leaching and favouring the continuous plant absorption of the nutritional requirements (Grilli et al. 2011).

3.2 Pozzolana

3.2.1 Hydraulic Mortars

If in the distant and recent past, both NYT and LYT were simply considered good dimension stones, today, following the results of researches on their technological applications, it can be stated that they, although still representing a significant economic resource in the building industry, are important for the preparation of hydraulic binders (Sersale 1991, 1995; Fursenko et al. 2000).

The possibility to obtain hydraulic mortars, namely a binder which harden also underwater, was already known to Romans who utilised mixtures of lime and *pozzolana* as described by Marcus Vitruvius Pollio (15 BC), a Roman writer, architect and engineer, in his II volume of "De Architectura" ("On Architecture"), "*Est etiam genus pulveris, quod efficit naturaliter res admirandas. Nascitur in regionibus Bajanis,............ quod commixtum cum calce et caemento, non modo caeteris aedificiis praestat firmitates, sed etiam moles quae construuntur in mari, sub aqua solidesc unt.*" (It is a sort of powder with marvellous natural properties, close to Baia .... mixed with lime and stones it hardens any construction, particularly those underwater).

Actually, the Romans used, mostly in the construction of ports, what today is called a mix design composed, further than the two basic components, water and lime, of *pozzolana*, tuff in powder and fragments, and the usual sedimentary aggregates. Firmly convinced of the excellent results secured by these two Phlegraean rocks, Romans used them even to build ports in distant lands (*Caesarea* in Israel, *Alexandria* in Egypt, *Chersonisos* in Crete and *Pompeiopolis* in Turkey; Stanislao et al. 2011; Rispoli et al. 2019).

It is well known that *pozzolana* has a capacity to react with lime producing, in presence of water, Calcium Silicate Hydrate (C-S-H) and Calcium Aluminate Hydrate (C-A-H), as it usually occurs in the hydration of modern cements:

$$\begin{aligned}&Ca(OH)_2 + SiO_2 + Al_2O_3 + H_2O\\&= C\text{-}S\text{-}H + C\text{-}A\text{-}H.\end{aligned}$$

This reaction permits to obtain a binder with a mechanical resistance much higher than that of aerial mortars and can harden even under water and resists the flushing action of rainwater or that of wave motion in the case of maritime works. If the mortar of lime and *pozzolana* is exposed to air, a part of the $Ca(OH)_2$ transforms into $CaCO_3$, as occurs in the hardening of aerial mortar via carbonatation:

$$Ca(OH)_2 + CO_2 = CaCO_3.$$

The discovery of hydraulic mortars based on lime and *pozzolana*, enabled the formulation of the so-called "Roman concrete", mixing lime, pozzolanic sand, water and scrap from bricks or, more often, from tuff.

3.2.2 Roman Concrete

The discovery of *pozzolana*'s properties marked a revolutionary progress in ancient constructions for which concrete was utilised. The etymology of the Italian term for concrete, *calcestruzzo*, derives from the Latin *calcis structi*, that means "based on lime". In the majority of Roman works, concrete was actually used as a filler between external brick or stone wall facings that functioned as permanent formworks. Vitruvius also suggested the ratios in the mixture for preparation of the concrete: pieces of tuff, two parts of *pozzolana* and one of lime; for paving: three parts of crushed brick and one part of lime; five parts of stone in pieces of a volume inferior to the content of one hand and two parts of lime; three parts of stone in pieces, unused, and a part of lime; two parts of stone in pieces, a part of brick dust and a part in lime. Both Vitruvius and Pliny the Elder, the latter in his "Naturalis Historia" (AD 77), recommended beating and compressing the mortars with an iron body, especially in building foundations. This aspect evidences how important was a thorough compactness of the conglomerate that, along with the lowest but necessary amount of added water, allowed to carry out a solid and a long-lasting construction.

Today, the reason for the strength of such a concrete is well known to the scientific community: the extraordinary adherence at the *pozzolana* and/or tuff/lime interface is due to the great reactivity of lime towards the glass component, in case of *pozzolana*, or zeolitic minerals, in the case of tuffs. Recently, Jackson et al. (2017 and references therein) have outlined the importance of pozzolanic reaction of volcanic ash with hydrated lime in the cementing processes and durability of 2,000-year-old Roman harbour concrete. Pozzolanic crystallisation of Al-tobermorite, a rare, hydrothermal, calcium-silicate-hydrate mineral with cation exchange capabilities, has been previously recognised in relict lime clasts of the concrete.

4 Conclusive Remarks

Based on what discussed in the previous paragraphs, it is evident the important role played by the Phlegraean lithoid rocks and the *pozzolana* as raw materials in the building industry and more recently in technological sectors. Presently, the question that needs to be answered is which should be the future of the resources taking into account their availability in the territory. It should also be considered the possible coexistence of the mining activity and the urban development, often chaotic, in an area with a clear commitment to tourism.

For some lithotypes (i.e., NYT and CI), it is definitely unthinkable to propose a renewal of the mining activity with the same intensity and technology that characterised more than half a century, from the end of World War II with the construction boom, in the city of Naples and its surroundings, until the beginning of this century. On the other hand, the Regional Plan for Mining Activities of the Campania Regional Government (2006) posed very precise limits to the exploitation of those stone materials defined as historical. These limits represent a significant contribution to the conservation of the resource in regard to the environment, although it might be more appropriate to redefine these limits for each lithotype based on their final use and the type of mining procedures.

With regards to PI and lavas, it is worth highlighting that the available materials are extremely scarce and hence a systematic use of these resources is clearly not appropriate. At most, they could be merely used for restorations.

Obviously, all the aspects concerning the use of zeolitised tuffs in technological sectors of elevated interest that exploit their main properties, should not be disregarded. In fact, some recent studies foresee good perspectives to the use of natural zeolites as pozzolanic activity additives, in substitution of the *pozzolana strictu* sensu in cement and lime-based binders (Chulines Domínguez et al. 2018 and references therein). It is worth to be mentioned also the possible use of zeolites in the traditional ceramic processes in partial replacement of the more expensive feldspathic fluxes or in the preparation of structural and insulating lightweight aggregates (de Gennaro et al. 2004, 2005, 2007, 2008).

Finally, good perspectives are linked to the use of modified and non-modified natural zeolites in the biomedical field as active carriers in the formulation of laboratory products for the treatment of skin or digestive system diseases.

Acknowledgements Data reported in the present chapter are the result of several researches carried out in the last decades or in progress, granted by the following projects: Italian Research Project of National Interest (PRIN) 2005, 2008, 2010–2011; Campania Regional Law n° 5 2002, 2003 as well as by research activities granted by mining and technological companies of the Campania region. We wish to warmly thank Diego Di Martire for drawing Fig. 1. Thanks are also due to the reviewers (Emilio Galan and an anonymous one).

References

Aveta A (1987) Materiali e tecniche tradizionali nel napoletano. Note per il restauro architettonico. Arte Tipografica, Napoli, Italy, pp 1–224

Calcaterra D, Cappelletti P, Langella A, Morra V, de Gennaro R, Colella A (2000) The building stones of the ancient centre of Naples (Italy): the Piperno from Phlegrean Fields. Contributions to the knowledge of features of a long-time used stone. J Cult Herit 1:415–428

Calcaterra D, Cappelletti P, de Gennaro M, de Gennaro R, de Sanctis F, Flora A, Langella A (2007) The rediscovery of an ancient exploitation site of Piperno, a valuable historical stone from Phlegraean Fields (Italy). In: Přikryl R, Smith BJ (eds) Building stone decay: from diagnosis to conservation, vol 271. Geological Society London Special Publications, London, pp 23–31

Calcaterra D, Langella A, Morra V, Cappelletti P, Colella A, de Gennaro R, de Gennaro M (2013) Il Piperno. In: de Gennaro M, Calcaterra D, Langella A (eds) Le Pietre Storiche della Campania - dall'oblio alla riscoperta. Luciano, Napoli, Italy, pp 179–197

Calcaterra D, D'Amore M, Di Martire D, de Gennaro M, Langella A (2018) Le risorse lapidee della Campania: riscoperta e valorizzazione dei siti estrattivi. In: Aveta A, Marino BG, Amore R (eds) La Baia di Napoli. ArtstudioPaparo, Napoli, Italy, pp 28–34

Cappelletti P, Cerri G, Colella A, de Gennaro M, Langella A, Perrotta A, Scarpati C (2003) Post-eruptive processes in the Campanian Ignimbrite. Miner Petrol 79:79–97

Cappelletti P, Rapisardo G, de Gennaro B, Colella A, Langella A, Graziano SF, Bish DL, de Gennaro M (2011) Immobilization of Cs and Sr in aluminosilicate matrices derived from natural zeolites. J Nucl Mater 414:451–457

Cappelletti P, Colella A, Langella A, Mercurio M, Catalanotti L, Monetti V, de Gennaro B (2017) Use of surface modified natural zeolite (SMNZ) in pharmaceutical preparations Part 1. Mineralogical and technological characterization of some industrial zeolite-rich rocks. Micropor Mesopor Mat 250:232–244

Caputo D, de Gennaro B, Pansini M, Colella C (1999) Chromium removal from water by ion exchange using zeolites and solidification of the resulting sludge in a cement matrix. In: Kiricsi I, Pál-Borbély G, Nagy JB, Karge HG (eds) Porous Materials in Environmentally Friendly Processes. Stud Surf Sci Catal 125:723–730

Cardone V, Papa L (1993) L'identità dei Campi Flegrei. CUEN, Napoli, Italy, pp 1–299

Chulines Domínguez IA, Abdellaoui Y, Abatal M, Patiño-Carachure C (2018) Characterizations and industrial applications for cement and concrete incorporated natural zeolite. In: Rashed MN, Palanisamy PN (eds), Zeolites and their applications. https://doi.org/10.5772/intechopen.70980. ISBN 978-1-78923-343-8

Colella C (1996) Ion Exchange equilibria in zeolite minerals. Miner Dep 31:554–562

Colella A, de Gennaro B (2001) Evaluation of Italian phillipsite and chabazite as cation exchangers for Ba^{2+} and Co^{2+}. Stud Surf Sci Catal 140:153–162

Colella C, de Gennaro M, Langella A, Pansini M (1998) Evaluation of natural phillipsite and chabazite as cation exchangers for copper and zinc. Sep Sci Technol 33:467–481

Colella C, de Gennaro M, Aiello R (2001) Use of zeolitic tuff in building industry. In: Bish DL, Ming DW (eds) Reviews in mineralogy 45:551–587. Mineralogical Society of America. Washington DC, USA

Colella A, Calcaterra D, Cappelletti P, Langella A, Papa L, Perrotta A, Scarpati C, de Gennaro M (2013) Il Tufo Giallo Napoletano (TGN). In: de Gennaro M, Calcaterra D, Langella A (eds) Le Pietre Storiche della Campania - dall'oblio alla riscoperta. Luciano, Napoli, Italy, pp 129–154

Colella A, Di Benedetto C, Calcaterra D, Cappelletti P, D'Amore M, Di Martire D, Graziano SF, Papa L, de Gennaro M, Langella A (2017) The Neapolitan Yellow Tuff: an outstanding example of heterogeneity. Constr Build Mater 136:361–373

Coppola E, Battaglia G, Bucci M, Ceglie D, Colella A, Langella A, Buondonno A, Colella C (2003) Remediation of Cd- and Pb-polluted soil by treatment with organo-zeolite conditioner. Clays Clay Miner 51:609–615

de Gennaro M, Colella C (1991) The critical role of temperature in the natural zeolitization of volcanic glass. Neues Jahrb Mineral Monat 8:355–362

de Gennaro M, Langella A (1996) Italian zeolitized rocks of technological interest. Miner Dep 31:451–472

de Gennaro M, Colella C, Franco E, Stanzione D (1988) Hydrothermal conversion of trachytic glass into zeolite. 1. Reactions with deionized water. Neues Jahrb Mineral Monat 4:149–158

de Gennaro M, Colella C, Pansini M, Langella A (1992) Reconstruction of a natural zeolitization process through laboratory simulations. In: von Ballmoos R, Higgins JB, Treacy MMJ (eds) Ninth international zeolite conference. Butterworth-Heinemann, Boston, USA, pp 207–214

de Gennaro M, Colella C, Pansini M (1993) Hydrothermal conversion of trachytic glass into zeolite. 2. Reactions with high-salinity waters. Neues Jb Miner Monat 3:97–110

de Gennaro M, Langella A, Cappelletti P, Colella C (1999) Hydrothermal conversion of trachytic glass to zeolite. 3. Monocationic model glasses. Clays Clay Miner 47:348–357

de Gennaro M, Calcaterra D, Cappelletti P, Langella A, Morra V (2000a) Building stones and related weathering in the architecture of the ancient centre of Naples. J Cult Herit 1:399–414

de Gennaro M, Cappelletti P, Langella A, Perrotta A, Scarpati C (2000b) Genesis of zeolites in the Neapolitan Yellow Tuff: geological, volcanological and mineralogical evidence. Contrib Mineral Petrol 139:17–35

de Gennaro R, Calcaterra D, Di Girolamo P, Langella A, de Gennaro M (2003) Discovering the stone heritage of Southern Italy: the Mondragone Marble from Campania region. Environ Geol 44:266–276

de Gennaro R, Cappelletti P, Cerri G, de Gennaro M, Dondi M, Langella A (2004) Zeolitic tuffs as raw materials for lightweight aggregates. Appl Clay Sci 25:71–81

de Gennaro R, Cappelletti P, Cerri G, de Gennaro M, Dondi M, Langella A (2005) Neapolitan yellow tuff as raw material for lightweight aggregates in lightweight structural concrete pro-duction. Appl Clay Sci 28(1–4):309–319

de Gennaro R, Dondi M, Cappelletti P, Cerri G, de Gennaro M, Guarini G, Langella A, Parlato L, Zanelli C (2007) Zeolite-feldspar epiclastic rocks as flux in ceramic tile manufacturing. Micropor Mesop Mat 105:273–278

de Gennaro R, Langella A, D'Amore M, Dondi M, Colella A, Cappelletti P, de Gennaro M (2008) Use of zeolite-rich rocks and waste materials for the production of structural light-weight concretes. Appl Clay Sci 41:61–72

de Gennaro R, Graziano SF, Cappelletti P, Colella A, Dondi M, Langella A, de Gennaro M (2009) Structural concretes with waste-based lightweight aggregates: from landfill to engineered materials. Environ Sci Technol 43:7123–7129

de Gennaro B, Catalanotti L, Cappelletti P, Langella A, Mercurio M, Serri C, Biondi M, Mayol L (2015) Colloid Surface B 130:101–109

de Gennaro B, Mercurio M, Cappelletti P, Catalanotti L, Dakovic A, De Bonis A, Grifa C, Izzo F, Krakovic M, Monetti V, Langella A (2016) Use of surface modified natural zeolite (SMNZ) in pharmaceutical preparations. Part 2. A new approach for a fast functionalization of zeolite-rich carriers. Micropor Mesopor Mat 235:42–49

de Gennaro B, Izzo F, Catalanotti L, Langella A, Mercurio M (2017) Surface modified phillipsite as a potential carrier for NSAIDs release. Adv Sci Lett 23 (6):5941–5943

Deere UD, Miller RP (1966) Engineering classification and index properties of intact rocks. Air force lab technical report AFNL-TR-65-116, Albuquerque, New Mexico, USA, pp 1–302

Dell'Erba L (1923) Il Tufo Giallo Napolitano – studio scientifico-tecnico esteso alle cave e frane. Pironti, Napoli, Italy, pp 1–288

De Lorenzo (1904) History of volcanic action in the Phlegraean Fields. Q J Geol Soc Lond 60:296–315

Di Vito MA, Isaia R, Orsi G, Southon J, De Vita S, D'Antonio M, Pappalardo L, Piochi M (1999) Volcanism and deformation since 12,000 years at the Campi Flegrei caldera (Italy). J Volcanol Geotherm Res 91:221–246

Fursenko BA, Katazantseva LK Belitsky IA (2000) Recent advances in the use of zeolitic tuffs in Russia for manufacturing building Materials. In: Colella C, Mumpton FA (eds) Natural zeolites for the third millennium, De Frede, Napoli, Italy, pp 337–349

Galante D (1971) Relazione sulle condizioni del sottosuolo del vecchio centro urbano. Comune di Frattamaggiore, Napoli, Italy, pp 1–26

Grilli E, Colella A, Coppola E, Langella A, Buondonno A (2011) Modelling pedogenization of zeolitized tuff: effects of water and phenolic substances on weathering rates of the Campanian Ignimbrite (yellow facies). Clay Miner 46:311–327

Hay RL, Sheppard RA (2001) Occurrence of zeolites in sedimentary rocks: an overview. In: Bish DL, Ming DW (eds) Reviews in mineralogy, vol 45. Mineralogical Society of America, Washington DC, USA, pp 217–234

Izzo F, Mercurio M, de Gennaro B, Aprea P, Cappelletti P, Dakovic A, Germinario C, Grifa C, Smiljanic D, Langella A (2019) Surface modified natural zeolites (SMNZs) as nanocomposite versatile materials for health and environment. Coll Surf B 182:110380

Jackson MD, Mulcahy SR, Chen H, Li Y, Li Q, Cappelletti P, Wenk HR (2017) Phillipsite and Al-Tobermorite produced by cementitious water-rock reaction in Roman marine concrete. Am Mineral 102:1435–1450

Johnston-Lavis HJ (1888) Excavation near naples. Report of the committee appointed for the investigation of the volcanic phenomena of vesuvius and its neighbourhood. Spottiswoode and Co., London, UK, pp 1–7

Langella A, Calcaterra D, Cappelletti P, Colella A, d'Albora M, Morra V, de Gennaro M (2009) Lava stone from Neapolitan volcanic districts in the Campanian architecture. Environ Earth Sci 59:145–160

Langella A, Bish DL, Calcaterra D, Cappelletti P, Colella A, Graziano SF, Papa L, Perrotta A, Scarpati C, de Gennaro M (2013a) L'Ignimbrite Campana (IC) In: de Gennaro M, Calcaterra D, Langella A (eds) Le Pietre Storiche della Campania – dall'oblio alla riscoperta. Luciano, Napoli, Italy, pp 155–177

Langella A, Bish DL, Cappelletti P, Cerri G, Colella A, de Gennaro R, Graziano SF, Perrotta A, Scarpati C, de Gennaro M (2013b) New insights into the mineralogical facies distribution of Campanian Ignimbrite, a relevant Italian industrial material. App Clay Sci 72:55–73

Markovic M, Dakovic A, Krajisnik D, Kragovic M, Milic J, Langella A, de Gennaro B, Cappelletti P, Mercurio M (2016) Evaluation of the surfactant/phillipsite composites as carriers for diclofenac sodium. J Mol Liq 222:711–716

Mercurio M, Mercurio V, de Gennaro B, de Gennaro M, Grifa C, Langella A, Morra V (2010) Natural zeolites and white wines from Campania region (southern Italy): a new contribution for solving some oenological problems. Period Mineral 79:95–112

Mercurio M, Langella A, Cappelletti P, de Gennaro B, Monetti V, de Gennaro M (2012) May the use of Italian volcanic zeolite-rich tuffs as additives in animal diet represent a risk for the human health? Period Mineral 81:393–407

Mercurio M, Cappelletti P, de Gennaro B, de Gennaro M, Bovera F, Iannaccone F, Grifa C, Langella A, Monetti V, Esposito L (2016) The effect of digestive activity of pig gastro-intestinal tract on zeolite-rich rocks: an in vitro study. Micropor Mesopor Mat 225:133–136
Mercurio M, Izzo F, Langella A, Grifa C, Germinario C, Dakovic A, Aprea P, Pasquino R, Cappelletti P, Graziano FS, de Gennaro B (2018) Surface-modified phillipsite-rich tuff from the Campania region (southern Italy) as a promising drug carrier: an ibuprofen sodium salt trial. Am Mineral 103(5):700–710
Morra M, Calcaterra D, Cappelletti P, Colella A, Fedele L, de Gennaro R, Langella A, Mercurio M, de Gennaro M (2010) Urban geology: relationships between geological setting and architectural heritage of the Neapolitan area. In: Beltrando M, Peccerillo A, Mattei M, Conticelli S, Doglioni C (eds) The geology of Italy, vol 36, paper 26, Electronic edition. Journal of the Virtual Explorer, pp 1–60. ISSN 1441-8142
Orsi G, de Vita S, Di Vito M (1996) The restless, resurgent Campi Flegrei nested caldera (Italy): constraints on its evolution and configuration. J Volcanol Geotherm Res 74:179–214
Orsi G, Civetta L, Valentine G (1999) Special issue—Volcanism in the Campi Flegrei—Introduction. J Volcanol Geotherm Res 91(2–4):121
Pansini M (1996) Natural zeolites as cation exchangers for environment protection. Miner Dep 31:563–575
Pansini M, Colella C, de Gennaro M (1991) Chromium removal from water by ion exchange using zeolite. Desalination 83(1–3):145–157
Pansini M, Colella C, Caputo D, de Gennaro M, Langella A (1996) Evaluation of phillipsite as cation exchanger in lead removal from water. Micropor Mat 5:357–364
Pansini M, de Gennaro R, Parlato L, de Gennaro M, Langella A, Marocco A, Cappelletti P, Mercurio M (2010) Use of sawing waste form zeolitic tuffs in the manufacture of ceramics. Adv Mater Sci Eng 1:1–9
Penta F (1935) I materiali da costruzione dell'Italia meridionale. Fondazione Politecnica del Mezzogiorno, 2 vol in 8°, Napoli, Italy, pp 1–672
Pliny the Elder (77) Naturalis Historia
PRAE (2006) Piano Regionale delle Attività Estrattive della Campania. http://www.sito.regione.campania.it/lavoripubblici/Elaborati_PRAE_2006
Primavori P (1999) Planet stone. Zusi Editore, Verona, Italy, pp 1–326
Rispoli C, De Bonis A, Guarino V, Graziano SF, Di Benedetto C, Esposito R, Morra V, Cappelletti P (2019) The ancient pozzolanic mortars of the Thermal complex of Baia (Campi Flegrei, Italy). J Cult Her 40:143–154
Scherillo A. (1955) Petrografia chimica dei tufi Flegrei. II. Tufo Giallo, Mappamonte, Pozzolana (Italia). Rend Accad Sci Fis Mat 22:317-330
Serri C, de Gennaro B, Catalanotti L, Cappelletti P, Langella A, Mercurio M, Mayol L, Biondi M (2016) Surfactant-modified phillipsite and chabazite as novel excipients for pharmaceutical applications? Micropor Mesopor Mat 224:143–148
Sersale R (1991) La storia della calce dall'antichità ai nostri giorni. L'industria Italiana Del Cemento 651:56–62
Sersale R. (1995) Zeolitic tuffs as pozzolanic addition in the manufacture of blended cements. In Ming DW, Mumpton FA (eds) Natural zeolites'93: occurrence, properties, use. Brockport, New York, USA, pp 603–612
Sheppard RA, Hay RL (2001) Formation of zeolites in open hydrologic systems. In: Bish DL, Ming DW (eds) Reviews in Mineralogy, vol 45. Mineralogical Society of America. Washington DC, USA, pp 261–275
Stanislao C, Rispoli C, Vola G, Cappelletti P, Morra V, de Gennaro M (2011) Contribution to the knowledge of ancient Roman seawater concretes: Phlegrean pozzolan adopted in the construction of the harbor at Soli-Pompeiopolis (Mersin, Turkey). Per Mineral 80:471–488
Vitruvio Pollione M (15 AC) De Architectura, libro II, C sextum
Zezza F, Baldassarre R, Pagliarulo R, Racaniello F (1985) Applicazioni del point-load test per la classificazione della resistenza e la correlazione con le prove a compressione monoassiale delle pietre da costruzione dei bacini di Trani e di Apricena. In: Quarry and construction, Atti III Convegno Naz. Attività Estrattiva dei Minerali di 2ª Categoria, Bari, Italy, pp 178–181

The Urban Development of Campi Flegrei, Italy

Ugo Leone, Massimo D'Antonio, and Giovanni Orsi

Abstract

Since the Late Pleistocene, Campi Flegrei have been inhabited by humans because of the region's soil fertility, mild climatic conditions, and both commercially and strategically advantageous position in the Mediterranean basin. After a period of frequent eruptions that affected the environment and forced indigenous farmers temporarily to abandon the area after each event, in 740 BC Campi Flegrei were chosen by Chalcidian Greeks from Ischia to establish the settlement of Cuma, which was the first Greek colony in what is today the Italian mainland. Since its earliest colonisation, the Phlegraean area, because of its location and environmental characteristics, has been a cause of conflicts over its control, and has also witnessed a significant increase in population and urban development. Even the threat of ongoing bradyseismic activity, which has caused seismicity and ground deformation, and the AD 1538 Monte Nuovo eruption did not discourage people from living in the area. Campi Flegrei have undergone socio-economic development which for most of its history has been sustainable. Only with industrialisation over recent decades, has it become unsustainable. This unsustainable development has also caused an increase in the vulnerability of the area in terms of its volcanic risk. Knowledge of the history of urbanisation in Campi Flegrei may, therefore, assist the civil authorities in effecting socio-economic policies aimed at disaster risk reduction.

1 Introduction

The Neapolitan area, in southern Italy, is dominated by the presence of four volcanoes, of which Somma-Vesuvio, Campi Flegrei and Ischia are still active, while Procida volcanic field last erupted at ~23 ka (Morabito et al. 2014) (Fig. 1). Somma-Vesuvio is a strato-volcano, whereas Campi Flegrei and Ischia are resurgent calderas (Orsi et al. 2003; Santacroce et al. 2003; see Chap. Volcanic and Deformation History of the Campi Flegrei Volcanic Field, Italy). Despite the activity of these volcanoes (Fig. 2), the area has been the site of human settlement over millennia because of its soil fertility, mild climate conditions and advantageous position both commercially and strategically within the Mediterranean Basin (Orsi et al. 2003 and references therein). From the Neolithic to the Bronze Age, many eruptions of these volcanoes

U. Leone
Olim Professor of Environmental Policy, Università degli Studi di Napoli Federico II, Napoli, Italy

M. D'Antonio · G. Orsi (✉)
Dipartimento di Scienze della Terra, dell'Ambiente e delle Risorse, Università degli Studi di Napoli Federico II, Napoli, Italy
e-mail: giovanni.orsi@unina.it

© Springer-Verlag GmbH Germany, part of Springer Nature 2022

G. Orsi et al. (eds.), *Campi Flegrei*, Active Volcanoes of the World,
https://doi.org/10.1007/978-3-642-37060-1_15

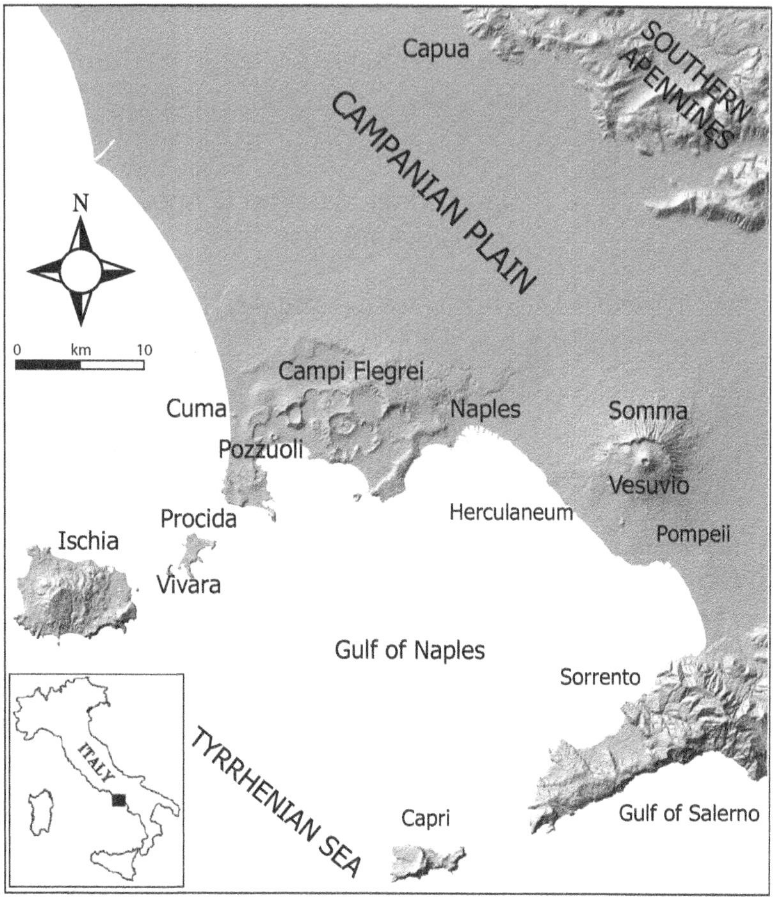

Fig. 1 Digital Terrain Model of the area including the Neapolitan volcanoes within the Campanian Plain

have affected both the environment and human populations (Marzocchella 1998; Saccoccio et al. 2013). Until the Late Copper Age, a series of Phlegraean eruptions caused modifications to the environment and forced farmers temporarily to abandon the area (Marzocchella 2000).

In the Phlegraean area, magmatism and volcanism are among the natural elements that have influenced the relationships between humans and environment. In particular, these relationships have been affected by volcanic eruptions (Chap. Volcanic and Deformation History of the Campi Flegrei Volcanic Field, Italy), hydrothermal activity with generation of fumaroles and thermal springs (Chap. The Hydrothermal System of the Campi Flegrei Caldera, Italy), and ground deformation (Chaps. The Permanent Monitoring System of the Campi Flegrei Caldera, Italy; Historic Unrest of the Campi Flegrei Caldera, Italy; Source Modelling from Ground Deformation and Gravity Changes at the Campi Flegrei Caldera, Italy). The Campi Flegrei caldera (CFc) is a complex and nested structure that results from two major collapses related to the Campanian Ignimbrite (Fisher et al. 1993; Rosi et al. 1996; Civetta et al. 1997; Ort et al. 2003) ($\sim$40 ka; Giaccio et al. 2017) and the Neapolitan Yellow Tuff (Orsi et al. 1992, 1995; Scarpati et al. 1993; Wohletz et al. 1995) ($\sim$15 ka; Deino et al. 2004) catastrophic eruptions (Orsi et al. 1992, 1996; Chap. Volcanic and Deformation History of the Campi Flegrei Volcanic Field, Italy). Over the past 15 kyrs, the CFc has been the site of intense and intimately linked volcanic and tectonic activities (Chap. Volcanic and Deformation History of the Campi Flegrei Volcanic Field, Italy). Volcanism has generated not less than 70 eruptions grouped in three epochs of activity (Fig. 2). The largest

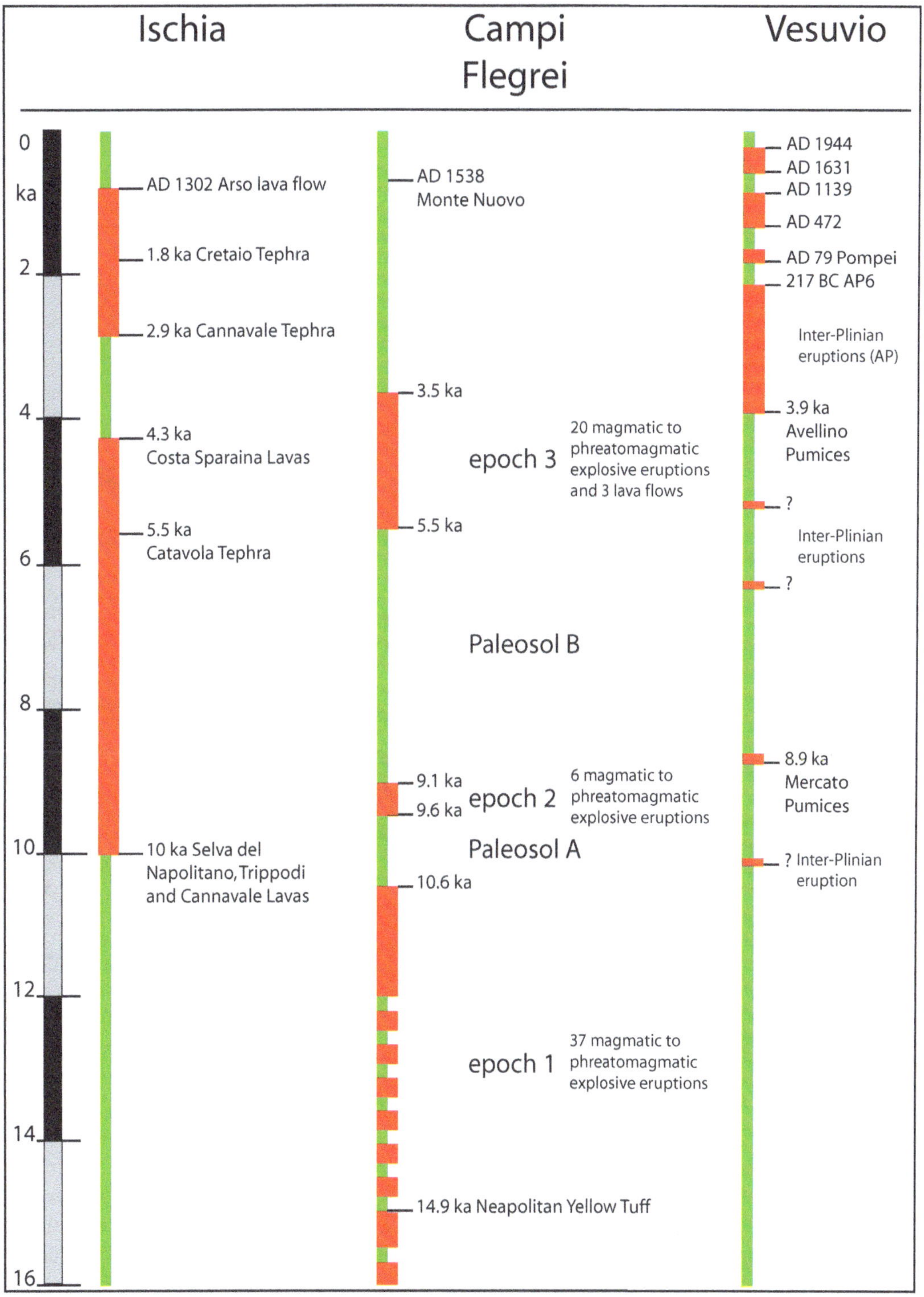

Fig. 2 Chronograms of the activity of the Neapolitan active volcanoes in the past 15 kyrs. Figure modified after Orsi et al. (2013) and Chap. Volcanic and Deformation History of the Campi Flegrei Volcanic Field, Italy

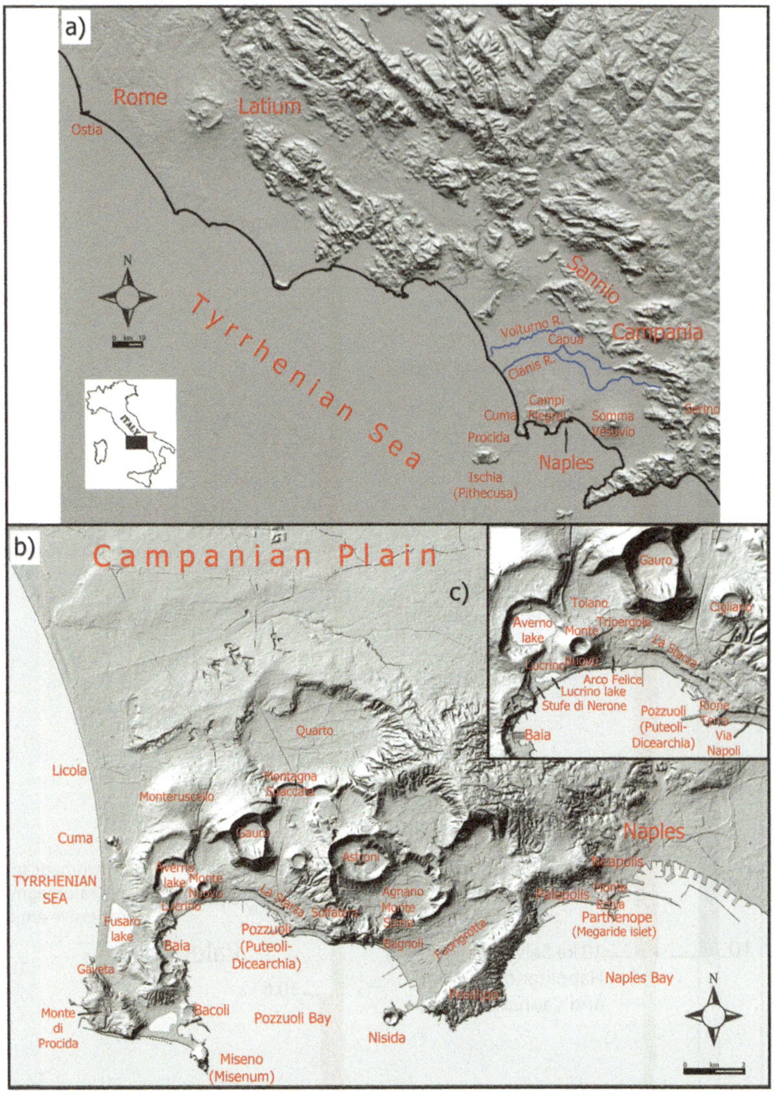

Fig. 3 **a** Digital Terrain Model of the area between Rome and Naples, Southern Italy; **b** Digital Terrain Model of the Phlegraean area; **c** Digital Terrain Model of the Pozzuoli area with the location of the place-names mentioned in the text

eruption of the past 5 kyrs was the Agnano-Monte Spina event (de Vita et al. 1999) (4,482–4,625 cal. years BP, Smith et al. 2011). The last eruption has occurred in AD 1538 (Guidoboni and Ciuccarelli 2011; Di Vito et al. 2016 and references therein), after about 3,000 years of quiescence, and formed the Monte Nuovo cone next to the town of Pozzuoli (Fig. 3). Tectonism has resulted in ground deformation that has produced a net uplift of 90 m on the La Starza marine terrace in the central part of the caldera (Cinque et al. 1984). This ongoing long-term deformation has been interpreted (Orsi et al. 1999) as being the result of many transient episodes of short-term deformation (i.e., ground uplift and subsidence), which has been known since the beginning of the last century in the volcanological literature of Campi Flegrei as bradyseism. This word (from the Greek *bradi*, slow and *seism*, movement) was coined by Issel (1883) to indicate slow subsidence or uplift of the ground. Vertical ground movement that has occurred in historical times in the CFc has been investigated by many authors (e.g., Lyell 1830; Parascandola 1947; Dvorak and Mastrolorenzo 1991; Morhange et al. 2006; Del Gaudio et al. 2010; Ricco et al. 2019). More recently, two major bradyseismic events have generated ground uplift in 1969–1972 (170 cm) and 1982–1984 (180 cm). These events have been followed by some minor ones, the last of those began in late 2004 and is still ongoing. Reviews of the caldera current state and hypotheses on its causes are also reported in Chaps. Volcanic and Deformation History of the Campi Flegrei Volcanic Field, Italy; The Permanent Monitoring System of the Campi Flegrei Caldera, Italy; The Hydrothermal System of the Campi Flegrei Caldera, Italy; Historic Unrest at the Campi Flegrei Caldera, Italy; Source Modelling from Ground Deformation and Gravity Changes at the Campi Flegrei Caldera, Italy.

Physiography of the Phlegraean area results either from natural events related to both endogenous and exogenous processes, or from anthropogenic activity. The latter has generated a significant degradation of the landscape (Maiuri 1958, 1970, 1977, 1991; Farneti et al. 1971).

Reconstructing the urban development of a densely inhabited area is a very important task, as it has significant effects on both humankind and environment. This goal can be achieved by means of the concept of "sustainable development". The latter may be defined as achieving development without reducing options for the future generations. Therefore, the impact of an economic development on the environment and the effects it produces on the landscape, need to be understood. In particular, the issue of whether a collective loss of memory could increase the proneness to volcanic hazards has to be considered in planning the urban development of an active volcanic area. The increase in population, building and infrastructures increases the vulnerability of such an area and, consequently, the volcanic risk. Furthermore, knowledge of the history of urbanisation should help stimulate civil authorities into developing policies aimed at risk reduction.

Ongoing magmatic and volcanic activity as well as human inhabitation over millennia make the CFc an ideal site to investigate the relationships among human settlement, urban development and natural environment. This paper reports the results of such an investigation since the earliest human settlement in the area, highlighting types of urban development and its modification over time.

2 The Urban Development of Campi Flegrei

2.1 From the Earliest Settlement to the Roman Empire

At the time of the catastrophic Campanian Ignimbrite eruption ($\sim$40 ka), hominids had already settled in the Neapolitan area (Lowe et al. 2012 and references therein). Since the time of the Neapolitan Yellow Tuff eruption ($\sim$15 ka), the Campanian Plain has been permanently inhabited by organised communities. During the Eneolithic and Bronze Age, roughly spanning the period between the fourth and the end of the second millennium BC, the active Neapolitan volcanoes generated effusive and explosive eruptions of varying magnitudes (Fig. 2).

Long-lasting human occupation means that the area became an important and well documented centre of urbanisation within the Mediterranean region (Maiuri 1970, 1977; Capasso 1978; Miccio and Potenza 1994; Annecchino 1996; Napoli 1996; Marzocchella 1998, 2000). The presence of natural harbours, the mild climate conditions, the availability of freshwater, the presence of defensible and sheltered sites, all favoured human settlement and encouraged re-population of the area following successive volcanic eruptions. The eruptions of the Neapolitan volcanoes followed each other with sufficient time to allow recovery of the area. Since the Neolithic, farmers had settled on low hills close to rivers on the plain north of Naples (Marzocchella 1998, 2000) and archaeological artefacts dating from the fourth millennium BC have been recovered from Ischia (Buchner 1969).

Since the sixteenth century BC, metal workers had settled on this island that was located along the routes from Greece to Tuscany and to the islands of Sardinia and Elba (Fig. 4) where a variety of metals were quarried (Buchner 1986). At the end of the ninth and at the beginning of the eighth centuries BC, there was a Rhodian commercial colony on the Megaride islet along the coast of Monte Echia in Naples (Fig. 3b).

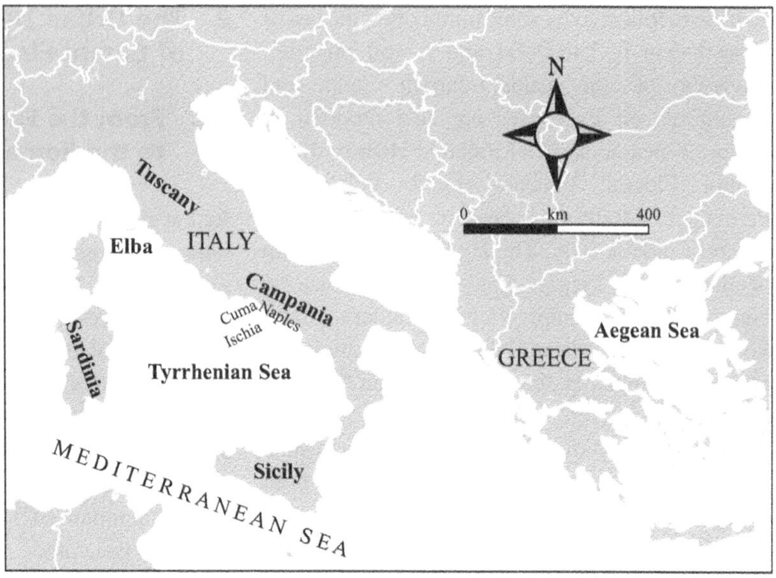

Fig. 4 Location of the Greek colonies of the seventh–sixth century BC. Figure modified after Orsi et al. (2003). *Map source* Wikimedia Commons, https://commons.wikimedia.org/wiki/File:Empty_map_of_Europe.svg

Stabilisation of the human settlement in Campi Flegrei, as witnessed by the prehistoric hamlets of Monte di Procida, Fuorigrotta and Montagna Spaccata (Fig. 3b), was eased by the onset of a quiescence period in Phlegraean eruptive activity during the second millennium BC (3.5 ka; see Chap. Volcanic and Deformation History of the Campi Flegrei Volcanic Field, Italy), while in the neighbouring Ischia and Somma-Vesuvio there was volcanic activity (Fig. 2) (Santacroce et al. 2008; de Vita et al. 2010; Di Renzo et al. 2011; Orsi et al. 2013). Intense and ongoing fumarolic and hydrothermal activity caused the earliest Greek settlers to name the area Campi Flegrei ("burning fields" in Greek). The quiescence in the Phlegraean volcanic activity enhanced the alluring peculiarities of the CFc, that became a site of more intense population growth and caused conflicts among the populations for controlling the area.

The development of an organised society in the Neapolitan area began with the settlement of Greek colonists along the commercial routes of the Mediterranean region (Fig. 4). The earliest settlers in Campi Flegrei were Chalcidian Greeks from the island of *Pithecusa*, the present Ischia, who were in search of a safer harbour on the mainland. Around 740 BC they selected a locality facing Ischia along the Tyrrhenian coast of the mainland in which they founded Cuma, the first Greek colony in what is today the Italian peninsula (Fig. 3b). The site of Cuma had all the peculiar characteristics of a Greek settlement; in particular a sheltered bay for landing vessels, and a hill on which to build an acropolis. The occurrence of marshes towards the mainland was an important element in its defence and acted as an obstacle to possible invading enemies. All these features permitted Cuma to rule over the coastal territory in the following seventh and sixth centuries BC. Cuma spread the Greek culture, by disseminating the Chalcidian alphabet that was adopted by both the Etruscans and Latins, becoming the alphabet of the language and literature of Rome and, indeed, of all the western culture.

In 680 BC, shortly after the establishment of their town, the Cumans founded *Parthenope* in honour of the homonymous siren of the Greek mythology (Fig. 5). At that time the coastline was different from that of the present-day and urban areas were all built on easily defensible promontories with sheltered landing places for vessels. The *Parthenope* settlement, established in a strategic position on the Megaride Islet, developed into a town that extended across the flat top of Monte Echia, a rocky promontory behind the islet (Fig. 3b). *Parthenope* hold,

Fig. 5 Statue of the *Parthenope* siren (carved by Onofrio Buccini in 1869) in Sannazzaro Square, Naples (Photo *source* Wikimedia Commons, https://it.wikipedia.org/wiki/Partenope_()

therefore, all the characteristics of the other Cuman settlements which had natural harbours and strongholds (Giampaola and D'Agostino 2005). In 530 BC the Cumans founded the town of *Neapolis* ("new town" in Greek), the present Naples, to the east of *Parthenope* (Fig. 3b). The town was located on a highland delimited by a scarp towards the sea and by deep gullies along the other sides, while the surrounding plains to the east and north were marshes. After foundation of *Neapolis*, *Parthenope* was renamed *Palepolis* ("old town" in Greek). The two towns, although not physically linked, were effectively a single settlement that soon replaced Cuma as the principal centre for both commercial shipping and control of trade along the coast.

In 528 BC the Cumans allowed new Greek people arriving from Samo to occupy the hill of the present town of Pozzuoli (Rione Terra; Fig. 3c) and established the settlement of *Dicearchia* ("town of the fair government" in Greek). The Etruscans of Capua (Fig. 3a), feeling their economic and political expansion seriously compromised, attacked Cuma in 524 BC but were defeated. Anyhow, their power was still more dramatically affected and their expansion in the Hellenic Italy halted by a defeat inflicted by Greeks from Sicily led by Jerone of Syracuse (478–466 BC) in the famous Cuma naval battle of 474 BC. The result of this battle increased the prestige of Cuma so much that the Greek historian Diodorus Siculus (90–27 BC), in his monumental universal history "Bibliotheca historica", referred to the entire Phlegraean region as "Cuma's Land".

In 421 BC Cuma and *Dicearchia* were invaded and occupied by the Osci from Sannio (Fig. 3a). To face the advance of this new enemy, Cuma entered into an alliance with Rome. Following this, Rome influence continuously expanded in southern Italy until it dominated the entire Campania region, including the Neapolitan-Phlegraean area. In 334 BC the Roman authorities designated Cuma as a town without right to vote, and in 194 BC they founded *Puteoli* (the present Pozzuoli; Fig. 3c) on the site of *Dicearchia*. The presence of this important harbour gave a new stimulus to the Phlegraean economy, mostly to its commercial trades, which produced a concomitant increase in the urban area. During the Roman civil wars (44–30 BC) that saw Octavius in opposition to Sextus Pompey and put an end to the Roman Republic, Cuma backed Octavius who, after his victory, became the first Roman emperor taking the name of Octavius Augustus (27 BC–AD 14). During

Augustan times, Cuma, together with the entire Phlegraean area, became the favourite site for Roman elites to relax and find amusement. Furthermore, the harbour of *Puteoli*, formerly the port of Cuma and centre of shipping, became vital in supplying provisions to Rome. Some Roman Patricians as well as Emperors, attracted by the scenic quality of the landscape, built houses on sites with fine views. The occurrence of large thermal spas and numerous hot springs whose healing powers were known over the entire Empire, attracted a large number of travellers from Rome causing a further rapid expansion in the urban fabric.

In the first century CE, under the Emperor Claudius (AD 41–54), an aqueduct was built and named after Claudius (Fig. 6a). The aqueduct collected water from the area of Serino and transported it through Naples to Baia over a distance of about 92 km. It ended in the

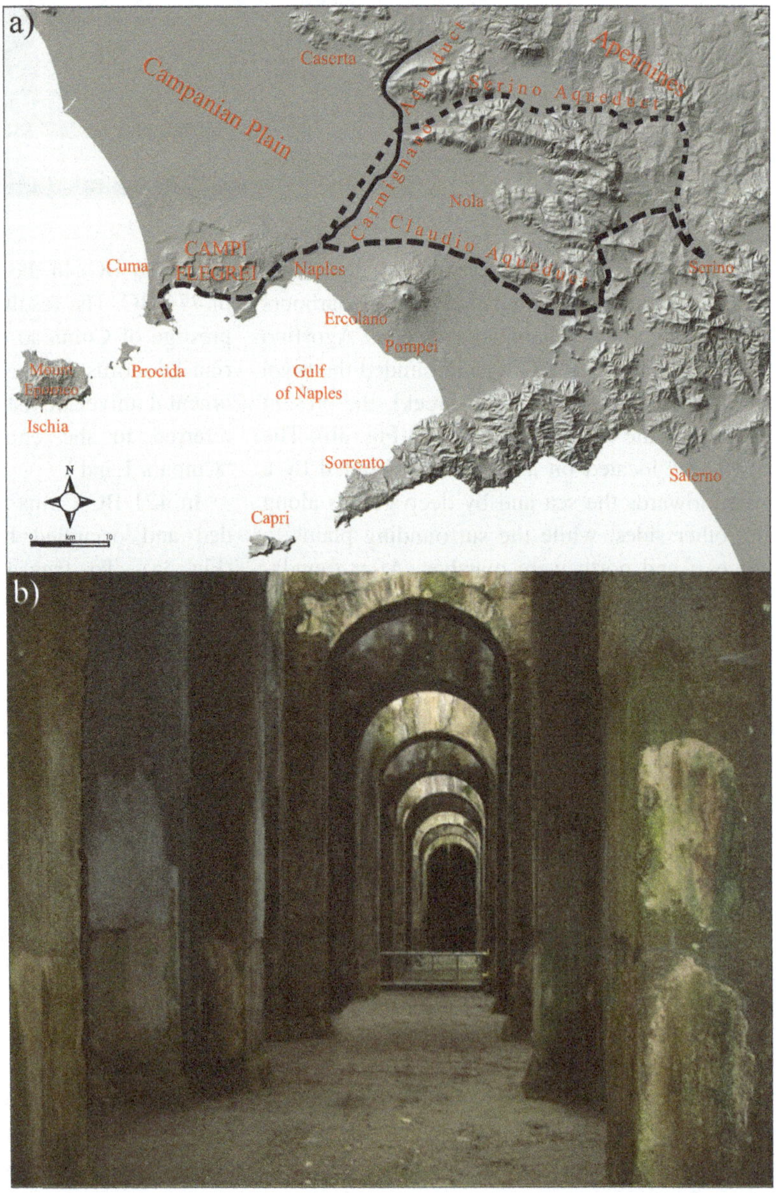

Fig. 6 **a** Plan of the main aqueducts, serving the town of Naples since Roman times until the ninenteenth century CE (modified after Orsi et al. 2003); **b** The huge reservoir, named Piscina Mirabilis, at the end of the Claudio aqueduct. *Photo source* Wikimedia Commons, https://upload.wikimedia.org/wikipedia/commons/2/23/Piscina_Mirabilis.jpg

underground Piscina Mirabilis (Fig. 6b, the largest water reservoir built by the Romans in order to supply fresh water to their fleet based in Miseno.

The development of harbour facilities further favoured the socio-economic development of the area, and remarkably increased trading of locally produced goods. These were mostly related to fishing and extraction of *pozzolana*, which is a loose pyroclastic material used for preparation of mortar (see Chap. Building Stones and Technological Materials of the Campi Flegrei Caldera, Italy). As a consequence, *Puteoli* witnessed both a significant demographic and building boom. It spread well beyond the acropolis (present Rione Terra; Fig. 3c) and population number reached 40,000.

The urbanisation of the territory was quite rapid with a phase of maximum demographic expansion corresponding to the imperial period of Rome (27 BC–AD 395), during which *Puteoli* became its principal harbour. This development did not, however, have a negative impact on the environment. Horace (65–8 BC) wrote that there is no place on Earth that outshines the agreeable Baia ("Nullus in orbe locus Baiis praelucet amoenis" in Latin) (Fig. 3b) (see Colletta 1984).

The thermal springs and fumaroles of Agnano, Baia, Averno and Stufe di Nerone next to the Lucrino lake (Fig. 3b), although used in pre-Roman times, were intensively exploited by the Romans with the development of thermal spas (Maiuri 1970, 1977). In Republican times, luxurious villas were built on the hills along the coast of Baia which became the most important thermal centre in ancient Italy. Marcus Vicinius Crassus, Gaius Marius, Caesar, Pompey, Antonii, Marcus Terentius Varro, Cicero and Quintus Hortensius were among the owners of the villas. After the civil wars, Baia, due to medical quality of its many thermal springs, became one of the principal residential and tourist sites in the entire Roman Empire. It never became a city, either from an administrative or from an urban point of view, its core being represented solely by the thermal spas (Colletta 1988). This land-use contributed to the establishment of pisciculture, with fish and oyster farms being developed. The deformation of the Neapolitan Yellow Tuff caldera has caused subsidence of a portion of its floor along the coast of Baia (Capuano et al. 2013; Chap. Volcanic and Deformation History of the Campi Flegrei Volcanic Field, Italy) and, therefore, monuments as well as thermal spas, that local tradition has named temples, still occur on land with many being at about 8–10 m b.s.l. (Cinque et al. 1988; Di Fraia 1993; Scognamiglio 2002; Lombardo 2009; Passaro et al. 2013; Bruno et al. 2015).

One of the most important Roman public works was produced between 38 and 36 BC with the construction of the harbour named *Portus Julius* at the behest of Augustus. At the time of its construction the morphology of the area was different from that of today. In particular, the coastline was nearer to open sea (Cinque et al. 1988) and the harbour was built by connecting the Lucrino and Averno lakes to the sea (Fig. 3c). A channel was also excavated to connect Averno Lake to Cuma. The harbour also operated as a military base during the civil wars. The construction of this harbour marked the beginning of further urbanisation that in a short time spread over the entire area from *Puteoli* to *Misenum* (the present Miseno) (Fig. 3b). About 20 years after its construction, *Portus Julius* could not be used by the military fleet because of silting up of Lucrino Lake as a result of ground deformation related to bradyseismic movements. Therefore, in 12 BC the Roman fleet was transferred to *Misenum* harbour, which was already used by the Cumans. This gave birth to the very powerful naval base of *Misenum*, from which the fleet dominated the known seas in the name of the Roman Empire. However, *Portus Julius* was active as a commercial harbour at least until the fourth century BC. During Augustan times, urbanisation related to the harbour flourished and extended towards *Puteoli* with the construction of two new suburbs.

Around AD 100 two reasons concurred in the progressive decline in the importance of *Puteoli* which ceased to be the principal commercial harbour within the Mediterranean. The first reason was the development of Ostia under the Emperor Trajan (AD 98–117) as a more convenient harbour, being much nearer to Rome (Fig. 3a), and the second was the bradyseismic deformation. The urban area progressively shrunk and became focussed on the fortified castrum (Rione Terra; Fig. 3c). Human activities were reduced to farming and fishing solely to support the need of the local population and the infrastructure (e.g., roads and aqueducts) quickly deteriorated. Nevertheless, as late as AD 530 the Roman senator and historian Cassiodorus (~AD 485 to ~AD 585) described Campi Flegrei as an oasis of peace where the nature provided waters full of fish, good harbours and hot water springs.

2.2 From the Decline of the Roman Empire to the Monte Nuovo Eruption

The decline of the Roman Empire (from around AD 395) was marked by a reduction in the importance of Phlegraean towns and the abandonment of farmland. Cuma, which had become one of the major Campanian Christian centres because of its location on a fortified and inaccessible hill, was not particularly badly affected and held out against barbarian raids for much longer than Puteoli. Military vicissitudes and Saracen raids in the first half of ninth century CE, however, caused its progressive decay and population decline. Following the silting up of the Clanis and Volturno rivers, the land at the mouths of these rivers became a wide marshland (Amorosi et al. 2012). The entire littoral area of Licola (Fig. 3b) was a marsh until the beginning of twentieth century CE, when land reclamation (Ciasca 1928; De Magistris et al. 2004) and archaeological excavations (Maiuri 1970, 1977) began.

Although relative obscurity characterised the region, from the sixth century CE, the thermal spas were continuously patronised by important people including Popes John VIII (AD 872–882) and Boniface IX (AD 1389–1404), and the Emperors of the Holy Roman Empire Louis II (AD 850–875) and Frederick II (AD 1220–1250). Frederick II gave a new stimulus to the thermal spas after he benefited from treatments undertaken in AD 1226.

After two millennia, Cuma period of prominence was drawing to a close, with roads disappearing under marshes and brushwood, and its territory becoming unhealthy (Maiuri 1970, 1977). Nevertheless, the strategic importance of Campi Flegrei remained unchanged. This was well understood by the Aragoneses who during their rule (AD 1442–1501) built important fortifications, the best example of which is the Castle of Baia (Fig. 3c). In this period, the population of Pozzuoli constantly grew until the plagues of AD 1656, and the urban development was characterised by building of magnificent villas and country houses. In the area encompassed by the present Arco Felice, Lucrino and Toiano (Fig. 3c), was the famous village of Tripergole. The name of the village means "three rooms", namely cool room, warm room and hot room of a thermal spa (frigidarium, tepidarium, and calidarium in Latin, respectively). Such a name accurately describes the intended use of the village and here the courts of Angevine (AD 1266–1442) and Aragonese (AD 1442–1503) rulers came to rest and relax. In the Tripergole area, which also included the Tritoli hill with Stufe di Nerone, there were 18 active thermal spas that were described in AD 1197 by *Petrus de Ebulo* (in English, Peter of Eboli) in his book "De Balneis Puteolanis" (Various Authors 1995).

Between September 29 and October 17, 1538, after a quiescence of about 3,000 years, CFc produced its last eruption that built the Monte Nuovo cone (Fig. 3b). A detailed reconstruction of the various phases of the eruption, as well as of the precursory and subsequent phenomena (e.g., seismicity, ground deformation and hydrothermal activity) are presented by Guidoboni and Ciuccarelli (2011) and Di Vito et al. (2016). This eruption caused more than 24 casualties and devastated the morphology of the territory surrounding the vent area. In particular, it destroyed the entire village of Tripergole that was uninhabited at the time due to a long period of seismic precursory events. The eruption also caused the loss of the majority of the thermal springs. An attempt to re-launch the thermal activity of Pozzuoli and Tripergole was made by Don Pedro Antonio de Aragon who was Viceroy of Naples

(AD 1666–1671), a province of the Spanish Kingdom from AD 1504 to 1707.

After about 20 years of control by the Austrian Royal House of Hapsburg (1707–1734), the Kingdom of Naples was conquered by the Bourbons who governed until the unification of Italy in 1860. The Bourbons, following the example of the Romans, discovered that the Phlegraean area was ideal for leisure activities. Therefore, they turned wide areas (e.g., the Astroni and Fusaro lakes) (Fig. 3b) into royal estates. Such a policy re-established the use of Campi Flegrei as a site for entertainment, although only for a short time. In the eighth century CE the area became a focus of visits by foreign travellers undertaking the Grand Tour. This involved a journey through Europe of aristocratic rich young men with the aim of increasing their learning through exploring the roots of the western civilisation.

Since the ninth century CE new activities have been set up in the Campi Flegrei area. These have included the establishment of the Armstrong industrial plants for production of armaments between Pozzuoli and Arco Felice (in 1885), and of the Ilva (later renamed Italsider) ironworks at Bagnoli (in 1907) (Fig. 3b). Recently, the decision to close these factories has generated a breakdown of the established socio-economical setting and has raised the issue of finding new uses for the area.

2.3 From the Monte Nuovo Eruption to the Beginning of the Industrialisation

Just after the last Phlegraean eruption in AD 1538, what may be termed modern urbanisation began. The inhabitants of Pozzuoli, terrified by precursory phenomena (e.g., intense seismicity and ground deformation) and by the eruption itself, fled towards Naples (Guidoboni and Ciuccarelli 2011). They returned to their homes the following year, when Pedro Álvarez de Toledo y Zúñiga (Spanish Viceroy of Naples in AD 1532–1553), in order to calm the fears of the population, located his residence in a newly constructed palace in Pozzuoli. The viceroy also decreed fiscal measures and credit facilities for the Phlegraean area that brought forth a growth in building activity and a revival in the area's infrastructure. During this period, a new coastal road (presently Via Napoli; Fig. 3c) was built to connect Naples to Pozzuoli. This road was damaged during the eruption but was restored and a programme of new construction began at the foot of the Rione Terra Hill. All existing constructions were completely destroyed by the eruption and the new urban expansion had to face an environment dramatically modified by the effects of the eruption, especially in the vicinity of the vent area. The planning and restyling of the town of Pozzuoli, enclosed as it was in a fortified fence with the turreted palace of the Viceroy at its summit, and the construction of a system of watchtowers testify to a desire to re-fortify the Phlegraean area in order to improve its defence (Colletta 1988). The importance of defence was stressed by the Viceroy of Naples, Ramiro Núñez de Guzmán, who in AD 1643 granted Pozzuoli the privilege of not having to be sold together with the other state-owned towns. This was because Pozzuoli controlled four harbours (Baia, Miseno, Nisida, Pozzuoli), two fortresses that were among the best in the Kingdom (Baia, Nisida), and three watch towers (Miseno, Monte di Procida, Gaveta) (Fig. 3b).

Post-1538 residential settlements did not modify the original Greek-Roman urban setting until the onset of industrialisation. The effects of the urban development of the Phlegraean area in this period are highlighted by abundant iconographic and literary sources. In particular, tour guides and diaries were published, which reflected the status of the area as one of the destinations of the Grand Tour. There was also a huge increase in urban iconography and cartography during the sixteenth and seventeenth centuries CE. Although the natural sites and monumental buildings of Campi Flegrei were already portrayed in the Classical Age, the first artistic view of Pozzuoli and its gulf was drawn in AD 1539 by the Portuguese artist Francisco de

Fig. 7 View of the Campi Flegrei from Monte Epomeo at Ischia, illustrated by P. Fabris, in Hamilton (1776–79). *Photo source* Wikimedia Commons, https://upload.wikimedia.org/wikipedia/commons/6/64/The_Phlegraean_plain%3B_birdseye_view_from_the_top_of_the_San_Wellcome_V0025276.jpg?uselang=it

Hollanda (AD 1517–1585) and is preserved at the Real Biblioteca del Monasterio de San Lorenzo in El Escorial (Madrid, Spain).

2.4 From Beginning of Industrial Times to Present

At the end of the nineteenth and the beginning of the twentieth centuries CE, strengthening of the capitalist enterprise generated a change in urbanisation that was associated with large industrial plants and specialised infrastructure. Since then, the Phlegraean territory has progressively lost its peculiar character as a site for leisure. This is because urban growth occurred without any control and/or land-use planning, which disrupted the traditional character of many Phlegraean towns. Bacoli, Monte di Procida, Quarto and Pozzuoli (Fig. 3b) have outgrown their natural boundaries represented by morphostructural features. Furthermore, increase in building construction has caused the progressive loss of the Mediterranean maquis vegetation and terrace-cultivation which were once typical of the local landscape (Colletta 1988). In particular, urban development has modified the morphology of the Pozzuoli area (e.g., Solfatara, INA Arco Felice, Toiano, Monteruscello districts; Fig. 3c). The chosen locations of the industrial plants (e.g., Selenia at Fusaro Lake, ILVA-Italsider at Bagnoli, shipyards at Baia; Olivetti in Pozzuoli; Fig. 3b, c) often fail, either to allow plant efficiency to be optimised, or preserve and develop natural resources. In particular, the shipyards at Baia (Fig. 3b, c) have produced a discontinuity in the development of the littoral area that once was the site of Roman imperial villas and imposing thermal spas that were in contact with the sea.

Views of Campi Flegrei depicted by P. Fabris and illustrating a book by Sir William Hamilton (1776–79) (Fig. 7) bear witness to a landscape in harmony with its natural features. The description of localities such as Capo Miseno, Baia and Pozzuoli (Fig. 3b) made by Peyrefitte (1991), in his account of Campi Flegrei during a trip made in 1950s from Vesuvio to Etna (Sicily), points out an environment not yet significantly affected by signs of an unsustainable development. Maiuri (1991), in a transcription of a lecture delivered in 1961, draws striking contrasts in the life in Pozzuoli: between immeasurable riches

Table 1 Variation of the population in the towns of Campi Flegrei according to the censuses taken from 1951 to 2011

Town	1951	1961	1971	1981	1991	2001	2011
Bacoli	15,090	17,395	20,749	23,558	26,475	26,507	26,648
Monte di Procida	9419	10,265	11,140	11,803	12,490	12,838	12,975
Pozzuoli	41,457	51,308	59,813	69,861	75,142	78,754	80,357
Quarto	5,250	6,686	8,295	18,741	30,587	36,543	39,221
Total	71,216	85,654	99,997	123,963	144,694	154,642	159,201

Data from Italian Istituto Nazionale di Statistica (ISTAT); http://www.istat.it/en/

versus great poverty and sumptuous homes on the hills, versus evil-smelling slums near to the harbour. According to Maiuri (1958), about 60 years ago Bacoli was still a "picturesque town" and there were no visible or obvious signs of degradation of the environment from the Virgil tomb (Posillipo) to the Sibilla Grotto (Cuma) (Fig. 3b). Maiuri (1958) highlighted that only a few remnants of the Stufe di Nerone thermal spa were preserved, because of the construction of a road and *pozzolana* quarrying. He also stated that the quarrying was carried out in an uncontrolled way and was, thus, a major threat to the beauties of the area. Farneti et al. (1971) pointed out that Campi Flegrei, despite their environmental importance, were the object of a phase of recent degradation that was developing rapidly.

The total permanent population of Campi Flegrei has more than doubled in sixty years, from 1951 to 2011 (Table 1). In particular, in the 30 years following the 1969–1972 bradyseismic event, the population of Pozzuoli has increased by ~11,000 people, while that of Quarto has more than doubled. This increase in population occurred also in the Vesuvian area, which is also exposed to volcanic hazards. In both cases many migrants were from Naples. Population increase has caused a dramatic increase in building and infrastructure that, because of their location and construction type, has had a negative impact on the quality of the landscape. Such degradation has continued despite the recent and ongoing bradyseismic events also felt by the population (see Chaps. Volcanic and Deformation History of the Campi Flegrei Volcanic Field, Italy; The Permanent Monitoring System of the Campi Flegrei Caldera, Italy; The Hydrothermal System of the Campi Flegrei Caldera, Italy; Historic Unrest at the Campi Flegrei Caldera, Italy; Source Modelling from Ground Deformation and Gravity Changes at the Campi Flegrei Caldera, Italy). In particular, after the two major bradyseismic events recently occurred, this negative impact has become worse with the abandonment of the Rione Terra and urbanisation of the Toiano quarter (1969–1972 event), and the construction of a town at Monteruscello (1982–1984 event) (Fig. 3b). Uncontrolled and disordered urbanisation that has characterised the Phlegraean area, has also adversely affected urban green space, which is now represented only by small town parks, some formerly private gardens that now have public access and few green spaces in within areas of public housing (Di Caterino 2000).

3 Conclusive Remarks

In discussing the development of Campi Flegrei it is necessary to refer to the landscape, both natural and human-formed, which has given the area its distinctiveness. In the light of the historic, demographic and urban development discussed so far, both type of economic development and its impact on the landscape have to be addressed. According to the previously given definition of sustainable development, it is fundamental to understand the impact that the economic development of Campi Flegrei has had on its environment and landscape. Therefore, it is useful to understand whether the type of economic development of Campi Flegrei through time has been sustainable, or when and why it has become unsustainable. The economy of the Phlegraean area before the beginning of industrialisation was based essentially on thermal (i.e., spa) tourism, extraction of building materials, agriculture, fishing, fish farming and shipping business focussed on the harbour of Pozzuoli. In Roman times the way of life and villas of patricians, men of letters and local middle-class inhabitants, the thermal spas, fish farming and harbour activities, had little significant impact on the environment, and its high quality was preserved.

A reconstruction of the long-term sustainable character of the development of the area and dating the beginning of the present-day manifest degradation as a consequence of an unsustainable development, may make use of iconographic and literary source materials. According to the paintings of Fabris (Fig. 7), it can be established that the degradation of both landscape and quality of the environment has taken place during the past two hundred years. In fact, this period of major change from sustainable to unsustainable development is no longer than 60 years, as highlighted by Peyrefitte (1991). According to Maiuri (1958), the beginning of the degradation may be dated to the 1960s, when also large portions of the city of Naples and the Vesuvian area were irreparably undermined. Furthermore, the notation of Maiuri (1991) on the striking social contrasts holds the answer to the previously posed question about who and how many people would have benefited from a sustainable development. Environmental degradation may be ascribed to variable causes, including largely uncontrolled industrialisation, farming and urban development.

Campi Flegrei is an excellent example of co-existence between humans and active volcanoes over millennia. Human history also highlights the succession of two opposite ways in which co-existence may be achieved. Sustainable versus unsustainable development, is based on the socio-economical setting and development of the region at different times in its history. From the earliest settlements until recent industrial times, a time span of about three thousand years, the development of the area was sustainable, whilst in the past few decades it has become unsustainable.

Development has been sustainable for millennia despite significant ground uplift and subsidence and the Monte Nuovo eruption in the sixteenth century CE. Modern urban development in Campi Flegrei has largely disregarded the earlier intimate and positive relationships between humankind and natural environment. Only with the rapid industrialisation, in particular in the second half of the twetieth century CE, has development become unsustainable. In the last sixty years, the number of inhabitants, buildings and infrastructure has dramatically increased. This process has generated a loss of natural landscapes, a decrease of the material and non-material living standard of the people of the area and an increase in the value exposed to future volcanic events.

Acknowledgements The authors gratefully acknowledge D. Chester and D. De Rita for thoughtful revisions of an early version of the manuscript that have greatly improved its quality.

References

Amorosi A, Pacifico A, Rossi V, Ruberti D (2012) Late Quaternary incision and deposition in an active volcanic setting: The Volturno valley fill, southern Italy. Sediment Geol 282:307–320

Annecchino R (1996) Storia di Pozzuoli e della zona flegrea. Adriano Gallina Editore, Naples, Italy, pp 1–414

Bruno F, Lagudi A, Gallo A, Muzzupappa M, Davidde Petriaggi B, Passaro S (2015) 3D documentation of archaeological remains in the underwater park of Baiae. Int Arch Photogramm Rem Sens Spatial Inf Sci XL-5/W5:41–46. https://doi.org/10.5194/isprsarchives-XL-5-W5-41-2015

Buchner G (1969) Mostra degli scavi di Pithecusa. Dialoghi Di Archeologia 3:85–101

Buchner G (1986) Eruzioni vulcaniche e fenomeni vulcanotettonici di età preistorica e storica nell'isola d'Ischia. In: Albore Livadie C (ed) Tremblements de Terre, Eruptions Volcaniques et Vie des Hommes dans la Campanie Antique, vol 7. Centre Jean Bérard, Institut Français de Naples, Naples, Italy, pp 145–188

Capasso B (1978) Napoli Greco-Romana. Arturo Berisio, Naples, Italy, first published in 1905 by Società Napoletana di Storia Patria, Pierro, Naples, Italy, pp 1–225

Capuano P, Russo G, Civetta L, Orsi G, D'Antonio M, Moretti R (2013) The active portion of the Campi Flegrei caldera structure imaged by 3D inversion of gravity data. Geochem Geophys Geosyst 14:4681–4697. https://doi.org/10.1002/ggge.20276

Ciasca R (1928) Storia delle bonifiche del regno di Napoli. Laterza, Bari, Italy, pp 1–253

Cinque A, Rolandi G, Zamparelli V (1984) L'estensione dei depositi marini olocenici nei Campi Flegrei in relazione alla vulcano-tettonica. Boll Soc Geol It 104 (2):327–348

Cinque A, Laureti L, Russo F (1988) Le variazioni della linea di costa lungo il litorale campano durante il Quaternario. TERRA 5:19–25

Civetta L, Orsi G, Pappalardo L, Fisher RV, Heiken GH, Ort M (1997) Geochemical zoning, mixing, eruptive dynamics and depositional processes—the Campanian Ignimbrite, Campi Flegrei, Italy. J Volcanol Geotherm Res 75:183–219

Colletta T (1984) Atlanti di città del Cinquecento. ESI, Naples, Italy, pp 1–142

Colletta T (1988) Pozzuoli città fortificata in epoca vicereale: una mappa inedita conservata alla Biblioteca nazionale di Parigi. In: Storia dell'urbanistica/Campania. Edizioni Kappa, Rome, Italy, pp 7–40

De Magistris A (2004) High-Rise: percorsi nella storia dell'architettura e dell'urbanistica del XIX e del XX secolo attraverso la dimensione verticale. UTET, Turin, Italy, pp 1–106. ISBN 9788877508041

Deino AL, Orsi G, de Vita S, Piochi M (2004) The age of the Neapolitan Yellow Tuff caldera-forming eruption (Campi Flegrei caldera—Italy) assessed by 40Ar/39Ar dating method. J Volcanol Geotherm Res 133:157–170

Del Gaudio C, Aquino I, Ricciardi GP, Ricco C, Scandone R (2010) Unrest episodes at Campi Flegrei: a reconstruction of vertical ground movements during 1905–2009. J Volcanol Geotherm Res 195:48–56

Di Caterino C (2000) Paesaggio e verde urbano nei Campi Flegrei. In: Piano di Tutela delle Aree - natura 2000. Parco Regionale dei Campi Flegrei, Naples, Italy

Di Fraia G (1993) Baia sommersa. Nuove evidenze topografiche e monumentali. Archeologia Subacquea. Studi Ricerche Documenti 1:21–48

Di Renzo V, Arienzo I, Civetta L, D'Antonio M, Tonarini S, Di Vito MA, Orsi G (2011) The magmatic feeding system of the Campi Flegrei caldera: architecture and temporal evolution. Chem Geol 281:227–241

de Vita S, Orsi G, Civetta L, Carandente A, D'Antonio M, Deino A, di Cesare T, Di Vito MA, Fisher RV, Isaia R, Marotta E, Necco A, Ort M, Pappalardo L, Piochi M, Southon J (1999) The Agnano-Monte Spina eruption (4100 years BP) in the restless Campi Flegrei caldera (Italy). J Volcanol Geotherm Res 91:269–301

de Vita S, Sansivero F, Orsi G, Marotta E, Piochi M (2010) Volcanological and structural evolution of the Ischia resurgent caldera (Italy) over the past 10 k.y. In: Groppelli G, Viereck-Goette L (eds) Stratigraphy and geology of volcanic areas. Geol Soc Am Spec Pap, Boulder (CO), USA, vol 464, pp 193–241

Di Vito MA, Acocella V, Aiello G, Barra D, Battaglia M, Carandente A, Del Gaudio C, de Vita S, Ricciardi GP, Ricco C, Scandone R, Terrasi F (2016) Magma transfer at Campi Flegrei caldera (Italy) before the 1538 AD eruption. Sci Rep 6:32245. https://doi.org/10.1038/srep32245

Dvorak JJ, Mastrolorenzo G (1991) The mechanisms of recent vertical crustal movements in Campi Flegrei caldera, southern Italy. In: Geol Soc Am Spec Pap 263:1–47. Boulder (CO), USA

Farneti G, Pratesi F, Tassi F (1971) Guida alla natura d'Italia. Mondadori, Milan, Italy, pp 1–556

Fisher RV, Orsi G, Ort M, Heiken G (1993) Mobility of a large volume pyroclastic flow emplacement of the Campanian Ignimbrite, Italy. J Volcanol Geotherm Res 56:205–220

Giaccio B, Hajdas I, Isaia R, Deino A, Nomade S (2017) High-precision ^{14}C dating and $^{40}Ar/^{39}Ar$ dating of the Campanian Ignimbrite (Y-5) reconciles the time-scales of climatic-cultural processes at 40 ka. Sci Rep 7:45940

Giampaola D, D'Agostino B (2005) Osservazioni storiche e archeologiche sulla fondazione di Neapolis. In: Harris W, Lo Cascio E (eds) Noctes Campanae. Studi di storia antica e archeologia dell'Italia preromana e romana. In memoria di H. W. Frederiksen, Luciano Editore, Naples, Italy, pp 49–80

Guidoboni E, Ciuccarelli C (2011) The Campi Flegrei caldera: historical revision and new data on seismic crises, bradyseisms, the Monte Nuovo eruption and ensuing earthquakes (twelfth century 1582 AD). Bull Volcanol 73:655–677

Hamilton W (1776–79) Campi Phlegraei. Observations on the Volcanoes of the Two Sicilies, as they have been communicated to the Royal Society of London, 3 vols, Naples, Italy

Issel A (1883) Le oscillazioni lente del suolo o bradisismi. Atti R Univ Genova, Genoa, Italy IV, pp 1–210

Lombardo N (2009) Baia: le terme sommerse a Punta dell'Epitaffio. Ipotesi di ricostruzione volumetrica e creazione di un modello digitale. Archeologia e Calcolatori 20:373–396

Lowe L, Barton N, Blockley S, Ramsey CB, Cullen VL, Davies W, Gamble C, Grant K, Hardiman M, Housley R, Lane CS, Lee S, Lewis M, MacLeod A, Menzies M, Müller W, Pollard M, Price C, Roberts AP, Rohling EJ, Satow C, Smith VC, Stringer CB, Tomlinson EL, White D, Albert P, Arienzo I, Barker G, Boric´ D, Carandente A, Civetta L, Ferrier C, Guadelli JL, Karkanas P, Koumouzelis M, Müller UC, Orsi G, Pross J, Rosi M, Shalamanov-Korobar L, Sirakov N, Tzedakis PC (2012) Volcanic ash layers demonstrate resilience of Neanderthal and early modern humans to natural hazards. Proc Natl Acad Sci 109 (34):13532–13537

Lyell C (1830) Principles of geology; an attempt to explain the former changes of the earth's surface by reference to causes now in operation. John Murray, London, UK

Maiuri A (1958) I Campi Flegrei, dal sepolcro di Virgilio all'antro di Cuma. 1a edizione, Istituto Poligrafico dello Stato, Rome, Italy, pp 1–168

Maiuri A (1970) I Campi Flegrei, dal sepolcro di Virgilio all'antro di Cuma. 5a edizione, Istituto Poligrafico dello Stato, Rome, Italy, pp 1–168

Maiuri A (1977) The Phlegraean Fields, from Virgil's Tomb to the Grotto of the Cumaean Sibyl. 5th edn. Istituto Poligrafico dello Stato, Rome, Italy, pp 1–168

Maiuri A (1991) La Campania al tempo dell'approdo di S. Paolo. Franco Di Mauro Editore, Naples, Italy, pp 1–72

Marzocchella A (1998) Tutela archeologica e preistoria nella pianura campana. In: Guzzo PG, Peroni R (eds), Archeologia e vulcanologia in Campania. Atti del convegno, Pompei, December 21st 1996, Naples, Italy, pp 97–133

Marzocchella A (2000) Storie di contadini alle falde del Vesuvio. Archeo 182:36–45

Miccio B, Potenza U (1994) Gli acquedotti di Napoli. A. M.A.N.: Azienda Municipalizzata Acquedotto di Napoli, Naples, Italy, pp 1–188

Morabito S, Petrosino P, Milia A, Sprovieri M, Tamburrino S (2014) A multidisciplinary approach for reconstructing the stratigraphic framework of the last 40 ka in a bathyal area of the eastern Tyrrhenian Sea. Global Planet Change 123(A):121–138

Morhange C, Marrimer N, Laborel J, Todesco M, Oberlin C (2006) Rapid sea-level movements and noneruptive crustal deformations in the Phlegrean Fields caldera, Italy. Geology 34(2):93–96. https://doi.org/10.1130/G21894.1

Napoli M (1996) Napoli Greco-Romana. Colonnese, Naples, Italy, pp 1–237

Orsi G, D'Antonio M, de Vita S, Gallo G (1992) The Neapolitan Yellow Tuff, a large-magnitude trachytic phreatoplinian eruption: eruptive dynamics, magma withdrawal and caldera collapse. J Volcanol Geotherm Res 53:275–287

Orsi G, Civetta L, D'Antonio M, Di Girolamo P, Piochi M (1995) Step-filling and development of a three-layers magma chamber: the Neapolitan Yellow Tuff case history. J Volcanol Geotherm Res 67:291–312

Orsi G, de Vita S, Di Vito M (1996) The restless, resurgent Campi Flegrei nested caldera (Italy): constraints on its evolution and configuration. J Volcanol Geotherm Res 74:179–214. https://doi.org/10.1016/S0377-0273(96)00063-7

Orsi G, Civetta L, Del Gaudio C, de Vita S, Di Vito MA, Isaia R, Petrazzuoli S, Ricciardi G, Ricco C (1999) Short-term ground deformations and seismicity in the nested Campi Flegrei Caldera (Italy): an example of active block resurgence in a densely populated area. J Volcanol Geotherm Res 91:415–451

Orsi G, de Vita S, Di Vito M, Nave R, Heiken G (2003) Facing volcanic and related hazards in the Neapolitan area. In: Heiken G, Fakundiny R, Sutter J (eds) Earth sciences in the cities: a reader. AGU Special Publication Series 56, Washington, pp 121–170

Orsi G, Cioni R, Di Renzo V (2013) The Campanian plain during the bronze age: development of volcanism and impact of the Vesuvius Avellino eruption in a densely populated area. In: Proceedings of 4th Archaeol conference of Central Germany "1600-Cultural Change in the shadow of the Thera-Eruption?" pp 117–134

Ort M, Orsi G, Pappalardo L, Fisher RV (2003) Anisotropy of magnetic susceptibility studies of depositional processes in the Campanian Ignimbrite Italy. Bull Volcanol 65:55–72

Parascandola A (1947) I fenomeni bradisismici del Serapeo di Pozzuoli. Stabilimento Tipografico G Genovese, Naples, Italy, pp 1–156

Passaro S, Barra M, Saggiomo R, Di Giacomo S, Leotta A, Uhlen H, Mazzola S (2013) Multi-resolution morpho-bathymetric survey results at the Pozzuoli-Baia underwater archaeological site (Naples, Italy). J Archaeol Sci 40:1268–1278

Peyrefitte R (1991) Dal Vesuvio all'Etna. Campania. Ediprint, Siracusa, Italy, pp 1–171

Ricco C, Petrosino S, Aquino I, Del Gaudio C, Falanga F (2019) Some investigations on a possible relationship between ground deformation and seismic activity at Campi Flegrei and Ischia volcanic areas (southern Italy). Geosciences 9(5):222

Rosi M, Vezzoli L, Aleotti P, De Censi M (1996) Interaction between caldera collapse and eruptive dynamics during the Campanian Ignimbrite eruption, Phlegraean Fields, Italy. Bull Volcanol 57:541–554

Saccoccio F, Marzocchella A, Vanzetti A (2013) The field system of Gricignano d'Aversa (Southern Italy) and the agrarian impact in the Piana Campana, ca. 3900 cal BP. Quatern Internat 303:82–92. https://doi.org/10.1016/j.quaint.2013.02.021

Santacroce R, Cristofolini R, La Volpe L, Orsi G, Rosi M (2003) Italian active volcanoes. Episodes 26(3):227–234

Santacroce R, Cioni R, Marianelli P, Sbrana A, Sulpizio R, Zanchetta G, Donahue DJ, Joron JJ (2008) Age and whole rock-glass compositions of proximal pyroclastics from the major explosive eruptions of Somma-Vesuvius: a review as a tool for distal tephrostratigraphy. J Volcanol Geotherm Res 177:1–18

Scarpati C, Cole P, Perrotta A (1993) The Neapolitan Yellow Tuff—A large volume multiphase eruption from Campi Flegrei, Southern Italy. Bull Volcanol 55:343–356

Scognamiglio E (2002) Nuovi dati su Baia sommersa. Archeologia Subacquea. Studi, Ricerche, Documenti 3:47–55

Smith VC, Isaia R, Pearce NJG (2011) Tephrostratigraphy and glass compositions of post-15 kyr Campi Flegrei eruptions: implications for eruption history and chronostratigraphic markers. Quat Sci Rev 30:3638–3660

Various Authors (1995) Le terme puteolane e Salerno nei codici miniati di Pietro da Eboli. Fausto Fiorentino, Naples, Italy, pp 1–184

Wohletz K, Orsi G, de Vita S (1995) Eruptive mechanisms of the Neapolitan Yellow Tuff interpreted from stratigraphic, chemical and granulometricdata. J Volcanol Geotherm Res 67:263–290. https://doi.org/10.1016/0377-0273(95)00002-C

The manufacturer's authorised representative in the EU is Springer Nature Customer Service Centre GmbH, Europaplatz 3, 69115 Heidelberg, Germany. If you have any concerns regarding our products, please contact ProductSafety@springernature.com

Printed and bound by CPI Group (UK) Ltd, Croydon, CR0 4YY
22/09/2026
02220954-0019